3rd edition

physiology

National Medical Series

In the basic sciences

anatomy, 2nd edition
behavioral sciences
 in psychiatry, 3rd edition
biochemistry, 3rd edition
clinical epidemiology and
 biostatistics
genetics
hematology
histology and cell biology,
 2nd edition

human developmental anatomy
immunology, 3rd edition
introduction to clinical medicine
microbiology, 2nd edition
neuroanatomy
pathology, 3rd edition
pharmacology, 3rd edition
physiology, 3rd edition
radiographic anatomy

In the clinical sciences

medicine, 2nd edition
obstetrics and gynecology,
 3rd edition
pediatrics, 2nd edition
preventive medicine and
 public health, 2nd edition
psychiatry, 2nd edition
surgery, 2nd edition

In the exam series

review for USMLE Step 1,
 3rd edition
review for USMLE Step 2
geriatrics

The National Medical Series for Independent Study

3rd edition
physiology

John Bullock, M.S., Ph.D.

Adjunct Associate Professor of Physiology
New Jersey Medical School
University of Medicine and Dentistry
  of New Jersey
Newark, New Jersey

Joseph Boyle, III, M.D.

Associate Professor of Physiology
Clinical Assistant Professor of Medicine
New Jersey Medical School
University of Medicine and Dentistry
  of New Jersey
Newark, New Jersey

Michael B. Wang, Ph.D.

Professor of Physiology
Temple University
  School of Medicine
Philadelphia, Pennsylvania

Williams & Wilkins

Philadelphia • Baltimore • Hong Kong • London • Munich • Sydney • Tokyo

A Waverly Company

Williams & Wilkins

Managing Editor: Jane Velker
Development Editor: Melanie Cann
Production: Laurie Forsyth
Illustration: Patricia MacAllen

Copyright © 1995
Williams & Wilkins
Rose Tree Corporate Center, Building II
1400 North Providence Road, Suite 5025
Media, PA 19063-2043 USA

Library of Congress Cataloging-in-Publication Data
Bullock, John, 1932-
    Physiology / John Bullock, Joseph Boyle III, Michael B. Wang.—
3rd ed.
       p. cm.—(The National medical series for independent study)
    Includes index.
    ISBN 0-683-06259-X (alk. paper)
    1. Human physiology—Outlines, syllabi, etc. I. Boyle, Joseph.
II. Wang, Michael B. III. Title. IV. Series.
    [DNLM: 1. Physiology—examination questions. 2. Physiology—
outlines.   QT 18 B938p 1994]
QP41.P492   1994
612—dc20
DNLM/DLC
for Library of Congress   93-23683
                                     CIP

ISBN 0-683-06259-X

©1995 Williams & Wilkins, Malvern, PA

10 9 8 7 6 5 4 3 2

Contents

Preface

Since its publication in 1984, NMS *Physiology* has garnered considerable response from medical students around the world. Based on these written and oral commentaries, the authors have revised the text to conform better to the needs of students preparing for course examinations and standardized medical or other health science examinations. Accordingly, this new edition of NMS *Physiology* contains a well-defined fund of knowledge the authors consider essential for medical practitioners. Although this book does not focus specifically on pathology, it does provide an insight into the pathophysiologic aspects of disease by comparing normal processes in the body with the abnormal.

The outline format has allowed the authors to emphasize the relative importance of the facts presented and to provide students with a source of fundamental information about physiology. Such a core guide to the discipline should prove especially valuable now that medical students are expected to assume more and more responsibility for their own learning.

John Bullock
Joseph Boyle, III
Michael B. Wang

Acknowledgments

We collectively wish to express our thanks to Melanie Cann (Development Editor) for her personal support, perseverance, understanding, and expertise. We also recognize the special contribution of Patricia MacAllen as medical illustrator. Our greatest measure of gratitude goes to Debra Dreger, who has earned our utmost respect for her energy and commitment to the NMS series.

The authors

To Barbara, who, through her love and willing sacrifice, encouraged me to complete two graduate degrees in physiology and biochemistry. Also to Laura, John, and Katherine—all important well-springs of inspiration. Not to be forgotten are the many medical and dental students together with the foreign medical graduates who have become friends and colleagues over the years.

John Bullock

To Patti, for her love and support over many years. To many colleagues who have made the study of physiology an exciting and stimulating career. And, to the many students who have asked the questions which have enlightened me.

Joseph Boyle, III

My thanks to family, friends, colleagues, and students for their encouragement and support over many years.

Michael B. Wang

To the Reader

Since 1984, the National Medical Series for Independent Study has been helping medical students meet the challenge of education and clinical training. In this climate of burgeoning knowledge and complex clinical issues, a medical career is more demanding than ever. Increasingly, medical training must prepare physicians to seek and synthesize necessary information and to apply that information successfully.

The National Medical Series is designed to provide a logical framework for organizing, learning, reviewing, and applying the conceptual and factual information covered in basic and clinical studies. Each book includes a concise but comprehensive outline of the essential content of a discipline, with up to 500 study questions. The combination of distilled, outlined text and tools for self-evaluation allows easy retrieval and enhanced comprehension of salient information. Each question is accompanied by the correct answer, a paragraph-length explanation, and specific reference to the text where the topic is discussed. Study questions that follow each chapter use current National Board formats to reinforce the chapter content. Study questions appearing at the end of the text in the Comprehensive Exam vary in format depending on the book; the unifying goal of this exam, however, is to challenge the student to synthesize and expand on information presented throughout the book. Wherever possible, Comprehensive Exam questions are presented in the context of a clinical case or scenario intended to simulate real-life application of medical knowledge

Each book in the National Medical Series is constantly being updated and revised to remain current with the discipline and with subtle changes in educational philosophy. The authors and editors devote considerable time and effort to ensuring that the information required by all medical school curricula is included and presented in the most logical, comprehensible manner. Strict editorial attention to accuracy, organization, and consistency also is maintained. Further shaping of the series occurs in response to biannual discussions held with a panel of medical student advisors drawn from schools throughout the United States. At these meetings, the editorial staff considers the complicated needs of medical students to learn how the National Medical Series can better serve them. In this regard, the staff at Williams & Wilkins welcomes all comments and suggestions. Let us hear from you.

PART I

GENERAL PHYSIOLOGY
Michael B. Wang

Chapter 1

Membrane and Cellular Physiology

I. **HOMEOSTASIS.** The compositions of the extracellular fluid (ECF) and the intracellular fluid (ICF) differ from each other (Table 1-1; see also Chapter 22) and are maintained in a steady-state condition by a variety of regulatory processes called **homeostatic mechanisms**.

A. **Extracellular fluid** consists of the blood plasma and interstitial fluid. The composition of the ECF is maintained by the cardiovascular, pulmonary, renal, gastrointestinal (GI), endocrine, and nervous systems acting in a coordinated fashion.

B. **Intracellular fluid.** The composition of the ICF is maintained by the cell membrane, which mediates the transport of material between the ICF and ECF by diffusion, osmosis, active transport, and vesicular transport.

II. **CELL MEMBRANE.** All animal cells are enveloped by a cell membrane composed of **lipids** and **proteins**.

A. The **lipids** form the basic structure of the membrane.

TABLE 1-1. Major Ionic Components of the Intracellular and Extracellular Fluids

Ion	Intracellular Concentration		Extracellular Concentration	
	(mOsm/L)	(mEq/L)	(mOsm/L)	(mEq/L)
Na^+	15	15	142	142
K^+	135	135	4	4
Ca^{2+}	10^{-4}	2×10^{-4}	2	4
Mg^{2+}	20	40	1	2
Cl^-	4	4	106	106
HCO_3^-	10	10	24	24
HPO_4^{2-}	10	20	2	4
SO_4^{2-}	2	4	0.5	1
Proteins$^-$, amino acids$^-$, urea, etc.	99	152	0.5	1

The osmolarity of the intracellular fluid (ICF) is the same as that of the extracellular fluid (ECF). Also, the milliequivalents of cations and anions are equal in the ICF as well as in the ECF.

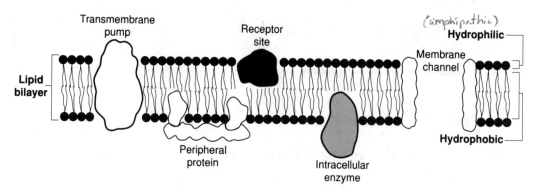

FIGURE 1-1. Some of the functions performed by proteins within the lipid bilayer of cell membranes.

1. The lipid molecules (primarily phospholipids, cholesterol, and glycolipids) are **amphipathic;** that is, they have a hydrophilic polar region at one end of the molecule and a hydrophobic hydrocarbon tail at the other end (Figure 1-1).

2. The lipid molecules arrange themselves into **bilayers** when they are placed in an aqueous solution (see Figure 1-1).

   a. The hydrophilic ends of the lipid molecules line up facing the ICF and ECF.
   b. The hydrophobic tails of the molecules face each other in the interior of the bilayer.

B. The **proteins** assist the membrane in carrying out a large variety of physiologic processes (see Figure 1-1).

1. According to the **fluid mosaic model** of membrane structure, membrane proteins are embedded in the lipid bilayer.

2. Some proteins (called **integral** or **intrinsic proteins**) bind to the hydrophobic center of the lipid bilayer.
   a. Transmembrane proteins are integral proteins that **span the entire bilayer.** Transmembrane proteins serve as:
     (1) Channels, through which small, water-soluble substances can diffuse
     (2) Carriers, which actively transport materials across the bilayer
     (3) Pumps, which actively transport ions across the bilayer
     (4) Receptors, which, when activated, initiate intracellular reactions
   b. Integral proteins that are **present on only one side of the membrane** serve primarily as **enzymes** that activate or inactivate various metabolic processes.

3. Other proteins (called **peripheral** or **extrinsic proteins**) bind to the hydrophilic polar heads of the lipids or to the integral proteins.
   a. Peripheral proteins that bind to the **intracellular surface** of the membrane contribute to the **cytoskeleton.**
   b. Peripheral proteins that bind to the **extracellular surface** of the membrane contribute to the **glycocalyx** (i.e., a structure composed of glycolipids and glycoprotein that covers cell membranes).

III. **TRANSPORT ACROSS CELL MEMBRANES.** Substances move through cell membranes by **diffusion, osmosis, active transport,** and **vesicular transport.**

A. Diffusion

1. **Simple diffusion** is a passive process by which uncharged particles in solution flow down their concentration (chemical) gradient (i.e., particles move from areas of high concentration to areas of lower concentration).

  a. Process. No external source of energy (or driving force) is required to move parti-
    cles down a concentration gradient by diffusion. Simple diffusion occurs because
    the heat content of the solution keeps all of the solvent and solute particles of the
    solution in constant motion.

    (1) Each particle moves in an unpredictable (random) fashion; however, it is more
        likely that a particle will move from an area of high concentration to an area of
        lower concentration.

    (2) Net movement ceases when the concentration of the particles is equal every-
        where within the solution (**diffusional equilibrium**).

    (3) Although random movement of the particles continues after diffusional equilib-
        rium is achieved, the concentration of the particles throughout the solution re-
        mains the same.

  b. Rate. Fick's law of diffusion describes the rate at which a material diffuses through a
    membrane as a function of its concentration gradient. It has two forms:

    (1) $$\text{Flux} = -\frac{D \cdot A}{x}(C_{in} - C_{out}), \text{ where}$$

$$D = \text{the diffusion coefficient (cm}^2\text{/sec); see III A 1 c}$$
$$A = \text{the area of the membrane (cm}^2\text{)}$$
$$x = \text{the diffusion distance, or the thickness of the membrane (cm)}$$
$$C_{in} \text{ and } C_{out} = \text{the concentrations of the material on the inside and out-side of the membrane, respectively (mmol/L or mmol/1000 cm}^3\text{)}$$

    The negative sign indicates that the material is moving down its concentration
    gradient.

    (2) Fick's law can be simplified when biologic membranes are considered, because
        the thickness of the membrane always is approximately 10^{-6} cm. Dividing the
        diffusion coefficient (D) by 10^{-6} (x) yields the permeability coefficient of the
        membrane, and Fick's law becomes:

$$\text{Flux} = -P \cdot A \cdot (C_{in} - C_{out})$$

    where P is the permeability coefficient (cm/sec).

  c. Permeability. The **permeability coefficient** (and the **diffusion coefficient**) depend on
    the solute and the membrane through which diffusion is taking place.

    (1) Lipid-soluble particles diffuse through the lipid bilayer of the cell membrane.
        Thus, their permeability is proportional to their lipid solubility.

    (2) Water-soluble particles diffuse through the aqueous channels formed by trans-
        membrane proteins. Thus, their permeability is proportional to their molecular
        size, shape, and charge.

2. Facilitated diffusion is a carrier-mediated process that enables particles that are too
  large to flow through membrane channels by simple diffusion (e.g., many ions and nutrients) to pass through the membrane. For example, facilitated diffusion accomplishes
  the transport of glucose into red blood cells and muscle and into adipose tissue (when
  insulin is present).

  a. Process. No external source of energy (or driving force) is required to move parti-
    cles down a concentration gradient by facilitated diffusion.

    (1) The carrier protein undergoes repetitive spontaneous conformational changes
        during which the binding site for the substance is alternately exposed to the ICF
        and ECF (Figure 1-2A).

    (2) The substance is more likely to bind to the carrier where it is more highly con-
        centrated and dissociate from the carrier where it is less highly concentrated;
        therefore, the substance flows down its concentration gradient.

  b. Rate. Unlike simple diffusion, the rate of facilitated diffusion increases as the con-
    centration gradient increases until all of the binding sites are filled. At this point, the
    rate of diffusion can no longer increase with increasing particle concentration. This
    is called **saturation,** or **Michaelis-Menten, kinetics** (Figure 1-2B).

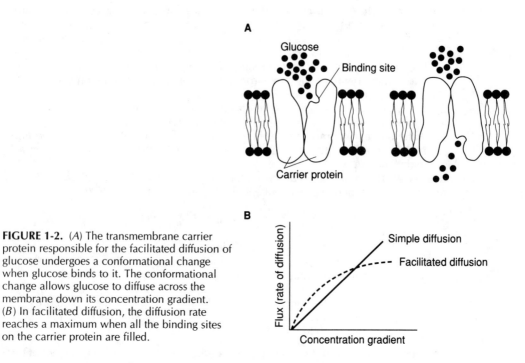

FIGURE 1-2. (*A*) The transmembrane carrier protein responsible for the facilitated diffusion of glucose undergoes a conformational change when glucose binds to it. The conformational change allows glucose to diffuse across the membrane down its concentration gradient. (*B*) In facilitated diffusion, the diffusion rate reaches a maximum when all the binding sites on the carrier protein are filled.

B. **Osmosis** is the passive flow of water across a selectively permeable membrane down an osmotic pressure gradient (Figure 1-3).

1. The **osmotic pressure gradient** is created by the presence of different concentrations of solutes in the solutions on either side of the membrane.
   a. Osmotic pressure is a **colligative property** of solutions; that is, it is related to the number of particles dissolved in the solution, not their size, molecular weight, or chemical constitution. Thus, the osmotic concentration difference (ΔC) of a substance reflects the difference in particle concentrations, not the difference in molar concentrations. For example, a 1 mmol/L glucose solution has an osmotic concentration of 1 mOsm/L, but a 1 mmol/L sodium chloride (NaCl) solution (which forms two particles when it is dissolved in water) has an osmotic concentration of 2 mOsm/L.
   b. **Calculating osmotic pressure**
      (1) Although it is not known how the particle concentration gradient creates an osmotic pressure, the pressure across the membrane can be calculated using the **van't Hoff equation:**

      $$\pi = \Delta C \cdot R \cdot T, \text{ where}$$

      π = the osmotic pressure (mm Hg)
      ΔC = the difference in the concentration of particles between the two solutions (mOsm/L)
      R = the natural gas constant (62 mm Hg $\cdot$ L/mmol $\cdot$ °K)
      T = absolute temperature (°K)

      (2) The osmotic pressure produced by a concentration difference can also be determined experimentally by measuring the hydrostatic pressure that must be applied to the solution with the higher particle concentration (i.e., the higher osmotic pressure) to prevent water from entering it from the solution with the lower particle concentration.

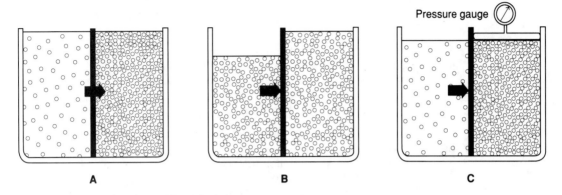

FIGURE 1-3. When a selectively permeable membrane separates two solutions of different osmolalities (*A*), water flows from the solution with the lower osmotic pressure (concentration) to the solution with the higher osmotic pressure (concentration). (*B*) Water will flow into the chamber until the pressure (i.e., hydrostatic and osmotic) difference between the two chambers is zero. Because the flow of water increases the height of the water in the chamber on the *right side*, the hydrostatic pressure in this chamber is higher than it is in the chamber on the *left*. Consequently, at equilibrium, the osmotic pressure (and thus, concentration) is somewhat less in the chamber on the *right* than in the chamber on the *left*. (*C*) The flow of water can also be prevented by applying pressure to the chamber that contains the higher solute concentration. The amount of pressure that must be applied to prevent the flow of water is a measure of the osmotic pressure difference between the two chambers.

 2. **Flow.** The flow of water caused by osmotic pressure differences can be calculated using the **osmotic flow equation:**

$$\text{Flow} = L \cdot A \cdot (\pi_1 - \pi_2), \text{ where}$$

$$
\begin{aligned}
L &= \text{ hydraulic conductivity of the membrane (L/sec/cm}^2\text{/mm Hg)} \\
A &= \text{ area of the membrane (cm}^2\text{)} \\
\pi_1 \text{ and } \pi_2 &= \text{ osmotic pressures on either side of the membrane (mm Hg)}
\end{aligned}
$$

   a. If the particles in the solution are permeable to the membrane, the osmotic pressure difference created by the concentration difference will be less than predicted by the van't Hoff equation. The reduction in osmotic pressure can be determined from the **reflection coefficient,** which is a measure of the relative membrane permeabilities of the solute and solvent.
   (1) If the particle is totally impermeable to the membrane (i.e., it is totally reflected from the membrane), it has a reflection coefficient of 1.
   (2) If the particle is as permeable to the membrane as the solvent (i.e., it is not at all reflected from the membrane), it has a reflection coefficient of 0.

   (3) The actual reflection coefficient can be calculated using the expression:

$$\sigma = 1 - \frac{P_{solute}}{P_{water}}$$

   where σ is the reflection coefficient and P_{solute} and P_{water} are the membrane permeabilities of the solute and solvent, respectively.
   b. The osmotic flow equation can be used to calculate the osmotic flow of water if it is modified to include the reflection coefficient:

$$\text{Flow} = \sigma \cdot L \cdot A \cdot (\pi_1 - \pi_2)$$

 3. **Clinical importance**
   a. Cells. Changes in plasma osmolarity cause cells to shrink or swell, as the osmotic pressure difference between the inside and outside of the cell causes water to flow out of or into the cells.

(1) Calculating cell volume change. The final volume of the cell subjected to a change in extracellular osmolality can be calculated using the expression:

$$\pi_i \cdot V_i = \pi_f \cdot V_f$$

where i and f refer to the initial and final osmolalities and volumes of the cell.

  (a) For example, if a red blood cell with an initial volume of 100 μm^3 and a normal osmolality of 285 mOsm/L is placed in a solution with an osmolality of 325 mOsm/L, its final volume (V_f) will be:

$$285 \cdot 100 = 325 \cdot V_f$$
$$V_f = 88 \ \mu m^3$$

  (b) The volume of water leaving the red cell is so small compared with the volume of the extracellular solution that the osmolarity of the extracellular solution is assumed to remain constant.

(2) The ability of a particle to cause a steady-state change in cell volume is referred to as its tonicity. Osmotic pressure differences can only cause a steady-state change in cellular volume if the particles creating the osmotic pressure difference are impermeable to the membrane. For example, because urea easily flows through cell membranes, a change in the extracellular concentration of urea does not cause a steady-state change in cell volume.

  (a) An extracellular solution that causes water to flow into the cell, making it swell, is called **hypotonic.**

  (b) An extracellular solution that causes water to flow out of the cell is called **hypertonic.** crenation

  (c) A solution causing no change in intracellular volume is called **isotonic.**

b. Capillaries. Almost all particles dissolved in blood plasma, except proteins, easily cross the capillary and, thus, do not cause water to flow between the capillary and the interstitial fluid. Plasma proteins have an osmolar concentration of approximately 1.2 mOsm/L and, therefore, create an osmotic pressure of approximately 23 mm Hg.

(1) The osmotic pressure produced by the plasma proteins, called the **colloid oncotic pressure,** draws water into the capillaries from the interstitial fluid and is counteracted by the hydrostatic pressure of the blood produced by the heart. The movement of water through the capillary walls as a result of hydrostatic or osmotic pressure differences is called **bulk flow** (see also Chapter 12 I C 3 b).

(2) Whether water flows into or out of the capillaries depends on whether the colloid osmotic pressure is greater or less than the hydrostatic pressure of the blood. When water flows into or out of capillaries, it carries dissolved particles with it. This is called solvent drag. in clotting also.

C. | **Active transport processes**

1. Primary active transport processes directly use the energy obtained from the hydrolysis of adenosine triphosphate (ATP) to transport material against an energy (e.g., concentration, electrical) gradient.

  a. The most common of these active transport systems is the **sodium-potassium (Na$^+$-K$^+$) pump (Na$^+$-K$^+$-ATPase),** which uses the membrane-bound ATPase as a carrier molecule.

  (1) Function. The Na$^+$-K$^+$ pump is responsible for maintaining the high K$^+$ and low Na$^+$ concentrations in the ICF.

  (2) Composition. ATPase is composed of four subunits (two α and two β). The α subunit has ATPase enzymatic activity [i.e., the ability to convert ATP to adenosine diphosphate (ADP), thereby releasing energy] and binding sites on its intracellular and extracellular faces.

   (a) The intracellular face contains binding sites for three Na$^+$ ions and an ATP molecule.

   (b) The extracellular face contains binding sites for two K$^+$ ions.

  (3) Process. The operation of the pump can be divided into two steps:

   (a) When three ions of Na$^+$ and an ATP molecule bind to the carrier on the inside of the cell, a phosphate group is transferred from the ATP molecule to

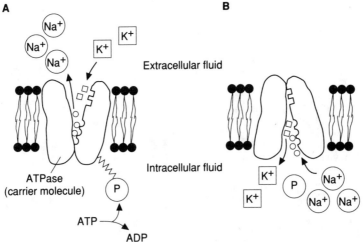

FIGURE 1-4. (*A*) The energy contained in the high-energy phosphate bond produces a conformational change in the carrier, during which three ions of Na$^+$ are transported out of the cell. A second conformational change occurs after two K$^+$ ions bind to the carrier on the outside of the cell, (*B*) transporting K$^+$ into the cell and causing the phosphate (*P*) to dissociate from the carrier.

an aspartic acid residue of the α subunit (Figure 1-4A). The addition of a high-energy phosphate group causes a conformational change in the protein, which results in the transport of the three Na$^+$ ions out of the cell.

  (b) When two ions of K$^+$ bind to the carrier on the outside of the cell, the aspartic acid–phosphate bond is hydrolyzed. The energy released by this dephosphorylation step results in a second conformational change during which the two K$^+$ ions are transported into the cell (Figure 1-4B).

(4) Inhibition of the pump

  (a) Digitalis, a drug used for the treatment of heart failure, produces its therapeutic effect by binding to the extracellular face of the α subunit and interfering with the dephosphorylation step of the transport process.

  (b) The pump requires binding by Na$^+$, K$^+$, and ATP for its operation; therefore, if the concentration of any of these substances is too low, the pump does not function.

b. Other primary active transport systems that directly rely on the hydrolysis of ATP to transport ions across membranes are:

  (1) The **calcium (Ca^{2+}) pump** on the sarcoplasmic reticulum (SR) of muscle cells, which maintains the intracellular ionic Ca^{2+} concentration below 0.1 μmol/L

  (2) The **potassium-hydrogen (K$^+$-H$^+$) pump** of the gastric mucosa cells, which affects the secretion of H$^+$ into the stomach during the digestive process

2. Secondary active transport processes use the energy stored in the Na$^+$ concentration gradient to transport material against an energy gradient (Figure 1-5).

  a. Function. The transport of many ions and nutrients against their electrochemical energy gradients is accomplished by Na$^+$-dependent secondary active transport. For example:

  (1) Glucose and amino acids are reabsorbed from the proximal tubule and absorbed from the intestinal lumen by Na$^+$-dependent secondary active transport mechanisms.

  (2) Calcium is removed from the cytoplasm of cardiac ventricular (and other muscle) cells by a Na$^+$-dependent secondary active transporter called the Na$^+$-Ca^{2+} exchanger. This mechanism [along with the SR Ca^{2+} pump described in III C 1 b (1)] causes muscle relaxation.

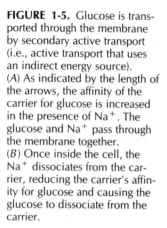 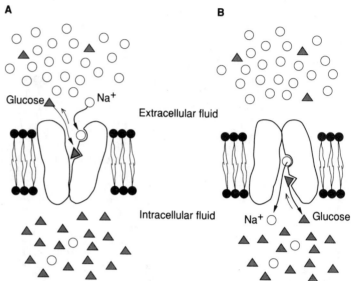

A

B

Glucose Na$^+$

Extracellular fluid

Intracellular fluid

Na$^+$ Glucose

FIGURE 1-5. Glucose is transported through the membrane by secondary active transport (i.e., active transport that uses an indirect energy source). (*A*) As indicated by the length of the arrows, the affinity of the carrier for glucose is increased in the presence of Na$^+$. The glucose and Na$^+$ pass through the membrane together. (*B*) Once inside the cell, the Na$^+$ dissociates from the carrier, reducing the carrier's affinity for glucose and causing the glucose to dissociate from the carrier.

    (3) H$^+$ produced by cellular metabolism is pumped out of cells by a Na$^+$-dependent secondary active transporter. This mechanism is important for maintaining normal intracellular pH in all cells, and in the proximal tubule of the kidney it is necessary for the reabsorption of bicarbonate.

  b. The energy stored in the Na$^+$ concentration is given by the expression:

$$\text{Energy} = z \bullet F \bullet E_m + R \bullet T \bullet \ln \frac{Na^+_{in}}{Na^+_{out}}$$

$$
\begin{aligned}
z &= \text{the valence of the ion}\\
F &= \text{the Faraday constant (96,500 coulombs/mol)}\\
E_m &= \text{the membrane potential}\\
R &= \text{the natural gas constant}\\
T &= \text{the absolute temperature (°K)}\\
Na^+_{in} \text{ and } Na^+_{out} &= \text{the intracellular and extracellular sodium concentrations, respectively}
\end{aligned}
$$

    (1) The term $z \bullet F \bullet E_m$ represents the energy produced by the resting membrane potential (see Chapter 2 II).

    (2) The term $R \bullet T \bullet \ln \dfrac{Na^+_{in}}{Na^+_{out}}$ represents the energy produced by the Na$^+$ concentration gradient.

    (3) Their sum is called the **electrochemical gradient** and is typically 14,000 joules.

  c. Process. The energy contained in the Na$^+$ electrochemical gradient causes a conformational change in the carrier molecule.

    (1) When Na$^+$ binds to the carrier molecule, the carrier molecule increases its affinity for the substance to be transported (see Figure 1-5A).

    (2) When both Na$^+$ and the substance are bound to the carrier molecule, the carrier undergoes a conformational change during which both molecules are transported across the membrane (see Figure 1-5B).

    (3) In some cases, both Na$^+$ and the substance are transported in the same direction, whereas in others they are transported in opposite directions.

 3. Carrier types

  a. Uniporters are carriers that transport a single particle in one direction, such as the facilitated diffusion of glucose.

      b. Symporters transport two particles in the same direction, such as the secondary active transport of glucose.

      c. Antiporters transport molecules in opposite directions, such as the Na^+-Ca^{2+} and Na^+-H^+ exchangers.

D. **Vesicular transport mechanisms.** Many substances are transported across the cell membrane by **endocytosis** and **exocytosis**.

    1. In **endocytosis,** extracellular material is trapped within vesicles that are formed by the invagination of the cell membrane.

      a. The endocytotic vesicle pinches off from the cell membrane and fuses with another intracellular vesicle (e.g., an endosome or lysosome), from which the ingested material is released into the ICF.

      b. In receptor-mediated endocytosis, the material to be transported first binds to a receptor, and then the receptor–substance complex is ingested by endocytosis. Iron and cholesterol are two important substances transported into cells by receptor-mediated endocytosis.

    2. In **exocytosis,** intracellular material is trapped within vesicles.

      a. The vesicles fuse with the cell membrane and release their contents to the ECF.

      b. Hormones, digestive enzymes, and synaptic transmitters (see Chapter 3 II B 3) are examples of materials transported out of the cell by exocytosis.

Chapter 2

Membrane Potentials

I. INTRODUCTION. All cells have membrane potentials.

A. **Definition.** A membrane potential (E_m) is the electrical energy difference between the inside and outside of the cell. It is measured in millivolts (mV).

B. **Origin.** The membrane potential is produced by the **separation of charge**. Charge separation can occur under several conditions.

1. When a **membrane** separating two solutions is **permeable to only one of the ions**, there is a tendency for ions to diffuse to the area of least concentration, creating an imbalance of charges.
   a. Therefore, the membrane potential **depends on the concentration gradient of the permeable ion**.
   b. The **Nernst potential** (also called the **diffusion potential** or **equilibrium potential**) is the electrical potential energy required to balance the energy in the concentration gradient (i.e., the chemical energy).
      (1) **Calculating the Nernst potential.** The Nernst potential can be calculated using the formula:

$$E_{ion} = - \frac{R \cdot T}{z \cdot F} \cdot \ln \frac{C_{in}}{C_{out}}, \text{ where}$$

$$
\begin{aligned}
E_{ion} &= \text{ the electrical potential (mV)} \\
R &= \text{ the natural gas constant} \\
T &= \text{ the absolute temperature (°K)} \\
z &= \text{ the valence of the ion} \\
F &= \text{ the Faraday constant (96,500 coulombs/mol)} \\
C_{in} &= \text{ the concentration of the ion inside the cell (mmol/L)} \\
C_{out} &= \text{ the concentration of the ion outside the cell (mmol/L)}
\end{aligned}
$$

      (2) **Simplification.** The Nernst equation can be simplified by substituting for the constants (R, T, and F) and converting to common logarithms, yielding:

$$E_{ion} = - 61 \cdot \log \frac{C_{in}}{C_{out}}$$

The valence (z) is omitted because it is $+1$ for both Na^+ and K^+. However, if the Nernst potential for Ca^{2+} is calculated, 61 must be divided by $+2$; similarly, if the Nernst potential for chloride (Cl^-) is calculated, 61 must be divided by -1. Table 2-1 gives the Nernst potentials for selected important electrolytes.

2. When a **membrane** separating two solutions is **permeable to several, but not all, of the ions,** a **Donnan equilibrium** is established.
   a. The membrane potential energy balances the energy in the concentration gradient for all of the ions permeable to the membrane.
   b. When the solutions contain two diffusible (permeable) ions and one nondiffusible (impermeable) anion, the diffusible ions arrange themselves so that their concentration ratios are equal:

$$\frac{Na^+_{\ in}}{Na^+_{\ out}} = \frac{Cl^-_{\ out}}{Cl^-_{\ in}}, \text{ where}$$

Na^+ and Cl^- represent the concentrations of diffusible ions on either side of the membrane (Figure 2-1).

TABLE 2-1. Nernst Potential for Ions Commonly Found in Nerve and Muscle Cells

Ion	Concentration (mmol/L)		Nernst Potential (mV)
	Intracellular	Extracellular	
Na^+	15	140	+58
K^+	135	4	−92
Ca^{2+}	10^{-4}	2	+129
H^+	10^{-4}	40×10^{-6}	−24
Cl^-	4	120	−89
HCO_3^-	10	24	−23

    c. Normal cells strive to achieve **osmotic equilibrium**. There are more particles inside the cell because of the presence of intracellular proteins (i.e., nondiffusible anions), which means that the osmotic gradient would cause water to flow into the cell.

        (1) Normal cells are prevented from reaching a Donnan equilibrium by the action of the Na^+-K^+ pump, which maintains an intracellular Na^+ concentration low enough to keep the inside and outside of the cell in osmotic equilibrium.

        (2) In brain ischemia, however, decreased activity of the Na^+-K^+ pump allows the Na^+ concentration to rise. The increase in intracellular osmolarity causes water to flow into the cell, producing neuronal swelling and damage.

II. RESTING (STEADY-STATE) MEMBRANE POTENTIALS

A. **Definition.** A **resting (steady-state) membrane potential** is established in normal cells when the membrane is permeable to several ions but the Na^+-K^+ pump prevents ions from reaching equilibrium.

B. **Calculating the resting membrane potential.** The magnitude of the membrane potential is determined by the concentration gradients and permeabilities of the ions that are able to diffuse across the membrane.

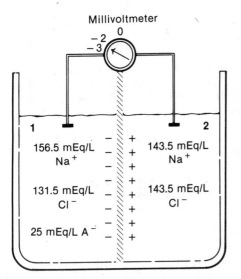

Millivoltmeter

1
156.5 mEq/L Na^+
131.5 mEq/L Cl^-
25 mEq/L A^-

2
143.5 mEq/L Na^+
143.5 mEq/L Cl^-

FIGURE 2-1. An established Donnan equilibrium. Compared with chamber 2, chamber 1 is negatively charged and has a larger solute concentration, a greater cation concentration (Na^+), and a smaller concentration of diffusible anions (Cl^-). Because the nondiffusible anion (A^-) is in chamber 1, the membrane potential is negative on the left side of the membrane and can be calculated by finding the Nernst potential for either of the diffusible ions. There are more particles in chamber 1 (because of the presence of the nondiffusible anion); therefore, the solutions are not in osmotic equilibrium.

1. Transference (chord conductance) equation

a. Variables. When a typical cell (e.g., a neuron) is in a steady-state, K^+ flows out of the cell and Na^+ flows in. The amount of leakage depends on the driving force acting on the ions and the ease with which the ions are able to cross the membrane.

(1) **Driving force.** The **driving force** is the difference between the electrical and concentration forces acting on the ion. The electrical force is represented by the membrane potential (E_m), and the concentration force is represented by the Nernst potential (E_{ion}). Thus, the driving force is expressed by the following:

$$E_m - E_{ion}$$

(2) **Conductance.** The ease with which the ion can cross the membrane is represented by the conductance (G) of the membrane to the ion. Although conductance and permeability represent similar properties of the membrane, they are not exactly the same. When describing the amount of current leaking across the membrane, the term conductance is used; however, when diffusional flux is measured, the term permeability is used.

(a) **Currents.** The currents (I) for Na^+ and K^+ can be calculated by applying **Ohm's law:**

$$I_{Na} = G_{Na} \cdot (E_m - E_{Na})$$
$$I_K = G_K \cdot (E_m - E_K)$$

(b) **Transferences.** In the steady-state, $I_{Na} = -I_K$. Setting these currents equal and opposite and solving for E_m yields:

$$E_m = E_K \cdot \frac{G_K}{G_K + G_{Na}} + E_{Na} \cdot \frac{G_{Na}}{G_K + G_{Na}}$$

The terms $G_K/(G_K + G_{Na})$ and $G_{Na}/(G_K + G_{Na})$ are the **relative membrane conductances** for K^+ and Na^+ respectively and are called the transferences for K^+ and Na^+ (T_K and T_{Na}).

b. Deriving the equation. Substituting the transference terms in the above equation yields the transference equation:

$$E_m = T_K \cdot E_K + T_{Na} \cdot E_{Na}$$

which can be used to calculate the resting membrane potential. Because the Na^+-K^+ pump is electrogenic (i.e., it pumps three Na^+ ions out of the cell while pumping two K^+ ions into the cell), it makes the cell slightly negative, and the transference equation should be written as:

$$E_m = T_K \cdot E_K + T_{Na} \cdot E_{Na} + E_{Pump}$$

However, in most cells, the magnitude of the pump potential is so small that it can be ignored.

(1) **Substituting for variables.** In a typical nerve or muscle cell, the conductance for K^+ is approximately nine times as great as the conductance for Na^+. Because the conductance for K^+ is so much higher than the conductance for Na^+, the membrane potential is primarily dependent on the concentration gradient for K^+ (E_K).

(a) **Increases in extracellular K^+,** which make the equilibrium potential for K^+ more positive, cause the membrane potential to **depolarize** (i.e., become more positive).

(b) **Decreases in extracellular K^+,** which make the equilibrium potential for K^+ more negative, cause the membrane potential to **hyperpolarize** (i.e., become more negative).

(2) **Solving the equation**

(a) Since G_K is nine times the value of G_{Na},

$$G_K = \frac{9}{9+1} = 0.9 \text{ and } G_{Na} = \frac{1}{9+1} = 0.1$$

(b) Therefore, the average resting membrane potential is:

$$E_m = (0.9 \cdot -92) + (0.1 \cdot +58) = -77 \text{ mV}$$

2. **Goldman-Hodgkin-Katz (GHK) equation.** The GHK equation is derived from the laws of diffusion rather than Ohm's law, and therefore uses permeability rather than conductance as a measure of the ease with which an ion passes through the membrane:

$$E_m = -\frac{R \cdot T}{F} \cdot \log \frac{P_K \cdot [K^+]_{in} + P_{Na} \cdot [Na^+]_{in}}{P_K \cdot [K^+]_{out} + P_{Na} \cdot [Na^+]_{out}}$$

   a. Note that if the permeability of Na^+ is zero, the GHK equation reduces to the Nernst equation for K^+.
   b. Like the transference equation, the GHK equation predicts that the membrane potential will be most influenced by the equilibrium potential for the ion to which the membrane is most permeable (or to which it has the highest conductance).

III. ACTION POTENTIALS

A. **Definition.** The action potential is a transient change in the membrane potential that conveys information within the nervous system.

B. **Origin.** Electrically excitable cells (e.g., nerve and muscle cells) generate action potentials when they are stimulated by a change in membrane potential (i.e., by the flow of current into and out of the cell).

C. **Recording action potentials**

1. **Intracellular recordings** of action potentials are made by inserting glass microelectrodes that have tip diameters of less than 0.5 μm through the cell membrane. The small tips prevent damage to the cell.
   a. **Nerve cells**
      (1) Figure 2-2A shows the recording made as a microelectrode is inserted into a nerve cell at rest. The electrical potential is 0 mV when the microelectrode is outside the cell and drops to −80 mV as soon as the microelectrode passes through the membrane and enters the intracellular fluid (ICF). Note that this is the **resting membrane potential**.
      (2) When the cell is stimulated, the microelectrode records the changes in membrane potential (i.e., the **action potential**).
   b. **Other cell types**
      (1) Figure 2-2B illustrates an action potential recorded from a cardiac ventricular muscle cell (see Chapter 10 II B).
      (2) Figure 2-2C illustrates an action potential recorded from a stomach smooth muscle cell (see Chapter 41 II C).
2. **Extracellular recordings** usually are made with metal electrodes that are placed on or near the nerve or muscle.
   a. Because these electrodes are outside the cell, they can record only changes in membrane potential (i.e., they can record action potentials but not resting potentials). They cannot record the exact magnitude or time-course of the action potentials.
   b. Extracellular recordings are useful in clinical situations when the electrical activity of excitable tissues must be monitored. For example, electroencephalograms (EEGs) are used to aid in the diagnosis of brain disease; electrocardiograms (EKGs) are used to detect damage to the heart; and electromyograms (EMGs), which are recordings from skeletal muscle, are used to aid in the diagnosis of neuropathies and myopathies.

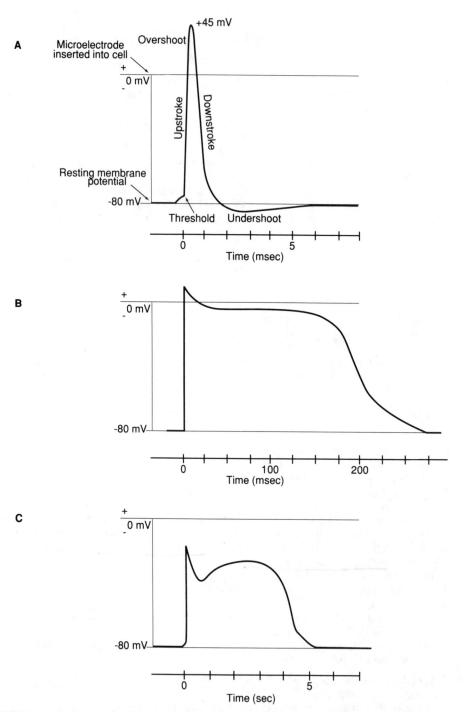

FIGURE 2-2. (*A*) Nerve cell. When a microelectrode is inserted into a nerve cell, a resting membrane potential of −80 mV is recorded. Stimulation produces an action potential, which is also recorded. Note that during upstroke, depolarization takes place as the conductance for sodium (G_{Na}) increases, and that during downstroke, repolarization occurs as the conductance for potassium (G_K) increases, allowing K^+ to flow out of the cell. (*B*) Cardiac ventricular muscle cell. As in the nerve cell, depolarization and repolarization take place as G_{Na} and G_K, respectively, increase; however, an increased conductance for calcium ($G_{Ca^{2+}}$) maintains the plateau phase that occurs during the cardiac action potential. (*C*) Gastric antrum smooth muscle cell. In these cells, the flow of Ca^{2+} into the cell causes the upstroke phase and maintains the plateau phase. As in the nerve and cardiac cells, an increased G_K and the flow of K^+ out of the cell cause the downstroke.

D. **Phases.** The phases of action potentials produced by various cell types differ slightly, but in all action potentials, the change in membrane potential is caused by the flow of current through ion-specific channels that open or close in response to changes in membrane potential. The phases of the **nerve cell** action potential are discussed here; detailed information about the cardiac and gastric action potentials is found in Chapters 10 and 41, respectively.

1. **Threshold.** Excitable cells undergo rapid depolarization if the membrane potential is reduced to a critical level (i.e., the threshold potential).
   a. Once the threshold potential is reached, the remainder of the depolarization is spontaneous.
   b. The action potential is an **all-or-none response** to a stimulus; that is, if the stimulus is strong enough to reach threshold, the changes in membrane potential that characterize the action potential are always the same.

2. **Upstroke.** The rapid depolarization of the membrane after threshold is reached is the **depolarization phase,** or upstroke, of the action potential. **The upstroke is produced by the flow of Na^+ into the cell.**

3. **Overshoot.** The portion of the action potential during which the membrane is positive is the overshoot. The peak of the action potential is the overshoot potential.

4. **Downstroke.** The rapid return of the membrane toward its resting potential is the **repolarization phase,** or downstroke, of the action potential. **The downstroke is produced by the flow of K^+ out of the cell.**

5. **Undershoot.** The membrane potential becomes more negative than its resting value at the end of the action potential. This is the **hyperpolarization phase,** or undershoot, of the action potential.

E. **Activation of the action potential**

1. **Ion-specific channels.** The all-or-none characteristic of the action potential is based on the behavior of the regulatory gates that cover the ion-specific channels present in electrically excitable cells. In nerve cells, there are two channels: one specific for Na^+ ions and the other specific for K^+ ions.
   a. **General composition.** The ion-specific channels are transmembrane integral proteins (Figure 2-3) that form an aqueous pore through the membrane. They are composed of several thousand amino acids. Each channel contains:
      (1) Selectivity filters, which give the channel its ion-selective characteristics
      (2) Gating particles, which open and close the channels
   b. **Regulation of the Na^+ channel.** The Na^+ channel has **two gating particles** (an **m gate** and an **h gate**).
      (1) The **m gate** covers the **extracellular side** of the Na^+ channel, and the **h gate** covers the **intracellular side** of the Na^+ channel.
      (2) **Both** the m and the h gates **must be open for Na^+ to flow through the Na^+ channel.**
         (a) When the **m gate** is **open,** the channel is said to be **activated.**
         (b) When the **h gate** is **closed,** the channel is said to be **inactivated.**
   c. **Regulation of the K^+ channel.** The K^+ channel has one gating particle, the **n gate.**
      (1) The **n gate** covers the **extracellular side** of the K^+ channel.
      (2) The n gate must be open for K^+ to flow through the K^+ channel.
         (a) When the **n gate** is **open,** the K^+ channel is **activated.**
         (b) The K^+ channel **does not have an inactivation gate.**
   d. **Recording channel activity.** The opening and closing of the channels can be recorded with **patch electrodes,** which are small (1–5 μm) glass capillary tubes with smooth edges that are attached to the membrane by applying suction to the electrode. One or more channels are typically contained in the patch of membrane covered by the patch electrode (Figure 2-4).

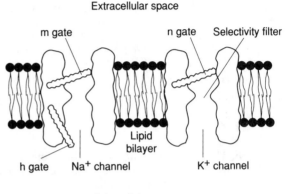

FIGURE 2-3. The gates that cover the Na$^+$ and K$^+$ channels. The negative resting potential tends to keep the m and n gates closed and the h gate opened.

2. **Voltage and time dependency.** Changes in the membrane potential cause the opening and closing of the gates in a time-dependent fashion. This phenomenon is called the **voltage and time dependency of the gating particles**.
   a. The **voltage- and time-dependent nature** of the gating particles produces an action potential when the nerve membrane is depolarized to threshold.
      (1) When the membrane is at rest (-70 to -90 mV), almost all of the channels are closed (see Figure 2-3).
         (a) The Na$^+$ channel is closed by the m gate. However, the h gate is in the open position. In this condition, the Na$^+$ channel is described as being closed but available for excitation.
         (b) The K$^+$ channel is closed by the n gate.
         (c) Although almost all of the channels are closed, there are approximately 10 times as many open K$^+$ channels as open Na$^+$ channels; therefore, the membrane potential is close to E$_K$.
      (2) When the membrane is depolarized to threshold (approximately -65 to -60 mV), a **positive feedback** (regenerative) **process** produces the all-or-none upstroke of the action potential. That is, the response of the membrane to the stimulus (the opening of the m gates) causes an effect (membrane depolarization) that produces an even greater response (Figure 2-5).
      (3) The downstroke follows the upstroke as part of the all-or-none response.
         (a) Depolarization of the membrane causes the inactivation of the Na$^+$ channels, and the flow of Na$^+$ into the cell stops.
         (b) Depolarization of the membrane activates the K$^+$ channels, and K$^+$ flows out of the cell, causing the membrane to repolarize.
      (4) The undershoot also occurs as part of the all-or-none response.
         (a) The n gates close slowly as the membrane repolarizes; therefore, the K$^+$ conductance (G$_K$) is higher at the end of the action potential than during the resting state. The high G$_K$ causes the membrane to hyperpolarize (i.e., become more negative than the resting membrane potential).
         (b) Eventually, the n gates close and the membrane potential returns to its resting level.
   b. The **time-dependent nature** of the gating particles is essential for the production of the all-or-none action potential.
      (1) If the h gates closed as rapidly as the m gates opened (i.e., if the Na$^+$ channel was inactivated and activated at the same time), Na$^+$ could not flow into the cell.
      (2) Similarly, if the n gates opened as fast as the m gates opened (i.e., if Na$^+$ and K$^+$ activation occurred at the same time), the upstroke could not occur.
   c. **The refractory period** (i.e., an interval during which it is more difficult to elicit an action potential) is produced by the behavior of the voltage and time-dependent gating particles (Figure 2-6). There are two refractory periods.

A

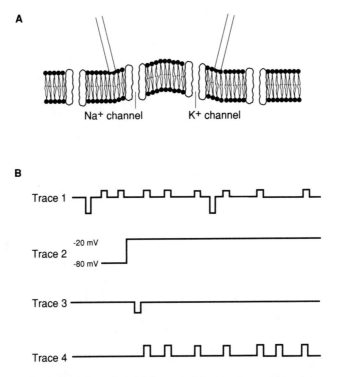

Na+ channel   K+ channel

B

Trace 1

Trace 2   -20 mV
           -80 mV

Trace 3

Trace 4

FIGURE 2-4. (*A*) The patch electrode is sealed tightly against the membrane by applying a partial vacuum to the electrode. (*B*) The currents passing through the Na$^+$ and K$^+$ channels underneath the patch electrode are illustrated in *trace 1*. The channels open rapidly, remain open for approximately 1 millisecond, and then close rapidly. The inward flow of Na$^+$ is shown as a downward current. It is larger than the current generated by the outward flow of K$^+$ because the driving force on Na$^+$ ($E_m - E_{Na} = -70 + -60$, or -130 mV) is larger than the driving force on K$^+$ ($E_m - E_K = -70 - -90$, or $+20$ mV). Because the K$^+$ channels open more often at the resting membrane potential, the resting K$^+$ conductance is greater than the resting Na$^+$ conductance. By alternately blocking the K$^+$ and Na$^+$ channels, the response of each channel to membrane depolarization can be studied in isolation. When the membrane potential is changed to -20 mV (*trace 2*), the Na$^+$ channel opens only once (*trace 3*) because the h gate inactivates the channel. The K$^+$ channel opens after the Na$^+$ channel (*trace 4*) closes. The K$^+$ channel is able to open repetitively because it does not have an h gate and, therefore, is not inactivated.

(1) **Absolute refractory period.** During this interval, another action potential cannot be elicited, regardless of the strength of the stimulus.
  (a) The absolute refractory period begins at the start of the upstroke and extends into the downstroke.
    (i) During the upstroke, a second action potential cannot occur because the m gates are opening as fast as possible (see Figure 2-6A).
    (ii) During the early portion of the downstroke, an action potential cannot occur because the Na$^+$ channels are inactivated by the h gates (see Figure 2-6B).

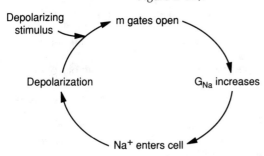

Depolarizing stimulus → m gates open

Depolarization G_{Na} increases

Na$^+$ enters cell

FIGURE 2-5. When the cell is depolarized to threshold, Na$^+$ channels open, causing an increase in the conductance for Na$^+$ (G_{Na}). This allows Na$^+$ to enter the cell, causing further depolarization. The positive feedback system represented by this cycle is responsible for the upstroke of the action potential.

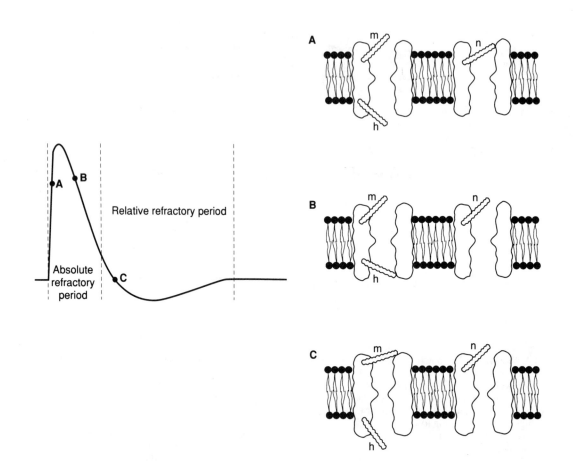

FIGURE 2-6. Diagram illustrating how the position of the Na$^+$ and K$^+$ gates changes during the action potential. (*A*) Both the m and h gates are opened during the upstroke of the action potential, allowing Na$^+$ to flow into the cell. (*B*) The closing of the h gates (inactivation) stops the flow of Na$^+$ into the cell. The opening of the n gates allows K$^+$ to flow out of the cell and initiates downstroke. (*C*) During the undershoot, the n gates are open, the m gates are closed, and the h gates are open. The high K$^+$ conductance causes the cell to hyperpolarize.

    (b) The absolute refractory period ends when the number of inactivated Na$^+$ channels is reduced by repolarization and another action potential can be initiated. The Na$^+$ channels are once again available for excitation when the h gates open during the downstroke.

  (2) Relative refractory period. During this interval, a second action potential can be elicited if the stimulus is sufficient. The stimulus must be greater than normal because some Na$^+$ channels are still inactivated and more K$^+$ channels than normal are still open (see Figure 2-6C).

    (a) The relative refractory period begins when the absolute refractory period ends.

    (b) The action potential elicited during this interval has a lower upstroke velocity and a lower overshoot potential than does the normal action potential.

       (i) These changes result from the increased number of inactivated Na$^+$ channels and activated K$^+$ channels that exist during the relative refractory period compared with the resting state.

(ii) The changes do not violate the all-or-none principle of the action potential but demand that the principle be revised: If a stimulus is sufficient to bring the membrane to threshold, the strength of the stimulus will not affect the magnitude and time-course of the action potential.

F. **Conduction of the action potential** occurs along the axon.

1. **Propagation.** The action potential generated at one location on the axon acts as a stimulus for the production of an action potential on the adjacent region of the axon (Figure 2-7). The magnitude of the action potential does not change as it is conducted along the axon because new action potentials are being generated constantly. This is different from the spread of an electrotonic potential (see III G), which diminishes in size along the axon.

2. The speed of propagation depends on the **type of fiber**. Propagation is faster in myelinated fibers than it is in unmyelinated fibers.
   a. In **myelinated fibers,** action potentials are generated only at the nodes of Ranvier, the region of the axonal cell membrane exposed to the extracellular fluid (ECF).
   (1) The nodes occur every 100–500 μm. In general, as the diameter of the axon increases, the internodal distance increases. The membrane area between the nodes is covered by an insulating sheath of myelin formed from Schwann cell membranes.
   (2) Because the action potential appears to jump from node to node, this type of propagation is called **saltatory** ("jumping" or "dancing") **propagation**.
   b. In **unmyelinated fibers,** new action potentials must be generated at each contiguous patch of membrane.

3. The speed of propagation is proportional to the **fiber diameter.**
   a. In **myelinated fibers,** the speed of propagation is approximately six times the fiber diameter. The diameter of myelinated fibers ranges from 1–20 μm; therefore, propagation velocities vary from **6–120 m/sec**.
   b. In **unmyelinated fibers,** the speed of propagation is proportional to the square root of the diameter. The largest unmyelinated fibers are approximately 1 μm in diameter; their action potentials propagate at a velocity of approximately **1 m/sec**.

G. **Electrotonic potentials** are passive changes in membrane potential that do not propagate.

1. **Types**
   a. A **local (subthreshold) response** is produced when a stimulus does not open enough m gates to elicit an action potential. Because of the small number of open m gates, the amount of Na$^+$ entering the cell is insufficient to initiate a positive feedback cycle.
   (1) The local response is **graded;** that is, the magnitude and duration of the response vary with the size and strength of the stimulus.
   (2) The local response is **nonpropagated;** that is, its magnitude is insufficient to generate another local response.

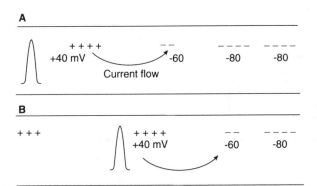

FIGURE 2-7. (*A*) The intracellular positive potential (+40 mV) produced during the overshoot of the action potential causes current to flow toward the negative, resting portion of the axon. The flow of current acts as a stimulus that depolarizes the axon toward threshold. (*B*) When threshold is achieved, an action potential is elicited and the entire process is repeated, causing propagation of the action potential along the axon.

b. **Other graded, nonpropagated responses** are produced on nerve and muscle membranes.
   (1) For example, the generator and receptor potentials produced by sensory stimuli on receptors (see Chapter 5 II C 2–3) and the excitatory and inhibitory potentials produced by neurotransmitters on synaptic membranes [see Chapter 3 II C 2 b (2)] are graded and nonpropagated.
   (2) The receptor and synaptic potentials differ from the local response in that the membrane channels producing them are not voltage- or time-dependent.

2. **Cable properties of the membrane.** The membrane changes produced by graded, nonpropagated responses spread passively, or electrotonically, along the membrane. Cable properties of the membrane determine the time-course and voltage changes of an electrotonic potential. The electrical equivalent of the cable properties of an axon is illustrated in Figure 2-8A.
   a. **Time constant.** When a current is applied to a membrane by a stimulus, a charge is added to the capacitor, causing a potential difference to develop across the membrane.
      (1) The potential difference, V_t, develops at a rate that is determined using the equation:

$$V_t = V_{max} \cdot (1 - e^{-t/\tau}), \text{ where}$$
$$V_t = \text{voltage at time t}$$
$$V_{max} = \text{voltage of the applied stimulus}$$
$$\tau = \text{time constant (seconds)}$$

τ is also equal to $r_m \cdot c_m$, where r_m = the membrane resistance (ohms) and c_m = the membrane capacitance (farads).

FIGURE 2-8. (*A*) Equivalent circuit of an axon. Each patch of axon contains a resistor representing the conductive pathways through the membrane (r_m), a capacitor representing the lipid bilayer of the membrane (c_m), and a resistor representing the intracellular pathway for the flow of ions along the axon (r_i). (*B*) When a stimulus is applied to a nerve membrane, the membrane depolarizes exponentially according to the equation $V_t = V_{max} \cdot (1 - e^{-t/\tau})$. When t = τ (the time constant), $V_\tau = 0.63 \cdot V_{max}$. (*C*) The magnitude of depolarization decreases as the distance from the stimulus increases according to the equation $V_x = V_{max} \cdot e^{-x/\lambda}$. When the distance from the stimulus equals λ (the space constant), $V_\lambda = 0.37 \cdot V_{max}$.

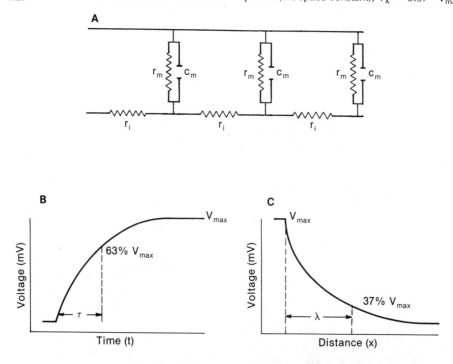

    (2) The time constant, τ, of the membrane is equal to the time required for the membrane voltage to reach approximately 63% $(1 - 1/e)$ of V_{max} (Figure 2-8B).

b. Space constant. The potential produced by the stimulus is spread passively along the membrane.

    (1) At any point along the membrane, the potential, V_x, is given by the equation:

$$V_x = V_{max} \cdot e^{-x/\lambda}$$

    where x is the distance (mm), and λ is the space constant.

    (2) The space constant, λ, is the distance from the stimulus to the point at which the applied voltage falls to approximately 37% $(1/e)$ of V_{max} (Figure 2-8C).

    (3) In an axon, the space constant is equal to r_m/r_i and, thus, increases if either of the following occurs:

        (a) The membrane resistance increases

        (b) The axoplasmic resistance decreases

c. The cable properties of a neuron play an important role in determining the ability of a stimulus to elicit an action potential and the propagation velocity of action potentials. Graded, nonpropagating responses (e.g., synaptic and receptor potentials) must spread passively from the patch of membrane where they are produced to a patch of membrane that is able to produce an action potential.

    (1) If the graded response is produced too far from the action potential–producing portion of the membrane (e.g., at the end of a long dendrite), it will decay too much to be able to depolarize the action potential–generating portion of the membrane to threshold.

    (2) If the membrane resistance is reduced (e.g., by an inhibitory synaptic transmitter), the space constant will be reduced, and the ability of an excitatory response to spread passively to the action potential–generating region of the membrane will be reduced.

    (3) If the time constant is increased, it will take longer for the action potential produced at one point along the axon to depolarize its adjacent region to threshold, and propagation velocity will slow.

        (a) In demyelinating diseases, such as multiple sclerosis, the loss of myelin increases the membrane capacitance, which increases the time constant.

        (b) The increase in the time constant causes action potential propagation to fail, producing the sensory and motor deficits characteristic of multiple sclerosis (e.g., paresis, diplopia, paresthesia).

Chapter 3

Synaptic Transmission

I. INTRODUCTION. This chapter describes synaptic transmission in the peripheral and central nervous systems (PNS and CNS).

A. **Synaptic transmission** is the process by which nerve cells communicate among themselves and with muscles and glands.

B. The **synapse** is the anatomic site where this communication occurs.

C. Most synaptic transmission is carried out by a chemical called a **neurotransmitter;** however, in some instances, synaptic transmission may be **electrical** and occur through **gap junctions**.

D. The neurotransmitter is released from a **neuron (the presynaptic cell)** and diffuses to its **target (the postsynaptic cell),** where it produces a **postsynaptic response**. The postsynaptic response may be **excitatory** or **inhibitory**.

   1. Excitation causes an action potential (if the target cell is another neuron), contraction (if the target cell is a muscle), or secretion (if the target cell is a gland).

   2. Inhibition reduces or blocks the activity of the postsynaptic cell.

II. NEUROMUSCULAR TRANSMISSION is synaptic transmission between an alpha motoneuron and a skeletal muscle fiber that occurs at the neuromuscular junction.

A. **Physiologic anatomy.** Figure 3-1 illustrates the anatomic structure of the neuromuscular junction.

   1. Light microscopic appearance (see Figure 3-1A)
      a. The **alpha motoneuron branches** as it approaches the muscle, sending axon terminals to several skeletal muscle fibers.
         (1) The number of skeletal muscle fibers innervated by an alpha motoneuron depends on the type of muscle fiber involved.
            (a) Alpha motoneurons innervating large muscles used primarily for strength or postural control innervate hundreds to thousands of skeletal muscle fibers.
            (b) Alpha motoneurons innervating muscles used for precision movements innervate just a few skeletal muscle fibers.
         (2) Each skeletal muscle fiber receives only one axon terminal.
      b. The **axon terminal** lies in a groove called the **synaptic trough,** which is formed by an invagination of the skeletal muscle fiber.
      c. Synaptic transmission occurs at the **end-plate region** (i.e., the postsynaptic membranes) of the skeletal muscle fiber.

   2. Electron microscopic appearance. Figure 3-1B illustrates the details of the presynaptic and postsynaptic (end-plate) membranes.
      a. **Synaptic vesicles** (approximately 50 nm in diameter) containing the neurotransmitter **acetylcholine (ACh)** are located in the presynaptic nerve terminal. They are concentrated around specialized presynaptic membrane structures called the **active zones**.
      b. The **synaptic cleft** (approximately 60 nm wide) is filled with an amorphous network of connective tissue called the basal lamina, in which the enzyme **acetylcholinesterase (AChEase)** is bound. AChEase degrades ACh after it has produced its effect on the postsynaptic membrane of the skeletal muscle fiber.

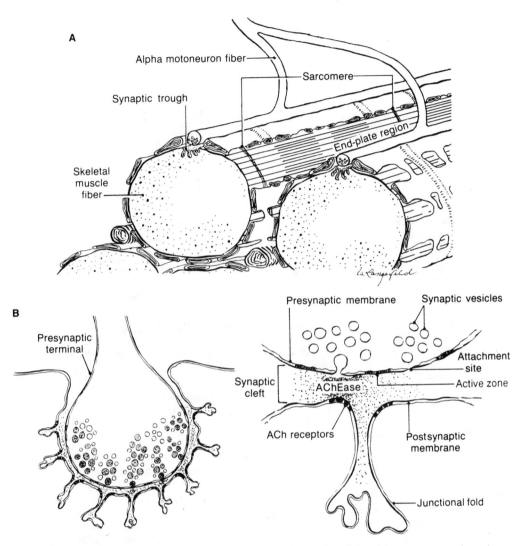

FIGURE 3-1. (*A*) Light microscopic view of a neuromuscular junction. The alpha motoneuron branches as it reaches the muscle, and each branch forms a synapse with a single muscle fiber at the end-plate region of the muscle. The axon terminal lies in the synaptic trough. Junctional folds increase the surface area of the postsynaptic membrane. (*B*) Electron microscopic view showing the synaptic vesicles within the presynaptic terminal, the acetylcholinesterase (*AChEase*) within the synaptic cleft, and the postsynaptic receptor sites. *ACh* = acetylcholine.

    c. The **postsynaptic membrane** contains numerous **junctional folds,** which are membrane invaginations located opposite the active zones. The receptor sites for ACh are found on the postsynaptic membanes near the junctional folds.

B. **Synthesis, storage, and release of ACh**

  1. Synthesis. ACh is synthesized in the presynaptic nerve terminal from **choline** and **acetyl coenzyme A (acetyl-CoA)** by **choline acetyltransferase (CAT).**

  2. Storage. Newly synthesized ACh is stored within the synaptic vesicles. Approximately 5000–10,000 molecules of ACh are stored within each vesicle.

  3. Release. Spontaneous release of ACh occurs by **exocytosis** (Figure 3-2) whenever a vesicle binds to an attachment site on one of the active zones.

    a. The vesicle fuses with the presynaptic membrane, exposing its contents to the extracellular fluid (ECF).

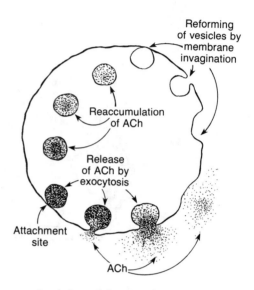

Reforming of vesicles by membrane invagination

Reaccumulation of ACh

Release of ACh by exocytosis

Attachment site

ACh

FIGURE 3-2. The life cycle of a synaptic vesicle. After binding to its attachment site, the vesicle releases its transmitter, acetylcholine (*ACh*), via exocytosis and then merges with the membrane. Later, a new vesicle is formed from invaginations of the presynaptic membrane. These vesicles are filled with transmitter and can be reused.

    b. ACh diffuses out of the vesicle into the synaptic cleft, and the vesicle merges with the presynaptic membrane.

    c. Later, new vesicles are formed from the presynaptic membrane by **endocytosis**.

        (1) The presynaptic membrane forms invaginations that eventually bud off to form new vesicles.

        (2) The new vesicles are refilled with ACh and, once again, are available to release their contents into the ECF.

C. **Events in synaptic transmission**

    1. Neurotransmitter release is triggered by the **depolarization** of the presynaptic nerve terminal. Depolarization occurs when an action potential propagates into the nerve terminal.

        a. Depolarization of the nerve terminal by the action potential causes Ca^{2+} channels (which are located in the active zones) to open, allowing Ca^{2+} to enter the cell down its electrochemical gradient.

        b. The increase in intracellular Ca^{2+} concentration causes approximately 200–300 vesicles to bind to attachment sites and release their contents into the synaptic cleft. The fusion of synaptic vesicles with the presynaptic terminal may be mediated by a protein called **synapsin I**.

            (1) When synapsin I binds to the synaptic vesicle, it inhibits exocytosis.

            (2) When Ca^{2+} enters the nerve terminal, it initiates the phosphorylation of synapsin I.

            (3) Phosphorylated synapsin I dissociates from the synaptic vesicle, allowing exocytosis to occur.

    2. Postsynaptic response. When ACh binds to the postsynaptic receptor, it causes a depolarization of the postsynaptic membrane called the **end-plate potential (EPP)**.

        a. The ACh receptor

            (1) ACh receptors in the end-plate region of skeletal muscle are called **nicotinic** (as opposed to muscarinic) receptors, because they are stimulated by nicotine (as well as ACh) and inhibited by curare.

            (2) The ACh receptor (Figure 3-3) is a transmembrane protein consisting of five subunits that form an aqueous channel within the lipid bilayer.

                (a) Two of the subunits, called α subunits, contain binding sites for ACh.

                (b) When the two α subunits are occupied by ACh, the proteins undergo a conformational change that opens a gate within the channel.

                (c) The channel is equally permeable to Na^+ and K^+. Unlike the Na^+ and K^+ channels found on electrically excitable membranes (which are activated by

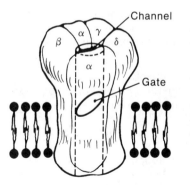

Channel

Gate

FIGURE 3-3. Acetylcholine (ACh) receptor. The receptor contains five subunits, two of which (the α subunits) contain binding sites for ACh. When both subunits are occupied, the channel's gate opens, allowing Na^+ and K^+ to pass through the membrane. (Adapted with permission from Anholdt R, Lindstrom J, Montal M: In *Enzymes of Biological Membranes.* Edited by Martonosi A. New York, Plenum Press, 1985, pp 335–401.)

changes in membrane voltage), the ACh-activated channel is **opened by the binding of the neurotransmitter to the receptor,** not by membrane depolarization; therefore, it is a **chemically activated channel**.

b. **Depolarization**

(1) Opening the channel causes the cell to depolarize.

 (a) When the channel is opened, Na^+ enters and K^+ leaves the cell down their respective electrochemical gradients.

 (b) The magnitude of the Na^+ current (I_{Na}) and the K^+ current (I_K) can be calculated using Ohm's law [see Chapter 2 II B 1 a (2) (a)]. Because the electrochemical gradient for Na^+ is greater than that for K^+ (and the conductances for Na^+ and K^+ are equal), the amount of Na^+ entering the cell exceeds the amount of K^+ leaving the cell, and the cell depolarizes.

(2) The EPP is a **graded response,** not an all-or-none response. The magnitude of depolarization is proportional to the number of open ACh channels.

 (a) If a **single ACh channel** is opened (as occurs when two ACh molecules bind to it—one to each α unit), the membrane will depolarize by only a few microvolts (μV).

 (b) If a **single vesicle** releases its contents of 5000–10,000 ACh molecules (i.e., a **quantum** of ACh), the membrane will depolarize by approximately 1 mV.

  (i) The small (1 mV) depolarization caused by the spontaneous release of one vesicle is called a **miniature end-plate potential (MEPP)**.

  (ii) Spontaneous release of vesicles occurs at a rate of approximately 1/sec. Thus, MEPPs occur every second or so.

  (iii) The MEPPs may be important in maintaining the integrity of the muscle fiber because denervation of skeletal muscle fibers leads to muscle atrophy.

 (c) The **200–300 vesicles** that release their contents when an action potential invades the presynaptic nerve terminal produce a depolarization of approximately 50 mV. This depolarization is the **EPP**.

(3) **The reversal potential** is the **maximum depolarization** that can occur on the end-plate membrane.

 (a) The reversal potential is the potential at which no net current flows through the channel (i.e., when the Na^+ and K^+ currents are equal and opposite to each other).

 (b) These currents are equal and opposite to each other when the membrane potential is −16 mV. [This can be verified by substituting the appropriate values of E_{Na} (+58 mV) and E_K (−92 mV) into the transference equation (see Chapter 2 II B 1 b).] Recall that the membrane potential calculated with the transference equation is the membrane potential at which inward and outward currents are equal and opposite.

 (c) The reversal potential is analogous to the equilibrium potential because it is the potential at which there is no net current through the channel.

  (i) At the reversal potential, no net current flows because the inward (in this case, Na^+) currents and the outward (in this case, K^+) currents are equal and opposite.

      (ii) At the equilibrium potential, however, no net current flows because the driving force for the ion down its electrical gradient is equal and opposite to the driving force for the ion down its concentration gradient [see Chapter 2 II B 1 a (1)].

  3. The EPP initiates an action potential on the muscle fiber membrane.

    a. There are no electrically excitable Na^+ and K^+ channels on the end-plate region; however, Na^+ and K^+ channels are located on the muscle membrane contiguous to the end-plate.

    b. The EPP depolarizes the contiguous membrane regions to threshold, and an action potential is generated.

    c. The action potential is propagated along the muscle membrane and is responsible for initiating a muscle contraction (see Chapter 4 II B).

  4. ACh is degraded rapidly.

    a. After binding to the ACh receptor, the ACh dissociates from the receptor and is hydrolyzed by the AChEase in the synaptic cleft.

    b. Degradation of ACh is necessary to prevent it from causing multiple muscle contractions.

    c. Enzymatic destruction is a unique method for inactivating the transmitter and occurs only at ACh synapses. Inactivation of the transmitter at all other synapses occurs when the transmitter diffuses out of the synaptic region or is actively transported back into the nerve terminal.

III. AUTONOMIC SYNAPTIC TRANSMISSION

A. **Organization of the autonomic nervous system (ANS)** [Figure 3-4]

  1. The ANS is divided into the **parasympathetic** and **sympathetic** systems, each of which has a **preganglionic neuron** in the **CNS** and a **postganglionic neuron** in the **PNS**.

  2. Synaptic transmission within the PNS involves several neurotransmitters (e.g., ACh, norepinephrine) and a variety of postsynaptic responses (e.g., inhibitory, excitatory), which are summarized in Table 3-1.

B. **Modes of transmission**

  1. Ganglionic transmission within the sympathetic and parasympathetic divisions of the ANS is essentially the same as that at the neuromuscular junction. ACh is released from the preganglionic (presynaptic) fiber, diffuses across the synaptic cleft, and binds to receptors on the postganglionic (postsynaptic) fiber, causing it to depolarize by opening channels that are equally permeable to K^+ and Na^+. Differences between neuromuscular transmission and ganglionic transmission include the following:

    a. Unlike the presynaptic fibers of neuromuscular transmission, a single preganglionic fiber does not release enough neurotransmitter to depolarize the postganglionic fiber to threshold. The postganglionic fiber can discharge an action potential only when there is a **summation** of the postsynaptic responses (see IV B 2).

    b. Although the ACh receptors at the synapses of the pre- and postganglionic fibers are nicotinic receptors, they are not identical to those on the skeletal muscle end-plate, because the pharmacologic agents needed to block the two receptors are different. The extreme toxicity of high concentrations of nicotine (it causes vomiting, diarrhea, diaphoresis, and high blood pressure) is primarily due to its action on the ANS.

      (1) Hexamethonium blocks ganglionic transmission.

      (2) Curare blocks neuromuscular transmission.

  2. Postganglionic (parasympathetic) transmission. ACh is the neurotransmitter released from most **parasympathetic postganglionic** fibers. The postganglionic receptors are **muscarinic** (i.e., they respond to muscarine but not to nicotine). Muscarinic receptors

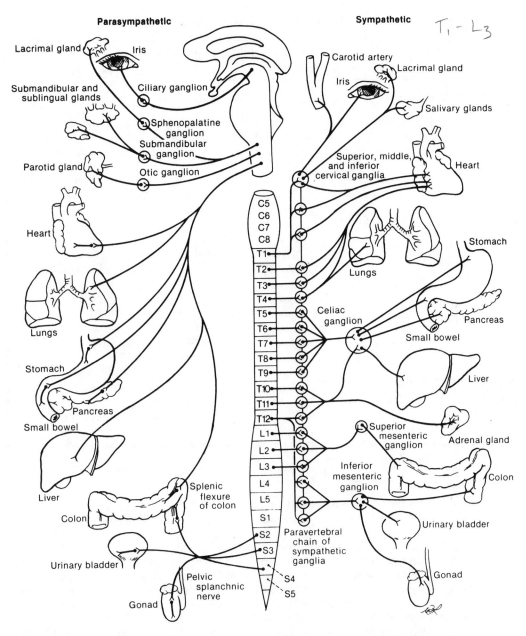

FIGURE 3-4. The autonomic nervous system (ANS). The parasympathetic division (*left*) arises from CN III, CN VII, CN IX, and CN X and from spinal cord segments S2 to S4. The sympathetic division (*right*) arises from spinal cord segments T1 to L3. (Reprinted with permission from NMS *Neuroanatomy*. Malvern, PA, Harwal Publishing, 1988, p 88.)

are blocked by atropine. ACh can have either an excitatory or inhibitory effect, depending on the postsynaptic receptor that is activated.

a. **Excitatory effects** of parasympathetic fibers are produced on a variety of **smooth muscles** (e.g., those within the stomach, intestine, bladder, and bronchi) and on **glands**. ACh can produce its excitatory effect by a variety of mechanisms.

   (1) ACh can bind to receptors that cause the membrane to depolarize, much in the **same way** ACh receptors effect the **depolarization of the skeletal muscle endplate**.

TABLE 3-1. Neurotransmitters Within the Peripheral Nervous System (PNS)

Transmitter	Receptor	Presynaptic Neuron	Postsynaptic Cell	Response
ACh	Nicotinic	Alpha motoneuron	Skeletal muscle fiber	Contraction
	Muscarinic	Postganglionic parasympathetic	Secretory cells	Secretion
			SA and AV nodes	Slow pacemaker activity and conduction
			Smooth muscle of GI tract	Relaxation of sphincters; contraction of other muscle
Norepinephrine, epinephrine	α	Postganglionic sympathetic	Smooth muscle of GI tract	Contraction of sphincters; relaxation of other muscle
			Vascular smooth muscle	Contraction
	β		Ventricular muscle	Contraction
			Bronchial and vascular smooth muscle	Relaxation
			Adipose tissue	Fatty acid mobilization

The table lists some of the most important actions of ACh and the catecholamines (e.g., norepinephrine and epinephrine) within the PNS. Although muscarinic, α, and β receptors all have several subtypes, these have not been listed in the table. ACh = acetylcholine; AV = atrioventricular; GI = gastrointestinal; SA = sinoatrial.

    (2) ACh can bind to receptors that **increase Ca^{2+} conductance**. The increase in Ca^{2+} conductance does not produce a major effect on membrane potential. However, the Ca^{2+} entering the cell through the channels opened by ACh can be used to initiate contraction in smooth muscles (see Chapter 4 IV B 2 b).

    (3) ACh can bind to receptors that activate the membrane-bound protein **guanosine triphosphate (GTP) binding protein,** or **G protein**. When the G protein is activated, it initiates a series of membrane and intracellular events that lead to muscle contraction (Figure 3-5; see also Chapter 4 II B).

  b. Inhibitory effects of parasympathetic fibers are produced on the **heart**.

    (1) ACh acts primarily on the pacemaker regions of the sinoatrial (SA) node to decrease the heart rate and the atrioventricular (AV) node to slow conduction of the action potential from the atria to the ventricles (see Chapter 10 III A 1).

    (2) When ACh binds to a receptor on the SA node, it slows the rate of pacemaker activity by two mechanisms:

      (a) Hyperpolarization. ACh causes a K^+ channel to open. K^+ flows out of the cell (down its electrochemical gradient) and the cell hyperpolarizes. The hyperpolarization acts to decrease the rate of pacemaker activity and to slow the conduction of the action potential through the AV node.

      (b) Direct inhibition. ACh acts directly to inhibit the pacemaker channel, slowing the rate of spontaneous depolarization.

  3. Postganglionic (sympathetic) transmission. Norepinephrine is the neurotransmitter released from most **sympathetic postganglionic** fibers. Norepinephrine can have an excitatory or inhibitory effect, depending on the type and location of the receptor activated.

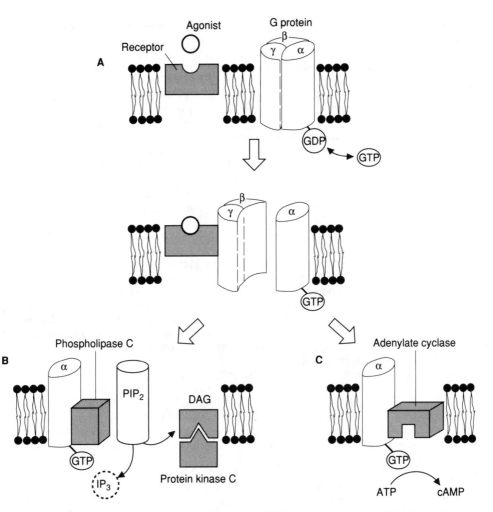

FIGURE 3-5. (A) When a neurotransmitter or other agonist binds to a receptor that is linked to a G protein–mediated second messenger system, the receptor diffuses within the membrane until it encounters a G protein complex. When the activated receptor binds to the G protein, it induces the G protein to exchange the guanosine diphosphate (*GDP*) molecule that is bound to the α subunit for a guanosine triphosphate (*GTP*) molecule. The presence of GTP causes the α subunit to separate from the G protein. The α subunit diffuses within the membrane until it encounters the enzyme that initiates the second messenger response. The G protein is inactivated when GTP is converted to GDP. (B) If the agonist binds to a muscarinic or α_1 receptor, the α subunit of the G protein activates phospholipase C, a membrane-bound lipase. The phospholipase C, in turn, hydrolyzes phosphatidylinositol diphosphate (*PIP$_2$*) into inositol triphosphate (*IP$_3$*) and diacylglycerol (*DAG*). IP$_3$ enters the cytoplasm, where it liberates Ca^{2+} from internal stores. The increase in intracellular Ca^{2+} leads to muscle contraction. DAG remains in the membrane and activates protein kinase C, a cytoplasmic enzyme that activates a series of cytoplasmic proteins by phosphorylating them. (C) If the agonist binds to a β receptor, the α subunit of the G protein activates adenylate cyclase. The adenylate cyclase catalyzes the conversion of adenosine triphosphate (*ATP*) to cyclic adenosine 3′,5′-monophosphate (*cAMP*). cAMP activates protein kinase A, which catalyzes the phosphorylation of cellular proteins.

    a. Norepinephrine receptors are either β or α receptors.
        (1) β-Adrenergic receptors. When norepinephrine binds to the β receptors, it activates a G protein similar to that activated by ACh (see Figure 3-5); however, in this case, the activated G protein activates a different membrane-bound protein: **adenylate cyclase**.
            (a) Adenylate cyclase stimulates the formation of **cyclic adenosine 3′,5′-monophosphate (cAMP)** from adenosine triphosphate (ATP). The G protein that stimulates the formation of cAMP is called the G$_s$ protein.

(b) cAMP activates the cytoplasmic enzyme **protein kinase A** (or **cAMP-dependent protein kinase**), which, in turn, produces a variety of physiologic responses by phosphorylating the amino acids serine and threonine on an assortment of intracellular proteins.

(i) **Heart**

The **phosphorylation of the Ca^{2+} channels** in cardiac ventricular muscle cells by **protein kinase A** increases the amount of Ca^{2+} entering the cell with each action potential and, thus, **increases the force of contraction** (see Chapter 4 III B 2 a).

The **phosphorylation of phospholamban** [a protein located on the sarcoplasmic reticulum (SR)] by **protein kinase A** enhances the activity of the sarcoplasmic reticular Ca^{2+} pump. Although phospholamban is normally inhibitory to the sarcoplasmic reticular Ca^{2+} pump, when phosphorylation occurs, the inhibition is removed. The increased activity of the sarcoplasmic reticular Ca^{2+} pump removes Ca^{2+} from the cytoplasm of the cardiac ventricular muscle cells, **reducing the duration of the contraction**.

The **phosphorylation of membrane channel proteins** increases the rate of pacemaker depolarization and shortens the duration of the action potential by activating the K^+ channels responsible for phase 3 repolarization (see Chapter 10 II B 1 d), thus **increasing the heart rate**.

(ii) **Lungs.** In bronchiole smooth muscle cells, the **phosphorylation of phospholamban** (and the resulting increase in sarcoplasmic reticular Ca^{2+} pump activity) **relaxes the bronchiole smooth muscle**.

(2) **α-Adrenergic receptors**

(a) When norepinephrine binds to the α_2 **receptor** (a subtype of the α-adrenergic receptors), it activates a G protein that inhibits adenylate cyclase, thus reducing the amount of cAMP in the cell. The G protein that inhibits the formation of cAMP is called the G_i **protein**.

(b) When norepinephrine binds to the α_1 **receptor** (another subtype of the α-adrenergic receptors), it activates a G protein that, similar to its action at the muscarinic receptor, results in the formation of IP_3 and diacylglycerol (DAG). Activation of the α_1 receptor by sympathetic nerve activity causes constriction of vascular smooth muscle.

b. Adrenergic receptors can be distinguished by the types of drugs that activate and inhibit them.

(1) **α-Adrenergic receptors** are **activated** preferentially **by epinephrine** and are **blocked by phenoxybenzamine**.

(2) **β-Adrenergic receptors** are **activated** preferentially **by isoproterenol** and are **blocked by propranolol**.

c. **Inactivation of norepinephrine** occurs by active transport into the nerve terminal and diffusion out of the synaptic cleft. In contrast to ACh, enzymatic degradation is not an important mechanism for inactivating norepinephrine.

C. The **enteric nervous system** is an independent component of the ANS. It is composed of the ganglia found within the wall of the gastrointestinal (GI) tract and is responsible for coordinating the activity of GI smooth muscle and gastric secretions.

1. The enteric nervous system **receives synaptic input from the sympathetic and parasympathetic postganglionic fibers**.

2. **Known active neurotransmitters** used by the enteric nervous system include serotonin, the enkephalins and endorphins, somatostatin, vasoactive intestinal peptides (VIPs), the purines ATP and adenosine, and nitric oxide; however, many active neurotransmitters have not yet been identified.

a. **Nitric oxide** is a gas that is synthesized from the amino acid L-arginine by **nitric oxide synthetase**.

(1) Nitric oxide synthetase activates guanylate cyclase, which stimulates the formation of cyclic guanosine monophosphate (cGMP) from GTP.

     (2) Nitric oxide is **unique** among neurotransmitters because, unlike other neuro-transmitters, which are stored in and released from synaptic vesicles, nitric oxide is **synthesized as needed** and **diffuses from the cytoplasm of the postsynaptic cell**.

    b. In the **GI system,** nitric oxide is found within the myenteric plexus and acts (along with the VIPs) as the **major relaxer of smooth muscle contraction**.

IV. CNS SYNAPTIC TRANSMISSION. An enormous variety of synaptic mechanisms occur within the CNS.

A. **Electrical synaptic transmission.** In a few locations (e.g., within the retina and olfactory bulb), synaptic transmission is accomplished by the **passive electrotonic spread of current between two cells**.

  1. Specialized junctions called **gap junctions** allow the spread of current between two cells.

    a. Only 2 nm separate the pre- and postsynaptic membranes at the site of gap junctions.

    b. Gap junctions are formed by **membrane bridges** that are constructed from integral membrane proteins called **connexin**.

      (1) An **aqueous channel** is formed in the membrane by six connexin molecules (Figure 3-6A).

      (2) The channel in one cell merges with a channel in the membrane of another cell to form the gap junction (Figure 3-6B), enabling small molecules and ions to pass from one cell to the other and establishing **cytoplasmic continuity**.

        (a) When an action potential propagating along the membrane in one cell reaches the gap junction, an electrical current flows passively through the gap from one cell to another.

        (b) Electrical current can pass through the gap in both directions, allowing either cell to serve as the pre- or postsynaptic cell.

FIGURE 3-6. A gap junction. (*A*) *Top view,* showing the six connexin molecules and the channel they form. The hexagonal unit is called a connexon. (*B*) *Side view,* showing how the connexin proteins from two cells align at the gap junction to form a channel that permits the passage of water-soluble molecules and electrical currents from cell to cell. The extracellular space between the two lipid bilayers measures approximately 2 nm; the channel itself is approximately 1.5 nm wide.

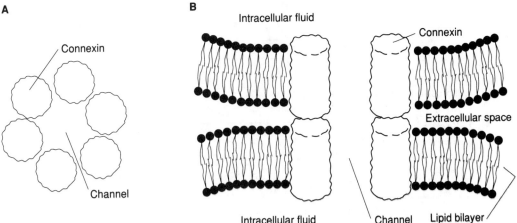

2. Although gap junctions are not common in CNS synaptic transmission, they play an important role in coordinating **muscle contraction in the heart and viscera**. Gap junctions rapidly transmit an action potential that is generated in one cell to all of the other cells within the organ, permitting the entire tissue to act as a **syncytium** and contract in a coordinated fashion.

B. **Chemical synaptic transmission**

1. Most synapses within the CNS use chemical neurotransmitters.

    a. The synaptic transmitter is synthesized in the nerve terminal, stored in vesicles, and released by exocytosis when an action potential invades the nerve terminal.

    b. After being released from the presynaptic terminal, the transmitter diffuses across the synaptic cleft, binds to a postsynaptic receptor, and causes the opening of channels through which ions can flow. Both **excitatory** and **inhibitory receptors** exist on the postsynaptic cell.

        (1) **Excitatory neurotransmitters** produce a depolarization of the postsynaptic membrane called the **excitatory postsynaptic potential (EPSP)**. The most common excitatory neurotransmitter within the CNS is **glutamate**.

        (2) **Inhibitory neurotransmitters** produce a hyperpolarization of the postsynaptic membrane called the **inhibitory postsynaptic potential (IPSP)**. The most common inhibitory neurotransmitters within the CNS are **glycine** and **γ-aminobutyric acid (GABA)**.

    c. The transmitter is inactivated in one of three ways.

        (1) It diffuses out of the synaptic cleft.

        (2) It is actively transported into the presynaptic terminal.

        (3) It is enzymatically degraded (if the transmitter is ACh).

2. **Summation.** The postsynaptic effects of the thousands of excitatory and inhibitory synapses that converge on a single neuron within the CNS (Figure 3-7) are integrated by the process of summation.

    a. **Postsynaptic response.** Because the presynaptic terminals are only a few micrometers in diameter (and therefore can only release a few synaptic vesicles at a time), summation is required.

FIGURE 3-7. Inhibitory and excitatory synapses are formed on an alpha motoneuron. (Although synaptic junctions cover most of the soma and proximal dendrites, only a few are shown here.) When the amplitude of the summated excitatory postsynaptic potentials (*EPSPs*) exceeds the amplitude of the summated inhibitory postsynaptic potentials (*IPSPs*), the axon hillock is depolarized to threshold and an action potential is generated.

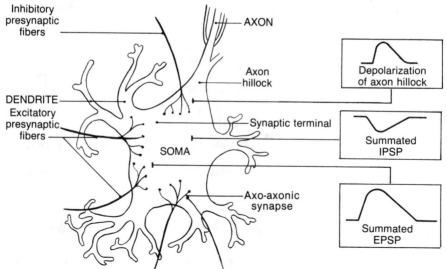

b. Summation occurs because the duration of the postsynaptic effect is relatively long and additional neurotransmitter can be released by the axon terminals before the effect of the first synaptic event has dissipated. Although the synaptic channel is open for only a few milliseconds, the membrane capacitance causes the postsynaptic response to decay more slowly.

   (1) Temporal summation occurs if another action potential invades the nerve terminal before the first postsynaptic potential has disappeared. The second postsynaptic potential adds to the first, producing a larger response.

   (2) Spatial summation occurs if several nerve terminals fire at approximately the same time.

c. The summated potentials produced by the excitatory and inhibitory neurotransmitters spread passively to the **axon hillock,** where the action potential is generated.

   (1) The threshold for producing an action potential at the axon hillock is approximately −65 mV.

   (2) If the summated EPSPs are large enough to depolarize the axon hillock to threshold, an action potential will be generated.

   (3) The summated IPSPs can prevent the axon hillock from being depolarized to threshold by hyperpolarizing the cell.

3. Presynaptic inhibition. The amount of neurotransmitter released from an axon terminal can be reduced by presynaptic inhibition.

   a. Presynaptic inhibition is produced by GABA, which is released from a neuron synapsing on the presynaptic nerve terminal. The synapse between the GABA-containing neuron and the presynaptic nerve terminal is called an **axo-axonic synapse**.

   b. GABA reduces the amount of neurotransmitter released from the nerve terminal by reducing the amount of Ca^{2+} entering the presynaptic nerve terminal during synaptic transmission. The mechanism by which the amount of Ca^{2+} is reduced depends on the type of GABA receptor present on the presynaptic nerve terminal.

   (1) When GABA binds to a receptor called the **$GABA_A$ receptor,** it opens Cl^- channels (Figure 3-8).

   (a) The opened Cl^- channels allow the negative Cl^- ion to flow into the cell.

   (b) When an action potential invades the presynaptic nerve terminal, the size of the action potential is reduced because of the increased Cl^- conductance.

   (c) Because the size of the action potential is smaller, less Ca^{2+} enters the nerve terminal, and the amount of neurotransmitter released is diminished.

   (2) When GABA binds to a receptor called a **$GABA_B$ receptor,** it activates a G protein. The G protein aids in reducing the amount of neurotransmitter that is released by acting in one of two ways.

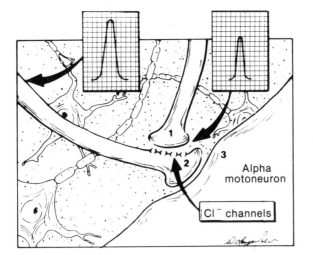

FIGURE 3-8. One way the axo-axonic synapse facilitates presynaptic inhibition. Neuron *1* releases γ-aminobutyric acid (GABA), which binds to a $GABA_A$ receptor on neuron *2*, causing Cl^- channels to open. The increased Cl^- conductance reduces the amplitude of the action potential as it approaches the nerve terminal. Because the action potential is smaller, less Ca^{2+} enters the nerve terminal, less transmitter is released from the nerve terminal, and the magnitude of the excitatory postsynaptic potential (EPSP) produced on the postsynaptic membrane of neuron *3* is reduced.

Alpha motoneuron

Cl^- channels

    (a) The G protein may open a K^+ channel that reduces the size of the action potential invading the nerve terminal by hyperpolarizing the presynaptic nerve terminal.

    (b) Alternatively, the G protein may directly block the opening of Ca^{2+} channels that normally occurs when an action potential invades the nerve terminal.

  c. In presynaptic inhibition, the excitability of the postsynaptic cell is not diminished, whereas an IPSP reduces the effectiveness of all excitatory input to a cell. Presynaptic inhibition allows a particular excitatory input to be inhibited without affecting the ability of other excitatory synapses to fire the cell.

Chapter 4

Muscle Contraction

I. INTRODUCTION

A. Muscle fibers are divided into **two types** based on their appearance in light micrographs.

1. **Striated muscle,** which includes **skeletal** and **cardiac** muscle, is characterized by alternating light and dark bands.

2. **Smooth muscle** has no distinguishing surface features.

B. Although all muscles function in a similar way, several important differences exist. In the following discussion, the structural and contractile properties of skeletal muscle are noted first, with the major differences in cardiac and smooth muscle noted afterward (Table 4-1).

II. SKELETAL MUSCLE contraction maintains posture and produces movement.

A. **Composition.** Skeletal muscle is composed of multinucleated skeletal muscle fibers (Figure 4-1). The fibers vary from approximately 10–100 μm in diameter and can be several centimeters in length.

1. **Fascicles.** The skeletal muscle fibers are grouped into fascicles of approximately 20 fibers by the **perimysium,** a connective tissue sheath that is continuous with the connective tissue surrounding the entire muscle.
   a. The perimysium is continuous with the **endomysium** surrounding each muscle fiber.
   b. The endomysium is continuous with the **sarcolemma,** a sheath that contains glycoprotein and closely envelops the true cell membrane of the muscle fiber.
   c. The tight connection between the cell membranes and the surrounding connective tissue structures enables the force developed by the muscle fibers to be transmitted effectively to the tendons.

2. **Myofibrils.** The individual skeletal muscle fibers are divided into myofibrils by a tubular network called the **sarcoplasmic reticulum (SR).**
   a. The myofibrils are approximately 1 μm in diameter and extend from one end of the muscle fiber to the other.
   b. The myofibrils are divided into functional units, or **sarcomeres,** by a transverse sheet of α-actinin protein called the **Z line.**
   c. The Z lines of neighboring myofibrils are aligned with each other so that in histologic slides, a Z line spans the entire width of the fiber.

3. **Filaments.** The myofibrils contain thick and thin filaments composed of contractile proteins.
   a. **Thick filaments** contain the protein **myosin** and are approximately 11 nm in diameter and 1.6 μm in length.
      (1) The thick filaments are interspersed between the thin filaments at the center of the sarcomere. They are attached to the Z line by a large filamentous protein called **titin.**
      (2) Projections from the thick filaments called **cross-bridges** extend toward the thin filaments. Cross-bridges play a fundamental role in muscle contraction.
   b. **Thin filaments** contain the proteins **actin, tropomyosin,** and **troponin** and are approximately 5 nm in diameter and 1 μm in length.
      (1) One end of the thin filament is attached to the Z line, so that filaments from opposing Z lines extend longitudinally toward each other into the center of the sarcomere.

TABLE 4-1. Comparison of Muscle Types

Muscle Type	Role of Ca^{2+}	Source of Ca^{2+}	Mechanism of Ca^{2+} Mobilization	Regulation of Force
Skeletal	Initiates contraction by binding to troponin	Intracellular from SR. Enough Ca^{2+} is released to activate all muscle protein	Depolarization of T-tubule	Summation, recruitment, and preload are varied to vary force
Cardiac	Initiates contraction by binding to troponin	Intracellular from SR. Extracellular through DHP receptor (L type) Ca^{2+} channels. Amount of Ca^{2+} released can be varied to vary contractile force	Ca^{2+}-induced Ca^{2+} release	Contractility and preload are varied to vary force; variations in contractility affect speed of contraction
Smooth	Activates calmodulin, which in turn activates MLCK	Intracellular from SR. Extracellular through voltage- and receptor-activated Ca^{2+} channels	IP_3 increases release of Ca^{2+}; protein kinase A increases uptake of Ca^{2+} by SR	Recruitment, summation, preload, and contractility are varied to vary force. Formation of latch-bridges reduces speed of contractility

DHP = dihydropyridine; IP_3 = inositol triphosphate; MLCK = myosin light-chain kinase; SR = sarcoplasmic reticulum.

      (2) The thin filaments appear to be held in place by a thin protein that connects their free ends.

  c. Banding pattern. The light and dark bands that characterize light microscopic views of skeletal muscle (see Figure 4-1) are created by the interdigitating thick and thin filaments.

      (1) The **light areas** on either side of the Z line are called **I bands**. They contain the **thin filaments**.

      (2) The **dark areas** in the center of the sarcomere are called **A bands**. The A bands contain the **thick filaments,** but the thin filaments may extend into the A band.

        (a) The area of the A band without any thin filaments is called the **H band**.

        (b) The **M line** vertically bisects the H band. Several proteins are located within the M line:

          (i) **Myomesin** is a structural protein that links neighboring thick filaments to each other.

          (ii) **Creatine phosphokinase (CPK)** is an important enzyme that helps maintain adequate adenosine triphosphate (ATP) concentrations in working striated muscle fibers.

      (3) The amount of overlap between thick and thin filaments varies with sarcomere length and determines how much force skeletal muscle will develop when it is stimulated (see II C). When the muscle is stretched or shortened, the thick and thin filaments slide past each other, and the I band increases or decreases in size.

  4. Tubules. Two tubular networks are present in skeletal muscle fibers.

    a. The **transverse (T) tubule** is formed as an invagination of the surface of the muscle membrane. In mammalian skeletal muscle, the T tubules are located at the junction of the A and I bands. (In frog muscle, the T tubules are at the Z line.)

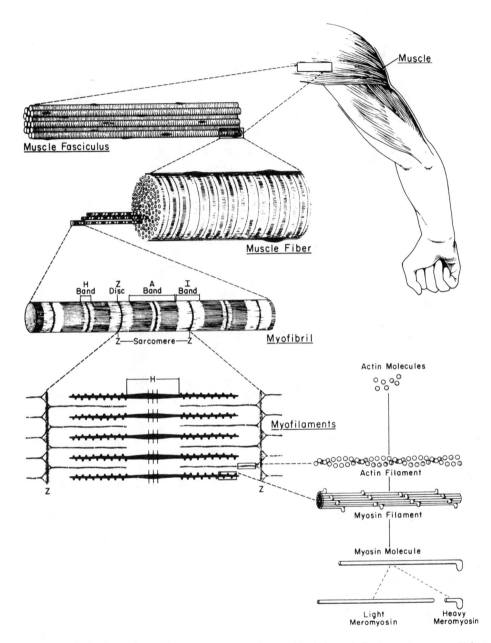

FIGURE 4-1. Skeletal muscle and its components as observed by light and electron microscopy. (Adapted from Fawcett DW: *Bloom and Fawcett's Textbook of Histology,* 11th ed. Philadelphia, WB Saunders, 1986, p 282.)

  b. An action potential spreading over the surface of the muscle membrane is propagated into the network of T tubules, which forms specialized contacts with the **SR,** the internal tubular structure that runs between the myofibrils.

   (1) The SR has a **high concentration of Ca^{2+},** which is used to initiate muscle contraction when the muscle is stimulated.

   (2) The ends of the SR expand to form **terminal cisternae (TC),** which make contact with the T tubule. Small projections, or **foot processes,** span the 20 nm separating the two tubular membranes (Figure 4-2).

   (a) The **SR membrane** contains a protein called the **ryanodine receptor** that contains the foot process and a Ca^{2+}-release channel.

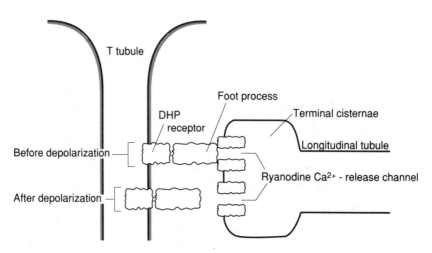

FIGURE 4-2. The dihydropyridine (*DHP*) receptor on the T tubule functions as a voltage sensor in skeletal muscle. When the T tubule is depolarized, the voltage sensor pulls the foot process away from the ryanodine Ca^{2+}-release channel in the sarcoplasmic reticulum (SR) membrane. The release of Ca^{2+} from the SR initiates muscle contraction. In cardiac muscle, the DHP receptor is associated with a Ca^{2+} channel. When Ca^{2+} enters the cell through the DHP Ca^{2+} channel, it opens the ryanodine Ca^{2+}-release channel in the SR. In smooth muscle, another receptor, called the inositol triphosphate (IP_3) receptor, is linked to the Ca^{2+}-release channel.

      (b) The **T tubule membrane** contains a **voltage-sensitive dihydropyridine (DHP) receptor** that opens the ryanodine Ca^{2+}-release channel on the SR membrane.

B. **Excitation–contraction (EC) coupling** is the process by which an action potential initiates the contractile process. EC coupling involves **four steps:** the propagation of the action potential into the T tubule and release of Ca^{2+} from the TC, the activation of the muscle proteins by Ca^{2+}, the generation of tension by the muscle proteins, and the relaxation of the muscle.

1. **Release of Ca^{2+} from the TC.** Depolarization of the T tubule causes the ryanodine Ca^{2+} channels to open, which leads to the release of Ca^{2+}. Ca^{2+} flows out of the TC and into the cytoplasm.

2. **Activation of muscle proteins.** For a muscle to contract, the thick and thin filaments must interact. When the cell is at rest, this interaction is inhibited; the influx of Ca^{2+} removes the inhibition.
   a. Ca^{2+} binds to **troponin,** one of two regulatory proteins located on the thin filament (Figure 4-3A).
   b. The troponin undergoes a conformational change that alters the position of the **tropomyosin,** the other regulatory protein located on the thin filament.
   c. The movement of tropomyosin (deeper into the groove of the thin filament) exposes the myosin binding sites on the actin, allowing the myosin cross-bridge on the thick filament to bind to actin on the thin filament (Figure 4-3B). A rise in Ca^{2+} concentration, from 0.1–10 μmol/L, is sufficient to activate all of the muscle protein within the skeletal muscle fiber.

3. **Generation of tension.** The activated muscle proteins undergo **repetitive cross-bridge cycling** during which the muscle uses the energy obtained from the hydrolysis of ATP to shorten and generate tension. The cross-bridge cycle can be divided into four steps.
   a. The **first step** is the binding of the cross-bridge to actin, as described in II B 2 c (see Figure 4-3B). Binding occurs spontaneously after Ca^{2+} binds to troponin.
   b. The **second step** is the bending of the cross-bridge (Figure 4-3C), which pulls the thin filament over the thick filament, generating tension (see II C).

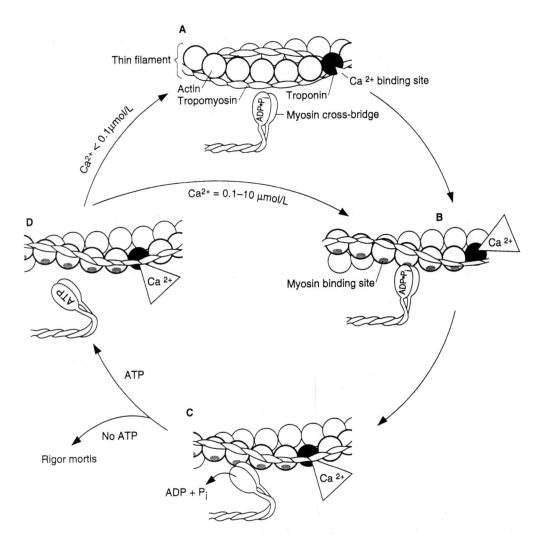

FIGURE 4-3. (*A*) Resting position. (*B*) When Ca^{2+} binds to troponin, the troponin undergoes a conformational change, pulling the tropomyosin away from the myosin binding site and allowing the cross-bridge cycle to begin. During this step of the cycle, the myosin binds to actin. (*C*) In the second step, the cross-bridge bends and the thin filament slides over the thick filament. During this step, the products of hydrolysis [i.e., adenosine diphosphate (*ADP*) and inorganic phosphate (*P_i*)] are released from the cross-bridge. (*D*) In the third step, a new molecule of adenosine triphosphate (*ATP*) binds to the cross-bridge and the cross-bridge detaches from the thin filament. The cross-bridge then stands up, and a new cycle begins if the intracellular Ca^{2+} concentration is sufficient to maintain troponin in its active state. Otherwise, the muscle relaxes.

  (1) The energy used to bend the cross-bridge is generated when ATPase splits the ATP molecule into **adenosine diphosphate (ADP)** and **inorganic phosphate (P_i)**.
  (2) Both the ATP molecule and the ATPase required for its hydrolysis are located on the cross-bridge; however, ATPase is activated only when myosin binds to actin. Therefore, hydrolysis occurs only during the cross-bridge cycle.
 c. The **third step** is the detachment of the cross-bridge from the thin filament. This occurs after the cross-bridge has bent (Figure 4-3D).
  (1) For detachment to occur, ADP and P_i must be removed from the cross-bridge and replaced with a new molecule of ATP. If no ATP is available, the thick and thin filaments cannot be separated (i.e., rigor mortis occurs).

   (2) As soon as myosin separates (or, perhaps, while it is separating) from actin, the initial steps of ATP hydrolysis occur, producing myosin • ADP • P$_i$, a high-energy ATP intermediate. Complete dissociation of the phosphate from the ADP does not occur until after myosin has bound to actin and completed its bending cycle.

  d. In the **fourth step,** the cross-bridge returns to its original upright position. Once there, it can participate in another cycle. Cycling continues as long as Ca^{2+} is bound to troponin.

4. Relaxation of muscle occurs when the Ca^{2+} is removed from the cytoplasm by Ca^{2+} pumps (Ca^{2+}-ATPase) located on the SR membrane. When the intracellular Ca^{2+} concentration falls below 0.1 μmol/L, troponin returns to its original conformational state, tropomyosin inhibition of myosin–actin interaction is restored, and cross-bridge cycling stops.

C. **Shortening and force development** is produced by the sliding of thin filaments over thick filaments. The contractile properties of muscle can be studied in two types of mechanical conditions: **isometric** and **isotonic contractions**.

1. Mechanical basis. The thin filaments are drawn to the center of the sarcomere by the repetitive cycling of the cross-bridge (Figure 4-4).
  a. Each time an attached cross-bridge bends, it generates a force that pulls the Z line toward the center of the sarcomere.
  b. The force developed by the bending of the cross-bridge is transmitted through the thin filament to the Z line and then through the sarcolemma and tendinous insertions of the muscle to the bones.
  c. In Figure 4-5, the thick and thin filaments are represented by the **contractile component,** and the tendons and other compliant structures of the muscle are represented by the **series elastic component (SEC)**.

2. An **isometric contraction** occurs when the ends of the muscle (or bones) do not move during the contraction; therefore, the length of the muscle remains constant but the **tension changes**.

FIGURE 4-4. Shortening of the sarcomere. (A) Repetitive cycling of cross-bridges causes the thin filament to slide over the thick filament, dragging the Z lines toward the center of the sarcomere and causing the muscle to shorten. (B) During the cross-bridge cycle, the myosin head binds to actin and pulls the thin filament toward the center of the sarcomere, advancing the Z line 7–10 nm. The myosin then detaches, unbends, and reattaches to another actin molecule on the thin filament, and a second cross-bridge cycle begins.

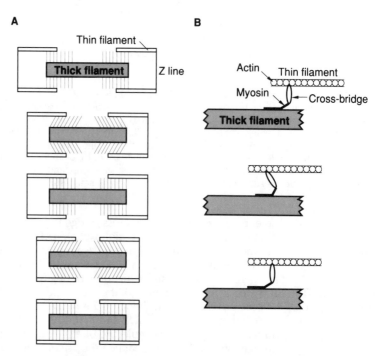

Relaxed

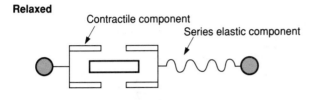

Contracted

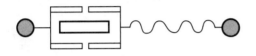

FIGURE 4-5. The contractile properties of a muscle can be explained using a mechanical analog consisting of a force-generating contractile component in series with an elastic element.

a. The intracellular Ca^{2+} concentration and the force developed during an isometric muscle **twitch** are illustrated in Figure 4-6.
   (1) Contraction begins when the Ca^{2+} released from the SR binds to troponin.
   (2) The muscle relaxes after the Ca^{2+} is resequestered into the SR by the Ca^{2+}-ATPase pump on the SR.
   (3) The duration of the twitch is longer than the duration of the **Ca^{2+} transient** (i.e., the period during which the intracellular Ca^{2+} concentration is above resting values and the cross-bridges are cycling) because the cross-bridges remain attached to actin for a period of time after the Ca^{2+} is removed from the sarcoplasm.
b. **Increasing the force of an isometric contraction.** The motor control system increases the force of an isometric contraction by either:
   (1) **Increasing the number of active alpha motoneurons.** When an alpha motoneuron fires, all the muscle fibers that it innervates will contract; therefore, recruitment of additional alpha motoneurons increases the number of active muscle fibers (and, consequently, the force of the contraction).
   (2) **Increasing the frequency of alpha motoneuron firings** (Figure 4-7)
       (a) The duration of the Ca^{2+} transient and, hence, the force of the isometric contraction, can be increased by increasing the frequency of alpha motoneuron firings because more Ca^{2+} is released from the SR each time the muscle is stimulated.
           (i) **Summation.** If the frequency increase is moderate, individual twitches "accumulate," increasing the force of contraction.
           (ii) **Tetanus.** If the frequency of stimulation is rapid, individual twitches become one continuous contraction (i.e., a maximal force is generated). The frequency required to produce a maximal force is called a tetanic frequency, and the resulting contraction is called tetanus.
       (b) The force generated by the cross-bridge can only be transmitted to the bones if the SEC is stretched.

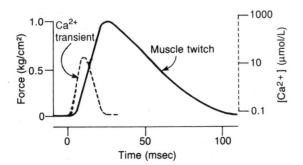

FIGURE 4-6. When intracellular Ca^{2+} concentration (*right ordinate*) increases to greater than 0.1 μmol/L, cross-bridge cycling causes an increase in muscle force (*left ordinate*). The resulting force development is the muscle twitch.

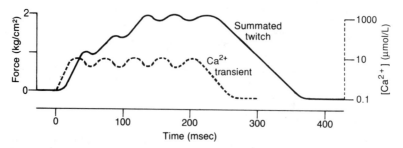

FIGURE 4-7. When the duration of the Ca^{2+} transient is increased by repetitive firing of the muscle, force development increases because there is sufficient time for the series elastic component (SEC) to be stretched completely.

      (i) Each time the cross-bridge bends, the Z line moves approximately 7.5–10 nm, and the SEC is stretched by the same amount. (Although the sarcomere is shortening, the contraction is still considered isometric because the total length of the muscle does not change.)

      (ii) The total shortening of the sarcomere (and the accompanying lengthening of the SEC) depends on the number of cross-bridge cycles that occur during the Ca^{2+} transient. The more the SEC is stretched, the more force is transmitted.

  c. Length–tension relationships. The force of an isometric contraction can be altered by altering the initial length of the muscle fiber. Figure 4-8 illustrates this relationship, which is called the length–tension relationship.

    (1) The **overlap** between the thick and thin filaments determines the number of cross-bridges that will bind to actin when the muscle is stimulated.

      (a) At an initial sarcomere length of 2.2 μm, each cross-bridge can bind to an actin molecule on the thin filament, and a maximum force is generated.

      (b) If the muscle is stretched to a sarcomere length of 3.5 μm, there is no overlap between the thick and thin filaments and, therefore, no force develops when the muscle is stimulated.

      (c) If the sarcomere shortens to lengths below 2.0 μm, the thin filaments from opposite sides of the sarcomere interfere with each other and the force of contraction decreases.

      (d) If the sarcomere shortens to 1.5 μm, the Z lines abut the thick filaments and no force can be generated.

    (2) Preload. Force must be applied to the skeletal muscle fiber to stretch it beyond 2.0 μm. The force is used to overcome elastic elements called **parallel elastic**

FIGURE 4-8. The length–tension relationship results from the overlap between thick and thin filaments. At a sarcomere length of 2.2 μm, overlap is optimal and force development is maximal. At lengths greater than 2.2 μm, force decreases because cross-bridge overlap is lessened. At lengths less than 2.2 μm, force is diminished because the thin filaments meet at the center of the sarcomere, causing an increased resistance to shortening.

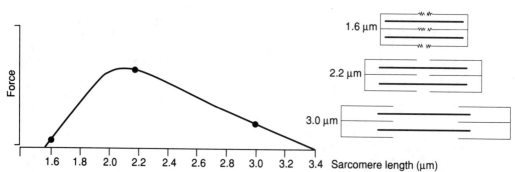

elements, one of which is most likely the protein titin, which attaches the thick filaments to the Z line. The force required to stretch the muscle to lengths beyond 2.0 μm is called the preload. The term preload is also used to indicate the length of the sarcomere or muscle before contraction.

(3) Varying the preload is not an important method of varying the contractile force of skeletal muscle. Often, the muscle length is determined by the particular motor task being performed. However, if muscle length is not constrained by the motor activity, more force can be obtained by holding the muscle at its optimal length (i.e., where the sarcomere length is 2.2 μm).

3. An **isotonic contraction** occurs when the muscle shortens; therefore, the tension remains constant but the **length changes**.
  a. The development of force and the change in muscle length that occurs during an isotonic contraction is illustrated in Figure 4-9.
    (1) The initial portion of the contraction is isometric, because the muscle only begins to shorten when the force developed by the muscle equals the load on the muscle. The weight that a muscle lifts during an isotonic contraction is called the **afterload**.
    (2) **Constants**
      (a) **Force.** While the muscle is shortening, the force remains equal to the afterload. The contraction is called isotonic because the force remains constant during the contraction.
      (b) The **velocity** of shortening remains constant.
  b. The **characteristics** of the contraction vary with the **magnitude of the afterload**. Increasing the afterload has the following effects.
    (1) The **duration of the isometric portion** of the contraction **increases,** because the SEC must stretch more to transmit the force required to lift the greater afterload.
    (2) The **velocity of shortening decreases** as the afterload increases.
      (a) Velocity decreases because each cross-bridge cycle takes longer.
      (b) The relationship between afterload and velocity is called the **load–velocity relationship** (Figure 4-10). The load–velocity relationship is an important characteristic of muscle because it indicates that the greatest velocity of

FIGURE 4-9. During an isotonic contraction, sufficient force must be generated for the muscle to shorten against the afterload on the muscle. Once sufficient force is developed, the muscle shortens at a constant velocity. Increasing the afterload increases the amount of force that must be developed before shortening can begin and decreases the velocity and extent of shortening.

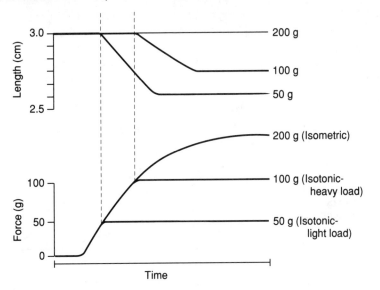

shortening is generated when the afterload on the muscle is zero. The peak velocity of shortening is an indication of the cross-bridge cycling speed.

(3) The **amount of shortening decreases** as the afterload increases (Figure 4-11).

    (a) As the muscle shortens below a sarcomere length of 2.2 μm, its ability to generate force decreases.

    (b) At some length, the maximal force the muscle can develop becomes slightly less than the afterload and shortening stops.

    (c) The greater the afterload, the longer it takes to reach the maximal force.

III. CARDIAC MUSCLE allows the heart to contract and propel blood through the circulatory system.

A. **Composition.** Although cardiac muscle fibers are striated, their structure differs somewhat from that of skeletal muscle.

1. Cardiac muscle fibers have a **single nucleus** and are smaller than skeletal muscles. Each fiber is approximately 15–20 μm wide, approximately 100 μm long, and only approximately 5 μm thick, making its shape more **ribbon-like** than cylindrical.

2. The T tubule is larger in cardiac muscle than in skeletal muscle and is located at the Z line rather than at the junction of the A and I bands. The SR makes contact with the T tubule and the cell membrane.

B. **EC coupling** of cardiac muscle differs in significant ways from that of skeletal muscle.

1. **Ca^{2+}-induced Ca^{2+} release.** Ca^{2+} release from the SR is triggered by Ca^{2+}, not by membrane depolarization. The Ca^{2+} responsible for Ca^{2+}-induced Ca^{2+} release enters the sarcoplasm during the plateau phase of the cardiac action potential.

    a. The T tubule DHP receptor in cardiac muscle, unlike the DHP receptor in skeletal muscle, contains a Ca^{2+} channel through which Ca^{2+} enters the cell during the action potential.

    b. The SR ryanodine receptor containing the Ca^{2+}-release channel is opened by the influx of Ca^{2+} from the T tubule. In contrast, the skeletal muscle Ca^{2+}-release channel is opened by a voltage-induced conformational change of the DHP receptor (see Figure 4-2).

FIGURE 4-10. The load–velocity relationship illustrates that velocity of shortening increases when the afterload on the muscle decreases. Velocity is zero when the afterload equals or exceeds the maximum force that the muscle is capable of generating (F_{max}). Maximum velocity of shortening (V_{max}) is achieved when the afterload is zero. In cardiac muscle, the intrinsic force-generating capability of the muscle (i.e., the muscle's contractility) can be increased by increasing the amount of Ca^{2+} that is released from the sarcoplasmic reticulum (SR). When contractility is increased, both F_{max} and V_{max} increase.

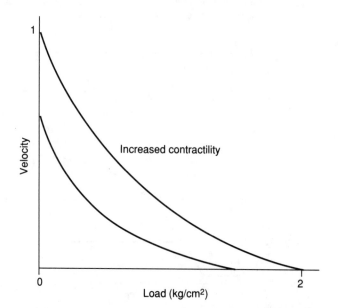

Increased contractility

Velocity

0   2

Load (kg/cm²)

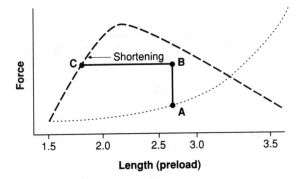

FIGURE 4-11. The relationship between force and length during an isotonic contraction (*solid line*). The isometric length–tension relationship (*dashed line*) indicates the maximum force the muscle can develop at any preload. The passive length–tension relationship (*dotted line*) indicates the amount of force required to stretch the resting muscle. When the muscle is stimulated (*point A*), it develops force but does not begin to shorten until its force equals the afterload (*point B*). At this point, the muscle shortens isotonically (i.e., its force remains equal to the afterload). As the muscle shortens to lengths below 2.2 μm, its ability to develop force (based on the isometric length–tension relationship) decreases. Eventually, depending on the afterload, the maximum force the muscle can develop equals the afterload and additional shortening cannot occur (*point C*).

    2. The amount of Ca^{2+} released from the SR is under physiologic control.

        a. The amount of Ca^{2+} entering the cell during the plateau phase of the cardiac action potential can be increased by norepinephrine and epinephrine. Increasing the amount of Ca^{2+} entering the cell increases the amount of Ca^{2+} released from the SR by Ca^{2+}-induced Ca^{2+} release.

        b. The amount of Ca^{2+} within the SR can be increased by catecholamine stimulation. When the Ca^{2+} concentration within the SR increases, the amount of Ca^{2+} released by Ca^{2+}-induced Ca^{2+} release increases.

        c. The amount of intracellular Ca^{2+} is regulated by the **Na^+-Ca^{2+} exchanger,** an antiport mechanism driven by the Na^+ gradient that transports one Ca^{2+} molecule out of the cell for every three Na^+ molecules that enter [see Chapter 1 III C 2 a (2)].

        (1) Reduced intracellular Ca^{2+}

            (a) During the plateau phase of the cardiac action potential, the driving force for Na^+ is reduced. Because the exchange rate is proportional to the Na^+ concentration gradient, the amount of Ca^{2+} leaving the cell is also reduced.

            (b) Cardiac glycosides. Ouabain and other cardiac glycosides can also affect Na^+-Ca^{2+} exchange. These drugs inhibit the Na^+-K^+ pump, causing Na^+ to accumulate in the cell. The increase in intracellular Na^+ reduces the driving force for Na^+ and thus decreases the ability of the Na^+-Ca^{2+} exchanger to remove Ca^{2+}.

        (2) Increased intracellular Ca^{2+}. Under certain circumstances, the driving force for Ca^{2+} entry may exceed the driving force for Na^+ entry. Under these conditions, Ca^{2+} is pumped into the cell by the Na^+-Ca^{2+} exchanger and Na^+ is pumped out of the cell.

C. **Shortening and force development** of cardiac muscle differs from that in skeletal muscle because of the duration of the action potential and the ability of cardiac muscle to regulate the amount of Ca^{2+} entering the cell.

    1. The action potential and the period of time that Ca^{2+} remains in the cytoplasm (the Ca^{2+} transient) are nearly equal in duration; thus, **summation and tetanus are not possible.** This is not a physiologic disadvantage for the heart, which must relax after each beat so blood can enter it.

2. The sarcomere length in a cardiac muscle before contraction (the preload) depends on how much blood has entered the heart. Because this amount is under physiologic control, it is an important regulator of the force of cardiac muscle contraction.

3. The force of contraction can vary at a given sarcomere length if the amount of Ca^{2+} entering the cell is changed. This also is under physiologic control and, thus, is an important regulator of cardiac muscle contractile force.

IV. SMOOTH MUSCLE plays a major role in the physiologic regulation of the airways, blood vessels, and gastrointestinal (GI) tract.

A. Composition. The structure of smooth muscle differs from that of striated muscle.

1. Fibers. Smooth muscle is composed of **elongated** (10–500 μm long), **thin** (5–10 μm wide) muscle fibers that contain a single nucleus.

2. Filaments
   a. Organization. Sarcomeres are absent in smooth muscle; instead, the thick and thin filaments are **dispersed throughout the cell.**
   (1) The thin filaments are attached to **dense bodies.** Some of the dense bodies are anchored to the cell membrane, but most float within the cytoplasm.
   (2) The dense bodies are composed of **α-actinin,** the same protein found in the Z lines.
   b. Proteins. The thick filaments contain myosin, and the thin filaments contain actin and tropomyosin; however, the thin filaments **lack troponin** in smooth muscle.

3. T tubules are absent and unnecessary in smooth muscle because the cell is small enough for a stimulus present on the cell surface to effectively activate the contractile machinery. Instead, smooth muscle fibers contain **caveolae,** small invaginations of the surface membrane that serve to increase the smooth muscle surface area and may (like T tubules) function to couple membrane potential changes to the SR.

B. EC coupling in smooth muscle is fundamentally different from that in striated muscle. In smooth muscle, cross-bridge cycling is regulated by Ca^{2+}**-induced phosphorylation of myosin** (Figure 4-12).

1. The myosin cross-bridges contain **four light chains,** two associated with each one of the head portions of the myosin molecule.

2. Myosin cannot bind to actin unless one of these light chains (called **LC_{20}** because it has a molecular weight of 20 kD) is phosphorylated.
   a. Phosphorylation of LC_{20} is catalyzed by the enzyme **myosin light-chain kinase (MLCK).** MLCK is activated by **calmodulin,** which in turn is activated by Ca^{2+}.
   b. Ca^{2+} can enter the cells in a variety of ways.
   (1) Stimulation by a neurotransmitter. When smooth muscle is stimulated by a neurotransmitter, a receptor-activated Ca^{2+} channel may open, allowing Ca^{2+} to enter the cell.
   (2) Voltage-operated Ca^{2+} channels. Ca^{2+} may also enter the cell through voltage-operated Ca^{2+} channels that open during the smooth muscle action potential.
   (3) Release from SR. Ca^{2+} may be released from the SR. The Ca^{2+}-release channel on smooth muscle SR is activated by inositol triphosphate (IP_3) and is called the **IP_3 receptor** to distinguish it from the ryanodine receptor found in striated muscle.

3. The light chains are **dephosphorylated** by the enzyme **myosin light-chain phosphatase (MLCP).**

C. **Shortening and force development** in smooth muscle differs from that in striated muscle.

1. The **speed of shortening** (i.e., the rate of cross-bridge cycling) is dependent on the phosphorylation of the myosin light chain. When the light chains are dephosphorylated by the phosphatase enzyme, the speed of shortening decreases.

2. **Latch-bridges.** The **dephosphorylated** cross-bridges remain **attached** to actin and are called latch-bridges. Latch-bridges provide smooth muscle with the ability to maintain tone with little energy consumption. Because the latch-bridges do not cycle, or cycle very slowly, they do not use much ATP.

FIGURE 4-12. Activated smooth muscle can exist in a phosphorylated and an unphosphorylated (latch) state. (*A*) Relaxed muscle. When myosin light-chain kinase (*MLCK*) catalyzes the phosphorylation (*P*) of light chain 20 (LC_{20}), rapid cross-bridge cycling ensues, enabling the muscle to develop force and shorten. (*B*) Each time the cross-bridge cycles, a molecule of adenosine triphosphate (*ATP*) is used. Actin–myosin–ATPase (*A–M–ATPase*) catalyzes the hydrolysis of ATP. (*C*) When the LC_{20} is dephosphorylated by myosin light-chain phosphatase (*MLCP*), a latch-bridge forms. In the latch-bridge state, cross-bridge cycling slows, enabling the muscle to maintain force with minimal energy consumption. (*D*) When the cytoplasmic Ca^{2+} concentration falls to resting levels, the latch-bridges dissociate and the muscle relaxes.

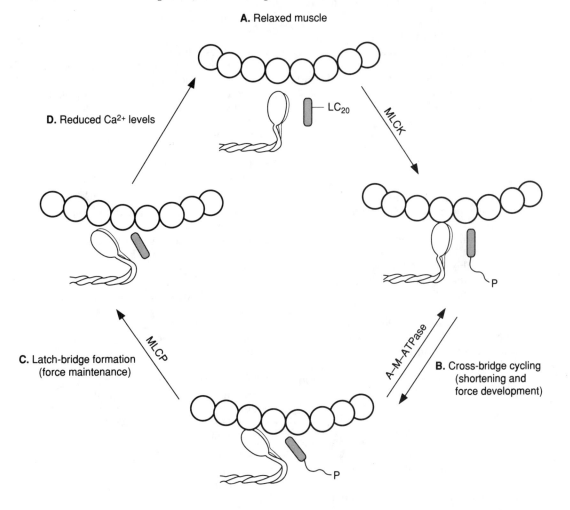

A. Relaxed muscle

D. Reduced Ca^{2+} levels

LC_{20}

MLCK

C. Latch-bridge formation (force maintenance)

MLCP

A–M–ATPase

B. Cross-bridge cycling (shortening and force development)

P

P

STUDY QUESTIONS

DIRECTIONS: Each of the numbered items or incomplete statements in this section is followed by answers or by completions of the statement. Select the ONE lettered answer or completion that is BEST in each case.

1. According to Fick's law of diffusion, particle flux will decrease if there is an increase in the

(A) particle's concentration difference across the membrane
(B) thickness of the membrane
(C) area of the membrane
(D) temperature of the solution
(E) diameter of the particle

2. When activated by β-adrenergic receptors, the G protein

(A) activates phospholipase C
(B) activates adenylate cyclase
(C) activates protein kinase C
(D) converts guanosine diphosphate (GDP) to guanosine triphosphate (GTP)
(E) stimulates the release of Ca^{2+} from the sarcoplasmic reticulum (SR)

3. Significantly decreasing the extracellular concentration of K^+ would

(A) increase the transport of Na^+ out of the cell by the Na^+-K^+ pump
(B) decrease the negativity of the membrane potential
(C) increase the conductance of the membrane to K^+ (G_K)
(D) increase the driving force on Na^+
(E) decrease the negativity of the equilibrium potential for K^+

4. Assuming that the extracellular concentration of Na^+ is 10 times the intracellular concentration, the equilibrium potential for Na^+ would be the same as the equilibrium potential for Ca^{2+} if the extracellular concentration of Ca^{2+} was

(A) 2 times the intracellular concentration of Ca^{2+}
(B) 10 times the intracellular concentration of Ca^{2+}
(C) 20 times the intracellular concentration of Ca^{2+}
(D) 100 times the intracellular concentration of Ca^{2+}
(E) the same as the intracellular concentration of Ca^{2+}

5. The synaptic channels on the end-plate of skeletal muscle are

(A) highly selective for Na^+
(B) opened when the cell membrane depolarizes
(C) activated by acetylcholine (ACh)
(D) inhibited by atropine
(E) responsible for the relative refractory period

6. Increasing the afterload on a skeletal muscle fiber

(A) increases the velocity of shortening
(B) decreases the force produced by the muscle during shortening
(C) decreases the interval between excitation and shortening
(D) increases the amount of shortening
(E) none of the above

7. Which one of the following statements regarding a Donnan equilibrium is true?

(A) The concentration of permeable anions is higher in the solution with the impermeable proteins
(B) The concentration of particles is greater in the solution that contains no impermeable proteins
(C) The membrane potential is more negative on the side with the impermeable proteins
(D) There is a net flux of cations into the solution that contains no impermeable proteins

8. The equilibrium potentials for K^+ and Na^+ are -90 mV and $+60$ mV, respectively. If the conductance for K^+ (G_K) is 4 times the conductance for Na^+ (G_{Na}), the resting membrane potential is

(A) -50 mV
(B) -60 mV
(C) -70 mV
(D) -80 mV
(E) -90 mV

9. In a nerve axon, which phase of the action potential is caused by the inactivation of the Na^+ channels?

(A) Upstroke
(B) Absolute refractory period
(C) Downstroke
(D) Undershoot
(E) Relative refractory period

10. The velocity of propagation along an axon will increase if there is a decrease in the

(A) membrane resistance (r_m)
(B) membrane capacitance (c_m)
(C) axon diameter
(D) refractory period
(E) axon excitability

11. Release of synaptic transmitter by exocytosis would be blocked most effectively by preventing the

(A) propagation of the action potential into the nerve terminal membrane
(B) depolarization of the nerve terminal membrane
(C) flow of Na^+ into the nerve terminal membrane
(D) flow of K^+ out of the nerve terminal membrane
(E) flow of Ca^{2+} into the nerve terminal membrane

12. In smooth muscle, Ca^{2+} is released from the sarcoplasmic reticulum (SR) by

(A) diacylglycerol (DAG)
(B) the guanosine triphosphate (GTP) binding protein (G protein)
(C) phospholipase C
(D) inositol triphosphate (IP_3)
(E) adenylate cyclase

13. The production of cyclic guanosine monophosphate (cGMP) is associated with the postsynaptic effect of

(A) acetylcholine (ACh)
(B) epinephrine
(C) γ-aminobutyric acid (GABA)
(D) glutamate
(E) nitric oxide

14. A red blood cell is placed in a saline (NaCl) solution, and the cell's volume increases to 1.5 times its original volume. This finding indicates that the Na^+ concentration in the saline solution is approximately

(A) 50 mEq/L
(B) 75 mEq/L
(C) 100 mEq/L
(D) 200 mEq/L
(E) 600 mEq/L

15. Which one of the following proteins is important for skeletal muscle contraction but not for smooth muscle contraction?

(A) Actin
(B) Myosin
(C) Troponin
(D) Myosin–adenosine triphosphatase (ATPase)
(E) Ca^{2+}-ATPase

16. The osmotic flow of water across a membrane will decrease if there is a decrease in

(A) the permeability of the membrane to the particles in solution
(B) the particle's concentration difference across the membrane
(C) both
(D) neither

17. Depolarization of the T tubule is directly linked to the opening of Ca^{2+} channels on the sarcoplasmic reticulum (SR) of

(A) skeletal muscle
(B) cardiac muscle
(C) both
(D) neither

DIRECTIONS: The incomplete statement in this section is negatively phrased, as indicated by a capitalized word such as NOT, LEAST, or EXCEPT. Select the ONE lettered answer or completion that is BEST.

18. All of the following transport processes display saturation (Michaelis-Menten) kinetics EXCEPT

(A) Na^+-coupled active transport
(B) primary active transport
(C) facilitated diffusion
(D) simple diffusion
(E) Na^+-Ca^{2+} exchanger

ANSWERS AND EXPLANATIONS

1. The answer is B [Chapter 1 III A 1 b]. According to Fick's law of diffusion, particle flux is inversely proportional to the thickness of the membrane and directly proportional to the area of the membrane, the concentration difference across the membrane, and the diffusion coefficient. The diffusion coefficient depends on the physical properties of the particle and the membrane through which it is diffusing. In general, the larger the diameter of the particle, the slower the rate of diffusion. Increasing temperature will increase the rate of diffusion.

2. The answer is B [Chapter 3 III B 3 a (1); Figure 3-5]. When norepinephrine binds to a β-adrenergic receptor, it activates a G protein [i.e., a guanosine triphosphate (GTP) binding protein], which, in turn, activates adenylate cyclase. Adenylate cyclase catalyzes the formation of cyclic adenosine 3′,5′-monophosphate (cAMP), which activates a variety of kinases. One of these kinases (protein kinase A) phosphorylates phospholamban, which reduces the inhibition of the sarcoplasmic reticular Ca^{2+} pump, increasing resequestration of Ca^{2+} from the cytoplasm.

3. The answer is D [Chapter 1 III C 1; Chapter 2 I B 1 b; II B 1 a (1), b (1) (b)]. Decreasing the extracellular K^+ concentration increases the equilibrium potential for K^+ (i.e., makes it more negative) and causes the membrane to hyperpolarize. Hyperpolarization increases the driving force on Na^+. The conductance of K^+ (G_K) will decrease when the cell hyperpolarizes because some of the K^+ activation gates close, and because there are fewer ions to carry current through the K^+ channels that remain open. The activity of the Na^+-K^+ pump decreases if the fall in K^+ is significant enough, because there is not enough K^+ to occupy all of the binding sites on the pump.

4. The answer is D [Chapter 2 I B 1 b (1)–(2)]. The equilibrium potential (E_{ion}) is calculated using the Nernst equation [$E_{ion} = -61 \cdot \log (C_{in}/C_{out})$], where C is the concentration of the ion. Because Ca^{2+} has a valence of 2, the log of the Ca^{2+} ratio would have to be twice the

log of the Na^+ ratio to produce the sameequilibrium potential. Since the log of 1/10 (the Na^+ ratio) $= -1$, a log of -2 would require a ratio of 1/100. Thus, the extracellular Ca^{2+} concentration would have to be 100 times the intracellular concentration of Ca^{2+}.

5. The answer is C [Chapter 3 II C 2; III B 2]. Acetylcholine (ACh) is released from the alpha motoneuron nerve terminal and activates the synaptic channels on the skeletal muscle endplate. These channels, unlike the channels that produce the action potential, are not affected by changes in the membrane potential. The ACh receptor is inhibited by curare; atropine blocks ACh receptors activated by postganglionic parasympathetic neurons. The channel opened by the ACh receptor is equally permeable to Na^+ and K^+.

6. The answer is E [Chapter 4 II C 3 b]. The afterload is the weight that a muscle lifts during an isotonic contraction. When the afterload on an isotonically contracting skeletal muscle is increased, the velocity of shortening slows, the amount of force produced by the muscle increases (because force must equal load for the muscle to shorten), the interval between excitation and shortening increases (because it takes longer for the muscle to build up enough force to lift the load), and the amount of shortening decreases.

7. The answer is C [Chapter 2 I B 2; Figure 2-1]. A Donnan equilibrium occurs when two solutions are separated by a membrane that is impermeable to one of the charged particles in solution. In biologic solutions, the impermeable particle is a negatively charged protein. The solution containing the protein will have a higher concentration of particles (and thus a greater osmotic pressure), a higher concentration of permeable cations, and a negative charge (compared to the solution without the proteins). Because the solutions are in equilibrium (i.e., the concentration differences between the permeable ions are balanced by the membrane potential differences), no net flux of particles occurs.

8. The answer is B [Chapter 2 II B 1]. The resting potential can be calculated using the transference equation ($E_m = E_K \cdot T_K + E_{Na} \cdot T_{Na}$). Since the conductance for K^+ (G_K) is 4 times the conductance for Na^+ (G_{Na}), T_K (the transference for K^+) is 0.8 and T_{Na} is 0.2. Substituting the appropriate values in the transference equation ($-90 \cdot 0.8 + 60 \cdot 0.2$) yields a membrane potential of -60 mV.

9. The answer is B [Chapter 2 III D, E 2 c; Figure 2-6]. The absolute refractory period is caused by the inactivation of Na^+ channels that occurs during the depolarization phase of the action potential. During this time, another action potential cannot be elicited because the closing of the h gates prevents Na^+ from entering the cell. The absolute refractory period ends when some of the h gates open during the downstroke. At this point, another action potential can be elicited but its threshold is above normal because some of the Na^+ channels are still inactivated and the conductance of the membrane to K^+ (G_K) is higher than at rest. This period of time is referred to as the relative refractory period. Although inactivation of the Na^+ channels contributes to the downstroke, the downstroke is caused primarily by K^+ channel activation. Na^+ channel inactivation that occurs during the upstroke opposes depolarization. Almost all Na^+ channel activation occurs before the undershoot phase.

10. The answer is B [Chapter 2 III G 2; Figure 2-8]. Propagation occurs because the current entering the axon during one action potential acts as a stimulus for the production of another action potential further along the axon. The velocity of propagation is proportional to the cable properties of the axon that modulate the passive flow of charge along the axon. The lower the membrane capacitance (c_m), the smaller the charge needed to depolarize the membrane to threshold and thus the greater the velocity of propagation. Decreasing the membrane resistance (r_m) causes charge to leak out of the cell, slowing propagation. Decreasing the axon diameter increases the axoplasmic resistance (r_a), decreasing the amount of charge that flows along the axon. Decreasing axon excitability makes it more difficult to elicit an action potential and, therefore, slows conduction velocity. The refractory period may affect the frequency of action potential firing, but not the propagation velocity.

11. The answer is E [Chapter 3 II B 3, C 1]. Preventing the flow of Ca^{2+} into the cell would prevent the release of transmitter, because Ca^{2+} initiates the intracellular events leading to the docking of the vesicle to its binding site on the active zone. Although Ca^{2+} normally enters the cell through voltage-operated channels that are opened by the depolarization of the nerve terminal that occurs as the action potential propagates along the nerve axon, release of transmitter will not occur if Ca^{2+} does not enter the nerve terminal. The flow of Na^+ into the nerve terminal would depolarize the membrane and open Ca^{2+} channels, leading to Ca^{2+} entry and exocytosis. However, Na^+ entry does not directly stimulate exocytosis. K^+ does not affect the nerve terminal membrane.

12. The answer is D [Figure 3-5; Chapter 4 IV B 2 b]. In smooth muscle, Ca^{2+} is released from the sarcoplasmic reticulum (SR) by an inositol triphosphate (IP_3)-activated channel. In striated muscle, Ca^{2+} is released from the SR through a ryanodine receptor that is activated by depolarization in skeletal muscle and by Ca^{2+} in cardiac muscle. Diacylglycerol (DAG), the guanosine triphosphate (GTP) binding protein (the G protein), and phospholipase C all play a role in excitation–contraction (EC) coupling but do not directly cause the release of Ca^{2+} into the cytoplasm. Adenylate cyclase catalyzes the conversion of adenosine triphosphate (ATP) to cyclic adenosine 3′,5′-monophosphate (cAMP). cAMP activates protein kinase A, which phosphorylates phospholamban, leading to an increase in Ca^{2+} sequestration by the sarcoplasmic reticular Ca^{2+} pump.

13. The answer is E [Chapter 3 III C 2 a]. Nitric oxide acts as a paracrine agent and as a neurotransmitter. It stimulates the action of guanylate cyclase which, in turn, promotes the formation of cyclic guanosine monophosphate (cGMP). The physiologic effects of nitric oxide are produced by cGMP-dependent kinases that phosphorylate a variety of intracellular proteins. In the gastrointestinal (GI) system, nitric oxide facilitates smooth muscle relaxation, and in the cardiovascular system, it relaxes blood vessels.

14. The answer is C [Chapter 1 III B 3 a (1)]. The steady-state volume of a cell placed in a solution containing an osmotic concentration different from the cell's can be found using the equation: $\pi_i \cdot V_i = \pi_f \cdot V_f$. The normal tonicity of red blood cells is approximately 285 mOsm/L. In this problem, a red blood cell is placed in a solution of unknown osmolality and swells to 1.5 times its original volume. Solving the equation for the osmolality of the extracellular fluid (π_f) yields 285/1.5, or 190 mOsm. A saline solution of this osmolality will have a Na^+ concentration of 95 mEq/L.

15. The answer is C [Chapter 4 II B 2; IV B]. In skeletal muscle, contraction is initiated when Ca^{2+} binds to troponin. Smooth muscle contraction is initiated by the phosphorylation of the myosin light-chain proteins. Both smooth and skeletal muscle rely on actin, myosin, and myosin–adenosine triphosphatase (ATPase) for cross-bridge cycling and on Ca^{2+}-ATPase for Ca^{2+} resequestration.

16. The answer is B [Chapter 1 III B 1]. Decreasing the membrane's permeability to the particle will increase the particle's effective osmolality and thus increase the osmotic flow of water. The osmotic flow of water is proportional to the osmotic pressure difference across a membrane, which is, in turn, proportional to the concentration difference across the membrane. However, the osmotic pressure actually produced depends on the particle's permeability to the membrane. The smaller the particle's permeability (or the greater its reflection coefficient), the greater the osmotic pressure produced for a given concentration difference.

17. The answer is A [Chapter 4 II B 1; III B 1]. Release of Ca^{2+} from the sarcoplasmic reticulum (SR) in both skeletal and cardiac muscle is activated by T tubule depolarization. However, in skeletal muscle, the Ca^{2+}-release channel in the SR is opened directly by the depolarization of the T tubule membrane. In cardiac muscle, Ca^{2+} is released from the SR by a Ca^{2+}-induced Ca^{2+} release process. The Ca^{2+} used in this process enters the cell through a voltage-activated Ca^{2+} channel [the dihydropyridine (DHP) receptor Ca^{2+} channel] associated with the T tubule membrane.

18. The answer is D. [Chapter 1 III A 1 b, 2 b]. Saturation, or Michaelis-Menten, kinetics occurs in all carrier-mediated transport processes because in these transport processes, the flux is proportional to the concentration of substrate–carrier complexes that are formed. Saturation (i.e., maximum flux) occurs when all of the carrier proteins are occupied. There is a limit to the number of particles that can pass through membrane channels by simple diffusion so, theoretically, saturation can also be achieved via this process. However, under physiologic conditions, simple diffusion obeys Fick's law, which indicates that flux increases linearly with increases in the concentration difference across a membrane.

PART II

NEUROPHYSIOLOGY
Michael B. Wang

Chapter 5

Cutaneous, Olfactory, and Gustatory Sensation

I. INTRODUCTION

A. **Sensory receptors transform (transduce) stimulus energy** into a local change in membrane potential called a **receptor (generator) potential**.

1. **The receptor potential** serves as a **stimulus for the generation of an action potential** or for the **release of neurotransmitter** by the sensory receptor.

2. In either case, information about the stimulus is **transmitted to the central nervous system (CNS),** where it is used to elicit a reflex response, alter behavior, or produce a conscious sensation.

B. This chapter first considers the general mechanisms used by the nervous system to encode sensory information, and then describes receptors that are stimulated by direct contact with the stimulus (i.e., those in the skin, nose, and tongue). Chapter 6 describes receptors that are stimulated by energy transmitted from a distance (i.e., those in the ear and eye).

II. GENERAL SENSORY MECHANISMS

A. **Receptor classification.** Receptors can be classified according to the:

1. **Source of the stimulus**
   a. **Exteroceptors,** such as the eye, ear, taste, and cutaneous receptors, receive stimuli from outside the body.
   b. **Enteroreceptors,** such as the chemoreceptors that measure blood gases, the baroreceptors that measure blood pressure, and the proprioceptors that measure the position of the limbs or the force of muscle contraction, receive stimuli from within the body.

2. **Type of stimulus energy**
   a. **Mechanoreceptors** detect skin deformation and sounds.
   b. **Thermoreceptors** detect environmental temperatures.
   c. **Photoreceptors** detect light.
   d. **Chemoreceptors** detect substances that produce the sensations of smell and taste.

3. **Type of sensation** (e.g., touch, heat, cold, pain, light, sound, taste, and smell receptors)

4. **Rate of adaptation** (see also II C 3)
   a. **Slowly adapting (tonic, static) receptors** fire action potentials continuously during stimulus application.
   b. **Rapidly adapting (phasic, dynamic) receptors** fire action potentials at a decreasing rate during stimulus application.

TABLE 5-1. Classification of Sensory Neurons

Classification				
Diameter	**Velocity**	**Fiber Type**	**Axon Diameter**	**Propagation Velocity**
I	Aα	Myelinated	12–20 μm	80–120 m/sec
II	Aβ	Myelinated	6–12 μm	35–75 m/sec
III	Aδ	Myelinated	1–6 μm	5–30 m/sec
IV	C	Unmyelinated	< 1 μm	0.5–2 m/sec

B. **Sensory neuron classification**

 1. Sensory neurons are classified on the basis of axon diameter or propagation velocity (Table 5-1). The information provided by the larger, more rapidly conducting fibers is more precise than that provided by the smaller, more slowly conducting fibers.
   a. **Roman numerals** are used to classify axons according to **size**.
   b. The **letters A** and **C** are used to classify axons according to **propagation velocity**.

 2. Examples of sensorimotor neurons and their classifications are given in Table 5-2.

C. **Energy transduction**

 1. **Adequate (appropriate) stimulus.** Each receptor is specialized to receive a particular type of stimulus.
   a. Although each receptor is exquisitely sensitive to its adequate stimulus, receptors can respond to other forms of energy if the intensity is high enough.
   b. For example, the retina can detect the presence of a single photon of light. However, a mechanical stimulus (e.g., rubbing the eyes) can produce a sensation of light.

 2. **Conversion of stimulus energy.** A **transducer region** and a **spike generation region** are contained within most receptor membranes (Figure 5-1). In some receptors (e.g., those in the eye or the ear) the transducer and spike generation functions are served by separate cells.
   a. The **transducer region** is responsible for converting the stimulus energy into an electrical signal [i.e., the receptor (generator) potential]. The mechanism used by the transducer region to produce the receptor potential varies depending on the type of receptor.
   b. The **spike generator region** of the receptor is responsible for converting the receptor potential into a train of action potentials.

TABLE 5-2. Examples of Sensorimotor Neuron Classification

Classification		**Efferent Nerve**	**Afferent Cutaneous Nerve**	**Afferent Muscle Nerve**
I	Aα	Motoneurons to skeletal muscle	. . .	Ia: All intrafusal (nuclear chain and nuclear bag) fibers Ib: Golgi tendon organs
II	Aβ	. . .	Mechanoreceptors	Nuclear chain fibers
III	Aδ, Aγ	Motoneurons to intrafusal muscle fibers	Nociceptors, thermoreceptors	. . .
IV	C	. . .	Nociceptors, thermoreceptors	. . .

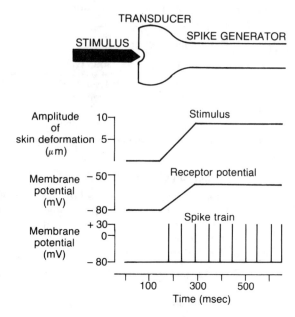

FIGURE 5-1. When a stimulus is applied to a sensory nerve ending, a receptor potential results. For example, in mechanoreceptors, the stimulus produces the receptor potential by deforming the nerve terminal. The deformation opens channels that are permeable to Na$^+$ and K$^+$, causing the membrane to depolarize. Note that the receptor potential follows the time-course of the mechanical stimulus. The magnitude of the receptor potential and the frequency of the action potential are proportional to the magnitude of the stimulus.

  (1) **The receptor potential spreads passively from the transducer region to the spike generator region.** If the spike generator is depolarized to threshold, an action potential is generated.
  (2) **At the end of one action potential, the receptor potential causes the spike generator membrane to depolarize toward threshold.**
     (a) If threshold is reached again, another action potential is generated.
     (b) Figure 5-1 illustrates the repetitive discharge of action potentials by the receptor axon in response to a receptor potential.
 3. **Adaptation** causes the receptor's response to decrease, despite the continued presence of a stimulus.
  a. **Functions**
     (1) Sensory adaptation functions primarily to **encode information about the rate of stimulus application**.
        (a) In **rapidly adapting receptors,** the discharge rate increases as the rate of stimulus application increases.
           (i) When a stimulus is applied too slowly, adaptation of the spike generator occurs before an action potential can be generated.
           (ii) When a stimulus is applied rapidly enough, excitation exceeds adaptation and a steady discharge of action potentials occurs.
        (b) In **slowly adapting receptors,** the discharge continues at a steady rate as long as the stimulus is applied.
     (2) In some cases, sensory adaptation is used to **decrease the amount of information reaching the brain**. Brain stem mechanisms responsible for consciousness and attention are capable of keeping the information flow to the brain within tolerable limits, however, so sensory adaptation is not often required for this purpose.
  b. **Mechanisms.** Sensory adaptation takes place via two major mechanisms.
     (1) In one, the **transducer mechanism fails to maintain a receptor potential** despite continued stimulus application.
     (2) In the other, the **spike generator fails to sustain a train of action potentials**. Although a receptor potential is present, the excitability of the spike generator membrane is diminished. An increase in the membrane conductance to K$^+$ or the activity of the electrogenic Na$^+$-K$^+$ pump, or the inactivation of Na$^+$ channels may be responsible for the decreased excitability of the membrane.

4. Encoding. The **intensity, location,** and **quality of a stimulus** are encoded by the sensory system.

   a. Stimulus intensity is encoded by the **firing frequency** of a sensory neuron. The firing frequency is proportional to the magnitude of the receptor potential.

      (1) As the intensity of a stimulus increases, the magnitude of the receptor potential increases (Figure 5-2). This relationship is expressed by the **Steven's power law function**

$$V = k \cdot [I - I_0]^n$$

or by the **Weber-Fechner law**

$$V = k \cdot \log \frac{I}{I_0} \text{ , where}$$

V = the magnitude of the receptor potential
k = a constant
I = the intensity of the stimulus
I_0 = the threshold
n = a constant

        (a) Assuming that the sensory system requires a receptor potential of at least 1 mV to detect the presence of a stimulus, the maximal response that can be detected is 50 times as great as the minimally detectable response.

        (b) The power law relationship between stimulus intensity and receptor potential magnitude enables the relatively small range of receptor potential amplitudes (i.e., 50 to 1) to encode a range of stimulus intensities that is greater than 1 million to 1.

      (2) The firing frequency of a sensory neuron, and, therefore, **sensory perception, are related to stimulus intensity** (see Figure 5-2). This relationship substantiates the concept that stimulus intensity is encoded by the rate of sensory nerve discharge.

FIGURE 5-2. The proportionality between the receptor potential and the stimulus can be expressed as a power function. This same relationship can be used to describe the proportionality between the stimulus magnitude and firing frequency, and between the magnitude of the stimulus and its perceived intensity. *I* = intensity.

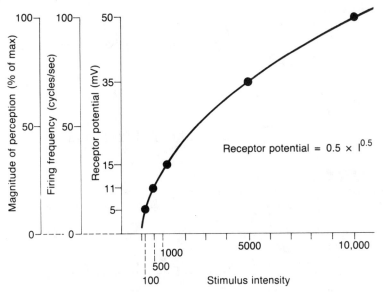

b. Stimulus location is encoded primarily by the location of the sensory projection in the cerebral cortex. This mechanism of encoding, called **topographic representation,** is used by the visual and somatosensory systems to localize the point of stimulus application. The auditory system uses **tonotopic representation** to encode the frequency of an auditory stimulus (see Chapter 6 II E 4 a).

    (1) Receptive field. Each sensory neuron receives information from a particular sensory area called its receptive field. The smaller the receptive field, the more precise the encoding of stimulus localization.

        (a) For example, the receptive fields of the fovea are the smallest within the eye.

        (b) Similarly, the receptive fields within the fingertips are much smaller than those on the hands and back, where it is difficult to precisely locate the point of a mechanical stimulus.

    (2) Lateral inhibition. Stimulus localization can be made more precise by lateral inhibition (Figure 5-3).

        (a) Two stimuli are applied to the skin, and the sensory task is to indicate whether one or two stimuli are present. The closest that two stimuli can be to each other and still be distinguished as separate stimuli is called the **two-point threshold**.

FIGURE 5-3. In order for two distinct stimuli to be perceived, they must stimulate receptive fields separated by an unstimulated receptive field. (*A*) Without lateral inhibition, the stimuli produce equal amounts of discharge in all three neurons. (*B*) With lateral inhibition, the neuron with the receptive field in the center is presynaptically inhibited by collaterals from the neurons with receptive fields located laterally. As a result, the receptive field in the center does not fire, and two stimuli are perceived.

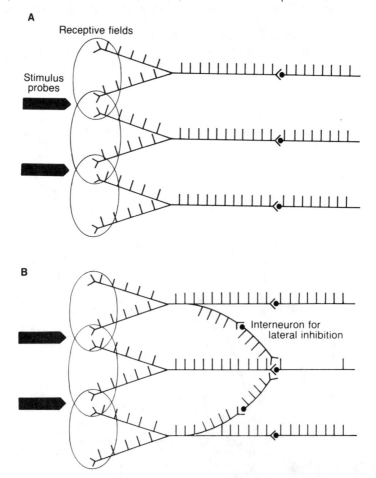

(b) Without lateral inhibition (see Figure 5-3A), the two stimuli are recognized as separate only if they are applied in receptive fields that are separated from each other by a nonstimulated receptive field.

(c) Lateral inhibition (see Figure 5-3B) can decrease the two-point threshold by reducing the discharge of the neuron innervating the receptive field in the center, thus making it apparent to the CNS that two stimuli are present.

c. **Stimulus quality** is encoded by a variety of mechanisms.

(1) The simplest mechanism uses a **labeled line,** in which the stimulus is encoded by the particular neural pathway that is stimulated.

(a) The basic sensory modalities are encoded in this way. Thus, the sensation of touch is elicited whether the receptors on the skin are excited by mechanical deformation or by electrical stimulation. Similarly, light sensation always is evoked no matter how the retina is activated, and sound sensation always results from stimulation of the cochlea.

(b) The same type of sensation results no matter where along the sensory pathway the stimulus is applied. For example, stimulating electrodes placed on the visual cortex evoke the sensation of light, and seizures within the olfactory cortex produce sensations of smell.

(2) A more complex mechanism of coding uses the **pattern of activity** within the neural pathway that is carrying information to the brain.

(a) In **temporal pattern coding,** the same neuron can carry two different types of sensory information depending on its pattern of activity. For example, cutaneous cold receptors indicate temperatures below and above 30°C by firing with or without bursts, respectively.

(b) In **spatial pattern coding,** the activity of several neurons is required to elicit a sensation. For example, three neurons may be required to encode different taste sensations. A sour taste may result if all three neurons are activated, whereas a salty taste may result if only two neurons fire.

(3) The most sophisticated mechanism of sensory coding uses **feature detectors**. These are neurons within the brain that integrate information from a variety of sensory fibers and fire to indicate the presence of a complex stimulus.

(a) For example, the location of an object in space can be encoded by cortical cells receiving information from a single eye. However, special feature detectors receiving information from both eyes are required to specify the depth of an object in space.

(b) Similarly, the location of a sound in space requires integration of information from both ears by feature detectors within the brain stem.

III. CUTANEOUS SENSATION. The skin contains receptors that are adapted to encode information about touch, pain, and temperature. Approximately 1 million sensory nerve fibers innervate the skin. Most of these are unmyelinated nerve fibers that are responsible for crude somatosensory mechanical sensation. Although far fewer in number, the large myelinated (group II) sensory fibers encode the important sensory qualities of touch, vibration, and pressure. The sensations of temperature and pain are encoded by small myelinated (Aδ) fibers and unmyelinated (C) fibers.

A. **Mechanoreceptors.** Precise information about mechanical stimulation from the hairless (glabrous) region of the skin is provided by four receptors (Table 5-3).

1. The **pacinian corpuscle** is a very rapidly adapting receptor with a large receptive field that is used to encode **vibratory sensation**.

   a. **Description.** The receptor is located on the end of a group II myelinated fiber, which is inserted into an onion-like lamellar capsule that is approximately 1 mm in diameter (see Table 5-3).

TABLE 5-3. Cutaneous Mechanoreceptors

Receptor		Receptive Field Size	Speed of Adaptation	Encoded Sensation
	Pacinian corpuscle	Large	Very rapid	Vibration
	Meissner's corpuscle	Small	Rapid	Speed of stimulus application
	Merkel's disk	Small	Slow	Location of stimulus
	Ruffini's corpuscle	Large	Slow	Magnitude and duration of stimulus

   b. Conversion of stimulus energy
   (1) When a mechanical stimulus deforms the outer lamellae of the capsule, the **deformation** is **transmitted through the capsule to the nerve terminal**.
      (a) The deformation of the nerve terminal increases the membrane permeability to Na^+ and K^+, producing a depolarizing receptor potential.
      (b) The size of the receptor potential increases in proportion to the magnitude of the deformation. However, because the pacinian corpuscle is a very rapidly adapting receptor, only a few action potentials are generated, regardless of stimulus intensity.
   (2) A **vibratory stimulus** produces a **steady discharge** of the pacinian corpuscle. Each time the stimulus is removed and reapplied, the pacinian corpuscle discharges another action potential.
   c. Adaptation occurs because the deformation of the nerve terminal is not maintained during continuous stimulus application.
   (1) Only rapid deformations are transmitted effectively to the core of the capsule.
   (2) If the stimulus is applied slowly or is left in place, the inner lamellae become rearranged so that they no longer deform the nerve terminal. This is an example of adaptation caused by failure to maintain the receptor potential.
   d. Encoding. The frequency of discharg by the pacinian corpuscle equals the frequency of a vibratory stimulus in the range of 50–500 Hz (cycles/sec).
   (1) This is about the same range of frequencies identifiable by humans, indicating that the **frequency of firing is encoding the frequency,** not the intensity, of vibration.
   (2) **Intensity** is encoded by the number of action potentials generated by each deformation (limited to two or three) and the number of pacinian corpuscles responding to the vibration.
2. Meissner's corpuscle is a rapidly adapting receptor with a small receptive field that is used to encode the **rate of stimulus application.**

  a. Description. The receptor is located at the end of a single group II afferent fiber that is inserted into a small capsule (see Table 5-3).
  b. Receptive field. Meissner's corpuscle can be stimulated only by **deformation of the small region of the skin lying just above the receptor;** therefore, it has a **small receptive field.**
  c. Adaptation. Meissner's corpuscle **rapidly adapts to a maintained or slowly applied stimulus** (Figure 5-4).
   (1) Because the frequency of action potentials is proportional to the magnitude of the receptor potential, a slowly applied stimulus produces a lower frequency of nerve discharge than a rapidly applied stimulus.
   (2) Thus, the CNS uses the frequency of firing in a neuron innervating a Meissner's corpuscle to detect the speed of stimulus application.
    (a) Rapid deformation of the skin occurs when the skin is jabbed quickly with a probe or when the fingers move over a rough object.
    (b) The ability to detect the rate of skin deformation when the skin is moved over an object is especially important to individuals using braille.

 3. Merkel's disk is a slowly adapting receptor with a small receptive field that is used to encode the **location of a stimulus.**
  a. Description. Merkel's disk is unique because the transducer is not on the nerve terminal but on the epithelial cells that make up the disk. The epithelial sensory cells form synaptic connections with branches of a single group II afferent fiber (see Table 5-3).
  b. Receptive field. The disk is about 0.25 mm in diameter and can be stimulated only if the stimulus is applied directly to the disk. The small receptive field of the Merkel's disk makes it an ideal receptor to encode information about the location of the stimulus.

 4. Ruffini's corpuscle is a slowly adapting receptor with a large receptive field that is used to encode the **magnitude of a stimulus.**
  a. Description. The receptor is located on the terminal of a group II axon that is covered by a liquid-filled collagen capsule. Collagen strands within the capsule make contact with the nerve fiber and the overlying skin (see Table 5-3).
  b. Receptive field. Any deformation or stretch of the skin causes the nerve terminal to depolarize and generate action potentials.

FIGURE 5-4. Meissner's corpuscle encodes the rate of stimulus application. If the stimulus is applied fairly slowly, adaptation occurs during the application of the stimulus and the magnitude of the receptor potential is reduced. If the stimulus is applied very rapidly, little adaptation occurs during the application of the stimulus and a large receptor potential is produced.

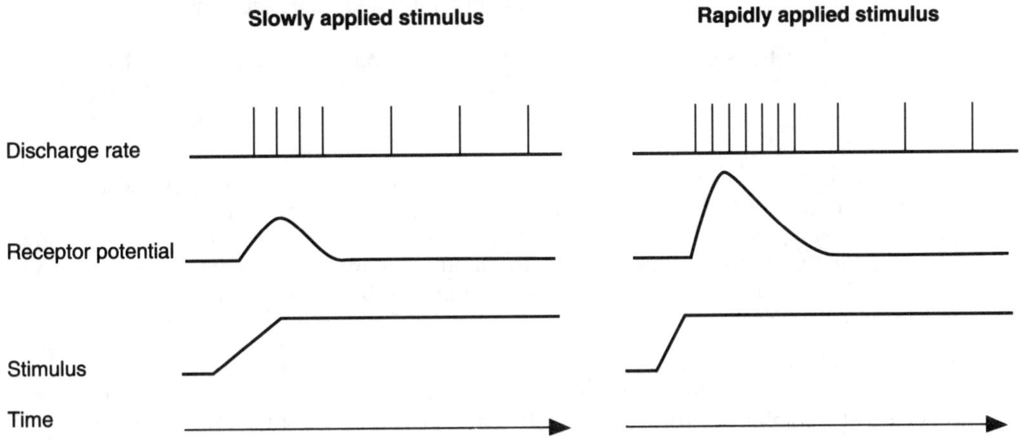

Slowly applied stimulus   **Rapidly applied stimulus**

Discharge rate

Receptor potential

Stimulus

Time

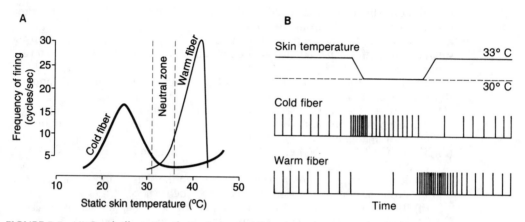

FIGURE 5-5. (*A*) Graph illustrating the tonic level of firing in both warm and cold fibers as a function of temperature. At temperatures within the neutral (comfort) zone, complete perceptual adaptation occurs (i.e., awareness of temperature disappears). (*B*) Spike trains illustrating the dynamic response of warm and cold fibers to a change in temperature. When the temperature decreases, cold fibers increase their rate of firing and then adapt to the firing rate indicated by the graph. Similarly, when the temperature increases, warm fibers phasically increase their firing rate before adapting to the rate indicated by the graph.

B. **Thermoreceptors.** Temperature sensation is encoded by thermoreceptors located on the free endings of small myelinated (Aδ) and unmyelinated (C) fibers. Separate receptors with discrete receptive fields exist for encoding warm and cold sensations.

1. **Warm fibers** are active when the skin temperature is between 30°C and 43°C (Figure 5-5A).
   a. The steady-state firing rate of warm fibers reaches a peak at temperatures of approximately 42°C.
   b. Warm fibers transiently increase their firing rate when skin temperatures increase, and decrease their firing rate when skin temperatures decrease (Figure 5-5B). Because the thermoreceptors respond transiently to the direction of a temperature change, the sensation produced by a small change in temperature depends on the current skin temperature. For example, a stimulus of 35°C will feel warm if the skin is at 30°C, and cool if the skin is at 40°C.

2. **Cold fibers** are active when the skin temperature is between 15°C and 38°C (see Figure 5-5A).
   a. The steady-state firing rate of cold fibers reaches a peak at temperatures between 23°C and 28°C. Paradoxically, temperatures between approximately 45°C and 49°C stimulate cold fibers as well as pain fibers, producing a mixed sensation of cold and pain.
   b. Cold fibers transiently increase their firing rate when skin temperatures decrease, and transiently decrease their firing rate when skin temperatures increase (see Figure 5-5B).

C. **Nociceptors.** Pain sensation is different from other sensations because its purpose is not to inform the brain about the quality of a stimulus, but rather to indicate that the stimulus is physically damaging. Although unpleasant, pain is a useful sensation if it leads to removal of the damaging stimulus. Much more information is required about the physiologic and psychologic mechanisms of pain before its elimination becomes a routine part of medical practice.

1. **Peripheral mechanisms**
   a. The **receptors for pain,** which are called nociceptors to indicate that they respond to noxious stimuli, are on the free nerve endings of small myelinated (Aδ) and unmyelinated (C) fibers.

       (1) Nociceptors are **specific for painful stimuli,** responding to damaging or poten-
tially damaging mechanical, chemical, and thermal stimuli. Cutaneous receptors
that respond to nonpainful levels of these stimuli do not elicit pain sensations no
matter how intense the stimulus.

       (2) Although the **adequate stimulus** for nociceptors is not known, it is assumed that
a chemical such as histamine or bradykinin is released from cells damaged by
the pain stimulus and that the chemical substance activates the nociceptors.

  b. Two types of pain sensation result from the application of a strong, noxious stimulus
to the skin.

       (1) Fast (initial) pain is a discrete, well-localized, pinprick sensation that results
from activating the nociceptors on the Aδ fibers.

       (2) Slow (delayed) pain is a poorly localized, dull, burning sensation that results
from activating the nociceptors on the C fibers.

  c. Pathways. Different pathways are used to reach the centers of consciousness in the
brain.

       (1) Somatic sensation

          (a) Fast pain. Action potentials that are propagated by the fast pain fibers travel
faster and, thus, reach the brain before those conducted by the slow pain
fibers. The sensory fibers for fast pain have small receptive fields, travel to
the cortex through the spinothalamic tract, and are topographically represented on the cortex—all factors that account for the ability of these fibers
to encode the location of the stimulus producing the fast pain.

          (b) Slow pain, which has a more diffuse pathway, travels to the brain through
the spinoreticulothalamic system. Collaterals of this system pass through the
reticular formation to activate fiber tracts that produce the emotional per-
ceptions accompanying pain sensations. These pathways account for the
intense unpleasantness associated with slow pain.

       (2) Visceral sensation. Referred pain (i.e., pain originating in visceral organs that is
referred to sites on the skin) most likely occurs because the visceral and somatic
pain fibers share a common pathway to the brain.

          (a) Because the **skin is topographically mapped and the viscera are not,** the
pain is identified as originating on the skin and not within the viscera.

          (b) Because of this anatomic relationship, **diagnosis of visceral disease can be
made based on the location of the referred pain**. For example, ischemic
heart pain is referred to the chest and the inside of the arm.

       (3) Projected pain. Referred pain should not be confused with projected pain,
which occurs as a result of directly stimulating fibers within a pain pathway.

          (a) Because a **labeled-line mechanism** is used to encode the location of the
pain, stimulation anywhere along the pathway will result in the same perception. For example, striking the elbow causes pain to be projected to the
hand.

          (b) Amputees often have sensations that appear to come from the severed limb.
This is known as **phantom limb sensation** and presumably results from activation of the sensory pathway either at the site of amputation or within the
CNS. Occasionally, the phantom sensation is one of pain, but with no obvious source of stimulation, it is difficult to eliminate the pain.

  d. Reflexes

       (1) Fast pain evokes a **withdrawal reflex** (see Chapter 7 III A) and a **sympathetic
response,** including an increase in blood pressure and a mobilization of body
energy supplies.

       (2) Slow pain produces nausea, profuse sweating, a lowering of blood pressure,
and a generalized reduction in skeletal muscle tone. (Pain sensations originating
in the muscles, blood vessels, and viscera produce similar reflexes.)

2. The **central mechanisms** of pain sensation are not well known.

  a. Generally, it is assumed that pain sensation is conveyed to the CNS through the **anterolateral quadrant**.

b. Chronic pain (i.e., a sensation of pain that endures long after the stimulus is removed and the injury is healed) is an extremely debilitating condition that is difficult to treat. Treatment is based on attempts to remove the pain area within the brain by surgery or to reduce the activity of the pain pathways by activating inhibitory pathways projecting to the pain areas; however, surgical section of the cord through the anterolateral quadrant is not very successful in relieving chronic pain.

    (1) Failure to relieve pain with this procedure may result from the existence of **parallel pain pathways** outside the anterolateral tracts.

    (2) Alternatively, chronic pain may result from the spontaneous activity of pain centers within the CNS.

        (a) Reverberating circuits that develop because of continuous pain input may fail to stop firing when the input is removed.

        (b) Denervation supersensitivity (i.e., an increased sensitivity to circulating neurotransmitters that occurs when the normal synaptic input is removed from a neuron) may develop in pain centers subsequent to the removal of pain fiber input.

IV. OLFACTORY SENSATION.

Olfactory stimuli are detected by specialized receptors located on the free nerve endings of the olfactory nerve (cranial nerve I) fibers.

A. **Olfactory (nasal) mucosa.** The olfactory receptors are located within the olfactory mucosa, which is distinguished from the surrounding respiratory mucosa by the presence of tubular **Bowman's glands,** the **absence of** the **rhythmic ciliary beating** that characterizes the respiratory mucosa, and a distinctive **yellowish-brown pigment**. A mucous layer covers the entire epithelium.

    1. Innervation. The olfactory mucosa is innervated by the olfactory nerve (cranial nerve I) and some branches of the trigeminal nerve (cranial nerve V). The irritative character of some odorants results from stimulation of the free nerve endings of the trigeminal nerve.

    2. Histologic structure. Three cell types comprise the olfactory mucosa (Figure 5-6).

        a. The **receptor cells** are **bipolar neurons**.

            (1) Their **dendrites** terminate in a knob. Cilia project from the knob into the mucous layer of the olfactory mucosa.

            (2) Their **axons** form the olfactory nerve.

FIGURE 5-6. The olfactory mucosa. Olfactory receptor cells are situated among supporting cells. The cilia lie within the mucous layer that covers the epithelium. New receptors are generated from basal cells. Bowman's glands, which contribute to mucus secretion, are not shown.

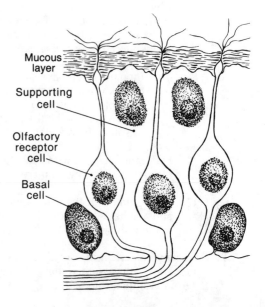

Mucous layer

Supporting cell

Olfactory receptor cell

Basal cell

    b. The **supporting cells** have a columnar shape. Microvilli extend from the surface of these cells into the mucous layer covering the nasal mucosa.

    c. The **basal cells** are **stem cells** from which new receptor cells are formed. There is a continuous replacement of receptor cells by mitosis of basal cells.

3. Location. The olfactory mucosa is located on the **superior nasal concha,** adjacent to the nasal septum. Because of its superior position in the nasal cavity, the olfactory mucosa is not directly exposed to the flow of inspired air entering the nose. Odorant molecules come in contact with the olfactory mucosa by **sniffing** (i.e., by short, forceful inspirations). Sniffing produces turbulence in the airflow and thereby transports molecules to the receptor cells.

B. **Stimuli. Odorant molecules** must dissolve in the mucous layer lining the nose before they can come in contact with the olfactory receptors. Therefore, to be effective, an odorant molecule must be:

1. Volatile, because the olfactory receptors respond to chemicals transported by the air into the nose

2. Water-soluble (to some degree), in order to penetrate the watery mucous layer lining the nasal epithelium to reach the receptor cell membrane

3. Lipid-soluble (to some degree), in order to penetrate the cell membranes of the olfactory receptor cells to stimulate those cells

C. **Receptors**

1. Sensitivity. The olfactory receptors are extremely sensitive. For many odorants, the sensitivity is so high that a few molecules interacting with receptors are sufficient to produce excitation.

    a. Number of odorants. The olfactory system can discriminate among a vast number of odorants. In some instances, the olfactory system can discriminate dextro- and levorotatory forms as well as *cis*- and *trans*- conformations of a molecule.

    b. Concentration. The olfactory system has limited ability to discriminate differences in odorant concentration in ambient air.

    c. Adaptation. Olfactory sensation adapts very rapidly with continued exposure to an odorant.

2. Receptor potential. The adsorption of odorant molecules to the plasma membrane of the cilia of the receptor cells generates a receptor potential in the receptor cell.

    a. A specific olfactory receptor does not respond to a particular compound or category of compounds; instead, an individual receptor responds to many odors. Furthermore, no two receptor cells have identical responses to a series of stimuli. **Sensory perception, therefore, is based on the pattern of receptors activated by the stimulus**.

    b. The electrical response recorded from the olfactory mucosa in response to an olfactory stimulus is called an **electro-olfactogram (EOG)**. The amplitude of the EOG increases when the intensity of the stimulus increases.

V. **GUSTATORY SENSATION.** Gustatory (taste) stimuli are detected by taste receptors within the tongue, mouth, and pharynx. Taste must be distinguished from flavor, which includes the olfactory, tactile, and thermal attributes of food in addition to taste.

A. **Stimuli. Sapid (taste-producing) substances** must dissolve in the saliva before they can stimulate the taste receptors. There are **four basic types** of taste sensations; all taste sensations are assumed to result from various combinations of these four primaries (Figure 5-7A).

1. A **sweet sensation** is produced by various classes of organic molecules including sugars, glycols, and aldehydes. The **tip of the tongue** is the area most sensitive to sweet stimuli.

A

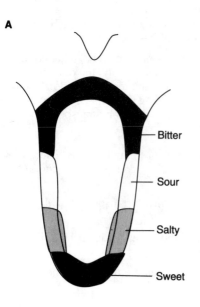

- Bitter
- Sour
- Salty
- Sweet

B

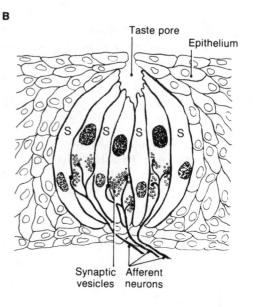

Taste pore

Epithelium

S   S   S   S

Synaptic   Afferent
vesicles   neurons

C

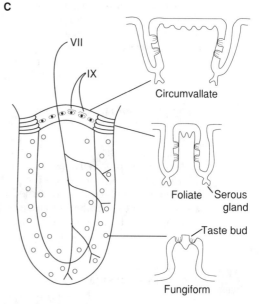

VII

IX

Circumvallate

Foliate   Serous
gland

Taste bud

Fungiform

FIGURE 5-7. (A) View of the dorsal lingual surface, showing the distribution of the four primary taste sensations. (B) Structure of a taste bud. Taste receptor cells contain synaptic vesicles and receive afferent nerve terminals. Supporting cells (S) are not innervated. (C) Distribution of the gustatory papillae. Innervation by the cranial nerves (*roman numerals*) is also indicated. (A and C redrawn and modified from Kandel EK, Schwartz JH, Jessell TM: *Principles of Neural Science,* 3rd ed. New York, Elsevier, 1991, p. 519.)

2. A **bitter sensation** is produced by alkaloids, such as quinine and caffeine. Many alkaloids are harmful when swallowed. The **back of the tongue** is the area most sensitive to bitter stimuli.

3. A **salty sensation** is produced by the anions of ionizable salts. The **front half of each side of the tongue** is the area most sensitive to salty stimuli.

4. A **sour sensation** is produced by acids; this sensation relates, to some degree, to the pH of applied stimulus solutions. The **posterior half of each side of the tongue** is the area most sensitive to sour stimuli.

B. **Receptors.** Taste cells are located within **taste buds,** which, in turn, are located within papillae.

1. **Taste buds**
   a. **Distribution.** Taste buds are located on the tongue papillae, hard and soft palate, epiglottis, and in the pharynx.
   b. **Structure.** Each taste bud is a cluster of 40–60 taste cells and numerous supporting cells (Figure 5-7B).
      (1) **Taste pore.** Each taste bud contains a taste pore that allows substances to reach the interior of the taste bud. The taste receptors are located on **microvilli,** which project from the taste cells into the taste pores.
      (2) **Taste cells** are modified epithelial cells that communicate with gustatory nerve endings by synaptic transmission.
         (a) **Innervation** (Figure 5-7C). Taste cells are innervated by branches of the facial, glossopharyngeal, and vagus nerves (cranial nerves VII, IX, and X, respectively). The tactile and temperature receptors of the mouth, tongue, and pharynx are innervated by the trigeminal nerve (cranial nerve V).
            (i) The taste buds in the **anterior two-thirds** of the tongue are innervated by lingual branches of the **facial nerve;** the lingual nerve, which branches from the chorda tympani, is part of the facial nerve. The cell bodies are located in the geniculate ganglion, and the nerve terminals end in the **nucleus solitarius** of the medulla.
            (ii) The taste buds in the **posterior third** of the tongue are innervated by the **glossopharyngeal nerve.** The cell bodies lie in the superior and inferior ganglia of this nerve. The fibers relating to taste sensation terminate in the nucleus solitarius.
            (iii) Taste receptors in the **pharyngeal aspect** of the tongue and on the hard palate, soft palate, and epiglottis are innervated by fibers of the **vagus nerve.** The cell bodies are located in the superior and inferior ganglia of the vagus nerve and terminate in the nucleus solitarius.
         (b) **Regeneration.** Each taste cell has a life cycle of only a few days.
            (i) The degenerating taste cell is replaced by a cell that arises from the supporting epithelial cells.
            (ii) Contact with the afferent neuron converts an epithelial cell into a taste cell. Conversely, nerve transection causes the disappearance of taste cells.

2. **Papillae.** There are four types of papillae; three contain taste buds and one contains mechanical receptors.
   a. **Gustatory papillae** (see Figure 5-7C)
      (1) **Fungiform papillae** are located in the anterior two-thirds of the tongue. There are 8–10 taste buds on each papilla.
      (2) **Circumvallate papillae** are arranged in a V-shaped row of 7–12 on the posterior part of the tongue. Each papilla has approximately 200 taste buds. These taste buds are located on the sides of these large structures.
      (3) **Foliate papillae,** located on the lateral border of the tongue anterior to the circumvallate papillae, have numerous taste buds.
   b. **Mechanical papillae.** Filiform papillae are not gustatory structures; however, they may play a role in breaking up food particles.

C. **Conversion of stimulus energy.** Taste receptors respond to taste stimuli in a variety of ways.

1. **Sweet-tasting substances** depolarize taste cells by:
   a. **Opening Na^+-selective channels.** These channels are blocked by amiloride and are similar to Na^+ channels found in renal and other epithelial cells.
   b. **Activating adenylate cyclase.** The cyclic adenosine 3′,5′-monophosphate (cAMP) produced by adenylate cyclase leads to the closing of K^+-selective channels.

2. **Bitter-tasting substances** stimulate the production of inositol triphosphate (IP_3) by taste cells. IP_3 increases intracellular Ca^{2+} levels, which leads to the release of synaptic transmitter and the activation of the gustatory nerve fiber.

3. **Salty-tasting substances** depolarize taste cells by activating an amiloride-sensitive Na^+ channel. No specific Na^+ receptor has been identified.

4. **Sour-tasting substances** (e.g., citric acid) depolarize taste cells directly by raising the intracellular H^+ ion concentration, which blocks K^+ channels.

D. **Encoding.** Each nerve fiber in the gustatory nerves responds to more than one taste stimulus. However, each fiber responds best to one of the four primary taste qualities. Therefore, the coding of a gustatory sensation is not a simple, labeled-line, chemical sensory system; instead, it **depends on the pattern of nerve fibers** activated by a particular stimulus.

Chapter 6

Vision and Audition

I. VISION

A. **Introduction.** A mental image of the external world is created by the visual system.

  1. An **image is formed on the retina** by the refractive surfaces of the eye.

  2. The **light energy is transduced into an electrical signal** by the rods and the cones.

  3. The **information needed to create the mental image is encoded** by the neurons within the retina.

  4. The information is used by the **visual cortex** to create the visual perception described as "seeing."

B. **Image formation**

  1. **Principles of image formation. Convex (converging) lenses** form real images of illuminated objects (Figure 6-1).
     a. **Object point.** The **light rays** emanating from each point on a luminous object (i.e., the object point) **diverge** from each other.
     b. **Image point.** When the **light rays** pass through the convex lens, the rays are **refracted** so that they bend toward each other and eventually **converge at a single point** (i.e., the image point).
     c. **Image distance.** The distance between the lens and the image point (i.e., the image distance) **depends on the converging power of the lens**.
        (1) **Determination of converging power.** The converging power of a lens can be calculated mathematically from the optical properties of the lens or determined experimentally by measuring the distance between the lens and the image when the rays of light entering the eye are parallel to each other.
           (a) **Focal distance.** When the rays of light entering the eye are parallel to each other, the image distance is called the focal distance of the lens.
           (b) The **converging power** of the lens **is the reciprocal of the focal distance:**

           $$P = \frac{1}{f} \text{ , where}$$

           P = the converging power in diopters (D)
           f = the focal distance in meters (m)

           (c) In practice, the converging power of the eye can be measured if the object is placed 20 ft (6 m) from the eye because, at this distance, the rays of light entering the eye from each point on the object are almost parallel to each other.
        (2) **Lens formula.** The relationship between the **object distance** (i.e., the distance from the lens to the object point), the **image distance,** and the **focal distance** is given by the lens formula:

        $$P = \frac{1}{o} + \frac{1}{i} = \frac{1}{f} \text{ , where}$$

        o, i, and f are the object, image, and focal distances, respectively (in meters) and P is the power of the lens (in diopters).

  2. The **cornea** and **lens provide the converging power of the eye**.
     a. The **cornea** is the **avascular, transparent outer surface** of the eye.
        (1) **Refractive power.** The cornea is responsible for approximately two-thirds of the refractive power of the eye, and its **refractive power cannot be altered physiologically**.

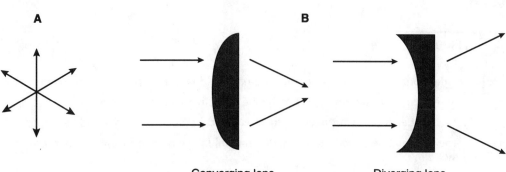

Converging lens   Diverging lens

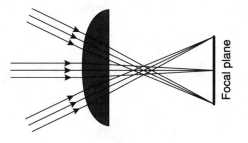

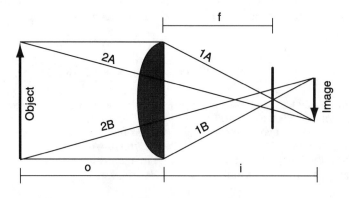

FIGURE 6-1. Fundamental optical principles. (*A*) Although light rays travel in all directions from a source of light, only those passing through the lens are used to form an image. (*B*) Converging lenses focus all rays of light coming from a point on an object onto a single image point. Diverging lenses cannot form an image. (*C*) If the rays of light entering a converging lens are parallel to each other, their image will be formed on the focal plane. (*D*) If the rays of light diverge when they enter the lens, the image will be formed behind the focal plane. Note that rays *1A* and *1B*, which are parallel, bend so that they pass through the focal point, but rays *2A* and *2B*, which are not parallel, do not bend. The relationship between the object distance (*o*), focal distance (*f*), and image distance (*i*) is given by the lens formula.

   (2) Transparency. The absence of blood vessels in the cornea allows light to pass
   through unhindered, but because it has no blood vessels, the cornea must receive O_2 and nutrients via diffusion through the **aqueous humor**.
   - **(a)** The aqueous humor is continuously secreted into the eye from the capillar-
     ies of the ciliary body in the posterior chamber, passes through the pupil
     into the anterior chamber, and returns to the circulation through the trabec-
     ular mesh located where the iris and cornea meet.
   - **(b)** Increased formation of aqueous humor or blockage of outflow can cause
     intraocular pressure to rise above its normal value of 10–20 mm Hg. In-
     creased intraocular pressure is called **glaucoma** and can lead to blindness if
     not treated.
 b. The **lens,** like the cornea, is avascular and transparent.
   - **(1) Refractive power.** Unlike the cornea, the **refractive power** of the lens is **under
     physiologic control** (see I B 3 b).
   - **(2) Transparency.** As an individual ages, the lens develops opacities called **cata-
     racts**. Normal vision can be restored by surgically removing the opaque lens
     and replacing it with a plastic lens.

3. **Focusing of image.** In a normal eye, an image of a distant object is focused on the retina by the cornea and lens.
   a. Reduced eye model (Figure 6-2). This diagram, which illustrates the optical properties of the eye, is called a reduced eye model because a single refractive surface
   represents the total converging power of the cornea and lens.
   - **(1)** The **axial length** (i.e., the distance between the single refractive surface and the
     retina in the reduced eye model) is 17 mm. In a normal eye, the distance from
     the cornea to the retina is approximately 25 mm.
   - **(2)** The **converging power** of the refractive surface is 58.8 D.
   - **(3)** The **image distance** of an object placed infinitely far from the eye is equal to the
     focal distance which, in this case, is $\frac{1}{P} = \frac{1}{58.8} = 17$ mm. That is, the image is
     focused on the retina.
   - **(4)** If the object is placed 6 m from the reduced eye model, the image distance (i),
     according to the lens formula, is 17.05 mm from the refractive surface:

$$\frac{1}{i} = \frac{1}{0.017} - \frac{1}{6}$$

   This is close enough to the axial length to be in focus on the retina.

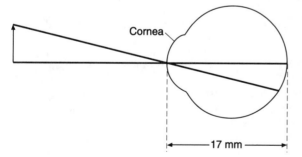

FIGURE 6-2. Reduced eye model. The converging power of the unaccommodated lens
(approximately 20 D) and the cornea (approximately 40 D) are combined into a single
refractive surface having a converging power
of 58.8 D and a focal distance of 17 mm.
The axial length of the reduced eye is 17
mm, so images of distant objects
(i.e., objects greater than 20 ft away) are
formed on the retina.

Cornea

17 mm

   b. Accommodation can increase the converging power of the eye in a young adult by
   approximately 12 D.
   (1) Accommodation reflex. Objects closer than 6 m (20 ft) can be focused on the
   retina by the accommodation reflex, which has three components.
   (a) Bulging of the lens. Contraction of the ciliary muscle releases tension on the
   zonular fibers. The elastic capsule surrounding the lens retracts, increasing
   the convexity (and thus the power) of the lens.
   (b) Pupillary constriction. By reducing the area through which light can enter
   the eye, **spherical aberration** is reduced. Reducing spherical aberration increases the **depth of focus** (i.e., the degree to which the image can form in
   front of or behind the retina and still appear to be in focus). **Squinting** also
   increases the depth of focus.
   (c) Convergence. The gaze of the two eyes shifts toward the center of the head
   in order to keep both eyes focused on the object.
   (2) The **near point** and **far point define the range of distances over which a clear
   image can be formed** by the eye.
   (a) The **near point** is the nearest point at which an object can be seen clearly.
   A typical young adult can increase his or her converging power by 12 D;
   therefore, at a maximum lens power of 70.8 D, the near point is approximately 8.3 cm from the eye.
   (b) The **far point** is the distance from the eye that an object must be placed so
   that it can be seen clearly without accommodation. For a normal person,
   this is 6 m.
   (3) Presbyopia is the loss of accommodative power that occurs with age. As an individual ages, his or her near point increases, making it difficult to see objects
   placed close to the eye. Reading glasses are converging lenses that compensate
   for the loss of accommodation.
   c. Refractive errors, in which distant objects cannot be seen clearly, are caused by
   variations in the converging power of the cornea, lens, or both.
   (1) Myopia (nearsightedness) results when the focal distance of the eye is less than
   the axial distance. In this case, the **focal point is in front of the retina;** thus,
   distant objects are not focused on the retina.
   (a) The object can be seen clearly if it is moved closer to the eye so that the
   image forms on the retina (i.e., behind the focal point). In severe myopia,
   the far point may be only 10–15 cm from the eye.
   (i) Because the **far point** is close to the eye in myopia, objects must be
   brought near to the eye to be seen clearly (hence the term "nearsightedness").
   (ii) Myopes have **near points** very close to the eye. Thus, they are able to
   do fine work without magnifying glasses.
   (b) A **diverging lens** can be placed in front of a myopic eye, reducing its converging power. When wearing such a lens, the myope can see distant objects clearly.
   (2) Hyperopia (farsightedness) results when the focal distance of the eye is greater
   than the axial distance. In this case, the **focal point is behind the retina;** thus,
   distant objects are not focused on the retina.
   (a) A distant object can be seen clearly if the hyperope increases his or her
   converging power by accommodating. If the degree of hyperopia is minimal, the person is usually unaware of the refractive error. However, if a
   large amount of **accommodation** is required to focus distant objects, the
   constant contraction of the ciliary muscle will cause **eye strain**.
   (b) Because some amount of accommodation is used to see distant objects,
   there is not so much available for near vision. Thus, the near point in hyperopes is farther from the retina than normal (i.e., objects must be placed
   farther from the eyes to be seen clearly, hence the term "farsightedness").
   (c) A **converging lens** can be placed in front of a hyperopic eye to increase its
   converging power. When wearing a converging lens, a hyperope can see
   distant objects clearly without accommodation.

(3) **Astigmatism** results from an **uneven cornea**. Normally, the cornea has a spherical surface; in astigmatism, it has more of an egg-shaped surface. As a result, the power of the lens is different in different axes.
  (a) **Images** formed with an astigmatic lens **are distorted**. For example, a point is seen as a line, and a line appears to have a halo on either side of it.
  (b) **Cylindrical lenses** can be worn by an astigmatic individual. These lenses increase converging power in only one axis, allowing the lens system to behave as a spherical surface.

C. **Energy transduction.** The **rods** and **cones** (Figure 6-3) are the photoreceptors of the eye.

  1. **Morphology.** Both cell types consist of:
    a. An **inner segment containing the nucleus, abundant mitochondria,** and **synaptic vesicles**
    b. An **outer segment containing membranous disks**
      (1) The membranous disks are **continuously formed at the base of the outer segment and migrate toward the apex,** where they are sloughed off.
      (2) The membranous disks **contain a visual pigment,** called **rhodopsin,** which absorbs light rays.
        (a) **Rhodopsin** consists of a protein called **opsin** and a light-absorbing analogue of vitamin A (retinol) called **11-*cis* retinal** (Figure 6-4).
        (b) The amino acid composition of opsin determines the wavelength of light absorbed by the photopigment.
          (i) **Rods** contain a **single type of opsin.** The gene encoding for rod opsin is located on chromosome 3.
          (ii) **Cones** contain **three types of opsins (blue, green,** or **red,** depending on the portion of the visual spectrum they absorb best).

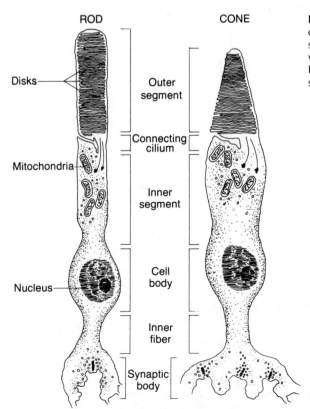

FIGURE 6-3. Morphology of rod and cone receptor cells. Cones, which are responsible for color perception and high visual acuity, are found in the fovea. Rods, which are responsible for night vision, are located in the peripheral retina.

A. Membrane depolarized

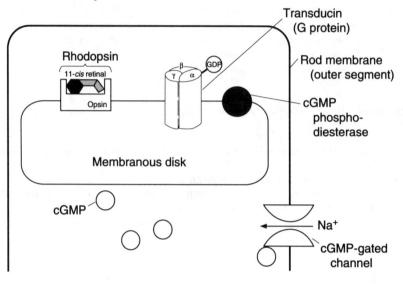

B. Membrane hyperpolarized

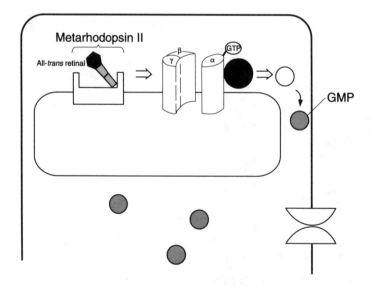

FIGURE 6-4. (*A*) In the dark, Na⁺ channels are kept open by cyclic guanosine monophosphate (*cGMP*) and the membrane is depolarized. (*B*) When rhodopsin absorbs light, the 11-*cis* retinal is converted to its more stable isomer, all-*trans* retinal. The photoisomerization of 11-*cis* retinal to all-*trans* retinal produces metarhodopsin II, the active form of rhodopsin. Metarhodopsin II activates transducin, a G protein located within the outer segment disk membrane. The activated transducin activates a cGMP phosphodiesterase that hydrolyzes cGMP, forming guanosine monophosphate (*GMP*). As the concentration of cGMP falls, cGMP dissociates from the Na⁺ channels, causing them to close. The reduced Na⁺ conductance causes the receptor to hyperpolarize. *GDP* = guanosine diphosphate; *GTP* = guanosine triphosphate.

2. Electrophysiology

    a. Darkness. Rods and cones are **depolarized** in the dark. Their resting membrane potential is low, approximately -40 mV.

        (1) The low resting membrane potential results from the **high Na$^+$ conductance of the outer segment** (see Figure 6-4A).

            (a) Na$^+$ flows into the cell through Na$^+$ channels in the outer segment and is transported out of the inner segment by Na$^+$-K$^+$ pumps.

                (i) Na$^+$ channels are **maintained in the open state** by **cyclic guanosine monophosphate (cGMP),** which is synthesized from guanosine triphosphate (GTP) by guanylate cyclase. When cGMP binds to the Na$^+$ channel, the channel opens. That is, in this case, cGMP acts by activating the channel directly, not by activating a protein kinase.

                (ii) The numerous mitochondria in the inner segment provide the large quantities of adenosine triphosphate (ATP) required to maintain the high Na$^+$-K$^+$ pump activity.

            (b) The large flow of current into the cell through the outer segment and out of the cell through the inner segment is called the **dark current**.

        (2) The low resting membrane potential allows **continuous release of synaptic transmitter**.

    b. Light. The photoreceptors **hyperpolarize** when stimulated by light. Absorption of light by rhodopsin initiates a series of reactions resulting in the hydrolysis of cGMP, the closing of the Na$^+$ channel, and the hyperpolarization of the cell (see Figure 6-4B).

3. Functions. The rods and cones perform different sensory functions.

    a. The **rods,** which are **more sensitive to light** than cones, are responsible for **night vision**.

        (1) Rods can **absorb more light** than cones.

            (a) Rods contain more rhodopsin in their outer segments.

            (b) Rods can detect light entering the eye from any direction whereas cones respond only to light directly along their axis.

        (2) Rods **produce a greater response for each photon of light absorbed**.

        (3) Rods **remain polarized for a longer time** than cones; therefore, the response from several photons of light can summate more easily in rods than in cones.

    b. The **cones,** which have **three different photopigments,** are responsible for **daylight, high acuity,** and **color vision**.

        (1) Cones can **respond to light over a large range of intensities** (e.g., from that produced by a light bulb to that produced by sunlight on a snow-covered field); however, more light, such as that provided by daylight or normal room lighting, is required to stimulate cones. In contrast, rods are saturated at fairly low light levels.

        (2) Cones achieve **high visual acuity** because they are **concentrated in the center of the retina** (where the clearest images are formed), they **do not respond to scattered light,** and their **response to light is brisk**. That is, the same properties that reduce the cones' sensitivity to light enhance their ability to produce an image of high quality.

        (3) **Color vision** is achieved by combining the information contained in cones, which absorb light in the red, green, or blue range of the visual spectrum (see I E 3).

D. **Encoding** of the visual stimulus is carried out by the cells in the **retina**.

1. Morphology of the retina (Figure 6-5). The retina is a thin sheet of cells consisting of three cellular and two synaptic layers. The retina is bounded at the back of the eyeball by the pigment epithelium and separated from the vitreous humor (on its inner surface) by the cell membranes of glial cells (Müller's fibers), which fill the spaces between the other retinal cells. Light has to pass through all of the retinal layers before it reaches the photoreceptors.

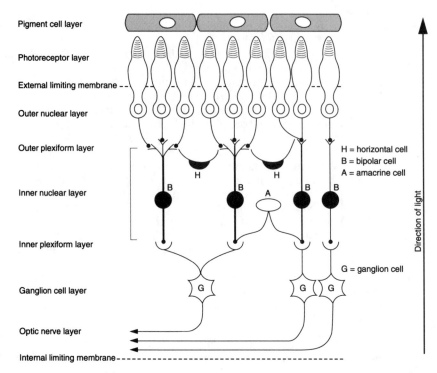

Pigment cell layer

Photoreceptor layer

External limiting membrane

Outer nuclear layer

Outer plexiform layer

H = horizontal cell
B = bipolar cell
A = amacrine cell

H

Inner nuclear layer

B

A

B

B

Inner plexiform layer

Direction of light

G = ganglion cell

Ganglion cell layer

G

G

G

Optic nerve layer

Internal limiting membrane

FIGURE 6-5. Organization of the retina. Photoreceptors converge on bipolar cells, which converge on ganglion cells. All of the receptors that convey information to a ganglion cell are part of that ganglion cell's receptive field.

  a. The **pigment epithelium** is a single sheet of melanin-containing epithelial cells.
    (1) **Functions**
      (a) **Light absorption.** Most of the light reaching the back of the eye is absorbed by the pigment epithelium so that light scattering does not degrade visual acuity.
      (b) **Phagocytosis.** The membranous disks and other debris sloughed from the photoreceptors is phagocytosed by the cells of the pigment epithelium.
      (c) **Vitamin A (retinol) storage.** The pigment epithelium serves as a repository for vitamin A, which is needed for the synthesis of rhodopsin.
    (2) **Clinical significance. Retinal detachment** (i.e., detachment of the rest of the retina from the pigment epithelium) can lead to **blindness;** however, reattachment can be accomplished by laser surgery.
  b. **Cellular layers**
    (1) **Outer nuclear layer (ONL).** The **nuclei of the photoreceptors** form the ONL, the **outermost cellular layer** of the retina.
    (2) **Inner nuclear layer (INL).** The **nuclei from neurons (bipolar, horizontal,** and **amacrine cells)** and **Müller's fibers** comprise the **second cellular layer,** the INL.
    (3) **Ganglion cell layer.** The **third cellular layer** of the retina is the ganglion cell layer. **Axons** of the ganglion cells **form the optic nerve**.
  c. **Synaptic layers**
    (1) **Outer plexiform layer (OPL).** Synaptic connections between the photoreceptors and horizontal and bipolar cells in the ONL are formed in the OPL.
    (2) **Inner plexiform layer (IPL).** Synaptic connections between the ganglion cells and bipolar and amacrine cells in the ONL occur in the IPL.

2. **Receptive fields.** Each ganglion cell collects information from a group of cones organized into a receptive field (Figure 6-6).
   a. **Circular region.** The receptive field of each ganglion cell occupies a circular region of the retina. In the fovea (where visual acuity is greatest), the receptive field may be only 10 μm in diameter. Near the periphery of the retina, receptive fields are much larger (up to 1 mm in diameter).
      (1) The cones in the center of the receptive field (the **field center**) convey information **directly** to the ganglion cell by synapsing with **bipolar cells.**
      (2) The cones at the periphery of the receptive field (the **field surround**) reach the ganglion cell **indirectly** through **horizontal cells.**
   b. **Receptive field properties**
      (1) **Organization.** The receptive fields of ganglion cells are organized into **center-surround antagonistic regions** (see Figure 6-6A).
         (a) In an **on-center, off-surround receptive field,** the ganglion cell is excited when light stimulates the cones in the center of the field, and inhibited when light stimulates the cones in the periphery of the field (see Figure 6-6B).
         (b) In an **off-center, on-surround receptive field,** the ganglion cell is inhibited when light strikes the cones in the center of the field, and excited when light stimulates the cones in the surround (see Figure 6-6C).
      (2) The properties of the center-surround receptive fields result from the synaptic organization of the retina. The behavior of the off-center receptive field is opposite that of the on-center receptive field.
         (a) **Illuminating the center of an on-center receptive field stimulates ganglion cell firing** (see Figure 6-6B).
         (b) **Illuminating the surround of an on-center receptive field inhibits ganglion cell firing** (see Figure 6-6D).
         (c) **Illuminating the center and surround of an on-center receptive field has no effect on ganglion cell firing** (see Figure 6-6E).
   c. **Neurotransmitters within the retina**
      (1) **Cones release glutamate.**
         (a) Glutamate's **inhibitory effect** on bipolar cells (in **on-center receptive fields**) is produced by **opening K^+-selective ion channels** or by **closing Na^+-selective ion channels**. Both mechanisms are thought to operate through a G protein–mediated second messenger system.
         (b) Glutamate's **excitatory effect** on bipolar cells (in **off-center receptive fields**) is produced by **opening cation-selective ion channels**.
      (2) **Horizontal cells release γ-aminobutyric acid (GABA).** GABA **produces both its inhibitory effect** (on on-center bipolar cells) **and its excitatory effect** (on off-center bipolar cells) **by depolarizing cones,** thus decreasing the amount of transmitter they release (i.e., GABA returns cones to the membrane potential they were at before being activated by light).

3. **Contrast and intensity.** The receptive field organization of the retina is used to transmit information about contrast and intensity of the visual image.
   a. **Contrast.** The change in a ganglion cell's firing rate is maximized when its receptive field is on the border between two different illumination levels (Figure 6-7).
   b. **Intensity.** Light intensity is encoded by the frequency of the ganglion cell firing rate.
      (1) **At moderate light levels, both cones and rods are involved** in the stimulation of ganglion cells.
         (a) Rods activate cones through gap junctions; therefore, they communicate with ganglion cells through cone bipolar cells.
         (b) The center-surround organization is preserved so the ability to see contrast and form can be maintained at fairly low light levels.
      (2) **At very low light levels, only rods are active.**
         (a) The gap junctions between rods and cones are closed at low light levels; therefore, the rods communicate with ganglion cells through rod bipolar cells.

A. Center-surround organization

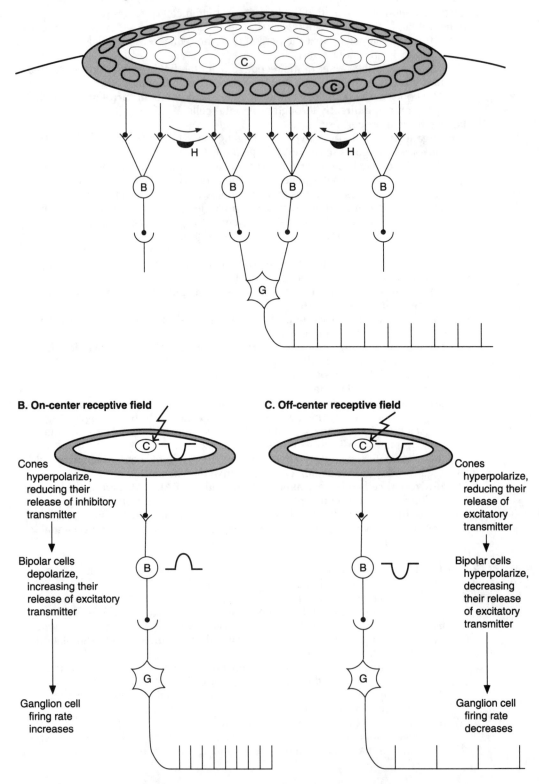

B. On-center receptive field

Cones
hyperpolarize,
reducing their
release of inhibitory
transmitter

Bipolar cells
depolarize,
increasing their
release of excitatory
transmitter

Ganglion cell
firing rate
increases

C. Off-center receptive field

Cones
hyperpolarize,
reducing their
release of
excitatory
transmitter

Bipolar cells
hyperpolarize,
decreasing
their release
of excitatory
transmitter

Ganglion cell
firing rate
decreases

FIGURE 6-6. (*A*) Center-surround organization of a ganglion cell. Those cones that converge on a ganglion cell (*unshaded circles*) are part of the field center. Those cones that affect ganglion cells via horizontal cells (*shaded circles*) are part of the field surround. The ganglion cell fires at a constant rate, even when the retina is not exposed to light. (*B,C*) On- and off-center receptive fields occupy overlapping regions of the retina so that light striking any region of the retina will cause on-center cones to increase their firing rate and off-center cones to decrease their firing rate. The intensity of the light stimulus is signalled by the difference in firing rate between on- and off-center receptive fields. (*D*) Illuminating cones in the field surround of an on-center receptive field will decrease the ganglion cell firing rate. (*E*) When light strikes the center and surround of an on-center receptive field, it produces opposite effects on the cones. The cones in the center of the field are hyperpolarized by light. At the same time, the horizontal cells that were activated by light striking the cones in the surround depolarize the cones in the center of the field. As a result, there is no change in transmitter release by the cones, no change in bipolar cell membrane potential, and no change in the ganglion cell firing rate. *B* = bipolar cell; *C* = cone; *G* = ganglion cell; *H* = horizontal cell.

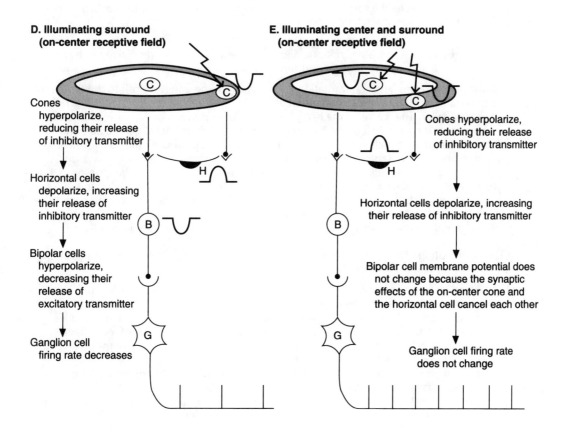

**D. Illuminating surround
(on-center receptive field)**

Cones
hyperpolarize,
reducing their release
of inhibitory transmitter

Horizontal cells
depolarize, increasing
their release of
inhibitory transmitter

Bipolar cells
hyperpolarize,
decreasing their
release of
excitatory transmitter

Ganglion cell
firing rate decreases

**E. Illuminating center and surround
(on-center receptive field)**

Cones hyperpolarize,
reducing their release
of inhibitory transmitter

Horizontal cells depolarize, increasing
their release of inhibitory transmitter

Bipolar cell membrane potential does
not change because the synaptic
effects of the on-center cone and
the horizontal cell cancel each other

Ganglion cell firing rate
does not change

A   B   C

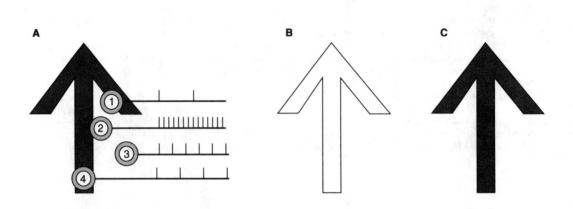

FIGURE 6-7. (A) Receptive fields of four ganglion cells are represented on an area of the retina covered by an image of a black arrow. When the center of an on-center receptive field is covered by a dimmer portion of the image than its surround (*receptive field 1*), the ganglion cell decreases its firing rate. When the center of an on-center receptive field is covered by a brighter portion of the image than its surround (*receptive field 2*), the ganglion cell increases its firing rate. If both the center and surround are illuminated equally (*receptive fields 3 and 4*), the ganglion cell's firing rate does not change. The ganglion cell associated with receptive field 3 has a slightly higher firing rate because it is in the light. (B) Because only the ganglion cells on the border of the arrow display a change in activity, only the outline of the arrow is transmitted to the cortex. (C) The brain creates the correct image of the arrow by assuming that the level of illumination does not change within or outside the borders of the arrow.

      (b) Because rods are not organized into center-surround receptive fields, the ability to see contrast and form is diminished.

    4. Information relay. Visual information is relayed from the ganglion cell to the cortex by cells within the **lateral geniculate nucleus**.
      a. Optic tract. The axons of the ganglion cells are rearranged at the **optic chiasm**. Axons from the right half of each retina project to the **right lateral geniculate,** and axons from the left half of each retina pass to the **left lateral geniculate**. This means that fibers of the **temporal** retina on each side pass through the chiasm without crossing, while fibers of the **nasal** half of each retina decussate at the chiasm.
      b. Lateral geniculate nucleus. The input from each eye remains segregated in the lateral geniculate; uncrossed fibers go to layers 2, 3, and 5, and crossed fibers go to layers 1, 4, and 6. Neurons in the lateral geniculate also retain the on-center, off-surround (or the off-center, on-surround) organization of the ganglion cells.

E.   **Visual perception.** The **primary visual cortex** (Brodmann's area 17 of the occipital lobe) decodes information about **contrast, color,** and **depth**.

    1. Contrast. Information about contrast is relayed from the lateral geniculate nucleus to cortical neurons called **simple cells**.
      a. Simple cells have **center-surround receptive fields**. Unlike the circular receptive fields of ganglion cells, simple cell receptive fields are **rectangularly organized**.
      b. Ganglion cells from each area of the retina project to a group of **columns**. Each column contains simple cells oriented in a particular direction.
        (1) The orientation of all of the simple cells in a small (approximately 50 μm wide) columnar region of the cortex is the same.
        (2) Adjacent columns have orientations that are approximately 10° apart; therefore, lines of any angle at any point on the retina are decoded by a specific column of cells in the cortex.

    2. Depth. Information about depth is relayed from the lateral geniculate nucleus to cortical neurons contained in **ocular dominance hypercolumns**.

    a. The **hypercolumn** contains simple cell columns that represent each angle of orientation.

    b. **Ocular-dominant cells** are **binocular** (i.e., they receive input from both eyes).

    c. **Depth perception** (i.e., **stereopsis**) is possible because each eye perceives a slightly different image.

        (1) The images are fused by the binocular ocular-dominant cells in the hypercolumns.

        (2) The amount of disparity between the images perceived by each eye is used to assign depth to the image.

            (a) Fixation point. Most binocular cells respond best when they receive stimuli from the **fovea of both eyes**.

            (b) Some binocular cells respond best when they receive stimuli from areas **medial** to the fixation point in both eyes. These images are interpreted as being **behind the fixation point**.

            (c) Some binocular cells respond best when they receive stimuli from areas **lateral** to the fixation point in both eyes. These images are interpreted as being **in front of the fixation point**.

3. Color. Information about color is relayed from the lateral geniculate nucleus to cortical cells organized into **oblong regions** called **blobs**. The blobs are located **within the hypercolumn** of cells that decode contrast and depth.

    a. **Ganglion cells** relaying information about color to the lateral geniculate nucleus respond oppositely to red and green light; therefore, they are called **color-opponent cells**.

        (1) **Red center, green surround** cells are stimulated by red cones in the center and inhibited by green cones in the surround.

        (2) **Green center, red surround** cells are stimulated by green cones in the center and inhibited by red cones in the surround.

    b. **Cortical cells** interpret the information received from the lateral geniculate to assign color to each region of the image formed on the retina. For example, if green center and red center cortical cells are equally excited, the color assigned that region of the cortex is yellow.

    c. **Color blindness** results from the absence of the gene that codes for either the red, green, or blue opsin.

        (1) **Red–green color blindness** results from the inability to synthesize either red or green pigment. The genes for these pigments are on the X chromosome (i.e., it is a sex-linked genetic defect). Approximately 9% of the male population has some sort of red or green color deficit.

        (2) **Blue color blindness** is very rare. The gene encoding for the blue pigment is on chromosome 7.

II. **AUDITION.** The detection of distant sounds may serve to warn of impending danger, or to localize friends. But most importantly, audition allows social communication.

A. **Structure of the ear.** Sound waves must travel through the **three divisions** of the ear, the **external, middle,** and **inner ear** (Figure 6-8), before reaching the auditory receptor cells within the organ of Corti.

    1. The **external ear** consists of the **pinna** (auricle) and the **external auditory canal**.

    a. **Pinna.** By transforming the sound field, the convoluted pinna **helps to identify sound sources**.

        (1) In **lower animals,** the cartilaginous pinna can be moved by muscular action in the direction of a sound source to collect sound waves for the receptor organ.

        (2) In **humans,** these muscles have little action, but the shape of the pinna aids in discerning the source of a sound (e.g., in front of versus behind the head).

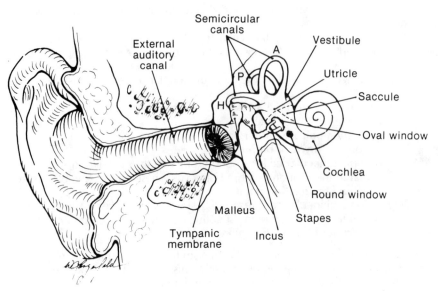

FIGURE 6-8. Major anatomic components of the human ear. The bony labyrinth consists of the saclike otolith organs (i.e., the saccule and utricle) and three semicircular canals: the horizontal (*H*), anterior (*A*), and posterior (*P*) canals.

    b. The **external auditory canal** extends from the pinna to the tympanic membrane of the middle ear. Its outer portion is cartilaginous and its inner portion is osseous.

2. The **middle ear (tympanic cavity)** is an **air-filled** cavity within the temporal bone that contains the **tympanic membrane (eardrum)** and three small bones called the **auditory ossicles (ossicular chain).**
    a. Tympanic membrane. Sound stimuli that pass through the pinna and external auditory canal strike the tympanic membrane, causing it to vibrate.
    b. Auditory ossicles. The vibrating tympanic membrane causes the middle ear bones to vibrate.
        (1) The **malleus** resembles a **mallet.** The manubrium (handle) of the malleus is connected to the inner surface of the tympanic membrane (see Figure 6-8).
        (2) The **incus,** which resembles an **anvil,** articulates with the head of the malleus (see Figure 6-8).
        (3) The **stapes** looks like a **stirrup.** The head of the stapes articulates with the incus, and the oval footplate contacts the membrane of the oval window of the cochlea (see Figure 6-8).

3. The **inner ear** is a **fluid-filled** cavity within the temporal bone that contains the **vestibular** and **auditory receptor apparatuses.**
    a. The **vestibular receptors** are contained within the **saccule, utricle,** and **semicircular canals** and are described in Chapter 7.
    b. The **auditory receptors** are contained within the **cochlea.**
        (1) Cochlear organization. The cochlea is a spiral tube which, in humans, has two and one-half turns (see Figure 6-8).
            (a) Two membranes divide the cochlea into **three compartments** (Figure 6-9).
                (i) Reissner's membrane separates the **scala vestibuli** from the **scala media.**
                (ii) The **basilar membrane** separates the **scala tympani** from the **scala media.**
            (b) Helicotrema. Reissner's membrane and the basilar membrane meet near the apex of the cochlea, allowing the scala tympani and scala vestibuli to join. The junction between the two scalae is called the helicotrema.

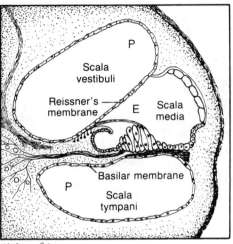

W. Langenfeld

FIGURE 6-9. The components of the cochlea shown in cross-section. P = perilymph; E = endolymph.

(c) **Fluid composition**
  (i) **Perilymph,** the fluid within the **scala tympani** and **scala vestibuli,** is **similar to extracellular fluid (ECF)** in that it is high in Na^+ and low in K^+.
  (ii) **Endolymph,** the fluid within the **scala media,** is more like **intracellular fluid (ICF)** in that it is high in K^+ and low in Na^+. Endolymph is secreted by the stria vascularis, which covers the lateral wall of the scala media.
(d) The vibrating middle ear bones cause the fluid within the cochlea to vibrate. The vibrating fluid ultimately stimulates the auditory receptors (i.e., the hair cells in the organ of Corti).
  (i) The footplate of the stapes contacts the cochlea at the **oval window,** which is the membrane-covered opening of the **scala vestibuli**.
  (ii) When the sound pressure wave impinges on the oval window, the **round window** (i.e., the membrane-covered opening of the **scala tympani**) bulges outward.
  (iii) As the sound energy passes from the scala vestibuli to the scala tympani, it causes the basilar membrane to vibrate, stimulating the organ of Corti.
(2) The **organ of Corti,** which is situated on top of the basilar membrane, contains **hair (auditory receptor) cells**.
  (a) **Two groups of hair cells** lie on the basilar membrane (see Figure 6-10).
    (i) The **inner layer** is a **single row** of hair cells. The inner hair cells are responsible for **fine auditory discrimination**. Over 90% of the auditory nerve fibers innervate these cells.
    (ii) The **outer layer** consists of **three rows** of hair cells. The outer hair cells are responsible for **detecting the presence of sound**. They receive approximately 10% of the auditory nerve fibers.
  (b) The hair cells contain **stereocilia,** which protrude into the overlying **tectorial membrane**.

B. **Stimuli. Sound waves** are produced by vibrating objects, which cause alternating phases of compression and rarefaction in the medium surrounding the object. The compressions and rarefactions spread out as a sound wave.

1. **Speed of sound**
  a. **In air,** sound travels at a rate of approximately **330 m/sec** (1100 ft/sec, 700 miles/hr).
  b. **In water,** sound travels much faster, at a rate of approximately **1500 m/sec**.

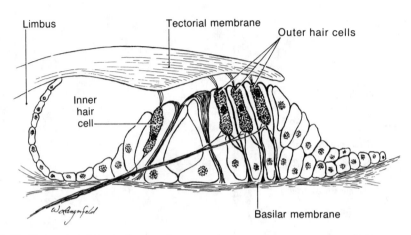

FIGURE 6-10. The organ of Corti, showing the connection between the tectorial membrane and the cilia of the hair cells.

   2. Frequency and **amplitude** characterize sound waves (Figure 6-11).
     a. The **frequency of sound** is measured in **hertz (Hz)**.
       (1) The **range for human hearing** is approximately 20–20,000 Hz.
       (2) The **range of the average speaking voice** is approximately 2000–5000 Hz.
     b. The **amplitude (intensity) of sound** is measured using a logarithmic scale. The unit of intensity is the **decibel (dB)**.
       (1) Decibels are measured using the formula:

$$dB = 20 \cdot \log \frac{P_{sound}}{P_{SPL}} \text{ , where}$$

$$\begin{aligned} dB &= \text{the number of decibels} \\ P_{sound} &= \text{the pressure of the sound stimulus} \\ P_{SPL} &= \text{the sound pressure level (SPL) at the threshold for} \\ &\quad \text{human hearing} \end{aligned}$$

          (a) Thus, sound intensities are measured on a ratio scale using a subjective intensity (i.e., the threshold), rather than an arbitrary intensity, as a base.
          (b) The actual SPL intensity is 0.0002 dynes/cm^2.
       (2) The formula can be used to calculate the decibels above threshold for any sound for which the amplitude is known. Sound pressures above 140 dB (10^7 times threshold) are painful and damaging to the auditory receptors.

FIGURE 6-11. The minimum audibility curve traces the threshold for hearing at different frequencies in the human hearing range. Maximum sensitivity occurs at about 4 kHz. The actual threshold at this frequency is 0.0002 dynes/cm^2 (or 0 dB).

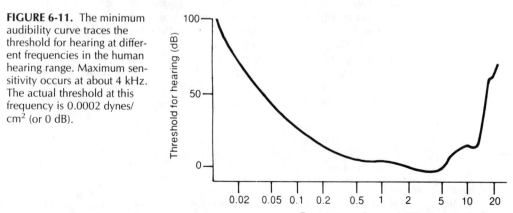

     (a) The sound pressures used during normal conversation are approximately 1000 times as great as threshold, or 60 dB.

     (b) An airplane produces a sound pressure of approximately 100,000 times that of threshold, or 100 dB.

C. **Conduction of stimulus energy.** The sound stimulus must be amplified if it is to be effectively transferred from the air-filled middle ear to the fluid-filled cochlea.

   1. **Impedance mismatching.** Effective transfer of sound energy from an air to a fluid medium is difficult because most of the sound is reflected as a result of the different mechanical (i.e., elastic, resistive, and inertial) properties of the two media.

   2. **Impedance matching.** The **middle ear** functions as an impedance matching device, primarily by amplifying the sound pressure.

     a. Amplification of sound pressure

       (1) Mechanisms. Together, the following effects increase the sound pressure 22-fold (i.e., by 27 dB).

         (a) Most amplification occurs because the area of the tympanic membrane (55 mm^2) is approximately 17 times greater than the **stapes–oval window surface area.** The size difference means that the force produced by the sound is concentrated over a smaller area, thus amplifying the pressure.

         (b) A small additional amount of amplification is obtained by the mechanical advantage that results from the leverage of the **auditory ossicles**.

       (2) The ability of the middle ear to amplify some sounds better than others accounts for the shape of the **minimum audibility curve** (see Figure 6-11).

         (a) Amplification is greatest for sounds between 2000 and 5000 Hz, the frequencies used for speech.

         (b) Sounds below 20 Hz or above 20,000 Hz are not amplified at all.

     b. Reduction of sound pressure. The **middle ear muscles** reduce sound pressure amplitude by affecting the mobility and transmission properties of the auditory ossicles, thereby reducing the pressure of sounds reaching the inner ear.

       (1) There are **two middle ear muscles.**

         (a) The **tensor tympani** is an elongated muscle that inserts on the manubrium of the malleus and is innervated by the **trigeminal nerve** (cranial nerve V).

         (b) The **stapedius** is a small muscle that inserts on the neck of the stapes and is innervated by the **facial nerve** (cranial nerve VII).

       (2) Significance. These muscles **contract reflexly in response to intense sounds**. Although the middle ear muscles may act to prevent receptors from being damaged by high-intensity sounds, contraction occurs too long after the stimulus to provide much protection. Muscle contraction also occurs just **prior to vocalization and chewing,** which suggests that the middle ear muscles may act to reduce the intensity of the sounds produced by these activities.

D. **Conversion of stimulus energy. Auditory transduction** occurs in the organ of Corti.

   1. **Movement of the organ of Corti**

     a. The in-and-out motion of the oval window produced by the pressure wave is converted into an up-and-down motion of the basilar membrane.

       (1) Sound waves entering the inner ear from the oval window spread along the scala vestibuli as a traveling wave.

       (2) Most of the sound energy is transferred directly from the scala vestibuli to the scala tympani. Very little of the sound wave ever reaches the helicotrema at the apex of the cochlea.

       (3) As the sound energy passes from the scala vestibuli to the scala tympani, it causes the basilar membrane to vibrate.

     b. The up-and-down motion of the basilar membrane causes the organ of Corti to vibrate up and down, which, in turn, causes the stereocilia to bend back and forth (Figure 6-12).

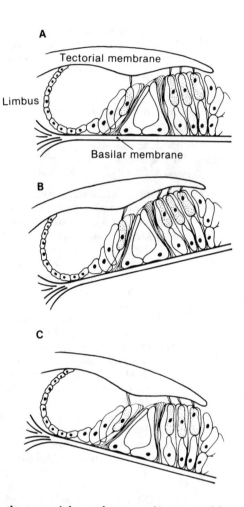

FIGURE 6-12. Up-and-down movement of the basilar and tectorial membrane causes the stereocilia extending from the hair cells to bend back and forth. (*A*) The tectorial and basilar membranes are attached to the limbus at different points. (*B*) When the membranes rotate upward, the tectorial membrane slides forward relative to the basilar membrane, bending the stereocilia away from the limbus. (*C*) When the membranes rotate downward, the stereocilia bend toward the limbus.

    (1) The **attachment of the stereocilia to the tectorial membrane** makes it possible for the up-and-down movement of the organ of Corti to be translated into the back-and-forth movement of the stereocilia.

        (a) The **bottoms of the hair cells** are anchored to the **basilar membrane,** and the **stereocilia** are connected to the overlying **tectorial membrane**.

        (b) Because the **tectorial and basilar membranes** are **attached at different points on the limbus** (see Figure 6-12A), they slide past each other as they vibrate up and down, causing the stereocilia on the hair cells to bend back and forth.

            (i) When the organ of Corti moves **up,** the **stereocilia bend away** from the limbus (see Figure 6-12B).

            (ii) When the organ of Corti moves **down,** the **stereocilia bend toward** the limbus (see Figure 6-12C).

    (2) **Polarization of the stereocilia**

        (a) When the organ of Corti moves **upward,** the stereocilia bend **away** from the limbus and they **depolarize**.

        (b) When the organ of Corti moves **downward,** the stereocilia bend **toward** the limbus and they **hyperpolarize**.

  2. **Receptor potential.** The hair cells are depolarized by the movement of K^+ into the cell.

    a. The electrochemical gradient for K^+ on the apical surface of the hair cells (where the stereocilia are located; see Figure 6-9) favors the movement of K^+ into the cell.

      (1) The **endolymph** contains a high concentration of K^+ (135 mEq/L) and is **electrically positive** in comparison to the perilymph.

    (2) Hair cells, like all other cells, contain a high concentration of K$^+$ and are **electrically negative** in comparison to the perilymph. The inside negativity is maintained by Na$^+$-K$^+$ pumps on the basal surface of the hair cell (i.e., the surface that faces the perilymph of the scala tympani).

    (3) Because the endolymph is electrically positive and the hair cell is electrically negative, a **very large potential difference** (in excess of -100 mV) exists across the hair cell membrane.

    (4) The large negative potential and lack of a K$^+$ concentration difference between the inside and outside of the hair cell create a **driving force, pushing K$^+$ into the cell**.

  b. The **gating of the K$^+$ channels** is controlled by the bending of the stereocilia.

    (1) When the **stereocilia bend away** from the limbus, they cause **K$^+$ channels** to **open**. K$^+$ then flows into the cell and the **hair cell depolarizes**.

    (2) When the **stereocilia bend toward** the limbus, they cause **K$^+$ channels** to **close** and the **hair cell hyperpolarizes**.

  c. Release of synaptic transmitter

    (1) When the hair cell depolarizes, a Ca^{2+} channel opens, allowing **Ca^{2+}** to enter the cell. Ca^{2+} initiates the release of a synaptic transmitter, which stimulates the auditory nerve fiber.

    (2) The cell bodies of the auditory nerve fibers are located within the **spiral ganglion**. Their axons join those from the vestibular apparatus to form the **vestibulocochlear nerve (cranial nerve VIII)**.

E. Encoding

1. Frequency

  a. Place principle of frequency determination. The frequency of sound that activates a particular hair cell depends on the location of the hair cell along the basilar membrane. The basilar membrane is narrowest and stiffest at the base of the cochlea (near the oval and round windows) and widest and most compliant at the apex of the cochlea (near the helicotrema).

    (1) The energy contained in high-frequency sounds passes through the organ of Corti near the base of the cochlea.

    (2) In contrast, the energy in low-frequency sounds passes through the organ of Corti near the apex of the cochlea.

  b. Encoding of frequency. The auditory nerve fiber activated by a particular sound frequency is similarly dependent on the location of the hair cell it innervates.

    (1) There are about 30,000 nerve fibers in each auditory nerve.

    (2) For low-frequency sounds, the auditory nerve fibers can fire at the same frequency as the sound wave. This mechanism of sensory encoding is called the **volley principle of frequency discrimination**.

    (3) The sound frequency producing the greatest response in an auditory nerve fiber is called the **characteristic frequency** of that nerve fiber.

2. Intensity. The frequency of firing in an auditory nerve fiber increases as the intensity of the sound wave increases. In addition, a larger portion of the basilar membrane is vibrated as the sound intensity increases so that more auditory nerve fibers are activated. Thus, sound intensity is encoded by the frequency of auditory nerve discharge and by the number of auditory nerve fibers that are active.

3. Inhibitory innervation. The hair cells receive a very prominent efferent innervation from the superior olivary nucleus via the olivocochlear bundle. Although such a large pathway probably plays an important role in auditory transduction, its purpose has not been deciphered.

4. Auditory pathways and cortex

  a. Tonotopic organization. Generally, any neuron of the auditory pathway can be tested with tones of different frequencies to determine a characteristic frequency for that cell. Neurons responding best to low-frequency tones will be located at one end of a nucleus, while neurons responding best to high-frequency tones will be represented at the opposite end of the nucleus. This orderly arrangement of frequency sensitivity, termed tonotopic organization, resembles the retinotopic organization of the visual system and the somatotopic organization of the somatosensory system. The tonotopic map reflects the methodical arrangement of frequency sensitivity along the length of the basilar membrane from base to apex. Tonotopic organization is prominent in the cochlear nuclei but becomes less precise in more rostral structures of the auditory pathway.

b. **Feature detection.** Higher auditory centers respond to particular features of sound stimuli. For example, cortical neurons may respond specifically to a shift from high- to low-frequency notes, which is why lesions of the auditory cortex may not impair the ability to discriminate frequency. Instead, lesions of the auditory cortex cause a loss of ability to recognize a patterned sequence of sounds. In addition, the ability to identify the position of a sound source is impaired.

c. **Localization of sound in space.** Detection of the position of a sound source depends on the ability of the central nervous system (CNS) to compare intensity differences and phase differences. A sound source located behind the head, closer to one ear than the other, produces a slightly more intense sound in the near ear than in the remote ear. Sound absorption by the tissues of the head attenuates sound intensity in the remote ear. In addition, at low frequencies there may be phase differences in the sound waves striking the two ears. These minute phase and intensity differences between the two ears are detected and discriminated by the auditory system. The extensive decussation and complex circuitry of the auditory pathway provide the CNS with the information necessary to identify a sound source.

F. **Hearing impairments**

1. **Tinnitus** is a ringing sensation in the ears caused by irritative stimulation of either the inner ear or the vestibulocochlear nerve.

2. **Deafness.** Two classes of deafness are distinguished based on the location of the abnormality or lesion.

    a. **Conduction deafness** is caused by interference with the transmission of sound to the sensory mechanism of the inner ear. Conduction deafness represents a defect of the external or middle ear.

    b. **Nerve deafness** results from defects of either the inner ear or the vestibulocochlear nerve. Deafness due to a lesion in the CNS structures generally is not found clinically because of the redundancy and bilateral nature of the central auditory pathways.

Chapter 7

Motor Control System

I. COMPONENTS OF THE MOTOR CONTROL SYSTEM (Figure 7-1).

A. **Cerebral cortex.** The cerebral cortex is responsible for generating the idea for voluntary movements and issuing the motor commands for their execution.

B. **Subcortical centers** are responsible for modulating and coordinating the motor commands so that tasks are properly carried out.

    1. The **basal ganglia** provide the motor patterns necessary to maintain the postural support required for motor commands to be carried out properly.

    2. The **cerebellum** receives information from the motor cortex about the nature of the intended movement and from the spinal cord about how well it is being performed. This information is used to adjust the motor command so that the intended movement is executed smoothly.

    3. The **brain stem** is the major relay station for all motor commands except those requiring the greatest precision, which are transferred directly to the spinal cord. In addition, the brain stem is responsible for maintaining normal body posture during motor activities.

C. **Spinal cord.** The spinal cord contains the final common pathways through which a movement is executed. By selecting the proper motoneurons for a particular task and by reflexly adjusting the amount of motoneuron activity, the spinal cord contributes to the proper performance of a motor task.

D. **Receptors** provide sensory feedback to the central nervous system (CNS) that can be used to adjust the motor commands during a movement.

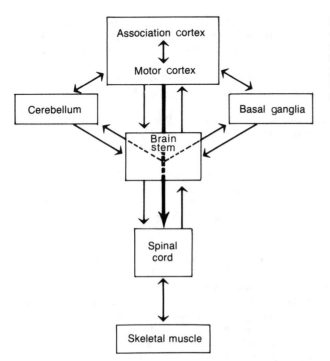

FIGURE 7-1. Diagram illustrating the extensive interconnections between the components of the motor control system. Note that all of the descending pathways except for the pyramidal tract (*thick arrow*) communicate with the spinal cord through the brain stem.

1. **Proprioceptive information** (i.e., **unconscious information** about the position of the body in space and the tension produced by the contracting skeletal muscles) is provided by the **joint, muscle,** and **vestibular systems**.

2. **Conscious information** about the position of the body and limbs in space is provided primarily by the **visual** and **cutaneous sensory organs**.

II. **MOTOR UNITS.** The motor unit is the functional module used by the motor control system to carry out a movement.

A. **Components.** A motor unit consists of a **motoneuron** and **all the muscle fibers it innervates**.

1. **Alpha motoneurons** are the final common pathway over which the motor control system coordinates the activity of skeletal muscle fibers.
   a. **Function.** If an alpha motoneuron is stimulated, skeletal muscle fibers contract; if the alpha motoneuron is not stimulated, the skeletal muscle fibers relax. Therefore, each component of the motor control system produces its effect on movement by altering the amount of excitation or inhibition impinging on the alpha motoneurons within the CNS.
   b. **Organization.** The alpha motoneurons are organized into **motoneuron pools**. Gamma motoneurons are randomly interspersed among alpha motoneurons within the pool, and some overlap among neurons from separate motoneuron pools takes place.
      (1) Motoneurons innervating distal muscle are located more laterally within the ventral horn than motoneurons innervating proximal muscles.
      (2) Motoneurons innervating extensor muscles are located more dorsally within the ventral horn than motoneurons innervating flexor muscles.
      (3) Usually, the motoneurons leave the spinal cord in several contiguous ventral roots and then combine into a single motor nerve containing the alpha and gamma motoneurons as well as the Ia, Ib, and II afferent fibers from the muscle [see III B 1 a (2)].

2. **Muscle fibers.** All of the muscle fibers in a motor unit are of the same physiologic type and are categorized according to their **histochemical** and **contractile characteristics**.
   a. **Fast-twitch fatigable fibers** contract quickly, fatigue easily, and have the following characteristics.
      (1) **Rapid contractile speeds** result from the high myosin–adenosine triphosphatase (ATPase) activity of the cross-bridges and the rapid sequestering of Ca^{2+} by the sarcoplasmic reticulum (SR).
      (2) **Rapid fatigue** occurs because fast-twitch fatigable fibers have few mitochondria and, thus, cannot make use of oxidative metabolism. They rely on glycolysis for their adenosine triphosphate (ATP) supply and fatigue when their glucose stores are depleted.
      (3) **Sparse capillary supply.** Because fast-twitch fatigable fibers do not make use of oxidative metabolism, the growth of surrounding capillaries is limited.
      (4) **Large size.** Although fast-twitch fatigable fibers cannot sustain activity for long periods of time, they can generate large contractile forces. Their large size is not a disadvantage from a diffusional point of view because they do not make use of oxidative metabolism.
   b. **Slow-twitch fibers** contract slowly, are virtually untiring, and have the following characteristics.
      (1) **Slow contractile speeds** result from the low myosin–ATPase activity of the cross-bridges and the slow sequestering of Ca^{2+} by the SR. The long contraction times make summation and tetanus possible at low frequencies of stimulation.
      (2) **Great resistance to fatigue** is a consequence of the ability of slow-twitch fibers to use oxidative metabolism as a primary source of ATP. The low myosin–ATPase activity and slow sequestering of Ca^{2+} by slow-twitch fibers reduce the amount of ATP they require.

(3) **Rich capillary supply.** O_2 needed by slow-twitch fibers during oxidative metabolism encourages the growth of surrounding capillaries.

(4) **Small size.** Although slow-twitch fibers cannot produce a large amount of force, they can sustain force for a long time. Their small size allows O_2 to diffuse into the center of the fiber and waste products to diffuse out of the fiber.

c. **Fast fatigue-resistant fibers** have characteristics of both fast-twitch fatigable and slow-twitch muscle fibers.

B. **Characteristics**

1. The motor units within a muscle vary in size from a few muscle fibers to several thousand muscle fibers. Muscles that perform precise movements (e.g., the extraocular muscles or those responsible for finger movements) have smaller motor units than those responsible for large body movements and for maintaining posture.

2. Whenever an alpha motoneuron fires, all of the muscle fibers in its motor unit are activated.

3. **Distribution of muscle fibers.** The muscle fibers belonging to a single motor unit are dispersed throughout the muscle so that the force they produce is distributed evenly.

C. **Recruitment.** The orderly recruitment of motor units during a contraction enhances the ability of the motor control system to carry out its task. During the performance of a motor task, the small motor units, because they are more excitable, are recruited before the large ones. This **size principle of motor unit selection** has significant physiologic advantages.

1. **Simplification of motor command structure.** Because of the size principle, the motor cortex does not need to specify the particular motoneuron to be recruited during a movement. Therefore, the number of cortical neurons that are involved in generating a movement is reduced.

a. **Force.** To perform a precision movement requiring small amounts of force, it is advantageous to use small motor units. When more force is required, larger motor units must be activated.

(1) When a small amount of force must be applied, the motor cortex provides a minimal amount of input to the motoneuron pool, activating only the smallest motor units.

(2) If more force is required, the motor cortex increases its input to the motoneuron pool, and larger motor units are recruited.

b. **Endurance.** To perform a task requiring endurance, the smaller, fatigue-resistant motor units must be recruited. When power is required, the larger motor units must be recruited.

(1) When an endurance movement must be performed, the motor cortex provides a minimal input to the spinal cord, and the smallest, most fatigue-resistant motor units are recruited.

(2) When the amount of force being generated by the muscle is not sufficient to execute the movement, the motor cortex increases its input and recruits more motor units. In all cases, the most fatigue-resistant fibers are recruited first without requiring that the motor cortex determine which motoneuron to activate.

2. **Development of fatigue resistance.** The relationship between fatigue resistance and muscle fiber size is a consequence of the size principle. A muscle fiber's fatigability can be altered by its activity. Small motor units are recruited first. Thus, they are involved in all movements and, consequently, develop fatigue resistance.

III. **SPINAL CORD REFLEXES** enhance the ability of the motor control system to produce a coordinated movement. A reflex is an automatic response to a stimulus carried out by a relatively simple neuronal network consisting of a receptor, an afferent pathway, and an effector organ. Spinal cord reflexes are categorized according to the receptor from which they originate.

A. **Cutaneous reflexes.** The most important of the cutaneous reflexes is the **withdrawal (flexor, pain) reflex,** which effects the removal of a body part from a painful stimulus.

1. **Receptors** for the withdrawal reflex are the **nociceptors** located on the free nerve endings of Aδ and C fibers.

2. **Effector organs** of the withdrawal reflex are the **skeletal muscles** that cause withdrawal of the limb. Although they are called flexors, these muscles are **flexors in the physiologic, not anatomic, sense**. For example, the muscles that cause the fingers to open in order to drop a hot coal, although anatomically referred to as extensors, are considered flexors, because they are involved in the withdrawal reflex.

3. **Polysynaptic (multisynaptic) pathway.** Limb withdrawal is produced by a polysynaptic pathway that begins with the stimulation of a cutaneous pain afferent and ends with the firing of the alpha motoneuron that excites flexor muscles.
   a. **Excitation of flexor muscles.** Upon entering the spinal cord, the pain fibers synapse on many **interneurons**. Some of these convey information to the CNS. Other interneurons contribute to reflex pathways that coordinate the withdrawal of the limb.
      (1) **Afterdischarge** refers to the continuation of reflex withdrawal even after the sensory receptor has stopped firing. Afterdischarge is produced by **reverberating circuits** (i.e., a branch from the axon of one interneuron in the reflex pathway feeds back onto previously excited neurons, reexciting them and prolonging alpha motoneuron firing).
      (2) **Local sign** refers to the ability of the reflex to confine the withdrawal to the portion of the body affected by the noxious stimulus. For example, if an individual accidentally touches a hot stove, it is likely that he or she will only jerk the hand away from the stove.
      (3) **Irradiation** refers to the activation of a large number of muscles when the noxious stimulus is strong enough. For example, if an individual picks up a hot coal, not only will the fingers open to drop it, but the entire arm will withdraw and the individual may even leap away from the fire.
      (4) **Crossed extensor reflex.** Interneurons form pathways that cross the spinal cord to innervate the extensor motoneurons on the contralateral side. In the lower limbs, the crossed extensor reflex allows one limb to support the body while the other is raised off the ground.
   b. **Inhibition of the antagonist to the flexor muscle** is produced by a polysynaptic pathway that begins with the stimulation of a cutaneous pain afferent and ends with the inhibition of the alpha motoneuron that innervates the antagonistic muscles. This type of neuronal organization, in which the reflex pathway activating one group of alpha motoneurons also inhibits its antagonistic motoneuron, is quite common within the spinal cord and is called **reciprocal innervation**. By inhibiting the motoneurons that innervate muscles antagonistic to those withdrawing the limb, reciprocal innervation **ensures that the flexion movement is not impeded by contraction of the extensors**.
   c. **Integration of the withdrawal reflex** occurs on the alpha motoneuron, the final common pathway through which all the afferent fibers act. If the excitatory pathways dominate, the alpha motoneuron discharges a train of action potentials; if the inhibitory pathways dominate, the neuron does not fire.
   d. **Characteristics of the withdrawal reflex**
      (1) **Long latency.** The withdrawal reflex has a relatively long latency because the afferent pathway uses small, slowly conducting fibers and involves many synapses.
      (2) **Response outlasts stimulus.** The afterdischarge that results from the parallel pathways and reverberating circuits causes the response to outlast the stimulus, keeping the affected limb away from the painful stimulus while the brain determines where to place it next.
      (3) **Patterned response.** The crossed extensor reflex produces a patterned response in which the affected limb flexes while the contralateral limb extends.

B. **Muscle reflexes.** Two important reflexes originate in the muscles: the stretch reflex and the lengthening reaction.

1. The **stretch reflex** causes the reflex contraction of a muscle that is stretched. For example, when the patellar tendon is tapped by a reflex hammer, it stretches the quadriceps muscle. The stretched muscle contracts reflexly, elevating the leg (i.e., the **knee jerk reflex** takes place).
   a. **Receptor.** The receptor for the stretch reflex is the **muscle spindle**. The number of spindles in each muscle depends on the task performed by the muscle. Muscles involved in precision movements contain many more spindles than muscles used to maintain posture. For example, hand muscles have approximately 80 spindles, which is 20% of the number of spindles contained in back muscles weighing 100 times as much. This complex, spindle-shaped, encapsulated receptor contains muscle fibers that have both sensory and motor innervation (Figure 7-2).
      (1) **Intrafusal muscle fibers.** The muscle fibers within the spindle are called **intrafusal muscle fibers,** in contrast to **extrafusal muscle fibers,** which are responsible for generating tension.
         (a) There are **two major types** of intrafusal muscle fibers.
            (i) The **nuclear bag fibers** are 30 μm in diameter and 7 mm in length. Their **nuclei** appear to be **gathered in the center of the cell** as if in a bag. Approximately **2–5** nuclear bag fibers exist in a typical spindle.
            (ii) The **nuclear chain fibers** are 15 μm in diameter and 4 mm in length. Their **nuclei** are **lined up** in a **single file** in the center of the fiber. Approximately **6–10** nuclear chain fibers exist in each spindle.
         (b) The connective tissue surrounding the intrafusal fibers is continuous with that of the extrafusal fibers. As a result, when the extrafusal fiber contracts, the intrafusal fiber is shortened, and when the extrafusal fiber stretches the intrafusal fiber is lengthened. The **muscle spindle,** thus, is **in parallel** with the **extrafusal muscle fibers**.
      (2) **Afferent sensory neurons**
         (a) There are **two types** of sensory neurons that emerge from the muscle spindle.
            (i) A **single large fiber,** called a **group Ia fiber (primary ending),** sends branches to **every intrafusal fiber** within the muscle spindle.
            (ii) **Several smaller neurons,** called **group II fibers (secondary endings),** innervate the **nuclear chain fibers.**

FIGURE 7-2. Diagram of an intrafusal muscle fiber, showing its nuclear bag and nuclear chain fibers. The afferent innervation (Ia and II fibers) and efferent innervation (gamma dynamic and gamma static fibers) of the intrafusal muscle fiber also are illustrated.

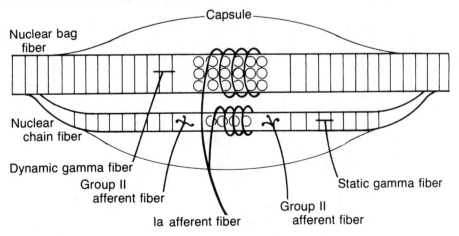

      (b) The **primary endings surround the center** of the intrafusal muscle fiber; the **secondary endings terminate on either side** of the primary endings.

    (3) Efferent fibers. The efferent fibers to the muscle spindle are called **gamma fibers** because their axons belong to the Aγ group of fibers. Gamma efferent fibers **control the sensitivity of the receptors to stretch** [see III B 1 d (2)].

      (a) There are **two types** of gamma fibers.

        (i) Dynamic gamma fibers primarily **innervate nuclear bag fibers** and **increase the sensitivity of the Ia afferent fiber to stretch**. That is, the Ia afferent firing rate for a given velocity of stretch is increased by the discharge of the dynamic gamma fibers.

        (ii) Static gamma fibers primarily **innervate nuclear chain fibers** and **increase the tonic activity** in the Ia afferent fibers at any given muscle length.

      (b) Some nuclear bag fibers have characteristics similar to those of nuclear chain fibers. These fibers, called static nuclear bag fibers, are innervated by the static gamma fibers.

b. Effector organs. Both **extensor** and **flexor muscles** exhibit stretch reflexes, which are elicited routinely during neurologic examinations to test for damage to either the spinal cord or the sensory or motor neurons.

c. Monosynaptic pathway. The Ia fiber enters the spinal cord through the dorsal root and sends branches to every alpha motoneuron that goes to the muscle from which the Ia originated.

    (1) Monosynaptic pathway. Because of its monosynaptic pathway, the stretch reflex **does not exhibit afterdischarge or radiation**.

    (2) Reciprocal innervation. Like the withdrawal reflex, the stretch reflex is characterized by reciprocal innervation. When a stretch reflex is elicited, the muscle antagonistic to the stretched muscle is inhibited, allowing the agonistic muscle to contract without interference.

    (3) Integration. Like the withdrawal reflex, the **alpha motoneuron** is the final common pathway, serving as both and **integrating center and efferent pathway**.

    (4) Characteristics of the stretch reflex. The rapidly conducting afferent fiber of the stretch reflex allows for a **short latency**.

d. Role. The stretch reflex is used by the motor control system to aid in the performance of a movement. During activity generated by the motor command center, the Ia fibers from the muscle spindle inform the motor control system about the changes in muscle length and provide the alpha motoneuron with a source of excitatory input in addition to that coming from higher centers.

    (1) Ia afferent discharge increases when the muscle stretches and decreases when the muscle contracts (Figure 7-3).

      (a) In parallel. Because the extrafusal and intrafusal muscle fibers are in parallel, stretching the extrafusal muscle fibers stretches the intrafusal muscle fibers as well.

      (b) Central region. Most of the stretch occurs in the central, more compliant region of the intrafusal fiber, which lacks sarcomeres.

      (c) Deformation. Stretching the central region deforms the primary endings of the Ia afferents, opening ion channels that cause the membrane to depolarize and generate a train of **action potentials**. The action potentials **discharge phasically** at a frequency proportional to the **velocity of stretch** and then adapt to **discharge statically** at a frequency proportional to the **amount of stretch**.

        (i) The central region of the **nuclear bag fiber** stretches rapidly when the muscle is rapidly stretched and then returns to its initial length. Thus, the nuclear bag fiber produces a **phasic discharge** of the Ia afferents that is proportional to the rate of stretch.

        (ii) The central region of the **nuclear chain fiber** stretches as the muscle stretches and then maintains its length. Thus, the nuclear chain fiber produces a **static discharge** that is proportional to actual muscle length.

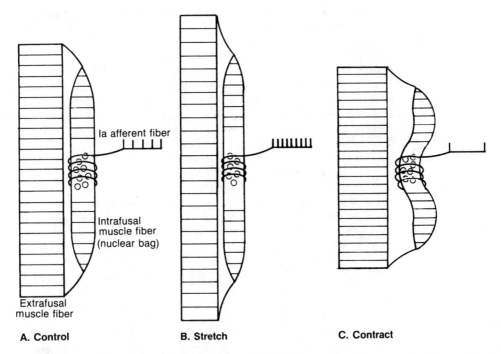

la afferent fiber

Intrafusal
muscle fiber
(nuclear bag)

Extrafusal
muscle fiber

A. Control   **B. Stretch**   **C. Contract**

FIGURE 7-3. (*A*) la afferent fiber. (*B*) The firing rate of the la afferent fiber increases when the extrafusal muscle fiber is stretched because the intrafusal muscle fiber also is stretched. Most of the stretch occurs at the center region of the intrafusal muscle fiber. (*C*) Contraction of the extrafusal muscle fiber compresses the central region of the intrafusal fiber, reducing the deformation of the la fiber terminal and reducing the firing rate of the la fiber.

(2) Gamma efferent discharge. The decreased rate of la afferent discharge that occurs during muscle contraction, called **unloading,** is functionally disadvantageous because the CNS stops receiving information about the rate and extent of muscle shortening. Unloading can be prevented by the activity of the gamma efferent motoneurons (Figure 7-4). The gamma motoneurons cause the sarcomeres of the intrafusal muscle fibers to shorten as the extrafusal muscle fiber shortens. As a result, the central region of the intrafusal muscle fiber remains stretched during muscle contraction, and unloading does not occur.

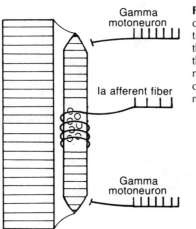

Gamma
motoneuron

la afferent fiber

Gamma
motoneuron

FIGURE 7-4. The firing rate of the la afferent fiber does not decrease if the gamma motoneuron discharge accompanies contraction of the extrafusal muscle fiber. Unloading is prevented because the intrafusal muscle fiber also shortens. Since the center region of the extrafusal muscle fiber does not contain sarcomeres, it does not shorten and, in fact, may lengthen if the gamma motoneuron causes sufficient shortening of the intrafusal muscle fiber sarcomeres.

    (a) When the **dynamic gamma motoneurons** are fired, only the **nuclear bag fibers shorten**. Because the nuclear bag fibers are responsible for the phasic (i.e., velocity-sensitive) portion of the Ia afferent response to stretch, stimulation of the dynamic gamma fibers increases phasic activity without affecting the static activity.

    (b) When the **static gamma motoneurons** are fired, only the **nuclear chain fibers shorten**. Because the nuclear chain fibers are responsible for the static (i.e., length-sensitive) component of the Ia afferent response to stretch, stimulation of the static gamma fibers increases static activity without affecting phasic activity.

(3) Coactivation of alpha and gamma motoneurons

    (a) During a **normal movement** (e.g., **lifting a weight**), the motor control system coactivates both the alpha and gamma motoneurons, diminishing the amount of unloading that occurs during muscle contraction and allowing the CNS to determine if its motor commands are being carried out.

        (i) As the extrafusal muscle fiber shortens, the intrafusal muscle fiber sarcomeres also shorten.

        (ii) If the two muscle fibers shorten at the same rate, then the central region of the intrafusal fiber is neither compressed nor lengthened, keeping Ia activity at a constant level.

        (iii) The constant level of Ia input to the CNS during a movement indicates that the motor command is being carried out.

    (b) If the **weight to be lifted is underestimated** by the CNS, the motor command system does not activate a sufficient number of alpha motoneurons to lift the weight and the extrafusal muscle fibers do not shorten.

        (i) The intrafusal muscle fibers do shorten, however, and because the tendon ends of the muscle cannot move, the central portion of the intrafusal fiber lengthens.

        (ii) Stretching the central region of the fiber causes Ia activity to increase, indicating that the motor command is not being carried out. The CNS uses this information to readjust its command to the spinal cord.

        (iii) Even before the CNS responds to the information provided by the Ia fibers, the Ia activity is used at the spinal cord level to adjust the alpha motoneuron activity to meet the unexpectedly high load. Because the Ia fiber synapses on the alpha motoneuron, its activity increases the excitability of the alpha motoneuron, leading to an increase in the frequency of action potential generation and an increase in muscle force development.

(4) Gamma loop. Theoretically, the CNS is capable of initiating movements directly by stimulating only the gamma motoneurons, using a pathway called the gamma loop (Figure 7-5).

    (a) Increasing gamma motoneuron activity causes the intrafusal muscle fiber sarcomeres to shorten, which, in turn, leads to stretching of the central portion of the intrafusal fiber and activation of the Ia fiber. Firing the Ia fibers causes alpha motoneuron activity to increase, which results in an increased amount and force of skeletal muscle activity.

    (b) Although the gamma loop can elicit movement on its own, it normally does not do so. However, because of coactivation, the gamma loop is activated during all movements and thus contributes to the excitability and firing rate of the alpha motoneurons.

2. The **lengthening reaction** causes inhibition of the alpha motoneurons that innervate muscles that are under tension, allowing them to lengthen.

  a. Receptor. The receptors for the lengthening reaction are the **Golgi tendon organs**. These small (0.5–1 mm long), encapsulated receptors are located in the tendons, between the muscles and tendon insertions.

    (1) The Golgi tendon organs have neither muscle fibers nor an efferent innervation.

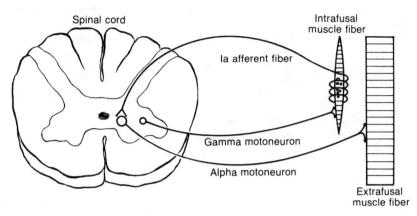

FIGURE 7-5. The gamma loop increases the firing of the alpha motoneuron during muscle contraction. The loop begins with the gamma motoneuron, which discharges to cause intrafusal muscle fiber contraction. This leads to an increase in Ia afferent fiber activity, which, in turn, causes increased alpha motoneuron discharge via a monosynaptic reflex.

      (2) The Golgi tendon organs are stretched whenever the muscle contracts and, thus, in contrast to the muscle spindle, are **in series** with the extrafusal muscle fibers.
   b. Effector organs. Both **extensor** and **flexor muscles** exhibit the lengthening reaction.
   c. Disynaptic pathway. The afferent fiber innervating the Golgi tendon organ is a **group Ib fiber**. It enters the dorsal root and forms a disynaptic pathway, which ends on the alpha motoneurons that send axons to the muscle from which the Ib fiber originated.
      (1) The disynaptic pathway of the Golgi tendon organ is **inhibitory to the alpha motoneuron.** The Ib fiber, like all sensory fibers, releases an excitatory transmitter. To produce inhibition, an **inhibitory interneuron** must be activated.
      (2) The lengthening reaction displays reciprocal innervation but lacks afterdischarge and irradiation.
   d. Role
      (1) Historically, the lengthening reaction has been described as a **protective reflex** in which a strong and potentially damaging muscle force reflexly inhibits the muscle, causing the muscle to lengthen instead of trying to maintain the force and risking damage.
      (2) Although the lengthening reaction is a protective reflex, it is now clear that the reflex plays a more important role in **regulating tension during normal muscle activity**. The lengthening reaction is described as **autogenic inhibition,** which indicates that the force generated when the muscle contracts is the stimulus for its own relaxation.

IV. **BRAIN STEM.** The brain stem contains the medulla, pons, midbrain, and parts of the diencephalon. Neuronal circuits within these areas control many physiologic functions (e.g., blood pressure, respiration, body temperature, sleep, and wakefulness). In addition, the **reticular formation** and **vestibular nuclei** are **important components of the motor control system**.

A. The **reticular formation** plays an important role in coordinating normal movements.

   1. The **motor control centers** within the reticular formation are a relay station for all descending motor commands, except those traveling directly to the spinal cord through the medullary pyramids (e.g., fine movements performed by the distal muscles of the fingers and hands).

a. The motor control centers **receive and modify the motor commands to the proximal and axial muscles** of the body.

b. These centers are **responsible for maintaining normal postural tone**.

(1) Neurons within the pontine reticular formation send axons to the spinal cord in the medial reticulospinal tract and are **excitatory to the alpha and gamma motoneurons that innervate the extensor antigravity muscles**.

(2) These neurons are prevented from firing too rapidly by **inhibitory input** derived from the cerebral and cerebellar components of the motor control system.

(a) The amount of inhibition is increased to reduce postural tone and is decreased to enhance postural tone.

(b) The withdrawal of inhibition, called **release of inhibition,** frequently is used to increase neuronal activity within the CNS.

2. Lesions

a. Lesions within the motor control centers of the reticular formation produce the syndrome of **spinal shock**.

(1) The **initial result** of removing the spinal cord from the control of the brain stem is the **complete loss of reflex activity,** which can last for days in cats and for months in humans.

(2) **When reflexes return,** they no longer are under the influence of the brain stem and, therefore, **do not follow their normal patterns**. For example, the local sign that characterizes the withdrawal reflex disappears. Instead, even a light touch on the foot can cause activation of all the flexor muscles in the body.

(3) **Ultimately,** the excitability of the motoneurons becomes excessive, causing particular groups of muscles to contract continuously.

b. Lesions within the cerebrum that interfere with inhibitory input to the motor control centers within the reticular formation cause **spasticity**. When discussing the effects of the brain stem on antigravity muscles, the terms spasticity and **rigidity** are used interchangeably. Clinically, however, spasticity and rigidity are not alike. Spasticity refers to the condition in which the stretch reflexes of the antigravity muscles are increased as a result of increased activity of the alpha or gamma motoneurons. Rigidity refers to the condition seen in Parkinson's disease [see V C 4 a (1)], in which there is increased activity in all of the muscles at a joint.

(1) Without the inhibitory input from the higher centers, the pontine reticular formation fires uncontrollably, subjecting both the alpha and gamma antigravity motoneurons to intense excitation.

(a) **Excessive firing of the alpha motoneurons** causes the antigravity muscles (i.e., the leg extensors and the arm flexors) to contract continuously.

(b) **Firing of the gamma motoneurons** activates the gamma loop, increasing discharge of the Ia afferent fibers and reflexly adding to the alpha motoneuron excitation produced by the reticulospinal tract. Because the gamma motoneurons are firing at higher than normal rates, the muscle spindles become more sensitive to stretch. Therefore, spasticity is increased further when the affected muscle is stretched.

(2) Cutting the dorsal roots reduces the amount of Ia input to the spinal cord and reduces spasticity by eliminating the reflex excitation of the alpha motoneurons.

c. Severing the pathways between the cerebrum and cerebellum and **the motor control centers** within the reticular formation causes **decorticate posturing** or **decerebrate rigidity**.

(1) **Decorticate posturing** is produced by lesions that involve the internal capsule or rostral cerebral peduncle. It is characterized by flexion of the arms and extension and internal rotation of the legs.

(2) **Decerebrate rigidity** occurs when the brain stem is severed above the pontine reticular formation. It is characterized by an increased tone in all the antigravity muscles, resulting in **opisthotonos** (arching of the back and neck), extension and hyperpronation of the arms, and extension and internal rotation of the legs.

B. The **vestibular nuclei,** located within the brain stem and cerebellum, receive information from **vestibular receptors** via **vestibular nerve fibers** (cranial nerve VIII). **Vestibular system reflexes** maintain tone in antigravity muscles and coordinate the adjustments made by the limbs and eyes in response to changes in body position.

1. **Vestibular receptors**
   a. **Location and role.** The vestibular receptors are located within a system of fluid-filled, membrane-bound structures called the **labyrinth** (see Figure 6-7). The labyrinth consists of **two otolith organs** (i.e., the **saccule** and **utricle**) and **three semicircular canals**.
      (1) **Otolith organs**
         (a) **Location.** The saccule communicates directly with the cochlea, which is beneath it, and the utricle, which is above it.
         (b) **Role.** The receptors within the otolith organs are responsible for detecting **linear acceleration** and the **static position of the head**.
            (i) The receptors within the **utricle** are oriented in a **horizontal** plane.
            (ii) The receptors within the **saccule** are oriented in a **vertical** plane.
      (2) **Semicircular canals**
         (a) **Location.** Both ends of all three semicircular canals emerge from the utricle.
         (b) **Role.** The receptors, which are located in the expanded end of the canal (i.e., the **ampulla**), are responsible for detecting **angular accelerations of the head**. Each canal is oriented in a different plane.
            (i) The plane of the **horizontal canal** is oriented along a line that is approximately **parallel to the earth** when the head is held in a normal upright position.
            (ii) The plane of the **anterior vertical canal** is oriented along a line from the **center of the head toward the eye**.
            (iii) The plane of the **posterior vertical canal** is oriented along a line from the **center of the head toward the ear**.
            (iv) The orientation of the posterior canal on one side of the head is roughly parallel to the orientation of the anterior canal on the other side of the head.
   b. **Receptor cells.** The receptor cells of the vestibular system, called **hair cells,** are polarized (Figure 7-6). A large cilium, called the **kinocilium,** is located at one end of the cell. When the stereocilia are bent toward the kinocilium, the cell depolarizes. When the stereocilia are bent away from the kinocilium, the cell hyperpolarizes.
      (1) **Otolith organs.** The hair cells of the utricle and saccule are located on a mass of tissue called the **macula.** The cilia are enmeshed in a gelatinous substance

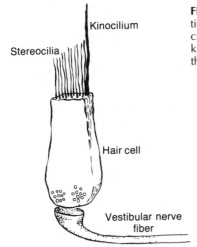

FIGURE 7-6. Diagram illustrating the polarized hair cells of the vestibular system. The large cilium is called the kinocilium. The smaller cilia are called stereocilia. When the stereocilia are bent toward the kinocilium, the cell depolarizes; when the stereocilia are bent away, the cell hyperpolarizes.

Kinocilium

Stereocilia

Hair cell

Vestibular nerve fiber

filled with small calcium carbonate crystals called **otoconia**. Because the otoconia are heavier than the fluid of the otolith organs, they **tend to sink and,** thus, can **bend the cilia of the hair cells**.

    (a) When the head deviates from the horizontal position, the hair cells bend. The kinocilium of each hair cell is oriented in a different plane so that, regardless of the direction in which the head is tilted, some of the hair cells are maximally stimulated while others are maximally inhibited.

    (b) Linear acceleration, (i.e., the type of movement experienced when one jumps down stairs or pulls away from a traffic light in a car) also displaces otoconia and stimulates the otolith organs.

  (2) Semicircular canals. The hair cells of the semicircular canals are located on the **crista,** a mass of tissue within the ampulla. The cilia are embedded in a gelatinous structure called the **cupula,** which completely fills the ampullar space.

    (a) General mechanism

      (i) When the head begins to move, the fluid within the semicircular canals lags behind and pushes the cupula backward, causing the cilia of the hair cells to bend. Depending on whether the stereocilia are pushed toward or away from the kinocilium, the hair cell depolarizes or hyperpolarizes.

      (ii) After 15–20 seconds of continuous movement at a constant velocity (e.g., the movement of a twirling dancer), the velocity of fluid movement catches up to that of the head, and the cupula returns to its resting position. The return of the cupula to the resting position causes the cilia to return to their upright position and the hair cell to return to its resting membrane potential. Thus, **the semicircular canals signal changes in motion (acceleration) but are insensitive to movements at a constant angular velocity**.

      (iii) When the head stops moving, the fluid within the semicircular canals continues to move, pushing the cupula forward, causing the cilia to bend in the opposite direction. Thus, if the original movement caused the hair cell to depolarize, the hair cell hyperpolarizes when the movement ceases.

    (b) Angular acceleration of the head. Because each of the three canals is oriented in a different plane, movement of the head in any direction generates a unique pattern of activity within the semicircular canals. This information is used by the CNS to interpret the speed and direction of head movement and to make the appropriate adjustments in posture and eye position.

      (i) In the **horizontal canals,** the kinocilium is located on the side of the hair cells closest to the utricle.

        Head movements that bend the stereocilia toward the utricle (**utriculopedal** movements) depolarize the hair cells; movements that bend the stereocilia away from the utricle (**utriculofugal** movements) hyperpolarize the hair cells.

        The horizontal canals of the left and right ear work opposite each other: when the left horizontal canal is stimulated, the right is inhibited, and vice versa (Figure 7-7).

      (ii) In the **vertical canals,** the kinocilium is located on the side of the hair cells away from the utricle. The vertical canals also work in pairs: when the anterior vertical canal on one side is stimulated, the posterior vertical canal on the other side is inhibited.

2. Vestibular nuclei. The vestibular nuclei send their axons into the spinal cord through a number of **vestibulospinal tracts**.

  a. The **input from the vestibular nuclei is excitatory to antigravity alpha motoneurons**.

  b. The vestibular nuclei, like the motor control centers within the reticular formation, **receive inhibitory input from the cerebrum and cerebellum**. If the inhibitory input from the cerebrum and cerebellum is removed, the vestibular nuclei greatly increase

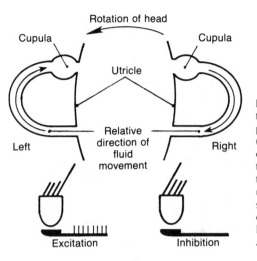

FIGURE 7-7. When the head is rotated toward the left, the endolymph in the left horizontal semicircular canal pushes the cupula and stereocilia toward the utricle (i.e., the fluid lags behind the head movement and causes the cupula to move to the right), depolarizing the hair cells. At the same time, the hair cells within the right horizontal canal are pushed away from the utricle, causing them to hyperpolarize. When the head stops rotating, the fluid continues to move, pushing the cupula in the opposite direction. This action causes the hair cells in the left horizontal canal to hyperpolarize and those in the right horizontal canal to depolarize.

their firing rate, leading to signs of spasticity similar to those observed after the inhibitory input to the reticular formation is severed.

(1) The spasticity produced by the vestibular nuclei differs from that produced by the reticular formation in that the vestibular nuclei primarily affect the alpha motoneurons, rather than both alpha and gamma motoneurons. Spasticity caused by the vestibular nuclei is called **alpha rigidity** to distinguish it from spasticity caused by the reticular formation, which is called gamma rigidity.

(2) Because gamma motoneurons are not involved in spasticity produced by the vestibular nuclei, alpha rigidity is not reduced greatly by cutting the dorsal roots and eliminating the Ia input.

3. **Vestibular reflexes**
   a. The **vestibulo-ocular reflex** maintains visual fixation during movements of the head. For example, if the head is rotated to the left, the eyes move toward the right in order to prevent an image from moving off the fovea. When the eyes have rotated as far as they can, they are rapidly returned to the center of the socket. If the rotation of the head continues, the eyes once again move in the direction opposite the head rotation. These movements of the eyes are called **nystagmus.**
      (1) The slow movement of the eyes to maintain visual fixation is initiated by the **activity within the semicircular canals**. When the head rotates to the left, the activity of receptors in the left horizontal canal causes the eyes to move toward the right.
      (2) After the body has been rotated and the movement ceases, the receptors within the right horizontal canal are stimulated, and these cause **postrotary** eye movements to occur in the opposite direction. That is, the eyes slowly move to the left until they reach the end of the socket, at which point they return quickly to the center. These movements continue until the cupula returns to its resting position.
      (3) **Lesions within the vestibular pathways** can cause nystagmus to occur spontaneously. **Damage to the vestibular receptors** can prevent nystagmus from occurring during rotation of the head.
   b. **Otolith reflexes.** The otolith organs initiate a reflex that prevents leg injuries when an individual walks down stairs or jumps from a platform. When making such a descent, the muscles in the leg begin to contract before the feet reach the ground in order to cushion the force of impact.
      (1) The otolith receptors responsible for this reflex are **stimulated by the linear acceleration of the head** that occurs during the descent.
      (2) Individuals lacking otolith reflexes are prone to leg injuries because of the large contact forces that occur during descent (e.g., stepping off a bus).

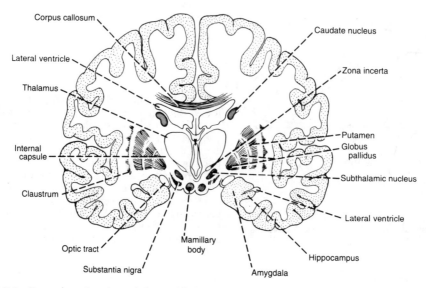

FIGURE 7-8. Coronal section through the midthalamus at the level of the mamillary bodies. The basal ganglia are prominent at this level and include the caudate nucleus, putamen, globus pallidus, subthalamic nucleus, and substantia nigra. (Reprinted from Fix JD: *BRS Neuroanatomy,* Malvern PA, Harwal Publishing, 1992, p 266.)

V. **BASAL GANGLIA.** The primary function of the basal ganglia is to **aid the motor cortex in planning and generating motor commands**. The basal ganglia also play a role in cognitive and affective behavior.

A. **Basal nuclei**

1. The basal ganglia consist of five nuclei (Figure 7-8):
   a. Caudate nucleus
   b. Putamen
   c. Globus pallidus
   d. Subthalamic nucleus
   e. Substantia nigra

2. Together, the **caudate nucleus** and **putamen** are considered the **striatum (neostriatum)**.

B. **Pathways.** The basal ganglia form extensive interconnections with the cortex and the thalamus (Figure 7-9). Because the basal ganglia do not make any direct sensory or motor connections with the spinal cord, their contribution to the control of movement is made indirectly through the sensorimotor cortex.

1. **Primary feedback loop**
   a. **Afferent fibers** from all areas of the cortex **project to the striatum**.
      (1) The **putamen** receives information related to **motor control**.
      (2) The **caudate nucleus** receives information related to the **control of eye movement**. In addition, the caudate nucleus receives information from the cortex that is related to cognitive and affective behavior, rather than to motor control.
   b. **Internuclear connections.** The **striatum sends most of its output to the globus pallidus** and to **the reticular nucleus of the substantia nigra**.
   c. **Efferent fibers** from the striatum and the reticular nucleus of the substantia nigra **project to the thalamus**. The **information received by the thalamus is conveyed back to the cortex,** completing the feedback loop.

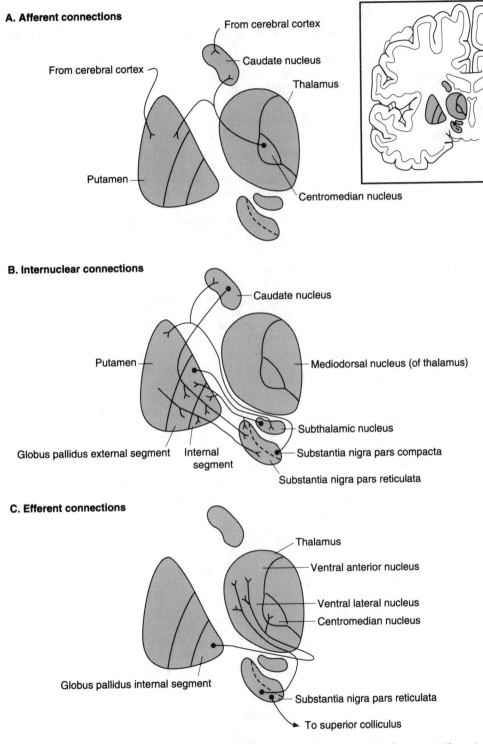

A. Afferent connections

From cerebral cortex

Caudate nucleus

From cerebral cortex

Thalamus

Putamen

Centromedian nucleus

B. Internuclear connections

Caudate nucleus

Putamen

Mediodorsal nucleus (of thalamus)

Subthalamic nucleus

Globus pallidus external segment

Internal segment

Substantia nigra pars compacta

Substantia nigra pars reticulata

C. Efferent connections

Thalamus

Ventral anterior nucleus

Ventral lateral nucleus

Centromedian nucleus

Globus pallidus internal segment

Substantia nigra pars reticulata

To superior colliculus

FIGURE 7-9. The basal ganglia form extensive connections with the cortex and thalamus. (*A*) The striatum (i.e., the caudate nucleus and putamen) receives afferent input from all areas of the cortex. (*B*) Extensive connections exist among the nuclei of the basal ganglia. (*C*) The thalamus receives efferent fibers from the basal ganglia. (Redrawn with permission from Kandel ER, Schwartz JH, Jessell TM: *Principles of Neural Science,* 3rd ed. New York, Elsevier, 1991, p 649.)

2. Additional pathways

a. Fibers from the pars compacta of the substantia nigra project to the putamen. This pathway utilizes dopamine as a neurotransmitter and damage to it produces **parkinsonism**.

b. Fibers from the subthalamic nucleus project to the globus pallidus. Because the subthalamic nucleus receives input from the motor cortex, lesions within the subthalamic nucleus lead to uncontrolled flinging movements of the limbs (i.e., **ballismus**).

C. **Lesions** in the basal ganglia produce characteristic deficits in motor behavior.

1. Lesions in the globus pallidus result in an inability of the trunk muscles to maintain postural support. The head bends forward so that the chin touches the chest, and the body bends at the waist.

    a. The motor deficits are not the result of muscular weakness or failure of voluntary control because individuals with these lesions can stand upright when requested to do so.

    b. Because the globus pallidus is the major outflow tract of the basal ganglia, it is possible that the motor deficits occur because the cortex is deprived of information it needs to automatically control the trunk muscles.

2. Lesions in the subthalamic nucleus cause spontaneous, wild, flinging movements of the limbs. This syndrome is called **hemiballismus**.

    a. The movements, which are **caused by a release of inhibition, appear on the side opposite the lesion**. The subthalamic nucleus inhibits the cortex indirectly.

      (1) Efferent fibers from the subthalamic nucleus release glutamate, which excites thalamic neurons.

      (2) The thalamic neurons use γ-aminobutyric acid (GABA) to inhibit the cortex.

      (3) Damage to the subthalamic nucleus reduces the excitation of inhibitory fibers in the thalamus and leads to excitation of the cortex.

    b. Because the movements of hemiballismus appear to be like those performed when an individual is thrown off balance, the subthalamic nucleus is believed to be involved with controlling the centers that issue the motor commands for **balance**.

      (1) Normally, the subthalamic nucleus responds to the need for initiating the balancing movement by momentarily withdrawing its inhibition from these centers.

      (2) When there is a lesion in the subthalamic nucleus, these centers no longer are under inhibitory control, and the movements are generated spontaneously.

3. Lesions within the striatum produce a variety of motor syndromes that are also related to a release of inhibition.

    a. Huntington's chorea, an inherited disorder, is characterized by continuous uncontrollable movements of the limbs that resemble the movements of hemiballism.

      (1) In Huntington's chorea, the **GABA-containing neurons** that originate **within the striatum** are **destroyed**. These GABA-ergic neurons normally inhibit neurons within the globus pallidus, which, in turn, inhibit neurons within the subthalamic nucleus.

      (2) When the striatal neurons are destroyed, the pallidal cells are released from inhibition and thus increase their inhibitory effect on the subthalamic nucleus.

      (3) Reduction in subthalamic nucleus activity may produce spontaneous movements resembling those observed when the subthalamic nucleus is destroyed.

    b. Athetosis is characterized by continuous, slow, irregular, twisting motions of the limbs, fingers, and hands.

    c. Dystonia is typified by twisting, tonic-type movements of the head and trunk.

4. Lesions within the pars compacta of the substantia nigra produce **Parkinson's disease**.

    a. Characteristics. Parkinson's disease is characterized by rigidity, hypokinesia (i.e., reduction in voluntary movement), and tremor.

      (1) The **rigidity** in Parkinson's disease involves all of the muscles at a joint and, thus, is different from spasticity that is associated with cortical lesions.

        (a) The rigidity of Parkinson's disease has been described as **lead-pipe rigidity** because the rigid limb, when moved, remains where it is placed.

(b) The rigidity seen in Parkinson's disease also has been described as **cogwheel rigidity**. When an examiner tries to move the limb, the limb periodically gives way and then reestablishes its resistance to movement like cogs on a wheel.

(2) The **hypokinesia** reduces the movement patterns normally associated with motor activity. For example, a Parkinson's disease patient may not swing his or her arms when walking, or display varied facial expressions during conversation.

   (a) The hypokinesia is **not related to a loss of muscle strength** or power, because normal movements occur under certain conditions.

   (b) The hypokinesia is **not caused by rigidity,** because hypokinesia may occur in the absence of rigidity.

(3) The **tremor** in Parkinson's disease occurs at rest and usually disappears during voluntary activity. It occurs at a frequency of approximately 4–7 cycles/sec. (A normal physiologic tremor has a frequency of approximately 10 cycles/sec.)

b. Therapy. Because the lesion of Parkinson's disease involves a pathway that uses dopamine as its neurotransmitter, some success has been achieved in treating the disease with L-dopa, a precursor of dopamine that can cross the blood–brain barrier. Dopamine inhibits the striatum sufficiently to reduce some of the clinical signs of the disease.

VI. **CEREBELLUM.** The cerebellum is intimately associated with control of the **timing, duration,** and **strength of a movement**. Its removal produces no deficits in emotional or intellectual function but causes profound disturbances in the ability to produce smooth, coordinated movements.

A. **Lobes.** Two transverse fissures divide the cerebellum into three lobes: the **anterior, posterior,** and **flocculonodular lobes** (Figure 7-10).

  1. Phylogenetic nomenclature. These lobes developed at different times during evolution.
   a. The **flocculonodular lobe** evolved early in the evolution of vertebrates and is therefore called the **archicerebellum**.
   b. The **anterior lobe** evolved next and is called the **paleocerebellum**.
   c. The **posterior lobe** was the last to evolve and is called the **neocerebellum**.

FIGURE 7-10. Diagrammatic dorsal view of the cerebellum, showing the anterior, posterior, and flocculonodular lobes. (Reprinted with permission from *NMS Neuroanatomy*. Malvern PA, Williams & Wilkins, 1988, p 189.)

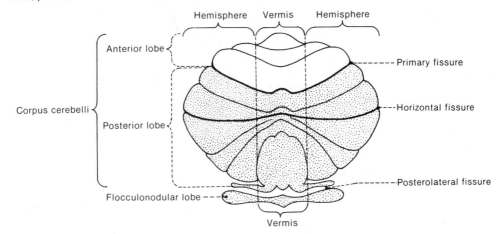

2. **Functional nomenclature.** These lobes are also named according to the connections they make with other components of the motor control system.
   a. The **flocculonodular lobe** is functionally related to the vestibular apparatus. Therefore, it is also called the **vestibulocerebellum**.
   b. The **entire anterior lobe,** and those parts of the posterior lobe that receive information from the spinal cord, are called the **spinocerebellum**. The spinocerebellum occupies the medial portion of the cerebellar cortex.
   c. The **remainder of the posterior lobe** receives input from the cerebral cortex and, thus, is called the **cerebrocerebellum**. The cerebrocerebellum occupies the more lateral regions of the cerebellar cortex.

B. **Pathways.** The organization of the synaptic connections within each lobe is the same.

1. **Layers.** The cerebellar cortex is divided into **three** layers (Figure 7-11).
   a. **Granule cell layer.** The **innermost layer** contains **10 billion granule cells,** as well as a smaller number of interneurons called **Golgi cells**.
   b. **Purkinje cell layer.** The **middle layer** contains **Purkinje cells**. The highly branched dendritic tree of the Purkinje cells extends vertically into the outer portion of the

FIGURE 7-11. Schematic diagram of the three layers of the cerebellar cortex, showing the neuronal elements and their connections. The *circular broken line* contains a cerebellar glomerulus. Climbing and mossy fibers provide excitatory input. Purkinje cell axons provide the sole output from the cerebellar cortex, which is inhibitory. (Reprinted from Fix JD: *BRS Neuroanatomy,* Malvern PA, Harwal Publishing, 1992, p 198.)

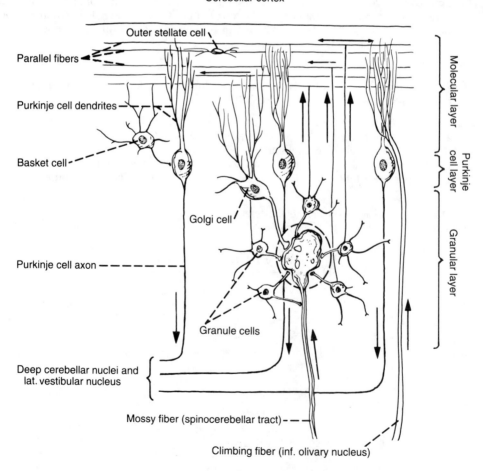

Cerebellar cortex

cortex, and the axon of the Purkinje fiber descends to the cerebellar nuclei (located in the granule cell layer).

   c. **Molecular layer.** The **outer layer** contains two types of **interneurons,** the **basket cells** and the **stellate cells**.

   (1) The axons of the granule cells project vertically into the molecular layer, where they form two branches (i.e., the **parallel fibers**), which extend parallel to the cortical surface.

   (2) The parallel fibers synapse with the dendrites of the Purkinje cells in the outer layer.

2. **Afferent fibers**

   a. **Excitatory.** The Purkinje cells receive **two types of excitatory input**.

   (1) Mossy fibers arise from cells within all levels of the nervous system that send projections to the cerebellum.

      (a) Mossy fibers **excite granule cells** which, in turn, **excite Purkinje cells** via the parallel fiber axons. Each Purkinje fiber receives axons from approximately 20,000 granule cells.

      (b) The Purkinje cells lie along a narrow path several millimeters in length.

   (2) Climbing fibers arise from cells within the inferior olivary nucleus. Each climbing fiber excites approximately ten Purkinje cells.

   b. **Inhibitory**

   (1) Basket cells inhibit Purkinje cells.

      (a) The basket cells, like Purkinje cells, are excited by parallel fibers.

      (b) The axons of the basket cells run perpendicular to the parallel fibers and act to inhibit the Purkinje cells on either side of the Purkinje cells activated by the parallel fibers. Thus, granule cell excitation produces a strip of excited Purkinje cells surrounded by parallel strips of inhibited Purkinje cells.

   (2) Golgi cells inhibit granule cells.

      (a) The dendrites of the Golgi cells ascend vertically from the granule cell layer to the molecular layer, where they receive excitatory input from the parallel fibers.

      (b) The axons of the Golgi cell inhibit the granule cells. Thus, excitation of the granule cell is rapidly extinguished by a **negative feedback loop:** Granule cell axons (parallel fibers) excite Golgi cell dendrites, whose axons inhibit the granule cells.

3. **Efferent fibers.** The **Purkinje cell axons,** which are **inhibitory,** provide the only output from the cerebellum.

   a. Axons originating in the **cerebrocerebellum** and **spinocerebellum project to the deep cerebellar nuclei**.

   b. Axons originating in the vestibulocerebellum project to the **vestibular nuclei**.

C. **Lesions** in the various lobes produce characteristic motor deficits.

1. **Lesions in the vestibulocerebellum** cause deficits related to the loss of vestibular function (e.g., loss of equilibrium, **ataxia**). Individuals with vestibulocerebellar lesions are unable to maintain their balance and tend to fall over when standing. When walking, these individuals tend to stagger and have a wide stance.

2. **Lesions in the spinocerebellum** have **no obvious effects** in humans, probably because spinocerebellar functions can be assumed by the cerebrocerebellum. In cats, lesions in the anterior lobe increase the tone of the antigravity muscles.

3. **Lesions in the cerebrocerebellum** cause **small motor deficits,** unless an extensive area of the cerebellar cortex is affected. **If the outflow pathways are damaged,** however, **the ability to produce smooth, coordinated movements is lost.**

   a. A major sign of cerebrocerebellar disease is **decomposition of movement**. Instead of acting in a coordinated way to produce a movement, the muscles act individually. For example, when reaching for an object, extension of the arm first takes place at the shoulder, followed by extension at the elbow, and finally by extension of the hand.

b. Dysmetria (i.e., the inability to stop a movement at the appropriate time or to direct it in the appropriate direction) is another sign of cerebrocerebellar disease. Dysmetria results from the loss of the neuronal circuitry required to control the duration and strength of a movement.

c. The **intention tremor** that results from cerebrocerebellar lesions also is related to the inability to time and sequence movements properly. The intention tremor is different from the resting (spontaneous) tremor of Parkinson's disease and appears to occur because an entire movement cannot be directed by a single motor command. Instead, the movement is partially directed and then halted. Several other motor commands are required before the movement is completed.

d. Adiadochokinesia is the inability to make rapidly alternating movements (e.g., turning the hands back and forth). This, too, appears related to the inability to time the duration of a movement.

Chapter 8

Cortical Control of Movement and Consciousness

I. **CORTICAL CONTROL OF MOVEMENT.** The cerebral cortex contains the neuronal circuits responsible for the conception, planning, and generation of motor commands. **Two parallel** systems of **descending pathways** originate in the cerebral cortex: the **pyramidal** and **extrapyramidal systems**. Clinically, these systems are considered together because lesions within the cortex almost always involve both of them; however, these systems are functionally different.

A. **Motor areas.** The cerebral cortex contains three motor areas.

1. The **primary motor cortex** is located within the precentral gyrus and corresponds to Brodmann's area 4.
   a. **Organization**
      (1) The primary motor cortex is **organized somatotopically** (i.e., each area of the body is controlled by a specific area of the primary motor cortex). For example, the head is controlled by cells located laterally and the feet are controlled by cells located on the medial surface of the cortex.
      (2) The **area of the cortex devoted to each part** of the body **is proportional to the amount of motor control that is exerted over that part**. Therefore, the hands and face have a much larger representation than the trunk and legs.
   b. **Role.** Neurons within the primary motor area are responsible for continuously **exciting spinal motoneurons** during the performance of a movement.
      (1) Although primary motor area neurons responsible for performing fine dextrous movements of the fingers may send axons to a group of motoneurons controlling a single muscle, most corticospinal tract neurons send axons to several synergistic motoneuron pools.
      (2) The firing frequency of the primary motor area neurons is proportional to the force that must be generated by the muscle to carry out its motor command.
      (3) The primary motor area neurons receive sensory input from the spinal cord. This information is used to modify ongoing motor commands.

2. The **supplementary motor cortex** is located on the lateral and medial surface of the cortex in Brodmann's area 6.
   a. **Organization.** The supplementary motor cortex is organized **somatotopically**. However, unlike the primary motor area, which controls muscles on the contralateral side of the body, the supplementary motor cortex controls muscles on both sides of the body.
   b. **Role.** Neurons within the supplementary motor area are responsible for **generating the plan for a movement;** therefore, they are activated before those of the primary motor area.
   c. **Lesions** within the supplementary motor area interfere with the coordination of complex motor tasks, particularly those requiring both limbs.

3. The **premotor motor cortex** is located on the lateral surface of the cortex (in Brodmann's area 6), just in front of the primary motor cortex.
   a. **Organization.** The premotor motor cortex is organized **somatotopically**.
   b. **Role.** The premotor motor cortex **coordinates the proximal and axial muscles** during a motor task.

B. **Pathways.** Two major descending pathways emerge from the motor areas.

1. **Corticospinal tract**
   a. **Organization.** The corticospinal tract originates in the motor cortex, crosses to the contralateral side within the pyramids on the ventral medulla, and terminates within

the spinal cord or the cranial nerves controlling facial muscles. Because its axons pass through the pyramids on their way to the spinal cord, the corticospinal tract is also referred to as the **pyramidal system**.

(1) **Axons.** The corticospinal tract contains approximately 1 million axons.

    (a) The cell bodies of these axons are in layer V of the cerebral cortex. A small number of the axons (approximately 30,000) **originate from** large pyramidal cells called **Betz cells**. The remainder come from **smaller pyramidal cells**.

    (b) Approximately half of the axons come from the primary motor cortex. A third come from the other motor areas, and the remainder come from the somatosensory cortex.

(2) **Collaterals** from the corticospinal tract travel to the basal ganglia and cerebellum, which use this information to modify and coordinate ongoing movements.

b. Role. The corticospinal tract is responsible for **controlling muscles that make precision movements** (e.g., the muscles that move the fingers and hands and the muscles that produce speech).

c. Lesions within the pyramidal tract system produce only **minor deficits** in motor control. Most movements are not affected because they can be adequately controlled by cortical fibers descending in the extrapyramidal system (see I B 2).

(1) Muscle weakness or paralysis, particularly of the distal muscles responsible for fine, highly coordinated movements, occurs with pure pyramidal tract lesions.

(2) A number of cutaneous reflexes (e.g., the cremasteric reflex) are abolished or more difficult to elicit and a positive Babinski sign is produced when the plantar surface of the foot is stroked.

2. Corticoreticular and corticovestibular tracts

a. Organization. These pathways are referred to as the **nonpyramidal or extrapyramidal system,** because their influence over motoneurons is not exerted through axons that travel through the pyramids. In the past, the basal ganglia were also referred to as the extrapyramidal system; the two usages of this term should not be confused.

(1) Axons from cells within the nuclei of the pontine reticular formation (the primary nucleus reticularis pontis oralis and caudalis) and the lateral vestibular nucleus pass ipsilaterally to the spinal cord, where they excite motoneurons controlling antigravity muscles of the limbs.

(2) Motoneurons controlling motoneurons going to antigravity muscles of the neck and back muscles are excited by axons originating within the medial and inferior vestibular nuclei.

b. Role. The corticoreticular and corticovestibular tracts provide inhibitory input to contralateral nuclei within the pontine reticular formation and the vestibular system. They are responsible for **maintaining postural tone** and for **directing voluntary movement**.

c. Lesions to the extrapyramidal system remove inhibition from the pontine reticular formation and lead to spasticity (see Chapter 7 IV A 2).

II. SLEEP AND CONSCIOUSNESS

A. **Diurnal (circadian) rhythms.** Many of the body's regulatory mechanisms vary in their activity during the day. For example, body temperature is approximately 1°C higher during the early evening than it is at dawn, and adrenocortical hormones are secreted at levels that are higher in the morning than they are at night. The cycle of these diurnal rhythms is roughly **24 hours**.

1. If an individual is isolated from the environmental stimuli that indicate the normal day–night periods, the cycle time lengthens, demonstrating that the circadian rhythms are not rigidly linked to the rotation of the earth but can be driven by an individual's internal biological clock.

2. It is necessary to understand normal variations in physiologic activities when evaluating pathologic functions. For example, it would be misleading to compare temperatures obtained at different times of the day.

B. **Sleep–wake cycle.** The most obvious, and probably most important, diurnal rhythm is the sleep–wake cycle. When awake, an individual is able to perform all activities that are required for individual and species survival. When asleep, an individual is not aware of the environment and is unable to perform activities that require consciousness.

1. **Assessment of sleep states.** The presence of sleep can be assessed by **behavioral analysis** (e.g., an individual who does not move and does not respond when spoken to or touched is often asleep). A more accurate assessment of sleep can be obtained from an **electroencephalogram (EEG)**.

   a. **Obtaining the EEG**

      (1) The EEG is obtained by placing electrodes on the scalp. The location of the electrodes and the amplification and paper speed of the polygraph used for recording the EEG are standardized.

      (2) The brain waves recorded by an EEG represent the summated activity of millions of cortical neurons. The inhibitory postsynaptic potentials (IPSPs), excitatory postsynaptic potentials (EPSPs), and the passive spread of electrical activity into the dendrites of these neurons, rather than their action potentials, form the basis for the EEG.

   b. **Variations in the EEG during sleep and wakefulness** (Figure 8-1). The two states of sleep—**slow-wave** and **fast-wave** sleep—have characteristic EEG patterns.

      (1) When an individual is **awake,** the electrical activity recorded from the brain is asynchronous and of low amplitude. This type of brain electrical recording is called a **beta wave**.

      (2) If an individual sits quietly for a while, the brain waves gradually become larger and highly synchronized. The **typical resting EEG pattern** has a frequency of 8–13 cycles/sec and is called an **alpha wave**. When the eyes open or when conscious mental activity is initiated, the EEG shifts from an alpha to a beta pattern (i.e., **alpha blocking** takes place).

      (3) As consciousness is reduced still further, an individual enters a state of sleep called **slow-wave sleep,** which progresses in an orderly way from **light to deep sleep**. Behaviorally, slow-wave sleep is characterized by a progressive reduction in consciousness and an increasing resistance to being awakened. Muscle tone is reduced, the heart and respiratory rates decrease, and, in general, body metabolism slows.

         (a) **Light sleep** is characterized by an EEG that shows high-amplitude waves of approximately 12–15 cycles/sec, called **sleep spindles,** which periodically interrupt the alpha rhythm.

FIGURE 8-1. As an individual passes from wakefulness to deep sleep, the electroencephalogram (EEG) wave increases in amplitude and decreases in frequency. Sleep spindles indicate the presence of light sleep.

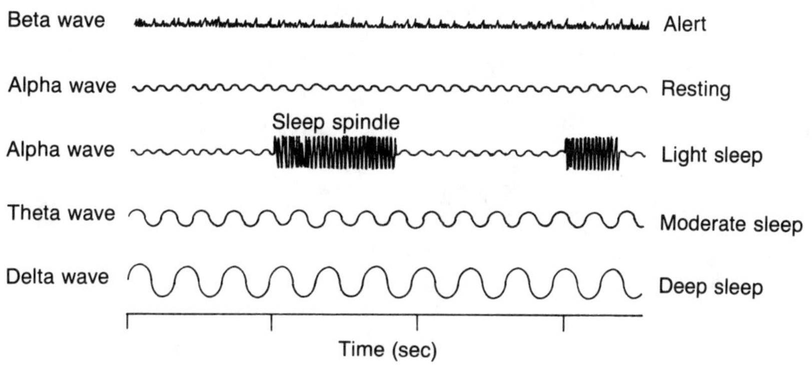

(b) Moderate sleep is characterized by an EEG that displays slower and larger waves called **theta waves**.

(c) Deep sleep produces an EEG pattern with very slow (4–7 cycles/sec), large waves called **delta waves**.

(4) Fast-wave (desynchronized) sleep is characterized by the same high-frequency and low-amplitude EEG pattern that is seen in the waking state; however, the individual clearly is unresponsive to environmental stimuli and, thus, is asleep. For this reason, fast-wave sleep also is called **paradoxical sleep**.

(a) Because this state of sleep is characterized by the presence of rapid eye movements, it also is called **rapid eye movement (REM) sleep**.

(b) Because dreaming occurs during REM sleep, it is also called **dream sleep**.

(c) Behaviorally, REM sleep is quite different from slow-wave sleep.

(i) It is as difficult to arouse an individual from REM sleep as it is from deep sleep. However, when awakened from REM sleep, the individual is immediately alert and aware of the environment.

(ii) The eyes are not the only organs that are active during REM sleep. The middle ear muscles are active, penile erection occurs, heart rate and respiration become irregular, and there are occasional twitches of the limb musculature. Because muscle tone is reduced tremendously during REM sleep, the frequency and intensity of muscle twitching do not produce injuries or awaken the individual.

2. The sleep cycle. There is an orderly progression of sleep stages and states during a typical sleep period (Figure 8-2).

a. When an individual falls asleep, the light stage of slow-wave sleep is entered first. During the next hour or so, the individual passes into progressively deeper stages of sleep until deep sleep is reached. After approximately 15 minutes of deep sleep, the depth of sleep starts to decrease and continues to do so until the individual reenters the light stage of sleep (about 90 minutes after the start of the first sleep cycle). At this point, the individual passes from slow-wave sleep to REM sleep.

b. This cycle repeats itself about five times during the night. However, as Figure 8-2 demonstrates, after the second cycle, the intervals between periods of REM sleep shorten and the duration of each period of REM sleep lengthens. As morning approaches, an individual spends less time in the deeper stages of slow-wave sleep and periodically awakens.

c. The sleep cycle shown in Figure 8-2 is typical of an adult. The cycle varies greatly with age.

FIGURE 8-2. Diagram indicating the pattern of sleep during one sleep cycle. As the evening progresses, the depth of slow-wave sleep decreases and the duration and frequency of rapid eye movement (*REM*) sleep episodes increase. Note that occasional periods of wakefulness occur during the night.

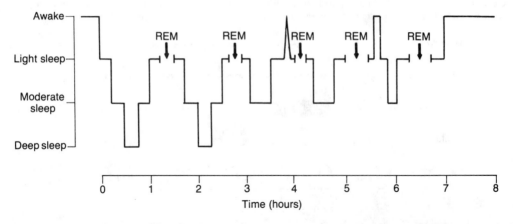

(1) During infancy, approximately 16 hours of every day are spent asleep. This figure drops to 10 hours during childhood and to 7 hours during adulthood. Elderly individuals spend less than 6 hours of each day sleeping.

(2) It is interesting to note that prematurely born infants spend approximately 80% of their sleep time in REM sleep, whereas full-term infants spend only 50% of their sleep time in REM sleep. The total time spent in REM sleep is reduced to about 1.5–2 hours by puberty and remains unchanged thereafter.

(3) During infancy and childhood, therefore, the reduction in sleep time from 16 hours to 10 hours occurs almost entirely by a reduction of the amount of time spent in REM sleep. In adulthood, the reduction in sleep time is caused by a reduction in the time spent in the deep stages of slow-wave sleep.

3. **Physiologic basis for sleep.** Areas throughout the entire brain participate in the sleep–wake cycle.
   a. The **waking state** is maintained by the **ascending reticular activating system (RAS),** a diffuse collection of neurons within the medulla, pons, midbrain, and diencephalon. Electrical stimulation anywhere within this area causes the EEG pattern to change abruptly from that of the sleep state to that of the waking state [i.e., a **cortical alerting (arousal) response** takes place].
   b. The **sleep state** does not result from the passive withdrawal of arousal. **Two sleep centers** exist in the brain stem; one is responsible for producing slow-wave sleep, and the other produces REM sleep.
      (1) The **slow-wave sleep center** is located in a midline area of the medulla containing the rapheal nuclei.
         (a) The neurons within these nuclei use serotonin (5-hydroxytryptamine) as a neurotransmitter.
         (b) Administration of serotonin directly into the cerebral ventricles of experimental animals induces a state of slow-wave sleep, whereas lesions in this region induce a permanent state of insomnia.
      (2) The **REM sleep center** is located in specific nuclei of the pontine reticular formation, including the locus ceruleus, which uses norepinephrine as a neurotransmitter. Lesions within this area eliminate the electrophysiologic and behavioral signs of REM sleep.

4. **Sleep disorders.** As noted previously, there is a cycling between slow-wave sleep and REM sleep during a normal sleep period.
   a. In **narcolepsy,** REM sleep is entered directly from the waking state.
      (1) Individuals suffering from narcolepsy often report an intense feeling of sleepiness just prior to an attack, although sleep sometimes occurs without warning.
      (2) In some narcoleptics, the profound reduction in muscle tone characteristic of REM sleep can occur without loss of consciousness. During such an attack, called **cataplexy,** the individual suddenly becomes paralyzed, falls to the ground, and is unable to move.
      (3) Another symptom associated with narcolepsy is the presence of a dream-like state during wakefulness, which narcoleptics describe as a hallucination.
   b. Most of the other symptoms of sleep disorders are associated with slow-wave sleep. These include **sleepwalking (somnambulism), bed-wetting (nocturnal enuresis), and nightmares (pavor nocturnus),** all of which occur during stages of slow-wave sleep.
      (1) During a nightmare that occurs in slow-wave sleep, the individual wakes up screaming and appears terrified. However, no reason for the acute anxiety is recalled. These episodes are called night terrors.
      (2) By contrast, terrifying dreams that occur during REM sleep are graphically remembered.

5. **Disturbances of consciousness**
   a. **Coma.** A lesion blocking the connection between the ascending RAS and the thalamus produces a permanent state of sleep, or coma. In this situation, stimulation of sensory pathways can cause a momentary desynchronization of the EEG but does not produce any behavioral signs of arousal.

    (1) Coma is not simply a deep sleep state. It is characterized by a loss of consciousness from which **arousal cannot be elicited**.

    (2) **O_2 consumption by the brain is reduced** during coma. This is in marked contrast to normal sleep, in which there is no change in brain O_2 consumption from the waking state.

  b. Syncope (fainting). A transient pathologic loss of consciousness is called syncope. More persistent losses of consciousness (from which arousal can be obtained) are called stupor.

  c. Brain death occurs when the brain no longer can achieve consciousness. Because of the desire to obtain organs for transplant operations and the desire to remove heroic life support systems, an objective standard for determining the presence of brain death has been developed.

    (1) Brain death is said to occur when a loss of consciousness is accompanied by a flat EEG (i.e., an EEG with no brain waves) and a loss of all brain stem regulatory systems (e.g., those systems that control respiration and blood pressure).

    (2) Moreover, these clinical signs must be due to traumatic or ischemic anoxia and not to hypothermia or metabolic poisons, from which later recovery is possible.

    (3) Finally, the criteria for brain death must be present for 6–12 hours.

STUDY QUESTIONS

DIRECTIONS: Each of the numbered items or incomplete statements in this section is followed by answers or by completions of the statement. Select the ONE lettered answer or completion that is BEST in each case.

1. Which one of the following stimuli normally activates a receptor that is located on the free nerve ending of a sensory neuron?

(A) Taste
(B) Gravity
(C) Light
(D) Sound
(E) Smell

2. Which characteristic of a sensory stimulus is encoded better by phasic receptors than by tonic receptors?

(A) How strong the stimulus is
(B) The type of energy producing the stimulus
(C) How rapidly the stimulus is applied
(D) The duration of the stimulus
(E) Where the stimulus is located

3. Which one of the following sensory systems uses unmyelinated fibers to convey information to the central nervous system (CNS)?

(A) Proprioception
(B) Vision
(C) Vibration
(D) Temperature
(E) Pressure

4. Which one of the following statements about pain sensation is correct?

(A) Painful sensations can be elicited by any sensory neuron if its firing frequency is high enough
(B) Painful sensations arising from a particular area of the skin occur only when pain fibers from that area of the skin are stimulated
(C) Cutting the anterolateral tract on both sides of the spinal cord will permanently eliminate painful sensations arising from skin regions innervated by sensory neurons located below the site of the lesion
(D) Pain fibers conduct impulses to the spinal cord and to skin regions surrounding the site of a painful stimulus

5. Which one of the following statements best describes cold receptors?

(A) Cold receptors produce a sensation of warmth when their firing frequency is very low
(B) Sudden decreases in temperature always increase the firing frequency of cold receptors
(C) Cold receptors are tonic receptors that slowly increase their firing rate when the temperature is decreased
(D) Cold receptors do not fire at skin temperatures above body temperature
(E) Cold receptors produce a sensation of pain when their firing frequency is very high

6. Which one of the following statements about the optical properties of a myopic eye is correct?

(A) A converging lens can be used to correct the optical defect
(B) The image of a distant object is formed in front of the retina
(C) The power of accommodation for near vision is greater than normal
(D) The refractive power of the lens is less than normal
(E) The far point is greater than normal

7. Which of the following refractive problems most closely resembles presbyopia?

(A) Hyperopia
(B) Myopia
(C) Astigmatism
(D) Cataract
(E) Diplopia

8. If the refractive power of the unaccommodated eye of an emmetropic woman is 60 diopters (D), the axial length of her eye is closest to

(A) 14.5 mm
(B) 15.5 mm
(C) 16.5 mm
(D) 17.5 mm
(E) 18.5 mm

9. Which of the following statements correctly describes the role played by transducin during the response of rods and cones to light?

(A) Transducin reduces membrane conductance by closing Na^+ channels
(B) Transducin stimulates synaptic transmitter release by opening Ca^{2+} channels
(C) Transducin initiates the photoreceptor response by converting 11-*cis* retinal to all-*trans* retinal
(D) Transducin enhances the action of rhodopsin by converting vitamin A to 11-*cis* retinal
(E) Transducin reduces the concentration of cyclic guanosine monophosphate (cGMP) by activating a phosphodiesterase enzyme

10. Which one of the following is more descriptive of rods than of cones?

(A) Not located within the fovea
(B) Provide information about the color of an object
(C) Recover their sensitivity more rapidly after exposure to bright light
(D) Responsible for the high visual acuity of the visual system
(E) Organized into on-center, off-surround receptor fields

11. A sound stimulus of 20 decibels (dB) is

(A) 10 times threshold
(B) 20 times threshold
(C) 50 times threshold
(D) 100 times threshold
(E) 200 times threshold

12. Which one of the following statements best describes the muscle fibers that are recruited first during a normal movement?

(A) They have a very limited capillary supply
(B) They store large quantities of glycogen
(C) They have low myosin–adenosine triphosphatase (ATPase) activity
(D) They fatigue easily
(E) They rapidly sequester Ca^{2+} into their sarcoplasmic reticulum (SR)

13. Which one of the following receptors is responsible for monitoring the rate of muscle stretch?

(A) Nuclear bag intrafusal fibers
(B) Nuclear chain intrafusal fibers
(C) Golgi tendon organs
(D) Pacinian corpuscles
(E) Ruffini's corpuscles

14. Which of the following will occur in a girl who suddenly stops spinning after several seconds of spinning to the left?

(A) The hair cells in the right semicircular canal will depolarize
(B) Her eyes will move slowly to the right
(C) When asked to point to a target, the girl will point to the right of the target
(D) The cupula in the right semicircular canal will move away from the utricle
(E) The objects in the visual field will appear to be spinning to the right

15. Which of the following is most closely related to slow-wave sleep?

(A) Dreaming
(B) Atonia
(C) Bed-wetting
(D) High-frequency electroencephalogram (EEG) waves
(E) Irregular heart rates

16. Movement disorders related to the removal of inhibition are produced by lesions to all of the following components of the motor control system EXCEPT the

(A) striatum
(B) internal capsule
(C) substantia nigra
(D) medullary pyramids
(E) subthalamic nucleus

Questions 17–18

Match each receptor potential with the sensory system it is associated with.

(A) Taste
(B) Olfaction
(C) Audition
(D) Touch
(E) Vision

17. The receptor potential is produced by the flow of K^+ into the cell

18. The receptor potential causes the photoreceptors to hyperpolarize when stimulated

Questions 19–20

Match each function with the component of the auditory system that performs it.

(A) Stria vascularis
(B) Scala media
(C) Auditory ossicles
(D) Oval window
(E) Basilar membrane

19. Production of endolymph

20. Amplification of sound stimuli

ANSWERS AND EXPLANATIONS

1. The answer is E [Chapter 5 IV A 1]. Olfactory receptors are on the free nerve endings of the olfactory nerve (cranial nerve I). Gravity is detected by hair cells in the utricle and saccule, sound by hair cells in the organ of Corti, light by rods and cones, and taste by epithelial cells within the taste buds. Hair cells, rods and cones, and epithelial taste cells then communicate with afferent nerves via synaptic transmission.

2. The answer is C [Chapter 5 II A 4, C 3 a (1)]. Phasic (rapidly adapting) receptors are better suited than tonic receptors to encode the rate of stimulus application. When stimulated, phasic receptors produce a high frequency discharge that declines rapidly after the stimulus reaches its peak amplitude. The rate of decline depends on the rate of stimulus application. If the stimulus reaches its peak rapidly, very little adaptation occurs during the stimulus application and the initial rate of firing is very high. However, if the stimulus rises slowly to its peak, adaption limits the maximum frequency that can be obtained. Therefore, the firing encodes the rate of stimulus application. Tonic receptors are required to encode stimulus strength. Both types of receptors can provide information about stimulus quality, location, and duration. Phasic receptors encode duration by firing when the stimulus is applied and when it is removed. Tonic receptors encode duration by firing for as long as the stimulus is applied.

3. The answer is D [Chapter 5 III B]. The thermoreceptors are on the free nerve endings of unmyelinated C fibers and small myelinated (Aδ) sensory fibers. Proprioception (i.e., the sense of muscle tension and length) is conveyed by Ia and Ib afferent fibers, which innervate muscle spindles and Golgi tendon organs. Vibration is detected by Pacinian corpuscles and pressure is detected by Ruffini's corpuscles. Both are innervated by large, myelinated (group II) afferent neurons.

4. The answer is D [Chapter 5 III C]. Small myelinated and unmyelinated nociceptive fibers that convey information to the spinal cord have collateral branches that innervate blood vessels in the region of the pain stimulus. Antidromically conducted action potentials cause the release of bradykinins from the blood vessels, which stimulates the release of pain-producing substances (e.g., histamine) from neighboring cells. The pain-producing substances enlarge the area from which painful sensations arise and produce hyperalgesia (an increase in the sensitivity to painful stimuli). Only stimulation of pain fibers produces the sensation of pain. Other sensory neurons do not elicit a painful stimulus. Visceral pain can be referred to areas of the skin where there are no painful stimuli to activate the pain fibers in that region. Although cutting the anterolateral tract eliminates the acute pain and temperature sensations arising from areas below the lesion, chronic pain is not eliminated and occasionally worsens.

5. The answer is B [Chapter 5 III B 2; Figure 5-5]. Specific cold receptors are located on the free nerve endings of small myelinated and unmyelinated neurons. These receptors are distinguished from warm receptors because they produce a phasic burst of action potentials whenever their temperature is decreased and reduce their firing rate when they are warmed. Their steady-state firing rate reaches a peak at skin temperatures of about 23°C–28°C and declines as the temperature is raised or lowered from this point. If the temperature is raised to about 45°C, the cold fibers start firing again. At the same time, pain fibers are activated, producing a burning cold sensation.

6. The answer is B [Chapter 6 I B 3 c (1)]. A myopic eye is one in which the power of the unaccommodated lens is too high for the axial length of the eye, so that the image is formed in front of the retina. The refractive error can be corrected by a diverging lens that reduces the strength of the high converging power that characterizes a myope's eye. The far point is the point at which the image can be clearly seen by the unaccommodated eye (i.e., the image is located on the retina). Myopes must bring the object close to the eye to be seen (hence the term "nearsighted"); therefore, the myope's far point is closer than the far point of a person with normal vision.

7. The answer is A [Chapter 6 I B 3 b (3), c (2)]. Presbyopia is the loss of accommodative power that occurs as a person ages. Because accommodation cannot occur, the near point moves away from the eye and images must be held at some distance from the eye to be clearly seen. This is similar to the situation in hyperopia where the converging power is too weak for the axial length of the eye. Hyperopes must accommodate to see distant objects clearly and thus have less accommodation available for near vision. Thus hyperopes, like presbyopes, must hold objects at a distance to see them clearly (hence the term "farsighted").

8. The answer is C [Chapter 6 I B 3]. The focal point of an optical system (in meters) is the reciprocal of the refractive power of the system. Therefore, the focal point is 1/60 diopter (D), or 16.7 mm. An emmetropic eye forms a focused image of a distant object on the retina without accommodation. If the axial length is 16.5 mm, the image will be focused on the retina.

9. The answer is E [Chapter 6 I C; Figure 6-4]. Transducin is a G protein that, when activated by the photoisomerization of 11-*cis* retinal to all-*trans* retinal, activates a cyclic guanosine monophosphate (cGMP) phosphodiesterase. The hydrolysis of cGMP by the phosphodiesterase leads to the closing of Na^+ channels, the hyperpolarization of the membrane, and a reduction in the release of synaptic transmitter. Formation of 11-*cis* retinal from vitamin A requires two steps; the vitamin A must be isomerized to its 11-*cis* isomer by an isomerase enzyme and then the 11-*cis* retinol must be oxidized to the aldehyde, 11-*cis* retinal.

10. The answer is A [Chapter 6 I C 3]. Rods are not located within the fovea; only cones are present. The three types of cones (red, green, and blue) make color vision possible. Because cones are smaller than rods, have a greater sensitivity to light, and are organized into smaller center-surround receptor fields, they are capable of high visual acuity. After being exposed to bright light, cones recover their sensitivity about five times faster than rods.

11. The answer is A [Chapter 6 II B 2 b (1)]. A decibel (dB) is a unit of sound intensity based on the formula:

$$dB = 20 \cdot \log \frac{I}{I_0}, \text{ where}$$

$$I = \text{stimulus intensity}$$
$$I_0 = \text{threshold stimulus}$$

According to this formula, a sound intensity of 20 dB is 10 times threshold:

$$20 = 20 \cdot \log \frac{10 \cdot I_0}{I_0}$$

12. The answer is C [Chapter 7 II A 2 b, C]. The first muscles to be recruited during a movement are the slow-twitch muscle fibers. Slow-twitch fibers are characterized by low myosin–adenosine triphosphatase (ATPase) activity associated with slow contractile speeds, slow sequestering of Ca^{2+} that facilitates summation, and high oxidative enzyme activity and a rich capillary supply that render them fatigue-resistant.

13. The answer is A [Chapter 7 III B 1 a (1) (a) (i); Figure 7-3]. The Ia afferent fibers innervating the intrafusal muscle fibers contained within muscle spindles convey proprioceptive information about muscle length to the central nervous system (CNS). When nuclear bag intrafusal fibers are stretched, the Ia afferents respond by generating a burst of action potentials with a frequency proportionate to the rate of muscle stretch. Stretching the nuclear chain fibers produces a steady discharge proportional to the amount of stretch. Golgi tendon organs respond to the force developed by muscles, pacinian corpuscles respond to vibratory stimuli, and Ruffini's corpuscles respond to pressure applied to the skin.

14. The answer is A [Chapter 7 IV B 1 b (2); Figure 7-7]. When the head spins to the left, the vestibular nerve innervating the left horizontal semicircular canal will be stimulated. When the girl stops spinning, the endolymph within the horizontal semicircular canals will continue to move toward the left, pushing the cupula in both horizontal semicircular canals toward the left. In the right horizontal semicircular canal, the cupula moves toward the utricle and bends the stereocilia toward the kinocilium, causing the hair cell to depolarize. Depolarization of the hair cells stimulates the right vestibular nerve, which cause the eyes to move slowly toward the left. Stimulation of the right vestibular nerve also causes the girl to feel that she is spinning to the right or that the world is spinning to the left. As a result, when she reaches for a target, she points to the left of the target (toward the direction she senses the target is moving).

15. The answer is C [Chapter 8 II B 1 b (3), 4 b]. A variety of sleep disturbances, including bed-wetting, sleepwalking, and night terrors all occur during slow-wave sleep. Dreaming occurs most often during rapid eye movement (REM) sleep, which is also characterized by high-frequency electroencephalogram (EEG) waves, irregular heart rates and breathing patterns, and total inhibition of alpha motoneurons.

16. The answer is D [Chapter 7 V C 2–4; Chapter 8 I B 1 b]. The medullary pyramids contain the axons of the corticospinal tract. These axons provide excitatory input directly to alpha motoneurons within the spinal cord or to interneurons that excite alpha motoneurons. Lesions to these axons, therefore, reduce excitability. The striatum, substantia nigra, and subthalamic nucleus are components of the basal ganglia, which normally inhibit motor activity. Lesions in these structures increase motor tone or spontaneous movements. The internal capsule contains axons of upper motor neurons that normally inhibit brain stem neurons that are, in turn, excitatory to antigravity muscles. When these axons are damaged, the brain stem neurons are released from inhibition, causing the antigravity muscles to become hyperexcitable.

17–18. The answers are: 17-C [Chapter 6 II D 2], **18-E** [Chapter 6 I C 2 b]. The stereocilia

and apical surface of hair cells in the organ of Corti are surrounded by the endolymph that fills the scala media. Endolymph contains high levels of K^+. When the vibration of the basilar membrane causes the stereocilia to bend toward the kinocilium, K^+ channels open, allowing K^+ to enter the cell. The flow of K^+ into the cell causes the hair cell to depolarize. K^+ is forced into the cell by the large potential difference between the endolymph (which is made positive by the secretion of K^+ from the scala vestibuli) and the intracellular fluid (ICF) of the hair cell (which is negative because the basolateral potions of the hair cell are bathed in the Na^+-containing perilymph of the scala vestibuli.

In the dark, rods and cones are depolarized by the flow of Na^+ into the cell through Na^+ channels that are kept open by cyclic guanosine monophosphate (cGMP). When light strikes the eye, rhodopsin is activated. Rhodopsin activates a G protein called transducin, which in turn activates a phosphodiesterase that hydrolyzes cGMP. When cGMP levels fall, cGMP is removed from its binding site on the Na^+ channels, closing the Na^+ channel and causing the photoreceptor to hyperpolarize.

19–20. The answers are: 19-A [Chapter 6 II A 3 b (1) (c) (ii), **20-C** [Chapter 6 II C 2]. The stria vascularis is a profusely perfused layer of epithelial cells that is responsible for the production of endolymph. Endolymph resembles intracellular fluid (ICF) in that it contains a high concentration of K^+ and a low concentration of Na^+. It is secreted into the scala media, where it surrounds the cells of the organ of Corti. The scala media is bounded above by Reissner's membrane and below by the basilar membrane.

Sound stimuli striking the tympanic membrane are amplified by the auditory ossicles before reaching the oval window. Amplification is necessary for sounds to pass from an air to a fluid environment.

PART III

CARDIOVASCULAR PHYSIOLOGY
Joseph Boyle, III

Chapter 9

Hemodynamics

I. CARDIOVASCULAR SYSTEM

A. **Function.** The primary function of the cardiovascular system is **convection** (i.e., the mass movement of fluid caused by a difference in pressure between two points). The cardiovascular system:

1. Distributes substrates and oxygen (O_2) to all body cells
2. Collects waste products and carbon dioxide (CO_2) for excretion
3. Controls blood flow to the skin and extremities to enhance or retard heat loss to the environment
4. Distributes hormones to distant sites
5. Aids in body defense mechanisms by delivering antibodies, platelets, and leukocytes to affected areas of the body

B. **Components**

1. The **heart** provides the driving force for the cardiovascular system.
2. The **arteries** serve as distribution channels to the organs.
3. The **microcirculation,** which includes the capillaries, serves as the exchange region.
4. The **veins** serve as blood reservoirs and collect the blood to return it to the heart.

II. THE HEART AS A PUMP. The heart is two pumps in series (i.e., the right and left sides) that are connected by the pulmonary and systemic circulations.

A. **Valves** (Figure 9-1). Each side of the heart is equipped with two valves that normally maintain a one-way flow of blood.

1. **Atrioventricular (AV) valves** separate the atria from the ventricles.
   a. The **right AV valve** is the **tricuspid valve.**
   b. The **left AV valve** is the **mitral valve.**
   c. These valves open during **ventricular relaxation (diastole)** to allow blood to fill the ventricles and close during **ventricular contraction (systole)** to prevent backflow (regurgitation) of blood from the ventricles into the atria.
2. **Semilunar valves (aortic and pulmonary)** open to allow the ventricles to eject blood into the arteries during systole and close to prevent backflow of blood into the ventricles during diastole.

A

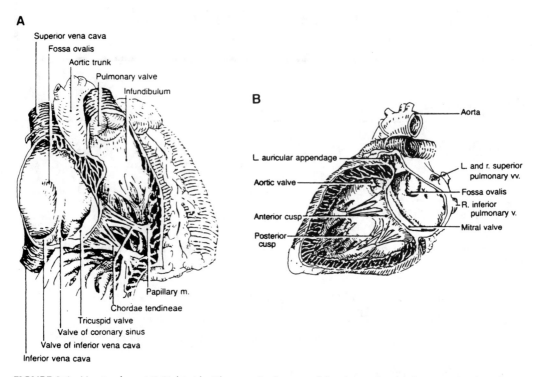

FIGURE 9-1. Heart valves. (A) Right side. The anterior heart wall has been dissected away, revealing the right atrium, tricuspid valve, right ventricle, and pulmonary valve. (B) Left side. The posterior heart wall has been dissected away, revealing the left atrium, mitral valve, left ventricle, and aortic valve. (Reprinted from *NMS Anatomy*, 2nd ed. Baltimore, MD, Williams & Wilkins, 1990, pp 142–143.)

B. **Ventricles**

1. The **right ventricle** pumps relatively large volumes of blood at low pressures through the pulmonary circulation.
    a. The right ventricle ejects blood by the concurrent shortening of the free wall and bulging of the interventricular septum into the right ventricle (the **bellows function**).
    b. The **normal** cross-section of the right ventricular chamber is **crescent-shaped**. If the right ventricle must eject blood against a high pressure for prolonged periods (as seen in certain pulmonary diseases), it assumes a much more **cylindrical** appearance and there is a thickening of the right ventricular free wall (**right ventricular hypertrophy**).

2. The **left ventricle** pumps blood through the systemic circulation. It is cylindrical in shape and normally has a thicker wall than does the right ventricle.
    a. The left ventricle ejects blood primarily by reducing the cross-sectional area of the cylinder. Thus, changes in volume are a function of changes in the radius squared (area $= \pi \cdot r^2$).
    b. The left ventricle works much harder than the right ventricle because of the higher pressures in the systemic circulation. Consequently, the left ventricle is more commonly affected by disease processes than is the right ventricle.

C. **Cardiac output** is the blood flow generated by each ventricle per minute. The cardiac output is equal to the volume of blood pumped by one ventricle per beat times the number of beats per minute:

$$Q = SV \cdot HR$$

where Q = cardiac output, SV = stroke volume, and HR = heart rate.

1. **Normal cardiac output.** The stroke volume for each ventricle averages 70 ml of blood, and a normal heart rate is approximately 70–75 beats/min; therefore, the cardiac output at rest is approximately 5 L/min.
   a. The heart rate is under neural control. Cardiac sympathetic efferent activity increases heart rate, whereas parasympathetic (vagal) efferent impulses decrease heart rate.
   b. The stroke volume varies with changes in the force of ventricular contraction, the arterial pressure, and the volume of blood in the ventricle at the onset of contraction.

2. **Systemic flow versus pulmonary flow.** Both ventricles must pump the same volume of blood during any significant time interval because of the series arrangement of the systemic and pulmonary circulations (Figure 9-2). In other words, pulmonary blood flow (right-sided output) must equal systemic blood flow (left-sided output).
   a. **Frank-Starling mechanism.** The balanced output is achieved by an intrinsic property of cardiac muscle known as the Frank-Starling mechanism (see Chapter 11 III A).
   b. **Congestive heart failure** occurs when the output of one or both ventricles is not sufficient to supply the needs of the body.
      (1) The cardiac output is limited to the smaller ventricular output (because the healthier ventricle can only pump as much blood as it receives), and a large volume of blood accumulates behind the lagging ventricle. The congestion impairs many of the exchange functions of the cardiovascular system and might be lethal.
      (2) Congestive heart failure can be caused by many different conditions that either decrease contractile function (e.g., myocarditis, valvular heart disease) or increase the pressure or volume work.

3. **Parallel distribution to the organs** (see Figure 9-2)
   a. Although the right and left sides of the heart are connected in series, the various systemic organs receive blood flow through parallel distribution channels.

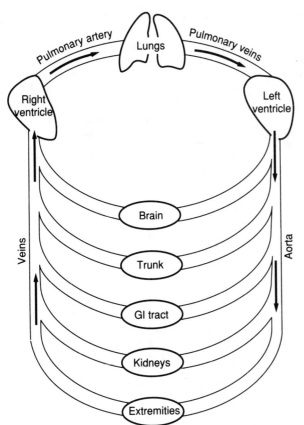

FIGURE 9-2. A schematic illustration of the organization of the cardiovascular system. Note that the right and left sides of the heart are connected in series, but the body organs receive blood through a parallel arrangement of vessels. GI = gastrointestinal.

b. The parallel arrangement supplies systemic organs with blood that has the same arterial composition (e.g., the same O_2 and CO_2 tensions, pH, glucose levels, and so on) and essentially the same arterial pressure.

III. **HEMODYNAMICS** is the study of the factors that determine blood flow and blood pressure in the body.

A. **Pressure** is a force per unit area (dynes/cm^2) and is usually expressed in terms of the height of a column of fluid that the pressure will support. The common units of pressure are mm Hg and cm H_2O (1 mm Hg = 1.36 cm H_2O = 1330 dynes/cm^2).

1. **Pressure gradient (ΔP)** is the difference in total energy (or pressure) between two locations within the system. Fluid always flows from an area of higher pressure to one of lower pressure; in other words, water (or blood) always flows downhill (Figure 9-3).

2. The **total energy** at any point equals the sum of the potential energy (pressure) and the kinetic energy.
   a. **Kinetic energy** is the momentum that blood gains because of its mass (m) and velocity (v):

   $$\text{kinetic energy} = \frac{m \cdot v^2}{2}$$

   Obviously, the greater the velocity of the blood flow, the greater the kinetic energy that is attained.
   b. **Potential energy** includes both hydrostatic and lateral pressure components.

FIGURE 9-3. Mean (average) lateral pressure in various components of the cardiovascular system. Note the progressive decrease in pressure from the left ventricle through the venous system until the blood enters the right ventricle. The right ventricle generates additional force to pump the blood through the pulmonary circulation. The resistance of the large vessels is minimal; therefore, the pressure drop is minimal until the smaller vessels are reached. The most drastic pressure drop occurs in the arterioles.

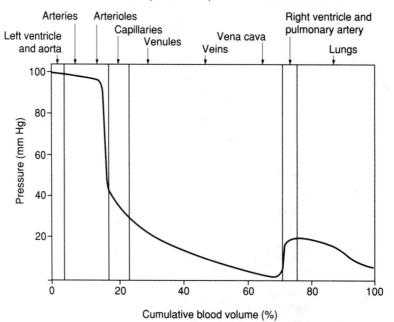

(1) Hydrostatic pressure (P_h) is caused by the effects of gravity on a fluid-filled system. Because of gravity, fluid has weight that generates force. The force is proportional to the vertical height of the column of fluid exposed to gravitational conditions (e.g., the pressure at the bottom of a lake is higher than it is at the surface):

$$P_h = \delta \cdot h \cdot g, \text{ where}$$

δ = the density of the fluid
h = the height of the fluid column above or below a reference level
g = the gravitational constant

(a) In the human vascular system, hydrostatic pressure occurs as a result of the weight of the blood in the vessels.
    (i) Assuming that in an upright person the foot is 150 cm below the heart, the pressure in the vessels of the foot would be 150 cm H_2O (110 mm Hg) higher than the pressure at the root of the aorta (Figure 9-4).
    (ii) In a supine subject, the hydrostatic effect is eliminated because the entire cardiovascular system is at essentially the same horizontal level. Therefore, it is important to eliminate the hydrostatic effect by measuring vascular pressures at the **zero reference (phlebostatic) level,** which is equivalent to the level of the right atrium.
(b) Both arteries and veins are affected equally by the hydrostatic column of blood so that the pressure gradient between arteries and veins is not altered (see Figure 9-4). However, the hydrostatic pressure can cause distention of dependent blood vessels (i.e., vessels that are located below the level of the heart), altering flow indirectly. This distention is especially significant in very compliant vascular beds (e.g., the veins or the pulmonary vessels).
(2) Lateral (static) pressure represents the pressure in the cardiovascular system that is usually measured with a gauge or transducer after eliminating the hydrostatic pressure effect. It does not include kinetic energy.
c. Conversion of energy. Energy can be converted between potential and kinetic energy.
    (1) Figure 9-5 shows a system of pitot tubes and standpipes. A pitot tube with its opening facing upstream measures both lateral pressure (static energy) and kinetic energy, whereas a standpipe measures only the lateral pressure.
    (2) The difference in total pressure between any two points in the system represents the **pressure gradient** that produces flow. If there is no flow, then the fluid level (pressure) in all the columns equals the fluid level at the source.
    (3) There is a progressive loss in total energy as the fluid flows through the pipe.
        (a) The progressive loss of energy along the pipe when there is flow is a result of frictional or resistive factors (see III B 2 b).

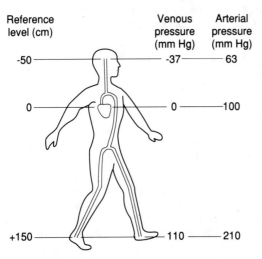

Reference level (cm)

	Venous pressure (mm Hg)	Arterial pressure (mm Hg)
-50	-37	63
0	0	100
+150	110	210

FIGURE 9-4. Effects of hydrostatic column of blood on the arterial and venous pressures. The zero reference level is at the right atrium.

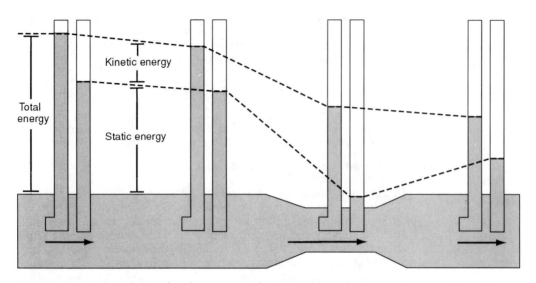

FIGURE 9-5. A system of pitot tubes demonstrating the variation in total energy (pressure), potential energy, and kinetic energy in relation to flow velocity. The *arrows* represent relative velocities in the different segments of the system. The narrowed section, a high-resistance segment, causes a large pressure drop as a result of the conversion of potential energy to kinetic energy. In addition, resistance (frictional) forces contribute to the drop in pressure. In the right segment, potential energy is gained as the velocity of flow slows again.

      (b) The narrowed section of the pipe is a high-resistance segment. The greater resistance leads to a greater loss of total energy per unit length of the pipe.

B. **Flow** (Q) is the mass movement of a volume of fluid per unit time, which is usually expressed in ml/sec or L/min. It can be represented as the average velocity of movement (cm/sec) times the cross-sectional area of the tube (cm²).

    1. Continuity principle. In any system arranged in series, the flow through each vascular component (e.g., the right and left ventricles, the aorta, the arteries, the capillaries, and the veins) must be equal to the flow through every other, unless one segment becomes progressively distended. Obviously, the limit for distention of the vascular system is rapidly reached in the body.

      a. The cross-sectional area of various vascular segments varies in the body.

      b. To keep the flow rate equal, the velocity of flow must vary inversely with the cross-sectional area for each vascular segment (Figure 9-6).

      c. In Figure 9-5, the narrowed segment also represents an area of increased velocity because this area has a decreased cross-sectional area. An increase in velocity results in a **Bernoulli effect** [i.e., a large conversion of potential energy (pressure) into kinetic energy (velocity)].

    2. Poiseuille's law for laminar flow is expressed as:

$$Q = \frac{\Delta P \cdot \pi \cdot r^4}{8 \cdot \eta \cdot L}, \text{ where}$$

          ΔP = the pressure gradient
            r = the radius of the tube
            η = the viscosity of the fluid
            L = the length of the tube

      a. Flow, pressure gradient, and resistance. Poiseuille's law is valid for straight, rigid tubes that contain a fluid with constant flow rate and constant viscosity; therefore, it is not strictly accurate in regard to the vascular system. Nevertheless, important principles relating flow, pressure gradient, and resistance remain applicable.

        (1) If resistance (R) is defined as $\frac{8 \cdot \eta \cdot L}{\pi \cdot r^4}$, then Poiseuille's law simplifies to a relationship analogous to Ohm's law: Current (flow) = Voltage (pressure gradient) / R.

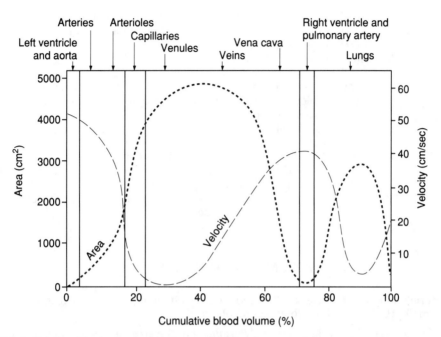

FIGURE 9-6. Relationships between velocity, cross-sectional area, and pressure in various segments of the cardiovascular system plotted as a function of the cumulative blood volume. Velocity of blood flow and area are inversely related.

Therefore, blood flow equals the difference in pressure between two points in the vascular system divided by the resistance to flow: $Q = \Delta P/R$.

(a) In the body, an increase in the pressure gradient is usually produced by elevating the arterial pressure.

(b) An increase in the pressure gradient causes an increase in blood flow, whereas an increase in resistance causes a decrease in blood flow.

(2) **Total peripheral resistance (TPR)** is usually expressed as the ratio of the difference in pressure to the flow: $TPR = \Delta P/Q$ (in mm Hg/L/min). An alternative unit is the peripheral resistance unit (PRU), which equals 1 mm Hg/ml/sec. The normal systemic circulation exhibits a resistance of approximately 1 PRU; in the pulmonary circulation, the resistance is 0.1–0.2 PRU.

b. **Factors that affect resistance.** The major factors that determine the resistance to blood flow are the **radius of the vessels** and the **viscosity of the blood**. Although an increase in the length of blood vessels would cause a proportional increase in resistance, blood vessels in the body generally have a fixed length.

(1) **Vessel radius.** Because the vessel radius (r) is raised to the fourth power, any change in the radius produces a change in resistance that markedly alters the flow. For instance, for a constant pressure gradient (ΔP), if the radius is reduced to one-half, the flow will be one-sixteenth of its previous value.

(a) **Arteriolar radius.** Control of blood flow to organs or tissues is regulated primarily by altering the radius of the arterioles. Because the arterioles control the flow to various vascular beds, they are considered the **stopcocks** of the circulation. Arteriolar radius is controlled by altering the number of sympathetic impulses to the blood vessels, by various drugs or hormones, or by the release of various chemicals from the tissues [e.g., histamine, prostaglandin, CO_2, hydrogen (H^+)].

(b) **Distensibility of vessels.** Blood vessels in the body are not rigid but distensible; the diameter of the vessels varies as a function of the internal and external pressures. Thus, increasing arterial pressure enlarges the vessel and increases the driving force for blood flow. As a result of the increased vascular radius, vascular resistance decreases so that blood flow in the body increases to a greater extent than would be predicted from the increased arterial pressure, per se.

    (2) Viscosity is the internal friction to flow in a fluid. Blood is a complex fluid because of the presence of cells and proteins, and its viscosity varies as a function of the flow rate and the vessel size. Variation in the **hematocrit,** which is the percentage of the blood that is occupied by red blood cells, is the major factor that changes the viscosity of blood.

        (a) The normal hematocrit for men is 40–45 whereas the normal value for premenopausal women is 35–40.

        (b) Blood with a hematocrit of 40 has a viscosity that is approximately three times that of water. Blood with a hematocrit of 60 has approximately twice the viscosity of blood with a hematocrit of 40.

        (c) An increase in hematocrit requires a greater pressure to produce a given flow rate because of the increased viscosity.

IV. ARTERIAL BLOOD VOLUME AND PRESSURE

A. **Arterial blood volume** refers to the amount of blood within the arterial vessels, which is normally **10%–15% of the total blood volume**.

    1. The **arterial blood volume is one determinant of the arterial pressure** because the blood in the arteries stretches the walls of the blood vessels, which then recoil on the contained blood, generating a pressure (Figure 9-7). This recoil force is termed **elastance**.

      a. The **unstressed (resting) volume** of the arterial tree is the volume of blood in the arteries when the internal pressure equals the external pressure (i.e., when the transmural pressure is zero).

      b. The **arterial elastance** represents the ratio of change in pressure for a change in volume.

        (1) The **slope** of the pressure–volume curves for the various age groups in Figure 9-8 represents the elastance.

        (2) Elastance is the **reciprocal of compliance** (i.e., distensibility, which is a change in volume for a change in pressure).

    2. Acute changes. The arterial pressure varies constantly due to changes in the arterial blood volume, which is altered by changes in heart rate, peripheral resistance, and stroke volume (Figure 9-9).

    3. Long-term changes in arterial pressure are caused by changes in the unstressed (resting) volume or in the elastance of the arterial system (i.e., because the arterial system enlarges and becomes more rigid with age).

B. Arterial blood pressure

    1. Factors that increase arterial pressure

      a. Increased heart rate

        (1) Increasing the heart rate increases arterial volume and pressure because the time the blood has to leave the arterial tree is reduced (i.e., the **length of the cardiac cycle is reduced**). In addition, the **cardiac output usually increases**.

        (2) After an increase in heart rate, the arterial pressure rises until the amount of blood exiting the arterial system equals the amount entering. At this time,

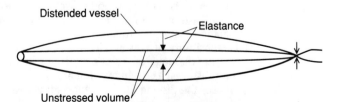

FIGURE 9-7. The vascular system exhibits elastic properties similar to a balloon. The blood volume distends the vessel, which then recoils and generates a pressure.

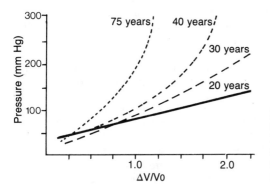

FIGURE 9-8. The average pressure–volume relationships of arterial systems in 20-, 30-, 40-, and 75-year-old individuals. Note the increasing elastance (decreased distensibility) that occurs with aging. ΔV = volume change, V_0 = unstressed arterial volume.

a new equilibrium is established (i.e., inflow equals outflow), but the transfer of blood from the venous to the arterial system raises the arterial pressure and lowers the venous pressure.

    b. Increased peripheral resistance

        (1) Peripheral resistance increases when **contraction of the smooth muscle in the walls of the arterioles narrows the lumina, raising the arterial volume** by reducing the amount of blood that leaves the arterial system per minute.

        (2) The arterial pressure rises until the new pressure is sufficient to overcome the additional resistance to flow, and outflow again equals inflow. The arterial pressure remains elevated as long as the volume of blood in the arteries remains high.

    c. Increased stroke volume. Increasing the stroke volume raises the arterial pressure by increasing the cardiac output (flow rate). The arterial pressure increases until outflow equals inflow and a new higher pressure is attained.

    d. Increased elastic constant. The elastic constant refers to stiffness of the arterial system, which progressively increases from birth until the time of death.

        (1) The slope of the pressure–volume relationship for any age group (see Figure 9-8) represents the elastic constant.

        (2) An increased stiffness or elastance of the arterial system causes an increase in pressure for any increase in arterial volume.

  2. Factors that decrease arterial pressure

    a. Obviously, **decreases in heart rate, peripheral resistance, stroke volume,** or the **elastic constant** will decrease the arterial pressure.

    b. Hemorrhage and **blood pooling** reduce the arterial pressure by **decreasing the circulating blood volume**. Gravity or dilation of vessels by neural, chemical, or mechanical factors can lead to blood pooling in dependent portions of the body.

FIGURE 9-9. A flow diagram showing the relationships among factors that determine cardiac output and the mean arterial pressure. Increases are indicated by *solid arrows;* decreases are indicated by *dashed arrows.*

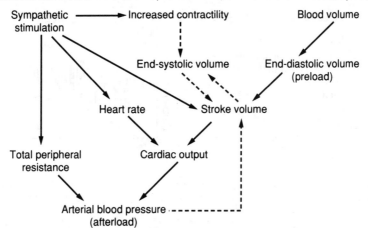

3. **Arterial pulse pressure.** The arterial pressure varies continuously because the heart ejects blood into the arteries intermittently but the blood flows out through the capillaries continuously.
   a. **Terminology**
      (1) **Systolic pressure** occurs during ventricular ejection and is the highest pressure in the arterial system. The normal arterial systolic pressure is approximately 120 mm Hg.
      (2) **Diastolic pressure** occurs just before the onset of ventricular ejection and is the lowest pressure in the arterial system. The normal arterial diastolic pressure is approximately 80 mm Hg.
   b. **Factors affecting pulse pressure.** Pulse pressure is the arithmetic difference between systolic and diastolic pressures (e.g., the normal pulse pressure is 40 mm Hg).
      (1) The pulse pressure depends on three factors: **age, stroke volume,** and the **arterial elastic constant**. The arterial elastic constant includes the arterial blood volume, which is proportional to the mean arterial pressure. The mean arterial pressure depends on cardiac output and peripheral resistance. Figure 9-10 depicts the relationship among these factors, which is integral to properly interpreting changes in a patient's heart rate and blood pressure.
      (2) The pulse pressure is increased by the effects of age and increased stroke volume, and decreased by increases in heart rate (HR) or total peripheral resistance.

FIGURE 9-10. Factors that alter pulse pressure. (*A*) An increased heart rate but constant stroke volume increase cardiac output and mean arterial pressure. Because the elastic constant remains unchanged, the pulse pressure must decline to maintain equality. (*B*) An increased stroke volume at a constant heart rate also increases cardiac output and mean arterial pressure, but pulse pressure increases. (*C*) An increased peripheral resistance transiently reduces capillary flow, and arterial volume and pressure increase. The result is a decrease in pulse pressure. (*D*) An increased elastic constant increases the pulse pressure, but the mean arterial pressure remains normal.

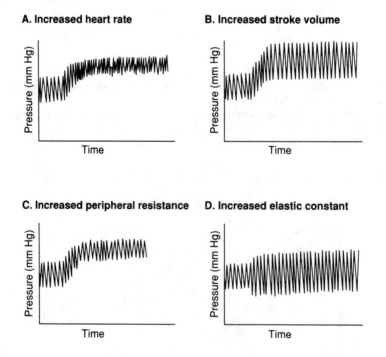

Chapter 10

Electrical Events

I. CONTRACTION

A. **Terminology**

1. **Systole** is the **contractile phase** and **diastole** is the **relaxation (filling) phase**.

2. The terms systole and diastole usually refer to ventricular events, but may be prefixed by "atrial" to refer to atrial contraction and relaxation, respectively.

B. **Rate.** The heart has an intrinsic contraction rate because it contains its own **pacemaker,** normally located in the **sinoatrial (SA) node**.

1. **Noradrenergic (sympathetic postganglionic) neurons** increase the pacemaker rate.

2. **Cholinergic (parasympathetic) neurons** decrease the pacemaker rate. The parasympathetic nerve to the heart is the **vagus nerve (cranial nerve X)**.

C. **Myocardial syncytium.** The heart muscle (myocardium) is composed of separate cardiac muscle cells that are electrically connected with one another by gap junctions (see Chapter 3 IV A 1–2).

1. Depolarization of one cardiac cell is transmitted to adjacent cells via the gap junctions; thus, the myocardium is a functional syncytium (a mass of cytoplasm with numerous nuclei).

2. The heart actually functions as two syncytia (i.e., the atria and the ventricles).
   a. These two masses of muscle are separated by the fibrous atrioventricular (AV) valve ring, which insulates the electrical events of the atria from those of the ventricles.
   b. Normally, there is only one functional electrical connection between the atria and the ventricles—the AV node and its continuation, the bundle of His.

II. ELECTRICAL ACTIVITY

A. **Resting membrane potential** (see Chapter 2 II). The myocardial cells maintain a voltage difference across their cell membranes of 60–90 mV.

B. **Cardiac action potential.** The action potential is the change in membrane potential that occurs after the cell receives an adequate stimulus, which depolarizes the membrane to the threshold potential (see Chapter 2 III D 1 a).

1. **Phases.** The action potential is divided into five phases (Figure 10-1).
   a. **Phase 0.** The resting membrane is relatively impermeable to Na^+, but once the threshold potential is reached, the Na^+ channels open and allow Na^+ to rush into the cell along its electrochemical gradient. This is a positive feedback (regenerative) process (see Chapter 2, Figure 2-5). The movement of Na^+ ions is a current-carrying process that results in the depolarization of the cell membrane (Phase 0).
      (1) This process is extremely rapid in fast fibers (see II B 2).
      (2) In slow fibers, Ca^{2+} is the current-carrying ion; therefore, phase 0 coincides with an increase in the Ca^{2+} conductance.
   b. **Phase 1.** During phase 1, the Na^+ conductance is rapidly reduced when the Na^+ channels close, but the membrane conductance for both Ca^{2+} and K^+ increases. The overall effect is a small change in the membrane potential toward the resting membrane potential (repolarization).

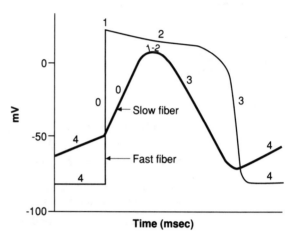

FIGURE 10-1. Action potentials from fast and slow cardiac fibers showing the different phases of the action potential. Note the diastolic depolarization (phase *4*) and the gradual phase *0* in the slow fiber.

    c. Phase 2. The plateau of the action potential in fast fibers coincides with an increased membrane conductance for Ca^{2+}. The inward movement of Ca^{2+} and the decreased efflux of K^+ maintain the membrane potential near zero during this phase of the action potential. In cardiac muscle, the influx of Ca^{2+} is important in the excitation–contraction (EC) process.

    d. Phase 3 is a rapid repolarization due to a reduction of the inward Na^+ and Ca^{2+} currents and a large increase in the outward K^+ current.

    e. Phase 4

        (1) Nonpacemaker cells (e.g., atrial and ventricular myocardium) exhibit a constant membrane potential during phase 4.

        (2) In **pacemaker tissues** (e.g., the SA and AV nodes and Purkinje fibers), there is a slow diastolic depolarization during phase 4, which indicates the presence of **automaticity** (i.e., the ability of the heart to generate its own beat).

           (a) The diastolic depolarization, also termed a **pacemaker potential,** brings the membrane potential toward threshold. The cell whose membrane potential first reaches threshold is the pacemaker of the heart, and the depolarization then spreads to the remainder of the heart.

           (b) The normal pacemaker of the heart is the SA node because its cells usually have the highest automaticity (i.e., the steepest slope of phase 4). Elimination of SA node activity or an increased slope of phase 4 in other areas of the heart causes a pacemaker outside of the SA node to initiate the depolarization process. This new pacemaker is termed an **ectopic pacemaker** because it is outside of the normal place.

  2. Slow and fast fibers. Myocardial cells may be slow or fast fibers, depending on the shape and conduction velocity of the action potential (Figure 10-2).

    a. Slow fibers are normally present only in the **SA and AV nodes;** however, the effects of hypoxia or certain drugs can convert fast fibers to slow fibers.

        (1) Resting membrane potential. Slow fibers have a resting membrane potential of **60–70 mV.**

        (2) Action potential. In slow fibers, the inward movement of positive ions (Na^+ and Ca^{2+}) neutralizes the normally negative intracellular charge, producing the action potential. During the action potential, the membrane potential approaches zero and the cell is said to be depolarized.

           (a) Slow channels. Ions apparently enter cells through "slow channels" in slow fibers.

               (i) These slow channels limit the rate of ion entry, slowing the rate of cell depolarization.

               (ii) The upstroke of the action potential requires approximately 100 milliseconds in slow fibers, compared with the 1 millisecond required by fast fibers.

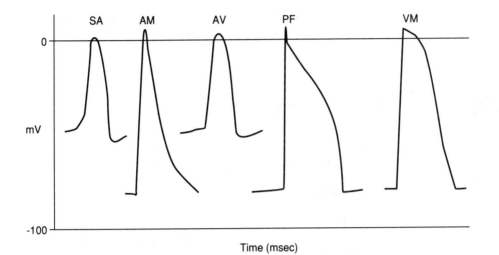

FIGURE 10-2. The form of the action potentials from various cardiac tissues. SA = sinoatrial node; AM = atrial muscle; AV = atrioventricular node; PF = Purkinje fiber; VM = ventricular muscle.

    (b) The **conduction velocity** of the action potential along the cells is directly related to the resting membrane potential and the rate of depolarization. Because the resting membrane potential is low and the rate of depolarization is slow, the conduction velocity of slow fibers is only 0.02–0.10 m/sec.

    (c) The **absolute refractory period** (see Chapter 2 III E 2 c) lasts the duration of the action potential in all myocardial fibers.

    (d) The **relative refractory period** may last for several seconds in slow fibers.

        (i) A stimulus applied during the relative refractory period must be stronger than normal to elicit an action potential. This action potential has a slower conduction velocity and lower amplitude than normal.

        (ii) A longer than normal refractory period frequently interferes with the conduction of impulses within the AV node, preventing all of the atrial depolarizations from reaching the ventricles and causing AV nodal block (see III D 3).

  b. Fast fibers. The normal atrial and ventricular myocardial cells and the specialized conducting tissues of the heart are fast fibers.

    (1) The **resting membrane potential** in fast fibers is **80–90 mV**. The inside of the cells are negatively charged with respect to the outside.

    (2) Action potentials

        (a) Na^+ channels. The action potential of fast fibers exhibits an extremely rapid rate of rise because once the threshold potential is reached, the cell membrane becomes extremely permeable to Na^+. Because the Na^+ concentration is much higher outside the cell and because of the negative charge inside the cell, there is a very large electrochemical gradient for Na^+, causing Na^+ to rush into the cell through channels that open once the threshold potential is reached.

            (i) The duration of the action potential varies in different fast fibers and is longest in the Purkinje and bundle of His fibers.

            (ii) The long action potential and absolute refractory period provide the heart some protection against certain arrhythmias by limiting the maximal ventricular rate.

        (b) The **conduction velocities** of fast fibers vary from 0.3–1.0 m/sec in myocardial cells to approximately 4 m/sec in Purkinje fibers. The fast conduction velocity ensures that the entire myocardium is depolarized almost instantaneously, which improves the effectiveness of myocardial contraction.

        (c) The **absolute refractory period** in fast fibers lasts until the repolarization has reached a membrane potential of 50–60 mV.

(d) The **relative refractory period** ends when the resting membrane potential of 80–90 mV is reestablished.

III. CONDUCTION PATHWAYS (Figure 10-3A&B)

A. **SA node.** Located near the junction of the superior vena cava and the right atrium, the SA node is the normal pacemaker of the heart. The isolated SA node has a firing rate of 90–120 beats/min; this rate is higher in young individuals and declines with advancing age. The firing rate is increased by a rise in temperature, thyroid hormone, epinephrine, or norepinephrine.

FIGURE 10-3. (*A*) The specialized conducting tissues of the heart. (*B*) Flow diagram summarizing the path of depolarization in the heart. SA = sinoatrial; AV = atrioventricular.

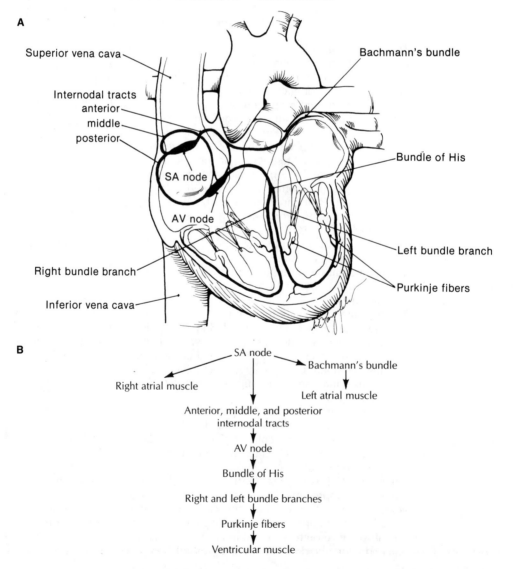

1. **Vagus nerve.** The right vagus nerve densely innervates the SA node and liberates **ace-tylcholine (ACh)** from its nerve endings when stimulated.
   a. Normally, vagal activity (vagal tone) slows the firing rate of the SA node from its automatic rate of 90–120 beats/min to the actual heart rate of approximately 70 beats/min.
   b. Strong vagal stimulation can completely eliminate SA node impulses (**sinus arrest**), leading to **asystole** (cardiac standstill) until an ectopic pacemaker begins to function.

2. **Sympathetic nerves.** Stimulation of cardiac sympathetic nerves or the injection of sympathetic-like drugs (e.g., epinephrine) **increases heart rate** (i.e., has a **positive chronotropic effect**) and increases the automaticity of ectopic sites. Stimulation of the cardiac sympathetic nerves markedly **increases the force of contraction** (i.e., has a **positive inotropic effect**).

B. **Interatrial tract (Bachmann's bundle).** This band of specialized muscle fibers runs from the SA node to the left atrium. The interatrial tract causes almost simultaneous depolarization and contraction of both atria because the conduction velocity through atrial muscle is very fast (Table 10-1).

C. **Internodal tracts.** Three bundles of specialized cells connect the SA and AV nodes: the **anterior, middle,** and **posterior** internodal tracts. The internodal tracts increase the likelihood that impulses from the SA node will reach the AV node and initiate ventricular depolarization, because the internodal tracts are more resistant than atrial muscle to the blockade of impulses.

D. **AV nodal.** Located just beneath the endocardium on the right side of the interatrial septum near the tricuspid valve, the AV node is normally the **only path for excitation** to proceed from the atria to the ventricles.

1. **Wolff-Parkinson-White syndrome.** Occasionally, an alternate depolarization pathway develops between the atria and ventricles during the formation of the heart. Its presence results in abnormal excitation of the ventricles and can be detected by an electrocardiogram (EKG).

2. **AV nodal delay.** The conduction velocity through the AV node is extremely slow (see Table 10-1), so that ventricular depolarization is delayed for 100–150 milliseconds after atrial depolarization.
   a. This delay provides time for atrial contraction to occur, which enhances ventricular filling, especially at fast heart rates.
   b. The AV node is richly supplied by fibers from both the sympathetic and vagal nerves, which affect the conduction of impulses.
      (1) **Sympathetic stimulation** increases the conduction rate through the AV node and enhances the transmission of impulses.
      (2) **Parasympathetic (vagal) stimulation** slows the conduction velocity and prolongs the refractory periods, making it more likely that AV block will occur.

TABLE 10-1. Cardiac Conductive Properties

Tissue	Fiber Diameter (μm)	Resting Membrane Potential (mV)	Conduction Velocity (m/sec)
Sinoatrial node	. . .	40–50	0.05
Atrial muscle	8–10	70–80	0.3–0.5
Internodal tracts	15–20	80–90	1.0
Atrioventricular node	Variable	50	0.02–0.05
Purkinje fibers	70–80	70	2.0–4.0
Ventricular muscle	10–16	80	< 1.0

3. AV nodal block. The AV node cells have a long refractory period and a very slow conduction velocity, thus limiting the number of impulses that can be conducted through the AV node to about 180 beats/min. The ventricles are protected from being driven at excessive rates, which can reduce the pumping effectiveness of the heart.

  a. Various diseases (e.g., coronary artery disease, myocarditis) can completely prevent atrial impulses from passing through the AV node; when this occurs, an ectopic ventricular pacemaker usually functions to drive the ventricles (but at a very slow rate).

  b. Under these conditions, it may be necessary to implant an **electronic pacemaker** in order to stimulate the ventricles to contract at an adequate rate.

E. **Ventricular conduction.** The impulses conducted through the AV node are distributed to the ventricles by specialized fibers.

  1. The **bundle of His** is the continuation of the AV node and is located beneath the endocardium on the right side of the interventricular septum. The bundle of His **divides into the right** and **left bundle branches**.

  2. The **right** and **left bundle branches** proceed on each side of the interventricular septum to their respective ventricles (see Figure 10-3A).

  a. Congenital or acquired lesions can lead to blockage of the depolarization process in either bundle branch. The affected ventricle is eventually depolarized by the spread of impulses through the regular ventricular muscle fibers.

  b. A right or left bundle branch block produces characteristic EKG changes (see IV C 4).

  3. Purkinje fibers arise from both bundle branches and branch out extensively just beneath the endocardium of both ventricles. These cells have the largest diameter in the heart and possess the highest action potential conduction velocity (see Table 10-1). The high conduction velocity ensures that both ventricles contract almost simultaneously, which increases the effectiveness of contraction.

F. **Ventricular muscle.** Depolarization of ventricular muscle occurs from the endocardial surface to the epicardium.

  1. Depolarization initially reaches the surface of the ventricle at the apex and spreads throughout the myocardium. The last area of the ventricles to be depolarized is the base of the left ventricle.

  2. The conduction velocity of the action potential through the ventricular myocardium is 0.3–0.4 m/sec.

IV. **ELECTROCARDIOGRAPHY.** The science of electrocardiography is approximately 100 years old. The techniques of recording electrical activity from the heart originally were developed by a Dutch physiologist named Willem Einthoven. Although the Anglicized acronym for electrocardiogram is ECG, the acronym EKG (from *elektrokardiogramm*) is used here in deference to Dr. Einthoven, who received a Nobel prize for his work. The EKG provides a method of evaluating excitation events, arrhythmias, cardiac hypertrophy, myocardial damage, and the presence of ischemia or necrosis. The EKG does not provide information on the mechanical performance of the heart, and at times even abnormal electrical activity may not be recordable.

A. **Volume conduction**

  1. Recording voltage changes. The body tissues function as electrical conductors because they contain electrolytes. The heart is assumed to lie centrally within the thorax, and its electrical activity is conducted to the body surface through the body fluids.

  a. Attaching electrodes to the body surface allows the voltage changes within the body to be recorded after adequate amplification of the signal. A **galvanometer** within the

EKG machine (electrocardiograph) is used as a recording device. Galvanometers record potential differences (voltages) between two electrodes.

b. EKGs are merely the recordings of differences in voltage between two electrodes on the body surface as a function of time.

   (1) Zero potentials. During diastole, the cardiac cells are polarized—positively charged on the outside and negatively charged on the inside. Electrodes on the skin do not detect voltage differences because all parts of the heart are equally polarized. Thus, the recording shows no deflection from the **zero potential line** (Figure 10-4A).

   (2) Action potentials

   (a) Depolarization. Excitation of a portion of the heart causes the cells to depolarize, producing a reversal of membrane potential in the area. The outside of the cells is now negatively charged with respect to ground. Thus, a potential difference exists between the depolarized cells and the neighboring, nonexcited cells (Figure 10-4B).

   (i) This potential difference can be recorded from surface electrodes, and the direction of its deflection will depend on the polarity of the electrodes.

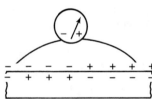

A. Resting

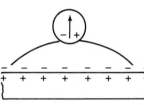

B. Depolarizing

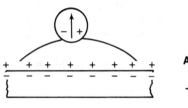

C. Depolarized

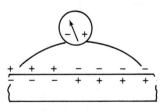

D. Repolarizing

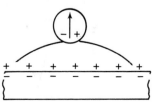

E. Repolarized

FIGURE 10-4. The effects of depolarization and repolarization on isolated, excitable tissue. The dial represents a galvanometer that is connected to the tissue bath using standard connections. Note that when depolarization and repolarization are in the same direction, the galvanometer deflections are oppositely directed. In the limb leads of an electrocardiogram (EKG), the QRS and T waves are in the same direction because depolarization and repolarization occur in opposite directions.

      (ii) When the entire heart has been depolarized, all of the cells are negatively charged outside, and both electrodes again "see" the same potential. The galvanometer reading returns to zero because the potential difference between the two electrodes is zero (Figure 10-4C).

    (b) Repolarization. Assuming that repolarization proceeds in the same direction as depolarization, the galvanometer will be deflected in the opposite direction (Figure 10-4D&E) during the repolarization process.

    (c) The resulting record of depolarization and repolarization is termed a **biphasic action potential** because there are two opposite waves.

  2. Equivalent dipole. The voltage differences among resting, depolarized, and repolarizing cells function as a battery. The various charges that are present are summated and are termed an **equivalent dipole**.

    a. The total charge that is present depends on the mass of tissue involved as well as the magnitude of the membrane potentials.

    b. The cardiac dipole is a **vector quantity** because it has both **magnitude** and **direction**. Vectors are represented as arrows with the arrowhead indicating the direction and the length of the arrow indicating the magnitude.

  3. The **surface potential,** or the magnitude of the voltage recorded at the body surface, is a function of electrode position and the orientation and magnitude of the dipole. Figure 10-5 depicts these relationships, which are essential to understand when analyzing EKG recordings.

    a. By convention, a wave of depolarization approaching the positive electrode results in an upward deflection of the EKG tracing.

    b. A wave of depolarization proceeding parallel to an electrode axis (the line connecting two electrodes) produces the maximal deflection for that dipole.

    c. A depolarization wave perpendicular to the electrode axis produces no net deflection of the tracing (i.e., the positive and negative waves are equal).

B. The **standard 12-lead EKG** consists of three bipolar limb leads (I, II, and III), three augmented limb leads (aVR, aVL, and aVF), and six chest leads (V$_1$–V$_6$). Unless there is a change in the sequence of excitation (such as would result from a transient block), each lead shows the same cardiac events as the other leads, but from a different view. Additional leads are used in special circumstances.

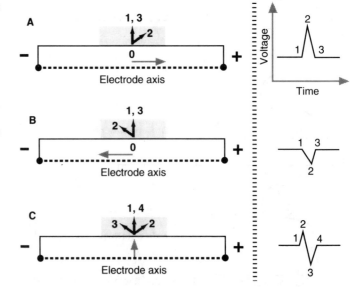

FIGURE 10-5. The deflection of a galvanometer needle using standard connections. (*A*) A wave of depolarization approaching a positive electrode causes an upward (positive) deflection. (*B*) A wave of depolarization approaching a negative electrode causes a downward deflection. (*C*) A wave of depolarization that is proceeding perpendicular to the electrode axis produces no net deflection.

1. **Einthoven's triangle.** The torso is considered to be an equilateral triangle with the right and left shoulders and the left leg as the apices. The right leg serves as a ground connector; all four electrodes must always be attached to the extremities, which merely serve as conductors from the torso.

    a. **Zero potential lines.** If lines are drawn perpendicularly from the center of each side of an equilateral triangle, they will meet at the center of the triangle. These lines represent the zero potential lines for the three sides of the triangle (Figure 10-6A).

    b. **Bipolar limb leads.** Three leads are formed by measuring the potential differences between any two of the active limb electrodes [i.e., the **right arm (RA), left arm (LA),** and **left leg (LL)**]. These leads are selected by a switch on all standard EKG machines.

    (1) **Lead I** (mV) = LA − RA.

    (2) **Lead II** (mV) = LL − RA.

    (3) **Lead III** (mV) = LL − LA.

2. **Unipolar (V) leads.** If the three limb leads are connected to a common terminal, the combined voltage from the three leads theoretically will be zero. This common terminal can be attached to the negative pole of a galvanometer and a fourth, or exploring, electrode can be attached to the positive pole. The galvanometer still can read only the potential difference between two points; however, if the common electrode is at zero volts, the exploring electrode will provide the actual or absolute voltage at the body surface. This arrangement of connections is termed a **unipolar electrode** and is used to record the **precordial** or **chest leads** from standardized sites. There are six precordial (V) leads in the standard EKG.

    a. **V_1** is located at the fourth intercostal space just to the right of the sternum.

    b. **V_2** is located at the fourth intercostal space just to the left of the sternum.

    c. **V_4** is located at the midclavicular line in the fifth intercostal space.

    d. **V_3** is halfway between V_2 and V_4.

    e. **V_5** is located at the anterior axillary line at the same level as V_4.

    f. **V_6** is located at the midaxillary line at the same level as V_4 and V_5.

3. **Augmented unipolar (aV) leads.** Alternatively, the unipolar exploring electrode can be placed on the limbs to record cardiac potentials; however, the deflections are small. The size of the recordings is increased by eliminating the electrode of interest from the common terminal, which does not seem to create any significant error. The potentials recorded in this manner from the **RA, LA,** and **LL** are termed **aVʀ, aVʟ,** and **aVғ,** respectively.

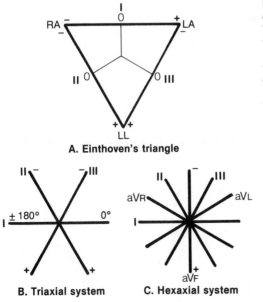

A. Einthoven's triangle

B. Triaxial system

C. Hexaxial system

FIGURE 10-6. (A) Einthoven's triangle showing connections, bipolar limb leads, and lead polarity. The electrical zero occurs at the center of each of the bipolar leads. (B) The triaxial reference system in which leads I, II, and III are collapsed onto their respective zero points. (C) The hexaxial reference system is obtained by adding the augmented unipolar limb leads (aVʀ, aVʟ, aVғ) to the triaxial system. RA = right arm; LA = left arm; LL = left leg.

    a. A **triaxial reference system** is obtained by moving the sides of Einthoven's triangle so that they intersect at the center of the triangle. Lead I then divides the system into an upper (negative) and a lower (positive) hemisphere (Figure 10-6B).

    b. Hexaxial reference system. Superimposing the axes of the aV leads on the triaxial system provides a hexaxial reference system with axes every 30°. The aV lead axes bisect the angles of Einthoven's triangle (Figure 10-6C). In this system, leads I and aVF divide the system into quadrants with the **zero potential point** for all leads at the central intersection.

C. **EKGs** are the tracings of the surface cardiac potentials recorded against time. These tracings usually are made at a standard recording speed (25 mm/sec) and amplification (1 mV = 1 cm deflection). Using standard EKG paper, each small horizontal division represents 0.04 second and each large division 0.2 second. Each small vertical division represents 0.1 mV.

1. EKG waves and intervals (Figure 10-7)

    a. A **P wave** results from **atrial depolarization** and is normally positive (upright) in the standard limb leads and inverted in aVR.

    b. The **QRS complex** is caused by **ventricular depolarization,** and its duration is normally less than 0.08 second.

        (1) Prolongation of the QRS duration indicates either an intraventricular conduction block caused by blockage of one of the bundle branches or the presence of an ectopic ventricular pacemaker.

        (2) Terminology. The **ventricular depolarization complex** is termed the **QRS complex** regardless of whether all three components are present.

            (a) A **Q wave** is a **negative wave before a positive wave**.

            (b) An **R wave** is a **positive wave**. If there are two positive waves, the second is designated "**R'**" and the smaller of the two is designated by "**r.**" For example, an rSR' complex would consist of a small, initial R wave, an S wave, and a large, final R wave (see Figure 10-18B).

FIGURE 10-7. An electrocardiogram (EKG) showing the various waves and intervals as well as the calibration times and voltages.

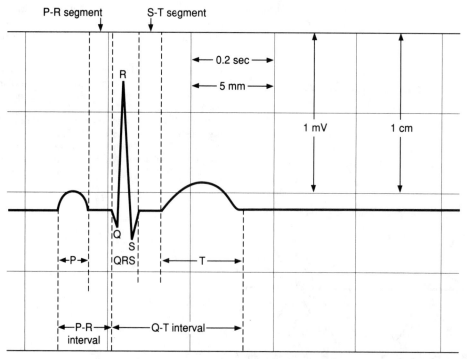

(c) An **S wave** is a negative wave following an R wave. If no R wave is present, a **completely negative wave** termed a **QS complex** results.

c. The **T wave** is caused by **ventricular repolarization** and normally deflects in the same direction as the QRS complex, because ventricular repolarization follows a path opposite to depolarization.

d. The **P-R interval** is measured from the onset of the P wave to the onset of the QRS complex and normally varies between 0.12 and 0.21 second, depending on heart rate. The P-R interval is a measure of the **AV conduction time,** including the delay through the AV node.

e. The **R-R interval** is the time between successive QRS complexes and is usually used to measure the **cardiac cycle length**. The heart rate is equal to 60 divided by the R-R interval in seconds.

2. The **mean electrical axis (MEA)** is the average vector produced by any given wave in the EKG. An electrical axis can be derived for the P, QRS, and T waves.

a. **Derivation.** The MEA can be derived using any two standard (i.e., bipolar) limb leads or any two augmented limb leads. An augmented limb lead and a bipolar limb lead cannot be used together because of the difference in amplification of the two leads.

b. **Method of measurement.** Clinically, the MEA of the QRS complex is determined algebraically by summing the heights of the Q, R, and S waves in each of two leads. The resultant sum represents the magnitude of the vector for the respective lead. The direction of the vector is toward either the positive or negative electrode depending on the value of the summation. Figure 10-8 summarizes this method of determining the MEA.

(1) **Plotting lead vectors.** The vector from each lead is plotted, on an appropriate scale, using either the Einthoven triangle, the triaxial, or the hexaxial reference system.

(2) **Plotting the MEA.** The tail of the vector for each lead is placed on the zero potential point, and the vector is drawn along the respective lead with the head of the arrow pointing toward the correct pole (positive or negative). Perpendiculars are then drawn from the heads of the two vectors until they intersect.

(3) **Drawing the mean vector.** The head of an arrow is drawn at the intersection of the two perpendiculars, and the tail is drawn to the center of the triangle or the zero potential point. The length of this arrow represents the magnitude of the MEA, and its direction (in degrees) represents the electrical axis in the frontal plane.

FIGURE 10-8. Mean electrical axis (MEA) determination by vector analysis. The Q and S waves (negative values) are added algebraically to the R waves (positive values) for each of two leads. The result gives the magnitude and the direction of the vector (+ or −) in each lead. The QRS magnitude is plotted along the respective leads from the zero point toward the appropriate polarity. Perpendiculars to each axis are drawn through the arrowheads, and the intersection marks the head of the mean electrical vector. The MEA is drawn from the center of the triangle (electrical zero for the system) to the perpendiculars' intersection. The MEA (in degrees) and the magnitude are given by the direction and the length of the mean electrical vector. *RA* = right arm; *LA* = left arm; *LL* = left leg.

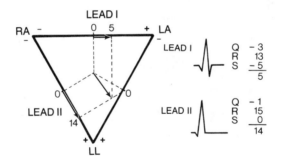

c. **Electrical axis in the frontal plane** (Figure 10-9A). The MEA of the QRS complex determines if axis deviation is present.
  (1) A **normal axis** is present if the MEA lies between −30° and +120° when plotted on the hexaxial reference system (or 2 and 7 o'clock, respectively, if the hexaxial system is considered a clock face).
  (2) **Right axis deviation (RAD)** is present when the MEA lies between +120° and +180° (or 7 and 9 o'clock).
    (a) RAD is caused by **right ventricular hypertrophy** secondary to chronic lung disease or pulmonary valve stenosis, or by **delayed activation of the right ventricle** (as occurs in right bundle branch block).
    (b) Figure 10-9B shows QRS complexes characteristic of RAD.
  (3) **Left axis deviation (LAD)** is present when the MEA lies between −30° and −90° (or 2 and 12 o'clock).
    (a) LAD is associated with **obesity, left ventricular hypertrophy,** or **left bundle branch block**.
    (b) Figure 10-9B shows typical QRS complexes that occur with LAD.
  (4) An **indeterminate axis** is present if the MEA lies between −90° and −180° (or 12 and 9 o'clock). Indeterminate axis may result from either **extreme right or extreme left axis deviation**.
d. **Electrical axis in the horizontal plane.** The MEA in the horizontal plane is determined using the precordial (chest) leads.

D. **Vector loops.** The instantaneous cardiac vector represents the electrical vector generated by the cardiac dipole during the depolarization process. This vector begins at the zero isopotential point and inscribes a loop as the tissues are depolarized. Three loops can be recorded during one cardiac cycle. The **P loop** is caused by atrial depolarization; the **QRS loop** is caused by ventricular depolarization; and the **T loop** results from ventricular repolarization (Figure 10-10). Atrial repolarization cannot be recorded with standard techniques because of the prolonged time course and the small voltages involved.

FIGURE 10-9. (A) Axis definitions in the frontal plane (using the hexaxial reference system). Lead I divides the frontal plane into upper (negative) and lower (positive) hemispheres and the aVF lead divides the frontal plane into quadrants. Note the degree designations and the division into various axis designations. (B) QRS configuration as seen in leads I, II, and III for a normal axis and right and left axis deviations. *RAD* = right axis deviation; *LAD* = left axis deviation.

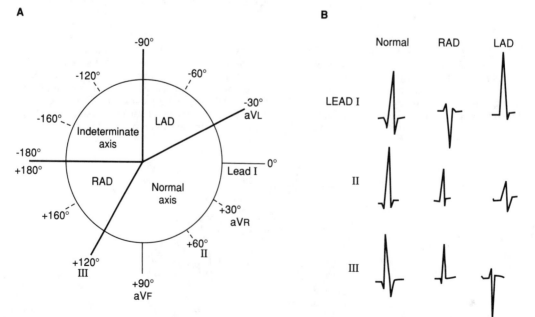

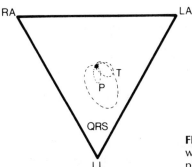

FIGURE 10-10. The vectorcardiographic loops P, QRS, and T, which can be recorded with appropriate equipment and electrode placement. *RA* = right arm; *LA* = left arm; *LL* = left leg.

1. The **P loop** is small and is directed leftward and inferiorly, resulting in a positive P wave in the three bipolar limb leads.

2. The normal **QRS loop** is inscribed counterclockwise and is directed leftward, inferior and posterior.

   a. **Stages.** Ventricular depolarization is a continuous process, but it can be divided into stages for discussion (Figure 10-11A).

      (1) **Stage 1** is the initial phase of ventricular depolarization and involves the left endocardial surface of the interventricular septum. Depolarization spreads superiorly and to the right resulting in a small vector directed toward the right shoulder. This initial vector produces small negative waves in leads I and II and small positive waves in the right precordial leads (Figure 10-11B).

      (2) **Stage 2** represents depolarization of the remainder of the interventricular septum and the subendocardial areas of both ventricles. The resultant vector is directed inferiorly and produces large positive waves in leads II, III, and aVF (see Figure 10-11B).

      (3) **Stage 3** depolarization involves the remainder of the right ventricle and a large portion of the left ventricle. Because the left ventricle normally has a larger muscle mass than the right, the stage 3 vector is directed toward the left. This vector produces R waves in lead I and causes a progressive increase in the R waves in the left precordial leads V_2–V_6.

      (4) **Stage 4** represents the last portion of the ventricle to depolarize (i.e., the posterior base of the left ventricle). This vector is directed toward the left shoulder and produces small S waves in leads II, III, and aVF (see Figure 10-11B).

FIGURE 10-11. (*A*) Arbitrary stages of ventricular depolarization. (*B*) Reconstruction of the QRS complex from the QRS loop for the three bipolar limb leads. Arrows indicate the direction of the loop recording. *RA* = right arm; *LA* = left arm; *LL* = left leg.

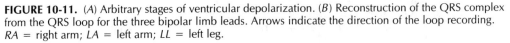

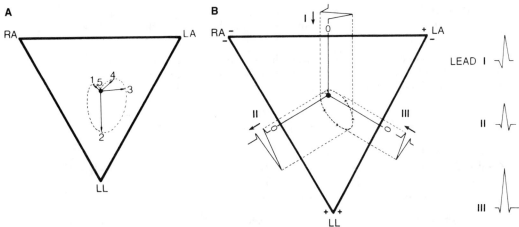

(5) **Stage 5** is the return of the cardiac potential to the zero potential point and the end of the depolarization process.

b. **Relationship between the QRS loop and QRS complex.** Depolarization of the cardiac tissue generates the vector loop.

(1) A **scalar EKG** records the depolarization process by plotting voltage as a function of time. Figure 10-11B depicts how the vector loop generates the QRS complex on the different leads.

(2) It is possible to reconstruct the vector loops from scalar EKGs (and vice versa) if basic electrical principles are understood.

3. The **T loop** represents the repolarization process, which is roughly opposite in direction to that of depolarization. This occurs because the inner layers of the myocardium are slightly hypoxic as a result of vascular compression during systole, and hypoxia prolongs the relative refractory period and the onset of repolarization. The opposite direction of repolarization compared with depolarization results in T waves that normally are in the same direction as the QRS complex.

V. CARDIAC RATE, RHYTHMS, AND CONDUCTION DISTURBANCES

A. Cardiac rate

1. **Normocardia** is a normal heart rate that ranges between 60 and 100 beats/min.

2. **Tachycardia** is a heart rate in excess of 100 beats/min.

3. **Bradycardia** is a heart rate that is less than 60 beats/min.

B. Cardiac rhythms

1. **Sinus rhythm** is present when the SA node is the pacemaker. Sinus rhythm can be assumed if each P wave is followed by a normal QRS complex, the P-R and Q-T intervals are normal, and the R-R interval is regular.

a. **Sinus arrhythmia** (Figure 10-12) is characterized by a normal QRS complex, P-R interval, and Q-T interval, but the R-R interval (cardiac rate) varies in a set pattern.

(1) Sinus arrhythmia is usually, but not always, synchronized with respiration. Usually, heart rate increases during inspiration (note in Figure 10-12 the shorter cycle lengths during inspiration) and slows during expiration as a result of variations in vagal tone that affect the SA node.

(2) Sinus arrhythmia is common in children and in endurance athletes with slow heart rates.

b. **Sinus tachycardia** is a normal response to exercise and also occurs in the presence of fever, hyperthyroidism, and as a reflex response to low arterial pressures (Figure 10-13).

c. **Sinus bradycardia** represents a heart rate of less than 60 beats/min and may be abnormal, but is more commonly seen in highly trained endurance athletes (see Figure 10-13).

2. **Atrial rhythms** are generated when there is an ectopic atrial pacemaker.

a. **Atrial tachycardia** occurs when there is an ectopic atrial pacemaker and the heart rate exceeds 100 beats/min.

FIGURE 10-12. Sinus arrhythmia. The heart rate increases (i.e., the cycle length decreases) during inspiration. Note the differences in the T-P intervals during inspiration and expiration.

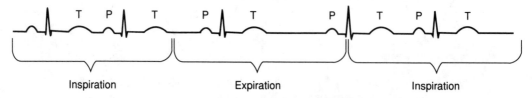

Inspiration   Expiration   Inspiration

Normal

Tachycardia

Bradycardia

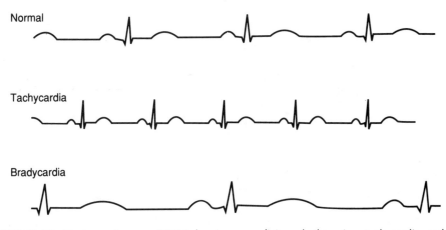

FIGURE 10-13. Electrocardiograms (EKGs) showing normal sinus rhythm, sinus tachycardia, and sinus bradycardia.

    (1) Characteristics. Atrial tachycardias are characterized by very regular rates ranging from 140–220 beats/min.
        (a) Usually, atrial tachycardias have a rapid onset and last only for seconds or minutes, which is why they are termed **paroxysmal atrial tachycardia (PAT).**
        (b) Occasionally, these tachycardias may last for hours or days. A prolonged bout of atrial tachycardia is termed a **supraventricular tachycardia** because it is usually not possible to determine the actual pacemaker site.
    (2) Causes
        (a) These attacks may be precipitated by overindulgence in caffeine, nicotine, or alcohol, or they may occur during anxiety attacks.
        (b) PAT may be caused by discharge of a single ectopic site or it may result from a **reentry phenomenon,** which occurs when an action potential reexcites an area of myocardium that had been depolarized previously.
            (i) Reentry requires at least two paths of depolarization that have different conduction velocities, refractory periods, or path lengths.
            (ii) Cells that have regained excitability are reexcited from impulses arising from the initial impulse carried over the slower conducting paths.
  b. Premature atrial contractions (PACs, atrial extrasystoles)
    (1) Cause. PACs occur if an atrial ectopic site fires and becomes the pacemaker for one beat.
    (2) Diagnosis (Figure 10-14A)
        (a) EKG appearance. A premature P wave is present, followed by a normal QRS complex and T wave. The premature P wave may have an aberrant configuration because the impulse arises outside of the sinus node. The P-R interval may also be altered because of the different path of atrial depolarization that occurs.
        (b) Bedside diagnosis. The condition can usually be diagnosed at the bedside because the premature impulse discharges the SA node, which then must repolarize and fire after the normal interval, resulting in a shift in cardiac rhythm (see Figure 10-14A).
    (3) Significance. These beats are entirely normal. The patient may note an occasional irregularity in the cardiac rhythm.
  c. Atrial flutter (Figure 10-15A) occurs when atrial rates are 220–350 beats/min. During atrial flutter, the **AV node is unable to transmit all of the atrial impulses**. A physiologic AV block develops, and the ventricular rate is one-half, one-third, or one-fourth of the atrial rate. Changes in the AV block from 4:1 to 3:1, for example, can cause rapid shifts in the ventricular rate. Atrial flutter differs from atrial tachycardia only in the atrial rate.
    (1) Cause. Like atrial tachycardia, atrial flutter may be caused by a single ectopic focus or by a reentry phenomenon.

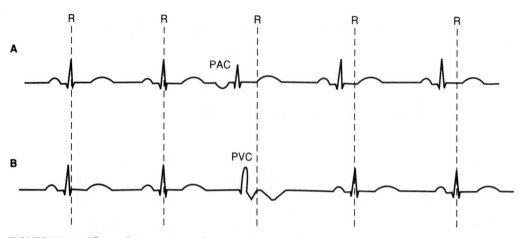

FIGURE 10-14. Effects of premature atrial (PAC) and ventricular (PVC) contractions on rhythm. The vertical lines represent the normal R-R interval. Note that the rhythm shifts with a PAC because the sinus node is reset, but the rhythm does not shift after a PVC because the following beat occurs after a compensatory pause.

        (2) EKG appearance. The P waves produce a saw-toothed appearance on the EKG that is virtually diagnostic of atrial flutter.

   d. Atrial fibrillation (Figure 10-15B) is an irregular, rapid atrial rate in which there is **contraction of only small portions of the atrial musculature at any one time because large portions of the atria remain refractory to depolarization**. The ventricular rate is completely (irregularly) irregular because only a fraction of the atrial impulses that reach the AV node are transmitted to the ventricles.

        (1) EKG appearance. The baseline of the EKG shows **small, irregular oscillations (F waves)** that result from the depolarization of small units of the atrial musculature. There are **no recognizable P waves** and the **R-R interval is irregularly irregular**. The QRS and T waves are normal because the impulses that are transmitted through the AV node are conducted normally through the ventricles.

        (2) Significance. Atrial fibrillation represents an even higher rate of atrial activity than does atrial flutter.

           (a) Atrial fibrillation frequently is associated with **enlarged atria** secondary to AV valve disease. Cardiac output may be reduced as a result of the valve disease and because of the loss of an effective atrial contraction.

           (b) Long-term atrial fibrillation is associated with **thrombi** in the atrial appendages. Atrial thrombi may be the source of pulmonary emboli (right atrium) or systemic emboli (left atrium).

FIGURE 10-15. (A) Atrial flutter exhibits a characteristic saw-toothed appearance. (B) Atrial fibrillation shows very small baseline fluctuations on the electrocardiogram (EKG) as small portions of the atria are depolarized in an uncoordinated manner.

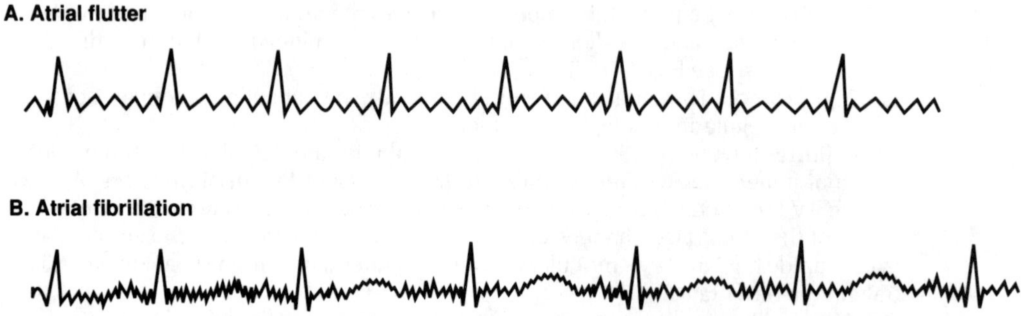

A. Atrial flutter

B. Atrial fibrillation

3. **AV junctional (nodal) rhythms.** Pacemaker cells have been found in the AV node. The term "junctional rhythms" is used if there is an ectopic pacemaker located in the AV node.
   a. **Junctional premature beats**
      (1) **EKG appearance.** Junctional premature beats are characterized by an **inverted P wave** (as the result of retrograde transmission) and **normal QRS complexes**. They have been subdivided into **high or low** junctional beats depending on whether the P wave precedes or follows the onset of the QRS complex.
      (2) **Significance.** Junctional beats have the same significance as PACs. If the AV nodal tissues recover excitability, a retrograde impulse may return and cause a second ventricular depolarization (known as a **reciprocal beat**).
   b. **Transient or permanent junctional rhythms** may arise in otherwise healthy individuals if the SA node is suppressed. Permanent junctional rhythms may arise secondary to a large number of organic heart diseases.
   c. **Junctional tachycardias** are similar to atrial tachycardias and may be indistinguishable on EKG; therefore, they should be considered supraventricular tachycardias.

4. **Ventricular rhythms**
   a. **Premature ventricular contractions (PVCs)** can arise from any portion of the ventricular myocardium and occasionally occur in otherwise healthy individuals.
      (1) **Cause.** Frequent PVCs occur with many forms of heart disease, especially coronary heart disease (because ischemia increases the irritability of the myocardium).
      (2) **EKG appearance.** PVCs are characterized on the EKG by a **prolonged** (greater than 0.1 second), **bizarre QRS complex that does not have a preceding P wave**.
         (a) The T wave is usually oppositely directed from the QRS complex (Figure 10-14B).
         (b) The origin of the PVC can be determined by vectorial analysis using the initial portion of the QRS complex.
      (3) **Compensatory pause.** Retrograde transmission of depolarization to the atria usually does not occur with PVCs; thus, the atrial rate remains unaltered.
         (a) The atrial depolarization that follows a PVC usually arrives while the AV node is still refractory and, therefore, is not conducted to the ventricles, creating a pause in the ventricular rhythm.
         (b) This pause is usually fully compensatory so that the R-R interval of the beat preceding the PVC and the PVC interval together equal two normal cycle lengths (see Figure 10-14B). The **beat following the PVC** is **stronger than normal** because of the added stroke volume and is usually **detectable by the patient**.
      (4) **Interpolated beats.** If the sinus rhythm is slow, a PVC may occur without altering the normal R-R interval. The premature beat in this case is termed an interpolated beat.
   b. **Ventricular tachycardias** result from a rapid, repetitive discharge from a ventricular site, usually as a result of a reentry phenomenon. Alterations in vagal tone (carotid sinus massage, Valsalva maneuver) do not affect ventricular tachycardias because the ventricles do not receive any efferent vagal innervation.
      (1) **Causes**
         (a) Ventricular tachycardias are almost always associated with **serious heart disease** or **drug toxicity**.
         (b) Ventricular tachycardias also occur when **catheters** (e.g., Swan-Ganz) stimulate the endocardial surface of the ventricles. A slight movement or withdrawal of the catheter may terminate the tachycardia.
      (2) **EKG appearance** (Figure 10-16A). The EKG reveals **wide, bizarre QRS complexes** occurring at a rapid rate. The P waves are usually indistinguishable, although the SA node activity continues independently of the ventricles.
      (3) **Significance.** Sustained ventricular tachycardia can be a life-threatening arrhythmia if it degenerates into ventricular fibrillation.

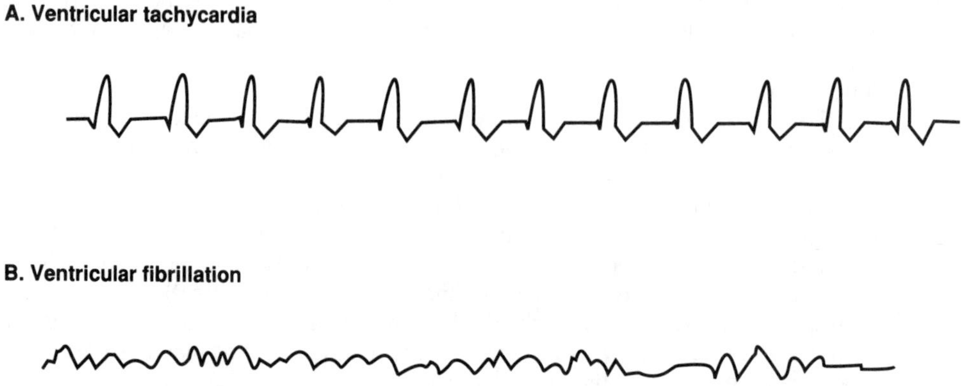

A. Ventricular tachycardia

B. Ventricular fibrillation

FIGURE 10-16. (*A*) Ventricular tachycardia is characterized by wide, bizarre QRS complexes that are caused by an ectopic ventricular pacemaker. (*B*) Ventricular fibrillation, like atrial fibrillation, results from uncoordinated depolarization of small segments of the myocardium.

> **(a)** Cardiac output is reduced during ventricular tachycardias because ventricular filling time is reduced, causing a decrease in stroke volume.
> **(b)** Cardiac output may also be reduced because the effectiveness of ventricular contraction is reduced by **asynchrony** of the contractile process. Because action potentials travel over the slowly conducting ventricular muscle (rather than the Purkinje fibers) in ventricular tachycardia, depolarization is prolonged and asynchrony occurs.

  c. Ventricular fibrillation results from rapid, irregular, ineffective contractions of small segments of the ventricular myocardium. The peripheral pulse is absent and **cardiac output is zero**. The electrocardiograph must be used to **distinguish** ventricular fibrillation **from cardiac standstill**.

> **(1) EKG appearance** (Figure 10-16B)
> > **(a)** The EKG shows **undulating waves of varying frequency and amplitude**.
> > **(b)** Ventricular fibrillation is **often precipitated by** one or more **PVCs,** one of which falls on the **vulnerable interval** of the T wave (i.e., the interval near the peak of the T wave when the ventricle is partially repolarized). A PVC during the vulnerable interval often produces fibrillation because the likelihood of reentry occurring is increased.
> **(2) Significance.** Cardiopulmonary resuscitation must be started immediately to prevent tissue death, and continued until cardioversion (defibrillation) can be performed.

C. | **Conduction disturbances**

  1. SA nodal block (sick sinus syndrome) consists of the disappearance of the P wave for several seconds while an ectopic pacemaker, usually in a junctional or ventricular site, drives the ventricles.

> **a.** SA block results in a **bradycardia** and is seen especially in the elderly and in patients recovering from a coronary occlusion.
> **b.** Various drugs and implanted pacemakers may be used to increase the heart rate to normal levels.

  2. AV nodal block (Figure 10-17)

> **a. First-degree AV nodal block** is a prolongation of the P-R interval beyond 0.21 second that results from slowed conduction through the AV node (see Figure 10-17A).
> > **(1) Causes.** Increased vagal tone and many systemic diseases may cause first-degree AV nodal block.
> > **(2) Significance.** This condition has no effect on the pumping ability of the ventricles but indicates that some conduction disturbance is present.
> **b. Second-degree AV nodal block** occurs when the AV node fails to transmit all of the atrial impulses. It is usually associated with organic heart disease.

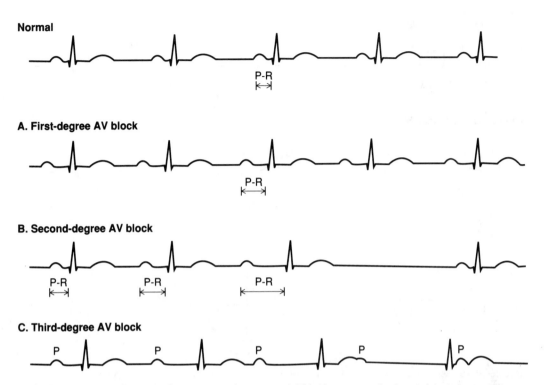

FIGURE 10-17. Various types of atrioventricular (AV) nodal block. (*A*) First-degree AV block. Note the prolonged P-R interval, which is diagnostic. (*B*) Second-degree AV block exhibits occasional dropped beats. Wenckebach block (Mobitz type I) is characterized as a series of progressively longer P-R intervals ending with a dropped beat. (*C*) Third-degree (complete) AV block represents a condition in which the atria and ventricles beat independently at their own rates.

        (1) Wenckebach block (Mobitz type I) is characterized by a progressive lengthening of the P-R interval in successive beats and finally a failure of one impulse to be transmitted (see Figure 10-17B).

        (2) Periodic block (Mobitz type II) is characterized by an occasional failure of conduction that results in an atrial-to-ventricular rate of, for example, 6:5 or 8:7. The P-R interval is constant in this condition.

        (3) Constant block is a form of second-degree block and represents a higher degree of block than the periodic type. The atrial-to-ventricular rate is a constant small number ratio (e.g., 2:1 or 3:1).

    c. Third-degree (complete) AV nodal block occurs when AV node conduction is completely interrupted, causing the atria and ventricles to beat at independent rates (see Figure 10-17C).

        (1) Cause. As in second-degree AV nodal block, organic heart disease is often the cause.

        (2) Significance. Third-degree AV nodal block may be associated with prolonged ventricular standstill until a ventricular focus begins firing. The cardiac arrest, if prolonged, can result in cerebral ischemia, syncope, or death; the condition is termed **Stokes-Adams syndrome**.

3. Wolff-Parkinson-White syndrome, also termed **ventricular preexcitation** or **accelerated conduction,** results from an aberrant conduction pathway between the atria and the ventricles. The aberrant pathway eliminates or reduces the normal delay between atrial and ventricular activation.

    a. EKG appearance (Figure 10-18A). The EKG shows a shortened P-R interval, usually less than 0.1 second, and a widened QRS complex with a slurred initial upstroke (delta wave). The slurred upstroke is caused by the slow conduction through the ventricular muscle.

A. Wolff-Parkinson-White syndrome

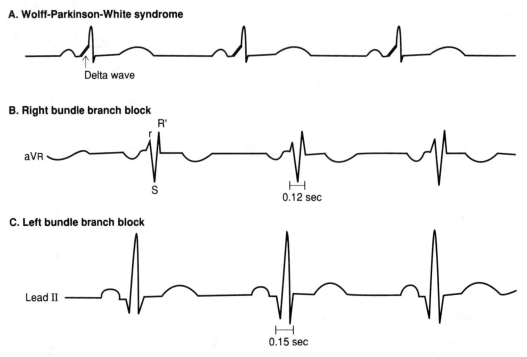

Delta wave

B. Right bundle branch block

R'

r

aVR

S

0.12 sec

C. Left bundle branch block

Lead II

0.15 sec

FIGURE 10-18. (*A*) The characteristic deformity of the QRS complex that results from an aberrant connection between the atria and ventricles, as seen in Wolff-Parkinson-White syndrome. The aberrant connection causes early activation of the ventricles and results in a slurring of the QRS upstroke (delta wave). (*B*) Right bundle branch block is caused by an interruption in the right bundle branch. The right ventricle is depolarized last, leading to a widening of the QRS complex. (*C*) Left bundle branch block causes late activation of the left ventricle and a widening of the QRS complex.

  b. **Significance.** The **accessory bundle** (i.e., the aberrant pathway) predisposes to paroxysmal tachycardias because of a reentry phenomenon. The atrial impulse is conducted to the ventricles through the AV node and returns to the atria via the accessory bundle. The circular depolarization pathway triggers the reentry mechanism.
  c. **Treatment.** Surgical ablation of the accessory bundle is possible if it can be adequately localized.
 4. **Bundle branch block.** Blockage of the right or left common bundles or the left anterior or posterior fascicles results in an abnormal sequence of ventricular excitation.
  a. **Right bundle branch block** delays the activation of the right ventricle.
   (1) **Causes.** Right bundle branch block can occur in an otherwise healthy individual as a transient or a permanent manifestation. It also occurs secondary to chronic pulmonary disease and may appear acutely as a consequence of pulmonary embolism.
   (2) **EKG appearance**
    (a) The **QRS duration** is prolonged beyond 0.12 second, except in **incomplete right bundle branch block,** in which the QRS duration is only 0.08–0.10 second.
    (b) The initial QRS vectors shift to the left because of the unopposed activation of the left ventricle. The final vectors shift to the right as excitation spreads to the right ventricle through the myocardium. This sequence of excitation produces a typical QRS pattern in the right precordial leads, a wide S wave in lead I, and a right axis deviation of the QRS.
    (c) Repolarization of the right ventricle follows the same path as depolarization, causing an inversion of the T wave in the right precordial leads (see Figure 10-18B).

b. Left bundle branch block

   (1) Causes. The occurrence of left bundle branch block is **rare in the absence of organic heart disease**. It is common in association with **coronary artery disease** or conditions leading to **left ventricular hypertrophy** (e.g., hypertension, aortic valve stenosis).

   (2) EKG appearance. Left bundle branch block is best diagnosed using the left precordial leads.

      (a) The QRS duration is greater than 0.12 second (see Figure 10-18C).

      (b) The initial and final QRS vectors are directed toward the left. Septal depolarization from right to left produces the initial vector, and the final vector results from delayed depolarization of the left ventricle.

D. **Other sources of cardiac dysfunction**

1. Narrowing or **occlusion of coronary arteries** leads to the reduction or absence of blood flow to an area of the myocardium. The most common causes of coronary artery narrowing and occlusion are atherosclerotic plaque and arteriosclerotic coronary thrombosis, respectively. Reduction of coronary blood flow results in **myocardial ischemia** or **infarction**.

a. Myocardial ischemia

   (1) Symptoms. The cardinal sign of coronary artery insufficiency is **angina** related either to exertion or emotional stress. The pain may be precordial or it may radiate to the shoulder, arm, or jaw. The pain is typically described as a tight band or weight on the chest that results in a **strangling sensation;** thus, the term **angina pectoris** is truly apt. The pain of coronary insufficiency typically lasts 1–10 minutes and is relieved by rest.

   (2) EKG appearance. The ischemic area of the myocardium has a reduced membrane potential compared with normal regions of the heart, which leads to a current flow from the normal regions to the ischemic area.

      (a) This current of injury produces **S-T segment elevation** from leads overlying the ischemic area. **S-T segment depression** occurs in leads on the opposite side of the heart.

      (b) Ischemia alters the path of repolarization, usually resulting in **T wave inversion**.

   (3) Diagnosis. The diagnosis is usually made from the **patient history,** because the resting EKG may be within normal limits.

      (a) Twenty-four hour EKG (Holter) monitoring may provide evidence of ischemic changes during anginal attacks.

      (b) Stress (exercise) testing with EKG monitoring is used to precipitate ischemic episodes for diagnostic purposes.

b. Myocardial infarction occurs when coronary blood flow ceases or is reduced below a critical level. The left ventricle is almost always the site of a myocardial infarction.

   (1) Symptoms of myocardial infarction are similar to those of coronary insufficiency, except that the pain is more intense, it generally lasts for more than 15–20 minutes, it may not be related to exertion or exercise, and it is not relieved by nitrates.

   (2) EKG appearance. The EKG undergoes a series of changes following a myocardial infarction. These changes must be recorded with daily EKG tracings for diagnostic purposes.

      (a) A **transmural infarct** results in **wide, deep Q waves** from leads overlying the infarcted area, because these leads essentially record an intracavitary potential.

      (b) These same leads demonstrate an **S-T segment elevation,** and, early after the infarct, there is an **inversion of the T wave**.

   (3) Diagnosis. The EKG provides supportive evidence, but **patient history** is critical in making the diagnosis of a myocardial infarct. Although measurement of **serum enzyme levels** [creatine phosphokinase (CPK), serum glutamic-oxaloacetic transaminase (SGOT), and lactic acid dehydrogenase (LDH)] may aid in the diagnosis, no enzyme is specific for the myocardium.

2. **Ventricular hypertrophy** occurs if the work of one or both ventricles is increased sufficiently (a common complication of pulmonary or systemic hypertension). In ventricular hypertrophy, the number of myocardial cells remains the same but the diameters of the individual cells increase, increasing the diffusion distance for O_2 and other metabolites. The increased diffusion distance may lead to borderline ischemic conditions.

   a. **EKG appearance**

      (1) **R wave.** There is a direct correlation between the thickness of the ventricular wall and the height of the R wave in the overlying leads. The increased height of the R wave is a reflection of the increased magnitude of the depolarization vector.

      (2) **QRS duration** is increased slightly because of the increased muscle mass that is present. The duration of the QRS complex is usually less than 0.12 second but occasionally it may exceed this value.

   b. **Left ventricular hypertrophy**

      (1) **Axis.** The MEA is shifted toward the left and superiorly. The transition zone of the precordial leads is also shifted to the left.

      (2) **QRS configuration.** Left ventricular hypertrophy is usually present if the R wave in V_5 or V_6 together with the S wave in V_1 are greater than 35 mm.

   c. **Right ventricular hypertrophy**

      (1) **Axis.** The MEA is usually vertical or greater than $+110°$ in the frontal plane.

      (2) **QRS configuration.** Right precordial leads generally show tall R waves, rather than normal S waves. The QRS is usually prolonged, but less than 0.12 second; the S-T segment is depressed and the T wave is inverted in the right precordial leads.

Chapter 11

Cardiodynamics

I. **INTRODUCTION.** The term **cardiodynamics** refers to the mechanical events that are associated with the contraction of the heart.

A. **The cardiac cycle** includes both electrical (e.g., myocardial cellular depolarization) and mechanical events (e.g., myocardial cell shortening leading to pressure generation and, thus, volume changes). The electrical events precede and initiate the mechanical events.

B. **Excitation–contraction (EC) coupling** is the term used to define the events that connect the depolarization of the cell membrane to the contraction of the muscle fibers (see also Chapter 4 III A–C). The action potential causes Ca^{2+} release from the sarcoplasmic reticulum (SR), and Ca^{2+} diffuses into the cell across the T tubules (i.e., there is a slow inward Ca^{2+} current). The action potential raises the sarcoplasmic Ca^{2+} concentration approximately tenfold, compared with resting conditions.

1. At high Ca^{2+} concentrations, troponin binds **four calcium ions per molecule**.
   a. When troponin is saturated with Ca^{2+}, the troponin changes shape and uncovers the cross-bridge sites on the tropomyosin molecule.
   b. Cross-bridges form between actin and myosin.

2. Normally, all of the troponin is not saturated with Ca^{2+}; therefore, there is a **potential for additional cross-bridge formation**.
   a. Factors that increase the intracellular Ca^{2+} concentration provide more sites for cross-bridge formation.
   b. **Increasing the number of cross-bridges that form increases the force of the contraction.**

3. **Contractility.** Cardiac muscle has a **graded response** on the force of contraction, which depends on the intracellular Ca^{2+} concentration. The increased force that results from an increased Ca^{2+} concentration is termed **positive inotropism** or **increased contractility**.

II. **THE CARDIAC CYCLE**

A. **Sequence of events** (Figure 11-1). The electrical and mechanical events of the cardiac cycle result in the generation of pressure within the cardiovascular system. The pressure (potential energy) is converted into flow (kinetic energy) to supply all of the body cells with nutrients and to serve other circulatory functions. A full understanding of the events of the cardiac cycle is essential to understanding the physical signs of cardiovascular disease that are elicited during the physical examination of a patient.

1. **Electrocardiogram (EKG).** The electrical events precede and initiate the corresponding mechanical events; thus, the P wave precedes atrial contraction, and the QRS complex precedes ventricular contraction.

2. **Atrial contraction**
   a. **Electrical events.** The **P wave** on the EKG precedes atrial contraction. The specialized **interatrial tract** that carries the excitation between the atria **allows the two atria to contract almost simultaneously**.
   b. **Function.** Atrial contraction increases the volume of blood in the ventricles.
      (1) The effects of atrial contraction are especially important when the heart rate is rapid and the ventricular filling time is reduced.

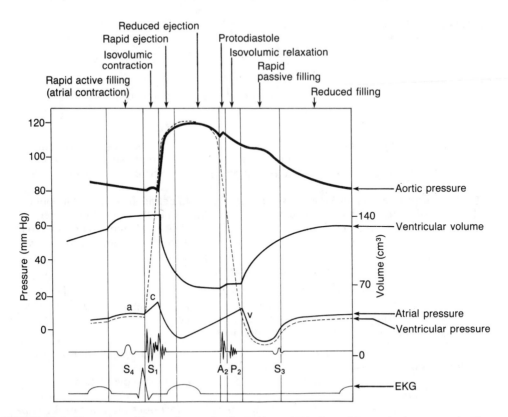

FIGURE 11-1. The events of the cardiac cycle. S_1, S_2 (A_2 and P_2), S_3, and S_4 are the heart sounds. The a wave (*a*) is caused by atrial contraction. The c wave (*c*) is caused by the atrioventricular (AV) valves ballooning back into the atria during isovolumic contraction. The peak of the v wave (*v*) results from the opening of the AV valves.

      (2) The **ventricular end-diastolic pressure (VEDP)** is the pressure of the blood in the ventricles at the termination of the atrial contraction. The VEDP is normally less than 12 mm Hg in the left ventricle and 5 mm Hg in the right ventricle.

    c. Clinical significance. The increase in atrial pressure that results from atrial contraction (i.e., the a wave) is reflected back into the large veins, where it can be recorded from the jugular vein with appropriate transducers.

      (1) The **right ventricular filling pressure** can be estimated from the level of blood in the jugular veins.

        (a) Normally, the jugular veins are collapsed when an individual is in an upright position, because right atrial pressure is only approximately 6–7 cm H_2O.

        (b) The jugular veins do not become distended unless the normal individual assumes a horizontal position, causing the jugular veins to fill with blood because they are at the same level as the heart.

      (2) Distended jugular veins indicate excessive pressures in the right atrium. Such excessive pressures can be caused by **depressed ventricular function** (i.e., congestive heart failure) or by a **narrowed atrioventricular (AV) valve (i.e., tricuspid stenosis),** which obstructs the blood flow.

  3. Ventricular contraction (systole)

    a. Terminology. Ventricular volume (i.e., the volume of blood in the ventricles) is difficult to measure clinically; however, **echocardiography** can be used to measure ventricular dimensions as well as ventricular wall and valve motion during the cardiac cycle.

      (1) The **ventricular end-diastolic volume (VEDV)** is the volume of blood in the ventricle just before the onset of ventricular contraction. The normal left VEDV is 120–140 ml.

(2) The **ventricular end-systolic volume (VESV)** is the volume of blood remaining in the ventricle at the end of ejection. The normal left VESV is 40–70 ml.

(3) The **stroke volume** is the volume of blood that is ejected with each beat. It is equal to the VEDV minus the VESV (e.g., 75–80 ml).

(4) The **ejection fraction** equals the stroke volume divided by the VEDV. The ejection fraction normally is 60%–70%, but declines markedly in patients with ventricular dysfunction.

b. Stages

(1) AV valve closure. Shortly after the QRS complex begins, the pressure in the ventricular cavities rises.

(a) The rising ventricular pressure exceeds the atrial pressure and causes the AV valves to close.

(b) Closure of the AV valves is the major component in generating the **first heart sound (S_1).** The S_1 is a low frequency sound that occurs just after the onset of ventricular contraction and **signals the onset of ventricular systole** (see Figure 11-1).

(2) Isovolumic (isovolumetric) contraction. Once the AV valves close, the ventricular chambers are sealed from both the atria and the arteries. The ventricular volume remains constant (isovolumic) until the ventricular pressure exceeds the pressure in the respective artery (i.e., the aorta or pulmonary artery).

(a) When the ventricular pressure exceeds the pressure in the respective artery, **ventricular ejection begins** (see Figure 11-1).

(b) Right ventricular ejection occurs before left ventricular ejection, because the pressure in the pulmonary artery is low compared with that in the aorta.

(3) Rapid ejection. As soon as ventricular ejection begins, the arterial pressure increases because the arterial volume rises.

(a) The initial part of ejection is rapid; approximately two-thirds of the stroke volume is ejected during the first third of systole (see the ventricular volume tracing, Figure 11-1).

(b) The end of the rapid ejection phase occurs at approximately the peak of ventricular and arterial systolic pressures.

(4) Reduced ejection. During the last two-thirds of systole, the rate of ejection declines and the ventricles begin to relax. Both ventricular and arterial pressures begin to decrease as the rate of blood flow through the peripheral vessels exceeds the rate of blood flow from the ventricles.

(5) Semilunar valve closure. Closure of the semilunar (i.e., the aortic and pulmonic) valves prevents the movement of blood back into the ventricles and produces the **second heart sound (S_2),** a relatively high frequency sound that indicates the end of systole and the onset of ventricular diastole.

(a) Normal splitting of S_2. The aortic and pulmonic valves do not close simultaneously under normal conditions.

(i) Normally, the aortic valve closes, producing the aortic component of S_2 (A_2) before the pulmonic valve closes, producing P_2. The aortic valve closes first because the rate of ejection from the left ventricle is higher than that from the right ventricle.

(ii) During inspiration, the pressure in the thorax becomes more negative, increasing venous return to the right ventricle. The increased venous return increases the stroke volume and prolongs the ejection period for the right ventricle, delaying pulmonic valve closure even more and enhancing the separation of A_2 and P_2. The separation between these components narrows during expiration as the increased blood flow reaches the left ventricle.

(b) Paradoxical splitting of S_2. Paradoxical splitting is present if the splitting of S_2 decreases during inspiration, indicating that pulmonic valve closure precedes aortic closure.

(i) Paradoxical splitting of S_2 is usually caused by **delayed closure of the aortic valve.**

             (ii) Delayed aortic valve closure indicates a **disease process affecting the left ventricle** [e.g., left bundle branch block (delayed activation), myocardial depression leading to prolongation of systole, or prolonged ejection caused by aortic stenosis].

  4. Ventricular diastole is the **relaxation** or **filling phase** of the cardiac cycle.

    a. Stages

      (1) Isovolumic (isovolumetric) relaxation begins with the closure of the semilunar valve.

        (a) Ventricular pressure falls rapidly as the ventricles relax; isovolumic relaxation lasts only approximately 0.04 second.

        (b) The ventricular volume remains constant from the time the semilunar valves close until the AV valves open, which occurs when the ventricular pressure falls below the atrial pressure (as indicated by the peak of the **v wave** on the atrial pressure tracing; see Figure 11-1).

      (2) Rapid passive filling. During ventricular systole, the venous return continues so that the atrial pressure is high when the AV valves first open.

        (a) The high atrial pressure causes a rapid, initial flow of blood into the ventricles (see Figure 11-1).

        (b) Once the AV valves open, the atria and ventricles are a common chamber and the pressure in both cavities falls as ventricular relaxation continues.

      (3) Reduced filling and diastasis. Toward the end of diastole, the pressure in the atria and ventricles rises slowly as blood continues to return to the heart.

        (a) With very slow heart rates, ventricular filling virtually ceases because the ventricles reach their volume limit; this phase is termed **diastasis**.

        (b) When the heart rate increases, the length of the cardiac cycle obviously shortens.

          (i) Diastasis and the reduced filling phase are most affected by shortening of the cycle.

          (ii) Increases in heart rate of up to 140–150 beats/min do not cause a significant decrease in ventricular filling. At these levels, cardiac output increases in almost direct proportion to the increased heart rate.

          (iii) Heart rates in excess of 180–200 beats/min do result in a decreased cardiac output, because the ventricular filling time is markedly reduced. The reduced ventricular filling time decreases the VEDV and, consequently, the stroke volume.

    b. Third and **fourth heart sounds** (S_3 and S_4)

      (1) The S_3 occurs during the **rapid passive filling stage** and is not normally audible in adults, but may be heard in children.

      (2) The S_4 occurs during **atrial contraction** and is occasionally heard in people with congestive heart failure. In these patients, a triple sound called a **gallop rhythm** is audible.

B. **Aortic and pulmonary arterial pressures.** The recordings of the pulsatile pressure changes in the aorta and pulmonary artery are termed **pressure pulses**.

  1. Systole. During the period of rapid ventricular ejection, the pressure in the aorta normally is slightly less than that in the ventricle. The peak arterial pressure is the arterial systolic pressure, which occurs at the end of the rapid ejection phase.

    a. The normal aortic systolic pressure is approximately 120 mm Hg; pulmonary artery systolic pressure normally is 15–18 mm Hg. In the elderly, the decreased arterial compliance that results from aging causes systolic pressure to increase.

    b. The **incisura** (i.e., the notch in the downslope of the arterial pressure pulse produced by the closure of the semilunar valves) indicates the end of ventricular systole.

  2. Diastole. During diastole, the arterial pressure declines as blood continues to flow from the arteries, through the capillaries, and into the veins.

    a. The arterial diastolic pressure is the lowest arterial pressure during a cardiac cycle, and it occurs just before the onset of the next ventricular ejection.

b. The normal diastolic pressure in the aorta is approximately 80 mm Hg; pulmonary artery diastolic pressure normally is 8–10 mm Hg.

3. The **pulse pressure** is the difference between the systolic and the diastolic pressures (e.g., the normal aortic pulse pressure is 120 − 80, or 40 mm Hg).

4. The **recorded blood pressure** is normally written as the arterial systolic pressure/diastolic pressure (e.g., 120/80 mm Hg).

C. **Valve lesions and murmurs.** The cardiac valves may be abnormal as a result of either congenital or acquired heart disease.

1. Abnormal cardiac valves lead to **abnormal blood flow and pressure gradients,** which, in turn, produce signs of dysfunction that are **recognizable during the physical examination**. Valve lesions affect the circulation by:
   a. Reducing cardiac output
   b. Imposing additional work on the heart by creating an extra pressure or volume load
   c. Producing the backup of blood

2. **Murmurs** are important physical signs of valve lesions.
   a. **Causes.** Murmurs are caused by turbulent blood flow through a narrowed valve or by changes in the direction of the blood flow.
   (1) **Laminar (normal) flow** is streamlined and silent.
   (2) **Turbulent flow** produces vibrations in the tissues that are heard as murmurs.
       (a) Turbulent blood flow occurs when the Reynold's number exceeds 2000.
       (b) The Reynold's number is a dimensionless value that is given by:

$$\frac{D \bullet \delta \bullet v}{\eta}$$

   where D = vessel diameter, δ = blood density, v = velocity, and η = blood viscosity.
   b. **Timing.** Murmurs may be **systolic, diastolic,** or **continuous** throughout the cardiac cycle. Murmurs may also be more specifically classified (e.g., presystolic, pansystolic, early, or middiastolic). The timing of the murmurs is associated with specific valve lesions.
   c. **Location and radiation of murmurs.** Murmurs are best heard on the chest wall closest to their origin and in the downstream direction of blood flow. For example, the murmur of aortic stenosis is best heard at the right sternal border in the second interspace, and the murmur radiates to, and is also clearly heard over, the vessels of the neck.

3. **Valvular stenosis** is a narrowing of the valve orifice that impedes the forward flow of blood.
   a. **Stenosis of the semilunar (aortic or pulmonary) valves** forces the left and right ventricles, respectively, to generate high pressures in order to eject blood through the narrowed orifice (i.e., the narrowed orifice acts as a high-resistance segment of the vascular system).
   (1) **Diagnosis.** A ventricular systolic pressure that is much higher than the systolic pressure in the respective artery is pathognomonic of semilunar valve stenosis (Figure 11-2A).
       (a) The pressure gradient can be detected during **cardiac catheterization.**
       (b) The murmur of semilunar valve stenosis is an **ejection-type (crescendo–decrescendo) systolic murmur**. It produces a diamond shape on the **phonocardiogram,** reflecting the pressure gradient during ejection, which peaks at mid-ejection.
   (2) **Clinical significance.** Stenotic lesions can lead to **ventricular hypertrophy** and may eventually produce ventricular failure.
   b. **Stenosis of the AV (mitral or tricuspid) valves** impedes the filling of the ventricles so that a pressure gradient develops between the atria and the ventricles during diastole (Figure 11-2B). Therefore, atrial (and, consequently, venous) pressure tends to be high.

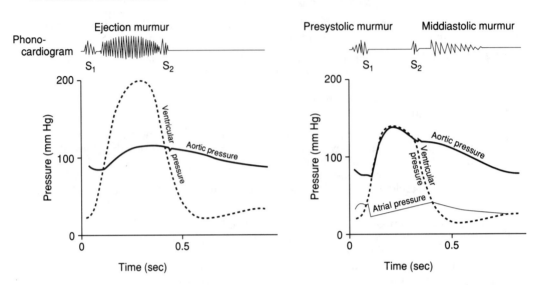

A. Semilunar valve stenosis

B. Atrioventricular valve stenosis

FIGURE 11-2. Valvular stenosis. (*A*) Semilunar (aortic or pulmonic) valve stenosis. The large pressure gradient between the ventricular and aortic pressures during ventricular ejection is pathognomonic. (*B*) Atrioventricular (AV; mitral or tricuspid) valve stenosis results in a sizeable pressure gradient between the atria and the ventricles during diastole. The pressure builds up in the veins behind the ventricle and may lead to edema. Presystolic and middiastolic murmurs are usually audible.

      (1) Diagnosis. AV valve stenosis produces diastolic murmurs that occur during atrial contraction (presystolic murmur) and the rapid passive filling stage (middiastolic murmur). These two murmurs are described as crescendo and decrescendo murmurs, respectively. As in semilunar valve stenosis, the pressure gradients produce a characteristic sound profile on the phonocardiogram.

      (2) Clinical significance

          (a) Mitral valve stenosis may lead to pulmonary edema (because of the high pulmonary venous pressures), enlargement of the left atrium, or atrial fibrillation.

          (b) Tricuspid stenosis elevates systemic venous pressure. It may also produce giant a waves that are visible as pulsations in the jugular veins.

   4. Valvular insufficiency (regurgitation) allows the backward flow of blood from the ventricles to the atria or from the aorta to the left ventricle.

     a. Insufficiency of the semilunar valves reduces the effective cardiac output because blood flows back into the ventricular cavities during diastole.

       (1) Pandiastolic murmur. Semilunar valve insufficiency results in a pandiastolic murmur that is relatively constant in intensity (Figure 11-3A).

       (2) Arterial pressures. The backward flow of blood causes a rapid fall in arterial pressure, producing **very low diastolic arterial pressures**. In addition, the backflow augments the next stroke volume, leading to **high systolic pressures**.

       (3) Clinical significance

          (a) Extremely large pulse pressures, called **Corrigan's (water-hammer) pulses,** are often seen in patients with **aortic insufficiency**. The large pulse pressure can be transmitted to the capillaries and is therefore **detectable in the nail beds**.

          (b) Semilunar valve insufficiency produces **marked ventricular dilation,** which eventually leads to **ventricular failure**.

          (c) A blood pressure of approximately 180/50 mm Hg may be diagnostic of aortic insufficiency in some patients.

     b. Insufficiency of the AV valves allows blood to flow from the ventricles to the atria during systole.

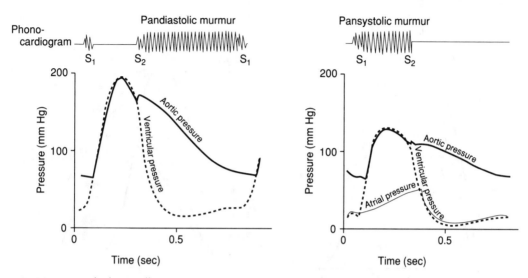

A. Semilunar valve insufficiency

B. Atrioventricular valve insufficiency

FIGURE 11-3. Valvular insufficiency (regurgitation). (*A*) Semilunar (aortic or pulmonic) valve insufficiency is caused by incomplete closure of the semilunar valves during diastole. The murmur is termed pandiastolic because it lasts throughout diastole, as the pressure gradient would lead one to predict. (*B*) AV (mitral or tricuspid) valve insufficiency causes a reflux of blood from the ventricles into the atria during systole. The regurgitant flow of blood creates a pansystolic murmur.

      (1) **Diagnosis.** The backflow of blood results in very high atrial pressures during systole (Figure 11-3B).

      (2) **Clinical significance.** Dilation of one or more heart chambers occurs to compensate for the regurgitant blood and can eventually lead to heart failure.

III. CARDIAC CONTRACTILE FORCE

A. **Frank-Starling effect. Starling's law of the heart** states that the energy produced by the heart when it contracts is a function of the end-diastolic length of its muscle fibers.

    1. **Factors.** One of the major factors that controls the force of cardiac contraction is the **initial length** (i.e., the **preload**) of the muscle fibers.

      a. In the intact heart, the **preload depends on the volume of blood in the ventricles just before contraction** begins (i.e., the **VEDV**).

        (1) The VEDV is considered to be proportional to both the VEDP and the central venous pressure (which are essentially equivalent because of the low resistance of the venous system).

        (2) An increase in VEDV or VEDP lengthens the cardiac fibers, which, up to an optimum level, exposes more of the myosin heads (cross-bridge sites) to the actin.

      b. **Cross-bridges.** Changes in preload alter the number of cross-bridges that are available by changing the overlap of the thick and thin filaments.

        (1) The additional cross-bridges allow the heart to generate more force with each contraction.

        (2) Thus, an increase in preload (or venous pressure) produces a stronger contraction, allowing the heart to pump more blood with each beat (Figures 11-4A, 11-5).

    2. The Frank-Starling mechanism is extremely important in **balancing the output of the two sides of the heart** over extended periods of time.

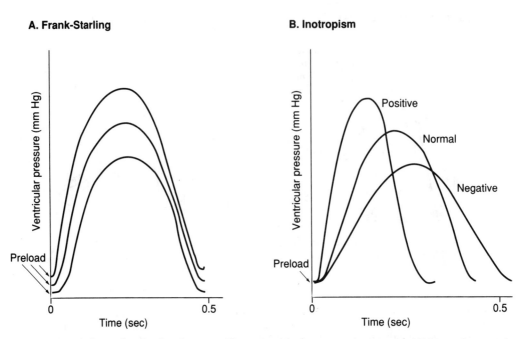

A. Frank-Starling

B. Inotropism

FIGURE 11-4. Effects of preload and contractility on ventricular pressure generated. (*A*) Increasing ventricular end-diastolic volume (VEDV) and pressure (VEDP) [i.e., the preload] increases the force of contraction by means of the Frank-Starling mechanism; the time required to generate peak pressure is unchanged. (*B*) Changes in contractility alter the peak force developed and the duration of the contractile process.

B. **Contractility.** In the intact heart, changes in stroke volume also depend on the contractility of the ventricles.

    1. Increased contractility (positive inotropism) is an increased contractile force at a constant preload or ventricular volume. Positive inotropism allows the ventricles to eject more blood from the same diastolic volume; consequently, the stroke volume increases.

        a. Positive inotropic mechanisms

            (1) Activation of the β_1 receptors [via **sympathetic nerve stimulation** or the **administration of drugs** (e.g., epinephrine, norepinephrine)] increases the concentration of Ca^{2+} within the cell and results in a more rapid and forceful contraction. In addition, the Ca^{2+} is taken up more rapidly by the SR, which shortens the duration of both the action potential and the contraction (Figure 11-4B).

            (2) Digitalis or a digitalis derivative (e.g., **ouabain, digoxin**) also increases the force of cardiac contraction by raising the intracellular Ca^{2+} concentration. Digitalis has been used for centuries to treat heart failure and was initially prescribed as a tea prepared from the leaves of a flowering plant, *Digitalis purpurea*.

        b. Positive inotropic effects. Figure 11-5 shows a family of ventricular function (Frank-Starling) curves and depicts the ventricular performance as a function of the preload. Positive inotropic effects produce a shift upward or toward the left (i.e., more cardiac force can be generated at a constant ventricular volume, or the same force can be generated from a smaller VEDV).

        c. Maximal velocity of shortening. Positive inotropism also can be defined as an increase in the maximal velocity of shortening (V_{max}) when it is plotted as a function of afterload. For the ventricles, the afterload is the systolic pressure during the ejection phase (Figure 11-6).

            (1) During any muscle contraction, the velocity of shortening and the force developed are inversely related.

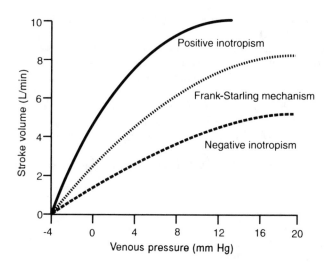

FIGURE 11-5. A family of Frank-Starling curves. Changes in stroke volume can be accomplished by alterations in the venous pressure [i.e., the ventricular end-diastolic volume (VEDV)], causing movement along one of the curves (Frank-Starling mechanism) or by shifting from one curve to another. The latter mechanism indicates changes in ventricular contractility (inotropism).

  (2) In Figure 11-6, the intercept of each line with the x-axis represents the load that prevents shortening of the muscle fibers. This condition is referred to as an **isometric** or **isovolumic contraction** because there is no shortening in an isolated muscle and no ejection of blood from the intact ventricle.
  (3) Figure 11-4B demonstrates that an increased contractility allows the ventricle to generate a greater amount of force from the same preload (venous pressure or VEDV). This is in contrast to the Frank-Starling mechanism, where an increase in the force of contraction is produced by an increase in fiber length or preload (see Figure 11-4A).

 2. Decreased contractility (negative inotropism) represents a decrease in the force of contraction at any fiber length or ventricular volume.

FIGURE 11-6. The velocity of shortening as a function of afterload. Changes in preload [i.e., venous pressure, ventricular end-diastolic pressure or volume (VEDP, VEDV)] do not cause an increase in the maximal velocity of shortening (V_{max}). Sympathomimetic substances (e.g., epinephrine) increase contractility and result in an increased V_{max}. L_0 = initial length of muscle.

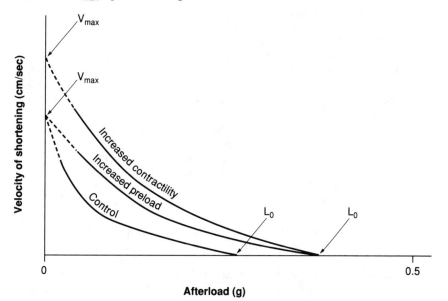

a. **Negative inotropic mechanisms.** Decreased contractility is caused by **hypoxia, acidosis, myocardial ischemia,** or **infarct.**
   (1) The damaged or dead tissue leads to a decrease in stroke volume, because less tissue is available to do the same amount of work. The ventricle's decreased pumping ability causes an additional increase in ventricular size as the venous pressure increases, dilating the ventricle.
   (2) The increased ventricular volume invokes the Frank-Starling mechanism, increasing the force of contraction and returning the stroke volume toward normal. However, as the ventricle dilates, it faces a double disadvantage because of the **Laplace effect.**
b. **Law of Laplace.** The **law of Laplace** for a thick-walled structure such as the ventricle can be expressed as:

$$T = \frac{(P \cdot r)}{h}, \text{ where}$$

   T = the wall tension
   P = the pressure difference across the ventricular wall
   r = the radius of the ventricle
   h = the ventricular wall thickness

   (1) **Wall tension.** Any hollow organ, such as the ventricle, generates pressure by increasing wall tension, which squeezes down on the contained volume.
   (2) **Wall thickness.** Ventricular dilation causes a thinning of the ventricular wall because a given ventricular muscle mass must enclose a larger volume of blood.
      (a) Because there are fewer myocytes per cross-sectional area of the ventricular wall, each myocyte must develop more force than was previously required to accomplish the same amount of work.
      (b) Eventually the ability of the myofibrils to generate force is exceeded and the amount of shortening or force generated is reduced. This represents a negative inotropic effect; less work is performed by the heart from any given VEDV.
   (3) **Ventricular radius.** Because ventricular dilation is equivalent to an increased ventricular radius, the ventricle must also generate a greater wall tension to maintain any given force generation or systolic pressure according to the Laplace relationship; therefore, a larger heart must work harder to generate the same stroke volume.
c. **Clinical significance**
   (1) **Diminished cardiac reserve.** Because the heart's ability to increase the stroke volume is limited, cardiac output is primarily maintained by **increasing the heart rate.** Both of these compensatory mechanisms reduce the ability of the ventricle to increase cardiac output in response to stress, causing patients to tire easily. Eventually, these compensatory mechanisms fail and cardiac output is reduced significantly.
   (2) **Edema.** In **severe ventricular failure,** both cardiac output and stroke volume are decreased, and the ventricular diastolic pressure is high. The increased ventricular diastolic pressure raises venous pressure, which is reflected back to the capillaries and produces leakage of fluid into the tissues (edema).
      (a) In **left-sided heart failure,** fluid accumulates in the lungs, causing difficulty breathing (dyspnea) and the inability to breathe when lying down (orthopnea).
      (b) In **right-sided heart failure,** edema occurs in the feet and ankles (dependent edema) because the high venous pressures are reflected into the systemic capillaries.
   (3) **Hypertrophy.** Any increase in the work of the heart causes the ventricular muscle to enlarge, just like any other muscle that is exercised sufficiently.

(a) The ventricles first dilate, producing thinning of the wall as noted previously; later the muscle hypertrophies. Myocytes have a limited ability to multiply; therefore, hypertrophy entails an increase in the diameter of the muscle fibers.
(b) The increased muscle dimensions may cause ischemia because the diffusion distance between capillaries increases. The hypertrophy also may be harmful if the blood supply that provides O_2 to the muscle does not increase proportionately.

IV. PRESSURE–VOLUME LOOPS. It is possible to record changes in ventricular pressure and volume during a cardiac cycle. These records result in a **loop** (Figure 11-7A), which provides a great deal of information regarding the **function and integrity of the ventricles**.

A. **Afterload.** The afterload represents an impediment to the shortening of muscle fibers or to ejection in the heart.

1. The afterload for the ventricles is the **arterial pressure,** which normally increases progressively throughout most of the ejection phase of the cardiac cycle.

2. The normal ventricle is able to maintain a relatively normal stroke volume in response to an increased afterload, but the failing ventricle is markedly affected by an increased afterload (Figure 11-7B).

B. **Ventricular end-diastolic volume (VEDV).** The VEDV affects the stroke volume because of the Frank-Starling mechanism. The greater the VEDV, the stronger the contraction and the larger the stroke volume at any one level of contractility (Figure 11-7C).

1. The VEDV increases when the filling pressure in the atria or the central veins increases, because the higher pressure stretches the ventricular walls. (This is similar to blowing up a balloon.)

2. The VEDV also increases if the distensibility of the ventricles increases or if the heart rate slows, allowing more time for the blood to enter the ventricles during diastole.

C. **Cardiac work** is reflected by the size of the pressure–volume loop (see Figure 11-7) and is proportional to both the pressure generated by the ventricular contraction and the stroke volume. Because both ventricles pump the same average stroke volume, and because the systolic pressure of the right ventricle is approximately one-seventh that of the left ventricle, the work of the right ventricle is approximately one-seventh that of the left ventricle.

1. **Myocardial oxygen consumption.** The amount of O_2 that is used by the heart depends on the amount of work performed. The difference in O_2 content between the coronary arterial and venous blood ($Cao_2 - Cvo_2$) and the coronary blood flow is an accurate reflection of O_2 consumption by the heart.

2. **Factors affecting cardiac work**
   a. Hypertension. The work of the ventricles is markedly increased if the blood pressure is high. Hypertension can occur in either the right or left side of the circulation.
   (1) **Pulmonary (right-sided) hypertension** almost invariably is caused by chronic lung disease, whereas **systemic (essential) hypertension** most commonly is caused by an unknown mechanism.
   (2) Hypertension can also be caused by endocrine, renal, or congenital conditions.
   b. Myocardial oxygen supply. Normal cardiac function requires a continuous delivery of O_2 to the heart.
   (1) The O_2 delivery (to any tissue) is equal to the arterial oxygen content (the volume of oxygen carried per 100 ml blood) multiplied by the blood flow.

A. Normal pressure–volume curve

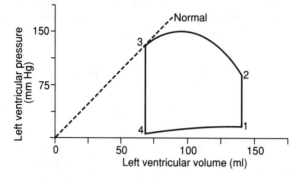

B. Effects of afterload

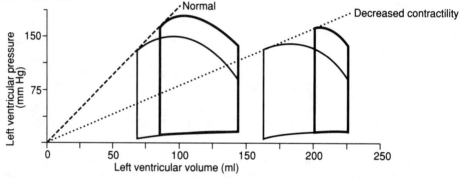

C. Increases in stroke volume

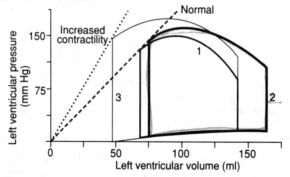

FIGURE 11-7. The effects of changes in afterload and ventricular end-diastolic volume (VEDV) on the ventricular pressure–volume loop. (*A*) Normal pressure–volume loop. *Point 1* represents the beginning of isovolumic contraction (or the VEDV). *Point 2* marks the onset of ejection. *Point 3* is the beginning of isovolumic relaxation, and *Point 4* is the onset of ventricular filling [or the ventricular end-systolic volume (VESV)]. The stroke volume equals the VEDV minus the VESV. The *dashed line* represents the end-systolic pressure–volume relationship, which is determined by the level of contractility. (*B*) In the normal heart (*curve 1*) an increased afterload (*thick line*) reduces the stroke volume because the increased afterload abbreviates myocardial shortening. The VESV is greater than normal, indicating that the stroke volume has been slightly reduced. If the heart's contractility decreases (*curve 2*), the VEDV increases as a Frank-Starling response to depressed ventricular function. The slope of the maximal pressure–volume relationship is reduced. An increased afterload markedly decreases stroke volume in a depressed ventricle. (*C*) *Curve 1* is the control curve. An increase in VEDV (*curve 2*) results in an increased stroke volume (i.e., the Frank-Starling effect), but the maximal pressure–volume relationship is unchanged. The slight increase in VESV is caused by the increased aortic pressure (i.e., afterload) that results from the larger stroke volume. An increased contractility (*curve 3*) increases the stroke volume because of the greater myocardial force and shortening that occurs (i.e., the VESV is reduced).

    (2) Impaired O_2 delivery to the heart is almost always caused by the partial or complete obstruction of a coronary artery, secondary to atherosclerosis.

D. **Increased contractility.** Positive inotropic mechanisms cause the end-systolic pressure–volume relationships to shift toward the left (see Figure 11-7C).

    1. Positive inotropic mechanisms allow the ventricle to achieve greater stroke volumes in the face of greater aortic pressures or afterloads.

    2. The resultant increased stroke volume is an important mechanism of increasing cardiac output during exercise.

Chapter 12

The Vascular System

I. VESSELS

A. Arteries

1. **Elastic arteries** are the **aorta** and the **carotid, iliac, and axillary arteries**.
   a. **Structure.** Elastic arteries are composed of a **thick medial layer** containing **large amounts of elastin** and **some smooth muscle cells**.
   b. **Functions**
      (1) **Windkessel.** The distensibility (compliance) of the elastic arteries allows them to accommodate the stroke volume of the heart with only a moderate increase in pressure.
      (2) **Elastic recoil.** In addition to being distensible, the vessels also generate elastic recoil during diastole. Elastic recoil creates potential energy (pressure), which helps to maintain blood flow during the diastolic phase of the cardiac cycle.

2. **Muscular arteries** comprise most of the named arteries in the body.
   a. **Structure.** The amount of smooth muscle contained in the medial layer of these vessels increases as the distance from the heart increases.
   b. **Function.** The muscular arteries serve as the **distributing channels** to the organs and have a relatively large lumen-to-wall thickness ratio. The large lumen minimizes the pressure drop that occurs as a result of resistance (i.e., friction; see Chapter 9, Figure 9-3).

B. Arterioles are the site of the major pressure drop in the cardiovascular system (see Chapter 9, Figure 9-3).

1. **Structure.** The arterioles contain a **thick layer of smooth muscle** within the medial layer and have a **relatively narrow lumen**.

2. **Functions**
   a. **Control flow.** Arterioles are the **stopcocks (valves) of the circulation**. The smooth muscle in the arteriole walls permits a fine control over the distribution of the cardiac output to the organs and tissues. The activity of the sympathetic nerves, the arterial pressure, and the local concentration of metabolites, various hormones, and many other mediators (e.g., prostaglandins, thromboxanes, and histamine) all influence the smooth muscle tone.
      (1) **Autoregulation** is the mechanism that allows many organs and tissues to adjust their vascular resistance and maintain a relatively constant blood flow in the presence of changes in arterial pressure.
         (a) Autoregulation is well developed in the kidneys, brain, heart, skeletal muscle, and mesentery.
         (b) Figure 12-1 schematically depicts autoregulation. Several theories have been proposed to explain its mechanism.
            (i) The **metabolic theory** suggests that an increase in arterial blood pressure initially increases blood flow to a tissue or organ. This increased blood flow washes out vasodilator substances in the area; vascular resistance increases; and blood flow returns to normal. Many vasodilator substances have been suggested, including CO_2, H^+, adenosine, prostaglandins, K^+, phosphate ions, and low O_2 levels; however, autoregulation cannot be explained in all instances by the changes in concentration of any of these substances.
            (ii) The **myogenic theory** suggests that vascular smooth muscle constricts when it is stretched, which would occur when the blood pressure

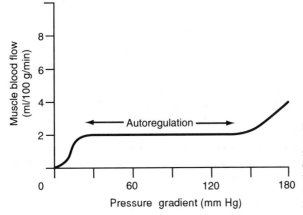

FIGURE 12-1. Autoregulation of blood flow. Note that blood flow remains relatively constant over a wide range of pressures. This can only be accomplished by changing the resistance to flow through the particular vascular bed in proportion to the change in pressure gradient.

inside the vessel increases. Vascular smooth muscle responds to wall tension, which is a function of pressure and wall radius according to the law of Laplace (see Chapter 11 III B 2 b). Therefore, an increase in arterial pressure raises wall tension, initially stretching the smooth muscle fibers. In response to this initial stretch, the vascular smooth muscle contracts, returning the wall tension to control levels. The narrowed lumen represents an increased resistance and compensates for the increased arterial pressure, returning the blood flow to control levels.

    (2) Reactive hyperemia is a phenomenon that occurs after the occlusion of the arteries to an organ or tissue.

        (a) When the occlusion is resolved, the blood flow exceeds the control level with a magnitude and duration that depends on the duration of the occlusion.

        (b) Reactive hyperemia seems to be caused by a metabolic mechanism that controls blood flow to tissues.

  b. Dampen pulsations. The arterioles convert the pulsatile flow in the arteries to a steady flow in the capillaries.

C. The **microcirculation** is a **meshwork** of vessels smaller than 100 μm in diameter.

  1. Organization

    a. Components. Metarterioles, arterioles, capillaries, and **postcapillary venules** comprise the microcirculation. In some tissues (especially in the skin, where they are involved in thermoregulation), there are short, low-resistance connections between the arterioles and the veins called **arteriovenous shunts**.

      (1) Metarterioles are relatively high-resistance conduits between arterioles and veins.

      (2) Capillaries arise directly from arterioles or metarterioles. A cuff of smooth muscle cells surrounds the origin of capillaries in some tissues and is termed the **precapillary sphincter**.

      (3) Postcapillary venules measure 20–60 μm in diameter and represent the most permeable portion of the microcirculation.

    b. Cross-sectional area. The many capillaries that arise from each arteriole and metarteriole provide the microcirculation with a total cross-sectional area of 0.4–0.5 m².

      (1) Blood flow velocity. This large area results in an average blood flow velocity of 0.3–0.4 mm/sec in the capillaries. The velocity varies widely—within short periods of time, it can range from 0 mm/sec to 1 mm/sec within the same capillary.

      (2) Vasomotion refers to the changes in capillary blood flow caused by constriction and dilation of regional arterioles.

(a) Not all of the capillaries in a tissue are functional at one time; therefore, blood flow alternates between different capillaries from moment to moment.

(b) Under basal conditions, 1%–10% of the capillaries are functional at any time. During high metabolic activities, significantly more capillaries carry blood simultaneously, enhancing the delivery of O_2 and metabolites to the tissues.

2. **Structure.** The **endothelial lining** of the vessels can enhance or impede the transport of substances between the vascular and the interstitial fluid compartments.

  a. The endothelial structure of the capillaries varies in different organs depending on the function of the particular tissue.

    (1) Fenestrations. In organs where transport of fluids is paramount [e.g., the renal glomeruli and the gastrointestinal (GI) tract], **large** (2000–20,000 nm) **transcellular openings** exist in the endothelial cells.

      (a) These fenestrations are coupled with an **incomplete basement membrane,** and a thin membrane may be present over the fenestrations.

      (b) Anatomically, there is minimal hindrance for fluid transport across these membranes.

    (2) Blood–brain barrier. The microcirculatory endothelium of the brain does not exhibit fenestrations and possesses a complete basement membrane. These structures are morphological evidence that the blood–brain barrier retards or prevents the transfer of many substances between the blood and the brain.

    (3) Gap junctions. Large gaps occur between the endothelial cells of the microcirculation in the bone marrow, liver, and spleen. Certain substances (e.g., histamine and bradykinin) promote the leakage of large molecules through the gap junctions.

    (4) Tight junctions. In most body tissues, the clefts between endothelial cells appear fused, but anatomic and physiologic evidence indicates that there are pores approximately 400 nm in diameter between endothelial capillary cells. These pores are absent in cerebral capillaries, providing further evidence of a blood–brain barrier.

  b. Endothelial functions. It has become apparent that the endothelium of the vascular system is an **active metabolic tissue**.

    (1) The endothelium plays a major role in the **control of blood flow** in many organs.

      (a) The endothelium responds to substances called **kinins** (e.g., **bradykinin, lysylbradykinin**) that **produce arteriolar dilation** and **increase capillary permeability**.

        (i) The kinins are formed from a precursor called a **kininogen** by a protease called **plasma kallikrein**.

        (ii) The kinins are enzymatically inactivated by **kininase I** and **kininase II**. Kininase II is identical to **angiotensin converting enzyme (ACE),** which converts angiotensin I to angiotensin II. (Angiotensin II is a very potent vasoconstrictor.)

      (b) Endothelium-derived relaxing factor (EDRF). EDRF has been identified as **nitric oxide** (see Chapter 3 III C 2 a).

        (i) Nitric oxide has a **continuous vasodilating effect** in blood vessels with an intact endothelium.

        (ii) Nitric oxide is **continually released** from the endothelium by a number of compounds, including acetylcholine (ACh), and a number of polypeptides, including bradykinin. Note that ACh causes contraction of vascular smooth muscle when endothelial tissue is not present.

      (c) Endothelins represent a family of polypeptides that have diverse roles in many different organs.

        (i) In the vascular system, they tend to produce an initial vasodilation followed by a potent and long-lasting vasoconstriction. They also have positive inotropic and chronotropic effects on the myocardium.

(ii) The endothelins appear to exert their effects by activating specific receptors.

(2) The endothelium also assists in **defense reactions** (e.g., hemostasis, inflammation).

3. **Function.** The microcirculation provides a total surface area of approximately 700 m^2 for exchange between the circulatory system and the interstitial compartment. Exchange of substances occurs primarily in the capillaries and postcapillary venules. **Diffusion, filtration (bulk flow),** and **cytopempis** are the **three mechanisms of exchange**.

a. **Diffusion** (see Chapter 1 III A) is the **principal mechanism** of microvascular exchange.

(1) The **rate** of diffusion is dependent on the solubility of the substance in the tissues, the temperature, and the available surface area, and is inversely related to molecular size and the distance over which diffusion must occur.

(a) Molecules such as O_2, CO_2, H_2O, and glucose rapidly achieve equilibrium across the microvascular endothelium.

(b) Larger molecules (e.g., albumin) cross the endothelial barrier very slowly or not at all.

(2) **Diffusion path.** The route of transport of most substances is controversial. Transport may occur through gap junctions, fenestrae, intracellular vesicles, or across the cells themselves.

b. **Filtration (bulk flow).** A **hydrostatic pressure difference** across the endothelium leads to filtration (i.e., the passage of water and solutes from the capillaries into the tissues). The presence of large molecules (e.g., proteins, especially albumin) in the blood creates **oncotic pressure,** which **counteracts filtration**.

(1) The **Starling hypothesis,** developed by Ernest Starling, holds that the balance between filtration and reabsorption of water depends on the difference in hydrostatic and oncotic pressures between the blood and the tissues, and on the permeability of the vessel.

(2) Figure 12-2 shows how intra- and extravascular factors affect filtration and reabsorption of fluid across the microcirculation. The **bulk flow of water (Q_w)** is expressed as:

$$Q_w = K \cdot [(P_c - P_i) - (\pi_c - \pi_i)], \text{ where}$$
$$K = \text{the permeability–surface area coefficient}$$
$$P_c = \text{the hydrostatic capillary pressure}$$
$$P_i = \text{the hydrostatic interstitial pressure}$$
$$\pi_c = \text{the oncotic blood pressure}$$
$$\pi_i = \text{the oncotic interstitial pressure}$$

(a) The **permeability–surface area coefficient (K)** varies as the number of microvessels receiving blood flow and the vascular permeability change. Hypercapnia, hypoxia, increased H^+ and K^+ concentrations, histamine, bradykinin, adenosine, and many other substances can result in vasodilation and an increase in the number of perfused capillaries.

(i) Permeability is increased when the vessels are injured by hypoxia or inflammatory agents (e.g., histamine).

(ii) Dilation of the arterioles, especially of the precapillary sphincters, increases the number of functional capillaries because dilation transmits a greater fraction of the arterial pressure into the capillary. The greater capillary pressure dilates vessels, so that more vessels are perfused and flow increases.

(b) The **hydrostatic capillary pressure (P_c)** tends to force fluid into the interstitium.

(i) Capillary pressure varies widely but ranges between 30–45 mm Hg in most tissues. The hydrostatic capillary pressure depends on the arterial blood pressure, the arteriolar resistance, the state of the precapillary sphincters, and, most importantly, the venous pressure.

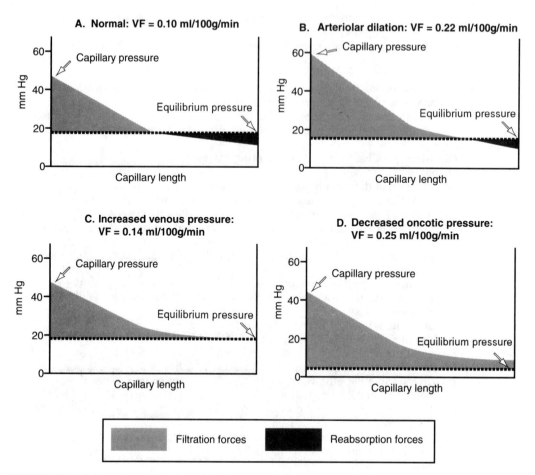

FIGURE 12-2. Schematic presentation of the intravascular pressures involved in determining the filtration of fluid across the microcirculation endothelium. The equilibrium pressure represents the pressure needed to balance the filtration and reabsorption forces. *VF* = volume filtered in ml/100 g tissue/min.

(ii) In the presence of blood flow, the capillary pressure must always exceed the venous pressure because there must be a driving force to produce blood flow. A low arteriolar resistance allows a greater proportion of the arterial pressure to be transmitted to the capillaries. For example, in the kidneys, the glomerular capillary pressure is approximately 50–60 mm Hg, producing the large force necessary for glomerular filtration.

(iii) Contraction of the arteriolar smooth muscle or precapillary sphincter can occlude the vascular lumen, causing the hydrostatic capillary pressure to equilibrate with the local venous pressure. Under these conditions, fluid may be reabsorbed along the entire length of the capillary. A decrease in the pressure within the capillaries (e.g., as in shock) tends to draw fluid from the tissues into the vascular system.

(iv) An increase in the blood pressure within the microvessels (i.e., an increase in hydrostatic capillary pressure) results in increased filtration of fluid from the capillaries and an accumulation of excess fluid in the tissues (edema). However, a loss of plasma proteins (i.e., a decreased hydrostatic capillary pressure), as occurs in starvation, can also cause edema because of the decrease in oncotic pressure that normally retards filtration. Injury to vessels or the presence of infection causes the vessels to become more permeable (i.e., the permeability–surface area coefficient increases), leading to the accumulation of tissue fluids.

(c) The **hydrostatic interstitial pressure (P_i)** increases if the interstitial fluid volume increases (e.g., in the presence of edema). Because the hydrostatic interstitial pressure is markedly increased in inflamed tissues, these tissues tend to be edematous. The increased hydrostatic interstitial pressure hinders additional fluid filtration from the capillaries. The magnitude of the hydrostatic interstitial pressure in normal tissue is controversial because different experimental techniques have shown it to be either slightly positive or slightly negative.

(d) The **oncotic pressure of blood (π_c)** results from the osmotic pressure of blood proteins, which represent only a small fraction (0.005%) of the total osmotic pressure of blood. The normal oncotic pressure of blood is 25–27 mm Hg. The blood oncotic pressure provides a force that tends to reabsorb fluid into the capillaries.

(e) **Oncotic pressure of the interstitium (π_i).** Significant amounts of plasma proteins cross the endothelial barrier through either intercellular pores or fenestrae, depending on the type of tissue. Protein in the interstitial space produces a force that tends to pull fluid out of the vascular system. The effective oncotic pressure in the interstitium is estimated to range between 5–10 mm Hg in most tissues.

c. **Cytopempis.** Some substances are transported across the endothelium by vesicular transport (see Chapter 1 III D). It has been proposed that this is how large, lipid-insoluble molecules are transported; however, the transport of water and small solutes would be negligible by this mechanism.

D. The **lymph vessels** originate as closed endothelial tubes that are permeable to fluid and high–molecular-weight compounds.

1. **Function.** The lymphatic system represents the only mechanism of returning albumin and other interstitial macromolecules to the circulatory system. (The higher concentration of albumin in the blood compared with the interstitium makes a net back-diffusion of albumin into the capillaries impossible.) In addition, excess fluid is removed from the interstitium to maintain a **gel state**.

2. **Mechanism**
   a. Lymphatic fluid is pumped out of the tissues by the contraction of the large lymph vessels and the contiguous skeletal muscles. The lymph vessels possess an extensive system of one-way valves that maintain flow toward the heart.
   b. The lymph collected from the peripheral tissues is returned to the cardiovascular system via the thoracic duct, which empties into the left subclavian vein near its junction with the left internal jugular vein.
   c. The **rate** of lymph flow varies in different organs and depends on tissue activity and capillary permeability. The GI tract and the liver produce the majority of the lymph; normal lymph flow is approximately 2 L/day for the entire body. The lymphatic system recovers approximately 200 g of protein daily that has been lost from the microcirculation.

E. **Veins.** The venous system is a low-resistance, low-pressure, highly distensible part of the vascular system. At normal pressures, the veins are approximately 20 times more distensible (compliant) than the arteries. In comparable vessels, the veins provide a larger cross-sectional area than do the arteries; therefore, the resistance to flow and the velocity of flow are less in the veins than in the arteries.

1. **Structure**
   a. The veins are relatively thin-walled structures that contain **small amounts of elastic tissue** and **smooth muscle**.
   b. **Valves**
      (1) The veins of the **dependent parts of the body** are equipped with a system of valves that **prevent the backflow of venous blood**.

    (2) These valves also **support the column of blood** so that increases in capillary pressure in the dependent parts of the body (caused by hydrostatic pressure) are minimized.

        (a) Venous valvular insufficiency (backward leaking) allows the column of blood to exert the full hydrostatic pressure on the capillaries, raising capillary pressure and leading to the formation of edema in the dependent parts.

        (b) Varicose veins represent a marked dilation and valvular insufficiency of the dependent veins as a result of a familial or occupational predisposition, or both.

2. Functions

    a. Reservoirs. The veins serve as a fluid reservoir: 65%–75% of the circulating blood volume is in the veins at any one time. Normally, the veins are not fully distended; therefore, small changes in pressure can cause large changes in volume (Figure 12-3).

    b. Conduits. The systemic veins carry blood from the tissues to the right atrium; the pulmonary veins collect blood from the lungs and return it to the left atrium.

      (1) Factors that enhance venous return

        (a) Venoconstriction is an extremely important mechanism related to increasing cardiac output. The smooth muscle of the veins constricts in response to sympathetic stimulation, displacing blood toward the heart and raising the ventricular filling pressure. The increased ventricular filling pressure increases cardiac output by increasing the ventricular end-diastolic volume (VEDV) and the ventricular stroke volume.

        (b) Skeletal muscle pump. Contraction of skeletal muscle aids in venous return by compressing the veins between the contracting skeletal muscles. **Competent venous valves** prevent backflow during relaxation of the skeletal muscles.

        (c) Respiration. The normal negative intrathoracic (interpleural) pressure **tends to enhance venous return** by increasing the pressure gradient between the heart and the peripheral vessels. **Inspiration** further amplifies the effect.

      (2) Factors that impede venous return

        (a) Peripheral pooling of blood results when an individual rises from a supine to an erect position.

          (i) Standing causes approximately 500 ml of blood to shift from the pulmonary circulation to the dependent veins of the legs. The shift is caused by the hydrostatic pressure generated by the erect position, which causes the leg veins to dilate and leads to a peripheral pooling of blood.

          (ii) Peripheral pooling of blood decreases venous return (consequently lowering cardiac output and, possibly, arterial blood pressure as well),

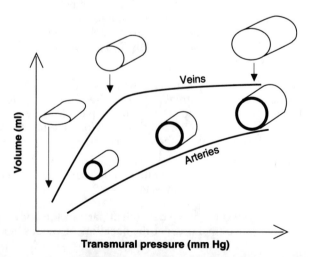

FIGURE 12-3. Pressure–volume relationship of arteries and veins. The slope of each line at any point represents compliance (distensibility). Note that the veins are much more compliant than the arteries at low pressures, because the veins are not completely distended at low pressures.

but a reflex increase of sympathetic tone to the veins and heart usually compensates for the lowered arterial blood pressure.

(b) Hydrostatic effects. The arterial pressure is approximately 30 mm Hg lower in the head than at heart level because of gravitational effects on the column of blood from the heart to the brain. A reflex increase in the smooth muscle tone in the veins, an increased heart rate, and increased peripheral vascular resistance help maintain an adequate arterial pressure.

(c) Both **positive end-expiratory pressure (PEEP)** and the **Valsalva maneuver** impede venous return and produce a complex set of direct and reflex changes in the cardiovascular system (see also Chapter 20 IV A 2 a).

II. SPECIAL CIRCULATIONS

A. Coronary circulation

1. **Organization.** The heart is supplied by the right and left coronary arteries; the left coronary artery divides after approximately 1 cm into the anterior descending and circumflex branches.

   a. Arteries

   (1) **Superficial vessels.** The major coronary vessels travel in the epicardium of the heart and subdivide, sending penetrating branches through the myocardium. It is the superficial vessels that are corrected by coronary bypass operations when atherosclerotic narrowing develops.

   (2) The **penetrating branches** subdivide into **arcades** that distribute blood to the myocardium. Normally, the coronary arteries appear to function as **end arteries;** however, the presence of an arterial plaque or occlusion allows **anastomoses** between vessels to become functional.

   b. Veins

   (1) **Left ventricular venous drainage** occurs primarily through the **coronary sinus**. The **thebesian veins** and **coronary–luminal vessels** (connections between the coronary vessels and the lumen of the heart) also return small amounts of blood to the left ventricle, creating an **anatomic shunt** [see Chapter 18 V A 2 b].

   (2) **Right ventricular venous drainage** occurs through the multiple **anterior coronary veins**. The coronary–luminal connections carry a larger proportion of the flow in the right ventricle than in the left ventricle.

2. **Blood flow.** A continuous flow of blood to the heart is essential to maintain an adequate supply of O_2 and nutrients.

   a. Normal coronary blood flow represents approximately 5% of the resting cardiac output (250 ml/min), or approximately 60–80 ml blood/100 g tissue/min.

   b. Coronary flow pattern. During ventricular systole, myocardial wall tension increases and compresses the vessels, thus increasing the resistance to flow.

   (1) This extravascular compression produces a complex coronary flow pattern (Figure 12-4).

   (2) Maximal flow in the left coronary vessels usually occurs during isovolumic relaxation while the arterial pressure is still relatively high and the myocardium is relaxed.

   c. Factors influencing coronary flow. The coronary blood flow is characterized by a **supply–demand** relationship. The heart metabolizes all substrates in approximate proportion to their vascular concentration.

   (1) **Myocardial O_2 consumption** averages 6–8 ml O_2/100 g of tissue/min.

   (a) Normally, hemoglobin releases approximately 50% of its arterial O_2 content to the myocardium; therefore, coronary venous hemoglobin is approximately 50% saturated with O_2. This is in contrast to the remainder of the body at rest, where hemoglobin releases only approximately 25% of its O_2 content as it passes through the systemic capillaries.

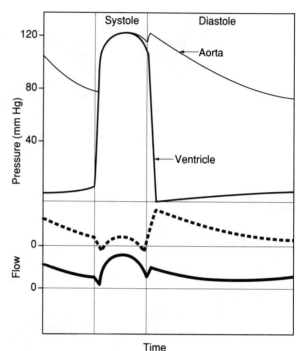

FIGURE 12-4. Diagram of the right coronary artery flow (*bold line*) and the left coronary artery flow (*dashed line*) associated with the left ventricular and aortic pressure pulses. The peak left coronary flow occurs at the end of the isovolumic relaxation, when extravascular coronary compression is low and the driving force (i.e., the aortic pressure) is high.

(b) The coronary venous partial pressure of oxygen (PO_2) is approximately 25–30 mm Hg under resting conditions and decreases during exercise or stress. During exercise, the cardiac O_2 consumption may increase five- to sixfold in well-trained athletes, partly because coronary flow increases, and partly because the arteriovenous O_2 content difference increases (i.e., the heart extracts more O_2 from the blood).

(2) **Adenosine.** Even at rest, the myocardium extracts large amounts of O_2 from the coronary blood. Therefore, it is critical during exercise that coronary flow increases to maintain an adequate O_2 tension in the myocardium. Recent evidence indicates that adenosine is a major factor in producing coronary vasodilation during hypoxic states.

(a) Adenosine is derived from the degradation of adenosine monophosphate by 5'-nucleotidase.

(b) Release of adenosine into the myocardium produces an extremely strong vasodilator response.

(3) **Sympathetic stimulation** increases the cardiac rate and contractility.

(a) The resultant **increase in myocardial metabolic activity** leads to coronary vasodilation. An increase in the myocardial metabolism without a compensatory increase in coronary flow would cause the breakdown of adenosine triphosphate (ATP) and, eventually, the release of adenosine.

(b) The coronary vessels contain both α and β receptors (see Chapter 3 III B 3). The α vasoconstrictor activity is rather weak, allowing the vasodilator (β) response to predominate. There are no sympathetic cholinergic vasodilator fibers in the myocardium.

B. **Cerebral circulation**

1. **Organization.** The **carotid arteries** are the major vessels supplying the brain in humans. The **carotid sinus** provides a reflex input that maintains blood pressure at a level to ensure adequate blood flow to the brain.

2. **Blood flow.** The brain, like the heart, relies on a continuous blood flow for adequate function. Interruption of blood flow for only 5–10 seconds causes a loss of consciousness; circulatory arrest for only 3–4 minutes results in irreversible brain damage.
    a. **Normal cerebral blood flow** represents approximately 15% of the resting cardiac output (750 ml/min), or about 50–55 ml blood/100 g tissue/min.
    b. **Control of cerebral blood flow.** The brain is surrounded by the rigid skull, but vascular diameter can be altered in response to changing blood flow needs by a narrowing of the venous bed, which compensates for vasodilation on the arterial side.
        (1) **Cerebral autoregulation** is an effective mechanism for controlling blood flow over a wide range (i.e., from 80–180 mm Hg) of arterial pressures. Moderate increases or decreases in CO_2 **tension** can respectively double or halve the normal cerebral blood flow.
            (a) **Hypocapnia.** The reduction in blood flow that accompanies hyperventilation can lead to cerebral hypoxia, which causes dizziness and, occasionally, fainting.
            (b) **Acidosis.** The H^+ that is derived from lactic or carbonic acid as a result of either hypoxia or hypercapnia causes cerebral vasodilation.
        (2) **Neural control** of cerebral blood flow seems to be limited to the larger vessels and those located within the pia mater. The cerebral vessels are supplied by both sympathetic and parasympathetic neurons, but the role of these neurons remains unknown.
        (3) **Cushing reflex** is a systemic vasoconstriction in response to an increase in cerebrospinal fluid (CSF) pressure. The marked systemic hypertension helps maintain adequate blood flow to the brain. The reflex is initiated by the medullary centers, which become activated when they become hypoxic as a result of limited blood flow.

C. **Skeletal muscle** comprises 40%–50% of the total body weight in an average adult man. Because of this large mass, even resting skeletal muscle utilizes approximately 20% of the total resting O_2 consumption.

1. **Resting skeletal muscle** receives a relatively large blood flow compared with its metabolic rate; therefore, the venous O_2 content is approximately 17–18 vol% (i.e., the hemoglobin saturation is 85%–90%).
    a. **Blood flow** in resting skeletal muscle varies from 1.5–6 ml/100 g tissue/min.
    b. **Factors affecting blood flow.** Blood flow to resting skeletal muscle is largely determined by the **type of muscle.** Muscles comprised mainly of red fibers receive a larger blood flow than those comprised mainly of white fibers.

2. **Exercising skeletal muscle**
    a. **Blood flow.** Exercising skeletal muscle receives approximately 80 ml blood/100 g tissue/min (a fifteen- to twentyfold increase over the resting state) and extracts approximately 80% of the O_2 from the arterial blood.
    b. **Factors affecting blood flow**
        (1) The **release of metabolic products** from the muscle tissue during exercise is responsible for the large increase in blood flow over resting skeletal muscle. These breakdown products include increased levels of CO_2, H^+, K^+, lactate, and adenosine, and decreased levels of O_2.
        (2) In some species, **cholinergic sympathetic vasodilator fibers** increase muscle blood flow before the actual onset of exercise.
    c. **Hemoglobin desaturation.** Even with the marked vasodilation that occurs, the blood flow is inadequate to maintain adequate O_2 delivery to the cells during moderate to strenuous exercise.
    d. The **anaerobic threshold** occurs at an O_2 consumption that is approximately 60% of the maximal consumption. The anaerobic threshold is defined as the level of exercise that produces a significant rise in the level of lactic acid in the bloodstream.

D. **Fetal circulation** differs in a number of significant ways from neonatal circulation.

1. Figure 12-5 shows the **circulatory pattern, blood flow,** and **hemoglobin saturation** in the fetus and the neonate.

2. The **placenta** is a major organ in the fetus and **receives more than 50% of the combined output of both ventricles**.

3. At birth, marked changes occur in both the systemic and pulmonary circulations as the placental blood supply is lost.
   a. **Pulmonary circulation changes** are primarily the result of a **marked decrease in pulmonary vascular resistance** as the lungs inflate after birth. After inflation of the lungs, the pulmonary vascular resistance decreases by approximately 90%, causing a **large drop in the afterload for the right ventricle** and **lowering** the **right atrial and ventricular pressures**.
      (1) **Foramen ovale.** The drop in the right-sided pressures causes pressure in the left atrium to exceed the pressure in the right atrium. (The opposite is true during fetal life.) The pressure reversal causes the flap-like valve to close over the foramen ovale. The valve then normally fuses to the interatrial septum over the next few days.
      (2) The **ductus arteriosus** conducts blood from the pulmonary artery to the aorta during fetal life.
         (a) The flow through this vessel reverses after birth as the pulmonary resistance is reduced and the pulmonary artery pressure becomes less than the aortic pressure.
         (b) The ductus arteriosus begins to constrict shortly after birth (perhaps in response to the higher O_2 content of the blood).

FIGURE 12-5. Fetal and neonatal circulatory patterns. Percentages within the organs indicate the percent of the total blood flow. (A) During fetal life, the two ventricles function in parallel as a result of the foramen ovale (*FO*) and the ductus arteriosus (*DA*). (B) Elimination of the placental circulation at birth almost doubles the systemic resistance, which increases the left atrial and ventricular pressures. Inflation of the lungs markedly reduces the pulmonary vascular resistance, lowering the right atrial and ventricular pressures. These pressure changes cause the foramen ovale to close and reverse the flow through the ductus arteriosus. *LA* = left atrium; *LV* = left ventricle; *RA* = right atrium; *RV* = right ventricle; *SAT* = hemoglobin saturation.

A. Fetal circulation

B. Neonatal circulation

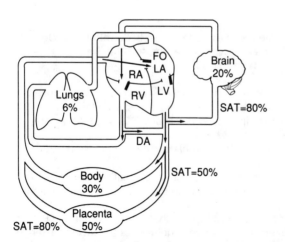

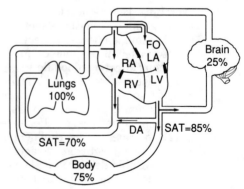

(i) There is evidence that the patency of the ductus arteriosus is maintained by a dilator prostaglandin; therefore, closure of the ductus arteriosus may be facilitated by the administration of aspirin, which inhibits prostaglandin formation.

(ii) The ductus arteriosus may remain patent for several days following birth. When this occurs, a continuous murmur may be heard over the upper chest as a result of the turbulent blood flow through the narrowed structure.

(3) The **thickness of the right ventricular wall** is approximately equal to the thickness of the left ventricular wall during fetal life. After birth, the thickness of the right ventricle diminishes over several months and, over a period of several years, the electrical axis of the heart swings toward the left until it lies within the normal adult quadrant.

b. **Systemic circulatory changes** occur because of the loss of the placental circulation after birth, which diverts more blood flow to the other body organs. The umbilical arteries constrict in response to cooling, trauma, circulating catecholamines, and high O_2 tensions.

(1) The loss of the placental circulatory bed after birth causes a **large increase in the systemic vascular resistance**.

(2) The increase in resistance **raises the arterial pressure,** consequently **increasing the afterload of the left ventricle**. The higher afterload leads to increases in left ventricular and left atrial pressures, which contribute to the closure of the foramen ovale.

(3) Closure of the foramen ovale prevents the right-to-left flow of venous blood and improves the oxygenation of the systemic arterial blood.

E. | Cutaneous circulation

1. **Organization**
   a. **Cutaneous arterioles** form a dense network just under the dermis layer of the skin. These arterioles give rise to **metarterioles,** which subdivide into **capillary loops**.
      (1) The **capillary loops** are located in the papillae of the dermis and provide a large surface area for heat exchange.
      (2) The outflow from several capillary loops forms a collecting venule that anastomoses with other venules to form an extensive **subpapillary venous plexus**.
   b. **Arteriovenous anastomoses** are located in the distal parts of the extremities and the nose, lips, and ears. These vessels are wide, low-resistance connections that serve as shunts and allow blood to bypass the superficial capillary loops.

2. **Innervation** of the cutaneous vessels is through the sympathetic adrenergic system, which provides a tonic vasoconstrictor effect.
   a. The cutaneous arterioles and metarterioles have both α and β receptors, but the arteriovenous anastomoses have only α receptors.
   b. The venous plexus has separate neural connections and can undergo marked venoconstriction, which minimizes the amount of blood in the skin.

3. **Functions**
   a. **Nutrient supply.** The metabolic rate of the skin is relatively small so that a minimal amount of blood flow to the skin can supply the nutritive functions.
   b. **Temperature control.** The major function of the cutaneous blood flow is to aid in the regulation of body temperature.
      (1) **Heat conservation** is accomplished by markedly diminishing the rate of blood flow to the skin. Reduction of superficial blood flow converts the skin and subcutaneous tissues into a layer of insulation. Thus, body heat is preserved because less heat is dissipated to the environment.
      (2) **Heat dissipation.** When body temperature rises as a result of increased metabolic activity, the added heat load must be eliminated from the body.

    (a) **Vasodilation.** The increased blood temperature stimulates centers in the hypothalamus that control the distribution of blood flow, inhibiting the normal tonic vasoconstrictor outflow to the cutaneous vessels and increasing blood flow to the skin.

    (b) **Radiation, conduction, and evaporation.** The increased flow carries heat to the surface of the body, where it is dissipated by radiation, evaporation, and conduction to the environment. If the environmental temperature is higher than the body temperature, heat can only be dissipated by means of **evaporation of sweat;** under these conditions, radiation and conduction would cause the body to gain heat.

4. **Skin color.** In light-skinned races, the skin color is dependent on the rate of blood flow.

    a. **High rates of cutaneous blood flow** and well-oxygenated hemoglobin impart a reddish color to the skin.

    b. **Marked vasoconstriction** of the cutaneous vessels, which occurs during strong sympathetic stimulation (as in the **fight-or-flight reaction**), produces paleness that is described as **ashen**.

    c. **Abrupt withdrawal of the vasoconstrictor input** to the vessels of the face and upper body causes vascular engorgement and the reddish coloration known as **blushing**.

    d. The **presence of methemoglobin or hemoglobin with a low O_2 saturation** (secondary to hypoxia) can impart a bluish tint to the skin and mucous membranes, termed **cyanosis**.

Chapter 13

Cardiac Output and Venous Return

I. **RELATIONSHIP BETWEEN CARDIAC OUTPUT AND VENOUS RETURN.** The heart can pump only as much blood as it receives, so that over any significant period of time the venous return must equal the cardiac output. In the normal, resting human, the cardiac output and the venous return are approximately 5 L/min. A complicated interaction of neural, humoral, and physical factors determines the flow rate.

A. **Vascular function curves** reflect the relationship between **venous return** and **venous pressure** in the circulatory system. With each contraction, the heart removes blood from the venous side of the circulation and ejects it under higher pressure into the arteries, causing the arterial pressure to increase and the venous pressure to decrease. Thus, an increased venous return, associated with an increased cardiac output, results in a decreased venous pressure (Figure 13-1A).

1. **Mean circulatory pressure (MCP).** If the heart stops, blood continues to flow from the arteries to the veins for a few more seconds until the pressures throughout the circulatory system equalize. Because the veins now contain more blood than they did when the heart was beating, the venous pressure is highest when the cardiac output (or venous return) is zero. This venous pressure is the MCP (see Figure 13-1A). The MCP varies with blood volume and the activity of the sympathetic nervous system (Figure 13-1B).
   a. An **increased blood volume** increases the MCP and shifts the entire vascular function curve to the right, reflecting the higher venous pressure. The additional blood volume distends the vascular system and increases the pressure in the same way that the pressure in a tire increases if more air is added.
   b. **Sympathetic stimulation** causes the smooth muscle in the veins and arteries to contract so that the vessels squeeze down on the blood with more force, raising the MCP and shifting the vascular function curve to the right.

2. **Increased resistance to flow** occurs when the vessels (especially the arterioles) constrict, creating more friction between the blood and the vessels. Because of the increased friction, there is a greater pressure drop between the arteries and the veins. This results in a lower venous pressure at any venous return and shifts the curve counterclockwise, as shown in Figure 13-1C.

B. **Cardiac function curves** reflect the relationship between **venous pressure** and **cardiac output (Starling's law of the heart;** see Chapter 11 III A).

1. To summarize, an increase in venous pressure increases the preload (i.e., the ventricular volume), which produces a stronger contraction and a greater ejection of blood during systole. Therefore, an increased venous pressure produces an increased cardiac output.

2. Figure 13-1D depicts the normal cardiac function curve, as well as the effects of sympathetic stimulation and ventricular failure (i.e., increased and decreased contractility, respectively) on the heart.

C. **Cardiovascular function curves.** Because both venous return and cardiac output depend on venous pressure, the normal curves from Figure 13-1A&D can be combined into one graph (Figure 13-2). The intersection of the venous return (vascular function) and the cardiac function (Frank-Starling) curves is the only point where this system can function, because this is where the two flow rates are equal. Any change in cardiac contractility, blood volume, or vascular resistance will cause the operating point to shift, but **cardiac output and venous return will always be equal**. Exercise and congestive heart failure are just two of many conditions that can alter the operating point of the system.

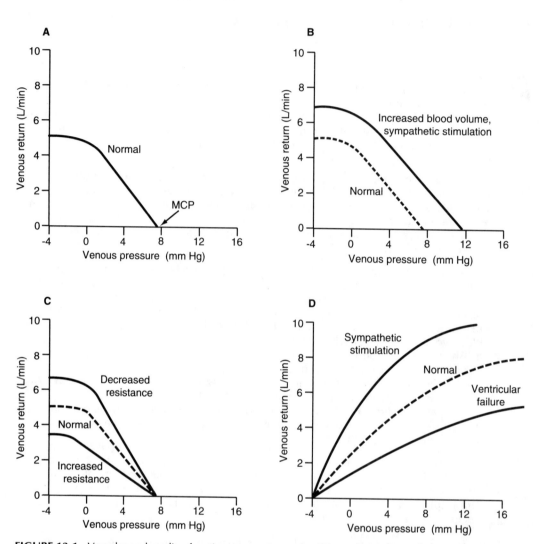

FIGURE 13-1. Vascular and cardiac function curves. In graphs A through C, venous pressure is the dependent variable but is plotted on the x-axis in contrast to the usual method of plotting the dependent variable along the y-axis. (A) Normal curve. Venous pressure decreases as venous return increases because the transfer of blood from the venous to the arterial system increases flow. When venous return (cardiac output) is zero, venous pressure is maximal and is referred to as the mean circulatory pressure (MCP). (B) An increased blood volume or sympathetic stimulation (which causes venous constriction) shifts the vascular function curve to the right and increases the MCP. (C) Changes in arteriolar resistance alter the slope of the vascular function curve. An increased resistance reduces the pressure transmitted through the microcirculation so that venous pressure is lower for any given flow rate. (D) Three cardiac function (Frank-Starling) curves demonstrate the effects of changes in contractility on the relationship between venous pressure and cardiac output.

    1. Exercise (Figure 13-3A). The overall effect of exercise is an **increase in cardiac output that is proportional to the exercise intensity**. Sympathetic stimulation of the heart and vascular system, as well as the release of vasodilators from the skeletal muscle beds, increases cardiac output.

        a. Activation of the sympathetic system increases ventricular contractility and heart rate, causing the cardiac output curve to shift to the left (i.e., to a higher level). Sympathetic stimulation also causes constriction of the venous smooth muscle, which increases the MCP, shifting the vascular function curve to the right.

        b. Vasodilation occurs in the skeletal muscles to increase the blood flow and provide

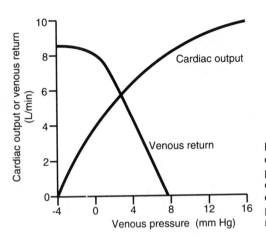

FIGURE 13-2. The vascular and cardiac function curves are superimposed. The only valid operating point on this graph is the intersection of the two curves, because this is the point where venous return equals cardiac output. In this graph, the cardiac output is 5 L/min and the venous pressure is approximately 4 mm Hg.

more O_2 and metabolites. The vasodilation reduces the total peripheral resistance, which causes the vascular function curve to become steeper (see Figure 13-1C).

2. **Congestive heart failure** (Figure 13-3B)
   a. **Decreased contractility** of the heart is the primary defect in congestive heart failure. The cardiac function curve shifts downward, and, initially, **cardiac output is reduced**.
   b. **Increased blood volume.** Because of the reduced cardiac output, the renin-angiotensin-aldosterone axis causes the kidneys to retain fluids and electrolytes, which causes the blood volume to increase over a period of several weeks or months.
      (1) The increased blood volume shifts the vascular function curve to the right, raising venous pressure, and **cardiac output returns toward normal**.
      (2) In addition, the slope of the vascular function curve decreases, reflecting the vasoconstriction that occurs as an attempt to maintain the arterial pressure in the face of a reduced cardiac output.
   c. **High venous pressure** invariably accompanies congestive heart failure because the cardiac output is reduced and there is an increase in the total blood volume. The high venous pressure results in an increased filtration of fluid from the vascular system into the tissues (edema). Edema may occur in the lungs (pulmonary edema) or in the feet (swollen ankles), depending on which ventricle is limited in its pumping ability.

II. MEASUREMENT OF CARDIAC OUTPUT.
The cardiac output (SV • HR) averages 5 L/min in a normal man and is approximately 20% less in a woman. The cardiac output is carefully controlled to provide an adequate mean blood pressure for delivery of nutrients and O_2. The cardiac output can be measured using either of the following techniques.

A. **The Fick principle** is an example of the law of conservation of mass, a widely applied principle in physiology. The Fick principle can be stated as:

$$Q = \frac{\dot{V}_{O_2}}{C_{aO_2} - C_{\bar{v}O_2}}, \text{ where}$$

$$\begin{aligned}
Q &= \text{cardiac output} \\
\dot{V}_{O_2} &= O_2 \text{ consumption (ml/min)} \\
C_{aO_2} &= \text{arterial } O_2 \text{ content} \\
C_{\bar{v}O_2} &= \text{mixed venous } O_2 \text{ content}
\end{aligned}$$

1. **Measuring cardiac output**
   a. The O_2 consumption for the whole body can be determined using a spirometer to measure O_2 uptake from the lungs.

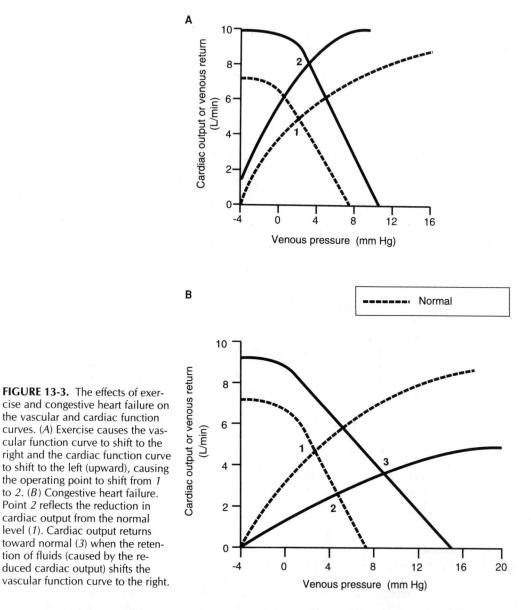

FIGURE 13-3. The effects of exercise and congestive heart failure on the vascular and cardiac function curves. (*A*) Exercise causes the vascular function curve to shift to the right and the cardiac function curve to shift to the left (upward), causing the operating point to shift from *1* to *2*. (*B*) Congestive heart failure. Point *2* reflects the reduction in cardiac output from the normal level (*1*). Cardiac output returns toward normal (*3*) when the retention of fluids (caused by the reduced cardiac output) shifts the vascular function curve to the right.

    b. The **arterial blood composition** is the same throughout the body.

    c. **Mixed venous blood,** which is representative in O_2 content of the entire body, must be obtained from the right ventricle or pulmonary artery.

  2. Measuring O_2 consumption. When used to measure the O_2 consumption of individual organs, the Fick principle is expressed as $\dot{V}o_2 = Q\,(Cao_2 - C\bar{v}o_2)$. The venous drainage from the particular organ must be sampled, and an independent measure of the arterial blood flow to the organ must be obtained. The arterial flow can be obtained by means of electromagnetic or Doppler flow probes.

B. Indicator dilution method

  1. Theory

    a. Volume. If a suitable indicator is injected into an unknown volume of distribution, the volume can be estimated from the resultant indicator concentration:

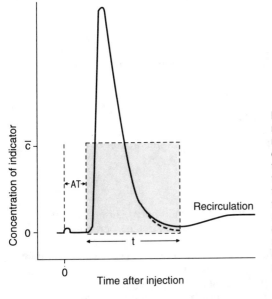

FIGURE 13-4. A dye concentration curve is used to measure cardiac output by the indicator dilution method. *AT* = appearance time (i.e., the time required by the dye to reach the sampling site); $\bar{c}$ = the mean dye concentration after one passage of the dye; t = the time required for the dye to pass the sampling site once. Because some of the dye will recirculate before the dye concentration reaches zero, the dye concentration curve must be extrapolated to the zero point (*dashed line*). The area of the *shaded rectangle* is equal to the area enclosed by the dye concentration curve. If $\bar{c}$ = 1 mg/L, t = 0.5 min, and a = 2.5 mg, then flow (Q) = 2.5 / (1 • 0.5), or 5 L/min.

$$V = A/c, \text{ where}$$
$$V = \text{the volume of distribution}$$
$$A = \text{the injected amount of indicator}$$
$$c = \text{the resultant indicator concentration}$$

   b. Flow. Similarly, the flow of a fluid (Q) can be measured by determining the time (t) required for the mean concentration ($\bar{c}$) of the indicator to pass a given site: $Q = A/(\bar{c} • t)$.

2. **Technique.** In the body, the indicator is injected instantaneously but is dispersed over time because of the different transit times between the injection and sampling sites. Usually, some recirculation of indicator occurs before passage of the indicator is complete (i.e., before the concentration reaches zero). After the peak concentration is reached, the indicator concentration follows an exponential time-course, so that the curve can be extrapolated to zero concentration, and the duration of the indicator passage (t) can be estimated (Figure 13-4).
   a. **Indicators** used for cardiac output measurement include Evans blue, Cardio-Green, hypertonic or hypotonic saline, ascorbate, and cold saline. Cold saline, used in thermodilution, is the most commonly used indicator in human cardiac output studies.
   b. **Thermodilution** uses a multilumened balloon catheter with a thermistor near the distal end (termed a **Swan-Ganz catheter** after its inventors).
      (1) The catheter is inserted into a peripheral vein and advanced into the central veins; the balloon is then inflated with air so that it acts as a sail to carry the catheter along with the blood through the right ventricle. The catheter is then advanced into the pulmonary artery.
      (2) **Cold saline** is injected through a proximal opening of the catheter. The opening lies within the right atrium or ventricle. The saline mixes with the blood in the right ventricle, and the temperature change of the blood is measured by the thermistor, which is downstream in the pulmonary artery.
      (3) Thermodilution is a preferred technique because little or no recirculation of indicator occurs; withdrawal of blood is unnecessary for analysis of the indicator concentration; cold saline is relatively innocuous; and little temperature change occurs in the surrounding tissues.

Chapter 14

Cardiovascular Control

I. **INTRODUCTION.** Control of blood volume and arterial pressure ensures that all of the organs receive sufficient blood flow.

II. **CONTROL OF BLOOD VOLUME** represents a complex interaction of neural, renal, and endocrine mechanisms.

A. **Neural regulation**

1. **Arterial (high-pressure) baroreceptors** are highly branched nerve endings located in the carotid sinus and the aortic arch that serve as **sensors** to detect increases in blood pressure.

   a. The arterial baroreceptors are stimulated by distention of the vessel walls; resultant nerve impulses are carried to the central nervous system (CNS) over the sinus and aortic nerves (branches of the glossopharyngeal and vagus nerves, respectively).

      (1) The carotid sinus represents the most distensible area of the arterial system, which indicates that these receptors respond to stretch of the arterial walls rather than to pressure per se.

      (2) Impulses from both high- and low-pressure receptors (see II A 2) are transmitted to the hypothalamus, where the neurohypophyseal cells are inhibited from releasing antidiuretic hormone (ADH) (Figure 14-1).

FIGURE 14-1. Major factors involved in cardiovascular control. *Solid lines* indicate an increase in the parameter; *dashed lines* indicate a decrease. Note that there are a large number of negative feedback loops that control both blood volume and arterial blood pressure. *ACE* = angiotensin converting enzyme; *ANP* = atrial natriuretic peptide; *GFR* = glomerular filtration rate.

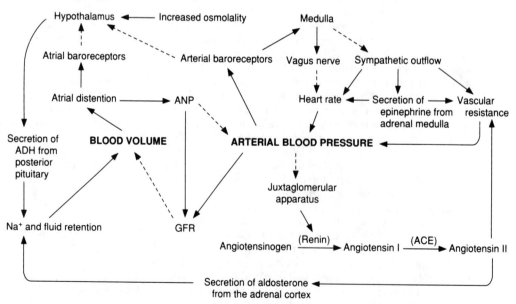

    b. Because the arterial baroreceptors are within the thoracic and cervical vascular systems, they detect changes in central vascular volume but not changes in the volume of the extracellular space. Standing causes fluid to shift from the thorax to the legs, which is interpreted by these receptors as a decrease in the vascular volume.

  2. Atrial (low-pressure, stretch) baroreceptors are located at the venoatrial junctions.

B. **Endocrine component**

  1. ADH. When blood with an increased osmolality perfuses the brain, osmoreceptors located in the anterior hypothalamus detect the increased osmolality, prompting the release of ADH from the posterior pituitary (see Figure 14-1). ADH acts on the kidneys to increase the reabsorption of water from the distal convoluted tubule and especially from the collecting duct. Water retention occurs, diluting the solutes and returning the osmolality of the blood to normal.

  2. The **renin-angiotensin-aldosterone system** (see Figure 14-1) is the primary regulator of electrolytes within the body. A decrease in the mean blood pressure or a decrease in the pulse pressure within the kidney causes the juxtaglomerular cells of the nephron to release the **enzyme renin** into the plasma.

    a. Renin reacts with a plasma protein called **angiotensinogen** to form the decapeptide **angiotensin I. Angiotensin converting enzyme** (ACE), located on the endothelial surface of many organs (particularly the pulmonary vasculature), converts angiotensin I to angiotensin II (an octapeptide).

    b. Angiotensin II is one of the most **potent vasoconstrictive agents** known.

      (1) Angiotensin II acts directly on the arterioles by **constricting vascular smooth muscle**. The vasoconstriction increases vascular resistance, which raises arterial blood pressure toward normal.

      (2) In addition, angiotensin II **causes the release of aldosterone** from the zona glomerulosa of the adrenals.

    c. Aldosterone causes the distal tubule of the kidney to reabsorb electrolytes (primarily Na^+) and obligated water. The increased fluid reexpands circulating blood volume, which also tends to increase blood pressure (see Figure 14-1).

  3. Atrial natriuretic peptide has been found as storage granules in certain myocardial cells of the left and right atria. Atrial natriuretic peptide, along with natriuretic factors from other sources, is released into the circulation when the atria are distended by increases in blood volume. As the blood volume decreases, the atrial distention is reduced and the rate of release is slowed.

    a. Natriuresis. Atrial natriuretic peptide dilates afferent arterioles in the kidney and may relax mesangial cells (increasing the filtration area of the glomerulus). Both actions increase the glomerular filtration rate (GFR), enhancing the excretion of Na^+ and water and lowering blood volume and pressure in the arterial system (see Figure 14-1).

    b. Vasodilation. In addition to being a generalized vasodilator, atrial natriuretic peptide reduces the response to various vasoconstrictors.

III. CONTROL OF ARTERIAL PRESSURE

A. **Baroreceptor response.** Input from the baroreceptors provides information regarding the heart rate, the rate of change in blood pressure (related to ventricular contractility), and pulse pressure to the cardiovascular control centers in the CNS.

  1. The arterial baroreceptors generate a **receptor potential** in response to pressure-induced distention of the vascular walls. The receptor potential has **two distinct components**.

    a. The **dynamic receptor response** is the initial receptor potential, which is proportional to the rate of change in arterial pressure.

    b. The **static receptor response** is proportional to the new steady pressure and continues without adaptation as long as the pressure is maintained.

2. The receptor potential, in turn, produces **action potentials in the afferent nerves**. Afferent neural activity over the sinus and aortic nerves produces the **baroreceptor (moderator) reflex** (i.e., a reflex slowing of the heart via the vagus nerve and a withdrawal of sympathetic tone to the arterioles, which results in vasodilation).
   a. The action potential frequency is **linearly related** to the increase in arterial pressure over a wide range.
   b. The **carotid sinus nerve** exhibits the **greatest sensitivity;** it responds to pressures that vary from approximately 50–160 mm Hg (Figure 14-2).
      (1) **Carotid sinus massage** is used clinically to interrupt paroxysmal atrial tachycardia by inducing a vagally mediated slowing of the heart.
      (2) **Stokes-Adams syndrome.** An increased sensitivity of the carotid sinus is seen in some elderly individuals who experience syncope as a result of vagally mediated sinus arrest, which leads to a prolonged period of ventricular asystole.

B. Central nervous system (CNS)

   1. **Medulla.** Diffuse collections of cells involved in the integration of cardiovascular function are located within the reticular formation of the brain stem.
      a. **Pressor** and **depressor areas** are located (respectively) in the lateral and medial portions of the reticular formation. The **depressor response** results from the **withdrawal of vasoconstrictor influence,** rather than an active vasodilation.
      b. The **cardioaccelerator** and **cardioinhibitory centers** are also located in the reticular formation. These centers send impulses to the heart over the cardiac sympathetic and vagal nerves, respectively.

   2. **Hypothalamus.** The medullary cardiovascular centers receive input from higher levels of the brain.
      a. The effects of **temperature changes** on the hypothalamic centers are relayed to the medulla, which causes the vessels of the skin to constrict (to conserve heat) or to dilate (to dissipate heat).
      b. **Emotional stresses** influence heart rate and blood pressure. These effects are relayed from the higher centers to stimulate or inhibit the medullary centers.
      c. In some animals, but apparently not in humans, the skeletal muscles receive a sympathetic cholinergic innervation as well as the α-adrenergic innervation that normally controls blood flow. Stimulation of these cholinergic fibers causes a vasodilation that occurs in anticipation of exercise. The vasodilation opens the thoroughfare channels in skeletal muscle, but once exercise begins the nutrient channels are dilated by the release of the local vasodilator substances.

   3. **Cortex.** Stimulation of various areas of the motor cortex leads to complex motor responses that include appropriate cardiovascular adjustments. These complex motor patterns involve autonomic responses.

FIGURE 14-2. The effect of altering arterial blood pressure on the activity of the carotid sinus nerve. The increased sinus nerve activity in response to increasing blood pressure causes central reflexes to inhibit vasoconstriction (by withdrawing sympathetic tone to the vessels) and promote bradycardia (by increasing vagal tone).

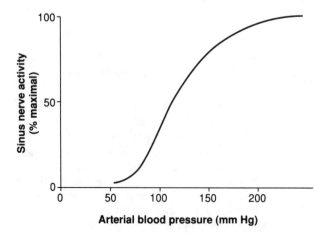

a. With increasing frequency, physicians are training patients in the use of **biofeedback therapy** in attempts to control heart rate and blood pressure. This trend emphasizes the control the higher centers can exert on the autonomic nervous system (ANS).

b. The sympathetic cholinergic outflow to the skeletal muscle arises from cerebral areas in lower animals and follows a separate descending tract from the sympathetic adrenergic outflow.

C. **Autonomic nervous system (ANS)**

1. **Sympathetic nervous system.** Most of the sympathetic nerves to blood vessels exhibit some neural activity that induces approximately 50% of the maximal **vasoconstrictor activity** in most of the vascular beds. Thus, vasoconstriction or vasodilation is produced by increasing or decreasing the sympathetic neural activity.

a. **Innervation.** The organs are supplied by sympathetic fibers at their segmental level.

  (1) Sympathetic **preganglionic fibers** arise from cell bodies in the intermediolateral gray columns of the spinal cord of the thoracodorsal regions. Axons of these cells exit the CNS through the ventral root and proceed to the paravertebral chain of ganglia, where they synapse with the **postganglionic cells** of the sympathetic system.

  (2) The **stellate ganglion** supplies **postganglionic sympathetic nerve fibers** to the heart. Sympathetic postganglionic fibers also innervate some exocrine glands and sweat glands, and are the vasodilator fibers in the skeletal muscle vascular beds of certain species.

  (3) **All** of the **preganglionic sympathetic fibers** are **cholinergic,** but **postganglionic sympathetic fibers** may be **either adrenergic or cholinergic.**

b. **Receptors.** The sympathetic nervous system contains α and β receptors. In the cardiovascular system, the α and β receptors have the following effects:

  (1) α **Receptors** are stimulated most strongly by epinephrine and norepinephrine. Stimulation of the α receptors causes **vasoconstriction.**

  (2) β **Receptors** are strongly stimulated by isoproterenol and less stimulated by norepinephrine. Pharmacologic agents are available that can stimulate or block the actions of both types of β receptors.

    (a) The β_1 **receptors** are located in the **myocardium** (not the coronary arteries); activation of these receptors results in positive inotropic or chronotropic responses.

    (b) The β_2 **receptors** are located in other body tissues. Stimulation of these receptors leads to inhibition of bronchoconstriction or vasoconstriction (i.e., bronchodilation or vasodilation).

  (3) **Denervation hypersensitivity** occurs following interruption of the sympathetic nerve supply to vascular smooth muscle and certain glands. Normally, receptors are located in the subjacent areas of the nerve terminals. Following denervation, however, receptors develop over the entire membrane and the end organs become extremely sensitive to any circulating catecholamines.

2. The **parasympathetic nervous system** rises from the cranial and sacral outflows of the CNS. The primary effect of the parasympathetic system on cardiovascular function is to slow the heart rate. Impulses conducted by the vagus nerve affect the sinoatrial (SA) and atrioventricular (AV) nodes and directly reduce atrial contractility. Marked efferent activity over the vagus nerve can result in extensive slowing of the heart, sinus arrest, or AV nodal block, or a combination of the three.

Chapter 15

Cardiovascular Responses to Stress

I. **EXERCISE.** The effects of exercise on the cardiovascular system are discussed in tandem with the effects of exercise on the respiratory system (see Chapter 20 I).

II. **SHOCK** is a condition characterized by inadequate delivery of O_2 and nutrients to critical organs [e.g., the heart, brain, liver, kidneys, gastrointestinal (GI) tract].

A. **Types of shock**

1. **Low resistance shock** occurs when neural reflexes or toxic substances cause excessive vasodilation within the vascular system.
   a. **Primary shock (syncope, fainting, loss of consciousness)** is usually caused by a reflex that produces arteriolar and venous dilation, which causes peripheral pooling of blood or cardiac slowing.
      (1) Without an adequate venous return, the cardiac output declines, arterial blood pressure decreases, and **blood flow to the brain decreases**.
      (2) Fainting is most commonly provoked by **pain, stress, fright, or heat**.
   b. **Anaphylactic shock** occurs when a generalized antigen–antibody reaction takes place in patients who are sensitized to certain antigens, resulting in severe vasodilation.
   c. **Other causes of low resistance shock** include endocrine failure (e.g., as in Addison's disease or myxedema), severe hypoxia, drug overdose, and trauma to the nervous system.

2. **Hypovolemic shock** is caused by a low blood volume and may result from a number of disorders.
   a. **Septic shock** occurs when infections in the bloodstream (especially gram-negative bacterial infections) or bacterial toxins released from an infected site cause hypotension. Gram-negative sepsis increases the permeability of the microvasculature, facilitating the movement of fluid into the tissues and intensifying the hypotension.
   b. **Hemorrhagic shock** occurs as a result of external or internal blood loss caused by ruptured vessels (e.g., from trauma). The typical sequence of changes that occurs during acute blood loss is summarized in Figure 14-1.
   c. **Dehydration shock.** Fluid loss from the GI tract (diarrhea or vomiting), kidneys (diabetes mellitus, diabetes insipidus, or excessive use of diuretics), or skin (burns, sweating, or exudation) can dehydrate the body and reduce the circulating blood volume.

3. **Cardiogenic shock** results when **cardiac output is markedly decreased**.
   a. **Inadequate cardiac function** (e.g., from infarction, myopathy, or mechanical disorders) can lead to cardiogenic shock.
   b. An **abnormally slow or fast ventricular rate** may cause cardiogenic shock.

4. **Obstructive shock** occurs when cardiac output is reduced as a result of a **mechanical impediment to left or right ventricular filling**. Massive pulmonary emboli, tension pneumothorax, positive end-expiratory pressure (PEEP) respiration, and cardiac tamponade can all lead to obstructive shock.

B. **Symptoms of shock** initially are agitation and restlessness, which later progress to lethargy, confusion, and coma.

C. **Physical signs of shock** depend on the stage of shock and the severity of the dysfunction.

1. The **initial stage** of shock is characterized by a moderate reduction in cardiac output secondary to venous pooling, fluid loss, or a negative inotropic effect on the heart.

   a. Generally, **compensatory reflexes** maintain the blood pressure at normal or low-normal levels.

   b. These reflex adjustments result in the following:

   (1) An **increased heart rate**

   (2) An **increase in sympathetic discharge,** as evidenced by decreased blood flow to the skin, which produces **pallor** and a **"cold sweat"**

   (3) A **marked increase in peripheral resistance,** in an attempt to maintain blood pressure. The increased resistance is **selective** in that the renal, skeletal muscle, and visceral beds undergo vasoconstriction, but the blood flow to essential organs (e.g., the brain and heart) is maintained.

2. The **second stage** of shock occurs after a loss of blood volume of 15%–25%. It is characterized by **intense arteriolar vasoconstriction**. Because of the moderately reduced cardiac output, the arteriolar vasoconstriction is usually not adequate for maintaining a normal blood pressure.

3. The **third stage** of shock occurs when small additional losses of blood or cardiac function produce a rapid deterioration of the circulation with inadequate blood flow to critical body organs.

   a. If this state persists, the inadequate blood flow causes **widespread tissue injury** (i.e., from hypoxia).

   b. The third stage of shock rapidly progresses to what is termed **irreversible shock,** which is **fatal**.

D. **Treatment of shock** requires establishing an adequate blood flow to all bodily tissues.

1. **Approaches** include raising the circulating blood volume (if this is depleted), reestablishing normal rhythm to the heart, increasing the inotropic state of the ventricles with digitalis or β-sympathetic agonists, or administering vasodilators.

2. **Evaluation.** The **pulmonary wedge pressure** is the most useful tool for evaluating the extracellular fluid (ECF) volume and the function of the left ventricle.

   a. **Method.** The pulmonary wedge pressure is obtained through a Swan-Ganz catheter that is inserted into a systemic vein and advanced through the right ventricle.

   (1) The catheter is wedged into a small pulmonary vessel; the pulmonary wedge pressure represents the back-pressure from the left atrium.

   (2) The Swan-Ganz catheter can also be used to measure the cardiac output by the thermal dilution method (see Chapter 13 II B 2 b).

   b. **Interpretation.** The wedge pressure provides an estimate of the filling pressure of the left ventricle. A low pressure indicates a depletion of the ECF; excessive pressures are associated with large fluid volumes, inadequate function of the left ventricle, or both.

III. **HYPERTENSION** refers to an **increase in arterial pressure;** the increased afterload causes damage to the heart. Damage to other organs occurs because of the added stress to the blood vessels.

A. **Pulmonary artery hypertension** occurs when **pulmonary vascular resistance increases**.

1. **Etiology.** Exposure to conditions associated with alveolar hypoxia (e.g., high altitudes or hypoventilation) or disease processes that destroy portions of the pulmonary vascular bed can cause pulmonary hypertension.

2. **Idiopathic pulmonary hypertension** is pulmonary artery hypertension from an unknown cause.

B. **Systemic hypertension** is generally diagnosed when the systolic pressure exceeds 140 mm Hg and the diastolic pressure exceeds 90 mm Hg; however, a diastolic pressure greater than 85 mm Hg may prove to be a viable threshold for diagnosis of systemic hypertension—evidence indicates that morbidity and mortality rates are increased in these patients.

1. **Etiology of hypertension**
   a. **Essential hypertension** comprises more than 90% of all cases of hypertension. The **cause** of essential hypertension is **unknown**. Excessive salt intake, resetting of the baroreceptors, increased salt and fluid in the arteriolar walls, and long-standing exposure to stress are some of the theories that have been proposed to explain the increased systemic arterial pressures.
   b. **Systolic hypertension** with a wide pulse pressure can be caused by various malformations or conditions that decrease the compliance of the aorta (e.g., aging) or increase the cardiac output or stroke volume of the left ventricle. These latter conditions include thyrotoxicosis, patent ductus arteriosus, arteriovenous fistula, fever, and aortic valvular insufficiency.
   c. **Secondary (systolic and diastolic) hypertension** can arise from a variety of causes.
      (1) **Renal disease,** if severe, results in systemic hypertension because of the release of renin and the formation of angiotensin II. Destruction of the kidney or narrowing of the renal arteries can result in hypertension.
      (2) **Endocrine diseases** such as Cushing's syndrome, primary hyperaldosteronism, pheochromocytoma, and acromegaly may result in hypertension.
      (3) **Miscellaneous causes** of hypertension include psychogenic causes, coarctation of the aorta (important to diagnose because it can be cured surgically), polyarteritis nodosa, and excessive transfusion.

2. **Symptoms of hypertension** are generally absent until the disease is far progressed and the hypertension has resulted in cardiac, renal, neural, or ocular damage. The disease is usually diagnosed during a routine physical examination. **Malignant hypertension** occurs when arterial pressure is markedly increased and damage occurs over weeks or months instead of years.
   a. **Cardiac symptoms** are caused by the increased work required of the left ventricle. The ventricle initially hypertrophies and (if the hypertension is severe) eventually fails.
      (1) **Failure of the left ventricle** results in pulmonary congestion and pulmonary edema. Both of these conditions can produce **dyspnea** (i.e., difficulty breathing).
      (2) Hypertension also increases the rate of **arteriosclerosis** (i.e., "hardening of the arteries"), which reduces blood flow to critical vessels in the heart. The reduced blood flow can cause **angina** or a heart attack.
   b. **Neural symptoms** include **occipital headaches** (especially in the morning), **dizziness, ringing in the ears, dimmed vision,** and **syncope.** Permanent brain damage can result from complete blockage of a blood vessel by arteriosclerosis or by hemorrhage secondary to the stress imposed on the vessels by the hypertension.
   c. **Renal effects** result from atherosclerotic lesions that affect the afferent and efferent arterioles and the glomerular tufts. These lesions decrease the glomerular filtration rate (GFR) and lead to **renal tubular dysfunction.** These pathological alterations eventually cause renal failure.

3. **Treatment of hypertension** is critically important to prevent or minimize damage to cerebral, ocular, renal, and cardiac tissues.
   a. **Treatment of essential hypertension** relies on prescribing diuretics, β-adrenergic blocking drugs, Ca^{2+}-channel blockers, or an angiotensin converting enzyme (ACE) inhibitor to lower the arterial blood pressure. The use of ACE inhibitors has proven very effective because these agents block the formation of angiotensin II and prevent the breakdown of bradykinin, a potent vasodilator.
   b. **Treatment of secondary hypertension** depends on treating the underlying disease.

■ STUDY QUESTIONS

DIRECTIONS: Each of the numbered items or incomplete statements in this section is followed by answers or by completions of the statement. Select the ONE lettered answer or completion that is BEST in each case.

1. In a recumbent person, the greatest difference in blood pressure would exist between the

(A) ascending aorta and brachial artery
(B) saphenous vein and right atrium
(C) femoral artery and femoral vein
(D) pulmonary artery and left atrium
(E) arteriolar and venous ends of a capillary

2. A patient with coronary artery disease undergoes coronary arteriography, which reveals a 50% decrease in the lumen diameter of the left anterior descending coronary artery. For any arteriovenous pressure gradient, the flow through this artery (compared with normal) will decrease by a factor of

(A) 2
(B) 4
(C) 8
(D) 12
(E) 16

3. Ventricular end-diastolic pressure (VEDP) can be used instead of the usual variable on the abscissa (x-axis) when plotting a Frank-Starling curve. What variable does the x-axis usually represent on a Frank-Starling curve?

(A) Cardiac output
(B) Ventricular stroke volume
(C) Cardiac work
(D) Ventricular contractility
(E) Ventricular end-diastolic volume (VEDV)

4. Which of the following events is specifically seen on an electrocardiogram (EKG)?

(A) Sinoatrial (SA) node depolarization
(B) Atrioventricular (AV) node depolarization
(C) Bundle of His depolarization
(D) Bachmann's bundle depolarization
(E) Atrial muscle depolarization

5. The greatest resting arteriovenous difference in O_2 content is found in the

(A) liver
(B) skeletal muscle
(C) heart
(D) kidney
(E) lung

6. Stimulation of the high-pressure baroreceptors is associated with

(A) an increase in cardiac contractility
(B) an increase in heart rate
(C) an increase in the discharge rate of vagal efferent cardiac neurons
(D) a decrease in systemic blood pressure
(E) stimulation of the vasopressor center

7. The rate of lymph flow in humans is approximately

(A) 10–20 L/day
(B) 100–200 ml/day
(C) 10–20 ml/hr
(D) 1–2 L/day
(E) 1–2 ml/hr

8. An increase in systemic blood pressure leads to which one of the following effects?

(A) An increase in the velocity at which blood is ejected from the left ventricle
(B) An increase in cardiac output
(C) An increase in the residual volume of blood in the left ventricle
(D) A decrease in the time it takes for the left ventricular wall to develop peak tension
(E) A decrease in the maximal wall tension developed in the left ventricular muscle

9. Distribution of blood flow is regulated mainly by the

(A) capillaries
(B) arterioles
(C) venules
(D) arteriovenous anastomoses
(E) postcapillary venules

10. A decrease in heart rate (while stroke volume and peripheral resistance remain constant) will cause an increase in

(A) arterial diastolic pressure
(B) arterial systolic pressure
(C) cardiac output
(D) arterial pulse pressure
(E) mean arterial pressure

11. The time from the upstroke of the carotid artery pulse to the incisura (dicrotic notch) is a measure of the period of

(A) atrial diastole
(B) ventricular ejection
(C) reduced ventricular filling
(D) rapid ventricular filling
(E) ventricular isovolumic relaxation

12. An electrocardiogram (EKG) reveals no P waves in any lead. This would indicate that impulses from which one of the following structures are being blocked?

(A) Sinoatrial (SA) node
(B) Bundle of His
(C) Purkinje fibers
(D) Left bundle branch
(E) Ventricular muscle

13. The pulmonic valve normally closes after the aortic valve because the

(A) diameter of the pulmonary artery is less than that of the aorta
(B) right ventricular contraction begins after left ventricular contraction
(C) velocity of ejection in the right ventricle is less than that in the left ventricle
(D) diastolic pressure in the pulmonary artery is less than that in the aorta
(E) leaflets of the pulmonic valve are stiffer and harder to close, compared with those of the aortic valve

14. Which one of the following mechanisms is most important for maintaining an increased blood flow to skeletal muscle during exercise?

(A) An increase in aortic pressure
(B) An increase in α-adrenergic impulses
(C) An increase in β-adrenergic impulses
(D) Vasoconstriction in the splanchnic and renal areas
(E) Vasodilation secondary to the effect of local metabolites

15. During diastole, blood flow into the ventricles sometimes produces

(A) a first heart sound (S_1)
(B) a second heart sound (S_2)
(C) a third heart sound (S_3)
(D) an ejection click
(E) an ejection-type murmur

16. Increasing the preload of cardiac muscle will

(A) reduce the ventricular end-diastolic pressure (VEDP)
(B) reduce the peak tension of the muscle
(C) decrease the initial velocity of shortening
(D) decrease the time it takes the muscle to reach peak tension
(E) increase the ventricular wall tension

DIRECTIONS: Each of the numbered items or incomplete statements in this section is negatively phrased, as indicated by a capitalized word such as NOT, LEAST, or EXCEPT. Select the ONE lettered answer or completion that is BEST in each case.

17. Venous return is enhanced during exercise by all of the following factors EXCEPT

(A) increased depth of respiration
(B) pumping action of skeletal muscles
(C) venoconstriction
(D) reduced arteriolar resistance
(E) an erect position

18. The increased circulating fluid volume in chronic congestive heart failure results from all of the following factors EXCEPT

(A) increased sympathetic discharge to the kidney
(B) decreased rate of firing of atrial volume receptors
(C) decreased renal perfusion
(D) aldosterone activity
(E) stimulation of the arterial baroreceptors

DIRECTIONS: Each set of matching questions in this section consists of a list of four to twenty-six lettered options (some of which may be in figures) followed by several numbered items. For each numbered item, select the ONE lettered option that is most closely associated with it. To avoid spending too much time on matching sets with large numbers of options, it is generally advisable to begin each set by reading the list of options. Then, for each item in the set, try to generate the correct answer and locate it in the option list, rather than evaluating each option individually. Each lettered option may be selected once, more than once, or not at all.

Questions 19–21

For each event, select the mechanism that is associated with it.

(A) Increase in contractility
(B) Increase in fiber length
(C) Both
(D) Neither

19. An increase in the maximal velocity of shortening (V_{max})

20. A shift to a new Frank-Starling curve

21. An increase in intracellular Ca^{2+}

Questions 22–26

For each condition, select the most appropriate cardiovascular response.

(A) Increased ventricular end-diastolic pressure (VEDP)
(B) Increased aortic diastolic pressure
(C) Both
(D) Neither

22. Myocardial tissue damage

23. Increased arteriolar resistance

24. Decreased ventricular contractility

25. Increased heart rate

26. Increased central venous pressure

Questions 27–30

Match the cardiac event with the interval of the cardiac cycle in which it occurs.

(A) Atrial contraction
(B) Isovolumic contraction
(C) Rapid ventricular ejection
(D) Reduced ventricular ejection
(E) Isovolumic relaxation

27. Second heart sound (S_2)

28. Achievement of maximal ventricular volume

29. Closure of the atrioventricular (AV) valves

30. Opening of the aortic valve

Questions 31–35

Match each cardiovascular adjustment with the factor that is responsible for it.

(A) Functional hyperemia
(B) Histamine
(C) Hypertension
(D) CO_2 tension (P_{CO_2})
(E) Capillary pressure

31. Regulation of capillary filtration rate

32. Elevation of arterial diastolic pressure

33. Metabolic regulation of blood flow

34. Regulation of cerebral blood flow

35. Increase in microvascular permeability

■ ANSWERS AND EXPLANATIONS

1. The answer is C [Figure 9-3; Chapter 12 I C 1 b]. In a recumbent person, the greatest difference in blood pressure would exist between the femoral artery and vein, because the largest pressure drop in the vascular system (40–50 mm Hg) occurs as blood passes through the systemic arterioles (e.g., the region between the femoral artery and femoral vein). Because the large arteries provide little resistance to flow, there is practically no pressure loss from the aorta to any of the distributing arteries. The pressure loss across the venous system amounts to approximately 10 mm Hg, because the pressure in the large veins may be 8–10 mm Hg, while the pressure in the right atrium is near zero. The atrium is distended because the pressure outside the atrium (i.e., the intrathoracic pressure) averages approximately -5 mm Hg, which means that the pressure across the wall of the atrium (i.e., the transmural pressure) is 5 mm Hg. The pulmonary circulation is a low-resistance circuit, and the gradient across this vascular segment also is only approximately 8–10 mm Hg. The pressure at the arteriolar end of a capillary averages approximately 40 mm Hg, whereas the pressure at the venous end is 25–30 mm Hg. Thus, even the smallest vessels in the body, which should have the highest resistance, exhibit only a 10–15 mm Hg pressure gradient. This seeming paradox occurs because of the parallel circuit arrangement of the tremendous number of capillaries. The total resistance to flow in a parallel circuit is the sum of the reciprocals of the resistance in each vessel, so that the greater the number of vessels, the lower the total resistance.

2. The answer is E [Chapter 9 III B 2]. The effect of vessel radius on flow is so powerful because the flow rate is directly proportional to the fourth power of the radius (Poiseuille's law). Thus, decreasing the radius to half reduces the flow to one-sixteenth of the normal value.

3. The answer is E [Chapter 11 III A 1]. The x-axis usually represents the independent variable, so that, for the Frank-Starling mechanism, this would be the factor that defines the length of the sarcomere. In the intact heart, the sarcomere varies in length as a function of the ventricular end-diastolic volume (VEDV). Assuming that there is no change in compliance of the ventricular muscle, the sarcomere will also vary as a function of the ventricular end-

diastolic pressure (VEDP). The factor plotted on the y-axis of such a graph represents the output of the heart, which can include functions such as cardiac output, stroke volume, and cardiac work (i.e., pressure times volume).

4. The answer is E [Chapter 10 IV C 1 a]. Atrial muscle depolarization is seen as the P wave on a standard electrocardiogram (EKG). An EKG does not reveal depolarization of the sinoatrial (SA) node, atrioventricular (AV) node, bundle of His, or Bachmann's bundle, because the mass of tissue involved is too small to cause a deflection on the surface electrograms that comprise the EKG. Depolarization of the bundle of His can be recorded from electrode catheters placed in the esophagus. Esophageal catheterization is frequently used to characterize supraventricular arrhythmias, which cannot be diagnosed using the standard EKG.

5. The answer is C [Chapter 12 II A 2 c (1)]. The greatest resting arteriovenous difference in O_2 content is found in the heart. Under resting conditions, the heart removes approximately half of the O_2 from the blood, which results in an O_2 tension of approximately 27 mm Hg. In contrast, blood flow through other tissues is much greater in proportion to the metabolic rate and results in an arteriovenous O_2 difference of approximately 5 ml/dl blood. The mixed venous O_2 content, which represents the weighted average of blood flow to all parts of the body, is 15 ml/dl blood, which represents a hemoglobin saturation of 75% and an O_2 tension of 40 mm Hg.

6. The answer is C [Chapter 14 II A 1 a; Figure 14-1].The high-pressure baroreceptors are stimulated by an increased arterial blood pressure. The afferent impulses are carried over the glossopharyngeal and vagus nerves, where they stimulate the cardioinhibitory and depressor centers. One portion of the efferent limb of the reflex causes an increased number of vagal impulses to the heart, which results in cardiac slowing leading to a decrease in cardiac output. This depressor response causes a decrease of sympathetic tone to the vascular system, resulting in decreased peripheral resistance and a pooling of blood in the venous system. Both of these peripheral effects also lead to a lowering of the arterial blood pressure.

7. The answer is D [Chapter 12 I D 2 c]. The rate of lymph flow is approximately 1–2 L/day. Although this rate seems high, it is quite low compared with the flow through the entire vascular system. The rate of blood flow is 5–6 L/min, or approximately 8000 L/day. Thus, approximately 99.98% of the blood flow returns to the heart via the veins, leaving only 0.02% of the blood flow to be returned to the heart via the lymphatics.

8. The answer is C [Chapter 11 IV A; Figure 11-7B]. An increase in the systemic pressure (i.e., the afterload that the ventricle must overcome) reduces the velocity of shortening of the contractile elements in the muscle. An increase in afterload also reduces the extent of muscle shortening, which results in a decreased stroke volume and, consequently, an increased residual volume (i.e., the volume of blood left in the ventricles at the end of ejection). The time that it takes for the left ventricular wall to develop peak tension is not reduced, because the velocity of contraction would be reduced, and the peak tension would be increased. Wall tension is a function of the product of ventricular pressure and volume. Because pressure and volume are increased, wall tension also is increased in the presence of an increased afterload.

9. The answer is B [Chapter 12 I B 2 a]. The arterioles often are referred to as the stopcocks of the circulation because they act as valves to restrict or enhance blood flow to different tissues and organs. Small changes in the diameter of these vessels can markedly alter the resistance to flow because resistance varies as the fourth power of the radius in the arterioles (as in any other vessel). However, the arterioles have a thick muscular wall compared with the lumen, so that a relatively small amount of smooth muscle contraction can cause a marked change in lumen size.

10. The answer is D [Chapter 9 IV B 3 b (1)]. The arterial pulse pressure increases when the heart rate decreases but the stroke volume and peripheral resistance remain constant. A constant stroke volume coupled with a decreased heart rate produces a decrease in cardiac output. The reduced flow, in tandem with the constant peripheral resistance, means that the mean arterial pressure must decline. The fact that the pulse pressure increases is obtained

from an analysis based on the vascular elastic modulus:

$$E = V \cdot dP/dV, \text{ where}$$
E = the elastic modulus
V = the arterial volume (which is determined by the arterial pressure)
dP = the pulse pressure
dV = the arterial uptake

When the cardiac output is reduced, arterial pressure, and therefore arterial volume (V), decrease. The arterial uptake (dV) remains constant because it is primarily related to the stroke volume. To maintain the elastic modulus, which varies only as a function of age, as a constant, the pulse pressure (dP) must increase to counteract the decreased arterial volume (V).

11. The answer is B (Chapter 11 II A 3 b (3), B 1 b]. The upstroke of the carotid artery pulse and the incisura (dicrotic notch) indicate the beginning and end, respectively, of ventricular ejection. The dicrotic notch is caused by the closure of the semilunar valve and coincides with the second heart sound (S_2). Atrial diastole represents the entire cardiac cycle, except for the period of atrial systole that is concomitant with the a wave in the atrial pressure pulse. The periods of ventricular filling represent diastolic events that can be determined from an atrial pressure pulse or the jugular volume pulse. The period of isovolumic relaxation occurs from the incisura, or S_2, to the peak of the atrial v wave, which represents the opening of the atrioventricular (AV) valves.

12. The answer is A [Chapter 10 III A; IV C 1 a]. The sinoatrial (SA) node is the normal pacemaker of the heart. The depolarization of the SA node cannot be seen on the electrocardiogram (EKG) because of the small mass of tissue involved, but normally, SA node depolarization spreads to the atrial muscle. The depolarization of the atrial muscle is seen on the EKG as the P wave. The bundle of His, Purkinje fibers, left bundle branch, and ventricular muscle lie below the atria and normally do not cause depolarization of the atria.

13. The answer is C [Chapter 11 II A 3 b (5) (a) (i)]. The onset of right ventricular ejection precedes ejection in the left ventricle and right ventricular ejection continues after left ventricular ejection ends; therefore, the aortic valve must close first. The major components of the second heart sound (S_2) are the A_2 (produced by the closure of the aortic valve) and the P_2 (produced by the closure of the pulmonic valve). During inspiration, the P_2 is delayed and the A_2 occurs slightly earlier (i.e., the interval between the closure of the aortic and pulmonic valves is lengthened). This sequence can be appreciated during auscultation of the heart and is referred to as normal splitting of the S_2. If there is a delay in left ventricular activation, as in left bundle branch block, the A_2 may follow the P_2 and, during inspiration, the interval between these events is shortened. This is termed paradoxical splitting of the S_2.

14. The answer is E [Chapter 12 II C 2 b (1)]. Vasodilation in active skeletal muscles, prompted by the release of metabolic products from the muscle tissue, reduces the resistance to flow and results in an increased blood flow to the involved tissues. The increase in blood flow is dependent on maintaining the mean arterial blood pressure, which normally rises only slightly with exercise. Thus, the slight rise in arterial pressure coupled with a significant rise in blood flow indicates that the total vascular resistance must decline during exercise. The decrease in vascular resistance occurs because skeletal muscle represents approximately 50% of total body weight in the normal adult. Therefore, a decrease in this tissue can more than compensate for the increased resistance in tissues such as the gastrointestinal (GI) tract, kidneys, and skin during periods of mild to moderate exercise.

15. The answer is C [Chapter 11 II A 3 b (1) (b), (5), 4 b]. The third heart sound (S_3) occurs during middiastole when the ventricular wall becomes tense toward the end of the rapid filling phase. An S_3 is not normally heard in adults but can be heard in children with thin chest walls. The first and second heart sounds (S_1 and S_2) indicate the beginning and end of systole, respectively. An ejection click and an ejection-type murmur are systolic sounds that are caused by blood leaving the ventricle.

16. The answer is E [Chapter 11 III A 1 a, B 2 b (1)]. Increasing the cardiac muscle preload increases the ventricular wall tension. An increase in both end-diastolic pressure and volume (VEDP and VEDV, respectively) is synonymous with an increased preload. The increased preload causes a more forceful ventricular contraction, which results in an in-

crease in the peak pressure generated by the ventricle as well as an increase in the stroke volume. This intrinsic property of the myocardium is termed the Frank-Starling mechanism. The increased preload results in an increased velocity of shortening, but because the peak pressure increases, the time to peak pressure usually remains constant under these conditions.

17. The answer is E [Chapter 9 III A 2 b (1); Chapter 12 I E 2 b (1)]. Standing erect exposes a column of blood to the effects of gravity, which raises the pressure in the dependent veins and diminishes venous return (compared with that in the supine position). Because the venous system is very compliant, the increase in pressure causes an increase in venous volume, which reduces venous return. Venoconstriction, which is caused by sympathetic outflow during exercise, partially compensates for the peripheral pooling caused by the effects of gravity. An increased depth of respiration causes a more negative pressure in the thorax, which increases the pressure gradient along the venous system and improves venous return. Contraction of the skeletal muscles compresses the veins and displaces the blood toward the heart; this effect is enhanced if the venous valves are competent and prevent a backflow of blood between contractions. A reduced arteriolar resistance in the skeletal muscles raises capillary and venous pressures and increases the flow rate through the systemic circulation.

18. The answer is E [Chapter 13 I C 2; Chapter 14 III A]. The arterial baroreceptors are stimulated by an increase in arterial pressure, not a decrease, as is likely in cardiac failure. In congestive heart failure, a generalized increase in sympathetic discharge and reduced renal perfusion cause the juxtaglomerular cells of the kidney to release renin. Renin, in turn, causes the adrenal glands to release aldosterone, which increases the fluid volume. There is evidence that the atrial volume receptors "accommodate" (i.e., their firing rate slows) when they are exposed for prolonged periods of time to vascular distention.

19–21. The answers are 19-A [Chapter 11 III B 1 c], **20-A** [Chapter 11 III A, B 1 b], **21-A** [Chapter 11 III B 1 a (1)]. An increased maximal velocity of shortening (V_{max}) is one definition of increased contractility, or positive inotropism. Increased contractility may also be defined by an increase in stroke volume at a constant ventricular end-diastolic pressure or volume (VEDP or VEDV).

Shifting from one Frank-Starling curve to another implies a change in contractility. A new curve above and to the left of the control curve indicates an increased contractility. Changes along any one curve indicate a change in ventricular fiber length (e.g., as a result of changing ventricular volume or pressure). A change in ventricular fiber length invokes the Frank-Starling mechanism of altering ventricular output (i.e., either by increasing cardiac work or stroke volume).

Positive inotropism is associated with an increase in the intracellular Ca^{2+} concentration. The increased Ca^{2+} allows saturation of a greater number of troponin molecules, which allows for the formation of more cross-bridges during the contraction process. The increased number of cross-bridges increases the force of contraction and the velocity of contractile element shortening at any load compared with the control condition.

22–26. The answers are: 22-A [Chapter 11 III B 2 a (1)], **23-C** [Chapter 9 IV B 1 b, 3 b; Figure 11-7], **24-A** [Chapter 11 III B 2], **25-B** [Chapter 11 II A 4 a (3) (b) (ii)], **26-C** [Chapter 9 III B 2 a (2)]. Myocardial tissue damage causes ventricular force development to decline, leading to a reduction in stroke volume. Transiently, the volume of blood pumped by the ventricle is less than the venous return, so the ventricular end-diastolic pressure (VEDP) and volume (VEDV) increase. The increase in VEDV increases the preload so that the remaining muscle generates a greater tension than before, as long as the ventricular pressure remains on the ascending limb of the Frank-Starling curve. The Frank-Starling compensatory mechanism is limited, however, because the ventricular pressure eventually exceeds the peak of the Frank-Starling curve, at which point there is a decline in output as the preload increases further.

Increased arteriolar resistance increases arterial (aortic) pressure, which, in turn, increases the VEDP. If the heart rate and stroke volume remain constant, the pulse pressure must decline in accordance with the principles determining arterial pressure. The increased aortic diastolic pressure represents an increase in the ventricular afterload (which slightly reduces stroke volume, further decreasing the pulse pressure).

Decreased ventricular contractility results in a reduction in cardiac output. As cardiac output decreases, the added inflow into the ventricle causes venous pressure to rise, increasing the VEDP and VEDV. As long as the ventricular pressure remains on the ascending limb of the Frank-Starling curve, cardiac output will be restored to some extent. However, the larger ventricular radius requires the myocardium to generate a higher wall tension in accordance with the Laplace equation. The one benefit of ventricular dilation is that the increased radius allows the myocardium to produce a given stroke volume with a smaller amount of myocardial shortening.

An increased heart rate increases cardiac output and reduces VEDP, because both ventricular filling time and central venous pressure are reduced. An increase in cardiac output as the primary event decreases central venous pressure and increases aortic diastolic pressure. Because the heart pumps blood from the venous to the arterial side of the circulation, an increased cardiac output means that an additional volume of blood is transferred from the venous to the arterial system, which raises the pressure in the arteries and lowers the pressure in the veins.

An increase in central venous pressure raises the gradient for ventricular filling, which increases the VEDP and VEDV. This represents a Frank-Starling mechanism that increases stroke volume, and consequently, cardiac output. As a result of the increased cardiac output, aortic diastolic and systolic pressures increase in accordance with the relationship between pressure (P) flow (Q), and resistance (TPR): $\Delta P = Q \cdot TPR$.

27–30. The answers are: 27-E [Chapter 11 II A 3 b (5), 4 a (1)], **28-A** [Chapter 11 II A 2 b], **29-B** [Chapter 11 II A 3 b (2)], **30-C** [Chapter 11 II A 3 b (3)]. The second heart sound (S_2) signals the beginning of ventricular relaxation (diastole), the first stage of which is isovolumic relaxation. The S_2 occurs when the semilunar valves close, stopping the backward flow of blood in the aorta and producing vibrations in the tissues.

Contraction of the atria, late in diastole, increases ventricular volume so that it reaches a maximum. Ventricular filling begins as soon as the atrioventricular (AV) valves open in diastole and depends on the rate of venous return. The increased filling caused by atrial contraction is particularly important at fast heart rates, because the time for ventricular filling is abbreviated as a result of the shortened diastolic interval. Closure of the AV valves isolates the left ventricular chamber, because the aortic valves are closed also as a result of the diastolic aortic pressure. Because both inflow and outflow valves are closed, there is no change in ventricular volume for a time; hence, the interval is called isovolumic (equal or constant volume) contraction.

Opening of the aortic valve indicates the onset of rapid ventricular ejection and occurs when the pressure in the left ventricle exceeds the pressure in the aorta, and blood begins to flow out of the ventricle. Because the rate of ventricular contraction is initially high, the blood leaves the ventricle rapidly during this interval, causing aortic pressure to increase rapidly in association with the pressure in the ventricle.

31–35. The answers are: 31-E [Chapter 12 I C 3 b], **32-C** [Chapter 15 III B], **33-A** [Chapter 12 II C 2 b], **34-D** [Chapter 12 II B 2 b], **35-B** [Chapter 12 I C 2 a (3)]. Of the factors listed, capillary pressure is most closely associated with the capillary filtration rate. According to the Starling hypothesis, the capillary pressure and interstitial fluid oncotic pressure are direct determinants of the capillary filtration rate. Capillary pressure is counteracted by the oncotic pressure of the blood and the tissue pressure.

An elevated arterial diastolic pressure reflects increased resistance in the systemic circulation and therefore is a more important indicator of hypertension (or the mean arterial pressure) than is systolic pressure, which is largely determined by the stroke volume. High arterial diastolic pressure leads to vascular changes in certain organs. The kidneys and eyes, especially, are characteristically affected by arterial hypertension.

An increased blood flow to skeletal muscles (i.e., a functional hyperemia) occurs during exercise in response to metabolic factors. The increased flow results from a local decrease in vascular resistance brought about by an increase in tissue temperature, osmolality, pH and CO_2 tension (P_{CO_2}), and a decrease in the local O_2 tension (P_{O_2}). None of these factors alone seems to be able to duplicate the effects of exercise.

P_{CO_2} is the most important factor that regulates resistance in the cerebral vessels. Hyperventilation, which reduces the P_{CO_2}, produces a marked vasoconstriction of cerebral vessels that, at times, can lead to hypoxia sufficiently severe to cause dizziness and even fainting.

Histamine, released from mast cells in areas of inflammation, results in an increased permeability of microvessels, primarily the postcapillary venules.

PART IV

RESPIRATORY PHYSIOLOGY
Joseph Boyle, III

Chapter 16

Respiratory Mechanics

I. INTRODUCTION

A. Functions of the respiratory system

1. The **primary role** of the respiratory system is to maintain a constant internal environment by providing O_2 for metabolic needs and excreting CO_2.
   a. **External respiration** involves the exchange of gas between the environment and the lungs, the transfer of gas across the respiratory membrane, and the transport of gas by the blood to and from the body cells. This chapter deals primarily with the process of external respiration.
   b. **Internal respiration** is concerned with intracellular oxygen utilization through metabolic transformations and is generally considered the province of biochemistry.

2. **Secondary roles** of the respiratory system include:
   a. Aiding in acid–base balance
   b. Defending the body against inhaled particles (e.g., bacteria and pollen)
   c. Acting as a filter to prevent clots from entering the systemic circulation
   d. Regulating various hormonal and humoral concentrations by means of the pulmonary capillary endothelium

B. The **study of respiratory mechanics** is the study of factors that determine lung volume. These factors include the forces generated by the respiratory muscles to inflate and deflate the respiratory system, the forces that impede volume changes (i.e., resistance and elastance), and the determinants of lung volume and its distribution within the lungs.

II. VOLUME CHANGES

A. **Pleural space.** The **visceral** and **parietal pleura** are continuous membranes that, together, line the **pleural sac**.

1. The **visceral pleura** covers the **outer surface of the lungs**.

2. The **parietal pleura** lines the **inner surface of the chest wall** and **diaphragm**.

3. The **pleural sac** is only a potential space, because normally the two pleural membranes abut each other and enclose only a small amount of pleural fluid.
   a. The **pleural fluid** acts as a lubricant and causes the visceral and parietal pleura to adhere to one another, much in the same way water behaves between two glass slides: The slides can readily move back and forth but are difficult to separate. Thus, **any movement of the chest wall causes similar changes in lung volume.**

b. Normally, no gas exists in the pleural space, but gas can enter the pleural space from a rupture of the lung or a penetrating wound of the chest wall. The presence of gas in the pleural space is termed a **pneumothorax** (see Chapter 20 VI).

B. **Muscles influencing gas flow.** The lungs are passive structures that follow movements of the chest wall. Because the respiratory system always returns to its equilibrium (resting) position when the muscles are relaxed and the airways are open, a force must be applied to the respiratory system to increase or decrease lung volume from the equilibrium position.

1. Inspiration is an **increase in lung volume** accomplished by applying an expansion force to the respiratory system. Normally, **inspiratory muscles** provide the necessary force (negative-pressure breathing).

a. Inspiratory muscles

(1) The **respiratory diaphragm,** the major muscle of inspiration, is a dome-shaped sheet of muscle that separates the thoracic and abdominal cavities.

(a) The diaphragm acts similarly to a piston or a syringe: Contraction of the muscle causes the dome of the diaphragm to descend, enlarging the volume of the thoracic cavity inferiorly (Figure 16-1).

(b) The dome of the diaphragm is supported by the abdominal organs; therefore, diaphragmatic contraction also lifts the lower ribs because the diaphragm originates from these structures. Because the ribs are angled downward, elevation of the ribs causes thoracic expansion laterally and anteriorly. This is referred to as a **bucket** or **pump handle effect**.

(2) The **external intercostal muscles** originate from an upper rib and insert on the next lower rib more anteriorly (see Figure 16-1). Contraction of these muscles also elevates the ribs, causing lateral and anterior enlargement of the thorax.

FIGURE 16-1. (*A*) The thorax and respiratory muscle relationships are shown during relaxation. (*B*) Maximal inspiration. Note the flattening of the diaphragm and the more horizontal position of the ribs denoting the increased volume of the thorax.

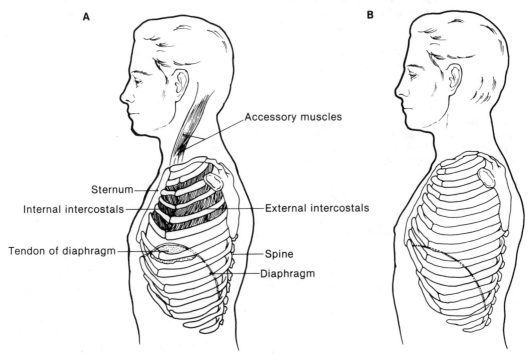

(3) The **accessory muscles** include the sternomastoid and other strap muscles of the neck. These muscles originate from the neck and skull and lift the clavicles and the sternum, helping to elevate the ribs and enlarge the thorax. The accessory muscles are not utilized during normal respiration but are activated when there is an increased respiratory demand (during exercise).

b. Inspiratory force is generated by contraction of the inspiratory muscles, which **expands the gas volume** within the respiratory system.

(1) Boyle's law states that the product of pressure and volume is a constant. Thus, if the volume of a structure containing a constant number of gas molecules is increased, the pressure will decrease so that the product of pressure and volume remains constant.

(a) At the beginning of inspiration, the lungs contain gas that has a pressure equal to the atmospheric pressure. Contraction of the inspiratory muscles expands the gas in the respiratory system, causing the gas pressure in the lungs to decrease. If the airways are open, gas will flow from an area of higher pressure (i.e., the atmosphere) to an area of lower pressure (i.e., the alveoli), renewing the O_2 concentration and diluting the CO_2 levels in the alveoli.

(b) Maximal inspiratory pressure is achieved by fully contracting the inspiratory muscles at a low lung volume when the airways are closed. This procedure is called a **Müller maneuver**.

(i) Normally, the maximal inspiratory pressure that can be achieved is about -80 to -100 cm H_2O (i.e., 80–100 cm H_2O below atmospheric pressure; Figure 16-2).

(ii) When one performs a Müller maneuver, the chest does not expand much but the negative pressure causes a suction effect that can be felt in the ears.

(2) Normal inspiration is **active** but requires alveolar pressures of only -3 to -5 cm H_2O to produce an adequate gas flow into the respiratory system (see Figure 16-2). Therefore, a tremendous reserve of muscle force is available in normal individuals, which can be used to increase respiration during exercise or stress, or to counteract the forces that impede inspiration (these forces may be increased by disease or injury).

(3) Impaired inspiratory force can be caused by many neurologic and muscular diseases (e.g., muscular dystrophy, poliomyelitis) that interfere with the ability of muscles to contract. An inadequate inspiratory force decreases O_2 levels and causes CO_2 retention in the body (i.e., **respiratory failure**). Patients in respiratory failure require mechanical respirators, which provide positive-pressure ventilation to maintain an adequate inspiratory volume (see Chapter 20 IV).

FIGURE 16-2. The *outer loop* represents the maximal pressure–volume relationships of the respiratory system. A subject varies the lung volume from minimal to maximal, and then inhales or exhales as forcefully as possible into a gauge. No gas can escape, so there is no airflow. Therefore, the pressure that is measured in the gauge is the same as the pressure in the alveoli. Note that maximal inspiratory pressures occur at low lung volumes, when the inspiratory muscles are at optimal length, and maximal expiratory pressures occur at maximal lung volumes (*dotted lines*). The *inner loop* represents the pressure and volume changes that occur during a normal respiratory cycle.

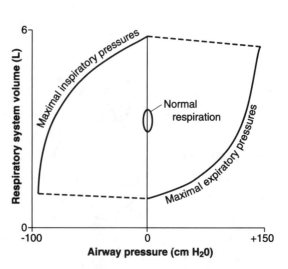

2. **Expiration** refers to a **decrease in lung volume,** which is a **passive process** during quiet, normal breathing (**eupnea**). During inspiration, the respiratory system is inflated above its resting volume, just like blowing up a balloon. During relaxation (releasing the neck of the balloon), the **elastic forces** generated by the inflation compress the gas in the respiratory system, which raises its pressure (in accordance with Boyle's law), and causes the gas to flow out until the system returns to its **resting volume.**

   a. **Expiratory muscles.** Expiratory muscle force is needed when respiration is increased during exercise or in the presence of severe respiratory disease. In addition, expiratory muscle contraction is necessary to achieve lung volumes below the normal resting volume.

   (1) The **rectus abdominus muscle** decreases lung volume by pulling down on the lower ribs. Contraction of the abdominal wall muscles also compresses the abdominal contents and forces the diaphragm upward, decreasing the volume of the thoracic cavity.

   (2) The **internal intercostal muscles** originate from the lower ribs posteriorly and insert on the upper ribs more anteriorly (see Figure 16-1); therefore, contraction of these muscles lowers the ribs.

   b. Expiratory muscle force compresses the gas in the respiratory system, increasing the gas pressure.

   (1) **Maximal expiratory pressure** is achieved by fully contracting the expiratory muscles with the lungs fully inflated and the glottis or airway closed. Forced expiration against a closed airway is termed a **Valsalva maneuver** and is commonly performed when lifting heavy objects or when defecating.

   (a) **Measurement.** The maximal expiratory pressure can be measured with a mercury column or a blood pressure (aneroid) gauge.

   (i) The subject takes in a maximal inspiration, holds the nose closed, and blows as forcefully as possible into the tubing to the gauge without allowing any air to escape. The maximal pressure that can be sustained for 5 seconds is read.

   (ii) Because these gauges are calibrated in mm Hg, the reading must be multiplied by 1.36 to convert to cm H_2O.

   (b) Normally, the maximal expiratory pressure that can be achieved is 100–150 cm H_2O greater than the atmospheric pressure. As lung volume decreases, the maximal achievable expiratory pressure decreases as well (see Figure 16-2).

   (2) **Normal expiration** is **passive,** because the elastic forces stored during inspiration compress the alveolar gases. During eupneic expiration, the inspiratory muscles gradually relax to provide a slow and controlled expiration.

   (3) **Impaired expiratory force** can be caused by many neuromuscular diseases; however, since expiration is normally passive, a weakness of the expiratory muscles is not as critical as a weakness of the inspiratory muscles.

C. **Pressures influencing gas flow** (Figure 16-3)

   1. **Alveolar pressure (PA).** The **rate of gas flow** into or out of the lungs depends primarily on the pressure gradient between the alveoli and the atmosphere (i.e., the **transairway pressure**).

   a. The highest gas flow rates occur when the pressure difference between the alveoli and the atmosphere is greatest. Conversely, when the alveolar pressure is zero (i.e., equals atmospheric pressure), the gas flow is also zero (see Figure 16-3).

   b. Because the alveolar pressure is negative during inspiration and positive during expiration, the gas flow changes direction accordingly.

   2. **Interpleural (pleural) pressure (PPL)**

   a. The interpleural pressure can be estimated in humans by measuring the intrathoracic esophageal pressure using a balloon-tipped catheter connected to a sensitive pressure gauge. The esophageal pressure is equivalent to the interpleural pressure because the normal esophagus is merely a floppy tube traversing the thorax that has the same pressure inside and outside.

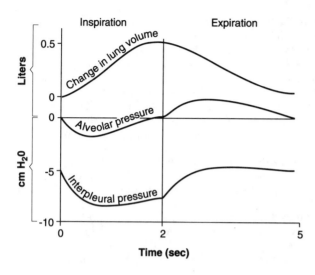

FIGURE 16-3. Pressure changes during the respiratory cycle. The alveolar pressure is negative during inspiration, creating the gradient that produces gas flow. At the end of inspiration, when gas flow ceases, the alveolar pressure is zero. During expiration, the elastic forces or contraction of the expiratory muscles compresses the gas and the alveolar pressure becomes positive, creating an outward gradient for flow. The interpleural pressure becomes progressively more negative during inspiration as a result of the increased transpulmonary pressure that is necessary to inflate the lungs to accommodate the tidal volume. During expiration, the interpleural pressure returns to its initial level.

 b. The interpleural pressure is less than the atmospheric pressure (i.e., it is "negative") and varies with the volume of the respiratory system and the action of the respiratory muscles.

  (1) During complete relaxation of the respiratory muscles, the interpleural pressure is about -5 cm H_2O.

   (a) When the respiratory muscles are completely relaxed, the lungs contain 2–2.5 L of gas. The lungs are distended to this degree because the chest wall exerts an inspiratory force that tends to increase lung volume. In turn, the elastic forces of the lungs try to compress the gas in the lungs and cause them to deflate. At the resting lung volume these two opposing forces are equal but in opposite directions and are exerted across the pleural space, generating a negative pressure in the interpleural fluid.

   (b) Because of the balanced forces, the volume of gas in the lungs when the respiratory muscles are relaxed is termed the **equilibrium volume,** or more commonly, the **functional residual capacity (FRC;** see VII C 4). The FRC is the lung volume at the end of a normal (eupneic), relaxed inspiration.

  (2) During inspiration, the interpleural pressure becomes progressively more negative (see Figure 16-3), because the inspiratory muscles must stretch the lungs to take up the additional volume of gas inspired with each breath. In addition, the negative alveolar pressure during inspiration is transmitted through the lungs into the interpleural space. At the end of inspiration, the stretched lungs recoil away from the rib cage with additional force, increasing the negativity of the interpleural pressure. The change in interpleural pressure, from the beginning to the end of inspiration, parallels the change in the elastance forces for that breath (see Figure 16-15).

  (3) During expiration, the interpleural pressure returns to its resting level. Normally, the interpleural pressure remains negative, but with forced expirations it can become positive, compressing the airways and slowing the rate of expiration.

D. **Recording events of the respiratory cycle.** The volume changes that occur in the lungs during various breathing maneuvers can be measured using a spirometer.

 1. A **spirometer** (Figure 16-4) consists of an inverted, airtight bell that is counterbalanced over a pulley so that it moves freely. The patient is connected to the spirometer by a mouthpiece and two tubes that contain one-way valves to minimize the dead space of the equipment. Expired gas is collected under the bell and inspired gas is withdrawn from the spirometer; usually, the expired gas is freed of CO_2 by means of a chemical absorbent. Records of the respiratory volume changes are made on a rotating drum or kymograph.

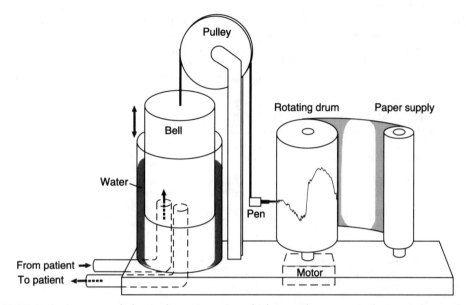

FIGURE 16-4. A water-sealed recording spirometer, which is used to measure changes in lung volume.

2. **Conversion to BTPS conditions (Table 16-1).** Most spirometers use a water seal to prevent the escape of gas, which means that the gas in the spirometer is at room temperature and saturated with water vapor [i.e., it is at **ATPS conditions** (see Table 16-1)].
   a. Gas expired into the spirometer occupies a different volume than it did in the body (i.e., at BTPS conditions) because of the change in the temperature and the amount of water vapor that is present.
   b. Any gas volume within the body should be converted to BTPS conditions. The volume that the gas occupied in the body can be calculated using the **general gas law,** which is a combination of Boyle's and Charles' laws, as shown in Table 16-1.

TABLE 16-1. The General Gas Law

The general gas law is stated as

$$\frac{P_1 \cdot V_1}{T_1} = \frac{P_2 \cdot V_2}{T_2}$$

To convert gas volumes from spirometer to body conditions, the general gas law is used in the following form:

$$V_1 = \frac{V_2 (P_2 - P_{H_2O}) T_1}{(P_1 - 47) T_2}$$

Where V_1 = gas volume in the body; V_2 = gas volume in the spirometer; P_1 = barometric pressure; P_2 = standard pressure (760 mm Hg); T_1 = body temperature (310° K); T_2 = room temperature (° K); P_{H_2O} = water vapor pressure at room temperature; and 47 = water vapor pressure at body temperature.

To identify the conditions for different lung volumes, the following definitions are used:
   ATPS = atmospheric temperature, pressure, saturated (conditions in a spirometer)
   BTPS = body temperature, pressure, saturated (gas volumes in the body)
   STPD = standard temperature (0° C), pressure (760 mm Hg), dry (used to express O_2 and CO_2 volumes for metabolic equivalence)

III. PRESSURE–VOLUME RELATIONSHIPS.

III. **PRESSURE–VOLUME RELATIONSHIPS.** The **elastic properties** of the respiratory system and its components are defined by the relationship between the structures' **transmural pressures** and the **gas volume** within the respiratory system. Elastic properties must be studied under static conditions when the respiratory muscles are completely relaxed so that gas flow, and therefore, resistance forces (see V), are eliminated.

A. The **transmural pressure (PTM)** (literally, "the pressure across the wall") is the difference in pressure between the inside (P_{in}) and the outside (P_{out}) of any structure: $PTM = P_{in} - P_{out}$ (Figure 16-5). The **equilibrium volume** of a structure is defined as the volume it contains when the transmural pressure is zero (i.e., when $P_{in} = P_{out}$).

1. A **positive transmural pressure ($P_{in} > P_{out}$)** is a distending force that tends to expand a structure (i.e., the volume exceeds equilibrium volume).

2. A **negative transmural pressure ($P_{in} < P_{out}$)** tends to deflate a structure (i.e., the volume is less than equilibrium volume).

B. The **transpulmonary pressure (PL)** is the transmural pressure **across the lungs**. The inside pressure is the **alveolar pressure (PA)** and the pressure just outside the lungs is the **interpleural pressure** (i.e., the pressure in the interpleural space, **PPL**): $PL = PA - PPL$.

1. **Lung volume.** The lung is a passive structure whose volume is determined by its transmural pressure (i.e., the transpulmonary pressure).
   a. **Limiting (maximal) volume.** The lung volume increases curvilinearly as the transpulmonary pressure increases until a limiting volume is reached (at a transpulmonary pressure of approximately 20–30 cm H_2O; Figure 16-6A). The limiting volume is thought to be determined by the complex arrangement of collagen fibers in the interstitium of the lung.
   b. The **equilibrium volume** is normally less than 10% of the maximal lung volume. The lung's equilibrium volume occurs where the transpulmonary pressure–volume curve crosses the 0 axis on the graph (see Figure 16-6A).
   c. **Minimal volume.** At volumes less than equilibrium volume, the transpulmonary pressure becomes negative and the airways collapse (i.e., airway closure takes place), trapping a small volume of gas in the lungs that cannot be removed. This volume is termed the **minimal volume**.

FIGURE 16-5. Schematic of the components of the respiratory system and the anatomic locations used to define the transmural pressures of the different structures. *PA* = alveolar pressure; *PPL* = interpleural pressure; *PBS* = pressure at the body surface (usually the atmospheric pressure); *PL* = transpulmonary pressure; *PCW* = transthoracic pressure; *PRS* = transrespiratory pressure (relaxation pressure).

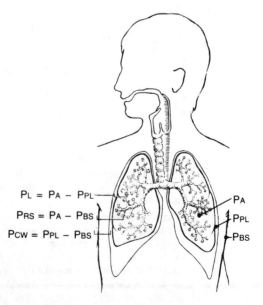

$PL = PA - PPL$

$PRS = PA - PBS$

$PCW = PPL - PBS$

2. Inflation and deflation of the lungs

    a. Compliance (distensibility) is the change in volume of a structure (ΔV) for each unit change in pressure (ΔP). Compliance is an indication of how easily a structure can be stretched or inflated (see also IV A).

    b. Elastance is the retractive (recoil) force that the distention of any structure generates (see also IV B).

    c. Relationship between compliance, elastance, and transmural pressure. If a rubber band is stretched between two hands, the force needed to lengthen the rubber band is the transmural pressure, and the recoiling of the rubber band against the hands is the elastance. The thicker the rubber band (i.e., the less compliant the structure), the greater the recoil force generated by any increase in length. Therefore, **compliance and elastance are inversely related ($C = 1/E$).**

        (1) Inflation of the lungs occurs when the transpulmonary pressure exceeds the recoil forces and an open airway is present. The lungs increase in volume until the elastance force once again balances the transpulmonary pressure and the volume becomes constant at a new equilibrium point.

            (a) Negative-pressure respiration (normal breathing) occurs when the external pressure (i.e., the interpleural pressure) decreases.

            (b) Positive-pressure respiration occurs when the internal pressure (i.e., the alveolar pressure) increases.

        (2) Equilibrium. When a distensible structure has a constant volume or length, the elastance force and the transmural pressure are equal but in opposite directions; in other words, the system is at an equilibrium.

C. The **transthoracic pressure (P_{CW})** is the transmural pressure **across the chest wall**. The inside pressure is the **interpleural pressure** and the outside pressure is the **pressure at the body surface (P_{BS});** therefore, $P_{CW} = P_{PL} - P_{BS}$. Normally, the pressure at the body surface is the atmospheric pressure.

    1. Expansion force. In the adult, the chest wall functions like a bellows that contains compression springs; the chest wall tends to expand because of the "springs." Because of this tendency, the equilibrium volume of the chest wall is near maximal volume (Figure 16-6B). The expansion force exerted by the relaxed chest wall helps to create the negative interpleural pressure that keeps the lungs expanded.

    2. If the lung volume is less than the chest wall equilibrium volume, then the transthoracic pressure is negative ($P_{PL} < P_{BS}$), which indicates that the chest wall is trying to expand.

D. The **transrespiratory pressure (P_{RS})** is the transmural pressure **across the entire respiratory system**. The inside pressure is the **alveolar pressure** and the outside pressure is the **pressure at the body surface,** so $P_{RS} = P_A - P_{BS}$.

    1. The **respiratory system pressure–volume curve** is often referred to as a **relaxation pressure–volume curve,** because of the method used to obtain it.

        a. Method. The subject is connected to a spirometer with a shut-off valve and a side arm connected to a pressure gauge. The spirometer measures changes in lung volume and the pressure gauge provides the airway pressure. The subject inflates the lungs to a predetermined volume, the valve is closed, the subject relaxes the respiratory muscles completely, and the airway pressure is recorded. Multiple pressure and volume coordinates are recorded, which are used to graph the pressure–volume curve.

            (1) The airway pressure equals the alveolar pressure because the lung volume is constant (i.e., there is no airflow) and resistance forces are absent.

            (2) If the subject does not relax completely, the pressure values will be distorted—contraction of the inspiratory muscles decreases the recorded pressures, and contraction of the expiratory muscles increases the airway pressure.

        b. Interpretation. The relaxation pressure–volume curve is sigmoid shaped, steepest in the middle and almost flat at both high and low lung volumes (Figure 16-6C).

            (1) Normally, the respiratory system functions near the middle of the curve, where the system is most distensible.

A. Lungs

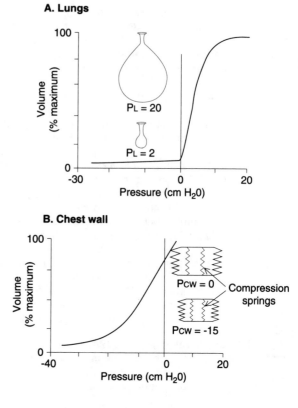

B. Chest wall

C. Lungs and chest wall

FIGURE 16-6. (*A*) Pressure–volume curve for the lungs, which are represented as a simple balloon whose volume depends on the transmural pressure. The equilibrium volume is defined as the volume when the transpulmonary pressure (P_L) equals zero. (*B*) Pressure–volume curve for the chest wall (P_{CW}), which is represented as a bellows with internal compression springs. Note that the equilibrium volume for the chest wall is about 80% of the maximal volume, and that the curve is located largely on the negative side of the pressure axis. A negative transmural pressure is necessary to lower the volume in this system. (*C*) Pressure–volume curve for the entire respiratory system, which is represented by the combination of the balloon and bellows. Algebraic summation of the transmural pressures for the lungs and chest wall will yield this curve. Note that the curve is sigmoid-shaped and that the equilibrium volume lies near mid-lung volume. This equilibrium volume represents the normal end-expiratory volume for the respiratory system. P_A = alveolar pressure; P_{PL} = interpleural pressure. (*D*) Superimposed pressure–volume curves for the lungs (P_L), chest wall (P_{CW}), and respiratory system (P_{RS}). The transmural pressure for each structure is used to plot the respective curve; alternatively, the transmural pressures for the lungs and chest wall may be algebraically summed to generate the respiratory system pressure–volume curve. The relative size and direction of the *arrows* indicates the forces at work at various points on the respiratory system pressure–volume curve.

D. Lungs, chest wall, and respiratory system

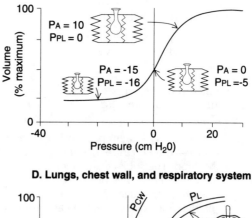

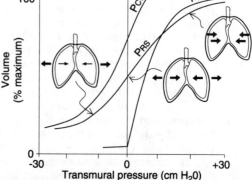

(2) The respiratory system consists of the chest wall and lungs; therefore, the relaxation pressure–volume curve results from summation of the elastic properties of these structures. In Figure 16-6D the pressure–volume curve for the respiratory system (PRS) was obtained by algebraically adding the pressures for the lungs and the chest wall at any given lung volume: PRS = PL + PCW.

2. Respiratory system volumes

a. The **equilibrium volume (FRC) of the respiratory system** occurs at about 40% of the maximal lung volume (see Figure 16-6D). This volume is also referred to as the **functional residual capacity (FRC)** and represents the normal end-expiratory volume of the respiratory system.

b. The **maximal volume** that the respiratory system can achieve is termed the **total lung capacity (TLC)**. Compliance, which is decreased at high lung volumes, and the strength of the inspiratory muscles affect the TLC. The maximal volume of the respiratory system will be reduced if either lung compliance or the strength of the inspiratory muscles is decreased.

c. The **residual volume (RV)** is the **smallest volume** of the intact respiratory system.

(1) Reduced compliance at low volumes and the strength of the expiratory muscles determine the RV.

(2) Compliance may be reduced by closure of the airways and subsequent trappings of gas in the lungs, or by increasing the rigidity of the chest wall. Early closure of the airways occurs when airways are narrowed by disease (e.g., bronchitis, emphysema) and also when lung recoil is diminished as a result of the aging process or emphysema.

IV. COMPLIANCE AND ELASTANCE

A. **Compliance.** In Figure 16-6, the slopes of the pressure–volume curves reflect the compliance of the various structures.

1. The **normal compliance** of the human lungs is about **0.2 L/cm H$_2$O** (i.e., for each cm H$_2$O increase in transmural pressure, the gas volume increases by 0.2 L). The amount of gas inspired with each breath (i.e., the **tidal volume**) is about 500 ml (0.5 L). Thus, the transpulmonary pressure must increase by about 2.5 cm H$_2$O in order to "stretch" the lungs to take up the normal tidal volume. This change in the transmural pressure normally is generated by the inspiratory muscles.

2. **Decreased compliance** of the lungs can be caused by various lung diseases [e.g., tuberculosis, adult respiratory distress syndrome (ARDS)] that produce scarring or fibrosis of the lungs, destroy functional lung tissue, or both.

a. The decreased compliance produces a condition termed **restrictive lung disease**. Patients with restrictive lung disease must generate significantly greater than normal forces to expand the lungs. In other words, they must overcome very high elastance forces.

b. **Role of functional lung tissue.** The compliance of any system is dependent on its size (e.g., the amount of functional lung tissue). Because compliance varies with the amount of functional lung tissue, it is not a good measure of absolute distensibility.

(1) **Example.** If a patient's lung compliance is 0.2 L/cm H$_2$O, then both lungs together are able to take up to 0.2 L of gas for each cm H$_2$O change in transpulmonary pressure. In other words, assuming equal compliance in both lungs, each lung will take up 0.1 L of gas. If the patient undergoes a pneumonectomy (i.e., the removal of one lung), the compliance will equal 0.1 L/cm H$_2$O.

(2) **Specific compliance** is the compliance adjusted for the volume of the structure: $C_{SP} = \Delta V/(\Delta P \cdot V)$, which is the fractional change in volume per unit change in pressure.

(a) Specific compliance allows comparison of the compliance properties of different structures or of subjects of different sizes. For instance, the specific lung compliances of mice and elephants are nearly identical, but the lung compliances vary by several orders of magnitude.

(b) If the patient in IV A 2 b (1) had an initial lung volume of 2 L, then the specific compliance prior to the pneumonectomy would equal 0.1 cm H_2O^{-1} (0.2/2). After the pneumonectomy, the lung volume is reduced by one half but the specific compliance remains the same: 0.1/1, or 0.1 cm H_2O^{-1}.

3. **Increased compliance** is produced by the pathologic processes that occur in **emphysema** and also as a result of the **aging process**.

   a. Alveolar septa, which provide some of the retractive force in the lungs, are destroyed in both conditions but emphysema causes a much more extensive loss of septa than the normal aging process.

   b. In emphysema, there is a marked increase in airspace size (due to the loss of many contiguous alveolar walls), a loss of alveolar surface area, and a reduction in the retractive forces of the lungs. These changes lead to abnormalities in gas exchange, in the pulmonary circulation, and in ventilation of the lungs (see Chapter 18 III C 3).

B. **Elastance.** The **recoil (retractive) forces** in the lungs are generated by both **tissue** (e.g., smooth muscle, elastin, collagen) **forces** and **surface forces**.

1. **Tissue forces.** The lung contains large amounts of **collagen** and **elastin,** but the role of these connective tissue fibers in generating elastic (retractile) forces in the lungs is not fully understood.

   a. It is thought that elastin fibers are stretched at low and medium lung volumes, and that collagen prevents overdistention of the lungs at high lung volumes.

   b. **Contractile fibers** in the lung parenchyma may also contribute to the tissue forces.

2. **Surface forces** (Figure 16-7) are generated at the interface between two different phases (e.g., liquid and gas).

   a. **Alveolar septa.** An interface between the alveolar septum, which is thought to be lined with a thin layer of fluid, and the alveolar gas generates a surface force (tension) because of the unbalanced attraction of the surface molecules by the molecules below the surface. A sphere has the smallest surface area for its volume and the surface force tends to minimize the surface area of the interface.

   b. **Pulmonary recoil.** Surface forces contribute to the total pulmonary recoil forces and increase the tendency of the lungs to deflate. An increased transmural pressure is necessary to counteract the effects of the surface forces.

      (1) Filling the lungs with a liquid (e.g., saline) will, theoretically, eliminate the surface forces so that only the tissue forces produce recoil. Figure 16-8 shows the results of such an experiment. Compared with air-filled lungs, fluid-filled lungs have markedly reduced recoil force, leading to the conclusion that surface forces play a major role in generating elastance forces in the lungs.

      (2) **Law of Young-Laplace.** The transmural pressure of a structure depends on both radius and surface (wall) tension (see Figure 16-7).

         (a) **Variable wall tension** exists in an **elastic structure** such as an inflated balloon (Figure 16-9).

            (i) The pressure inside the balloon is equal throughout, but the wall tension varies directly with the radius of curvature. For example, the neck of the balloon (which has a small radius) is very collapsible (i.e., the wall tension is low), but the body of the balloon is more rigid (i.e., the wall tension is higher).

            (ii) The balloon is **stable** because the wall tension varies to counteract the differences in the radii at different points.

         (b) **Constant surface tension.** Most **liquids** have a constant surface tension.

            (i) Figure 16-10 shows the pressure–volume relationships of a soap bubble. Once the bubble radius exceeds a minimum, the compliance of the bubble has a negative slope (i.e., as the volume increases, the distending pressure decreases). The negative slope indicates that this system is **unstable** (i.e., it tends to become overdistended and rupture easily).

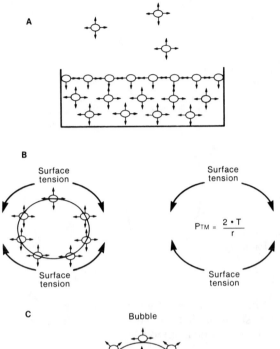

FIGURE 16-7. (*A*) Schematic of the inter-molecular forces for molecules in the gas phase, at the surface, and in the bulk phase of a liquid. Note that surface molecules have an unbalanced force tending to pull them into the liquid. (*B*) Surface forces in a hollow sphere tend to reduce the volume of the sphere. The relationship between surface tension and the surface component of transmural pressure is defined by Laplace's law. (*C*) To maintain a constant volume in a spherical structure, the transmural pressure must counterbalance the surface tension forces. P_{TM} = transmural pressure; T = tension; r = radius.

$$P_{TM} = \frac{2 \cdot T}{r}$$

(ii) **Significance.** The alveoli of the lungs are like millions of intercon-nected soap bubbles of varying size. If two alveoli of different sizes are interconnected and have the same surface tension, then the pressure in the smaller alveolus will be higher than the pressure in the larger alveolus. Since the alveoli are interconnected, the gas will flow from the smaller to the larger alveolus, because of the pressure gradient.

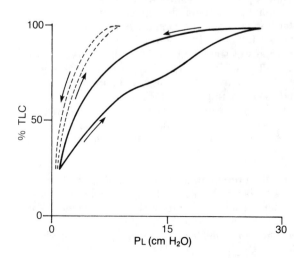

FIGURE 16-8. Pressure–volume loops for iso-lated lungs when filled with saline (*dotted line*) or air (*solid line*), showing the volume change as a percent of the total lung capacity (*TLC*). The *arrows* indicate the direction of volume change. Note the significant increase in the pressure that is required during inflation with air. P_L = transpulmonary pressure.

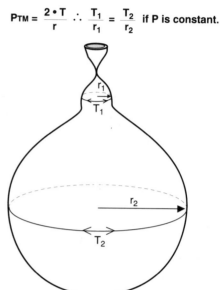

$$P_{TM} = \frac{2 \cdot T}{r} \quad \therefore \quad \frac{T_1}{r_1} = \frac{T_2}{r_2} \quad \text{if P is constant.}$$

FIGURE 16-9. In an inflated balloon, the internal pressure is constant throughout. The structure is stable (i.e., unlikely to burst) because there is a variation in the wall tension in areas with different radii; thus, the ratio of the wall tension to the radius is a constant value. P_{TM} = transmural pressure; r = radius; T = wall tension.

Thus, the smaller alveolus will eventually collapse (i.e., become **atelectatic**). Pulmonary surfactant helps to counter this phenomenon by stabilizing the alveoli and maintaining inflation.

3. **Alveolar stabilization**
    a. **Pulmonary surfactant** is a complex substance that lines the alveolar surface and markedly decreases surface tension. It is stored in **lamellar bodies**.
       (1) **Composition** (Table 16-2). Pulmonary surfactant is formed by type II alveolar cells, which are cuboidal cells located in the corners of the alveoli.
          (a) **Four unique proteins** have been identified in surfactant: SP-A, SP-B, SP-C, and SP-D.
             (i) SP-A and SP-D are hydrophilic proteins.
             (ii) SP-B and SP-C are strongly hydrophobic proteins.
          (b) **Phospholipids,** particularly **dipalmitylphosphotidylcholine (DPPC),** are strong **surface-active agents,** especially when they are compressed in the surface of an air–fluid interface.
             (i) A **surface balance** is an instrument that is used to measure surface tension; it allows the surface area to be varied while preventing any loss of surface-active material (i.e., material that concentrates at the surface). The surface tension at an air–fluid interface decreases when surface-active agents are initially added. Reducing the surface area of a surface balance compresses surfactants and further lowers the surface tension (Figure 16-11). Whether such changes in surface tension occur in the alveoli during normal breathing is still controversial.
             (ii) Surface balance studies comparing pure DPPC and pulmonary surfactant (obtained by lung lavage) reveal significant differences between the two materials—apparently, the combination of materials in surfactant produces complex surface properties that are not yet fully understood.
       (2) **Functions**
          (a) **Reducing surface tension** is the primary function of pulmonary surfactant.
             (i) **Adsorption of surface-active molecules** at the alveolar–air interface reduces the alveolar surface tension, increasing the compliance of the lungs and, thereby, decreasing the work of respiration.
             (ii) The low surface tension also facilitates the reopening of collapsed airways and alveoli.

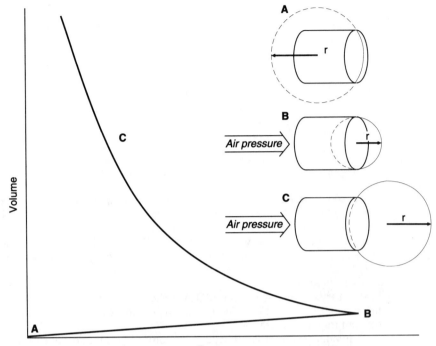

FIGURE 16-10. The pressure–volume characteristics of a soap bubble, whose wall tension equals the surface tension of the soap solution. (*A*) When the soap bubble has a volume of zero, the soap film has no curvature and, therefore, a radius (*r*) of infinity. Thus, the transmural pressure at this point is also zero. (*B*) As the bubble is inflated, the radius decreases from infinity (the starting point). The curve segment from point A to point B has a positive slope because the radius of the bubble decreases until it reaches a minimum at point B, which is defined by the radius of the tube. (*C*) As the bubble volume continues to increase, the radius also increases, causing the transmural pressure to decrease in accordance to the law of Young-Laplace. Segment C of the curve represents a negative compliance, which is an unstable condition because the transmural pressure decreases as the bubble increases in volume.

  (b) **Increasing alveolar radius.** Surfactant fills irregularities in the alveolar surface, which increases the mean alveolar radius. The increased radius reduces the transmural pressure required to maintain the alveolar inflation (per the law of Young-Laplace), increasing lung compliance and reducing the work of breathing.
  (c) **Reducing pulmonary capillary filtration.** Normal surfactant, by lowering the alveolar surface tension, reduces the retraction forces of the lungs. The

TABLE 16-2. Composition of Pulmonary Surfactant

Component		Percent Composition
Lipids		85%
DPPC	63.75%	
Neutral lipid	7.65%	
Cholesterol	5.95%	
Phosphatidylethanolamine	5.10%	
Sphingomyelin, lecithin	2.55%	
Proteins		13%
Other		2%

DPPC = dipalmitylphosphotidylcholine.

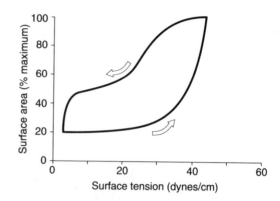

FIGURE 16-11. Surface tension as a function of surface area, obtained from bronchoalveolar lavage fluid. Note the large hysteresis and that the minimal surface tension approaches zero. The *arrows* indicate the direction of the recording.

decreased elastance makes the pulmonary interstitial pressure less negative, which reduces the filtration forces across the pulmonary capillary (Starling hypothesis; see Chapter 12 I C 3 b). Thus, normal surfactant helps to prevent pulmonary edema, which interferes with gas exchange by either increasing the thickness of the alveolar–capillary membrane or by flooding the alveoli.

(3) Clinical significance

(a) Respiratory distress syndrome (RDS) of the newborn (see also Chapter 20 III). Normal surfactant is formed by the type II alveolar cells only after the seventh month of gestation.

(i) Infants born prematurely have immature lungs that produce inadequate or abnormal surfactant (e.g., the minimal surface tension in lung lavage fluid taken from a premature infant is 20–25 dynes/cm, as compared with a minimal surface tension that is close to 0 dynes/cm in lung lavage fluid taken from a full-term infant).

(ii) Gas exchange is impaired in infants with RDS because of alveolar instability (atelectasis) and areas of edema and hemorrhage within the lungs, leading to a high mortality rate.

(iii) **Therapy** includes the application of **positive end-expiratory pressure (PEEP;** see Chapter 20 IV A 2) to prevent atelectasis, reduce edema, and enhance gas exchange; and the administration of **exogenous surfactant.** The exogenous surfactant, which is applied endotracheally and distributed by positional changes of the infant, somehow has an effect at the alveolar level.

(b) Adult respiratory distress syndrome (ARDS) can be caused by shock, systemic infection, or trauma that leads to atelectasis, pulmonary hemorrhage, and edema. Clinical trials are now underway using exogenous surfactant in an attempt to improve the outcome of this syndrome.

b. Tissue interdependence is a **mechanical method** of alveolar stabilization.

(1) The alveolar septa form a continous network with one another and are under tension because of lung inflation. If one tries to collapse one of the holes of a tennis net stretched across a court, the adjoining strings develop more tension to counteract the collapse forces. Likewise, the tendency of a lung segment to collapse is counteracted by tension that is generated in adjacent alveolar walls.

(2) Thus, alveoli are tethered together by their adjoining walls and cannot change volume independently of the neighboring lung tissue.

V. **RESISTANCE.** Changing lung volume creates resistance forces; therefore, resistance is a **dynamic property** (as opposed to compliance, which is a static property). Resistance is caused by the **friction** of gas molecules between each other and the walls of the airways **(airway resistance),** as well as by the friction of the tissues as the lung volume changes **(tissue resistance).**

A. **Airway resistance** is responsible for approximately **80%–90% of the total resistance forces** during breathing.

1. Airway resistance can be calculated by dividing the alveolar pressure during respiration by the rate of gas flow in or out of the lungs. The usual units of airway resistance are cm H_2O/L/sec.

2. **Factors determining airway resistance.** Airway resistance depends on the **rate of gas flow** and the **diameter and length of the airways** (although airway length changes relatively little during respiration and disease processes, and thus, is not a significant factor that changes airway resistance).
   a. The **rate of gas flow** is measured in liters per second (L/sec) and equals the product of the **velocity** of the gas molecules as they pass through the airways and the **cross-sectional area** of the airways.
      (1) The **velocity** of the gas molecules depends on the pressure difference between the alveoli and the mouth (i.e., the **transairway pressure**). This pressure gradient represents the **driving force** that determines airflow.
      (2) **Cross-sectional area.** The airway system is organized so that the total cross-sectional area of the airways increases about 1000-fold as the gas proceeds from the trachea through approximately 23 generations of branching to the alveoli. As cross-sectional area increases, the velocity of gas slows in accordance with the continuity principle. The **highest resistance** to flow occurs in the **intermediate-sized bronchi** because of the high airflow velocity in these segments. Normally, the **airways less than 2 mm in diameter** provide only about **10% of the total airway resistance,** because the large number of these airways provides a large total cross-sectional diameter.
   b. The **airway diameter (radius)** is the most powerful determinant of resistance (Poiseuille's law; see Chapter 9 III B 2). The smaller the airway, the higher the resistance, for any flow rate.
      (1) **Clinical significance**
         (a) **Chronic obstructive pulmonary disease (COPD),** such as **emphysema, asthma,** or **bronchitis,** narrows the small airways, creating a very high airway resistance. In order to maintain normal inspiratory and expiratory flow rates, COPD patients must generate much greater inspiratory and expiratory forces.
         (b) **Early airway disease.** The **small, terminal airways** that normally generate only 10%–20% of the total airways resistance (i.e., the **"silent region"**) are the ones primarily affected by COPD, making early airway disease difficult to detect.
            (i) For example, a bronchitic process that produces a three- to fourfold increase in the resistance of 50% of the small airways will result in only a 15%–20% increase in total airways resistance, which would still be within the normal range.
            (ii) Great effort has gone into developing pulmonary function tests to detect early airway disease, but so far success has been limited.
      (2) **Control of airway diameter**
         (a) **Sympathetic nerve (adrenergic) stimulation** to the airways **causes bronchodilation**.
            (i) **Direct sympathetic nerve activation** in humans causes only slight bronchodilation, because the adrenergic nerves do not actually innervate the bronchial smooth muscle.
            (ii) However, **circulating adrenergic substances** (e.g., epinephrine, isoproterenol, or adrenergic-like drugs) cause marked bronchodilation because bronchial smooth muscle contains a large number of β_2-receptors that produce smooth muscle relaxation in response to these substances. Many β_2-adrenergic drugs are now available via nebulizers for the **treatment of asthma.**

(b) **Vagus nerve (cholinergic) stimulation** to the lungs **causes bronchoconstriction** and **increases the formation of mucus,** narrowing the airways and increasing the resistance to airflow. Cholinergic activity is a major component in the pathogenesis of asthma in some individuals.

(c) **Nonadrenergic, noncholinergic activity.** The lungs are derived embryologically from the foregut. Recently, it has been recognized that many of the **neuropeptides** that control intestinal activity also control airway smooth muscle.

    (i) **Noncholinergic excitatory compounds,** such as **substance P** and **neurokinins A and B, cause bronchoconstriction, mucus secretion,** and **increased vascular permeability.** These compounds seem to be especially involved in the pathogenesis of asthma in certain individuals.

    (ii) The **nonadrenergic inhibitory system** consists of postganglionic neurons that release **vasoactive intestinal peptide (VIP)** and other mediators that **cause smooth muscle relaxation** and **inhibit mucus production**.

B. **Tissue resistance** plays only a minor role in determining the forces required to breathe.

C. **Effort-independent flow** occurs during forced expiration when interpleural pressures are positive and lung volumes are less than 80% of the TLC (Figure 16-12). Under these conditions, interpleural pressure is used as a measure of the effort exerted during expiration.

    1. **Driving force for flow.** In Figure 16-12, note that the expiratory flow rate remains constant at each lung volume, but the interpleural pressure varies over a large range (i.e., flow is independent of effort). The pressure gradient that produces flow under these conditions is no longer the transairway pressure (i.e., the alveolar pressure minus

FIGURE 16-12. (*A*) Isovolume flow diagram. At volumes less than 80% of total lung capacity (TLC), the expiratory flow rate is limited, and the flow rate is independent of effort as measured by the interpleural pressure. *VC* = vital capacity. (*B*) Flow–volume loop. Much information can be obtained about pulmonary mechanics by recording flow rate as a function of lung volume. The flow curve labeled *maximal effort* represents maximal flow rate and cannot be exceeded. The maximal flow rate at each volume depends on the state of the airways and the lung parenchyma and provides important diagnostic information. *% VC* = percent vital capacity.

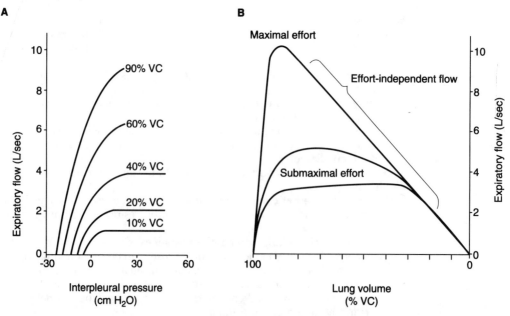

the mouth pressure)—instead, it is the **transpulmonary pressure,** which is a function of lung volume. This occurs because of the presence of a flow-limiting segment in the airways that is caused by the positive interpleural pressure.

2. **Dynamic compression of the airways** (Figure 16-13) occurs during forced expiration and limits the flow rate during the terminal 80% of expiration. The amount of dynamic compression depends on several variables that determine the limitation of the expiratory gas flow (Figure 16-14).

   a. **Lung volume** is the most powerful modulator of expiratory gas flow. When lung volume is increased, **radial traction** (i.e., a dilating force exerted on the outer walls of the airways by the alveolar septa) is enhanced, causing the airway diameter to expand. In addition, at high lung volumes, expiratory muscle force increases because the expiratory muscle length is optimal.

   b. **Elastic recoil.** Radial traction on the airway walls alters the diameter of the intrapulmonary airways.

      (1) **Increased elastic recoil** (as occurs in restrictive lung disease) increases radial traction and dilates the airways. Thus, patients with restrictive lung disease are able to achieve relatively high flow rates for their lung volumes.

      (2) **Decreased elastic recoil** (as occurs as a result of aging or emphysema) leads to very compressible airways and marked increases in airways resistance.

   c. **Pressure drop.** Increased airway resistance, caused by obstructive lung disease, increases the pressure drop along the small peripheral airways. This loss of pressure produces a greater compressive effect on the airways, leading to further narrowing and even greater reduction in the rate of gas flow at any given lung volume.

VI. **THE WORK OF RESPIRATION** equals the change in pressure that is needed to inflate the lungs multiplied by the change in volume (Figure 16-15).

A. **Work (pressure–volume) loops.** The work of breathing can be evaluated by plotting the change in lung volume versus the interpleural pressure (Figure 16-16A–C). The **area** of the rectangle ($\Delta P \cdot \Delta V$) has the units of work (kg $\cdot$ m) and is proportional to the O_2 utilized by the respiratory muscles. If either the resistance increases or the compliance decreases, the respiratory muscles must generate more force (and therefore, use more O_2) to overcome the added load.

Eupneic expiration

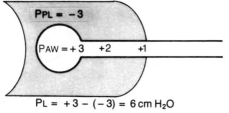

$PL = +3 - (-3) = 6 \, cm \, H_2O$

Forced expiration

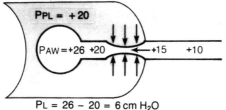

$PL = 26 - 20 = 6 \, cm \, H_2O$

FIGURE 16-13. Schematic representation of the flow-limiting segment in an airway during forced expiration. Note that the pressures within the airways decrease because of resistance forces during flow. During the forced expiration, the pressure decrement is steeper because of the increased flow rate. During eupneic expiration, the transmural pressure of the airways is a distending force, which dilates the airways. During a forced expiration, the interpleural pressure is greater than the pressure in the bronchi (as indicated by the *arrows*), which narrows the airways, increases airway resistance, and limits the expiratory flow rate. *PPL* = interpleural pressure; *PAW* = airway pressure; and *PL* = transpulmonary pressure.

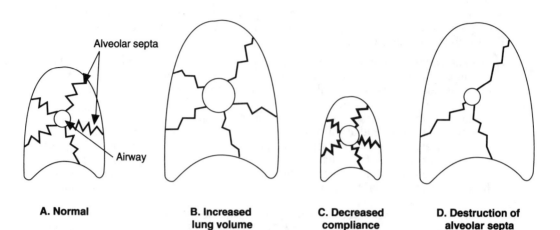

A. Normal

B. Increased lung volume

C. Decreased compliance

D. Destruction of alveolar septa

FIGURE 16-14. Various factors can alter the radial traction forces on the airway walls. (*A*) Normal. (*B*) Increased lung volume is the most powerful factor that increases airway diameter, leading to higher flow rates. (*C*) Restrictive lung disease reduces lung volume but the increased elastance forces support the airways and dilate them, compared with normal lungs of the same volume. (*D*) Destruction of alveolar septa (as occurs in emphysema) diminishes airway wall support, narrowing the lumen and creating high airway resistance.

B. **Normal work of breathing.** Normally, the work of breathing represents **2%–3% of the resting O_2 consumption**. Since expiration is a passive process, all of the work of breathing is done during inspiration.

C. **Minimal work of breathing.** The minimal work for respiration is a function of airway resistance and elastance in the respiratory system (Figure 16-17A). Recall that resistance and elastance are largely determined by the diameter of the airways, the lung compliance, and the respiratory rate. All species of animals seem to use the combination of tidal volume and respiratory rate that requires the minimal work of respiration, although it is unknown how the brain establishes this combination.

  1. **Restrictive lung disease.** Patients with restrictive lung disease (i.e., reduced compliance) must increase their elastance work in order to distend the lungs. The elastance work is proportional to the tidal volume and to the elastance of the respiratory system.
    a. **Work loop** (Figure 16-16B). In patients with restrictive lung disease, the interpleural pressure becomes more negative at the end of inspiration, the stored elastic energy is increased, and the total work of breathing is increased.

FIGURE 16-15. A graphic of the events of the respiratory cycle with the airway resistance and elastance forces identified. Increased resistance or increased elastance causes proportional increases in these areas. Thus, patients with obstructive lung disease (i.e., high airway resistance) must generate much higher gradients during inspiration and expiration, and patients with restrictive disease (i.e., increased elastance) must generate more negative interpleural pressures at the end of inspiration.

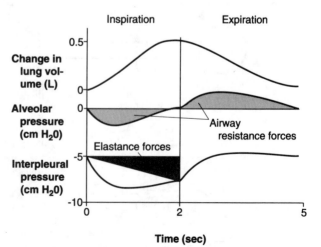

Respiratory Mechanics | **219**

A. Normal

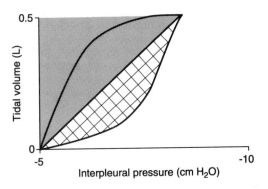

B. Restrictive lung disease

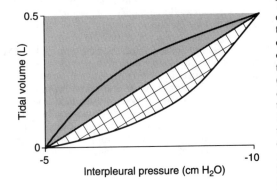

C. Obstructive lung disease

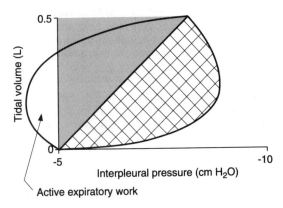

FIGURE 16-16. The work of ventilating the lungs can be evaluated by plotting the change in lung volume (i.e., the tidal volume) against the change in interpleural pressure. The resultant work loop is inscribed counter-clockwise; the peak of the loop represents the end of inspiration. (*A*) Normal. The slope of a line drawn from the beginning to the end of inspiration represents the compliance of the lungs and bisects the loop into inspiratory (lower) and expiratory (upper) segments. The *crosshatched area* represents inspiratory resistance, or the resistance work required during inspiration. The *shaded triangle* represents the stored elastic energy in the system that compresses the alveolar gas and creates expiratory gas flow. The total area represented by the tidal volume multiplied by the change in interpleural pressure ($V_T \cdot \Delta P_{PL}$) is proportional to the work that must be performed by the respiratory muscles. (*B*) Restrictive lung disease. Patients with restrictive lung disease must overcome significantly higher elastance forces (as indicated by the *shaded triangle*). If the tidal volume remains in the normal range, then contraction of the inspiratory muscles produces much more negative interpleural pressures. (*C*) Obstructive lung disease. Patients with obstructive lung disease rely on increased pressure gradients to generate adequate air flows. The more negative alveolar pressure is transmitted to the interpleural pressure, causing the work loop to broaden significantly. The work loop extends beyond the y-axis, which is indicative of the active contraction of the expiratory muscles that is necessary to supplement the stored elastic energy.

   b. **Compensatory mechanisms.** Elastance work can be minimized by **breathing rapidly and shallowly** (Figure 16-17B). The tidal volume is decreased, but the increased respiratory rate ensures adequate ventilation of the lungs.

2. **Obstructive lung disease.** Patients with obstructive lung disease must increase their resistance work in order to overcome increased airways resistance forces. If airway resistance is so high that contraction of the expiratory muscles is necessary to produce adequate expiratory gas flow, then the interpleural pressure may increase and even become positive, leading to airway compression and further increasing airway resistance.

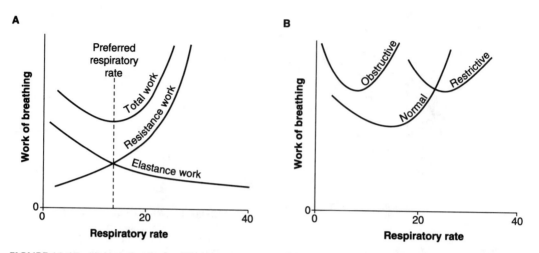

FIGURE 16-17. (*A*) Variation in the different components of the work of breathing at a constant level of alveolar ventilation. At low respiratory rates, the tidal volume is increased to maintain a constant ventilation. The larger tidal volumes require greater distention of the lungs and the elastance component of work is increased. Higher respiratory rates increase the velocity of flow through the airways, increasing the resistance component of work. Thus, for a given ventilation, there is a minimal work of breathing that is determined by a specific combination of tidal volume and respiratory rate. Normally, this optimal combination for humans occurs at a rate of 12–15 breaths per minute and a tidal volume of about 0.5 L. (*B*) Total work of breathing curves reflect the compensatory mechanisms employed by patients with obstructive or restrictive lung disease. Patients with obstructive pulmonary disease minimize their work of breathing by decreasing the respiratory rate, which minimizes the resistance component of respiration. Patients with restrictive disease minimize their respiratory work by decreasing tidal volume (to minimize the elastance component) and increasing their respiratory rate (to maintain a normal alveolar ventilation). The minimal respiratory work for patients with obstructive or restrictive lung disease is higher than normal but it is optimal for the pathophysiologic conditions that are present.

    a. Work loop (Figure 16-16C). The width of the work loop is increased in patients with obstructive lung disease because increased pressures are necessary to generate inspiratory and expiratory gas flow.

    b. Compensatory mechanisms. Resistance work can be minimized by **breathing more slowly and deeply** (see Figure 16-17B), because this reduces the flow rate of gas, which reduces the pressure gradient necessary to generate gas flow. The increased tidal volume compensates for the decreased respiratory rate so that a normal alveolar ventilation is maintained.

D. **Respiratory failure.** The respiratory muscles, just like any skeletal muscles, can become fatigued if exposed to a high work load for a long time. If the respiratory muscles cannot generate sufficient force, then pulmonary ventilation decreases, reducing the supply of O_2 to the body and impairing the excretion of CO_2.

VII. LUNG VOLUMES AND CAPACITIES

A. **Introduction**

    1. The amount of gas in the lungs can be divided into various fractions with specific functions.

        a. There are four lung **volumes** that are **nonoverlapping fractions** of the gas content and four **capacities** that are **combinations of** two or more of the **lung volumes**.

        b. Figure 16-18 shows how these volumes and capacities are organized and how they are dependent on the upper, lower, and equilibrium limits of the respiratory system.

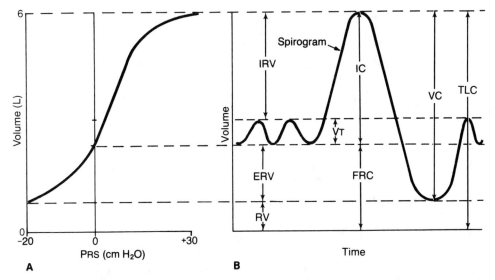

FIGURE 16-18. (A) A pressure–volume curve of the respiratory system. Residual volume (*RV*) and total lung capacity (*TLC*) are limited by the decreased compliance of the system, as indicated by the reduced slope of the curve. Functional residual capacity (*FRC*) is determined by the volume at which the transrespiratory pressure (*PRS*) is zero. (B) A spirogram showing several normal tidal volumes (*VT*); a maximal inspiratory effort followed by a maximal expiratory effort, termed a vital capacity (*VC*) maneuver; and the return to a normal tidal volume. This tracing can be used to define all of the subdivisions of lung volume. *ERV* = expiratory reserve volume; *IRV* = inspiratory reserve volume; and *IC* = inspiratory capacity.

2. The pressure–volume relationships of the respiratory system determine the sizes of the various lung volumes. Diseases alter lung volumes in a predictable fashion depending on how they alter the pressure–volume relationships of the respiratory structures (Figure 16-19).

B. | **Lung volumes**

1. The **tidal volume (VT)** is the volume of gas inspired or expired with each breath. The normal tidal volume is **500 ml (0.5 L)**.

2. The **inspiratory reserve volume (IRV)** is the additional volume of gas that can be inspired above the tidal volume, normally about 3.0 L. The IRV is "in reserve" and can be used if the tidal volume must be increased (e.g., during exercise).

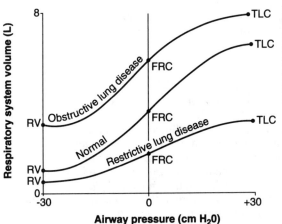

FIGURE 16-19. Typical changes in pressure–volume relationships and resultant lung volumes and capacities that occur in obstructive and restrictive lung diseases. The variations in the relaxation pressure–volume curves are produced by diseases that change the shape and position of the pressure–volume curves for the lungs or the chest wall.

3. The **expiratory reserve volume (ERV)** is the volume of gas that can be forcefully expired after a normal expiration, normally about 1.3 L. The ERV may be encroached upon when the tidal volume is increased.

4. The **residual volume (RV)** is the volume of gas that remains in the lungs after a maximal expiration, normally about 1.2 L. This gas prevents complete collapse of the alveoli; reopening collapsed alveoli requires extremely large forces. Because the RV cannot be expired into a spirometer, it must be measured using either a dilution method or a plethysmographic technique (see VII C 4).

C. **Lung capacities**

1. The **total lung capacity (TLC)** is the volume of gas in the lungs after a maximal inspiration. It equals the sum of the four lung volumes.

2. The **vital capacity (VC)** is the maximal volume of gas that can be expired after a maximal inspiration.

3. The **inspiratory capacity (IC)** is the sum of the tidal volume and the IRV and is the maximal volume that can be inspired after a normal expiration.

4. The **functional residual capacity (FRC)** is a reservoir of gas that remains in the lungs after a normal expiration and allows oxygenation of the blood between breaths. The FRC (or any lung capacity containing the RV) must be measured using either a dilution or a plethysmographic method.
   a. The **dilution method** is widely applied in medicine to measure gas volumes, body fluid compartment volumes, blood flow or cardiac output (i.e., via thermal dilution), and renal clearance. **To determine the FRC,** an indicator gas that can be readily detected (e.g., helium) is placed in a spirometer.
      (1) If the total volume of gas in the spirometer is 3 L and 20% of the gas is helium, then there is 0.6 L of helium in the spirometer.
      (2) If a person begins to breathe from this spirometer at the end of a normal expiration, then the volume of gas in the lungs at the beginning of the test (i.e., the **FRC**) is measured. The spirometer volume is maintained at 3 L by the addition of O_2.
      (3) **Equilibration.** After the subject breathes in and out of the spirometer for several minutes, the helium becomes equally distributed between the lungs and the spirometer. After equilibration, the fraction of helium is found to be 10% of the total gas volume because the helium in the spirometer has been diluted by the gas in the lungs. Although a small amount of helium dissolves in the pulmonary capillary blood, the total volume of helium can be considered approximately equal under the two conditions:

$$F_1 \cdot V_{sp} = F_2 \cdot (V_{sp} + FRC), \text{ where}$$

$$
\begin{aligned}
F_1 &= \text{the initial fraction of helium} \\
V_{sp} &= \text{the total volume of gas in the spirometer} \\
F_2 &= \text{the fraction of helium following equilibration} \\
FRC &= \text{functional residual capacity}
\end{aligned}
$$

Solving for the FRC yields:

$$FRC = F_1 \cdot V_{sp}/F_2 - V_{sp} = 0.2 \cdot 3/0.1 - 3 = 3 \text{ L.}$$

      (4) **Determination of RV.** If the ERV is measured from the spirometer tracing and found to be 2 L, then the RV is 1 L (RV = FRC − ERV).
   b. The **plethysmographic method** uses the principles of **Boyle's law** and an airtight chamber or plethysmograph to measure lung volume. The pressures within the chamber and the airways are measured while the subject (who is inside the chamber) breathes shallowly and rapidly at FRC against a closed airway (i.e., pants).
      (1) **Alveolar pressure measurement.** During the panting maneuver, gas does not flow in or out of the lungs; therefore, the pressure in the upper airways, which can be measured, is the same as the pressure in the alveoli (see also Chapter 9 III A 1, 2 c for a review of the principles underlying pressure gradients).

(2) **Changes in lung volume** caused by panting are proportional to changes in the gas pressure within the plethysmograph. The ratio of the change in airway pressure to the change in alveolar pressure is used to calculate the actual gas volume within the body by applying Boyle's law. This gas volume may differ slightly from the lung volume measured by the dilution principle because the dilution method only measures gas that is free to equilibrate with the spirometer.

VIII. VENTILATION OF THE LUNGS depends on the frequency and depth of breathing.

A. Properties of air

1. **Composition.** Air is a mixture of gases, primarily nitrogen (N_2) and O_2, with a variable amount of water vapor, a negligible quantity of CO_2, and a small amount of inert gases. In determining the composition of air, these inert gases usually are considered together with N_2, which is also inert in the body. **Dry atmospheric air** is approximately 79% N_2 and 21% O_2.

2. **Atmospheric (barometric) pressure** is the gas pressure exerted by the air.
   a. Gas molecules are in constant, rapid motion, which causes them to strike each other and the walls or surfaces of any container or structure. As temperature increases, so does the velocity of molecular movement, which increases the gas pressure if the volume occupied by the gas is constant (**Charles' law**). Because of this constant bombardment, gas molecules exert a pressure that can be measured using a **barometer**.
   b. Atmospheric pressure decreases as altitude increases. The **standard atmospheric pressure** is **760 mm Hg** at **sea level,** 380 mm Hg at 18,000 feet [0.5 atmospheres (atm)], and 190 mm Hg at 34,000 feet (0.25 atm).

3. **Partial pressures.** In a mixture of gases, each gas exerts its own partial pressure (tension). According to **Dalton's law,** partial pressure equals the fraction of gas present (i.e., the concentration) times the total pressure.
   a. The **partial pressure of N_2 (P_{N_2})** is 600 mm Hg at sea level ($760 \cdot 0.79 = 600$).
   b. The **partial pressure of O_2 (P_{O_2})** is 160 mm Hg at sea level ($760 \cdot 0.21 = 160$). Note that the sum of the partial pressures of all of the gases in a mixture equals the total barometric pressure (Dalton's law).

B. Pulmonary (total) ventilation

1. **Minute ventilation ($\dot{V}_E$)** is the volume of air inspired or expired per minute, which equals the tidal volume (V_T) multiplied by the respiratory rate or frequency (f): $\dot{V}_E = V_T \cdot f$. The tidal volume at rest averages 500 ml (0.5 L), and the normal respiratory rate is 12–15 breaths per minute; therefore, the normal minute ventilation is **6–7.5 L/min.**

2. **Dead space ventilation ($\dot{V}_D$)** is the portion of the minute ventilation that fails to reach the area of the lungs involved in gas exchange.
   a. The **anatomic dead space (V_D)** is the volume of gas that occupies the nose, mouth, pharynx, larynx, trachea, bronchi, and bronchioles (i.e., the upper airways). This area of the respiratory system (also called the conducting zone) does not participate in gas exchange because of the thickness of the walls of these structures.
      (1) $\dot{V}_D = V_D \cdot F$. The anatomic dead space contains approximately 150 ml (0.15 L) of gas; therefore, dead space ventilation equals approximately **2.25 L/min.**
      (2) **Functions** of the anatomic dead space
         (a) **Conditioning of air**
            (i) During nasal breathing, the inspired gas is **warmed to body temperature** and **saturated with water vapor** (i.e., it has 100% relative humidity) by the time it reaches the trachea. Breathing through the mouth or

a tracheostomy allows cool, dry air to reach the lower airways, which can lead to drying of the airway secretions and the formation of mucous plugs.

(ii) The addition of water vapor to the inspired air dilutes the O_2 and N_2 concentrations slightly.

(b) Removal of foreign material. Foreign particles are either filtered by the nose, impacted in the lower airways, or dissolved on the moist surface of the airways.

(i) Small particles (e.g., soot or pollen) impact on the surface of the airways and stick to the mucus lining, where they are carried in the mucus toward the mouth to be expectorated or swallowed. The mucus is propelled upward by the cilia of the respiratory epithelium that line the airways.

(ii) Foreign material in the inspired gas (e.g., cigarette smoke, smog) can stimulate irritant receptors in the airways, which causes coughing, increased secretion of mucus, and hypertrophy of the mucous glands. Prolonged breathing of air that contains foreign materials can cause chronic bronchitis with a resultant increase in airway resistance and difficulty in breathing.

b. Alveolar dead space. In healthy people, the anatomic dead space represents the entire dead space, but in patients with lung disease, some alveoli do not receive any blood flow and therefore do not participate in gas exchange. These alveoli form the alveolar dead space.

c. The **total (physiologic) dead space** includes the anatomic dead space and the alveolar dead space. There are several methods of **measuring the total dead space ventilation**.

(1) Bohr's method can be used to measure the total dead space by determining CO_2 tensions in the expired and alveolar gases: $\dot{V}_D = V_T \cdot (1 - P_{ECO_2}/P_{ACO_2})$, where P_{ECO_2} and P_{ACO_2} represent the CO_2 tensions in mixed expired and alveolar gases, respectively.

(a) This method requires measuring the CO_2 tensions in mixed gases and in end-tidal samples of expired gases, which, theoretically, represent pure alveolar gas.

(b) An increase in the dead space lowers the CO_2 tension in the mixed expired gas by dilution. Thus, the lower the mixed expired gas CO_2 tension, the greater the ratio of dead space to alveolar ventilation for any given level of alveolar CO_2 tension.

(2) Fowler's method of determining the dead space uses expired N_2 as an indicator. The subject inspires a VC of 100% O_2 and the expired N_2 is measured during the subsequent expiration [see Chapter 18 IV C 3 b (3)]. This method has limited applications, however; in people with severe pulmonary disease, the alveolar plateau (i.e., phase 3) is difficult to recognize.

C. **Alveolar ventilation ($\dot{V}_A$)** is the volume of gas that reaches the alveolar portion of the respiratory tract per minute and participates in the exchange of O_2 and CO_2. Alveolar ventilation equals the minute ventilation minus the dead space ventilation ($\dot{V}_A = \dot{V}_E - \dot{V}_D$) and averages **4.5–5 L/min.** An adequate alveolar ventilation is critical because it determines the O_2 and CO_2 tensions in the alveoli (see Chapter 18 II).

1. Alveolar CO_2 tension (P_{ACO_2})

a. The alveolar CO_2 tension is determined by the **ratio** of the **rate of CO_2 production** ($\dot{V}_{CO_2}$) to alveolar ventilation ($\dot{V}_A$): $P_{ACO_2} = K \cdot \dot{V}_{CO_2}/\dot{V}_A$, where K is a constant.

(1) Figure 16-20 depicts the effect of varying alveolar ventilation on the arterial CO_2 tension (P_{ACO_2}) at rest and at an increased metabolic rate.

(2) The alveolar and arterial CO_2 tensions are essentially equal because of the high diffusibility of CO_2, and they **depend on only two factors: the rate of CO_2 production** and the **alveolar ventilation** (assuming there is no significant CO_2 tension in the inspired gas). The presence or absence of O_2 has no effect on the alveolar CO_2 tension.

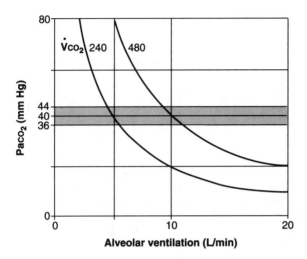

FIGURE 16-20. The graph depicts the effect of changing the alveolar ventilation on arterial CO_2 tension (*Paco₂*). Note that a doubling of the CO_2 production (*V̇co₂*) requires a doubling of the alveolar ventilation to maintain the normal arterial CO_2 tension (*shaded region*), and that a reduction of the alveolar ventilation by 50% will double the arterial CO_2 tension.

b. The alveolar CO_2 tension normally is very closely regulated to a value of 40 ± 4 mm Hg. Regulation of the alveolar CO_2 tension is important because changes in CO_2 tension alter pH, which affects the rates of many enzymatic reactions.

(1) Hypocapnia. An increase in alveolar ventilation (i.e., hyperventilation) with a constant rate of CO_2 production results in a decreased alveolar CO_2 tension because the CO_2 is washed out of the lungs by the increased ventilation. The decreased CO_2 tension in the body leads to **respiratory alkalosis** because CO_2 is an acid-generating molecule.

(2) Hypercapnia. An alveolar ventilation that is inadequate for the metabolic rate (i.e., hypoventilation) results in a rise in alveolar CO_2 tension, which, if it exceeds 45 mm Hg, is termed hypercapnia.

  (a) Hypercapnia usually is accompanied by **hypoxia** and results in an acidotic condition termed **respiratory acidosis**.

  (b) Hypercapnia can lower the O_2 tension in the alveoli by diluting the O_2 that is present.

2. Alveolar O_2 tension (PAO₂). O_2 is continually removed from the alveoli by diffusion into the pulmonary capillary blood. Inspiration brings fresh air into the alveoli, which normally maintains the alveolar O_2 tension at about 100 mm Hg.

a. The alveolar O_2 tension is calculated by means of the **alveolar gas equation**:

$$PAO_2 = (PB - 47) \cdot FIO_2 - PACO_2/R + K, \text{ where}$$

$$
\begin{aligned}
PB &= \text{the barometric pressure} \\
47 &= \text{the vapor pressure of water at body temperature} \\
FIO_2 &= \text{the fraction of oxygen in the inspired gas} \\
PACO_2 &= \text{the tension of } CO_2 \text{ in the alveolar gas} \\
R &= \text{the ratio of the volume of } CO_2 \text{ produced } (V̇CO_2) \text{ to the volume of } \\
&\quad\ O_2 \text{ consumed } (V̇O_2) \text{ per minute, generally assumed to be 0.8} \\
K &= \text{a constant that is relatively small and generally ignored}
\end{aligned}
$$

Therefore, at sea level, $PAO_2 = (760 - 47) \cdot 0.21 - 40/0.8$, or approximately 100 mm Hg.

b. The alveolar O_2 tension is affected by changes in the barometric pressure (e.g., mountain climbing or diving), the fraction of O_2 inspired (which depends on whether the subject is breathing air or a gas with supplemental O_2), and by the amount of CO_2 in the alveoli (i.e., hypoventilation or hyperventilation).

c. The alveolar O_2 tension at sea level can be raised above 600 mm Hg by administering 100% O_2 through a mask. Because prolonged use of 100% O_2 can lead to lung damage, high concentrations of O_2 should be used with caution.

IX. PULMONARY FUNCTION TESTS

A. **Vital capacity (VC).** Measurement of the VC under different conditions is the most commonly performed pulmonary function test.

1. The **slow VC** is used to evaluate the size of the lungs (see Figure 16-18).
   a. Many lung diseases decrease the VC by either reducing the TLC or increasing the RV.
   b. An abnormal VC is one that is more than two standard deviations below the predicted value for the patient's height, age, and sex.

2. The **forced vital capacity (FVC)** is used to evaluate the resistance properties of the airways and the strength of the expiratory muscles. The FVC is obtained by performing a VC maneuver as rapidly as possible into a recording spirometer (Figure 16-21).
   a. The **forced expiratory volume in 1 second (FEV$_1$)** is the most commonly used screening test for airway disease.
      (1) The FEV$_1$ is actually a flow rate that represents the volume of the FVC that is expired during the first second of expiration (see Figure 16-21).
      (2) Because flow equals the driving pressure divided by resistance, the FEV$_1$ varies as a function of the size of the VC, the airway resistance, and the maximal expiratory pressure.
      (3) The expiratory flow rate is directly proportional to the driving force and inversely related to the airway resistance. Therefore, patients with obstructive lung disease (i.e., a high airway resistance) will have a low FEV$_1$.
   b. The **FEV$_{1\%}$** is the percent of the VC that is expired in 1 second [i.e., FEV$_{1\%}$ = (FEV$_1$/FVC) • 100]. Normally, the FEV$_{1\%}$ is greater than 75%–80% of the FVC.
      (1) Patients with **restrictive lung disease** have a reduced VC but are able to achieve relatively high flow rates; therefore, their FEV$_{1\%}$ exceeds 80% (Table 16-3).
      (2) Patients with **obstructive lung disease** have low flow rates as a result of their high airway resistance; consequently, their FEV$_{1\%}$ is abnormally low (see Table 16-3).

B. **Maximal expiratory pressure.** Measuring the maximal expiratory pressure allows physicians to determine whether a low expiratory flow rate is caused by increased airway resistance or by weakness of the expiratory muscles.

1. **Method.** The maximal expiratory pressure is measured by having the patient expire as forcefully as possible into a pressure measuring device.
   a. Because there is no airflow under these conditions, the pressure generated in the alveoli is transmitted without loss to the airway opening (where the pressure is measured).

FIGURE 16-21. Spirogram of the forced vital capacity (FVC) maneuver, which consists of a maximal inspiration followed by a rapid maximal expiration. The FVC should have the same volume as the slow vital capacity (VC), but various measurements can be made on the FVC to evaluate the flow properties of the system. The most common measurement is the forced expiratory volume in 1 second (FEV$_1$), which is indicated by the *dotted portion* of the curve.

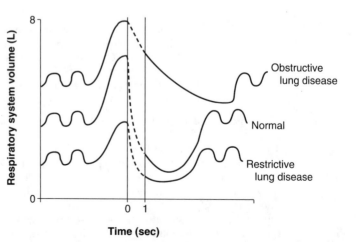

TABLE 16-3. Effects of Lung Disease on the Forced Vital Capacity

	Resistance	FVC	FEV$_1$	FEV$_1$ (%)
Normal	. . .	. . .	. . .	> 80%
Restrictive	↓	↓	↓	> 80%
Obstructive	↑	↓	↓	< 80%

FEV$_1$ = forced expiratory volume in 1 second; FEV$_1$ (%) = percent of forced vital capacity in 1 second; FVC = forced vital capacity.

    b. If flow occurs, the alveolar pressure is degraded by the resistance forces so that the maximal pressure cannot be recorded at the mouth.

2. Interpretation
    a. The patient with a **high airway resistance** can generate a normal expiratory pressure of approximately 100 cm H$_2$O.
    b. The patient with **expiratory muscle weakness** will generate a much lower expiratory pressure (see II B 2). Thus, the use of this simple test can determine the cause of the reduced expiratory flow, which is important to recognize because the treatment and prognosis of the two conditions are much different.

C. **Maximal voluntary ventilation (MVV).** The MVV results in the highest rate of ventilation for any condition and represents the maximal volume of gas that a patient can breathe. This test stresses all of the mechanical factors of breathing and becomes abnormal with increases in airway resistance, reduced compliance, or decreased respiratory muscle force. The largest drawback is that this fatiguing test depends on the patient's complete cooperation (thus the term ''voluntary'' in the name) and the results vary with the motivation of the patient.

    1. Method. The patient breathes as rapidly and forcefully as possible while the volume changes are recorded with a spirometer. The test is carried out for 12 seconds, and the results are expressed in L/min.

    2. Interpretation. Results should be compared to predicted values that are based on a patient's age, height, and sex. A normal value for a young adult man is approximately 200 L/min.

Chapter 17

Gas Transport

I. **INTRODUCTION.** In order for O_2 to cross the respiratory membrane, it must first dissolve in the tissues and then diffuse into the plasma of the pulmonary capillaries.

A. **Henry's law** defines the **volume of gas that will dissolve in any liquid:**

$$Cx = \alpha x \cdot Px, \text{ where}$$
$$Cx = \text{ the volume of gas "x"}$$
$$\alpha x = \text{ its solubility (Bunsen) coefficient}$$
$$Px = \text{ its partial pressure}$$

Gas molecules dissolve in liquids the same way that sugar dissolves in coffee. In order to know how much sugar is dissolved in coffee, one needs to know how many scoops were put in (e.g., the partial pressure) as well as the size of the scoops (e.g., the solubility coefficient).

1. **Partial pressure.** The amount of any **gas dissolved in a liquid increases as a function of the partial pressure**.
   a. Unlike sugar, which can cause a saturated solution, there are **no limits on the amount of gas that can be dissolved over physiologic ranges of gas tensions**.
   b. **Gas content versus partial pressure.** It is important to understand the difference between the partial pressure of a gas and the gas content of a liquid.
      (1) The **partial pressure** of the gas represents the pressure it would exert in the gas phase.
      (2) The **gas content** represents the volume of the gas per unit volume of liquid that is present; the units of gas content are ml gas/ml (or L) solvent.

2. **Solubility.** The solubility of a gas is a major determinant of the number of molecules that are present; therefore, solubility **affects the rate of diffusion**.
   a. The **solubility coefficient** is a **constant** for each gas "x" that equates gas content and partial pressure: $\alpha x = Cx/Px$.
   b. Increasing the temperature reduces the solubility of gases (Table 17-1).

B. **Equilibrium.** Liquids must be exposed to a gas tension for a finite time for the gases to dissolve in the liquid phase. If the exposure time is long enough, the gas tension in the liquid will become equal to that of the gas phase and an equilibrium will exist between the gas and the liquid phases.

1. The time required for this equilibrium to occur is a function of the area of contact (surface area) between the liquid and the gas, the solubility and diffusion properties of the gas, and the diffusion gradient.

2. In the lungs, the time available for this equilibrium to occur is the time that the blood spends in the pulmonary capillaries, normally about 0.75 second (see also Chapter 18 II C).

TABLE 17-1. Solubility Coefficients for Gases Important to Respiratory Physiology

Temperature (°C)	Solubility Coefficient (ml gas/ml saline/mm Hg gas tension)			
	O_2	N_2	CO_2	CO
0	0.049	0.024	1.71	0.035
20	0.032	0.016	0.90	0.023
37	0.024	0.012	0.58	0.019

II. O_2 TRANSPORT

A. Method

1. After diffusing across the respiratory membrane of the lungs, the **O_2 dissolves in the plasma of the pulmonary capillaries**.

2. From the plasma, the **O_2 diffuses into the red blood cell,** where it **combines reversibly** with the iron atoms of hemoglobin and **converts deoxyhemoglobin into oxyhemoglobin**.
   a. **Hemoglobin,** the O_2-carrying protein in red blood cells, increases the amount of O_2 that is carried in the blood (compared to plasma) almost 70-fold.
      (1) Each gram of hemoglobin can combine maximally with 1.34 ml of O_2; normally, there are 12–15 g hemoglobin/dl blood.
      (2) Therefore, the arterial blood carries approximately 20 ml O_2/dl; about 98.5% of this O_2 is carried by hemoglobin, and 1.5% is dissolved in the plasma.
   b. **Hemoglobin saturation** is the percent of hemoglobin that is combined with O_2.
      (1) Each hemoglobin molecule can combine with as many as **four O_2 molecules**. If all of the sites on the hemoglobin molecule are occupied by O_2, then the molecule is 100% saturated.
      (2) The amount of O_2 that combines with hemoglobin depends on the O_2 tension (Po_2), as indicated by the O_2–hemoglobin dissociation (association), or saturation, curve (Figure 17-1).

B. O_2–hemoglobin dissociation (association) curve. The **driving force** for the chemical reaction between hemoglobin and O_2 is the **O_2 tension in the pulmonary capillaries,** which is normally equal to the alveolar O_2 tension (Pao_2). Thus, the alveolar O_2 tension, which is usually about 100 mm Hg, represents the upper limit for the O_2 tension in the body.

1. The **loading (association) zone** of the hemoglobin saturation curve is the **plateau** that occurs above O_2 tensions of approximately 60 mm Hg (see Figure 17-1).
   a. **Arterial hemoglobin saturation.** The O_2 tension of arterial blood is normally 85–100 mm Hg, which results in an arterial hemoglobin saturation of 96%–98% (Figure 17-2).
   b. The loading zone provides a **margin of safety,** because the alveolar O_2 tension can be reduced substantially, yet the hemoglobin saturation remains quite high. This leeway allows people to climb mountains to moderate altitudes or to suffer from various pulmonary diseases without experiencing a significant decrease in the arterial **hemoglobin saturation,** even though the arterial O_2 tension decreases substantially under these conditions.

2. The **unloading (dissociation) zone** of the hemoglobin saturation curve is the **steep portion** of the curve that occurs at O_2 tensions below 60 mm Hg (see Figure 17-1).

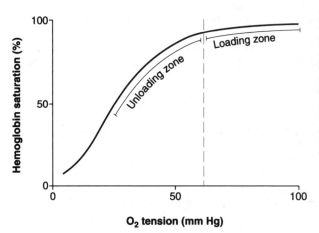

FIGURE 17-1. Hemoglobin saturation with O_2 as a function of the O_2 tension of the blood. In the loading zone (i.e., the pulmonary capillaries) the hemoglobin is exposed to an O_2 tension of approximately 100 mm Hg and the reversible, chemical combination between O_2 and hemoglobin occurs. When the hemoglobin reaches the unloading zone (i.e., the systemic capillaries), the low O_2 tension in the tissues causes O_2 to diffuse from the blood into the tissues.

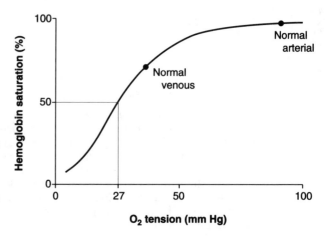

FIGURE 17-2. The normal arterial O_2 tension is about 95 mm Hg, which produces a hemoglobin saturation of approximately 98%. After O_2 is released to the tissues, the average venous O_2 tension is 40 mm Hg, which is equivalent to a hemoglobin saturation of 75%. The derivation of the normal P_{50} (i.e., the O_2 tension that produces a 50% saturation of hemoglobin with O_2) is shown.

a. **Establishment of an O_2 gradient**
   (1) The **O_2 tension in the tissues** is very low (because metabolism consumes O_2). An O_2 gradient is present between the capillary blood and the tissues, causing O_2 to diffuse from blood to tissue.
   (2) The **capillary O_2 tension** is determined by the ratio of the rate of local blood flow to the rate of O_2 consumption (i.e., a low blood flow or a high O_2 consumption will lead to low tissue and capillary O_2 tensions). As O_2 diffuses into the tissues, the capillary O_2 tension is reduced.
b. The steepness of the unloading zone indicates that **hemoglobin is able to release large amounts of O_2 in response to relatively small changes in O_2 tension.** This property of the dissociation curve tends to keep the O_2 tension in the capillary blood at a maximum level so that the diffusion gradient for O_2 is maintained.
c. **Venous hemoglobin saturation.** After the blood has transferred O_2 to the tissues via the microcirculation, the **mixed venous O_2 tension is about 40 mm Hg,** leading to a hemoglobin saturation of **approximately 75%** (see Figure 17-2). This is only an average value; the O_2 tension in the venous blood varies from organ to organ as a function of the ratio of blood flow to O_2 consumption in each tissue.

3. **Hemoglobin affinity** for O_2 is **inversely related** to the **P_{50},** which is the O_2 tension that produces a 50% saturation of hemoglobin with O_2.
   a. The **normal P_{50}** for arterial blood is **27 mm Hg** (see Figure 17-2). Hemoglobin is an allosteric enzyme that reacts with O_2; therefore, the affinity between hemoglobin and O_2 can be altered by various ligands.
      (1) **Physiologic alterations in P_{50}** (Figure 17-3) are produced by changes in temperature and by changes in the concentration of CO_2, H^+, and other ligands [e.g., diphosphoglycerate (DPG), adenosine triphosphate (ATP), adenosine diphosphate (ADP), and other organic phosphates].
      (2) **Bohr effect.** The Bohr effect refers to the displacement of the hemoglobin saturation curve, and the subsequent increase in P_{50}, that is caused by an increase in CO_2 tension (P_{CO_2}).
   b. A **decreased P_{50}** (i.e., an **increased hemoglobin affinity**) means that O_2 combines more readily with hemoglobin. A decreased P_{50} is an obvious benefit in the lungs because it enhances the hemoglobin saturation with O_2.
   c. An **increased P_{50}** (i.e., a **decreased hemoglobin affinity**) causes hemoglobin to release O_2 more readily at the tissue level. The increase in the P_{50} maintains the O_2 tension of the blood at higher levels, which improves the diffusion of O_2 from the blood to the tissues.

C. The **O_2 capacity** is the maximal amount of O_2 that can be carried in the blood by hemoglobin.

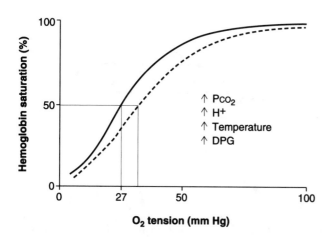

FIGURE 17-3. The effects of increasing the CO_2 tension (P_{CO_2}), H^+ concentration, temperature, and diphosphoglycerate (*DPG*) concentration on the P_{50} of the hemoglobin molecule (*dashed line*). The increase in P_{50} indicates a reduced affinity of hemoglobin for O_2, which raises the tissue O_2 tension. Decreases in these parameters cause a shift of the curve to the left, which represents an increased affinity of the hemoglobin for O_2.

1. Only the **functional hemoglobin concentration** (i.e., the hemoglobin available to react with O_2) determines the O_2 capacity.

2. The O_2 capacity equals the hemoglobin concentration (in g/dl of blood) multiplied by 1.34 ml O_2/g (Table 17-2).

3. Because the normal hemoglobin concentration is 15 g/dl, the **normal O_2 capacity** is **20.1 ml/dl** (or 20.1 vol%).

D. The **O_2 content** is the **total amount of O_2 actually being carried by the blood** (i.e., by the hemoglobin and dissolved in the plasma). Obviously, the amount of dissolved O_2 does not depend on the hemoglobin concentration, but on the O_2 tension in the blood. The **functional hemoglobin concentration,** the **O_2 tension in the blood,** and the **P_{50} of the hemoglobin** determine the O_2 content (Figure 17-4).

1. Hemoglobin saturation $= \dfrac{O_2 \text{ content} - \text{dissolved } O_2}{O_2 \text{ capacity}} \cdot 100$

2. **O_2 tensions and contents**
   a. **Normal values.** On average, approximately 4.5 ml of O_2 are released from every deciliter of blood in the systemic capillaries to supply tissue metabolism.
      (1) The **normal arterial O_2 content (Ca_{O_2})** is approximately 19.5 ml O_2/dl blood (or **19.5 vol%**) and has a tension of 90–95 mm Hg.
      (2) The **normal mixed venous O_2 content ($C\bar{v}_{O_2}$)** is approximately **15 vol%** at a tension of 40 mm Hg.
         (a) Mixed venous blood represents the **weighted average of blood flow and O_2 consumption throughout the body**.
            (i) The venous O_2 content (Cv_{O_2}) varies widely in different organs depending on organ blood flow and O_2 consumption.
            (ii) For example, the heart usually has a low venous O_2 content, but the kidney, which receives approximately 25% of the cardiac output, has a high venous O_2 content.

TABLE 17-2. Key Equations

O_2 capacity (ml/dl) $= [Hb] \cdot 1.34$

O_2 content (ml/dl) $= HbO_2 +$ dissolved O_2

Hb saturation (%)* $= \dfrac{O_2 \text{ content} - \text{dissolved } O_2}{O_2 \text{ capacity}} \cdot 100$

Hb = hemoglobin; [Hb] = hemoglobin concentration; HbO_2 = oxyhemoglobin.
*The P_{50} is the O_2 tension that produces 50% Hb saturation with O_2.

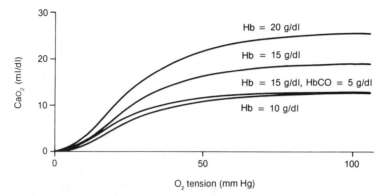

FIGURE 17-4. O_2 content of arterial blood (CaO_2) as a function of O_2 tension and hemoglobin (*Hb*) concentration. The arterial O_2 content varies according to the O_2 tension and the P_{50}, which have different values throughout the body. Note that carbon monoxide (*HbCO*) not only reduces O_2 content, it also reduces the P_{50}.

      (b) Mixed venous blood is obtained from the right ventricle or the pulmonary artery.

    b. The **Fick principle** (see Chapter 13 II A) can be used to determine the consumption of O_2 by the body. The normal cardiac output is 5–5.5 L/min (i.e., the blood flow equals 50–55 dl/min). If each deciliter of blood unloads 4.5 ml of O_2 in the systemic capillaries, then the body consumes 225–250 ml O_2/min.

3. O_2 delivery represents the **amount of O_2 that is presented to body cells per minute** and is equal to the arterial O_2 content times the blood flow. For the entire body, the average arterial O_2 content is 195 ml/L and the blood flow is the cardiac output (5–5.5 L/min). Thus, the **normal O_2 delivery to the entire body** is about **1 L/min**. Decreases in either the arterial O_2 content or cardiac output will reduce the O_2 delivery.

    a. Decreased arterial O_2 content

      (1) Causes

        (a) Decreased hemoglobin concentration

        (b) Reduced arterial O_2 tension

      (2) Effects. A reduced hemoglobin concentration or a reduced arterial O_2 tension can lead to inadequate O_2 delivery and tissue hypoxia (see Chapter 20 II C 1, 3).

      (3) Compensatory mechanisms. The body compensates for decreases in arterial O_2 by increasing cardiac output. A decreased arterial O_2 tension can be compensated by an increased hemoglobin concentration (i.e., an increased O_2 capacity) as well as an increased cardiac output.

    b. Decreased cardiac output

      (1) Causes. A variety of cardiac diseases can decrease cardiac output.

      (2) Compensatory mechanisms. Chronic reduction in cardiac output can lead to an **increased red cell production**. This is not an effective compensatory mechanism because the increased hematocrit (i.e., the percent of the blood occupied by red blood cells) increases viscosity, increasing resistance to blood flow and the ventricle's afterload. Both of these effects tend to reduce the cardiac output further.

4. Factors affecting O_2 content

    a. Carboxyhemoglobin is formed when carbon monoxide (CO) combines with hemoglobin; it imparts a typical **cherry-red color** to the skin and mucous membranes. Because CO binds to the same site on the hemoglobin molecule as O_2, each hemoglobin molecule can carry some combination of four molecules of either O_2 or CO. CO has about 200 times the affinity for hemoglobin as does O_2; therefore, carboxyhemoglobin is a rather stable molecule.

(1) Effects of CO

    (a) CO decreases the functional hemoglobin concentration—if a site is occupied by CO, it is unavailable for O_2 transport. Essentially, **CO poisoning** is a form of **acute-onset anemia**.

        (i) Because of the extreme affinity of CO for hemoglobin, a **carbon monoxide tension (Pco) of only 0.5 mm Hg causes 50% of the hemoglobin to react with CO** (Figure 17-5).

        (ii) The rapid loss of 50% of the functional hemoglobin can be fatal.

    (b) CO lowers the tissue O_2 tension by decreasing the O_2 content and the P_{50} for O_2 (i.e., it increases hemoglobin's affinity for O_2).

        (i) The CO combines with the most reactive hemoglobin molecules (i.e., those molecules that have the highest P_{50}), so that the remaining hemoglobin that is available for O_2 transport has a lower than normal P_{50} (i.e., a high O_2 affinity).

        (ii) The shift in the P_{50} moves the hemoglobin dissociation curve to the left and lowers the tissue O_2 tension even further than the decreased O_2 content would indicate (see Figure 17-4).

(2) Dissociation of CO from hemoglobin

    (a) CO is slowly displaced from hemoglobin at normal O_2 tensions; however, this process may require 8–12 hours.

    (b) Administration of 100% O_2 speeds the dissociation of CO from hemoglobin. In addition, significant amounts of O_2 (about 1.5 vol%) are dissolved in the plasma at this high O_2 tension, which helps to improve the O_2 delivery to the body tissues.

b. Methemoglobin is formed when **hemoglobin iron** is oxidized from the **ferrous (Fe^{2+})** to the **ferric (Fe^{3+}) state**. Methemoglobin is incapable of carrying O_2 and has a bluish color that can impart a cyanotic hue to tissues.

    (1) Because hemoglobin is exposed to O_2, it is oxidized to methemoglobin continuously at a slow rate within the red blood cell. Many drugs and foods (e.g., nitrites, phenacetin, fava beans) can produce methemoglobin, especially in people who have a genetic deficiency of glucose-6-phosphate dehydrogenase (G6PD).

    (2) Methemoglobin is converted back to functional hemoglobin by reducing compounds that are produced in the red blood cell by metabolic reactions.

c. Abnormal hemoglobin molecules occur as genetic variants and have either **abnormal physical characteristics** (as in sickle cell anemia) or **abnormal affinities for O_2**.

    (1) Sickle cell hemoglobin (hemoglobin S) becomes very insoluble in the deoxygenated state, causing these molecules to precipitate in the red blood cells and giving the red blood cells their characteristic, bizarre shape.

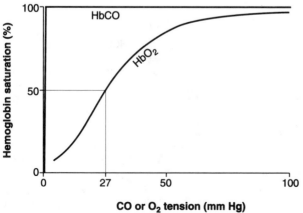

FIGURE 17-5. Hemoglobin saturation curves for both O_2 (*HbO$_2$*) and CO (*HbCO*). Note that the carboxyhemoglobin curve lies along the y-axis, which indicates the extremely high affinity of hemoglobin for CO. The P_{50} for O_2 is shown; the tension of CO required for 50% saturation of hemoglobin is about 0.5 mm Hg.

(2) The sickle cells are rigid and exhibit a higher viscosity to flow, causing them to lodge in capillaries. The absent or reduced blood flow results in ischemic damage to many different organ systems in affected individuals.

III. CO_2 **TRANSPORT.**
CO_2 is the usual end-product of oxidative metabolism and is formed in the tissues. CO_2 diffuses out of the cells and into the capillary blood, raising the CO_2 tensions of the capillary and venous blood to levels 5–6 mm Hg higher than the arterial CO_2 tension. CO_2 is transported in the blood in **several different forms**.

A. **Physically dissolved CO_2** accounts for approximately 5% of the total CO_2 in the blood.

B. **Carbaminohemoglobin** refers to CO_2 that combines with terminal amino (NH_3) groups on protein side chains to form NH_2-COO^- + H^+. Approximately 5% of the CO_2 is carried as carbamino compounds.

C. **Bicarbonate (HCO_3^-).** Approximately 90% of the CO_2 enters the red blood cell, where, in the presence of **carbonic anhydrase,** it rapidly reacts with water to form **carbonic acid** (Figure 17-6). The carbonic acid dissociates into bicarbonate (HCO_3^-) and one hydrogen ion (H^+). The HCO_3^- diffuses into the plasma, and the H^+ is buffered by the deoxygenated hemoglobin, which is a weaker (less dissociated) acid than oxyhemoglobin. Although HCO_3^- is formed within the red cell, most of the CO_2 is carried in the plasma as HCO_3^-.

1. **Chloride shift.** As the HCO_3^- is formed in red blood cells at the tissues, it diffuses out of the cells (see Figure 17-6). Since HCO_3^- is a negatively charged molecule, and the red blood cell membrane is relatively impermeable to cations, as HCO_3^- diffuses out of the cell it causes the inside of the cell to become less negatively charged. In order to neutralize this effect, the negatively charged chloride ion (Cl^-) diffuses from the plasma into the red cell to replace the HCO_3^-.

FIGURE 17-6. The chloride shift in a red blood cell. CO_2 tension increases in the blood as it passes through the systemic capillaries (*thin arrows*) causing CO_2 to diffuse into the red blood cell where it is quickly hydrated by carbonic anhydrase (*CA*). The hydrogen ion (H^+), derived from the dissociation of carbonic acid (H_2CO_3), is buffered by hemoglobin (*HHb*). CO_2 can also combine with the amino terminal (NH_2) of hemoglobin, giving rise to carboxyhemoglobin (*HHb - NHCOO$^-$*). Bicarbonate (HCO_3^-) diffuses out of the red blood cell and is replaced by chloride (*Cl$^-$*). Most of the CO_2 picked up from the tissues is transported in the plasma in the form of bicarbonate. In the pulmonary capillaries, these reactions are reversed (*thick arrows*).

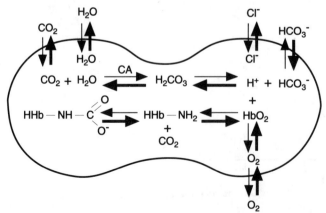

2. The **Haldane effect** refers to the increased CO_2 content of the blood at any CO_2 tension in the presence of deoxygenated (as opposed to oxygenated) hemoglobin.
   a. Deoxyhemoglobin is able to buffer (i.e., combine with) more H^+ than oxyhemoglobin, which allows more CO_2 to be transported as HCO_3^- at any CO_2 tension (Figure 17-7).
   b. The Haldane effect minimizes the increase in CO_2 tension that occurs in the venous blood, which, in turn, causes the venous blood to be less acidic than it would be otherwise. Thus, the Haldane effect is a way of buffering the CO_2 molecule, which has the potential to form carbonic acid after being hydrated.

FIGURE 17-7. CO_2 dissociation curves for the tissue capillaries (*dashed line*) and the lungs (bold line), which demonstrate the Haldane effect's role in the transport of CO_2. Note that the CO_2 content of blood varies depending on the hemoglobin saturation with O_2. Desaturated hemoglobin allows more CO_2 to be carried for the same CO_2 tension—its increased H^+ buffering ability allows the formation of more bicarbonate (HCO_3^-), which is the major transport form of CO_2. Thus, the release of O_2 by the hemoglobin in the capillaries causes CO_2 to bind to the hemoglobin at a greater rate, and in the lungs, the release of CO_2 is facilitated by the increase in O_2. Points A and B represent the normal CO_2 tensions in the tissues and arteries, respectively. Po_2 = oxygen tension.

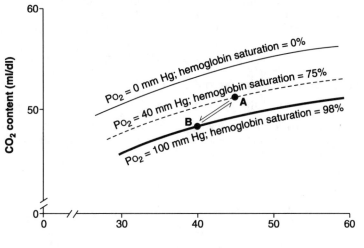

Chapter 18

Gas Exchange

I. INTRODUCTION. Alveolar ventilation and pulmonary blood flow (perfusion) are not evenly distributed throughout the lungs. This uneven distribution controls alveolar and pulmonary capillary gas tensions, which affect the rate of diffusion of O_2 and CO_2 across the alveolar–capillary membrane.

II. DIFFUSION is the movement of individual molecules from areas of higher concentration to areas of lower concentration by random motion (see also Chapter 1 III A). The rate of movement of the individual gas molecules is a direct function of the absolute temperature and is inversely related to their molecular weight.

A. **Factors affecting the rate of pulmonary gas diffusion** (Figure 18-1). The rate of pulmonary gas diffusion (i.e., the volume of gas per minute that crosses the alveolar–capillary membrane) is determined by several factors as defined by **Fick's Law** of diffusion:

$$\dot{V}\, gas = (P_C - P_A) \cdot \frac{D \cdot A}{d}$$

1. The **partial pressure of the gas in the alveoli** (P_A)

2. The **partial pressure of the gas in the pulmonary capillary** (P_C)

3. The **diffusion coefficient** of the gas (D), which is specific for each type of gas molecule and can be obtained from a table of physical constants. Physical constants that determine the diffusion coefficient include the molecular weight (because the diffusion rate

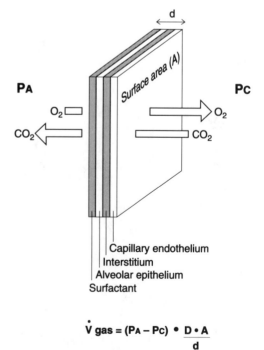

FIGURE 18-1. The alveolar–capillary membrane is multilayered but, on average, is only 0.5 μm (0.00002 inch) thick. The arrows indicate the direction of diffusion of O_2 and CO_2. Fick's equation, as it applies to pulmonary gas transfer, is shown. $\dot{V}$ = volume of gas transferred per minute; P_A = the alveolar gas tension; P_C = the pulmonary capillary gas tension; D = the diffusion coefficient; d = thickness of the membrane; and A = the surface area of the membrane.

P_A

P_C

O_2

O_2

CO_2

CO_2

Surface area (A)

d

Capillary endothelium
Interstitium
Alveolar epithelium
Surfactant

$$\dot{V}\, gas = (P_A - P_C) \cdot \frac{D \cdot A}{d}$$

is an inverse function of the square root of the molecular weight), solubility in a particular solvent, and absolute temperature (which is normally body temperature).

4. The **dimensions of the alveolar–capillary membrane** [i.e., the diffusion distance (d) and the total surface area (A) for exchange]
   a. The **diffusion distance** is the average thickness of the alveolar–capillary membrane, which consists of the surfactant layer, the alveolar epithelium (type I cells), the pulmonary interstitium, and the capillary endothelium (see Figure 18-1). Pulmonary edema or an increase in the amount of connective tissue in the interstitium (as occurs in some forms of pulmonary interstitial disease) will increase the diffusion distance.
   b. The **surface area** of the respiratory membrane is very large, averaging 80–100 m^2 in the normal adult. The large surface area and the small distance between the alveolar gas and the pulmonary capillary blood normally allow rapid equilibration of gas tension between the two.

B. **Diffusing capacity**

1. **Measurement of the diffusing capacity of the lungs (D_L)** permits evaluation of the diffusion properties of the alveolar–capillary membrane by measuring the rate of gas transfer (conductance) by the respiratory system. The diffusing capacity is defined as the **volume of gas absorbed by the lungs per minute ($\dot{V}$) per mm Hg partial pressure gradient** and its equation is derived from the Fick equation:

$$D_L = \frac{\dot{V}}{P_A - P_C}$$

Today, the diffusing capacity of the lungs is measured using a dilute concentration (about 0.01%) of CO because CO markedly simplifies the measurement.
   a. **Diffusing capacity of the lungs for CO (D_{LCO})**
      (1) The dilute CO mixture and the high affinity of hemoglobin for CO ensure that the pulmonary capillary CO tension (P_{CCO}) is close to zero during the test (unless the subject is a smoker or exposed to high atmospheric CO levels). Eliminating the pulmonary capillary gas tension simplifies the technique used to measure diffusing capacity.
      (2) The **normal value of the D_{LCO}** is approximately **30 ml/min/mm Hg.**
   b. The **diffusing capacity of the lungs for O_2 (D_{LO_2})** can be calculated from the D_{LCO} based on differences in the molecular weight and solubility of the two molecules. The **normal, resting value of the D_{LO_2}** is about **25 ml/min/mm Hg.**

2. **Factors affecting the diffusing capacity of the lungs.** Because the diffusion properties of the respiratory system are not the sole determinants of the diffusing capacity of the lungs, this test is called the **transfer factor** in England. Theoretically, the diffusing capacity of the lungs is proportional to the surface area of the lungs divided by the length of the diffusion path; however, in actuality, the transfer process is a function of both the diffusion properties of the system (i.e., the **membrane component**) and the rate of the chemical reaction with hemoglobin (i.e., the **blood component**). Each of these factors represents a resistance to gas transfer (recall that resistance = 1/conductance) and in normal individuals, these resistances are about equal.
   a. The **membrane component.** Pulmonary diseases affect the process of diffusion by reducing the exchange surface area (e.g., emphysema, which destroys the alveolar septa), increasing the diffusion distance (e.g., interstitial lung disease, pulmonary edema), or reducing the partial pressure gradient for the diffusion of gases (e.g., ventilation–perfusion abnormalities). All of these diseases reduce the rate of gas transfer across the respiratory membrane.
   b. The **blood component** depends on the **chemical reaction time** and also the **amount of hemoglobin** that is present in the pulmonary capillaries.
      (1) The chemical combination of gases with hemoglobin requires a finite time and represents a resistance to gas transfer.

(2) An increased amount of hemoglobin in the pulmonary capillaries, which occurs as the result of either an increased blood volume or an increased hemoglobin concentration, enhances the transfer of gases.

    (a) Thus, anemic patients have a reduced diffusing capacity but patients with pulmonary vascular congestion experience an enhanced diffusing capacity.

    (b) A supine position or an increased cardiac output also enhance the diffusing capacity. Exercise can increase the diffusing capacity to 60–75 ml O_2/min/mm Hg.

C. Equilibration

1. O_2

  a. **Diffusion of O_2 occurs from the alveolar gas to the pulmonary capillary blood** (recall that the normal alveolar O_2 tension is approximately 100 mm Hg, while the blood entering the pulmonary capillary normally has an O_2 tension of approximately 40 mm Hg). After crossing the alveolar–capillary membrane, the O_2 molecules dissolve in the plasma, raising the plasma O_2 tension and causing O_2 to diffuse into the red blood cell, where combination with hemoglobin takes place.

  b. **Equilibration time.** Normally, enough O_2 diffuses across the respiratory membrane so that the blood O_2 tension and the alveolar O_2 tension equalize in about **0.25 second** (Figure 18-2).

2. CO_2

  a. **Diffusion of CO_2 occurs from the pulmonary capillary blood to the alveoli.** The average CO_2 tension in the pulmonary capillary blood is 46 mm Hg, as opposed to 40 mm Hg in the alveoli. Although the CO_2 diffusion gradient is only one tenth of the O_2 diffusion gradient, CO_2 diffuses almost 20 times more rapidly than O_2 because it is about 25 times more soluble in body fluids.

  b. **Equilibration time.** It is estimated that the time required for the blood CO_2 tension and the alveolar CO_2 tension to equalize is also approximately 0.25 second.

3. Factors affecting gas exchange. The amount of gas exchanged across the respiratory membrane may be dependent on either the perfusion or diffusion properties of the system (Figure 18-3).

  a. **Perfusion-limited gas exchange** occurs when **equilibration takes place** between the alveolar gas and the pulmonary capillary blood (i.e., increased gas exchange can only be achieved by increasing blood flow). The transfer process is normally perfusion-limited.

    (1) As soon as the O_2 tension equilibrates, the net transfer of O_2 ceases; therefore, during perfusion-limited gas exchange, no additional O_2 uptake occurs until the capillary blood is replaced by new unsaturated venous blood.

FIGURE 18-2. Pulmonary capillary gas tensions as a function of the alveolar gas tension. The red blood cells spend an average of 0.75 second in the pulmonary capillaries at rest. A patient with a normal diffusing capacity of the lung for O_2 (D_{LO_2}) is able to achieve equilibration between alveolar gas and pulmonary capillary blood within 0.25 second. If the D_{LO_2} is diminished, progressively more time is required for O_2 transfer and the equilibration time is prolonged or never occurs. Carbon monoxide (*CO*) never equilibrates between the alveolar gas and the blood because it is administered in very low concentrations and the blood can combine with large volumes of the gas. Nitrous oxide (N_2O) equilibrates rapidly between alveolar gas and blood because it is relatively insoluble in the blood and only small amounts of N_2O produce a rapid rise in gas tension.

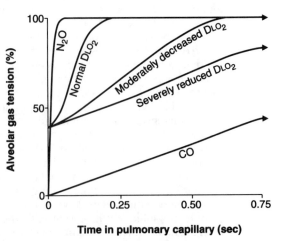

A. Perfusion-limited gas exchange   **B. Diffusion-limited gas exchange**

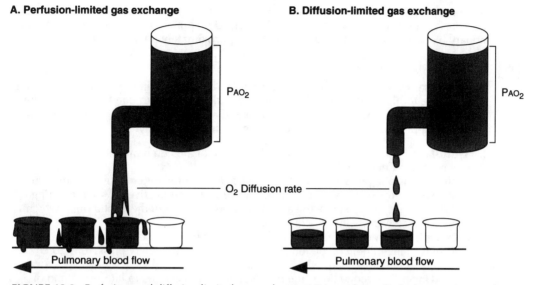

FIGURE 18-3. Perfusion- and diffusion-limited gas exchange. (*A*) In perfusion-limited gas exchange, the blood is equilibrated (as symbolized by the full beakers) so that the only method of increasing gas uptake is to increase pulmonary blood flow (by moving more beakers through the system). (*B*) In diffusion-limited gas exchange, the blood is not equilibrated (i.e., the beakers are not full) so that the diffusion rate must be increased to improve gas exchange. The diffusion rate can be increased by raising the alveolar O_2 tension (P_{AO_2}) and increasing the driving force for diffusion.

    (2) The average red blood cell spends approximately 0.75 second in the pulmonary capillary. If O_2 equilibration occurs in 0.25 second, then there is normally no increase in the O_2 content for the last 0.50 second of transit through the pulmonary capillary. This time provides a **safety margin** that ensures an adequate O_2 uptake during periods of stress (e.g., exercise, exposure to high altitudes) or impaired diffusion.

  b. Diffusion-limited gas exchange occurs whenever **equilibration does not occur** between the alveolar and pulmonary gas tensions. Increasing the diffusion gradient [e.g., by raising the fraction of inspired O_2 (F_{IO_2})] is one way to counteract diffusion-limited gas exchange.

    (1) CO exchange is diffusion-limited, which is why measurement of the D_{LCO} is an effective means of evaluating the diffusion properties of the respiratory system.

    (2) Clinical significance

      (a) Many pulmonary diseases reduce the rate of O_2 transfer, preventing equilibration from occurring between alveolar gas and pulmonary capillary blood (see Figure 18-2). Diffusion-limited gas exchange is one cause of **arterial hypoxia** (i.e., a reduced O_2 tension in the arterial blood).

      (b) Disease processes or **significant increases in altitude reduce the alveolar O_2 tension,** which in turn reduces the diffusion rate. Administration of a reduced F_{IO_2} was formerly used to evaluate the diffusion properties of the respiratory system—a reduced arterial O_2 tension indicated impairment of the lungs' diffusing capacity.

III. **DISTRIBUTION OF PULMONARY BLOOD FLOW.** Pulmonary blood flow varies throughout the lungs because of the low pressures that are present, the distensibility of the vasculature, and the hydrostatic effects of gravity.

A. **Pulmonary blood flow.** The entire blood flow from the right ventricle is distributed to the pulmonary blood vessels. Since right- and left-sided output are necessarily equal, the pulmonary blood flow **equals the cardiac output (approximately 5 L/min at rest)**.

B. **Pulmonary blood pressure**

1. **Pulmonary artery pressure** is much **lower than the pressure in the aorta, because the resistance to blood flow in the pulmonary circulation is about one tenth that of the systemic circulation**. The normal systolic and diastolic pressures in the pulmonary artery are approximately 20 mm Hg and 10 mm Hg, respectively.

2. **Hydrostatic pressure** (P_h) is caused by the weight of a column of fluid that is exposed to gravity [see Chapter 9 III A 2 b (1)].
   a. **The hydrostatic pressure adds to or subtracts from the potential energy (pressure) at levels below or above the zero reference plane, respectively.** When referring to the lungs, the zero reference plane is taken as the level of the right atrium, which is approximately at the middle of the lungs.
   b. **Example.** Assume that the lungs are 40 cm from base to apex, the zero reference level is in the middle of the lungs, the mean pulmonary artery pressure is 20 cm H_2O, and the blood density is 1 g/cm^3.
      (1) Under these conditions, the blood pressure is sufficient to lift the column of blood to a height of 20 cm above the right atrium.
      (2) If the pulmonary artery pressure decreases to 16 cm H_2O, then the top 4 cm of the lung will not be perfused, because the pressure is inadequate to raise the column of blood to that height.
      (3) In the supine or prone positions, the lung's exposure to gravity is reduced; therefore, hydrostatic effects are minimized in these positions.

3. **Perfusion zones of the vertical lung** have been defined that depend on the relationships between alveolar, pulmonary artery, and pulmonary venous pressures (Figure 18-4).
   a. **Zone 1** is the region at the lung apex that does not receive blood flow if the regional pulmonary artery pressure is less than the alveolar pressure (**Pa < PA**). Under these conditions, the pulmonary capillaries are collapsed, preventing blood flow (see Figure 18-4). Zone 1 is increased by factors that reduce the pulmonary artery pressure or increase the alveolar pressure [e.g., positive end-expiratory pressure (PEEP)]; it is decreased by factors that increase the pulmonary artery pressure or reduce the vertical height of the lung (e.g., lying down).

FIGURE 18-4. The effect of hydrostatic pressure and level on the distribution of pulmonary blood flow in a vertical lung. Pulmonary blood flow depends on the relationship between pulmonary artery pressure (*Pa*), alveolar pressure (*PA*), and pulmonary venous pressure (*Pv*). (Adapted from West JB, et al: Distribution of blood flow in the isolated lung: relation to vascular and alveolar pressure. *J Appl Physiol* 19:713, 1964.)

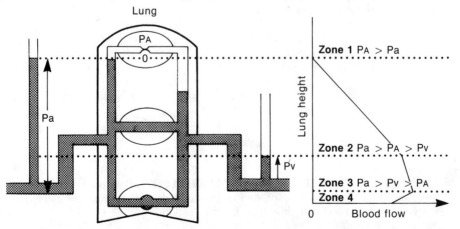

    b. Zone 2 is the region of the lung where the pulmonary artery pressure exceeds the alveolar pressure, which exceeds the pulmonary venous pressure (**Pa > P**A **> Pv**).

       (1) In this region, the arterial–alveolar pressure gradient produces flow, not the arteriovenous gradient, which is the usual mechanism.*

       (2) Since the pulmonary artery pressure increases 1 cm H_2O for each centimeter's descent from the apex of the lung, but the alveolar pressure remains constant, there is an increase in the driving force for blood flow in zone 2. Thus, flow increases linearly from top to bottom of zone 2 (see Figure 18-4).

    c. Zone 3 occurs in the lower portion of the lung, where the pulmonary artery pressure exceeds the pulmonary venous pressure, which exceeds the alveolar pressure (Pa > Pv > PA). The increased hydrostatic pressure in this zone causes distention and recruitment of both arteries and veins. Both distention and recruitment of blood vessels decrease the resistance to blood flow (Figure 18-5) so that blood flow increases down zone 3 (see Figure 18-4).

    d. Zone 4 impinges on zone 3 at the base of the lung and represents a zone of reduced blood flow (see Figure 18-4). Zone 4 is present only when there is an abnormally high pulmonary venous pressure (e.g., due to left ventricular failure or mitral stenosis), or when early-stage pulmonary edema produces edema in the interstitium around the blood vessels. This edema is termed **vascular cuffing;** it increases vascular resistance and reduces local blood flow.

C.   **Factors that alter pulmonary blood flow**

    1. Alveolar hypoxia (i.e., low alveolar O_2 tension) produces **hypoxic pulmonary vasoconstriction (HPV),** a local response (i.e., it occurs only in hypoxic areas) that is potentiated by hypercapnia and acidosis.

      a. Function. HPV is an important mechanism of balancing blood flow and ventilation. HPV results in an increased pulmonary vascular resistance in hypoxic areas of the lung, which are produced by a reduced alveolar ventilation.

       (1) The increased resistance reduces blood flow to the hypoxic area and shifts blood flow to pulmonary regions that are better ventilated and have a higher alveolar O_2 tension. Thus, HPV tends to match the amount of pulmonary blood flow to the alveolar ventilation in different areas of the lung; however, this matching is not perfect (see V A 1).

       (2) The **mechanism of HPV is unknown,** but it is important to remember that hypoxia, hypercapnia, and acidosis produce vasoconstriction in the pulmonary circulation, rather than vasodilation in the systemic circulation.

      b. Generalized alveolar hypoxia. If alveolar hypoxia is present throughout the lung (e.g., as occurs in hypoventilation or exposure to altitudes greater than 5,000–7,000 feet), then HPV increases the total pulmonary vascular resistance, which leads to **pulmonary hypertension.** Pulmonary hypertension increases the work of the right ventricle, resulting in right ventricular hypertrophy, right axis deviation (RAD) on the electrocardiogram (EKG), and, eventually, right ventricular failure.

    2. Changes in lung volume markedly alter the pulmonary vascular resistance.

      a. Pulmonary vascular resistance is minimal when the lung volume is close to functional residual capacity (FRC), and increases at both higher and lower lung volumes (Figure 18-6A).

---

*The flow conditions in zone 2 can be compared to a waterfall. The pulmonary artery pressure is the upstream height of the stream, the alveolar pressure represents the top of the waterfall, and the venous pressure is the height at the bottom of the waterfall. The flow rate depends on the difference in elevation (pressure) between the upstream site and the top of the waterfall. Changes in venous pressure alter the height of the waterfall but do not influence the overall flow rate through the system until the venous pressure exceeds the alveolar pressure; at this point, the flow conditions represent zone 3.

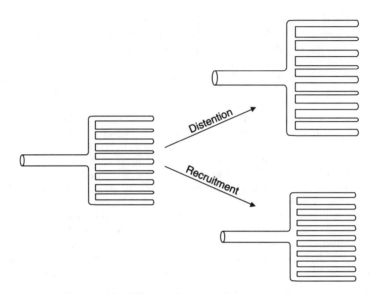

FIGURE 18-5. Schematic depicting the difference between distention and recruitment in the pulmonary vasculature. Both mechanisms reduce the pulmonary vascular resistance and increase blood flow in the affected areas. Distention and recruitment occur in the dependent portions of the lungs where there is an increase in the hydrostatic pressure.

    b. The effect of lung volume on pulmonary vascular resistance is complex, because two different categories of blood vessels are involved.

        (1) The **extraalveolar vessels** are the **larger distributing arteries and veins** in the lungs.

            (a) These vessels are dilated at high lung volumes because of the radial traction exerted on their walls by the alveolar septa (Figure 18-6B).

            (b) The increased diameter reduces the flow resistance in these vessels.

        (2) The **alveolar vessels,** which are **usually capillaries,** are located in the alveolar septa and are compressed by the alveoli at high lung volumes (see Figure 18-6B). During inspiration, the capillary vascular resistance increases while the larger vessels dilate, increasing their volume. Thus, much of the right ventricular output, which increases during inspiration, is stored in the pulmonary arteries until expiration, when the blood reaches the left ventricle. The fluctuation in the stroke volume of the right and left ventricles prolongs the respective ventricular ejection and represents the major factor that alters the splitting of the second heart sound [S_2, see Chapter 11 II A 3 b (5)].

  3. Pulmonary diseases (e.g., emphysema, pulmonary emboli) that obstruct or destroy pulmonary vessels can **alter the normal pattern of blood flow** (i.e., that shown in Figure 18-4) and **impair gas exchange**. The resultant uneven distribution of blood flow in adjacent areas of the lung alters the ventilation–perfusion ratio in the affected area.

IV. **DISTRIBUTION OF VENTILATION.** Alveolar ventilation, like pulmonary blood flow, is not evenly distributed throughout the lungs. The distribution of alveolar ventilation is **dependent on the interpleural pressure (Ppl) gradient, the time constant of the lung,** and **closure of the airways**.

A. The **interpleural pressure gradient** is caused by the effects of gravity on the lung within the intact thorax. Astronauts, exposed to zero-gravity conditions, have neither an interpleural pressure gradient nor regional variations in lung volume.

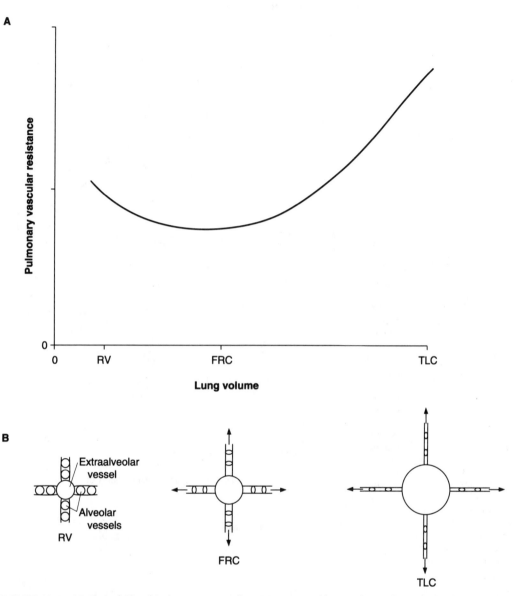

FIGURE 18-6. (*A*) The relationship between vascular resistance and lung volume shows a minimal value near the normal working lung volume, the functional residual capacity (*FRC*). (*B*) The effects of changes in lung volume on the alveolar vessels (i.e., the septal capillaries) and the extraalveolar vessels. When volume increases, radial traction (as indicated by the *arrows*) puts tension on the septa, which, in turn, dilates the extraalveolar vessels. The dilation decreases the flow resistance in these vessels. However, when the septa are stretched, the alveolar vessels (i.e., the septal capillaries) are compressed, which increases capillary flow resistance at high lung volumes.

1. **Cause.** The cause of the interpleural gradient is incompletely understood, but it is believed that the lung behaves as if it were a low-density fluid (because of its gas content) when it is exposed to gravity.
   a. The density of the normal lung is about 0.25–0.30 g/cm^3, which produces a hydrostatic pressure difference of 7.5–10 cm H_2O between the apex and the base of a vertical lung.
   b. The interpleural pressure in a standing adult increases from about -10 cm H_2O at the apex of the lung to about -2 cm H_2O at the base (Figure 18-7A).

A. Transpulmonary pressure

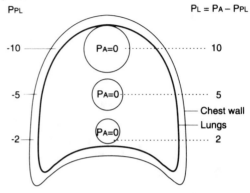

$$P_L = P_A - P_{PL}$$

B. End-expiratory phase

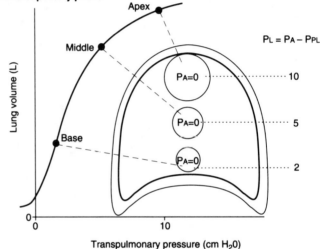

Transpulmonary pressure (cm H_2O)

FIGURE 18-7. (*A*) A gradient in the interpleural pressure (*P*PL) of about 8 cm H_2O exists between the apex and the base of the vertical lung. This gradient causes the regional transpulmonary pressure (*P*L) to vary from 2–10 cm H_2O between the base and the apex of the lung, respectively. This change in transmural pressure causes a marked difference in the extent of inflation between the base and apex of the lung. (*B*) End-expiratory phase. The correlation between the transpulmonary pressure (*P*L) and the degree of lung inflation is shown. The transmural pressure for the base and the apex of the lung places these regions on much different regions of the pressure–volume curve. The base of the lung is poorly distended but located on a very compliant segment. The apex is almost maximally distended, and therefore, has a low compliance. (*C*) End-inspiratory phase. The *points* on the pressure–volume curve indicate the pressure–volume coordinates at end-expiration, and the *arrowheads* indicate the pressure–volume coordinates at end-inspiration. The *thin arrows* along the y-axis show the volume change in the three regions that occurs with inspiration. Most of the tidal volume goes to the base of the lung because it is very distensible; conversely, very little of the tidal volume reaches the apex because of its low compliance.

C. End-inspiratory phase

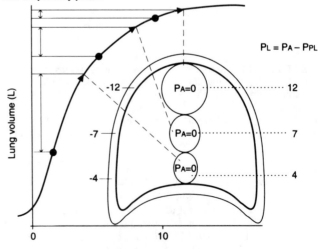

Transpulmonary pressure (cm H_2O)

2. Effects

a. The **regional lung volume** is affected by the local interpleural pressure, because the interpleural pressure is one of the determinants of the transpulmonary pressure $(P_L = P_A - P_{PL})$.

   (1) Because the alveolar pressure is zero throughout the lung under static conditions, the transpulmonary pressure varies from 2 cm H_2O at the base to 10 cm H_2O at the apex of the vertical lung (see Figure 18-7A).

   (2) The wide range between the transpulmonary pressures at the apex and base of the lung is reflected by their respective points on the pressure–volume curve (Figure 18-7B). The difference in lung volume causes **the apex and the base** to **function at very different compliances** (see Figure 18-7B).

b. The **tidal volume** is unevenly distributed in the vertical lung because of the variation in regional compliance.

   (1) In a vertical lung, most of the tidal volume goes to the base of the lung when inspiration starts at FRC (Figure 18-7C).

   (2) There is a linear reduction in regional tidal volume from base to apex under these conditions (Figure 18-8). The tidal volume is much more evenly distributed in a supine individual because the hydrostatic gradient of interpleural pressure is reduced.

B. The **time constant of the lung** is the **product of airway resistance (R)** and **lung compliance (C)**.

1. **Function.** The time constant establishes the rate of acinar volume change.

   a. The **acinus** is the functional unit of the lung; it consists of a respiratory bronchiole and all of the alveolar ducts and alveoli distal to it. The acinus **fills and empties exponentially.**

   b. **For each time constant, the volume changes 63% toward the equilibrium value** (Figure 18-9A).

2. **Effects of disease on the time constant.** In the healthy lung, all acini have the same time constant. In diseased lungs, some areas have low compliance and other areas have high airway resistance, which can cause a markedly unequal distribution of the tidal volume and, therefore, a markedly different alveolar ventilation throughout the lung. The variation in alveolar ventilation leads to wide differences in the alveolar O_2 tension in different areas of the diseased lung, which can markedly interfere with gas exchange in those areas of the lung.

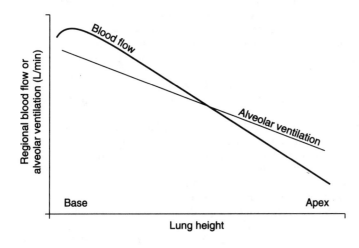

FIGURE 18-8. The distribution of pulmonary blood flow and alveolar ventilation is shown for different regions of the vertical lung. Note that blood flow (perfusion) exceeds ventilation at the base, but ventilation exceeds blood flow at the apex. Because of these relationships, the ventilation–perfusion ratio $(\dot{V}_A/\dot{Q}_C)$ is less than 1 at the base of the lungs and greater than 1 at the apex.

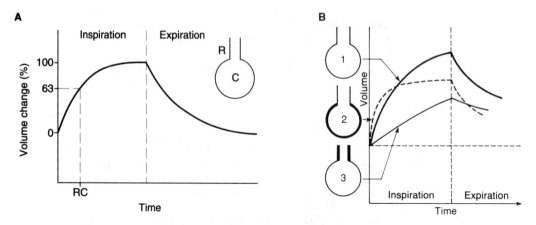

FIGURE 18-9. (A) The acinus can be modeled by a balloon and a tube; the balloon provides compliance (C) and the tube provides airway resistance (R). In such a unit, the volume change occurs exponentially and the rate of filling is determined by the time constant (R • C). The time constant (RC) is defined as the time required for the unit to reach 63% of its maximal volume. (B) Models of acini showing the effects of (1) normal compliance and resistance, (2) decreased compliance, and (3) increased resistance on the time course of filling. A decreased compliance results in rapid filling but a reduction in volume change. When airway resistance is increased, the acini fill slowly and the volume change depends on the duration of inspiration.

    a. Obstructive diseases, which **increase airway resistance, increase the time constant,** which results in a **slower rate of acinar filling and emptying** (Figure 18-9B). The high resistance (R) reduces the gas flow rate ($\dot{V}E$) for any driving pressure (ΔP): $\dot{V}E = \Delta P/R$).

      (1) The **dynamic lung compliance** is the ratio of the tidal volume to the change in interpleural pressure that occurs during the respiratory cycle.

        (a) In normal patients or those with restrictive lung disease, the dynamic lung compliance is equal to the static lung compliance; however, in patients with restrictive lung disease, these compliances are reduced. In patients with obstructive lung disease (i.e., high airway resistance), the dynamic lung compliance is reduced compared with the static lung compliance.

        (b) **Frequency-dependent compliance.** In obstructive lung disease, the presence of lung segments having an increased time constant results in the phenomenon of frequency-dependent compliance, because the extent of lung filling varies as a function of inspiratory time (see Figure 18-9B).

          (i) When the respiratory rate increases, the filling time is insufficient and the tidal volume decreases. Because the interpleural pressure change is constant, the dynamic lung compliance decreases. Tidal volume can be maintained at the normal level by increasing the pressure gradient for flow; however, this also results in a decrease in the dynamic lung compliance.

          (ii) Patients with frequency-dependent compliance can reduce respiratory work by breathing more slowly; the longer respiratory cycle provides additional time for pulmonary inflation and deflation to occur.

      (2) **Intrinsic PEEP.** Patients with high airway resistance frequently do not deflate their lungs to FRC at the end of expiration, and the additional lung volume that remains keeps the alveolar pressure from reaching zero. Thus, these patients function with a "built-in" PEEP that enhances the gradient for expiratory air flow [see also Chapter 20 IV A 2 a].

    b. Restrictive diseases, which **reduce compliance, reduce the time constant.** Areas of the lung that are scarred or fibrotic from the disease process change volume rapidly but to a smaller extent than normal (i.e., they have decreased compliance; see Figure 18-9B).

      (1) The decreased regional compliance results in an uneven distribution of tidal volume and increases the work of expanding the lungs.

      (2) In addition, the fibrosis may interfere with the diffusion process.

C. **Airway closure** (i.e., the collapse and apposition of the airway walls) alters regional alveolar ventilation (because the alveoli cannot be ventilated if the airways are collapsed), and increases the work of breathing because high pressures must be developed to reopen the closed airways. **Early airway closure** is considered to occur when airways close prematurely (i.e., at a higher volume than predicted for a person's age and sex).

1. **Cause.** Airway closure occurs when the regional transpulmonary pressure, which varies primarily as a function of lung volume, is reduced to a critical level.

    a. **Any process that narrows the airway lumen** leads to early airway closure and maldistribution of alveolar ventilation.

    b. **Inflammation** (as occurs in chronic bronchitis), **excess mucus production and contraction of bronchial smooth muscle** (as occurs in asthma), or **loss of radial traction on the airways due to destruction of alveolar septa** (as occurs in emphysema) are common causes of early airway closure.

2. **Terminology**

    a. The **closing volume** is the portion of the vital capacity (VC) that remains in the lungs when airway closure begins.

    b. The **closing capacity** is the closing volume plus the residual volume (RV).

3. **Detection of airway closure.** The **single-breath N_2 test** can be used to detect airway closure.

    a. **Method.** In this test, an individual expires to RV, inspires 100% O_2 to total lung capacity (TLC), and, finally, expires slowly and steadily to RV. The N_2 tension in the expired gas is measured during the second expiration.

      (1) At RV, the base of the lungs is almost completely deflated, but the apex is moderately inflated with gas containing 80% N_2.

      (2) After inspiring 100% O_2 to TLC, the base is maximally inflated by almost pure O_2. The apex is also maximally inflated, but contains significant amounts of N_2.

      (3) During the subsequent expiration, when basal airways begin to close, only gas from the apex continues to be expired, which has a higher N_2 tension.

    b. **Phases.** The expired N_2 tension shows **four distinct phases** (Figure 18-10).

      (1) **Phase 1** of the expired N_2 tension represents the first gas that is expired from the lungs, which is **pure dead space gas (100% O_2).**

      (2) **Phase 2** is a **mixture of gas from the dead space and alveoli** and represents a transition phase. The volume expired at the midpoint of this phase represents the volume of the anatomic dead space [as determined by Fowler's method; see Chapter 16 VIII B 2 c (2)].

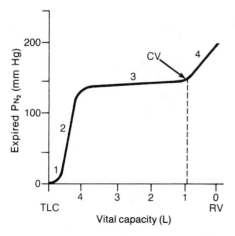

FIGURE 18-10. Record of a single-breath N_2 test. The tracing shows the expired N_2 tension (PN_2) following an inspiratory vital capacity (VC) of 100% O_2. CV = closing volume; TLC = total lung capacity; and RV = residual volume.

(3) **Phase 3** is **alveolar gas** and is referred to as the **alveolar plateau**. The slope of phase 3 varies as a function of the unevenness of tidal volume distribution.
(4) **Phase 4** is **gas from the apex of the lung**.
  (a) The **onset of phase 4 (terminal rise) indicates the closing volume** (see Figure 18-10).
  (b) The higher N_2 tension in phase 4 occurs because the dilution effect of gases coming from the base of the lung is absent.

V. THE VENTILATION–PERFUSION RATIO

($\dot{V}A/\dot{Q}c$) is the ratio of the alveolar ventilation to the pulmonary blood flow. The ventilation–perfusion ratio at the acinar level determines the alveolar O_2 and CO_2 tensions and is, therefore, critical in determining the gas exchange in each alveolus of the lung (Figure 18-11). Increased ventilation (i.e., inflow of O_2) raises the O_2 tension toward the inspired O_2 tension (normally 150 mm Hg), whereas increased blood flow (i.e., outflow of O_2) causes the alveolar O_2 tension to decline toward the mixed venous O_2 tension (normally 40 mm Hg). Ventilation–perfusion imbalances, which can result from many different pulmonary and cardiovascular diseases, are by far the most common cause of hypoxia.

A. Variations in the ventilation–perfusion ratio

1. **Normal ventilation–perfusion ratio.** The most efficient gas exchange occurs when the **ventilation–perfusion ratio is approximately equal to 1**. If the **alveolar ventilation is normally 4.5–5.0 L/min,** the **cardiac output is** approximately **5 L/min,** and the alveolar ventilation and pulmonary blood flow are both evenly distributed throughout the lungs, then the ventilation–perfusion ratio would be about 0.9.
   a. **Balanced conditions.** Under conditions of balanced ventilation and blood flow, the alveolar O_2 tension is about 100 mm Hg, as predicted by the alveolar gas equation (see Chapter 16 VIII C 2), and the alveolar CO_2 tension is about 40 mm Hg (Figure 18-12). The blood leaving this area of the lungs would have the same gas tensions as found in the alveoli.
   b. **Normal variations. Ventilation and blood flow vary in different regions of normal lungs** (i.e., they are low at the apex and high at the base of a vertical lung).
      (1) Because blood has a higher density than lung tissue, the change in blood flow is much greater than the change in ventilation per unit distance down the lung (see Figure 18-8). Thus, the lung **base is overperfused,** and the **apex is underperfused,** compared to the ventilation.

FIGURE 18-11. Changes in alveolar ventilation ($\dot{V}A$) and capillary perfusion ($\dot{Q}c$) determine the local O_2 tension in the lungs (P_{AO_2}). Increasing the alveolar ventilation raises the alveolar O_2 tension, which adds more O_2 to each unit of blood. Eventually, the alveolar O_2 tension stabilizes at a higher level when outflow of O_2 equals inflow. Increasing the blood flow will lower the O_2 tension so that less O_2 is added to each unit of blood. Eventually, the O_2 tension stabilizes at a new lower value. The alveolar O_2 tension is limited on the low end by the mixed venous blood and on the high end by the inspired O_2 tension.

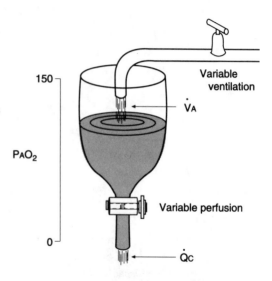

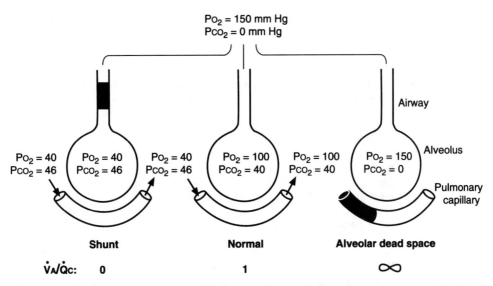

$Po_2 = 150$ mm Hg
$Pco_2 = 0$ mm Hg

Airway

Alveolus

| $Po_2 = 40$ $Pco_2 = 46$ | $Po_2 = 40$ $Pco_2 = 46$ | $Po_2 = 40$ $Pco_2 = 46$ | $Po_2 = 100$ $Pco_2 = 40$ | $Po_2 = 100$ $Pco_2 = 40$ | $Po_2 = 150$ $Pco_2 = 0$ |

Pulmonary capillary

Shunt   **Normal**   **Alveolar dead space**

$\dot{V}_A/\dot{Q}_C$:   0   1 ∞

FIGURE 18-12. The three-compartment model of gas exchange. The three lung units, comprised of a pulmonary capillary and an alveolus, all receive the same inspired gas and mixed venous blood gas concentrations. The *blackened areas* indicate blockage sites. The unit on the *left,* a shunt unit, receives no ventilation. Therefore, its ventilation–perfusion ratio ($\dot{V}_A/\dot{Q}_C$) equals zero. No gas exchange takes place; the blood leaving the unit has the same mixed venous concentration as that entering. The unit on the *right,* which represents alveolar dead space, receives no blood flow and therefore has a ventilation–perfusion ratio of infinity. Again, there is no gas exchange and the alveolar gas has the same composition as the inspired gas. The unit in the *center* receives ventilation and blood flow in equal proportions, so the alveolar gas and pulmonary venous blood have normal gas concentrations. Areas of the lung with ventilation–perfusion ratios less than one but greater than zero have gas concentrations between the normal and the shunt units. Lung regions with high ventilation–perfusion ratios have gas concentrations between the normal and the alveolar dead space units. Pco_2 = carbon dioxide tension; Po_2 = oxygen tension.

(2) **The ventilation–perfusion ratio varies from** about **0.6 at the base to** about **3 at the apex** of the normal lung. This relatively narrow range impairs gas exchange only slightly, but in the presence of pulmonary disease, the ventilation–perfusion ratio can vary from zero to infinity.

2. **A low ventilation–perfusion ratio** is caused either by **inadequate ventilation** or **excessive blood flow** to an area of the lung. Either condition causes the **alveolar O_2 tension to decrease** and the **CO_2 tension to increase,** because an inadequate amount of O_2 is brought into the lungs by ventilation, compared with the amount of O_2 that is carried away by the pulmonary capillary blood.
   a. **Physiologic shunts** occur when the **ventilation–perfusion ratio is greater than zero but less than one**.
      (1) The low alveolar O_2 tension means that even if equilibration occurs between the alveolar gas and the capillary blood, the blood leaving this area of the lungs will contain a low O_2 tension and a decreased O_2 content. When blood from areas with low ventilation–perfusion ratios mixes with blood from well-ventilated areas, the resultant mixture has an O_2 content dependent on the O_2 tension in the two areas as well as the relative amounts of blood flow from each. However, the arterial O_2 content will always be less than normal.
      (2) The decreased arterial O_2 tension and content caused by a low ventilation–perfusion ratio can lead to **arterial hypoxia** (see Chapter 20 II C 1 c).
   b. **Anatomic shunts** occur when the **ventilation–perfusion ratio equals zero** (see Figure 18-12).
      (1) When an area of the lungs receives no ventilation, the alveolar O_2 tension rapidly decreases until it equals the O_2 tension in the mixed venous blood. Therefore, additional gas exchange is impossible because there is no gradient for

diffusion. The blood leaving this area of the lungs has the same composition as the blood entering (i.e., the mixed venous blood from the pulmonary artery). The presence of an anatomic shunt means that true venous blood is mixing with oxygenated blood.

(2) An anatomic shunt may be **intra-** or **extrapulmonary**. Extrapulmonary anatomic shunts may result from **congenital cardiac malformations** (e.g., atrial or ventricular septal defects with right-to-left blood flow).

(3) **Diagnosis**

(a) **Administration of 100% O_2** at sea level to a patient with a significant anatomic shunt will not raise the arterial O_2 tension above 500 mm Hg, because the shunted blood is never exposed to the high alveolar O_2 tension. In contrast, when 100% O_2 is administered to a patient with a physiologic shunt (i.e., a ventilation–perfusion imbalance), the arterial O_2 tension may reach 673 mm Hg [see Chapter 20 II C 1 c (2)].

(b) **The shunt equation** can be used to measure the **fraction of shunted blood**. Because the relationship between arterial O_2 content and tension is nonlinear, it is necessary to use the O_2 content to calculate the amount of shunt. **Normal individuals** have an anatomic shunt consisting of **less than 5% of their cardiac output**.

(i) **Shunt equation**

$$\dot{Q}s/\dot{Q}T = (Cio_2 - Cao_2)/(Cio_2 - C\bar{v}o_2), \text{ where}$$

$\dot{Q}s/\dot{Q}T$ = the fraction of shunted blood (i.e., the shunt flow divided by the total flow)

Cio_2 = the ideal O_2 content (present in the pulmonary capillaries), which is dependent on the alveolar O_2 tension

Cao_2 = the arterial O_2 content

$C\bar{v}o_2$ = the mixed venous O_2 content

(ii) **Example calculation.** A patient is found to be cyanotic and subsequent studies reveal a hemoglobin concentration of 16 g/dl, an arterial O_2 content of 19 ml/dl, and a venous O_2 content of 14 ml/dl. Assuming that the alveolar O_2 tension is normal and that the pulmonary capillary blood has a normal saturation of 97%, then the $Cio_2 = 16 \cdot 1.34 \cdot 0.97$, or 20.8 ml/dl. Substituting in the equation gives: $\dot{Q}s/\dot{Q}T = (20.8 - 19)/(20.8 - 14)$, or 0.26. Thus, 26% of the patient's cardiac output is being shunted.

3. **A high ventilation–perfusion ratio** is caused either by **excessive ventilation** or **inadequate blood flow** to an area of the lung. A high ventilation–perfusion ratio does not cause arterial hypoxia per se; however, if the overall ventilation is normal, excessive ventilation in one section of the lung means there is inadequate ventilation (i.e., a low ventilation–perfusion ratio) in others.

a. In areas of the lung where the ventilation–perfusion ratio exceeds one, the **alveolar and capillary O_2 tensions** are **high** but the **O_2 uptake is poor** because there is little blood flow to carry the O_2 away. Because of the plateau of the O_2–hemoglobin dissociation curve (see Figure 17-1), O_2 exchange is inefficient for the level of ventilation because very little extra O_2 is added to the blood by the excess ventilation.

b. **Alveolar dead space**

(1) Alveolar dead space ventilation is the extra ventilation needed to raise the alveolar O_2 tension to above normal levels. It is considered wasted ventilation because it produces little gas exchange.

(2) **If the blood flow to an area of the lung is zero** (i.e., **the ventilation–perfusion ratio equals infinity**), then there is no gas exchange in these alveoli, and all of the ventilation represents alveolar dead space ventilation (see Figure 18-12). Alveolar dead space ventilation contributes, along with the anatomic dead space, to the total (physiologic) dead space.

(a) Unless the overall alveolar ventilation is increased, an increase in alveolar dead space results in CO_2 retention and hypoxia.

(b) In the typical patient with increased alveolar dead space, the initial hypercapnia stimulates the respiratory centers, increasing the minute ventilation. The CO_2 tension returns to normal, but the hypoxia remains due to the effects of a physiologic shunt.

4. The **O_2–CO_2 diagram** (Figure 18-13) is **one method of visualizing the effects of ventilation–perfusion alterations on the blood gas tensions.**
    a. **R** is the **ratio of the volume of CO_2 released per minute** ($\dot{V}_{CO_2}$) to the volume of O_2 absorbed per minute ($\dot{V}_{O_2}$). The **R lines** represent the O_2 and CO_2 tensions that are present when the rates of O_2 uptake and CO_2 release vary.
        (1) The **gas R line** is a straight line that connects the inspired gas point and the ideal gas point.
        (2) The **blood R line** connects the mixed venous blood point and the ideal gas point, curvilinearly. The curvature results from the nonlinear O_2–hemoglobin dissociation curve.
    b. The **alveolar dead space** is represented on the gas R line as a movement of the mixed expired gas point away from the ideal point (i.e., toward the inspired gas point). The ideal gas point represents the mean value of the O_2 and CO_2 tensions for all alveoli, so that the addition of alveolar dead space (which has a gas composition equal to the inspired gas) moves the mixed expired gas value toward the inspired gas point.
    c. A **physiologic shunt** causes the point representing the arterial blood gas composition to move along the blood R line toward the venous point.
        (1) The more severe the ventilation–perfusion abnormality, the closer the arterial point will move toward the venous point.
        (2) Note that there can be a marked reduction in O_2 tension with only a minimal rise in the CO_2 tension as one proceeds along the blood R line. This is typical of the blood gas values obtained from patients with ventilation–perfusion abnormalities.

B. **Distribution of ventilation–perfusion ratios.** The lungs are comprised of one or more populations of acini with different ventilation–perfusion ratios. The population of acini in the normal lungs can be represented by a log-normal distribution curve that has a range covering approximately 1 decade of ventilation–perfusion values (Figure 18-14A). Patients with a ventilation–perfusion imbalance may have a single mode of distribution with a much wider range of values than normal (Figure 18-14B), or they may have multiple modes of distribution (Figure 18-14C).

FIGURE 18-13. The O_2–CO_2 diagram. The ventilation–perfusion line (*solid line*) is defined by three points representing the mixed venous blood composition (*point v̄*), the average gas composition of the pulmonary capillary blood (i.e., the ideal alveolar gas composition, *point i*), and the inspired gas composition (*point I*). The arterial blood gas composition (*point a*) is determined by the ratio of shunt flow to cardiac output; the mixed-expired gas composition (*point E*) is determined by the ratio of the dead space to the tidal volume. The *R lines* respresent the ratio of CO_2 production to O_2 consumption. The blood R line (*dotted line*) and the gas R line (*dashed line*) intersect at the ventilation–perfusion line. P_{CO_2} = carbon dioxide tension; P_{O_2} = oxygen tension.

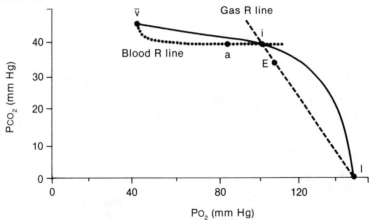

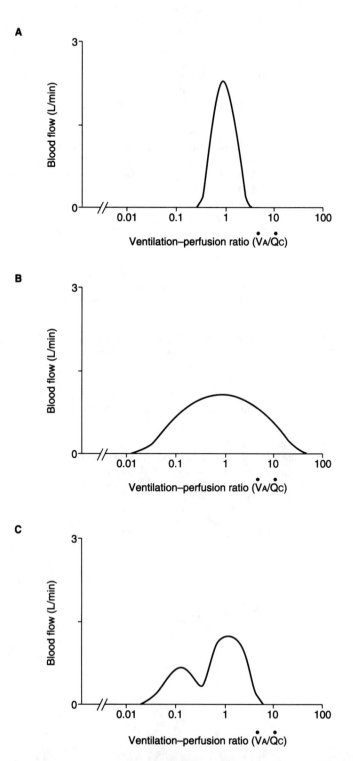

FIGURE 18-14. Varying distributions of ventilation–perfusion ratios. The *ordinate* represents the blood flow for each value of the ventilation–perfusion ratio, and the *abscissa* shows the ventilation–perfusion ratio on a log scale. (*A*) Normal. The mean ventilation–perfusion ratio is 1 and the standard deviation of the population is 0.2, giving a variation of 0.6–3 for the ventilation–perfusion ratios. (*B*) An increased standard deviation of the acinar population causes a wider distribution of values, which results in a shunt effect and increased alveolar dead space. (*C*) A bimodal distribution, which can be caused by a focal disease, indicates two separate populations of acini. A decrease in the overall mean ventilation–perfusion ratio would produce hypercapnia and hypoxia in the arterial blood.

1. **Single mode of distribution.** The mean ventilation–perfusion ratio for any lung is determined by the overall level of alveolar ventilation and the cardiac output.

   a. If all of the acini in the lung had the same ventilation–perfusion ratio, then there would be no physiologic shunting or alveolar dead space. This condition could be modeled by one acinus that received all of the ventilation and all of the blood flow; however, this is obviously an unrealistic model.

   b. A population of acini with varying ventilation–perfusion ratios (see Figure 18-14A) is a more realistic model. Thus, even in the normal lung, physiologic shunts and alveolar dead space are present.

      (1) Physiologic shunts produce a slight lowering of the arterial O_2 content and tension from the ideal gas point (see Figure 18-13).

      (2) The high ventilation–perfusion ratios increase the alveolar dead space, which is compensated for by increasing the minute ventilation.

2. **Multiple modes of distribution** represent subpopulations of acini with varying ventilation–perfusion ratios that are produced by focal disease processes. Each of these subpopulations may represent a physiologic shunt if it is below the normal mean ventilation–perfusion ratio, and alveolar dead space if it is above the normal ventilation–perfusion ratio.

Chapter 19

Respiratory Control

I. CENTRAL NERVOUS SYSTEM (CNS) MECHANISMS

A. **Introduction.** Unlike the heart, which automatically generates its own rhythm, the respiratory muscles are **typical skeletal muscles** requiring electrical stimulation from the CNS.

1. **Innervation.** Electrical stimuli are transmitted **via somatic nerves** to initiate contraction of the respiratory muscles. The major inspiratory muscle, **the diaphragm, is innervated by motor fibers in the phrenic nerve,** which arises from the upper cervical level of the spinal cord.

2. **Dual pathway.** Impulses reach the phrenic nerve from either **voluntary** (i.e., **corticospinal**) **or involuntary paths** in the CNS. This dual pathway allows voluntary control of breathing during activities such as talking, singing, and swimming, in addition to involuntary (i.e., unconscious) control, which allows humans to breathe automatically without conscious effort.

B. **Medullary centers**

1. **Function.** The automatic, involuntary, basic rhythm of respiration is generated within the medulla, but its exact source and mechanism of generation is unknown. Respiration continues as long as the medulla and the spinal cord are intact.

2. **Mechanism.** It appears as if there are two groups of neurons on either side of the medulla that act as oscillators to generate the basic rhythm. These groups of cells, called the **dorsal respiratory group (DRG)** and the **ventral respiratory group (VRG),** discharge rhythmically during either inspiration or expiration. Nervous activity in other areas of the brain [e.g., the pons, thalamus, reticular activating system (RAS), and the cerebral cortex] and afferent activity in the vagus, glossopharyngeal, and somatic nerves influence the VRG and DRG.

   a. The **DRG** is a bilateral group of cells that lies within the **nucleus of the tractus solitarius**. The DRG cells are primarily **inspiratory cells** (i.e., they discharge during inspiration).

      (1) **Function.** The DRG may be the **primary rhythm generator** for respiration, because the activity in these cells gradually increases during inspiration. The electrical activity of this center has been likened to a ramp, because the activity rises to a crescendo during inspiration and then rapidly disappears.

      (2) **Activity**

         (a) **Afferent input to the DRG** comes primarily from the **vagus and glossopharyngeal nerves,** which carry information from the peripheral chemoreceptors and mechanical receptors in the lungs (see II A–B; III A).

            (i) **Low O_2 tensions, high CO_2 tensions,** and **low pH levels** (i.e., an increased H^+ concentration) generate afferent impulses that **stimulate DRG activity.**

            (ii) **Increased electrical traffic within the RAS increases the neural activity of the DRG.** During sleep, the RAS activity diminishes, decreasing inspiratory drive and alveolar ventilation, and slightly increasing the arterial CO_2 tension.

            (iii) **Inflation of the lungs** produces afferent impulses from pulmonary stretch receptors that **inhibit DRG activity.**

         (b) **Efferent outflow from the DRG** goes to the **contralateral phrenic** and **intercostal motoneurons,** and to the **VRG.**

    b. The **VRG** is comprised of the **upper motor neurons of the vagus and the nerves to the accessory muscles of respiration**. The neurons of the VRG are active during both inspiration and expiration; however, expiratory activity does not activate the expiratory muscles during normal respiration (eupnea), because expiration is normally passive.

C. **The pontine centers** are areas of the brain stem that modify the activity of the medullary respiratory centers.

    1. The **apneustic center** lies in the **caudal area of the pons,** but it has not been identified with any specific collection of cells.

        a. **Efferent outflow** from the apneustic center **increases the duration of inspiration,** slowing the respiratory rate and resulting in a deeper and more prolonged inspiratory effort.

        b. **Inhibition.** The apneustic center is normally inhibited by **impulses carried by the vagus nerves,** and also by the activity of the **pneumotaxic center.** Bilateral vagotomy and destruction of the pneumotaxic center results in prolonged periods of inspiration (i.e., **apneusis**).

    2. The **pneumotaxic center** lies in the **upper part of the pons** and functions to inhibit the apneustic center. Stimulation of the pneumotaxic center shortens inspiration, leading to a shallower and more rapid respiratory pattern.

D. **Central chemoreceptors** are cells that **lie just beneath the ventral surface of the medulla; they respond to the H^+ concentration in the cerebrospinal fluid (CSF) and the surrounding interstitial fluid** (Figure 19-1).

    1. Mechanism

        a. **Charged ions** do not readily cross the **blood–brain barrier** (i.e., the endothelium of the blood vessels in the brain). Ionic composition of the CSF is regulated by ion pumps located in the choroid and pia arachnoid membranes.

        b. CO_2 readily crosses the blood–brain barrier, because it is a small, very soluble, uncharged molecule. The hydration and subsequent dissociation of CO_2 into H^+ and HCO_3^- alters the H^+ concentration in the CSF and brain tissues.

          (1) An **increase in CSF CO_2 (H^+)** causes the central chemoreceptors to **stimulate respiration.**

          (2) A **decrease in CSF CO_2 (H^+)** causes the central chemoreceptors to **inhibit respiration.**

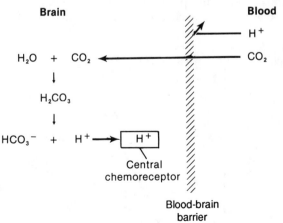

FIGURE 19-1. The effects of the blood–brain barrier on the transport of charged ions (e.g., H^+) and the diffusibility of CO_2. CO_2 can readily cross the blood–brain barrier; it serves as a source of H^+ to stimulate the central chemoreceptors.

2. **Approximately 85% of the resting (basal), chemical drive of respiration** results from the stimulatory effect of CO_2 (H^+) on the central chemoreceptors. Peripheral chemoreceptors are responsible for the remainder of the chemical drive.

II. PERIPHERAL CHEMORECEPTOR MECHANISMS

A. **Peripheral chemoreceptors** are located in the **carotid** and **aortic bodies**. These receptors **respond to lowered O_2 tensions, increased CO_2 tensions,** and **increased H^+ concentrations in the arterial blood**. Each of these chemical stimuli to respiration apparently affects the chemoreceptor individually, because the response to one chemical is augmented by the effects of the others. Figure 19-2 shows that the respiratory response at any CO_2 tension is markedly enhanced by a decreased O_2 tension.

1. **O_2 tension.** The peripheral chemoreceptors are the only sites in the body that detect changes in the O_2 tension of body fluids.
   a. **Mechanism**
      (1) The peripheral chemoreceptors receive a tremendous blood flow for their size and thus can be considered to **monitor the O_2 tension of arterial blood, rather than its content**. Decreases in the O_2 content of blood caused by anemia, methemoglobinemia, or CO poisoning do not stimulate the peripheral chemoreceptors, because the O_2 tension, which is determined by the amount of dissolved O_2, remains normal.
      (2) The peripheral chemoreceptors rapidly **increase their firing rate as the arterial O_2 tension falls below 60–80 mm Hg** (e.g., following exposure to moderate altitude). Afferent impulses from these receptors are carried to the brain from the aortic and carotid receptors via the vagal and glossopharyngeal nerves, respectively.
   b. The afferent impulses from the chemoreceptors **increase the rate and depth of respiration**. The increased alveolar ventilation that occurs secondary to hypoxia causes excessive elimination of CO_2 via the lungs, decreasing CO_2 concentration in the body fluids. The loss of CO_2, a volatile acid, produces **respiratory alkalosis,** which, in turn, inhibits the respiratory drive because of the reduced H^+ concentration in the CSF.
      (1) Over the course of several days, the ion pumps in the choroid and pia arachnoid transfer HCO_3^- from the CSF to the blood, returning the CSF pH toward normal. This removes the inhibitory alkalotic influence so that respiration continues to increase over the first several days of exposure to acute hypoxia.
      (2) The inhibitory effect of alkalemia on ventilation is also counteracted by the kidneys as they excrete HCO_3^-. It requires several days for the kidneys to return the pH of body fluids to normal. During this period, the respiratory response to hypoxia continues to increase as the braking effect of alkalosis is removed. As

FIGURE 19-2. CO_2 response curves obtained at alveolar O_2 tensions (P_{AO_2}) of 50, 100, and > 500 mm Hg. P_{ACO_2} = alveolar CO_2 tension.

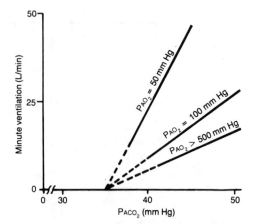

ventilation increases, increasing the alveolar O_2 tension and slightly decreasing the CO_2 tension, the hypoxia is ameliorated (see Chapter 16 VIII C 1).

2. **CO_2 tension.** Elevated CO_2 tensions also stimulate the peripheral chemoreceptors, but the major effect of CO_2 is on the central chemoreceptors.

3. **H^+ concentration.** An **increased H^+ concentration** (i.e., a **decreased pH**), which is caused by the addition of fixed acids (e.g., lactate, acetoacetate, butyrate) to the blood, **stimulates the peripheral chemoreceptors so that ventilation increases.** The increased alveolar ventilation lowers the CO_2 tension, and therefore, the acid content, of the arterial blood. The decrease in CO_2 tension tends to return the pH of the blood toward normal. Acutely, the fixed acid molecules cannot cross the blood–brain barrier, so that the resultant decrease in CO_2 tension results in an alkalosis in the CSF. This central alkalosis inhibits respiratory drive for several days until the CSF pH is returned to normal by ion transport by the meningeal tissues. The respiratory response to a low arterial pH caused by fixed acids results in a respiratory compensation that returns the arterial pH toward normal.

B. **Other receptors**

1. **Stretch receptors** located in the **small airways** are stimulated by inflation of the lungs. Stimulation of these receptors initiates the **Hering-Breuer (inspiratory inhibitory) reflex,** which **terminates inspiration** by sending impulses that inhibit the pontine and medullary respiratory centers via the vagus nerves. This reflex is not very powerful in adult humans, but functions in newborns and in some animal species.

2. **Irritant receptors** are located in the **large airways** and are stimulated by smoke, noxious gases, and particulates in the inspired air.
   a. These receptors initiate **reflexes** that cause **coughing, bronchoconstriction, mucus secretion,** and **breath holding** (i.e., **apnea**).
   b. **Chronic exposure** to inhaled irritants (e.g., from smog, industrial atmospheres, or cigarette smoke) can lead to chronic **bronchitis,** one of the diseases causing obstructive lung disease (which is characterized by high airway resistance and limitation of the respiratory system's ability to respond to stress).

3. **J receptors** are located in the **pulmonary interstitium** at the level of the pulmonary capillaries and are stimulated by distention of the pulmonary vessels (e.g., as caused by left ventricular failure, pulmonary embolization, and certain chemicals or drugs). These receptors initiate **reflexes** causing **rapid, shallow breathing** (i.e., **tachypnea**).

4. **Chest wall receptors** can detect the force generated by the respiratory muscles during breathing. If the force required to distend the lungs becomes excessive (either as a result of high airway resistance or low compliance), the information from these receptors gives rise to the sensation of **dyspnea (difficulty in breathing).**

III. DYSFUNCTION OF RESPIRATORY CONTROL

A. **Tests of respiratory control. Gases containing different amounts of O_2 or CO_2 can be administered to evaluate the respiratory drive of the respiratory system.**

1. Increasing the CO_2 tension of inspired gas raises the arterial CO_2 tension and should stimulate the chemoreceptors, leading to an increase in the minute ventilation (Figure 19-3). Hypercapnia is enhanced by a decrease in O_2 tension, so that the respiratory response is much greater than with either stimulus alone (see Figure 19-2).

2. Patients vary in their response to this test, depending on, for example, the sensitivity of their chemoreceptors, the activity of the RAS (which is reduced during sleep), the airway resistance, and the lung volume (Figure 19-4). The respiratory response is also blunted by drugs such as morphine, barbiturates, and cocaine.

FIGURE 19-3. Ventilatory response to several different respiratory stimuli. The highest level of ventilation is achieved by voluntary increases in respiratory rate and tidal volume [i.e., as takes place during maximal voluntary ventilation (*MVV*)]. This level of ventilation can only be maintained for a short time. The maximal level of ventilation that can be sustained for a significant period of time is approximately 66% of the MVV. Chemical factors that affect respiration (e.g., the O_2 and CO_2 tensions of the blood) are less powerful influences on ventilation.

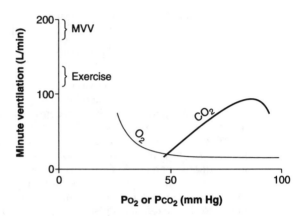

B. **Abnormal respiratory patterns.** Changes in the environment or diseases affecting the respiratory system, cardiovascular system, or brain can produce various respiratory patterns (Figure 19-5).

1. **Cheyne-Stokes respiration** is an abnormal respiratory pattern that occurs with depression of the brain due to disease, drug overdose, congestive heart failure, and hypoxia from other causes. Cheyne-Stokes respiration is characterized by periods of waxing and waning tidal volumes separated by periods of apnea.

2. **Biot's breathing** is another type of periodic breathing that consists of one or more large tidal volumes separated by periods of apnea. The condition occurs in many diseases producing brain damage (e.g., diseases that increase the intracranial pressure, meningitis).

3. **Kussmaul's respiration** is rapid, deep breathing often seen in patients suffering from diabetic ketoacidosis. It occurs as the body tries to compensate for metabolic acidosis by increasing the rate of CO_2 excretion.

4. **Ondine's curse**
   a. Ondine, a character from Greek mythology, was a water nymph who fell in love with a human. Ondine's father was king of the nymphs and placed a curse on the man that took away all of his automatic functions. Presumably, the man spent all of his time willing his heart to beat and his muscles to breathe and had little time left for his lover.
   b. Automatic respiratory control is lost when there is a **destruction of the involuntary neural pathways**. Patients with this condition can breathe only by conscious effort and, therefore, cannot sleep without the aid of a mechanical respirator.

5. **Sleep apnea syndromes** have been recognized as disorders of respiratory control that affect a large portion of the population, especially middle-aged and elderly men. Many sleep centers have been established to study patients by recording physiologic parameters (e.g., respiratory movements, airflow, electrical currents within the brain, and hemoglobin saturation) during sleep.

FIGURE 19-4. The effect of various conditions on CO_2 response curves. PCO_2 = carbon dioxide tension.

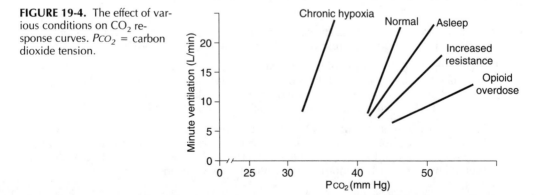

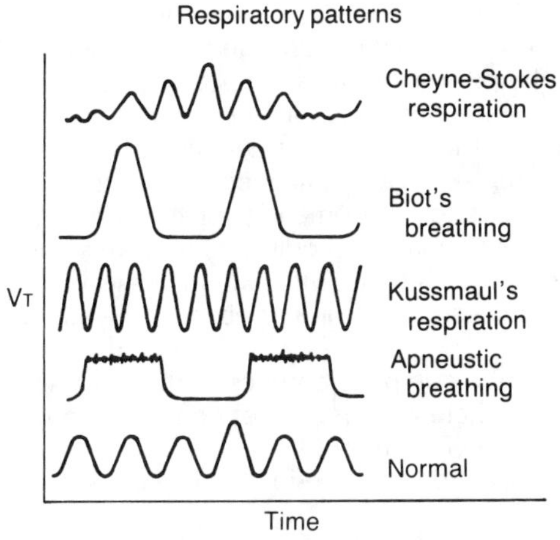

Respiratory patterns

Cheyne-Stokes respiration

Biot's breathing

V_T

Kussmaul's respiration

Apneustic breathing

Normal

Time

FIGURE 19-5. Illustration of various breathing patterns. V_T = tidal volume.

a. **Obstructive sleep apnea** occurs when inspiration is prevented by transient blockage of the airway.
   (1) **Causes**
       (a) **Loss of muscle tone.** Some individuals experience a marked loss of muscle tone in the **pharyngeal muscles** during **rapid eye movement (REM) sleep,** which causes partial or complete **obstruction of the oropharynx** by the soft tissues near the base of the tongue during inspiration.The negative pressures that are generated during inspiration cause the hypopharynx to collapse, preventing airflow even though strong inspiratory muscle contractions occur (Figure 19-6).
       (b) **Obesity.** When sleep apnea is associated with extreme obesity, it is referred to as the Pickwickian syndrome (after a character in Dickens' *Pickwick Papers*).
   (2) **Clinical signs**
       (a) **Partial airway obstruction causes snoring,** because the inspired air causes the soft tissues to vibrate.
       (b) In patients with **complete airway obstruction,** contraction of the respiratory muscles produces **movement of the chest** although no air can move because of the blocked airway.
       (c) Patients usually awaken because of the hypoxia that develops; muscle tone

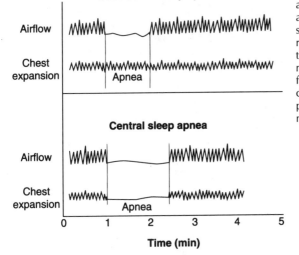

Obstructive sleep apnea

Airflow

Chest expansion

Apnea

Central sleep apnea

Airflow

Chest expansion

Apnea

Time (min)

FIGURE 19-6. Simulated records of air flow and chest movements in patients with sleep apnea syndrome. *Upper panel.* Obstructive sleep apnea is caused by occlusion of the pharyngeal airways. Although the patient contracts the inspiratory muscles, generating chest movement, the airway obstruction prevents air flow. *Lower panel.* Central sleep apnea is caused by the absence of phrenic nerve impulses; therefore, the inspiratory muscles do not contract and no chest movement is visible.

is reestablished, and sleep returns. These episodes may occur hundreds of times each night and are associated with a **marked drop in the hemoglobin saturation of the arterial blood**. The person awakens unrested because of the **poor sleep pattern**. Frequently, these people fall asleep at work, in lectures, or even while eating because of the lack of restful nighttime sleep.

(3) **Treatment** may involve the use of an **oropharyngeal appliance** to keep the airway open or a **positive-pressure nasal mask** to overcome the negative pressure in the hypopharynx. In severe cases, **surgery** may be required to remove some of the soft tissue to maintain an open airway. Patients with extreme cases of obstructive sleep apnea may require a **tracheostomy** in order to bypass the airway obstruction.

b. **Nonobstructive (central) sleep apnea** refers to a complete absence of rhythmic activity from the respiratory centers. Obviously, there is no respiratory muscle contraction because the neural impulses are absent (see Figure 19-6).

(1) These periods of apnea can last 30–60 seconds, and produce a marked drop in arterial hemoglobin saturation.

(2) Affected persons and some of their family members have been shown to have a **decreased chemoreceptor sensitivity to O_2 and CO_2.**

(3) Nonobstructive sleep apnea syndrome has been proposed as one of many possible causes of **sudden infant death syndrome (SIDS,** or **crib death)**.

Chapter 20

Respiratory Responses to Stress

I. EXERCISE

A. **Introduction.** The effects of exercise on the cardiorespiratory system have generated significant interest because of the current concern with **physical conditioning and prevention of cardiovascular disease**. Exercise may be used to evaluate the cardiovascular and respiratory systems (i.e., as a **stress test**).

1. **Specificity.** Endurance (aerobic) training produces different effects in the skeletal muscles than weight training. Training effects are specific for the particular muscle groups involved; **only endurance exercises produce cardiovascular conditioning**.

2. **Effects of endurance training.** Endurance training increases the availability of O_2 to the skeletal muscle cells.
   a. Muscles conditioned by endurance training exhibit increases in capillary density, myoglobin concentration, glycogen, and mitochondrial enzymes of the citric acid cycle.
   b. The mitochondrial enzymes allow the conservation of muscle glycogen by facilitating the breakdown of long-chain fatty acids, which serve as an alternate energy source during exercise.

B. **Cardiac responses**

1. **Heart rate.** Increases in heart rate cause a proportional increase in the cardiac output.
   a. The **heart rate** increases linearly with the work rate up to a maximum, which is determined by the subject's age. The **maximal heart rate (HR_{max})** is approximately equal to $210 - [0.65 \cdot$ age (years)], and is unaffected by conditioning.
   b. **Optimal cardiovascular conditioning** requires attaining a heart rate of 60%–70% of HR_{max} for 20–30 minutes, 3–4 times a week for at least 3 months.

2. **Stroke volume.** Endurance training increases the stroke volume of the heart by increasing the ventricular end-diastolic volume (VEDV). Thus, conditioned athletes can maintain any level of cardiac output at a lower heart rate than nonconditioned individuals.

3. **Maximal cardiac output.** The maximal cardiac output is greater in conditioned athletes than in deconditioned individuals (e.g., 30–35 L/min in olympic-class runners, as compared with 15 L/min in deconditioned adults.

4. **Anaerobic threshold.** The **anaerobic threshold** is higher in conditioned athletes than in deconditioned individuals. Training improves the absolute anaerobic threshold as well as the anaerobic threshold relative to the maximal O_2 consumption.
   a. The anaerobic threshold represents **the point when a limited O_2 supply causes the muscle metabolism to shift from aerobic to anaerobic,** producing a significant increase in the lactate concentration of the blood.
   b. The anaerobic threshold occurs at approximately **60% of the maximal exercise level,** regardless of the level of physical fitness.
   c. The onset of the anaerobic threshold is early in **cardiac patients,** because their cardiac output is low and they reach maximal heart rates at much lower levels of exercise than healthy individuals (Figure 20-1).

C. **Respiratory responses**

1. **Normal respiratory responses to exercise** (Figure 20-2)
   a. The **minute ventilation increases linearly** along with the work rate (O_2 consumption) until the anaerobic threshold is reached. Above the anaerobic threshold, minute ventilation increases more steeply as the work rate increases, because the lactic acid that is generated imposes an additional respiratory drive.

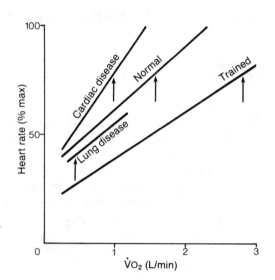

FIGURE 20-1. The effect of training, heart disease, and lung disease on the heart rate response to endurance (aerobic) exercise. *Vertical arrows* mark the onset of the anaerobic threshold, which occurs at about 60% of the maximal exercise level for every individual. $\dot{V}_{O_2} = O_2$ utilization.

    (1) Endurance (aerobic) training increases the maximal minute ventilation that may be achieved during exercise, but does not improve the maximal voluntary ventilation (MVV).

    (2) Respiratory muscle training. Recent evidence indicates that the maximal duration of exercise is limited by fatigue of the respiratory muscles, in both conditioned and nonconditioned individuals. Specific training of the respiratory muscles allows one to increase the duration and intensity of exercise.

  b. CO_2 output increases linearly with the work rate until the anaerobic threshold is reached. Above the anaerobic threshold, the CO_2 output increases more steeply because respiration increases. The arterial CO_2 tension declines as the body stores are depleted, because excretion exceeds production.

  c. O_2 consumption increases linearly with the work rate and is exactly dependent on the work performed.

    (1) The O_2 consumption does not decrease with training at any workload (i.e., training does not improve the body's efficiency unless muscle coordination is improved by practice).

FIGURE 20-2. The effects of exercise on respiratory parameters of gas exchange. The *dashed line* indicates the onset of the anaerobic threshold. $\dot{V}_E$ = minute ventilation; $\dot{V}_{CO_2}$ = CO_2 excreted in the expired gas (L/min); $\dot{V}_{O_2}$ = O_2 utilization (L/min); Pa_{CO_2} = arterial CO_2 tension; R = respiratory exchange ratio.

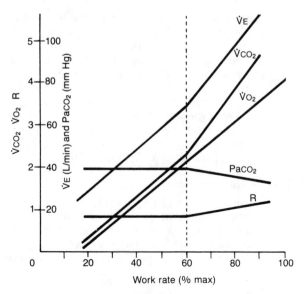

   (2) Conditioning produces a 5%–20% increase in the maximal O_2 consumption, because the cardiac output increases and the arteriovenous O_2 difference widens at the maximal exercise level.
   (3) The O_2 consumed by the entire body can be calculated from the **Fick principle,** which states that O_2 consumption equals the cardiac output times the difference in O_2 content between the arterial and mixed venous (i.e., pulmonary artery) blood.
   d. The **alveolar–arterial O_2 gradient** normally remains at 5–10 mm Hg during moderate levels of exercise, but widens slightly beyond the anaerobic threshold because the alveolar O_2 tension increases.
   e. The **respiratory exchange ratio** equals the volume of expired CO_2 per minute divided by the volume of O_2 that is consumed per minute ($\dot{V}CO_2/\dot{V}O_2$). At work levels above the anaerobic threshold, the respiratory exchange ratio exceeds 1, but normally never exceeds 1.25 even at maximal levels of exercise. The conversion of HCO_3^- to carbonic acid during the buffering of lactic acid increases the amount of CO_2 that is liberated by the lungs, and therefore, increases the respiratory exchange ratio.

2. Respiratory responses to exercise in patients with lung disease
   a. Patients with **chronic obstructive pulmonary disease (COPD)** are limited in the ability to exercise, mainly because of the onset of **severe dyspnea.**
   (1) In these patients, the heart rate does not reach the minimal level necessary to achieve benefits from training and the anaerobic threshold occurs at very low levels of exercise because O_2 uptake is impaired (see Figure 20-1).
   (2) Thus, exercise training in these patients has only slight benefits, which seem to relate primarily to desensitization of symptoms. It appears to be more worthwhile to train the respiratory muscles in these patients.
   b. In **patients with moderately or severely reduced diffusing capacities,** exercise may cause the arterial O_2 tension to decline, because the transit time of the red blood cells through the pulmonary capillaries is shortened as a result of the elevated cardiac output (Figure 20-3). These patients may **switch from perfusion–limited gas exchange at rest to diffusion-limited gas exchange during exercise.**

II. **HYPOXIA** is defined as an **inadequate O_2 supply to the body tissues**. Depending on the cause, the entire body or a localized region may be affected. Various disease processes can severely limit the O_2 supply anywhere between the atmosphere and the body's cells (Figure 20-4). Hypoxia occurs downstream of the limitation (i.e., toward the cells); a normal O_2 tension may be present upstream (i.e., toward the environment).

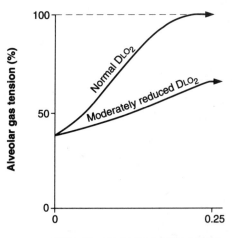

Time in pulmonary capillary (sec)

FIGURE 20-3. Effects of exercise on the diffusion process in the lungs. Normally, the red blood cell spends 0.75 second in the pulmonary capillary, but during exercise, this time is reduced to 0.25 second. During exercise, individuals with normal diffusing capacities are still able to achieve equilibration between the alveolar gas and the pulmonary capillary blood, in spite of the abbreviated transit time. In fact, in healthy people, exercise increases the diffusing capacity to approximately three times the resting value because the capillary blood volume increases and additional capillaries are perfused as a result of the increased pulmonary artery pressure and cardiac output. On the other hand, patients with a moderately reduced diffusing capacity may be unable to achieve equilibration during exercise (resulting in a decreased arterial O_2 tension) because of the reduced transit time. DLO_2 = diffusing capacity of the lungs for O_2.

A. **Symptoms of hypoxia** depend on the **rapidity** and the **severity of the decrease in O₂ tension,** the tissues that are involved, and the effectiveness of the body's compensatory mechanisms.

1. **Fulminant hypoxia** occurs within seconds after exposure to an **arterial O₂ tension of less than 20 mm Hg** (e.g., as would occur if an aircraft lost cabin pressure at altitudes above 30,000 feet and no supplemental O₂ was available, or the O₂ in a closed space was consumed by combustion or displaced by some other gas). **Unconsciousness** occurs in as few as 15–20 seconds, and **brain death** may follow in 4–5 minutes.

2. **Acute hyopxia** is produced by exposure to **arterial O₂ tensions of 25–40 mm Hg** (e.g., as would occur at altitudes 18,000–25,000 feet above sea level).
   a. **Symptoms** of acute hypoxia are very similar to the effects of ethyl alcohol; they **include incoordination, slowed reflexes, slurred speech, overconfidence,** and, eventually, **unconsciousness**.
   b. **Coma** and **death** can occur in minutes to hours if the compensatory mechanisms of the body are not adequate.

3. **Chronic hypoxia** is produced by exposure for extended periods of time to **arterial O₂ tensions of 40–60 mm Hg** (e.g., as would occur at altitudes of approximately 10,000–18,000 feet). Most clinical causes of hypoxia fall into this category. Patients with chronic hypoxia may be **bedridden** or limited to sitting in a chair, because respiratory or cardiac disease prevents them from increasing the O₂ supply to the tissues.
   a. **Symptoms** of chronic hypoxia are similar to those of **severe fatigue,** and include **dyspnea** and **shortness of breath**.
   b. **Respiratory arrhythmias** (e.g., **Cheyne-Stokes breathing**) can occur in patients with chronic hypoxia, especially during sleep, which can contribute to the hypoxic state.

B. **Signs of hypoxia**

1. **Cyanosis** is the **bluish color** of tissue caused by the presence of more than 5 g of deoxyhemoglobin/dl in the capillary blood. The coloration is most readily seen in the

FIGURE 20-4. The oxygen cascade. Normally, the transfer of O₂ along the pathway from the atmosphere to the tissues produces several small decrements in the O₂ tension. A large change in O₂ tension occurs when the blood in the systemic capillaries releases O₂ to the tissues; the amount of O₂ that is released depends on the tissues' metabolic rate and the rate of blood flow. The *thick lines* represent the normal O₂ tension for a subject at sea level. The *thin lines* represent an estimate of the changes in O₂ tension that can be caused by diseases. $\dot{V}_A/\dot{Q}_C$ = the ventilation–perfusion ratio; $\dot{Q}/\dot{V}_{O_2}$ = the blood flow–O₂ consumption ratio.

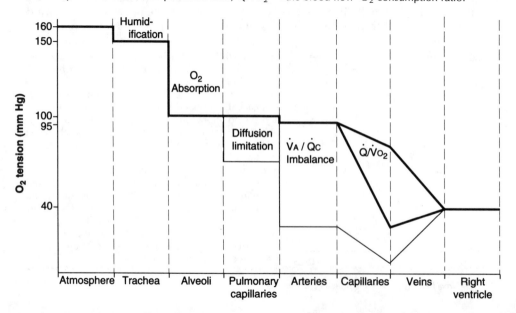

nail beds, lips, mucous membranes, and ear lobes, but it may not be recognized because of skin pigmentation or poor lighting. Cyanosis is **not a reliable sign of hypoxia**.

   a. **Anemic patients** may never develop cyanosis, even though they are extremely hypoxic because of their inadequate hemoglobin concentration. In contrast, **patients with polycythemia** [see II D 1 b (1)] may be cyanotic as a result of their high concentration of hemoglobin, even though their tissues are adequately oxygenated.

   b. **Methemoglobin,** because of its slate gray color, can also impart a bluish color to tissues.

2. **Tachycardia** (i.e., rapid heart rate) occurs as a reflex response to the low arterial O_2 tension. The hypoxia is detected by the aortic and carotid chemoreceptors, which then activate the sympathetic outflow to the heart. The increased heart rate increases the cardiac output, enhancing the O_2 delivery to the tissues (see also Chapter 17 II D 3).

3. **Tachypnea** (i.e., **rapid breathing**) and **hyperpnea** (i.e., **deep breathing**) are also reflex responses to hypoxia that are activated by the arterial chemoreceptors. The chemoreceptors prompt an increase in minute ventilation in an attempt to raise the alveolar O_2 tension toward its normal value of 100 mm Hg.

C. Types of hypoxia (Table 20-1)

1. **Arterial hypoxia** (Figure 20-5A) results from **inadequate oxygenation of the arterial blood,** which is caused by breathing gas with a low O_2 tension or by one of **four pathophysiologic mechanisms** (see Table 20-1) that impair gas exchange in the lungs. Arterial hypoxia is the only type of hypoxia in which there is a **decreased arterial O_2 tension**. Pulmonary gas exchange can be evaluated by calculating the **alveolar–arterial O_2 tension difference**. A large alveolar–arterial O_2 gradient indicates impaired pulmonary gas exchange.

   a. **Hypoventilation** occurs when the **rate** or **depth of respiration,** or both, is **inadequate**. The inadequate alveolar ventilation reduces both the alveolar and arterial O_2 tensions, and increases the alveolar and arterial CO_2 tensions. **Hypercapnia** (i.e., elevated CO_2 levels in the blood) is pathognomonic of hypoventilation.

   b. **Diffusion limitation** occurs when pulmonary disease destroys large portions of the alveolar membrane (decreasing the surface area) or significantly increases the diffusion distance across the alveolar–capillary membrane (see Chapter 18 II B 2 a).

   (1) These changes to the respiratory membrane reduce the diffusion rate of O_2 into the pulmonary capillary blood. Consequently, the O_2 tension in the pulmonary capillaries never equilibrates with the alveolar O_2 tension, which results in the abnormally low arterial O_2 tension.

   (2) Exercise decreases the arterial O_2 tension because it reduces the time that the blood remains in the pulmonary capillaries, thus even less O_2 is transferred into the pulmonary capillary blood.

TABLE 20-1. Differentiating Types of Hypoxia

Type of Hypoxia	PaO_2	$PaCO_2$	$P\bar{v}O_2$	PaO_2 During Exercise	Effect of 100% O_2
Arterial hypoxia					
Hypoventilation	↓	↑ *	↓	↓ ↑	↑ $PaCO_2$
Diffusion limitation	↓	Normal	↓	↓ ↓ *	PaO_2 > 600 mm Hg
Physiologic shunt	↓	↓ ↑	↓	↑ ↓	PaO_2 > 600 mm Hg
Anatomic shunt	↓	Normal	↓	↑ ↓	PaO_2 < 500 mm Hg*
Hypokinetic hypoxia	Normal	Normal	↓ *	↑ ↓	↑ dissolved O_2
Anemic hypoxia	Normal	Normal	↓ *	Normal	↑ dissolved O_2
Histotoxic hypoxia	Normal	Normal	↑ *	Normal	↑ dissolved O_2

$PaCO_2$ = arterial CO_2 tension; PaO_2 = arterial O_2 tension; $P\bar{v}O_2$ = mixed venous O_2 tension; ↑ = increased; ↓ = decreased; ↑ ↓ = variable; * = critical determinant.

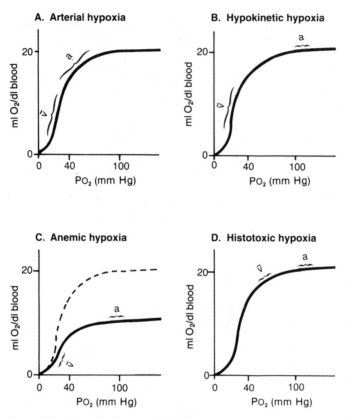

FIGURE 20-5. O_2–hemoglobin dissociation curves illustrating the four types of hypoxia. Note that the effects of 7.5 g/dl of hemoglobin are indicated on the anemic hypoxia curve (*curve C*) as the *solid line,* and the normal curve (as would be produced by 15 g/dl of hemoglobin) is indicated as the *dashed line.* The arterial O_2 tension is below normal only in arterial hypoxia (*curve A*), but the mixed venous blood point is below normal in all cases except in histotoxic hypoxia (*curve D*), where it is increased. a = arterial point; $\bar{v}$ = mixed venous point; $PO_2 = O_2$ tension.

    c. Physiologic shunts (ventilation–perfusion ratio imbalances) produce low O_2 tensions in areas of the lung with low ventilation–perfusion ratios.
        (1) When the blood leaves these areas of the lungs and mixes with better-oxygenated blood from other regions of the lung, the overall arterial O_2 content and tension are reduced (see Chapter 18 V A 2).
        (2) Administration of 100% O_2 to these patients can correct the hypoxia, because the O_2 flushes the N_2 from the alveoli, and the alveolar O_2 tension, even in low ventilation–perfusion areas, eventually exceeds 650 mm Hg at sea level.
    d. Anatomic shunts are characterized by the mixing of true venous blood and arterial (oxygenated) blood, which dilutes the normal O_2 concentration (see Chapter 18 V A 2 b). The blood that passes through an anatomic shunt is never exposed to ventilated alveoli, so the O_2 tension in this blood is equivalent to that of venous blood (i.e., the arterial O_2 tension is reduced in proportion to the fraction of the cardiac output that is shunted). The arterial blood in patients with significant anatomic shunts never achieves an O_2 tension in excess of 500 mm Hg.
        (1) Causes
            (a) Normal individuals have an **anatomic shunt** consisting of **less than 5% of their cardiac output**.
                (i) This shunted blood represents some of the coronary blood flow that enters the chambers of the left heart through the **thebesian veins** and **coronary–luminal connections.**

        (ii) In addition, much of the **bronchial venous blood** drains into the pulmonary veins.

    (b) Pathologically, an anatomic shunt may result from **congenital cardiac malformations** or from blood flow through **atelectatic areas** of the lungs.

  (2) Diagnosis

    (a) The **administration of 100% O_2** does not raise the arterial O_2 tension to its maximal level because the blood flow through anatomic shunts is never exposed to ventilated alveoli containing a high O_2 tension. At sea level, this maximal O_2 tension represents the difference between the atmospheric pressure and the sum of the water vapor and CO_2 tensions (i.e., 673 mm Hg).

    (b) The **shunt equation** $[\dot{Q}s/\dot{Q}_T = (Ci_{O_2} - Ca_{O_2})/(Ci_{O_2} - C\bar{v}_{O_2})]$ can be used to calculate the amount of shunt flow.

$$Ci_{O_2} = \text{the } O_2 \text{ content of the pulmonary capillary blood at the maximal alveolar } O_2 \text{ tension; normal value is 19.8}$$
$$Ca_{O_2} = \text{the arterial } O_2 \text{ content; normal value is 19.5}$$
$$C\bar{v}_{O_2} = \text{the mixed venous } O_2 \text{ content; normal value is 15}$$

Substituting average values for healthy individuals reveals that approximately 4%–5% of the cardiac output of a normal individual represents anatomic shunt flow.

  2. Hypokinetic (ischemic) hypoxia is caused by an **inadequate blood flow.**

    a. The arterial O_2 tension and content may be normal, but, because of inadequate blood flow, the tissues withdraw large amounts of O_2 from the capillary blood, so that the **venous O_2 content is markedly reduced** (Figure 20-5B).

    b. The reduced blood flow may involve the **whole body** (e.g., as in congestive heart failure), **or** it may involve only a **localized area of the body** (e.g., as in arteriosclerosis). **Arteriosclerosis,** the most common cause of arterial obstruction, occurs when deposits of cholesterol and other lipids in the endothelium narrow the vessel lumen. The narrowing increases the local vascular resistance, which severely reduces the blood flow.

    (1) If **coronary flow** is reduced, the resultant ischemia causes **anginal symptoms** (e.g., precordial pain and shortness of breath) or **myocardial infarction** (i.e., the actual death of tissues) because a portion of the heart muscle dies from lack of O_2.

    (2) Myocardial infarction may be fatal if a large portion of the heart muscle is affected, and cardiac output declines or arrhythmias occur.

    (3) Detection. The low O_2 tension decreases the amount of adenosine triphosphate (ATP) available for cellular processes, leading to cellular damage and the release of various cellular constituents, such as enzymes, into the general circulation. **Measurement of the isoenzyme levels** is frequently used clinically as an index of the type and extent of tissue damage.

  3. Anemic hypoxia is caused by an **insufficient amount of functional hemoglobin** (see Chapter 17 II C 1).

    a. The decrease in functional hemoglobin may be caused by deficiency of essential nutrients (e.g., as in iron deficiency anemia), or it may result from abnormal amounts of methemoglobin or carboxyhemoglobin.

    b. Patients with anemic hypoxia have a **reduced O_2 capacity** and, consequently, **a decreased O_2 content,** but the arterial O_2 tension remains normal (Figure 20-5C).

  4. Histotoxic hypoxia is caused by the **inactivation of certain metabolic enzymes** by chemicals such as **cyanide.** If these enzymes are not functioning, the tissues are unable to use O_2, even though there may be adequate O_2 in the tissues. In this condition, the O_2 delivery remains normal, but the venous O_2 tension and content are high because the O_2 is not consumed by the tissues (Figure 20-5D).

D. **Physiologic responses to chronic hypoxia**

  1. Compensatory mechanisms. The body has many ways to compensate for decreases in the O_2 supply to the body.

a. Accommodation refers to the **immediate reflex adjustments** of the **respiratory and cardiovascular systems** to hypoxia.

(1) Hyperventilation, produced by tachypnea and hyperpnea, occurs secondary to stimulation of the peripheral chemoreceptors by low O_2 tensions in the arterial blood.

(a) The increased ventilation **reduces the alveolar CO_2 tension, which raises the alveolar O_2 tension proportionately**.

(b) The reduced CO_2 tension causes **alkalosis,** which, in turn, lowers the respiratory drive. The alkalosis is slowly corrected by renal mechanisms over 1–2 weeks, so that the respiratory drive continues to increase as the alkalosis is corrected.

(2) Tachycardia is a reflex response to carotid body stimulation by hypoxia. Tachycardia increases O_2 delivery to the tissues by increasing cardiac output. In humans, the cardiac output returns to normal after spending several weeks at high altitude.

(3) The **diphosphoglycerate (DPG) concentration** increases in response to hypoxia and alkalosis. The increased DPG concentration increases the P_{50} of the hemoglobin, which helps to maintain the tissue O_2 tension at slightly higher levels than it would be otherwise [see Chapter 17 II B 3 a (1)].

b. Acclimatization refers to **changes in the body tissues** in response to **long-term exposure** to hypoxia.

(1) Polycythemia is an abnormally high number of red blood cells/μl of blood (> 5.5 million). This condition is also diagnosed if the hemoglobin level exceeds 18 g/dl in men or 16 g/dl in women, or if the hematocrit exceeds 50%.

(a) Mechanism. Polycythemia usually is secondary to tissue hypoxia, which prompts the release of **renal erythropoietic factor (REF)**. REF acts on a plasma globulin to form **erythropoietin**. Erythropoietin stimulates the production of erythrocytes by the bone marrow, which eventually increases the number of circulating erythrocytes, the hematocrit, and the hemoglobin concentration.

(b) Function. The increased number of red blood cells allows each unit of blood to carry additional O_2, which compensates for the decreased O_2 tension.

(2) Pulmonary hypertension occurs secondary to the generalized pulmonary vasoconstriction that is caused by the alveolar hypoxia [i.e., hypoxic pulmonary vasoconstriction (HPV)].

(a) Effect. The increased pulmonary artery pressure causes a more even distribution of the pulmonary blood flow, which can improve gas exchange by reducing the range of ventilation–perfusion values.

(b) Risks. The increased pulmonary artery pressure can induce right ventricular hypertrophy, right axis deviation (RAD) on the electrocardiogram (EKG), right bundle branch block, and right ventricular failure.

(3) Responses at the cellular and tissue level

(a) Oxidative enzyme concentrations increase within the mitochondria of many tissues, which allows more rapid generation of ATP via oxidative phosphorylation.

(b) Mitochondrial density increases within cells, which reduces the diffusion distance and provides more sites for O_2 utilization.

(c) Capillary density increases in skeletal and cardiac muscle, which reduces the diffusion distance from the blood into the cells.

(4) Life-long exposure to hypoxia causes additional alterations that seem to improve the body's tolerance of low O_2 levels.

(a) A **decreased respiratory drive** has been discovered in individuals exposed to hypoxia for prolonged periods. The reduced drive leads to higher CO_2 tensions and lower O_2 tensions, but it diminishes the work of respiration, which reserves more O_2 for use by other skeletal muscles.

(b) An **increased total lung capacity (TLC)** and **diffusing capacity** occur in high-altitude natives compared to their sea-level counterparts. The increase in TLC is evidenced by the enlarged chest that high-altitude natives develop.

2. Responses to hypoxia caused by high altitudes

a. Acute mountain sickness (AMS) occurs in many individuals who are unaccustomed to altitudes in excess of 9000–10,000 feet above sea level.

(1) Symptoms of AMS include fatigue, nausea, loss of appetite, headache, dyspnea, palpitations, and sleep disturbances.

(2) Risks

(a) In some individuals, **cerebral** or **pulmonary edema** occurs if the hypoxia is not treated.

(b) Exercise should be limited during the first several days at high altitude because it can lead to **severe pulmonary edema**.

(i) The pulmonary edema arises from the increased blood flow, which, when coupled with the HPV response, creates extremely high pulmonary artery pressures.

(ii) High altitude pulmonary edema is very unevenly distributed throughout the lungs and is thought to occur in areas of the lung that are not fully vasoconstricted. In these areas, the increased pulmonary artery pressure is transmitted to the pulmonary capillaries, increasing the transudation of fluid into the pulmonary interstitium and alveoli.

(3) Treatment relies on increasing the O_2 tension, either by administering supplemental O_2 or, preferably, by evacuating the individual to a lower altitude.

b. Chronic mountain sickness (Monge's disease) occurs in some long-term residents of high altitudes who develop extreme polycythemia, cyanosis, malaise, fatigue, and exercise intolerance. These individuals must be removed to a lower altitude to prevent fatal pulmonary edema from rapidly developing.

III. BIRTH is the most traumatic event that the respiratory system must withstand during the entire lifespan of the individual.

A. Perinatal respiration

1. In the **fetus,** the airways are filled with **pulmonary fluid,** which keeps the respiratory system at approximately functional residual capacity (FRC).

2. In the **neonate,** the fluid must be drained and absorbed from the airways while inflation of the airspaces is maintained. Because of the high viscosity of the pulmonary fluid, the newborn must generate very high interpleural pressures in order to initiate breathing (Figure 20-6).

B. Physiologic responses to birth

1. **Normal surfactant** is essential for initiating independent respiration, because surfactant lowers the surface tension in the lungs, which decreases the retraction (collapse) forces in the lung and enhances the absorption of fluid by the pulmonary capillaries [see Chapter 16 IV B 3 a (2)]. The low surface tension stabilizes the alveoli and allows the lung volume to increase steadily over the first few days and weeks of life (see Figure 20-6).

2. The transport mechanisms in the alveolar epithelium change from fluid secretion to fluid reabsorption mechanisms. The decreased fluid content of the lungs improves diffusion of gases by reducing the diffusion distance.

IV. MECHANICAL VENTILATION. Impaired function of the lungs, chest wall, or the neuromuscular system that controls the respiratory muscles can lead to either inadequate oxygenation of the blood (i.e., hypoxia) or inadequate ventilation of the lungs and CO_2 retention.

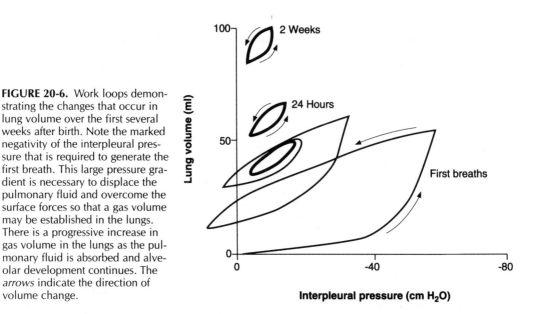

FIGURE 20-6. Work loops demonstrating the changes that occur in lung volume over the first several weeks after birth. Note the marked negativity of the interpleural pressure that is required to generate the first breath. This large pressure gradient is necessary to displace the pulmonary fluid and overcome the surface forces so that a gas volume may be established in the lungs. There is a progressive increase in gas volume in the lungs as the pulmonary fluid is absorbed and alveolar development continues. The *arrows* indicate the direction of volume change.

A. **Inadequate oxygenation** is usually produced by a ventilation–perfusion imbalance with excessive physiologic shunting, as commonly occurs in the **adult respiratory distress syndrome (ARDS)**. ARDS is characterized by atelectasis, pulmonary edema, ventilation–perfusion mismatch, and hemorrhage, which impair O_2 transfer across the alveolar–capillary membrane.

1. **Administration of supplemental O_2** (FIO_2 of 0.24–0.5) improves oxygenation of the blood.

2. **Mechanical ventilation** is used to provide positive pressure to the airways to maintain alveolar inflation, to administer supplemental O_2, and to ensure adequate ventilation because the work of breathing is increased in these patients. **Positive end-expiratory presssure (PEEP)** is the use of a ventilator to maintain alveolar pressure above zero at the end of expiration.

   a. **Method.** PEEP increases lung volume by increasing the transmural pressure of the respiratory system. The increase in alveolar volume helps to **stabilize the alveoli, reinflate atelectatic regions of the lung,** and **reduce pulmonary edema**. The reduction of edema occurs because the increased alveolar pressure opposes the filtration forces that tend to push fluid out of the pulmonary capillaries according to the Starling hypothesis [see Chapter 12 I C 3 b (1)].

   b. **Benefits.** PEEP is usually used to improve gas exchange in patients with respiratory distress syndrome, because these patients have disorders of surfactant function resulting in atelectasis, edema, and hemorrhage in different areas of their lungs. The judicious use of supplemental O_2 and PEEP enables some of these patients to survive without lung or heart–lung transplants. [Many ARDS patients develop **multiple organ failure syndrome** (i.e., the functional deterioration of several organ systems), which carries a very high mortality rate.]

   c. **Risks**

   (1) **Barotrauma.** The increased lung volume that the application of PEEP fosters can overdistend segments of the lung causing **laceration of tissues** and **leakage of air out of the alveolar spaces**.

   (a) Laceration of the visceral pleura with leakage of gas into the pleural space is termed **pneumothorax** and results in lung collapse (see VI).

   (b) The high pressures in the lungs can also force gas into the interstitium where it can dissect its way to the mediastinum and, from there, to the subcutaneous tissues, causing **subcutaneous emphysema**.

   (2) **Reduced cardiac output.** One of the biggest disadvantages of PEEP is that it raises the intrathoracic pressure, which reduces the vascular pressure gradient

responsible for venous return (remember that the interpleural pressure is normally negative). Thus, venous return and cardiac output decline because the driving force from the systemic capillaries to the right atrium is reduced.

    (a) In addition to the O_2 content of the blood, O_2 delivery to the tissues is a function of blood flow (i.e., cardiac output); therefore, the **reduced cardiac output is detrimental to tissue oxygenation**.

    (b) PEEP can be adjusted until the O_2 delivery (i.e., the arterial O_2 content times the cardiac output) is maximized. The **Swan-Ganz catheter,** which can be used to measure cardiac output, is an indispensible tool for managing patients with severe respiratory disease who require the use of PEEP.

B. **Inadequate ventilation** can result from depression of the respiratory centers by drugs, paralysis of the respiratory muscles, or respiratory muscle fatigue caused by excessive work of breathing. In all of these patients, **positive-pressure respiration** (i.e., mechanical ventilation) is used to maintain alveolar ventilation and normalize the arterial CO_2 tension.

    1. **Administration of positive-pressure respiration** is typically by a machine that delivers **regulated volumes of gas under positive pressure**. Inflation is produced by the positive pressure within the airways, and deflation occurs when the positive pressure is removed (Figure 20-7).

        a. **Endotracheal intubation.** The ventilation is usually administered through an endotracheal tube, made of flexible plastic, that is inserted into the trachea via the nose, mouth, or a tracheostomy, which provides better access to the airway. The tube contains an inflatable cuff that makes an airtight seal with the trachea.

        b. **Control adjustment.** Most modern ventilators are **volume-regulated rather than pressure-regulated**.

            (1) The **tidal volume** is set by the physician to maintain an adequate ventilation.

            (2) The **respiratory rate** of the ventilator can be adjusted over the physiologic range, as desired.

            (3) **Inspiratory–expiratory ratio.** The total duration of each respiratory cycle is determined by the respiratory rate, but the **ratio of inspiratory time to the total cycle duration** can be adjusted as well.

FIGURE 20-7. A record of airway pressure changes (*upper panel*) and the resultant change in lung volume (*lower panel*) that is produced by positive-pressure ventilation. The increase in lung volume is dependent on the peak airway pressure and the compliance of the respiratory system. Deflation (i.e., expiration) occurs when the positive airway pressure decreases. Note that the decrease in lung volume requires a significant time, which is dependent on the patient's airway resistance.

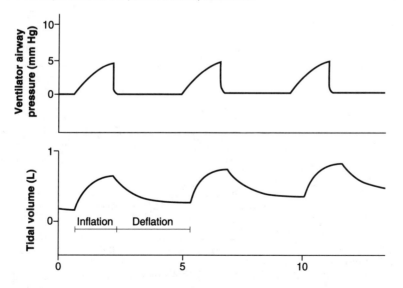

   (a) The duration of inspiration can be shortened by delivering the tidal volume at a higher flow rate, which provides a longer period of expiration for any given cycle length and allows maximal expiratory volume change to occur. An extended period for expiration is especially important in patients with a **high airway resistance** because expiration is generally passive.
   (b) **Intrinsic PEEP (auto-peep).** Patients with a high airway resistance may not deflate to their equilibrium volume (i.e., their normal FRC) because of low expiratory flow rates. Under these conditions, the alveolar pressure remains positive at the end of expiration. Increasing the respiratory rate abbreviates the time for expiration and prevents the lungs from deflating completely. Consequently, the lung volume increases until the added auto-PEEP is sufficient to expel the entire tidal volume (Figure 20-8).
(4) A **pressure-relief valve** is incorporated into the machine because very high pressures may develop if the compliance of the respiratory system is low or excessive tidal volumes are administered (Figure 20-9).

2. Common modes of positive-pressure respiration
   a. Controlled ventilation is required for patients who have lost all control over respiratory function (e.g., "brain dead" patients). The patient is connected to the ventilator, the respiratory rate and tidal volume are set by the physician, and the machine completely takes over the ventilatory functions of the respiratory system (Figure 20-10).
   b. Assisted ventilation is used in awake patients who are capable of initiating an inspiratory effort. An initial negative airway pressure, which is generated by the patient's inspiratory muscles, triggers the ventilator to deliver a tidal volume under positive pressure (see Figure 20-10)
      (1) This mode of ventilation is used to relieve the respiratory muscles of some of the work of breathing and is appropriate for patients with weakness or fatigue

FIGURE 20-8. Effect of increasing the respiratory rate on intrinsic positive end-expiratory pressure (PEEP). Note that the lung volume has not stabilized at the end of expiration, indicating that the respiratory system has not achieved its equilibrium volume [i.e., functional residual capacity (*FRC*)]. The increased lung volume creates PEEP. An increased respiratory rate shortens the time allotted for expiration (i.e., lung deflation), further increasing the lung volume and end-expiratory pressure.

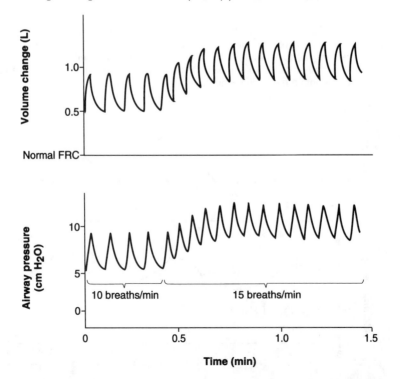

C_{RS} (L/cm H_2O) = 0.1   0.05   0.025

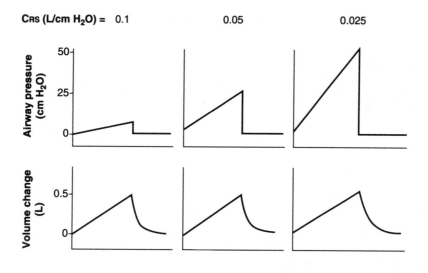

Time

FIGURE 20-9. Graphic depiction of the pressures generated during lung inflation in patients with varying respiratory system compliance (C_{RS}). As the compliance is reduced, the airway pressure progressively increases. The tidal volume (V_T) in each example is 0.5 L and the maximal pressure (assumed to be due only to elastance forces) is equal to V_T/C_{RS}. If the compliance varies in different areas of the lungs, those areas with high compliance may become overdistended. Overdistention can result in tissue damage, tearing or laceration of the tissues, and air leakage into the pleural space or interstitial lung tissues.

of the respiratory muscles. Patients with intrinsic PEEP (i.e., high airway resistance) must perform significantly more respiratory work than other patients because they must expand the lungs further to overcome the PEEP before they can generate the negative airway pressure that triggers the next tidal volume.

(2) Assisted ventilation usually has a minimal adjustable respiratory rate that ensures some minute ventilation even if spontaneous breathing ceases; however, for the most part, the patient's respiratory centers remain largely in control of determining the minute ventilation.

c. **Intermittent mandatory ventilation (IMV)** is used when respiratory rate and minute ventilation must be maintained above a patient's intrinsic breathing frequency. In this mode, the machine delivers a preset tidal volume at a respiratory rate set by the physician so that the patient is not required to make a respiratory effort. Even though the patient does not trigger the ventilator in this mode, he or she can take **independent breaths that exceed the preset ventilator rate and depth of respiration**.

V. **O_2 TOXICITY.** O_2 in high concentrations (> 50% at sea levels) can cause tissue damage by inactivating certain enzymes (via oxidation) and by forming O_2 radicals that cause tissue inflammation and edema. Two types of O_2 toxicity are recognized: **hyperbaric toxicity,** which affects the **central nervous system (CNS),** and **normobaric toxicity,** which affects the **lungs**.

A. **Hyperbaric (CNS) O_2 toxicity** requires exposure to a minimum of 1.5–2.0 atmospheres (atm) of O_2 tension for only 15 minutes (Figure 20-11).

1. **Causes.** Obviously, hyperbaric O_2 toxicity only occurs during diving or hyperbaric chamber operations (e.g., treatment for decompression sickness, gas gangrene).

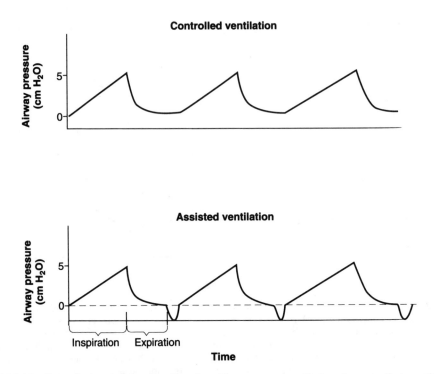

FIGURE 20-10. Controlled and assisted modes of positive-pressure ventilation. In controlled ventilation, the tidal volume and respiratory rate are set by the physician. Controlled ventilation is used to ventilate patients with no intrinsic function of the respiratory neuromuscular system. Assisted ventilation requires the patient to trigger the ventilator. The trigger is the negative airway pressure that is generated by the patient's initial effort to inspire. Assisted ventilation is used to augment the patient's inspiratory effort if the respiratory muscles are weakened or fatigued as the result of a disease process.

    2. Effects. CNS O_2 toxicity is evidenced by an increased irritability of nerves and muscles that eventually leads to grand mal convulsive seizures. Because of the danger of convulsions during scuba diving, it is forbidden to use 100% O_2 to refill scuba tanks for the general public.

B. **Normobaric (pulmonary) O_2 toxicity** requires exposure to a minimum of 0.5 atm of O_2 tension for 18–24 hours (see Figure 20-11).

    1. Causes. Administration of supplemental O_2 for long periods of time can produce pulmonary O_2 toxicity.

    2. Effects. The high O_2 tension destroys the type II cells that produce surfactant, leading to pulmonary edema, areas of atelectasis, and hemorrhage. The gas exchange functions of the lungs are impaired, and the arterial O_2 levels decline. In order to maintain arterial oxygenation at viable levels, the alveolar O_2 tension is increased, which leads to further damage of the lung tissue. This vicious cycle frequently occurs in patients with ARDS, who require external artificial oxygenators to sustain gas exchange until the lungs recover from the disease process.

VI. **PNEUMOTHORAX** is the presence of **air in the pleural space**. Humans have a complete mediastinum; therefore, a pneumothorax is typically **unilateral** because the two pleural sacs do not communicate.

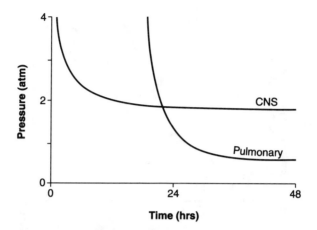

FIGURE 20-11. Typical dose–duration curves demonstrating the threshold for hyperbaric [central nervous system (*CNS*)] O_2 toxicity and normobaric (*pulmonary*) O_2 toxicity. The asymptotes for the CNS effects are 15 minutes and 1.5 atm, and the asymptotes for the pulmonary effects are 18 hours and 0.5 atm. The asymptotes represent the minimal values that produce the effects. Theoretically, the administration of 0.4 atm (40% FIO_2 at sea level) is safe for an infinite period of time, but administration of more than 50% FIO_2 at sea level will eventually produce O_2 toxicity in the lungs.

A. **Causes.** Laceration or rupture of a lung, traumatic penetration of the chest wall, or rupture of the esophagus can permit air to enter the pleural space.

B. **Effects.** The presence of air in the pleural space **uncouples the affected lung from the ipsilateral chest wall**. Normally, the interpleural pressure is negative, which means that any **communication between the pleural sac and an air source allows gas to enter the pleural space**.

1. As the volume of the pneumothorax increases, the interpleural pressure becomes less negative (i.e., it increases). The decreased negativity of the interpleural pressure causes the transmural pressure for both the chest wall and the lungs to decrease.

2. Figure 20-12 shows that the chest wall expands and the lungs deflate along their respective pressure–volume curves. The difference in volume between the chest wall and the lungs represents the volume of gas in the pleural space.

C. **Types of pneumothorax**

1. **Simple pneumothorax** occurs when the **air leak** into the pleural space **seals** and the patient is left with one partially collapsed lung and an expanded chest wall on the affected side. This condition can be detected by physical examination or by x-ray.
   a. **Needle aspiration.** A simple pneumothorax can be treated by needle aspiration of the air if the patient is in respiratory difficulty. A needle is inserted between two of the upper ribs and either the air is withdrawn by suction through a syringe, or the patient is connected to a series of suction bottles to maintain a negative pressure in the pleural space.
   b. **Resorption.** If the patient is not having respiratory difficulty, the pneumothorax can be allowed to resorb.
      (1) **Diffusion gradient.** Gas in the pleural space or in tissues is slowly absorbed because there is a diffusion gradient between the gas and the capillary blood. Recall that normally, the O_2 tension decreases from about 95 mm Hg in the arteries to an average of about 40 mm Hg in the veins, while the CO_2 tension increases from 40 mm Hg to 45 mm Hg as the blood traverses the capillaries. These changes result in a total gas tension in the veins that is approximately 50 mm Hg less than the atmospheric pressure. This diffusion gradient causes the slow absorption of gas pockets within the body.

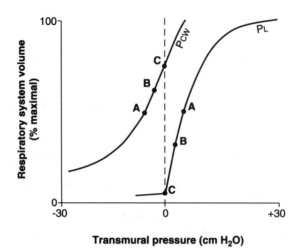

Transmural pressure (cm H₂O)

FIGURE 20-12. Pressure–volume curves for the chest wall (*Pcw*) and the lungs (*PL*). *Point A* on each curve represents the normal condition, where the interpleural pressure is −5 cm H_2O so that P_L = 5 cm H_2O and P_{CW} = −5 cm H_2O. Both points occur at the same respiratory system volume and all of the gas is contained within the lungs. The presence of a simple pneumothorax causes the interpleural pressure to decrease to about −3 cm H_2O, so that P_L = 3 cm H_2O and P_{CW} = −3 cm H_2O. The volume of gas in the lungs has decreased to *point B on the PL curve* but the total gas within the respiratory system is indicated by *point B on the Pcw curve*. The volume of gas in the interpleural space is given by the difference between points B on the two curves. *Point C* on each curve reflects the volume of gas within the chest wall and lungs, respectively, if the interpleural pressure increases to zero. A tension pneumothorax results in a positive interpleural pressure, which causes the chest wall to exceed its equilibrium volume and the lung to reach its minimal volume.

      (2) Rate of resorption. Normally, the pneumothorax is absorbed at the rate of about **1% per day**.

   2. Tension pneumothorax. At times, the site of gas entry into the pleural space acts as a **one-way valve,** letting gas in with each breath but not out. Under these conditions, the interpleural pressure slowly increases. Eventually, the interpleural pressure becomes positive and the vascular pressure gradient responsible for venous return is gradually eliminated.

     a. Diagnosis

      (1) The positive pressure causes **venous distention** that **can be noted in the jugular veins**.

      (2) A high pressure in one hemithorax causes the mediastinum to shift to the contralateral side. The **shifted mediastinum can be detected by physical examination**.

     b. Treatment. A tension pneumothorax presents an **acute emergency** because the cardiac output rapidly decreases and the patient goes into low cardiac output shock. Treatment requires that the positive intrathoracic pressure be relieved, which can be accomplished by **needle aspiration.**

STUDY QUESTIONS

DIRECTIONS: Each of the numbered items or incomplete statements in this section is followed by answers or by completions of the statement. Select the **one** lettered answer or completion that is **best** in each case.

1. A 62-year-old man is known to have chronic lung disease and hypercapnia. He needs a major operation to remove an intestinal tumor. To ensure that he has adequate alveolar ventilation while being anesthetized, which of the following should be available?

(A) Tank of 100% O_2
(B) Tank of 95% O_2, 5% CO_2
(C) Mechanical respirator
(D) Cardiac defibrillator
(E) Electrocardiograph

2. Maximal inspiratory gas flow occurs when the

(A) lung volume approaches total lung capacity (TLC)
(B) lung volume approaches residual volume (RV)
(C) alveolar pressure is most negative
(D) interpleural pressure is approximately -5 cm H_2O
(E) abdominal muscles are maximally contracted

3. The diffusion coefficient of O_2, as compared with that of CO_2, is

(A) greater because O_2 combines with hemoglobin
(B) less because O_2 is less soluble
(C) greater because of a higher pressure gradient
(D) less because of the lower molecular weight of O_2
(E) essentially the same

4. Which of the following is true regarding the transmural pressure for the lungs?

(A) It is always negative
(B) It is equal to the interpleural pressure minus the atmospheric pressure ($P_{PL} - P_B$)
(C) It is equal to the interpleural pressure minus the alveolar pressure ($P_{PL} - P_A$)
(D) It is equal to the alveolar pressure minus the interpleural pressure ($P_A - P_{PL}$)
(E) It is independent of lung volume when the muscles are relaxed

5. Airway resistance can be reduced by

(A) increasing vagal impulses to the lungs
(B) administering a β-adrenergic blocking drug
(C) decreasing the radial traction exerted by lung tissue
(D) performing a maximal forced expiration
(E) increasing lung volume

6. A reduction of arterial O_2 tension is typical of which one of the following?

(A) Anemia
(B) CO poisoning
(C) Moderate exercise
(D) Cyanide poisoning
(E) Hypoventilation

7. Which one of the following statements regarding the compliance of the respiratory system is true?

(A) It is greater than the compliance of the chest wall
(B) It is greater than the compliance of the lungs
(C) It is equal to the compliance of the chest wall
(D) It is equal to the compliance of the lungs
(E) It is less than the compliance of the chest wall

8. During the effort-independent portion of a forced vital capacity (FVC) maneuver, the expiratory flow rate

(A) varies as a function of the interpleural pressure
(B) is limited by compression of the airways
(C) depends on the alveolar pressure
(D) is maximal for that individual
(E) is constant

9. A lack of normal surfactant, as occurs in infants with respiratory distress syndrome (RDS), results in

(A) an increased lung compliance
(B) stabilization of alveolar volume
(C) an increased retractive force of the lungs
(D) a reduced alveolar–arterial O_2 tension difference
(E) a decrease in the filtration forces in the pulmonary capillaries

10. During inspiration, as the diaphragm contracts, the pressure in the interpleural space becomes

(A) equal to zero
(B) more positive
(C) more negative
(D) equal to the pressure in the alveoli
(E) equal to the pressure in the atmosphere

11. The volume of gas in the lungs at the end of a normal expiration is referred to as the

(A) residual volume (RV)
(B) expiratory reserve volume (ERV)
(C) functional residual capacity (FRC)
(D) inspiratory reserve volume (IRV)
(E) total lung capacity (TLC)

12. The major area of airway resistance during breathing is located in the

(A) oropharynx
(B) trachea and large bronchi
(C) intermediate-sized bronchi
(D) bronchioles less than 2 mm in diameter
(E) alveoli

13. A patient with restrictive lung disease typically has

(A) an increased forced expiratory volume in 1 second (FEV_1) and a normal lung compliance
(B) a decreased FEV_1 and an increased lung compliance
(C) a decreased FEV_1 and a decreased lung compliance
(D) an increased FEV_1 and an increased lung compliance
(E) an increased FEV_1 and a decreased lung compliance

14. The volume of N_2 dissolved in body fluids is greatest while breathing which of the following gas mixtures?

(A) Air at sea level
(B) Air at an altitude of 15,000 feet
(C) 20% O_2, 20% N_2, 60% He, while scuba diving at 2 atm of pressure
(D) 20% O_2, 30% N_2, 50% He, while scuba diving at 2 atm of pressure
(E) 20% O_2, 10% N_2, 70% He, while scuba diving at 5 atm of pressure

15. Increasing the tidal volume, while keeping everything else constant, will increase the

(A) dead space ventilation
(B) functional residual capacity (FRC)
(C) inspiratory capacity (IC)
(D) alveolar ventilation
(E) alveolar CO_2 tension

16. Which of the following statements is true regarding the fraction of O_2 in inspired (tracheal) gas?

(A) It equals 0.25 at sea level
(B) It decreases as a function of altitude
(C) It varies as a function of the weather
(D) It is less than the fraction of O_2 in the atmosphere
(E) It equals the fraction of O_2 in the alveoli

17. Which one of the following statements regarding the CO_2 tension in mixed expired gas is true?

(A) It is greater than the alveolar CO_2 tension
(B) It is less than the alveolar CO_2 tension
(C) It is equal to the alveolar CO_2 tension
(D) It is equal to the atmospheric CO_2 tension
(E) It is greater than the CO_2 tension in venous blood

18. Alveolar ventilation is equal to the

(A) dead space ventilation
(B) tidal volume times respiratory rate
(C) minute ventilation
(D) minute ventilation minus dead space ventilation
(E) CO_2 production/min

19. A reduction in local alveolar ventilation is associated with

(A) an increase in regional pulmonary blood flow
(B) a decrease in regional alveolar CO_2 tension
(C) a decrease in regional alveolar O_2 tension
(D) an increase in regional tissue pH
(E) an increase in capillary hemoglobin saturation

20. The following figure shows two ventilatory patterns: one normal and the other abnormal. Which experimental maneuver listed below will create the abnormal pattern?

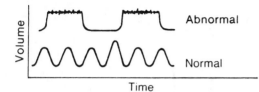

(A) Midpons transection with vagi intact
(B) Transection of the brain stem between the pons and medulla
(C) Midpons transection with vagi cut
(D) Transection rostral to the pons with vagi cut
(E) Transection rostral to the pons with vagi intact

21. The major sign of hypoventilation is

(A) cyanosis
(B) increased airway resistance
(C) hypercapnia
(D) dyspnea
(E) hypoxia

22. Which of the following statements regarding the normal alveolar CO_2 tension is true?

(A) It is equal in all alveoli
(B) It is highest at the base of vertical lungs
(C) It is directly proportional to the inspired O_2 tension
(D) It is directly proportional to the alveolar ventilation
(E) It is equal to 46 mm Hg

23. Hypercapnia affects respiration primarily by stimulating the

(A) carotid and aortic bodies
(B) receptors
(C) central (medullary) chemoreceptors
(D) arterial baroreceptors
(E) hypoglossal nerve

24. The vital capacity (VC) is the sum of the

(A) residual volume (RV), tidal volume, and expiratory reserve volume (ERV)
(B) RV, tidal volume, and inspiratory reserve volume (IRV)
(C) RV, ERV, and IRV
(D) ERV, IRV, and tidal volume
(E) functional residual capacity (FRC) and inspiratory capacity (IC)

25. The venous O_2 tension is higher than normal in which one of the following conditions?

(A) Cyanide poisoning
(B) Exercise
(C) Decreased cardiac output
(D) Anemia
(E) CO poisoning

26. The respiratory system is at the equilibrium position in all of the following conditions EXCEPT

(A) at the end of a normal expiration
(B) when the transrespiratory pressure is zero
(C) when lung recoil is balanced by chest wall expansion
(D) when lung volume is at residual volume (RV)
(E) when the respiratory muscles are relaxed and the airway is open

27. All of the following can reduce vital capacity (VC) EXCEPT

(A) a decreased total lung capacity (TLC)
(B) an increased residual volume (RV)
(C) a weakness of the inspiratory muscles
(D) a weakness of the expiratory muscles
(E) a decreased alveolar surface tension

Directions: Each group of items in this section consists of lettered options followed by a set of numbered items. For each item, select the **one** lettered option that is most closely associated with it. Each lettered option may be selected once, more than once, or not at all.

Questions 28–32

The following graph shows a normal respiratory cycle followed by a maximal inspiration, a maximal forced expiration, and another normal respiratory cycle. Match each of the lung volumes listed below to the appropriate lettered arrow on the graph.

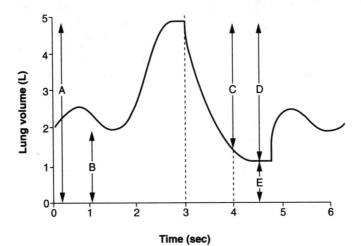

28. Vital capacity (VC)

29. Forced expiratory volume in 1 second (FEV$_1$)

30. Functional residual capacity (FRC)

31. Total lung capacity (TLC)

32. Residual volume (RV)

ANSWERS AND EXPLANATIONS

1. The answer is C [Chapter 16 VIII C 1 b 2; Chapter 20 II C 1 c (2), IV B]. The only treatment option that provides ventilation is the respirator. Administering supplemental O_2 may correct the hypoxia, but to correct hypercapnia, the lungs must be adequately ventilated. The patient certainly does not need extra CO_2, because he is hypercapnic, and neither the defibrillator nor the electrocardiograph will aid in ventilating the alveoli.

2. The answer is C [Chapter 16 II C 1 a, 2 b (1)]. The driving force for gas flow is the alveolar pressure; during normal breathing, negative pressures cause inspiratory flow, and positive pressures cause expiratory flow. It is the ratio of alveolar pressure to airway resistance that determines the actual flow of gas. As lung volume approaches either total lung capacity (TLC) or residual volume (RV), much of the muscle force is expended in overcoming the low compliance of the respiratory system. When the lung volume is in the mid-range, alveolar pressure may be either positive or negative depending on muscle activity. The abdominal muscles are expiratory muscles and, when contracted, generate positive alveolar pressures and expiratory flow, if the airways are open.

3. The answer is B [Chapter 18 II A 3; C 2 a]. O_2 is much less diffusible than CO_2 in the lungs. The diffusion coefficient concerns the movement of molecules in solution and is determined by the molecular weight and the solubility of the substance. O_2 has a lower molecular weight than CO_2, but is also about 25 times less soluble. The increased diffusibility of CO_2 is important because there is only a gradient of approximately 5 mm Hg to cause CO_2 to diffuse from the pulmonary capillary blood to the alveoli.

4. The answer is D [Chapter 16 III A, B]. In the lungs, the transmural pressure is the transpulmonary pressure, which equals the alveolar pressure minus the interpleural pressure ($P_A - P_{PL}$). The transmural pressure simply represents the pressure across the wall of a hollow organ and is defined as the inside pressure minus the outside pressure. In humans, the interpleural pressure is estimated by measuring the esophageal pressure, because the esophagus is a flaccid tube that essentially

traverses the pleural space. The interpleural pressure in animal experiments usually is measured by inserting a needle into an intercostal space and connecting it to a pressure gauge. The transmural pressure across the chest wall (i.e., the transthoracic pressure) equals the interpleural pressure minus the pressure at the body surface, which is usually equal to the atmospheric pressure ($P_{PL} - P_B$). The lung is a passive, elastic structure so that any change in lung volume produces a change in the transmural pressure.

5. The answer is E [Chapter 16 V A 2 b (2); C 2 a]. Increases in lung volume produce a mechanical force (i.e., radial traction) that acts on the airway walls, dilating the airways and reducing resistance. Radial traction is one of the most powerful factors that can alter airway resistance. The aging process can diminish radial traction, contributing to the higher airway resistance often seen in older people. Increasing vagal impulses to the lungs or administering a β-adrenergic blocking drug increases airway resistance by narrowing the airways: β-adrenergic blockers inhibit adrenergic substances (e.g., epinephrine) that act as bronchodilators, and stimulating the vagus nerve produces active contraction of bronchial smooth muscle. A forced expiration increases airway resistance, because the positive interpleural pressure compresses the large intrathoracic airways, creating a high-resistance, flow-limiting segment.

6. The answer is E [Chapter 17 II D 4 a; Chapter 20 II C 1 a, 4; Figure 20-5]. Hypoventilation can lead to arterial hypoxia, which is distinguished from other types of hypoxia by a low arterial O_2 tension, and hypercapnia. In anemia and CO poisoning, the O_2 capacity (i.e., the amount of O_2 carried by the hemoglobin) is decreased, but the O_2 tension (i.e., the amount of O_2 dissolved in the plasma) is unaffected. In these disorders, the tissue or venous O_2 tension is reduced as a consequence of the inadequate O_2 delivery. Moderate exercise should not affect the arterial O_2 tension in a normal person.

7. The answer is E [Chapter 16 III D 1 b (2); Figure 16-6D]. The compliance of the respiratory system (C_{RS}) is determined by the compliance of the lungs (C_L) and the chest wall (C_{CW})

and can be calculated as $1/C_{RS} = 1/C_L + 1/C_{CW}$. Because of the need to add reciprocals, the compliance of the respiratory system always is less than the compliance of either of its parts.

8. The answer is B [Chapter 16 V C 2]. Dynamic compression of the airways occurs during forced expiration and limits the flow rate during the terminal 80% of expiration [i.e., the effort-independent portion of a forced vital capacity (FVC) maneuver]. The driving force for expiration during the effort-independent period is the transpulmonary pressure, which is a function of lung volume. The term indicates that flow is independent of effort and that changes in expiratory force do not alter the flow rates. Therefore, the expiratory flow rate neither varies as a function of the interpleural pressure, nor depends on the alveolar pressure, because these values are altered by expiratory effort. The expiratory flow rate is not maximal for the individual because maximal flow can be achieved only when lung volume is just below total lung capacity (TLC).

9. The answer is C [Chapter 16 IV B 3 a (2)–(3)]. The lack of normal surfactant produces a high alveolar surface tension, which increases the retractile force of the lungs, resulting in a high transmural pressure. The high transmural pressure means that the lungs are less distensible, and the alveoli tend to collapse because of the increased surface forces. In addition, the increased alveolar surface tension decreases the interstitial pressure, which increases the filtration forces across the pulmonary capillaries and leads to edema. The edema and atelectasis cause an abnormal range of ventilation–perfusion ratios, which impairs gas exchange. The alveolar–arterial O_2 tension difference is a good measure of the gas exchange capabilities of the lungs; this difference increases in the presence of ventilation–perfusion or diffusion abnormalities.

10. The answer is C [Chapter 16 II C 2 b (2); Figure 16-3]. Contraction of the inspiratory muscles expands the chest wall, increasing the transpulmonary pressure and making the interpleural pressure more negative. The chest wall expansion also expands the gas in the lungs because the visceral and parietal pleurae are coupled together by the interpleural fluid. The expansion of the alveolar gases decreases the alveolar pressure, which causes the gas to

flow into the respiratory system through open airways.

11. The answer is C [Chapter 16 VII C 4]. The functional residual capacity (FRC) is the volume of gas in the lungs at the end of a normal expiration. Since expiration is passive, the lung volume decreases during expiration until the equilibrium volume (i.e., the FRC) is reached. The equilibrium volume represents the volume of a distensible structure when the transmural pressure (i.e., the pressure inside minus the pressure outside) is zero. The residual volume (RV) is the volume of gas in the lungs following a maximal expiration. The expiratory reserve volume (ERV) is the volume of gas that can be forcefully expired after a normal expiration, and the inspiratory reserve volume (IRV) is the additional volume of gas that can be inspired over the tidal volume. The total lung capacity (TLC) is the volume of gas in the lungs after a maximal inspiration.

12. The answer is C [Chapter 16 V A 2 a–b]. The highest resistance to airflow occurs in the intermediate-sized bronchi because of the high airflow velocity in these segments. The airway tree develops such that each generation of airways is only slightly smaller in diameter than the parent airways. Thus, there is almost an exponential increase in cross-sectional area proceeding toward the periphery of the lung. Because of this relationship, the linear velocity of gas molecules decreases markedly as these molecules approach the terminal bronchioles. Therefore, very little pressure is required to achieve this velocity (i.e., there is a low resistance). Direct measurements indicate that bronchioles less than 2 mm in diameter represent less than 10% of the total airway resistance.

13. The answer is C [Chapter 16 VI C 1; IX A 2 a–b; Figure 16-21]. By definition, a patient with restrictive lung disease has reduced lung compliance, and, typically, a reduced forced expiratory volume in 1 second (FEV_1) because of the reduced vital capacity (VC). These patients typically can expire a larger fraction of their own VC in 1 second because of the greater radial traction that results from the decreased compliance. Thus, these patients have a high $FEV_1\%$, which is the ratio of the FEV_1 to the VC. The increased radial traction reduces airway resistance in the lungs. These patients have no difficulty breathing at high frequencies because of the low airway resistance and usually choose a high rate of respiration coupled with a reduced tidal volume to minimize the work of breathing.

14. The answer is A [Chapter 17 I A]. Among these choices, the volume of N_2 dissolved in the body fluids is greatest when breathing air at sea level. Henry's law states that the volume of gas dissolved in a liquid equals the partial pressure of the gas times the solubility coefficient. Since the partial pressure of N_2 at an altitude of 15,000 feet would be less than the partial pressure of N_2 at sea level, the amount of N_2 dissolved in the tissues also would be less. The N_2 tension when breathing air at sea level is 0.79 atm. Gas equilibration in the body requires approximately 12 hours after any change in pressure and occurs at different rates in different tissues. For N_2, equilibration takes the longest in the fatty tissues because of the high N_2 solubility and the low blood flow to this type of tissue. Scuba diving at 2 atm while breathing a gas with 20% N_2 provides 0.4 atm of N_2 tension. The other two scuba conditions yield N_2 tensions less than 0.79 atm as well.

15. The answer is D [Chapter 16 VII B–C; VIII C]. If the respiratory rate, dead space, and ventilation–perfusion ratio remain constant, then an increase in tidal volume will increase the minute and alveolar ventilation. Since the dead space ventilation equals the dead space volume times the respiratory rate, increasing the tidal volume has no effect on the dead space ventilation. The functional residual capacity (FRC) is not altered, and the inspiratory capacity (IC) is reduced, by increases in tidal volume. Alveolar CO_2 tension is reduced by an increased alveolar ventilation because the increased ventilation washes out CO_2 from the alveoli.

16. The answer is D [Chapter 16 VIII A, C 2 a]. The fraction of O_2 in inspired gas is less than the fraction of O_2 in the atmosphere, because nasal breathing adds water vapor to inspired air, which decreases the O_2 content of the air by the time it reaches the trachea. The fraction of O_2 in the air is constant from sea level to several hundred thousand feet altitude and equals 0.21. However, the O_2 tension decreases with increased altitude, because the total pressure declines as one ascends. The partial pressure of any gas is given by the product of the mole fraction of the gas and the total or barometric pressure. Alveolar gas is

diluted by the addition of CO_2 from the blood, so that the fraction of O_2 in the alveoli is less than that in the trachea.

17. The answer is B [Chapter 16 VIII B 2 c]. Mixed expired gas has a lower CO_2 tension than alveolar gas because it is a mixture of alveolar and dead space gas. The composition of this gas varies depending on the ratio of dead space ventilation to alveolar ventilation. This fact can be used to determine the dead space–tidal volume ratio:

$$V_D/V_T = 1 - \frac{P_{ECO_2}}{P_{ACO_2}} \text{, where}$$

V_D/V_T = the dead space–tidal volume ratio
P_{ECO_2} = the CO_2 tension of mixed, expired gas
P_{ACO_2} = the CO_2 tension of alveolar gas

The CO_2 tension in venous blood is higher than that in alveolar gas, which is why CO_2 diffuses from the pulmonary capillaries into the alveoli.

18. The answer is D [Chapter 16 VIII C]. Alveolar ventilation equals the minute ventilation minus the dead space ventilation. Minute ventilation is the volume of gas expired per minute, which is equal to the product of tidal volume times respiratory rate. Not all of the minute ventilation reaches the gas exchange region of the lungs; some remains in the conducting system of the lungs.

19. The answer is C [Chapter 18 V A 2]. A decrease in local (i.e., regional) alveolar ventilation decreases the influx of gas to that region of the lung, so that, transiently, more O_2 is absorbed, and more CO_2 is released by the capillary blood. Consequently, in the affected area of the lung, the alveolar O_2 tension declines and the alveolar CO_2 tension increases. The hypoxia, hypercapnia, and resultant local acidosis cause hypoxic pulmonary vasoconstriction (HPV), which results in a decrease in pulmonary blood flow to the affected area of the lungs. The decrease in local O_2 tension reduces the hemoglobin saturation of the blood that leaves this area of the lung. These changes represent the effects of a low ventilation–perfusion ratio on gas exchange in a localized area of the lung.

20. The answer is C [Chapter 19 I C 1; Figure 19-5]. This pattern of breathing is called apneustic breathing, or inspiratory breath holding, which occurs when the inhibitory influences from both the periphery (via the vagus nerves) and the pneumotaxic center are interrupted. The pneumotaxic center lies in the rostral pons, and the apneustic center (which enhances inspiratory drive) lies in the caudal pons. The vagus nerve carries impulses from stretch receptors in the airways that tend to inhibit inspiration after a certain threshold of tidal volume is exceeded.

21. The answer is C [Chapter 16 VIII C 1 b (2); Chapter 20 II B 1, C 1 a]. Hypercapnia is pathognomonic of hypoventilation, because the alveolar CO_2 tension is determined primarily by the alveolar ventilation. Cyanosis may be present with hypoventilation, but it is a sign (and not a very good one) of hypoxia, not hypoventilation. Both an increased airway resistance and dyspnea may be symptoms of airway disease, but neither is characteristic of hypoventilation.

22. The answer is B [Chapter 18 Figure 18-8]. The normal alveolar CO_2 tension is highest at the base of the lungs, because it varies inversely as a function of the ventilation–perfusion ratio and the alveolar ventilation. The ventilation–perfusion ratio is low at the base of the lungs because blood flow exceeds ventilation. Therefore, CO_2 delivery to the lungs via the pulmonary blood flow is high and the alveolar ventilation is insufficient to lower it to normal. The O_2 tension varies reciprocally with the CO_2 tension. In a subject inspiring air at sea level, the sum of the two gas tensions cannot exceed 150 mm Hg, which is the inspired O_2 tension, because the remaining gas fraction is occupied by N_2, which is not utilized by the body. Normally, the average alveolar CO_2 tension is maintained at 40 mm Hg to maintain the pH of body fluids in the normal range.

23. The answer is C [Chapter 19 I D]. Approximately 85% of the effect of CO_2 on the respiratory drive is mediated through the central (medullary) chemoreceptors; only 15% of the effect comes from the carotid and aortic bodies (i.e., the peripheral chemoreceptors). CO_2 readily crosses the blood–brain barrier, but charged ions (e.g., H^+) do not. The hydration and subsequent dissociation of CO_2 into H^+ and HCO_3^- after it crosses the blood-brain barrier increases the H^+ concentration in the cerebrospinal fluid (CSF) and the brain tissues, stimulating respiraton. J receptors, arterial

baroreceptors, and the hypoglossal nerve are not affected by changes in CO_2.

24. The answer is D [Chapter 16 VII C; Figure 16-18]. The vital capacity (VC) consists of the expiratory reserve volume (ERV), the inspiratory reserve volume (IRV), and the tidal volume. It is the maximal amount of gas that can be expired after a maximal inspiration. Residual volume (RV), the fourth primary subdivision of lung volume, is not included in the VC; it is the small volume of gas that remains in the lungs after a maximal expiration. The RV prevents complete collapse of the alveoli.

25. The answer is A [Chapter 20 II C 4]. Cyanide poisoning causes the venous O_2 tension to be higher than normal, because the tissue oxidative enzymes are inactivated by cyanide and the cells utilize less O_2. Therefore, less O_2 diffuses out of the blood in the systemic capillaries, leaving a greater amount in the venous blood. During exercise, the tissues remove more O_2 from the capillary blood, which decreases the venous O_2 tension. A decreased cardiac output, anemia, and CO poisoning all reduce the O_2 delivery to tissues so more O_2 than normal is extracted from the capillary blood, and the venous O_2 tension declines.

26. The answer is D [Chapter 16 III C 1]. At residual volume (RV), the chest wall has a strong tendency to expand because it is far from its equilibrium position, which is about 80% of total lung capacity (TLC). At the same time, the recoil force of the lungs is reduced, because the RV is close to the equilibrium position of the lung. Due to these unequal forces, either the expiratory muscles must be contracting in order to hold the respiratory system at that level, or the glottis must be closed to prevent gas from entering the airways. If the glottis is closed and the respiratory muscles are relaxed, then the strong expansion force of the chest wall will cause the gas in the airways to expand, the alveolar gas pressure will become less than atmospheric, and the transrespiratory pressure will be negative.

27. The answer is E [Chapter 16 IV B 3 a (2) (a); VII B 4, C 1; IX A 1–2]. A decreased alveolar surface tension increases lung compliance, resulting in an increased total lung capacity (TLC) and an increased vital capacity (VC). The VC equals the TLC minus the residual volume (RV), so either a decrease in TLC or an increase in RV can reduce the VC. Expanding the lungs to the normal TLC requires a strong inspiratory muscle force. Thus, weakness of the inspiratory muscles will decrease the TLC. Similarly, expiratory muscle force is required to decrease the lung volume to the normal level of the RV. A decrease in expiratory muscle force can result in an increase in the RV.

28–32. The answers are: 28-D [Chapter 16 VII C 2; Figure 16-18], **29-C** [Chapter 16 IX A 2 a; Figure 16-21], **30-B** [Chapter 16 VII C 4; Figure 16-18, **31-A** [Chapter 16 VII C 1; Figure 16-18], **32-E** [Chapter 16 VII B 4; Figure 16-18]. The vital capacity (VC, indicated by the arrow labeled *D*) is the maximal volume of gas that can be expired after a maximal inspiration. The VC equals the sum of the tidal volume, the inspiratory reserve volume (IRV), and the expiratory reserve volume (ERV).

The forced expiratory volume in 1 second (FEV_1) is the volume of gas that can be expired in 1 second during a maximal forced expiration. The forced expiration begins at the third second on the graph; the volume expired 1 second later is indicated by the arrow labeled *C*.

The functional residual capacity (FRC, indicated by the arrow labeled *B*) is the reservoir of gas that remains in the lungs after a normal expiration. This gas buffers the changes in the O_2 and CO_2 tensions in the blood that traverses the capillaries between inspirations.

The total lung capacity (TLC, indicated by the arrow labeled *A*) represents the amount of gas in the lungs after a maximal inspiration.

The residual volume (RV, indicated by the arrow labeled *E*) is the amount of gas in the lungs after a maximal expiration. The RV cannot be removed from the lungs, because the chest wall becomes rigid, preventing further reduction of lung volume. In older individuals, the RV is set by closure of the airways, which prevents further emptying of the lungs.

PART V

RENAL PHYSIOLOGY
John Bullock

Chapter 21

Overview of Renal Function and Structure

I. FUNCTIONS

A. Maintenance of homeostasis

1. **Regulation of extracellular fluid (ECF) volume and composition.** The kidneys precisely balance the intake, production, excretion, and consumption of many organic and inorganic compounds via the conservation and excretion of water and solutes.
   a. **Intake of water and electrolytes**
      (1) **Ingested.** The gastrointestinal (GI) system is the primary source for the normal intake of water and electrolytes.
      (2) **Metabolically produced**
         (a) The **oxidation of food** provides a secondary but important source of **water**. The ordinary mixed diet will lead to the production of approximately 300 ml of metabolic water per day, mostly from the oxidation of fat.
         (b) Metabolism also produces **urea**, a nontoxic product of protein metabolism; **uric acid**, the end product of purine metabolism; and **creatinine**, an endogenous anhydride of muscle creatine.
   b. **Excretion of water and solutes**
      (1) **Excreted**
         (a) Normally, the kidneys excrete 1000–1500 ml of hypertonic urine in 24 hours.
         (b) Less than 100 ml of fluid together with 1 mEq each of sodium and chloride are excreted in the feces per day.
      (2) **Metabolically lost.** A nonperspiring young man in a basal metabolic state loses approximately 30 g of water/hr by insensible perspiration. This, together with water loss from the lungs, constitutes the **insensible water loss,** which can range between 800 and 1400 ml/day, depending on body surface area and metabolic rate. **Sensible sweat** is a hypotonic solution with a salt concentration ranging from 10 to 70 mEq/L of water.

2. **Acid-base homeostasis** (see Chapter 36)

B. Hormonogenesis

1. **Renin.** The formation and release of renin, which is a major component of the renin-angiotensin-aldosterone mechanism, allows the kidneys to regulate blood pressure by exerting control over fluid volume (see Chapter 14 II B 2).

2. **Renal erythropoietic factor (REF, erythropoietin).** The formation and release of REF increases the number of circulating erythrocytes.
   a. **Function.** REF is the primary regulator of red blood cell formation in the bone marrow and is often produced in response to arterial hypoxia, hypokinetic hypoxia, and anemic hypoxia (see Chapter 20 II D 1 b).

    b. Source. REF is a glycoprotein hormone synthesized in the epithelial cells of the renal cortical glomerular tuft and the juxtaglomerular cells. The liver is the primary source of REF in fetuses and neonates.

C. **Vitamin D₃ activation.** Dietary vitamin D_3 must undergo two hydroxylations in order to be useful to the body. The first step is performed by the liver. The final hydroxylation, performed by the kidneys, converts vitamin D_3 to its most biologically active form (1,25-dihydroxycholecalciferol) by the action of 1α-hydroxylase, a mitochondrial enzyme found in the cells of the proximal tubule.

D. **Gluconeogenesis.** The kidney acquires the important ability to synthesize and secrete glucose produced from noncarbohydrate sources (e.g., glutamine) only in unusual circumstances such as prolonged starvation and chronic respiratory acidosis.

II. **STRUCTURE** (Figure 21-1). The kidneys are paired organs that are located retroperitoneally in the upper dorsal region of the abdominal cavity. Each human kidney is composed of approximately 1 million nephrons, is about the size of a fist, and weighs approximately 150 g.

A. **Nephron** (Figure 21-2)

    1. Components. This basic functional unit of the kidney is composed of a glomerulus, with its associated afferent and efferent arterioles, and a renal tubule.
        a. Glomerulus (Figure 21-3). The glomerulus consists of an expanded, invaginated bulb (**Bowman's capsule**), which houses a **tuft of 20–40 capillary loops**.
        b. Renal tubule (see Figure 21-2)
            (1) Bowman's capsule forms the beginning of the renal tubule. Its epithelium is an attenuated layer that is about 4 μm thick.

FIGURE 21-1. Structure of the human kidney, cut away to show the various zones. (Redrawn with permission from Marsh DJ: *Renal Physiology*. New York, Raven, 1983, p 37.)

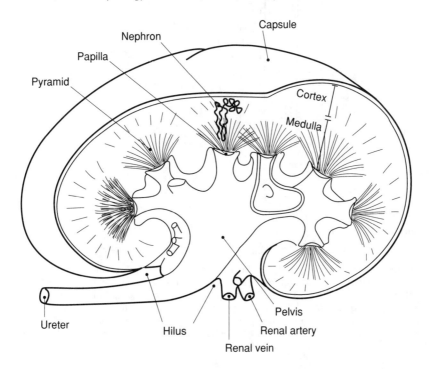

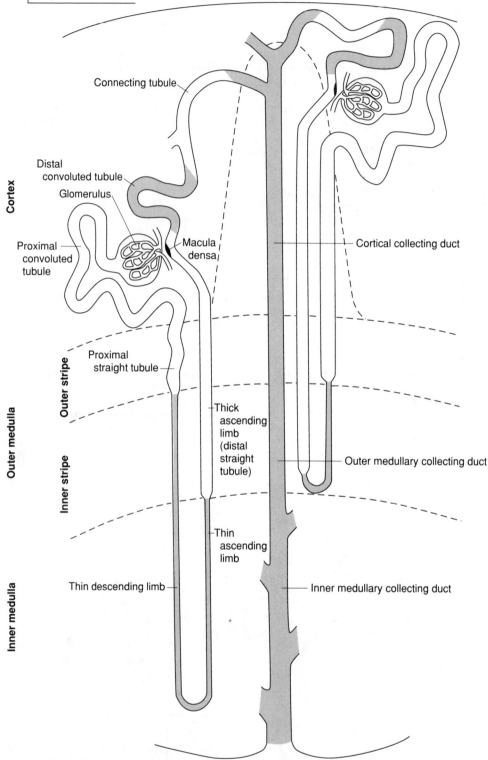

FIGURE 21-2. Detail of the functional unit of the kidney showing a cortical (superficial) nephron on the *right,* and a juxtamedullary nephron on the *left. Shaded areas* delineate tubule and collecting duct segments. The macula densa can be seen at the junction of the thick ascending limb and the distal convoluted tubule. Collectively, the distal convoluted tubule, connecting tubule, cortical collecting duct, and medullary collecting duct are known as the distal nephron. (Redrawn with permission from Windhager EE (ed): *Renal physiology VII.* In *Handbook of Physiology,* sect 8. Published for the American Physiological Society, New York, Oxford University Press, 1992, p 2443.)

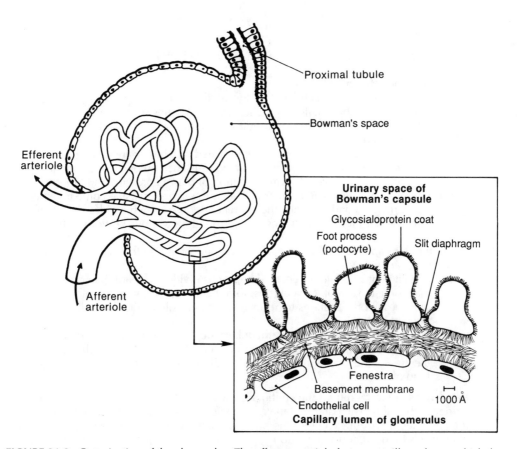

FIGURE 21-3. Organization of the glomerulus. The afferent arteriole forms a capillary plexus, which then fuses to form the efferent arteriole. The outer lining of the glomerulus, called Bowman's capsule, is continuous with the proximal tubule. (Reprinted from Marsh DJ: *Renal Physiology.* New York, Raven, 1983, p 41.) **Inset:** Glomerular capillary wall in cross-section. The luminal surface is covered by fenestrated endothelial cells. The basement membrane has a middle lamina densa surrounded by lamina rara interna and externa. Overlying this are the foot processes of epithelial cells, separated by small slit diaphragms. (Illustration by Nancy Lou Gahan Markris. Reprinted from Brenner BM, Beeuwkes R III: The renal circulations. *Hosp Pract* 13:35–46, 1978.)

    (2) Proximal tubule. The proximal tubule consists of a convoluted segment and a straight segment (pars recta).

    (3) Loop of Henle. The loop of Henle consists of the **thin descending limb** and the thin and thick segments of the ascending limb.

    (4) Cortical segments. The **distal convoluted tubule,** the **connecting tubule,** and the **cortical collecting duct** comprise the three cortical segments.

    (5) Medullary collecting duct. The medullary collecting duct carries the final urine to the renal pelvis and ureter.

2. Types of nephrons

    a. Cortical nephrons comprise approximately 85% of the nephrons in the kidney and have glomeruli located in the renal cortex. These nephrons have **short loops of Henle,** which descend only as far as the outer layer of the renal medulla.

    b. Juxtamedullary nephrons begin at the junction of the cortex and the medulla of the kidney. Juxtamedullary nephrons have **long loops of Henle,** which penetrate deep into the medulla and sometimes reach the tip of the renal papilla. These nephrons are important in the **countercurrent system,** by which the kidneys concentrate urine.

B. **Renal blood vessels**

1. Renal arteries. Each kidney receives a renal artery, which is a major branch from the aorta.
  a. Afferent and efferent arterioles
    (1) **Afferent arterioles.** Each renal artery subdivides into progressively smaller branches, and the smallest branches give off a series of afferent arterioles.
      (a) Each afferent arteriole contributes to the tuft of capillaries that protrudes into Bowman's capsule.
      (b) The capillary endothelium is **fenestrated** and has an **incomplete basement membrane**. These features minimize resistance (allowing plasma filtration) and act as a sieve (allowing retention of plasma proteins and blood cells).
    (2) **Efferent arterioles.** The capillaries within Bowman's capsule come together and form a second arteriole, the efferent arteriole, which divides shortly after to form the peritubular capillaries that surround the various portions of the renal tubule.
  b. Peritubular capillaries differ in organization depending on their association with different nephrons.
    (1) The efferent arterioles of **cortical nephrons** divide into peritubular capillaries that connect with other nephrons, forming a rich meshwork of microvessels. This meshwork functions to remove water and solutes that have diffused from the renal tubules.
    (2) The efferent arterioles of **juxtamedullary nephrons** form the **vasa recta**. The vasa recta descend with the long loops of Henle into the renal medulla and return to the area of the glomerulus, forming capillary beds at different levels along the loop of Henle.

2. Renal veins are formed from the confluence of the peritubular capillaries and exit the kidney at the **hilus**. The pattern of the renal venous system is similar to that found in the end arterial system except for the presence of multiple anastomoses between veins at all levels of the venous circulation.

III. RENAL BLOOD FLOW

A. **Rate.** Renal blood flow is approximately **1200 ml/min (420 ml/100 g tissue/min)**. Under basal conditions, the total renal blood flow is approximately **20% of the resting cardiac output**. During exercise, sympathetic tone to renal vessels increases and shunts renal blood flow to the skeletal muscles.

B. **Significance.** In addition to supplying O_2 and metabolic substrates to the kidneys, a large renal blood flow is required to produce a high glomerular filtration rate (GFR) for the excretion of metabolic by-products (e.g., urea, uric acid, creatinine).

1. O_2 requirements. In terms of O_2 consumption, the kidney is ranked second to the heart. Renal O_2 consumption is approximately **6 ml/100 g tissue/min**.
  a. Arteriovenous O_2 difference. The difference between the arterial and venous O_2 content in the human kidney is approximately 1.5 ml/dl of blood, which is the smallest arteriovenous O_2 difference of the major organ systems.
  b. Relationship to renal blood flow. The kidneys are unique in that changes in blood flow are accompanied by parallel changes in O_2 consumption. Renal O_2 consumption correlates best with the active reabsorption of Na^+, but a significant fraction is also required for H^+ secretion via the H^+-ATPase pumps.
    (1) A decline in blood flow is usually associated with a decrease in the GFR, which leads to a decrease in the filtered load of NaCl to be reabsorbed.

(2) Because tubular reabsorption of Na^+ is the major determinant of renal O_2 consumption, the metabolic demand for O_2 consumption is reduced when renal blood flow is lowered. Unlike other organs, where the blood flow is related to the O_2 requirements of the organ, in the kidney, the O_2 consumption is a function of blood flow.

2. GFR. The high blood flow to the kidney reflects the need to supply fluid (i.e., plasma) for filtration. The excretion of urea, the primary end-product of protein catabolism, is dependent on the GFR and renal blood flow.

Chapter 22

Body Fluids

I. **WATER CONTENT AND DISTRIBUTION.** The two major fluid compartments are the intracellular fluid (ICF) and extracellular fluid (ECF) volumes.

A. **Total body water (TBW)** constitutes 55%–60% of the body weight in young men and 45%–50% of the body weight in young women. (The lower percentage in women largely is due to the relatively greater amount of adipose tissue in women than in men.)

1. **Distribution.** Approximately one-third of the total body water is in the ECF compartment, and the remaining two-thirds is in the ICF compartment (Table 22-1). Distribution of TBW is as follows:
   a. Muscle (50%)
   b. Skin (20%)
   c. Other organs (20%)
   d. Blood (10%)

2. **Lean body mass (LBM).** Although the percentage of TBW declines with advancing age and with obesity, this percentage for any individual (regardless of gender) represents a constant 70% of that individual's LBM (fat-free mass).* Based on this constant relationship, the amount of body fat can be determined as:

$$\text{Body fat (\%)} = 100 - \frac{\text{percentage of TBW}}{0.7}$$

   and the LBM can be estimated as:

$$\text{LBM (kg)} = \frac{\text{TBW (L)}}{0.7}$$

B. **Extracellular fluid (ECF).** The ECF compartment includes several subcompartments.

1. **Plasma** volume, the fluid portion of the blood, represents approximately 25% of the ECF.
   a. **Blood volume,** which occupies approximately 80 ml/kg of body weight (8% of the total body weight) can be obtained from the plasma volume and the hematocrit:

$$\text{Blood volume (L)} = \text{plasma volume (L)} \cdot \frac{100}{(100 - \text{hematocrit})}$$

   b. **Plasma volume,** then, can be calculated from the blood volume and the hematocrit:

$$\text{Plasma volume (L)} = \text{blood volume (L)} \cdot \frac{(100 - \text{hematocrit})}{100}$$

2. **Interstitial fluid (milieu interieur)** surrounds all cells except blood cells and includes lymph, which constitutes 2%–3% of the total body weight. On average, the interstitial fluid represents approximately 15% of the total body weight and 75% of the ECF.

3. **Transcellular fluid** volume is about 1 L in most humans and occupies approximately 15 ml/kg of body weight (1.5% of body weight).

---

*LBM is defined as 15% bone, 10% fat, and 75% tissue.

TABLE 22-1. Distribution of Body Water in a Young, 70-kg Man

Compartment	Volume (L)	Percent Body Weight*	Percent Lean Body Mass	Percent Body Water
Total body water (TBW)	42[†]	60[‡]	70	100
Extracellular fluid (ECF)	14	20	24	33
Plasma	3.5	5	6	8
Interstitial fluid	10.5	15	18	25
Intracellular fluid (ICF)	28	40	46	67

*20–40–60 rule: ECF (20%) + ICF (40%) = TBW (60%).
[†]Total body water is 35 L in a young, 70-kg woman.
[‡]Body water accounts for 50% of the total body weight of a young, 70-kg woman.

    a. This ECF subcompartment represents fluid in the lumen of structures lined by epithe-
    lium and includes digestive secretions; sweat; cerebrospinal fluid (CSF); pleural,
    peritoneal, synovial, intraocular, and pericardial fluids; bile; and luminal fluids of
    the gut, thyroid, and cochlea.
      (1) Gastrointestinal luminal fluid constitutes about half of the transcellular fluid and
      occupies approximately 7.4 ml/kg of body weight.
      (2) CSF occupies approximately 2.8 ml/kg of body weight.
      (3) Biliary fluid volume is approximately 2.1 ml/kg of body weight.
    b. When the transcellular fluid compartment is unusually large, as in certain pathologic
    conditions (e.g., pleural effusions, ascites), it is referred to as the **third space** because
    this fluid is not readily exchangeable with the rest of the ECF.

C. **Intracellular fluid (ICF).** The volume of the ICF compartment varies but usually constitutes
30%–40% of the body weight. This is the larger of the two major fluid compartments,
containing about two-thirds of the TBW.

II. VOLUME MEASUREMENT IN THE MAJOR FLUID COMPARTMENTS

A. **Indicator dilution principle** (see also Chapter 13 II B). The volume of water in each fluid
compartment can be measured by the indicator dilution principle. This principle is based
on the relationship among the amount of a substance injected intravenously (A), the volume in which that substance is distributed (V), and the final concentration attained (c).

1. **Equation.** The equation for this relationship, based on the definition of concentration
(c), is

$$c = \frac{A}{V} \text{ or } V = \frac{A}{c}, \text{ where}$$

V is the volume (in ml or L) in which the quantity, A (in g, kg, or mEq), is distributed to
yield the concentration, c (in g/ml or L or in mEq/ml or L).

Example. If 25 mg of glucose are added to an unknown volume of distilled water and
the final concentration of glucose after mixing is 0.05 mg/ml, then the volume of solvent is

$$V = \frac{25 \text{ mg}}{0.05 \text{ mg/ml}} = 500 \text{ ml}$$

2. **Application.** Volume measurement by the dilution principle requires that the introduced substance be distributed evenly in the body fluid compartment being measured.

    a. The solute may leave the compartment through one of the following mechanisms:
        (1) Excretion in the urine or transfer to another compartment where it exists in a different concentration
        (2) Metabolism of the solute
        (3) Vaporization of the solute from the skin and respiratory tract
    b. The amount of substance lost from the fluid compartment, then, is substracted from the quantity administered:

$$V = \frac{A \text{ administered} - A \text{ removed}}{c}$$

Example. A 60-kg woman is infused with 1 millicurie (mCi) of tritium oxide (^3H_2O). After 2 hours, 0.4% of the administered dose is lost in the urine and by vaporization from the skin and respiratory tract. The radioactivity of a plasma sample is measured by liquid scintillation spectometry and indicates a concentration of 0.03 mCi/L of plasma water. Because the concentration of ^3H_2O throughout the body fluids should be the same as in plasma after the equilibration, the TBW can be calculated as

$$V = \frac{A \text{ infused} - A \text{ excreted}}{c}$$

$$= \frac{1 \text{ mCi} - (1 \text{ mCi} \cdot 0.004)}{0.03 \text{ mCi/L}}$$

$$= \frac{0.996}{0.03}$$

$$= 33.2 \text{ L}$$

  3. Markers. Regardless of compartment, desirable markers share **four qualities:**
    a. They are **measurable**.
    b. They **remain in the compartment being measured**.
    c. The **do not alter water distribution** in the compartment being measured.
    d. They are **nontoxic**.

B. **Extracellular fluid (ECF) volume**

  1. Plasma volume is measured using either of **two dilution methods**.
    a. The first method employs **substances that neither leave the vascular system nor penetrate the erythrocytes**. Such substances include:
        (1) Evans blue dye (T-1284)
        (2) Radioiodinated human serum albumin (RISA), which slowly leaks out of the circulation into the interstitial fluid
        (3) Radioiodinated gamma globulin and fibrinogen, which generally do not leak out of the bloodstream
    b. The second method is based on the fact that the radioisotopes of phosphorus (^{32}P), iron ($^{55,59}Fe$), and chromium (^{51}Cr) penetrate and bind to erythrocytes. The **tagged cells** are injected intravenously, and their volume of distribution is measured. Plasma volume is then calculated from the measured erythrocyte volume and hematocrit (see I B 1 b).

  2. Interstitial fluid volume cannot be measured directly, because no substance is distributed exclusively within this compartment. To measure the interstitial fluid volume, the capillary membranes must be permeable to a substance injected intravenously, which becomes distributed throughout the ECF and not exclusively in the interstitium. Therefore, the interstitial fluid volume is determined as the **difference between ECF volume and plasma volume**.

C. **Intracellular fluid (ICF) volume** cannot be measured directly by dilution, because no substance is confined exclusively to this compartment after intravenous injection of a marker substance. The ICF volume is obtained by subtracting the ECF volume from the TBW.

III. **DISTURBANCES OF VOLUME AND CONCENTRATION OF BODY FLUIDS**
(Figure 22-1, Table 22-2). Because aldosterone regulates the volume of body fluid compartments and antidiuretic hormone (ADH) regulates the concentration of the body fluids, these two hormones attempt to reestablish normal volumes and concentrations by increasing or decreasing their secretion.

A. **Terms and general concepts**

1. **Volume.** The general clinical terms for volume abnormalities are **dehydration** and **overhydration**. Both conditions are associated with a change in ECF volume.

2. **Concentration. Osmolarity** refers to the number of solute particles per liter of solution, and **osmolality** refers to the number of solute particles per kg of water.
   a. **Tonicity.** The tonicity of a solution is related to the effect of the concentration of the solution on the volume of a cell (e.g., erythrocytes).
      (1) **Isotonic** solutions do not change the volume of the cell.
      (2) **Hypotonic** solutions cause a cell to swell, and if sufficiently dilute, to burst (lyse).
      (3) **Hypertonic** solutions cause a cell to shrink (undergo crenation).
   b. The adjectives **isosmotic, hyperosmotic,** and **hyposmotic** refer to the osmolar concentration of the ECF in its new steady-state and are used to describe changes in volume (i.e., dehydration, overhydration).

3. **Osmosis** (see Chapter 1 III A 1) determines the distribution of body water in the ECF and ICF compartments.

B. **Dehydration**

1. **Definition.** Dehydration is an ambiguous term that does not distinguish between simple water loss and loss of Na^+. Both water loss and Na^+ loss are associated with a decrease in the **ECF volume,** which is **determined by** the amount of Na^+ in the body (i.e., **the Na^+ content**), not by the Na^+ concentration in the plasma.
   a. A simple water deficit reduces the ECF and ICF proportionately.
   b. A sodium chloride (NaCl) deficit always decreases the ECF volume.

2. **Detection.** Only a physical examination can diagnose dehydration (Table 22-3).

3. **Dehydration (volume contraction) states** (see Figure 22-1A)
   a. **Isosmotic dehydration**
      (1) **Causes** of isosmotic dehydration include hemorrhage, plasma exudation through burned skin, and gastrointestinal (GI) fluid loss (e.g., vomiting, diarrhea).
      (2) **Description**
         (a) Initially, fluid is lost from the plasma and then is repleted from the interstitial space. No major change occurs in the osmolality of the ECF; therefore, no fluid shifts into or out of the ICF compartment.
         (b) Finally, the volume of the ECF is reduced with no change in osmolality.
   b. **Hyperosmotic dehydration**
      (1) **Causes** of hyperosmotic dehydration include water deficits caused by decreased intake, diabetes insipidus (neurogenic or nephrogenic), diabetes mellitus, alcoholism, administration of lithium salts, fever, and excessive evaporation from the skin through heavy loss of sweat, which is hypotonic.
      (2) **Description**
         (a) Initially, fluid is lost from the plasma, which becomes hyperosmotic, causing a fluid shift from the interstitial fluid to the plasma.
         (b) The rise in interstitial fluid osmolality causes fluid to shift from the ICF back to the ECF compartment.
         (c) Finally, the ECF and ICF volumes both are decreased, and the osmolality of both major fluid compartments is increased.
   c. **Hyposmotic dehydration**
      (1) **Causes** of hyposmotic dehydration include renal loss of NaCl because of adrenal insufficiency [e.g., primary hypoadrenocorticalism (Addison's disease)].

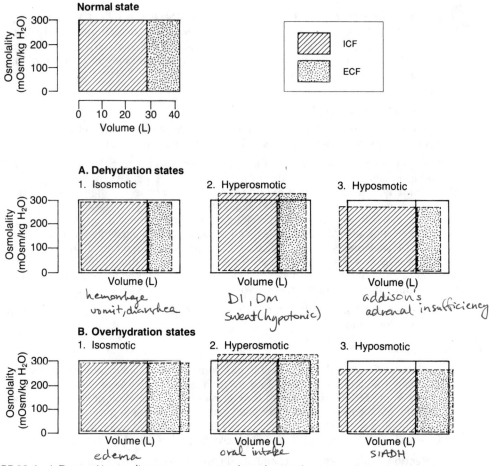

FIGURE 22-1. A Darrow-Yannet diagram representing the volume (*abscissa*) and osmolality (*ordinate*) of the intracellular and extracellular fluid compartments (*ICF* and *ECF*) in a 70-kg man. The combined plasma and interstitial fluid spaces are represented by the ECF compartment. This diagram is useful to simplify the clinical analysis of fluid balance. The area of each fluid compartment rectangle represents the total milliosmoles of solutes in that compartment. In all diagrams, the normal state is indicated by *solid lines* and the shifts from normality are indicated by *dashed lines*. (Reprinted from Valtin H: *Renal Function: Mechanism Preserving Fluid and Solute Balance in Health,* 2nd edition. Boston, Little, Brown, 1983, p 272.)

TABLE 22-2. Steady-State Changes in Volume and Osmolal Concentration of Body Fluids

	Volume (L)		Osmolality (mOsm/kg H_2O)	
Type of Change	ICF	ECF	ICF	ECF
Contraction (dehydration)				
Isosmotic	0	↓	0	0
Hyperosmotic	↓	↓	↑	↑
Hyposmotic	↑	↓	↓	↓
Expansion (overhydration)				
Isosmotic	0	↑	0	0
Hyperosmotic	↓	↑	↑	↑
Hyposmotic	↑	↑	↓	↓

The changes in volume and osmolality refer to the **ECF compartment** in the new steady-state.
ECF = extracellular fluid; ICF = intracellular fluid.

TABLE 22-3. Physical Signs of Dehydration

Decreases in Interstitial Volume	Decreases in Plasma Volume
Decreased skin turgor ("tenting")	Increased heart rate
Soft and sunken eyeballs	Flat neck veins (patient supine)
Dry mucous membranes	Increased arterial pulse
Gray, cool skin	Decreased blood pressure
Sunken fontanelles (in infants)	(severe cases)

C. **Overhydration (volume expansion) states** (see Figure 22-1B)

1. **Isosmotic overhydration**
   a. **Causes** of isosmotic overhydration are edema and oral or parenteral administration of a large volume of isotonic NaCl (150 mmol/L).
   b. **Description.** Isosmotic overhydration is characterized by an overall expansion of the ECF volume with no change in the osmolality of the ICF and ECF compartments.

2. **Hyperosmotic overhydration**
   a. **Cause.** Oral or parenteral intake of large amounts of hypertonic fluid causes hyperosmotic overhydration.
   b. **Description**
      (1) Oral intake of large amounts of salt or intravenous infusion of a hypertonic saline solution leads to an increase in the plasma osmolality.
      (2) The rise in plasma osmolality causes water to shift from the interstitium into the plasma, thereby initially increasing plasma volume.
      (3) Concomitantly, the increase in plasma salt concentration causes NaCl to diffuse into the interstitium. The net result is an increase in the osmolality of the ECF.
      (4) The increase in the osmolality of the ECF causes water to flow out of the ICF, which eventually decreases the volume of the ICF and increases the volume of the ECF. The osmolality of both major fluid compartments is increased.

3. **Hyposmotic overhydration**
   a. **Causes** of hyposmotic overhydration include ingestion of a large volume of water and renal retention of water due to the syndrome of inappropriate antidiuretic hormone secretion (SIADH).
   b. **Description**
      (1) Initially, water enters the plasma, causing a decline in the plasma osmolality, a shift of water into the interstitial space, and a decrease in the interstitial fluid osmolality.
      (2) The decrease in interstitial fluid osmolality causes water to shift from the ECF to the ICF compartment.
      (3) Finally, the ECF and ICF volumes increase and the osmolality of both major fluid compartments decreases.

IV. **IONIC COMPOSITION OF BODY FLUIDS** (Table 22-4)

A. **General considerations**

1. **Ions** constitute approximately 95% of the solutes in the body fluids.

2. The sum of the concentrations (in mEq/L) of the **cations** equals the sum of the concentrations (in mEq/L) of the **anions** in each compartment, making the fluid in each compartment **electrically neutral**.

TABLE 22-4. Electrolyte Concentration of Body Fluids

Ion	Plasma (mg/L)	Plasma (mEq/L H$_2$O*)	Plasma (mmol/L H$_2$O*)	Interstitial Fluid (mEq/L H$_2$O*)	ICF (mEq/L H$_2$O*)	Urine (mEq/L)
Cations						
Na$^+$	3266	153	153	147	10	50–130
K$^+$	156	5.4	5.4	4	145	20–170
Ca^{2+}	50	2.7	1.35	2.4	< 1	2–12
Mg^{2+}	27	1.9	0.95	1.8	27	5–18
NH$_4^+$	. . .	. . .	. . .	. . .	. . .	30–50
Total cations	3499	163	160.7	155.2	182	107–380
Anions						
Cl$^-$	3692	111	111	114	10	50–130
HCO$_3^-$	1464	26.2	26.2	30	10	. . .
Phosphate†	104	1	0.55	1	80	20–40
Sulfate$^{-‡}$	16	1.1	0.55	1	20	. . .
Proteinate$^-$	65,000	17.2	1.23	1	62	30–45
Organic acids	175	6.5	3.2	8.2	. . .	20–50§
Total anions	70,451	163	142.7	155.2	182	120–265
Total electrolytes	73,950	326	303.4	310.4		

At the pH of body fluids, the proteins have multiple charges per molecule (average valence of −15). Hence, the ICF has more total charges than does the ECF. The total concentration of cations in each compartment must equal the total concentration of anions in each compartment when expressed in mEq/L. Determining the solute concentration in urine is of limited value because of the high variability of urine volume.
ECF = extracellular fluid; ICF = intracellular fluid.
*Concentration per liter of water; to convert to concentration per liter of plasma, multiply by 0.93.
†HPO$_4^{2-}$ (ECF, ICF); H$_2$PO$_4^-$ (urine).
‡As free sulfur.
§Assuming average valence of −2.

> **3.** Physicians rely on the changes in electrolyte concentrations in the ECF compartment, particularly in the plasma, in the diagnosis and treatment of patients with fluid or electrolyte imbalances, or both.

B. | **Cations**

> **1.** The **monovalent cations** Na$^+$ and K$^+$ are the predominant cations of the ECF and ICF compartments, respectively. Essentially all of the body K$^+$ is in the exchangeable pool, whereas only 65%–70% of the body Na$^+$ is exchangeable. Only the exchangeable solutes are osmotically active.

> **2.** The **divalent cations** Mg^{2+} and Ca^{2+} exist in body fluids in relatively low concentrations. Almost all of the body Ca^{2+} (in bone) and most of the body Mg^{2+} (in bone and cells) is nonexchangeable. After K$^+$, Mg^{2+} is the main cation of the ICF. After Na$^+$, Ca^{2+} is the main cation of ECF.

C. | **Anions.** The chief anions of the body fluids are Cl$^-$, bicarbonate (HCO$_3^-$), phosphates, organic ions, and polyvalent proteins.

> **1.** Organic phosphates, proteins, and organic ions are the predominant anions in the ICF.

> **2.** Cl$^-$ and HCO$_3^-$ are the predominant anions in the ECF.

Chapter 23

Overview of Renal Tubular Function

 I. INTRODUCTION

A. The constancy of the body's internal environment is maintained, in large part, by the continuous functioning of its roughly 2 million nephrons. As blood passes through the kidneys, the nephrons clear the plasma of some substances (e.g., urea) while simultaneously retaining other, essential substances (i.e., water).

1. **Substances to be excreted** are removed by **glomerular filtration** and **renal tubular secretion** and passed into the urine.

2. **Substances that the body needs** are retained by **renal tubular absorption** (e.g., Na^+, HCO_3^-) and returned to the blood by **reabsorptive processes** (Figure 23-1).

B. **Glomerular filtration** is the initial step in urine formation.

1. The plasma that traverses the glomerular capillaries is filtered by the highly permeable **glomerular membrane,** and the resultant fluid, the **glomerular filtrate,** is passed into Bowman's capsule.

2. The term **glomerular filtration rate (GFR)** refers to the volume of glomerular filtrate formed each minute by all of the nephrons in both kidneys.

C. The terms **renal tubular secretion** and **renal tubular reabsorption** refer to the **direction of transport,** not to differences in the underlying mechanisms of transport.

1. **Secretion** refers to the transport of solutes from the peritubular capillaries into the tubular lumen (i.e., it is the **addition of a substance to the filtrate**).

2. **Reabsorption** denotes the active transport of solutes and the passive movement of water from the tubular lumen into the peritubular capillaries (i.e., it is the **removal of a substance from the filtrate**).

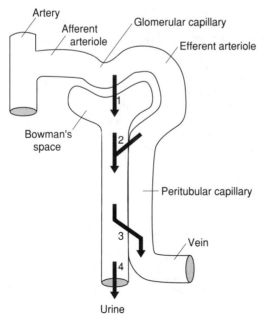

FIGURE 23-1. Filtration, secretion, and reabsorption. The amount excreted (*4*) equals the sum of the amounts filtered (*1*) and secreted (*2*) minus the amount reabsorbed (*3*).

I. RENAL CAPILLARY MEMBRANE TRANSPORT

A. **Filtration** is the bulk transport of a fluid with its dissolved small solutes across a membrane.

1. A **hydrostatic pressure difference** between the glomerular capillaries and Bowman's capsule promotes filtration (see also Chapter 12 I C 3 b).

2. Because there are **no concentration gradients for inorganic ions, glucose, amino acids,** and **inulin,** the concentrations in plasma and the ultrafiltrate in Bowman's space are the same.

3. The **filtered load** (i.e., the amount of solute transported across the glomerular membranes per unit time) is proportional only to the GFR and to the forces that affect filtration, namely hydrostatic and oncotic pressure differences. Mathematically, the filtered load is equal to the GFR times the plasma concentration of the solute.

B. **Simple diffusion** is the principal mode of transport in the interstitial space and within cells. For an ion, the driving force for diffusion consists of two gradients: a concentration gradient and an electric (voltage) gradient.

II. RENAL EPITHELIAL TRANSPORT.
The transport mechanisms of the proximal tubule are used as the basis for this discussion. Table 23-1 provides an overview of the transport mechanisms throughout the tubular segments of the nephron.

A. **Basolateral (abluminal) membrane transport systems** (Figure 23-2)

1. **Primary active transport.** Na^+ transport through the basolateral membrane is mediated principally by the Na^+-K^+-ATPase pump, which transports Na^+ against an electrochemical potential from the cell interior and **maintains a low intracellular Na^+ concentration** (see Figure 23-2B). Active transport of Na^+ (efflux) across all cells of the body consumes 30%–50% of the energy derived from metabolism in most cells.

2. **Facilitated diffusion,** which involves the passive transport of a substance by a protein carrier from a region of higher concentration to a region of lower concentration, translocates glucose from the intracellular fluid across the membrane to the interstitial fluid (see Figure 23-2B).

B. **Apical (adluminal) membrane transport systems.** The apical surface of the renal epithelium of the proximal tubule differs from the other cell membranes in that simple diffusion plays almost no role in the transport of Na^+, K^+, Cl^-, H^+, glucose, or amino acids. The movement of Na^+ across the luminal membrane (influx) is favored by the electrochemical gradient, but the bulk of Na^+ transport is mediated by specific membrane transport proteins and not by simple diffusion. The apical surface possesses two transport mechanisms for Na^+.

TABLE 23-1. Overview of Renal Transepithelial Transport

| | Carriers | | |
Pumps	Symporters	Antiporters	Channels
$3Na^+$-$2K^+$-ATPase	Na^+-Glucose	Na^+-H^+	Na^+
$3H^+$-ATPase	Na^+-Amino acid	Na^+-NH_4^+	K^+
H^+-K^+-ATPase	$2Na^+$-HPO_4^{2-}	Na^+-Ca^{2+}	Cl^-
Ca^{2+}-ATPase	Na^+-$3\ HCO_3^-$	Cl^--HCO_3^-	Ca^{2+}
	Na^+-$2Cl^-$-K^+		
	K^+-Cl^-		

A

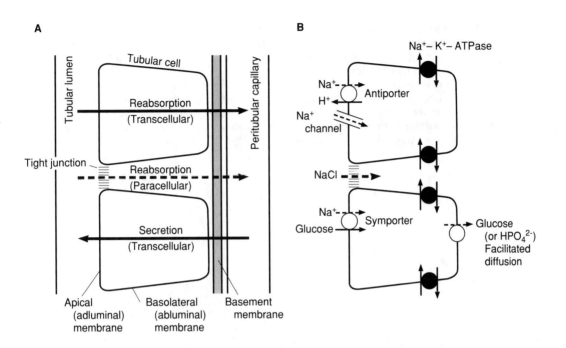

B

FIGURE 23-2. (*A*) Major modalities of renal transport. (*B*) Transport systems of the proximal tubule. Pumps (*solid circles*), carriers (*open circles*) and channels (*slanting parallel lines*) are important renal transport mechanisms. Diffusion takes place through channels, and facilitated diffusion is represented by *carriers with single dashed arrows. Dashed arrows* represent downhill transport (i.e., transport along a concentration gradient) and *solid arrows* represent uphill transport (i.e., transport against a concentration gradient).

1. A **diffusion** mechanism is available for Na^+ to pass through the tight junction in association with Cl^-. Most of the Cl^- that is reabsorbed never enters the cell (see Figure 23-2B).

2. **Carrier-mediated (secondary active) transport.** The transport of Na^+ down its gradient provides energy for active (uphill) transport of other solutes (e.g., glucose, amino acids). There are two types of Na^+-dependent transport mechanisms (see Figure 23-2B).
   a. **Cotransport (symport)** denotes the transport of two substances by a protein carrier in the **same direction**.
   b. **Countertransport (antiport)** defines the transport of two substances by a protein carrier in **opposite directions**.

C. **Transepithelial transport.** The differences between apical and basolateral membrane properties account for the transepithelial transport of all solutes, which can occur via the transcellular or the paracellular pathway (see Figure 23-2A).

1. The **transcellular pathway** is used for **active transepithelial transport**. Approximately two-thirds of the Na^+ transport is active and transcellular. There are three principal mechanisms of Na^+ transport in the proximal tubule. All involve protein carriers that also combine with Na^+ and thus are **secondary active transport processes**. These are:
   a. **Electrogenic Na^+ cotransport** (symport) with solutes such as sugars and amino acids
   b. **Electroneutral Na^+ cotransport** (symport) with phosphate and organic anions such as acetate, citrate, and lactate
   c. **Electroneutral Na^+ countertransport** (antiport) with H^+

2. The **paracellular (intercellular) pathway,** which follows the lateral intercellular spaces, is used for **passive transepithelial transport**. Approximately one-third of NaCl transport is passive and paracellular.

a. Paracellular pathway for Cl⁻ reabsorption in the proximal tubule
   (1) As water moves out of the tubule secondary to Na^+ reabsorption, the increase in luminal Cl^- concentration acts as a driving force for paracellular Cl^- reabsorption by diffusion.
   (2) Active transport of Na^+ leaves the luminal membrane negatively charged with respect to the interstitial fluid. This transtubular voltage gradient constitutes a second driving force for paracellular Cl^- reabsorption by diffusion.
b. A **paracellular pathway for K⁺** exists in the interspaces of the proximal tubule and the thick segment of the ascending limb of the loop of Henle.

Chapter 24

Renal Clearance

I. **INTRODUCTION. Clearance** is a measure of the volume of plasma completely freed of a given substance per minute by the kidneys. It is the **efficiency** with which the plasma is cleared of a given substance.

A. **Definitions**

1. **Renal clearance** of a given substance is the **ratio of the renal excretion rate of the substance to its concentration in the blood plasma**. This ratio is expressed as:

$$C_x = \frac{U_x \cdot \dot{V}}{P_x} \text{, where}$$

C_x = clearance of the substance (ml/min)
U_x = concentration of the substance in urine (mg/ml)
$P_x{}^*$ = concentration of the substance in plasma (mg/ml)
$\dot{V}$ = volume of urine output per minute (ml/min)

2. **Amount filtered and excreted.** Clearance is a comparison of the amount of substance removed from plasma per unit time ($C_x \cdot P_x$) with the amount of substance excreted in the urine per unit time ($U_x \cdot \dot{V}$), which is demonstrated by a simple rearrangement of the renal clearance ratio:

$$C_x \cdot P_x = U_x \cdot \dot{V}$$

B. **Units.** The unit for clearance is **ml/min**:

$$C_x = \frac{U_x \cdot \dot{V}}{P_x} = \frac{\text{(mg/ml) (ml/min)}}{\text{(mg/ml)}} = \text{ml/min}$$

C. **Mechanisms.** Substances can be cleared by **filtration, tubular secretion, or a combination** of these processes. The **magnitude of the tubular clearance** depends on the plasma concentration and the tubular transport capacity (Tm) for reabsorption or secretion. (The phenomenon of tubular transport capacity is discussed in more detail in Chapter 26 I).

II. **APPLICATIONS.** The concept of clearance can be applied to tubular function.

A. **Inulin clearance**

1. Inulin, a fructo-polysaccharide that does not occur naturally in the body, can be used in a **test for determining renal function**.
   a. **As a measure of glomerular filtration rate (GFR)**
      (1) **Inulin clearance** (C_{in}) is a measure of GFR because the volume of plasma completely cleared of inulin per unit time equals the volume of plasma filtered per unit time (i.e., $C_{in} = GFR$). The following characteristics of inulin account for this quality.
         (a) Inulin is only **freely filtered**. Because no inulin is secreted or reabsorbed, all excreted inulin must come from the plasma.

---

*Plasma concentrations rather than whole blood concentrations are used in calculations of renal clearance, because only the plasma is cleared by filtration.

FIGURE 24-1. Clearances of several substances plotted against their plasma concentrations. There are two types of substances with reference to the clearance of inulin: substances with clearances below that for inulin are filtered and reabsorbed (e.g., glucose), and substances with clearances above that for inulin are filtered and secreted [e.g., para-aminohippuric acid (PAH)]. Units of plasma concentrations vary over a wide range so that values on the abscissa are relative only. Note that the ordinate is interrupted, and PAH clearance at low plasma concentration is greater than 600 ml/min. (Reprinted from Bauman JW Jr, Chinard FP: Measurement of renal integrity. In *Renal Function: Physiological and Medical Aspects*. St. Louis, CV Mosby, 1975, p 43.)

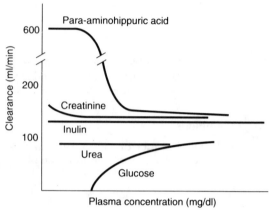

    (b) Inulin is **biologically inert** and **nontoxic**.
    (c) Inulin is **not bound to plasma proteins,** and it is **not metabolized to another substance**.
  (2) Although inulin is used most commonly, **other substances** can measure GFR. The most frequently used agents include mannitol, sorbitol, sucrose (intravenous), iothalamate, radio-active cobalt-labeled vitamin B_{12}, ^{51}Cr-labeled edetic acid (EDTA), and radioiodine-labeled Hypaque. (Endogenous creatinine clearance is used clinically as an estimate of GFR, as discussed in II B, because a small amount of creatinine is secreted in humans.)
 b. **As an indicator of plasma clearance mechanisms.** A comparison of the clearance of a given substance (C_x) with the clearance of inulin (C_{in}) provides information about the renal mechanisms used to remove the substance from plasma (Figure 24-1).
  (1) When C_x **equals C_{in},** excretion is by **filtration alone**. Therefore, the mass of the substance excreted in the urine per unit time equals the mass of the substance filtered during the same time:

$$U_x \bullet \dot{V} = C_{in} \bullet P_x *$$

  (2) When C_x **is less than C_{in},** excretion is by **filtration and reabsorption**. In this case, the mass of the substance excreted in the urine is less than the mass of the substance filtered during that time:

$$U_x \bullet \dot{V} < C_{in} \bullet P_x$$

  (3) When C_x **is greater than C_{in},** excretion is by **filtration and secretion**. In this case, the mass of the substance excreted in the urine is greater than the mass of the substance filtered during the same time.

$$U_x \bullet \dot{V} > C_{in} \bullet P_x$$

 2. **Clearance ratios.** The clearance ratio is the ratio of the clearance of any substance (C_x) to the clearance of inulin (C_{in}).
  a. **Interpretation of clearance ratios**
   (1) $C_x/C_{in} = 1$. A ratio of 1 indicates that the amount of the substance excreted per unit time equals the amount of the substance filtered in the same time:

$$U_x \bullet \dot{V} = C_{in} \bullet P_x$$

    (a) A clearance ratio of 1 indicates that, **on a net basis,** the substance is **neither reabsorbed nor secreted** and, therefore, is **only filtered**.

---

*The mass filtered per unit time (also called the **amount filtered** or **filtered load**) is equal to GFR $\bullet P_x$ (or $C_{in} \bullet P_x$) and is expressed in mg/min (see Chapter 26 I C).

    (b) Substances with clearance ratios close to 1 include mannitol, sorbitol, thiosulfate, ferricyanide, iothalamate, vitamin B_{12}, and sucrose (intravenous).

  (2) $C_x/C_{in} < 1$. A clearance ratio of less than 1 demonstrates that the amount of the substance excreted per unit time is less than the amount filtered per unit time:

$$U_x \cdot \dot{V} < C_{in} \cdot P_x$$

    (a) A clearance ratio of less than 1 indicates that, **on a net basis,** the substance undergoes **reabsorption**.

    (b) Substances with clearance ratios of less than 1 include glucose, xylose, and fructose.

  (3) $C_x/C_{in} > 1$. A clearance ratio that is greater than 1 indicates the net excretion of the substance over time is greater than the quantity filtered over time:

$$U_x \cdot \dot{V} > C_{in} \cdot P_x$$

    (a) A clearance ratio of more than 1 represents **net secretion** of the substance into the lumen; therefore, the substance is cleared by **filtration and secretion**.

    (b) Substances with clearance ratios greater than 1 include para-aminohippuric acid (PAH), phenol red, iodopyracet, certain penicillins, and creatinine (in humans).

 b. Tubular fluid-to-plasma (TF/P) concentration ratio. A comparison of the tubular fluid (TF) solute concentration to the plasma (P) solute concentration is called the TF/P concentration ratio. The TF/P concentration ratio is not compared to that of inulin and is, therefore, a single ratio.

  (1) This ratio **measures the tubular solute concentration along the nephron,** which is a function of solute transport (i.e., reabsorption or secretion) as well as water reabsorption (Figure 24-2).

  (2) TF/P solute concentration ratios also can represent **total solute concentration** rather than the concentration of a single solute. When measuring the total solute concentration, the TF/P or urine-to-plasma (U/P) ratio is expressed in terms of osmolality (osmolarity) and can be interpreted as follows.

    (a) When **$TF_{osm}/P_{osm} = 1$ or $U_{osm}/P_{osm} = 1$,** the tubular fluid or urine is **isosmotic** with respect to plasma.

    (b) When **TF_{osm}/P_{osm} or $U_{osm}/P_{osm} < 1$,** the tubular fluid or urine is **hyposmotic** with respect to plasma.

    (c) When **TF_{osm}/P_{osm} or $U_{osm}/P_{osm} > 1$,** the tubular fluid or urine is **hyperosmotic** with respect to plasma.

    (d) Example. States of hydropenia (dehydration) lead to antidiuretic hormone (ADH) secretion. This, in turn, leads to increased water reabsorption and an elevation of U_{osm}, which has an upper limit in humans of about 1400 mOsm/kg. With a P_{osm} of about 300 mOsm/kg, this is equivalent to a maximal U_{osm}/P_{osm} of approximately 4.7.

B. **Creatinine clearance.** Although inulin clearance can be used to measure GFR, in clinical practice it is more common to determine the 24-hour endogenous creatinine clearance as an estimate of GFR. Creatinine clearance determinations do not require administration of exogenous creatinine, as creatinine is a product of muscle metabolism.

1. Creatinine clearance has a normal range of 80–110 ml/min per 1.73 m^2 body surface area (estimated average body surface area of 25-year-old humans) and declines with age in healthy individuals. Because creatinine clearance is an index of GFR, it reflects the normal decline in GFR with age.

2. Plasma creatinine concentration remains remarkably constant through life, averaging 0.8–1.0 mg/dl in the absence of renal disease. The tendency to rely on serum creatinine concentration rather than creatinine clearance as an estimate of renal function is a serious error, because **plasma creatinine concentration overestimates renal function (i.e., GFR)**.

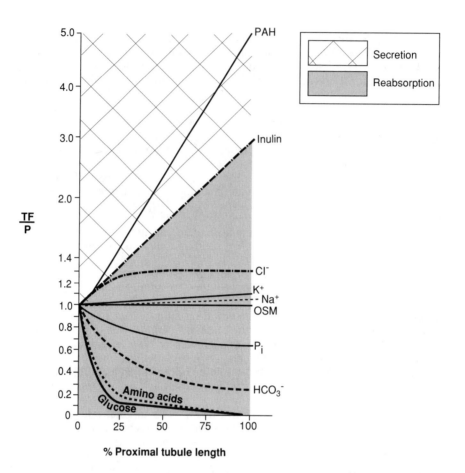

% Proximal tubule length

FIGURE 24-2. Transport of various solutes along the length of the proximal tubule. *TF/P* = tubular fluid-to-plasma concentration ratio for a particular solute. The TF/P concentration ratio for inulin rises to approximately 3 at the end of the proximal tubule, indicating water reabsorption (because inulin is not reabsorbed). Along the initial portion of the proximal tubule, there is little Cl^- reabsorption, so the TF/P concentration ratio rises as a result of the reabsorption of water. Bicarbonate (HCO_3^-) reabsorption lowers the TF/P concentration of ratio for HCO_3^- to approximately 0.2 at the end of the proximal tubule, indicating that the concentration of HCO_3^- in the tubular fluid is approximately 5 mmol/L, or 20% of its concentration in plasma. Glucose and amino acids are reabsorbed rapidly, so that at 25% of the proximal tubular length, their concentrations in the tubular fluid decline to approximately 10% of their concentrations in the glomerular filtrate (or plasma), as shown by the TF/P ratio of 0.1. *OSM* = osmolarity or osmolality (i.e., the total concentration of all solutes in the tubular fluid); *PAH* = para-aminohippuric acid; P_i = inorganic phosphate. (Modified from Brenner B, Coe FL, Rector FC: Transport functions of renal tubules. In *Renal Physiology in Health and Disease*. Philadelphia, WB Saunders, 1987, p 33.)

    a. Using serum creatinine concentration as a basis for drug dosages can lead to drug overdose, especially in elderly individuals. Overdose is a particularly serious risk with drugs that are cleared primarily by renal mechanisms (e.g., digoxin, aminoglycoside antibiotics). The dosages for these drugs frequently are based on serum creatinine concentration, although it is incorrect to do so.

    b. To avoid this error, patient age should be considered together with measurements of creatinine clearance and blood levels of the drug. Drug dosages should be adjusted to the 30%–40% decrease in GFR that normally occurs in individuals 30–80 years old.

  3. Urinary creatinine excretion. Because creatinine clearance normally declines with age and plasma creatinine concentration does not, the decline in clearance is attributed to a parallel decrease in the excretion rate of creatinine.

a. The decline in creatinine clearance is not due to a decreased tubular secretion of creatinine, because the clearance ratio of creatinine to inulin is a quite constant 1.2, about 120% of inulin clearance. Thus, urinary creatinine excretion exceeds the amount filtered, since approximately 20% of the urinary creatinine is derived from tubular secretion by the secretory pathway in the proximal tubule.

b. The decline in creatinine excretion with age is due to a primary decline in renal function and a secondary decline in muscle mass.

    (1) The decline in GFR is due to declines in renal plasma flow, cardiac output, and renal tissue mass. (Decreased GFR is the most clinically significant renal functional deficit occurring with age.)

    (2) The decline in creatinine excretion with a decline in muscle mass over time accounts for the relatively constant plasma creatinine concentration among healthy individuals.

Chapter 25

Glomerular Filtration and Renal Blood Flow

GLOMERULAR FILTRATION

A. **Factors affecting glomerular filtration**

1. **Role of hydrostatic and oncotic pressures.** Fluid movement is proportional to the permeability and to the balance between hydrostatic and oncotic (osmotic) forces across the glomerular membrane. When hydrostatic pressure exceeds oncotic pressure, filtration occurs. Conversely, when oncotic pressure exceeds hydrostatic pressure, reabsorption occurs.

   a. The term **effective filtration pressure (EFP)** refers to the **net driving forces** for water and solute transport across the glomerular membrane. EFP is a function of two variables: the **hydrostatic pressure gradient** driving fluid **out** of the glomerular capillary and into Bowman's capsule, and the **colloid osmotic pressure gradient** bringing fluid **into** the glomerular capillary (Figure 25-1). Using these variables, EFP is expressed as:

$$EFP = (P_{cap} - P_{Bow}) - COP_{cap}$$

FIGURE 25-1. Model of a glomerular capillary and the Starling forces across the filtration barrier. The effective filtration pressure (*EFP*) is the net outward force and is calculated as the difference between the outwardly and inwardly directed forces (*dark arrows*). When the hydrostatic pressure forces exceed the colloid oncotic (osmotic) pressure forces, outwardly directed filtration results. COP_{Bow} = Bowman's capsule oncotic pressure; COP_{cap} = capillary oncotic pressure; P_{Bow} = Bowman's capsule hydrostatic pressure; P_{cap} = capillary hydrostatic pressure; *light arrows* = direction of blood flow. (Modified and redrawn with permission from Berne RM and Levy MN: *Physiology*, 3rd ed. St. Louis, CV Mosby, 1993, p 735.)

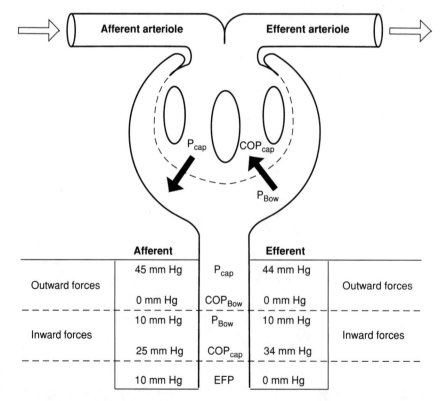

	Afferent		Efferent	
Outward forces	45 mm Hg	P_{cap}	44 mm Hg	Outward forces
	0 mm Hg	COP_{Bow}	0 mm Hg	
Inward forces	10 mm Hg	P_{Bow}	10 mm Hg	Inward forces
	25 mm Hg	COP_{cap}	34 mm Hg	
	10 mm Hg	EFP	0 mm Hg	

where $(P_{cap} - P_{Bow})$ = the outward forces promoting filtration, and COP_{cap} = the inward forces opposing filtration. The colloid osmotic pressure in Bowman's capsule (COP_{Bow}) is not expressed in this equation, although it is understood to be zero, because the glomerular filtrate normally contains little protein.

   b. The **filtration coefficient (K_f)** of the glomerular membrane is a function of the capillary surface area and the membrane permeability. The filtration coefficient is expressed in ml/min/mm Hg.

   c. The **glomerular filtration rate (GFR)** is equal to the EFP multiplied by the filtration coefficient: GFR = K_f • EFP.

      (1) The filtration coefficient normally equals approximately 12.5 ml/min/mm Hg.

      (2) The EFP normally equals about 10 mm Hg, which can be determined from the normal values for P_{cap} (45 mm Hg), P_{Bow} (10 mm Hg), and COP_{cap} (25 mm Hg).

      (3) The **normal GFR** for both kidneys is approximately 125 ml/min per 1.73 m^2 of body surface area.

2. Role of capillaries. Although the principle of opposing forces accounts for the magnitude and direction of flow, the main mechanism that alters intracapillary hydrostatic pressure probably is not the resistance along the length of the capillary but the contractile state of the **precapillary sphincters**.

3. Role of mesangial cells. Mesangial cells are intercapillary mononucleated, stellate cells embedded in an extracellular amorphous matrix and surrounded by glomerular capillaries.

   a. Certain mesangial cells **exhibit contractile responses** to angiotensin II, antidiuretic hormone (ADH, vasopressin), and norepinephrine. This type of mesangial cell possesses angiotensin II receptors and microfilaments similar to those found in smooth muscle cells.

   b. **Vasoactive substances** modulate the filtration coefficient by stimulating intrarenal angiotensin II production via the generation of cyclic adenosine 3',5'-monophosphate (cAMP). Locally, the glomerular angiotensin II causes mesangial contraction, reducing the glomerular filtration surface area and the filtration coefficient.

   c. The end result of the effect of angiotensin II on mesangial cells is a decline in glomerular capillary plasma flow.

B. **Comparison of glomerular filtration and systemic filtration.** The amount of filtration performed by the renal glomerular capillaries is better appreciated when this filtration is compared to that of the extrarenal (systemic) capillaries.

1. Capillary exchange area and filtration coefficient. The filtration coefficient of the glomerulus is 50–100 times greater than that of a muscle capillary.

   a. The **total glomerular capillary exchange area** is estimated to be 1.6 m^2, of which 2%–3% is available for filtration. Thus, the filtration surface measures between 320 and 480 cm^2.

   b. In the adult human, if the entire systemic capillary bed were patent at a given time, it would afford a total capillary exchange area of 1000 m^2. In a human at rest, approximately 25%–35% of the systemic capillaries are open at any time, and the **effective pulmonary capillary surface area** is about 60 m^2.

2. Filtration rate

   a. **Systemic filtration rate.** Approximately 20 L of fluid are filtered daily from the systemic capillaries. Of this, approximately 18 L/day are reabsorbed in the venular ends of the capillaries, and the remaining 2 L/day represent lymph flow. (**Diffusional exchange** across the entire systemic capillary bed, however, is about 80,000 L/day!)

   b. **GFR.** Approximately 180 L of fluid are filtered daily from the glomerular capillaries. Therefore, the transtubular flow of fluid out of the glomerular capillaries (GFR = 180 L/day) far exceeds the filtration from systemic capillaries (20 L/day).

II. RENAL BLOOD FLOW.

Although the major determinant of the GFR is the hydrostatic pressure within the glomerular capillaries, the renal blood flow through the glomeruli has an effect on the GFR. When the rate of renal blood flow increases, so does the GFR. The following factors determine the renal blood flow.

A. **Autoregulation of arterial pressure** [Figure 25-2; see also Chapter 12 I B 2 a (1)].

1. Over a wide range of renal arterial pressures (i.e., 90–190 mm Hg), the GFR and renal blood flow remain quite constant. This intrinsic phenomenon observed in the renal capillaries also occurs in the capillaries of muscles and is termed autoregulation. Intrarenal autoregulation is virtually absent, however, at mean arterial blood pressures below 70 mm Hg.

   a. Autoregulation has been observed to persist after renal denervation, in the isolated perfused kidney, in the transplanted kidney, after adrenal demedullation, and even in the absence of erythrocytes.

   b. Autoregulation of renal blood flow is necessary for the autoregulation of GFR.

2. Intrarenal autoregulation probably is mediated by changes in preglomerular (afferent) arteriolar resistance. The major factors that determine blood flow (Q) are a pressure gradient (ΔP) and resistance (R):

$$Q = \frac{\Delta P}{R}$$

It becomes clear that to maintain a constant blood flow with a concomitant increase in renal perfusion pressure (ΔP), there must be a commensurate increase in renal vascular resistance.

B. **Arteriolar resistance** (Figure 25-3)

1. **Effects of changes in arteriolar resistance** (Table 25-1)

   a. The **afferent arteriole** has a larger diameter than the efferent arteriole and is the major site of autoregulatory resistance. When resistance is altered in the afferent

FIGURE 25-2. Autoregulation of renal blood flow (*RBF*) and glomerular filtration rate (*GFR*) and the effect of arterial blood pressure on RBF, renal plasma flow (*RPF*), GFR, and the volume flow of urine (*V̇*). (Reprinted from Harth O: The function of the kidneys. In *Human Physiology*. Edited by Schmidt RF and Thews G. New York, Springer-Verlag, 1983, p 615.)

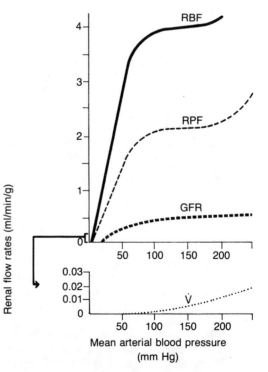

A. Constriction of a vessel

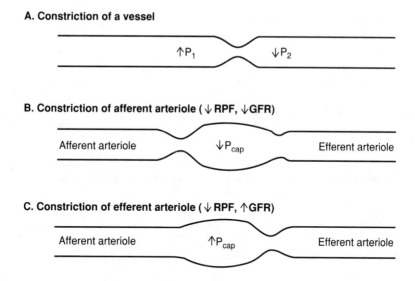

B. Constriction of afferent arteriole (↓ RPF, ↓GFR)

C. Constriction of efferent arteriole (↓ RPF, ↑GFR)

FIGURE 25-3. Relationship between arteriolar resistance, glomerular filtration rate (*GFR*), and renal plasma flow (*RPF*). (*A*) If flow is constant, vasoconstriction results in an increase in hydrostatic pressure proximally (*P₁*) and a fall in hydrostatic pressure distally (*P₂*). (*B*) Constriction of the afferent arteriole reduces both the renal plasma flow and the GFR without affecting the filtration fraction. (*C*) Constriction of the efferent arteriole also reduces the renal plasma flow, but the glomerular hydrostatic pressure (*P_{cap}*) increases, leading to an increase in filtration fraction.

arterioles, the GFR and renal blood flow change in the same direction. Therefore, **changes in afferent arteriolar resistance do not affect the filtration fraction** [i.e., the ratio of GFR to renal plasma flow].*

(1) **Constriction** of the afferent arteriole **decreases both** the renal plasma flow and the GFR.

  (a) An increase in vascular resistance proximal to the glomeruli leads to a decrease in renal plasma flow.

  (b) A decrease in hydrostatic pressure within the glomerular capillaries leads to a decrease in GFR.

(2) **Dilation** of the afferent arteriole increases both the renal plasma flow and the GFR.

  (a) A decrease in vascular resistance proximal to the glomeruli leads to an increase in renal plasma flow.

  (b) An increase in glomerular capillary perfusion pressure leads to an increase in GFR.

b. When resistance is altered in the **efferent arterioles,** the GFR and the renal plasma flow change in opposite directions. Therefore, **changes in efferent arteriolar resistance do affect the filtration fraction.**

(1) **Constriction** of the efferent arteriole **decreases the renal plasma flow** and **increases the GFR.**

  (a) An increase in vascular resistance distal to the glomeruli leads to a decrease in renal plasma flow.

  (b) An increase in the hydrostatic pressure within the glomerular capillaries leads to an increase in GFR.

  (c) Following an increase in efferent arteriolar resistance, the filtration fraction increases.

---

*The filtration fraction is a measure of the entering plasma volume (i.e., the renal plasma flow) that is removed through the glomeruli as filtrate. Filtration fraction refers to the bulk volume of fluid (not solutes) and normally measures about 20% of the renal plasma flow.

TABLE 25-1. Effects of Changes in Renal Vascular Resistance with a Constant Renal Perfusion

Renal Arteriolar Vascular Resistance		RPF (ml/min)	GFR (ml/min)	Filtration Fraction
Afferent	Efferent			
↑	. . .	↓	↓	no change
↓	. . .	↑	↑	no change
. . .	↑	↓	↑	↑
. . .	↓	↑	↓	↑↓

GRF = glomerular filtration rate, RPF = renal plasma flow.

Note that the GFR and renal plasma flow exhibit parallel shifts with changes in afferent arteriolar resistance but exhibit divergent shifts with changes in efferent arteriolar resistance. Increases in vascular resistance always lead to a decline in renal plasma flow, and decreases in arteriolar resistance always lead to an increase in renal plasma flow. In the first two columns, upward arrows denote the effect of vasoconstriction, and downward arrows denote the effect of vasodilation.

    (2) Dilation of the efferent arteriole **increases the renal plasma flow** and **decreases the GFR**.
        (a) A decrease in vascular resistance distal to the glomeruli leads to an increase in renal plasma flow.
        (b) A decrease in the hydrostatic pressure within the glomerular capillaries leads to a decrease in GFR.
        (c) Following a decrease in efferent arteriolar resistance, the filtration fraction decreases.

  2. Factors affecting arteriolar resistance
    a. Sympathetic stimulation
      (1) In addition to the autoregulatory response to hypotension (i.e., vasodilation), the reflex increase in sympathetic tone results in both afferent and efferent arteriolar vasoconstriction and an increase in renal vascular resistance together with a decrease in GFR.
      (2) Conversely, the autoregulatory response to increases in renal arterial blood pressure (i.e., vasoconstriction) causes the reflex decrease in sympathetic tone, which results in renal arteriolar vasodilation and a decrease in renal vascular resistance together with an increase in GFR.
    b. Hormonal regulation
      (1) Hormones that cause vasoconstriction, and thereby decrease renal blood flow and the GFR, include epinephrine, norepinephrine, angiotensin II, and adenosine. During hemorrhage, norepinephrine, epinephrine, and angiotensin II decrease the renal blood flow in an effort to conserve water.
        (a) Norepinephrine elicits an intense vasoconstriction of the afferent and efferent arterioles.
        (b) An increase in sympathetic activity increases the release of epinephrine and angiotensin II, enhancing vasoconstriction.
        (c) Low concentrations of angiotensin II cause a predominant constriction of the efferent arteriole. At high concentrations of this vasoactive hormone, constriction of both the afferent and efferent arterioles occurs.
      (2) Hormones that cause vasodilation, and thereby increase renal blood flow and GFR, include the prostaglandins PGE_2 and PGI_2 (prostacyclin).
        (a) Prostaglandins do not regulate renal blood flow or GFR in subjects who are in the basal state.
        (b) During hemorrhage, prostaglandins are produced locally within the kidneys in response to sympathetic nerve activity and angiotensin II. These prostaglandins, which enhance blood flow, vasodilate both the afferent and efferent arterioles.

III. INTRARENAL REGULATION.
There are three intrarenal mechanisms that act to prevent fluid delivery from exceeding the reabsorptive capacity of the collecting tubules.

A. **Autoregulation,** which keeps the GFR relatively constant despite changes in renal arterial pressure

B. **Tubuloglomerular feedback,** which lowers the GFR if the solute load to the macula densa is increased

1. **Basic mechanism.** Tubuloglomerular feedback is an important regulator of GFR, with changes in renal blood flow taking place as a secondary consequence.
    a. In response to increased tubular flow at the macula densa region, a signal is transmitted to the glomerular arterioles of the same nephron that reduces the GFR by producing vasoconstriction of the afferent arterioles.
    b. With increased tubular flow to the macula densa, the increased Na^+ and Cl^- concentrations in the area lead to increased reabsorption of Na^+ and Cl^- by the macula densa cells.

2. **Hormonal regulation**
    a. The **renin-angiotensin system** is the primary hormonal mediator of the tubuloglomerular feedback system.
        (1) Angiotensin II acts locally on the glomerular mesangium and the afferent and efferent arterioles to reduce the glomerular filtration coefficient and glomerular capillary plasma flow.
        (2) Increased transport of Cl^- by the macula densa serves to augment the production of renin and angiotensin II by the juxtaglomerular apparatus.
    b. Increased NaCl transport across the apical membrane of the macula densa cells leads to an increase in Na^+-K^+-ATPase activity in the basolateral membrane. The increased Na^+-K^+-ATPase activity causes an increase in **adenosine** concentration. Adenosine is a known vasoconstrictor in the kidney.

C. **Glomerulotubular balance,** an intrinsic function of the whole kidney whereby tubular reabsorption of a substance is adjusted in proportion to changes in the GFR

1. **Na^+ reabsorption.** Glomerulotubular balance applies to Na^+ reabsorption in the nephron. This reabsorption is integral to the role of Na^+ in maintaining the volume of the extracellular fluid (ECF). The net effect of GTB is to attenuate the large changes in Na^+ excretion that would result from changes in GFR.

2. The fraction of the filtered Na^+ and water reabsorbed in the proximal tubule tends to increase with volume depletion, and to decrease with volume expansion.

Chapter 26

Renal Tubular Transport: Reabsorption and Secretion

I. PARAMETERS OF RENAL ACTIVE TRANSPORT

A. **Renal tubular transport maximum (Tm)** refers to the maximal amount of a given solute that can be transported (reabsorbed or secreted) per minute by the renal tubules.

1. Definitions
    a. The highest attainable rate of reabsorption is called the **maximum tubular reabsorptive capacity** and is designated **Tm** or **Tr**. Substances that are reabsorbed by an active carrier-mediated process and that have a Tm include phosphate ion (HPO_4^{2-}), sulfate (SO_4^{2-}), glucose (and other monosaccharides), many amino acids, uric acid, and albumin, acetoacetate, β-hydroxybutyrate, and α-ketoglutarate.

    b. The highest attainable rate of secretion is called the **maximum tubular secretory capacity** and is designated **Tm** or **Ts**. Substances that are secreted by the kidneys and that have a Tm include penicillin, certain diuretics, salicylate, para-amino-hippuric acid (PAH), and thiamine (vitamin B_1).

    c. The **threshold concentration** (i.e., the plasma concentration at which a solute begins to appear in the urine), is characteristic for the substance.

2. Exceptions
    a. Uric acid is the only organic substance to be both reabsorbed and secreted by the kidney. K^+ is the only inorganic cation that is both reabsorbed and secreted by the kidney.

    b. Some solutes have no definite upper limit for unidirectional transport and, hence, have no transport maximum.

        (1) The reabsorption of Na^+ along the nephron has no transport maximum.

        (2) The secretion of K^+ by the distal tubules has no transport maximum.

B. **Filtered load and excretion rate.** Reabsorption and secretion are not directly measured variables but are derived from the measurements of the amount of solute filtered (filtered load) and the amount of solute excreted (excretion rate).

1. Definitions
    a. The **filtered load** is the amount of a substance entering the tubule by filtration per unit time and is mathematically equal to the product of the glomerular filtration rate (GFR) and the plasma concentration of the substance (P_x):

$$\text{Filtered load} = GFR \bullet P_x \text{ [(ml/min) (mg/ml)} = \text{mg/min]}$$

Because the GFR is equal to the clearance of inulin (C_{in}), the filtered load can be calculated as $\mathbf{C_{in} \bullet P_x}$.

    b. The **excretion rate** is the amount of a substance that appears in the urine per unit time and is mathematically equal to the product of urine flow rate ($\dot{V}$) and the urine concentration of the substance (U_x):

$$\text{Excretion rate} = U_x \bullet \dot{V} \text{ [(mg/ml) (ml/min)} = \text{mg/min]}$$

        (1) **Secretion.** If the secretion rate exceeds the filtered load [or if the clearance of the substance (C_x) is greater than C_{in}], net tubular secretion of that substance has occurred. Thus, secretion is expressed as:

$$U_x \bullet \dot{V} > C_{in} \bullet P_x, \text{ or } C_x > C_{in}$$

        (2) **Reabsorption.** If the filtered load exceeds the excretion rate (or if C_x is less than C_{in}), net reabsorption of that substance has occurred. Thus, reabsorption is expressed as:

$$C_{in} \bullet P_x > U_x \bullet \dot{V}, \text{ or } C_x < C_{in}$$

2. Calculation of transport maximum (Tr or Ts). Transport maximum is the difference between the filtered load ($C_{in} \cdot P_x$) and the excretion rate ($U_x \cdot \dot{V}$). Tr and Ts, then, are expressed as:

$$Tr = C_{in} \cdot P_x - U_x \cdot \dot{V} \text{ (in mg/min)}$$
$$Ts = U_x \cdot \dot{V} - C_{in} \cdot P_x \text{ (in mg/min)}$$

II. GRAPHIC REPRESENTATION OF RENAL TRANSPORT PROCESSES: RENAL TITRATION CURVES (Figure 26-1)

A. Construction of renal titration curves

1. A titration curve is constructed by plotting the following pairs of variables:
   a. The filtered load ($C_{in} \cdot P_x$) against the plasma concentration (P_x)
   b. The excretion rate ($U_x \cdot \dot{V}$) against P_x
   c. The difference between the filtered load and excretion rate (i.e., Tr or Ts) against P_x

2. The plotted renal titration curve, therefore, can be used to determine the plasma concentration (P_x) at which the renal tubular membrane carriers are fully saturated (Tm).

B. Glucose transport: reabsorption

1. **Characteristics of glucose transport**
   a. Glucose is an essential nutrient that is actively reabsorbed into the proximal tubule by a transport maximum-limited process. The transport maximum for glucose (Tm_G) is about 340 mg/min (or about 2 mmol/min) per 1.73 m^2 body surface area.
   b. Because the Tm_G is nearly constant and depends on the number of functional nephrons, it is used clinically to estimate the number of functional nephrons, or the tubular reabsorptive capacity (Tr).

2. **Glucose titration curve.** Figure 26-1A illustrates that glucose transport and excretion processes are functions of the plasma glucose concentration (P_G).
   a. Increasing the P_G results in a progressive linear increase in the filtered load ($C_{in} \cdot P_G$).
      (1) At a low P_G, the reabsorption of glucose is complete; hence, the clearance of glucose (C_G) is zero (see Figure 26-1C).
      (2) When P_G in humans reaches the renal threshold concentration of 180–200 mg/dl (10–11 mmol/L), glucose appears in the urine (glycosuria). Note that the threshold concentration is not synonymous with the P_G that completely saturates the transport mechanism, which is about 300 mg/dl. The appearance of glucose in the urine before the transport maximum is reached is termed **splay** and results from:
         (a) Heterogeneity in glomerular size and proximal tubular length within individual nephrons. For example, a nephron with a large glomerulus (i.e., a high filtered load), or a short proximal tubule (i.e., low reabsorptive capacity) will spill glucose into the urine at a lower P_G than predicted from the transport maximum for the whole kidney.
         (b) Variability in the transport maximum of the nephrons. For example, there is variability in the number of glucose carriers and in the transport rates of the carriers.
   b. As the Tm_G is approached, the urinary excretion rate increases linearly with increasing P_G (first-order reaction kinetics).
      (1) When the Tm_G is reached, the quantity of glucose reabsorbed per minute remains constant and is independent of P_G (zero-order reaction kinetics).
      (2) When the Tm_G is exceeded, the C_G becomes increasingly a function of glomerular filtration; therefore, the C_G approaches C_{in}, and the amount reabsorbed becomes a smaller fraction of the total amount excreted.

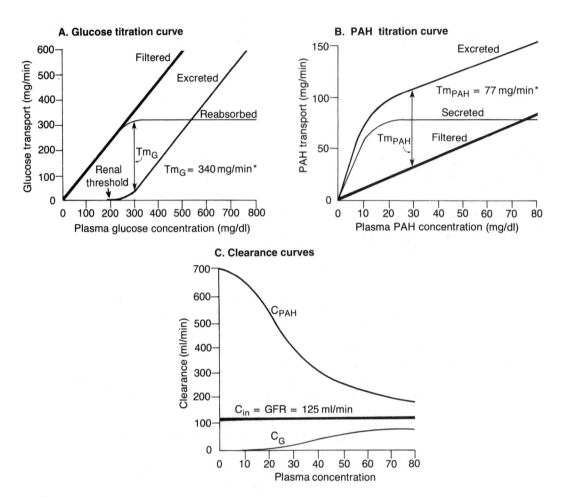

A. Glucose titration curve

B. PAH titration curve

C. Clearance curves

FIGURE 26-1. Relationships of renal titration curves and clearance curves to plasma concentration. The dependent variables [i.e., filtered load, excretion rate, maximum tubular reabsorptive capacity (Tm), and maximum tubular secretory capacity (Ts)] are plotted on the ordinate as a function of the independent variable [i.e., the plasma concentration of the substance (P_x)]. The splay (*curved portions*) observed in the glucose reabsorption and excretion curves (*graph A*) and the splay in the para-aminohippuric acid (*PAH*) secretion and excretion curves (*graph B*) are due to the kinetics of tubular transport and the heterogeneity of the nephron population in terms of number, length, and functional variations in transport. *Graph C* demonstrates that the clearance of glucose (C_G) increases and approaches the clearance of inulin (C_{in}) asymptotically as the plasma glucose concentration increases beyond the tubular transport maximum for glucose (Tm_G). Conversely, *graph C* shows that the clearance of PAH (C_{PAH}) declines and approaches the C_{in} asymptotically as the plasma PAH concentration increases above the tubular transport maximum for PAH (Tm_{PAH}). In *graph C*, the plasma PAH concentration is in mg/dl and the plasma glucose concentration is plotted as one-tenth of the actual concentration (i.e., mg/dl • 1.0^{-1}). * = per 1.73 m^2 body surface area. (Adapted from Selkurt EE: Renal function. In *Physiology*, 5th edition. Edited by Selkurt EE. Boston, Little, Brown, 1984, pp 424 and 429.)

C. **PAH transport: secretion**

1. Characteristics of PAH transport

    a. PAH is a weak organic acid that is actively secreted into the proximal tubule by a transport maximum–limited process. PAH is a foreign substance that is not stored or metabolized and is excreted virtually unchanged in the urine. By filtration and secretion, PAH is almost entirely cleared from the plasma in a single pass through the kidney, if the transport maximum for PAH (Tm_{PAH}) has not been reached. (The Tm_{PAH} in humans is about 80 mg/min per 1.73 m^2 body surface area.)

b. Because approximately 10% of the plasma PAH is bound to plasma proteins, PAH is not entirely freely filterable, and the concentration of PAH in the plasma (P_{PAH}) is greater than that in the glomerular filtrate. However, PAH binding does not significantly diminish the effectiveness of tubular secretion.

c. Because the Tm_{PAH} is nearly constant, it is used clinically to estimate tubular secretory capacity (Ts).

2. PAH titration curve. Figure 26-1B illustrates that the filtration and secretion of PAH are functions of the P_{PAH}. The amount of PAH excreted per minute ($U_{PAH} \cdot \dot{V}$) exceeds the amount filtered ($C_{in} \cdot P_{PAH}$) at any P_{PAH}.

a. When the P_{PAH} is low, virtually all of the PAH that is not filtered is secreted, and PAH is almost completely cleared from the plasma by the combined processes of glomerular filtration and tubular secretion.

b. When P_{PAH} is above 20 mg/dl, the transepithelial secretory mechanism becomes saturated, and the Tm_{PAH} is reached.

c. When the Tm_{PAH} is reached, the quantity of PAH secreted per minute remains constant and is independent of P_{PAH}.

d. When the Tm_{PAH} is exceeded, the C_{PAH} becomes progressively more a function of glomerular filtration; hence, the C_{PAH} approaches C_{in} (see Figure 26-1C), and the constant amount of PAH secreted becomes a smaller fraction of the total amount excreted.

3. PAH clearance (C_{PAH}) and renal plasma flow

a. Fick principle. According to the Fick principle, blood flow through an organ can be determined with the equation:

$$Q = \frac{R}{A_x - V_x}, \text{ where}$$

Q = the blood flow (ml/min)
R = the rate of removal (or addition) of a substance from (or to) the blood as it flows through an organ (in mg/min)
A_x = the concentration of the substance in the blood entering the organ (mg/ml)
V_x = the concentration of the substance in the blood leaving the organ (mg/ml)

b. Because **C_{PAH} can be used to measure renal plasma flow,** the Fick principle equation can be modified to:

$$RPF = \frac{U_{PAH} \cdot \dot{V}}{A_{PAH} - V_{PAH}}, \text{ where}$$

RPF = renal plasma flow (ml/min)
$U_{PAH} \cdot \dot{V}$ = the rate of PAH excretion (mg/min)
A_{PAH} and V_{PAH} = concentrations of PAH (in mg/ml) in the renal artery and the renal vein, respectively

(1) At low concentrations of PAH in arterial plasma, the renal clearance is nearly complete; therefore, as a first approximation, the PAH concentration in renal venous plasma may be taken to be zero.

(a) Approximately 10%–15% of the total renal plasma flow perfuses nonexcretory (nontubular) portions of the kidney, such as the renal capsule, perirenal fat, the renal medulla, and the renal pelvis; therefore, this plasma cannot be completely cleared of PAH by filtration and secretion.

(b) Because about 10% of the PAH remains in the renal venous plasma, the RPF calculated from C_{PAH} underestimates the actual flow by about 10%. Accordingly, the C_{PAH} actually measures the **effective renal plasma flow (ERPF):**

$$ERPF = \frac{U_{PAH} \cdot \dot{V}}{A_{PAH}} = C_{PAH}$$

(2) Because PAH is not metabolized or excreted by any organ other than the kidney, a sample from any peripheral vein can be used to measure arterial plasma PAH concentration, and the equation is written as:

$$ERPF = \frac{U_{PAH} \cdot \dot{V}}{P_{PAH}} = C_{PAH}$$

c. Effective renal blood flow (ERBF) is calculated from the relationship between plasma volume (PV), hematocrit (Hct), and blood volume (BV), as*:

$$\text{Blood volume} = \frac{\text{plasma volume} \cdot 100}{(100 - \text{hematocrit})}$$

and, therefore,

$$ERBF = \frac{ERPF \cdot 100}{(100 - \text{hematocrit})}$$

d. True renal plasma flow (TRPF) can be determined if the extraction ratio (E) is known, where:

$$E = \frac{A_{PAH} - V_{PAH}}{A_{PAH}}, \text{ and}$$

$$TRPF = \frac{ERPF}{E}; \quad TRBF = \frac{TRPF \cdot 100}{(100 - \text{hematocrit})}$$

III. RENAL TRANSPORT OF COMMON SOLUTES AND WATER (Table 26-1)

A. **General considerations** (see Figure 24-2)

1. **The proximal tubules reabsorb 70%–85% of the filtered Na^+, Cl^-, bicarbonate (HCO_3^-), and water and virtually all of the filtered K^+, HPO_4^{2-}, and amino acids.** In addition, glucose is reabsorbed almost completely by the proximal tubules and begins to appear in the urine when the renal threshold is exceeded (i.e., at approximately 200 mg glucose/dl arterial plasma, or 11 mmol/L).

2. Reabsorption of water is passive, and reabsorption of solutes can be passive or active; solute reabsorption generates an osmotic gradient, which causes the passive reabsorption of water (osmosis).

B. **Na^+ reabsorption**

1. **Proximal tubules.** Reabsorption of Na^+ across the proximal tubules occurs in two steps (see Figure 23-2B).
   a. **Step 1: Through the apical membrane.** Although Na^+ is transported through the apical membrane passively down an electrochemical gradient, the transport is mediated by specific membrane proteins and is not by simple diffusion. This first step involves two processes.
      (1) First, Na^+ undergoes **passive cotransport** with the active cotransport of glucose or amino acids. Thus, the energy for the uphill transport of glucose from lumen to cell is derived from the simultaneous downhill movement (influx) of Na^+ along the concentration gradient, which is maintained by the primary active Na^+-K^+-ATPase pump.

---

*In humans, the C_{PAH} is 600–700 ml/min and, when corrected to ERBF (assuming a hematocrit of 45%), is approximately 1100–1200 ml/min, which is about 20% of the resting cardiac output.

TABLE 26-1. Daily Renal Transport of Electrolytes, Nonelectrolytes, and Water in a Normal Adult

Substance	Unit	Filtered Load	Amount Reabsorbed	Amount Secreted	Amount Excreted	Filtered Load Reabsorbed (%)
Na^+	mEq/day	25,200	25,050	. . .	150	99.4
Cl^-	mEq/day	18,000	17,850	. . .	150	99.2
HCO_3^-	mEq/day	4320	4318	. . .	2	99.9+
K^+	mEq/day	720	620	50	100	86.1
Ca^{2+}	mEq/day	540	530	. . .	10	98.1
	mmol/day	270	265		5	
HPO_4^{2-}	mEq/day	260	234	. . .	26	90
	mmol/day	144	130		14	
Urea*	mmol/day	870	460	. . .	410	52.9
Glucose	mmol/day	800	799.5	. . .	0.5	99.9+
Uric acid	mmol/day	50	49	4	5	98.0
Creatinine	mmol/day	12	. . .	1	12	. . .
Total solute	mmol/day	50,386	49,541.5	55	848.5	98.1
Water	ml/day	180,000	179,000	. . .	1000	99.4

*Urea diffuses into and out of parts of the nephron (i.e., it is secreted and reabsorbed, respectively).
Amount filtered (filtered load) = $GFR \cdot P_x$
Amount reabsorbed = $(GFR \cdot P_x) - (U_x \cdot \dot{V})$
Amount secreted = $(U_x \cdot \dot{V}) - (GFR \cdot P_x)$
Amount excreted = $U_x \cdot \dot{V}$

(2) **Cation exchange.** Next, Na^+ reabsorption occurs, coupled with H^+ secretion across the apical membrane to maintain electroneutrality. (Na^+ transport in the proximal segment is not associated with K^+ secretion.) Because tubular H^+ is buffered by HCO_3^-, Na^+ influx must be electrically balanced by an anion.* This balance is accomplished by passive inward diffusion of Cl^- down its concentration gradient or by regeneration of HCO_3^- from cellular CO_2 and water in the presence of carbonic anhydrase.

b. **Step 2: Through the basolateral membrane**
   (1) Na^+ is actively transported through the basolateral membrane via the Na^+-K^+-ATPase pump (see Chapter 23 III A 1). Na^+ reabsorption is equivalent to Cl^- reabsorption + (H^+ + K^+) secretion. Approximately 67% of the proximal reabsorption of Na^+ is based on the active transport process.
   (2) The remaining percentage of Na^+ reabsorption across the basolateral membrane of the proximal tubule occurs passively by **solvent drag**.
       (a) The efflux of Na^+ into the intercellular spaces creates an electrical gradient for the efflux of Cl^- and HCO_3^-, which, in turn, generates an osmotic gradient for water transport into the interspaces. About 20% of the Na^+ movement is loosely coupled to the active secretion of H^+, resulting in the reabsorption of HCO_3^-.
       (b) The osmotic movement of water generates a small increment in hydrostatic pressure, and some bulk flow of water containing Na^+, Cl^-, and HCO_3^- proceeds into the peritubular capillaries.
   (3) As a result of these active and passive transport processes, Na^+ reabsorption together with anions along the proximal tubule allows for the isosmotic reabsorption of fluid.

*The transcellular transport of Na^+ is passive.

2. Distal and collecting tubules

   a. Na$^+$ transport via cation exchange. At the luminal membrane of the distal and collecting tubules, Na$^+$ is reabsorbed by two electrically linked cation-exchange processes.*

   (1) Na$^+$-H$^+$ exchange involves Na$^+$ reabsorption and H$^+$ secretion.

   (2) Na$^+$-K$^+$ exchange involves Na$^+$ reabsorption and K$^+$ secretion. The quantity of Na$^+$ reabsorbed distally is much greater than the amount of K$^+$ secreted.

   (3) These cation-exchange processes are competitive (i.e., K$^+$ competes with H$^+$ for Na$^+$) and, therefore, are not one-to-one. Both cation-exchange processes, however, are enhanced by aldosterone (see Chapter 29 II).

   b. Na$^+$ transport via Cl$^-$ reabsorption. In the thin segment of the ascending limb of the loop of Henle, NaCl is passively reabsorbed. In the thick segment, the reabsorption of Na$^+$ is mainly by a primary active process dependent on Na$^+$-K$^+$-ATPase pumps. The luminal entry of Na$^+$ in this segment is by cotransport with K$^+$ and Cl$^-$ via the Na$^+$-2Cl$^-$-K$^+$ symporter.

   (1) In the collecting duct, the apical transport of Na$^+$ is via Na$^+$ channels.

   (2) When the availability of either K$^+$ or H$^+$ is reduced, the exchange of the more available ion for Na$^+$ is increased.

   (a) Example 1. In hypokalemic alkalosis induced by chronic vomiting or hyperaldosteronemia, the exchange of H$^+$ for Na$^+$ is greater than the exchange of K$^+$ for Na$^+$, the urine becomes more acidic, and the plasma becomes more alkaline.

   (b) Example 2. In hypoaldosteronemia, the reduced secretion of K$^+$ and H$^+$ (due to the decreased rate of distal Na$^+$ reabsorption) causes hyperkalemia and acidosis.

   (3) The Na$^+$-H$^+$ exchange mechanism is not mandatory for H$^+$ secretion.

3. Nephron. Na$^+$ reabsorption along the nephron occurs in the following proportions:

   a. Proximal tubule (75%)

   b. Ascending limb of the loop of Henle (22%)†

   c. Distal tubule (4%–5%)

   d. Collecting tubule (2%–3%)

C. K$^+$ transport

1. General considerations

   a. K$^+$ is the only plasma electrolyte that is both reabsorbed and secreted into the renal tubules. Virtually all of the K$^+$ is actively reabsorbed by the proximal tubule, whereas K$^+$ secretion is largely a function of the distal tubule.

   b. The net secretion of K$^+$ by the nephron can cause the excretion of more K$^+$ than was filtered, as in hyperaldosteronemia and in metabolic alkalosis. (Elevated K$^+$ excretion also can occur, in part, as a result of decreased K$^+$ reabsorption, as observed in the osmotic diuresis seen with uncontrolled diabetes mellitus.)

   c. Changes in plasma [K$^+$] have important effects on acid-base balance, and, conversely, changes in acid-base balance have important effects on plasma [K$^+$] [see also III D 2 a (2)]. In the distal nephron, which has the capacity for both net secretion and net reabsorption of K$^+$, intracellular H$^+$ and K$^+$ behave as if they were competing with each other for secretion into the urine (Figure 26-2).

2. K$^+$ reabsorption. K$^+$ is completely filterable at the glomerulus, and tubular reabsorption occurs by active transport in the proximal tubule, ascending limb of the loop of Henle, distal tubule, and collecting duct.

3. K$^+$ secretion involves an active and a passive process and is influenced by aldosterone.

---

*Only a small amount of the filtered Na$^+$ that is reabsorbed in the distal and collecting tubules is by cation exchange with H$^+$ and K$^+$.
†Na$^+$ is passively reabsorbed in the thin segment of the ascending limb of the loop of Henle and is actively reabsorbed in the thick segment.

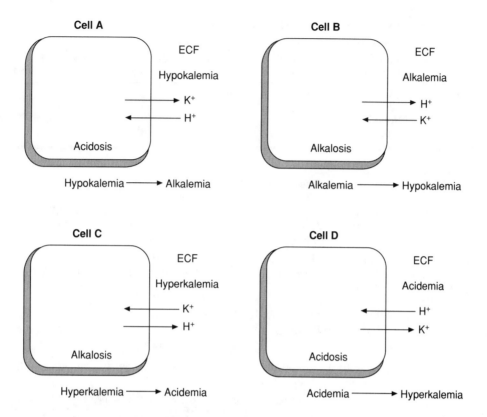

FIGURE 26-2. K^+/H^+ relationships. (*A*) Hypokalemia leads to alkalemia. During hypokalemic states, cellular K^+ enters the extracellular fluid (*ECF*) and H^+ enters the cell, leading to a state of relative intracellular acidosis and extracellular alkalemia. (*B*) Alkalemia leads to hypokalemia. During metabolic alkalosis, cellular H^+ enters the ECF and K^+ enters the cell, leading to a state of hypokalemia. (*C*) Hyperkalemia leads to acidemia. During hyperkalemic states, cellular H^+ enters the ECF and K^+ enters the cell, leading to a state of relative intracellular alkalosis and extracellular acidemia. (*D*) Acidemia leads to hyperkalemia. During a state of metabolic acidosis, cellular K^+ enters the ECF and H^+ enters the cell, leading to a state of hyperkalemia.

    a. The critical step is the active transport of K^+ from the interstitial fluid across the basolateral membrane into the distal tubular cell. This active transport step is associated with active Na^+ extrusion via the Na^+-K^+-ATPase system.

    b. The elevated intracellular $[K^+]$ achieved by the basolateral pump favors the net K^+ diffusion at the apical membrane down a concentration gradient into the tubular lumen. The transcellular transport of K^+ is passive.

    c. Aldosterone increases the activity of the basolateral Na^+-K^+-ATPase pump, which increases K^+ entry into the distal cells and the $[K^+]$ gradient from cell to lumen. Aldosterone also increases the permeability of the luminal membrane to K^+, further increasing K^+ diffusion (secretion).

D. **Renal regulation of H^+ balance** (Table 26-2). To maintain a normal plasma $[HCO_3^-]$, the kidney must secrete H^+ into the tubular lumen. The process of H^+ secretion and HCO_3^- reabsorption occurs throughout the nephron, except in the descending limb of the loop of Henle.

    1. **H^+ secretion** permits conservation of HCO_3^- via reclamation of filtered HCO_3^- (i.e., HCO_3^- reabsorption) and regeneration of HCO_3^- by urinary net acid excretion [see III D 2 b (1)].

        a. H^+ secretion into the tubular lumen occurs by active transport and is coupled to Na^+ reabsorption; for each H^+ secreted, one Na^+ and one HCO_3^- are reabsorbed.

        b. The renal epithelium secretes approximately 4300 mEq (mmol) of H^+ daily!

TABLE 26-2. Normal Renal Regulation of H^+ and HCO_3^-

Ion	Filtered (mEq/day)	Reabsorbed (mEq/day)	Secreted (mEq/day)	Excreted (mEq/day)	Filtered Load Reabsorbed (%)
HCO_3^-	4320	4318	None	2	99.9$^+$
HPO_4^{2-}	260	240	None	20	92
NH_4^+*	None	. . .	. . .	40	. . .
H^+	< 0.1	4315†	4375	58‡	. . .

*Secreted mainly as NH_3 and NH_4^+.
†98.6% of the secreted H^+ is reabsorbed daily.
‡Note that total H^+ excretion equals the sum of titratable acid (HPO_4^{2-}) and NH_4^+ minus HCO_3^-.

(1) Approximately 85% of the total H^+ secretion occurs in the proximal tubules.
(2) Approximately 10% of the total H^+ secretion occurs in the distal tubules; and about 5% occurs in the collecting ducts.

2. HCO_3^- conservation. Reabsorption of HCO_3^- alone cannot maintain normal acid-base balance. Therefore, the kidney is capable of generating new HCO_3^- as well.

a. HCO_3^- reabsorption
(1) Process (see Figure 36-1)
(a) H^+ secreted into the tubular fluid combines with filtered HCO_3^- to form carbonic acid (H_2CO_3):

$$H^+ + HCO_3^- \rightleftharpoons H_2CO_3$$

(b) Luminal (brush border) carbonic anhydrase catalyzes the dehydration of H_2CO_3 into CO_2 and H_2O:

$$H_2CO_3 \underset{\longleftarrow}{\xrightarrow{\text{carbonic anhydrase}}} CO_2 + H_2O$$

(i) The CO_2 that diffuses back across the luminal membrane adds to the intracellular CO_2 pool produced by metabolism.
(ii) This intracellular CO_2 pool serves as a substrate for the formation of H_2CO_3 via intracellular carbonic anhydrase.
(iii) The dissociation of this H_2CO_3 forms H^+ and HCO_3^-.
The H^+ is secreted again into the lumen.
The HCO_3^- is transported downhill across the basolateral membrane by a secondary active transport system.
(c) Buffering of secreted H^+ by filtered HCO_3^- does not contribute to the urinary excretion of H^+. The CO_2 formed in the lumen from secreted H^+ returns to the tubular cell to form another H^+ and, therefore, **no net H^+ secretion occurs**.
(2) Effect of plasma [K^+] on HCO_3^- reabsorption. Body K^+ stores influence HCO_3^- reabsorption. There is an inverse relationship between plasma [K^+] and proximal HCO_3^- reabsorption.
(a) Hypokalemia. A decrease in plasma [K^+] leads to an increase in HCO_3^- reabsorption.
(i) Hypokalemia is associated with **intracellular acidosis**.
(ii) Hypokalemia provides a concentration gradient for K^+ to move out of the cell (see Figure 26-2A). In response to the efflux of K^+, both H^+ and Na^+ enter the cell to maintain electroneutrality.
(iii) The increase in intracellular [H^+] favors HCO_3^- reabsorption because HCO_3^- reabsorption depends on H^+ secretion.
(b) Hyperkalemia. An elevated plasma [K^+] leads to a decreased HCO_3^- reabsorption.
(i) Hyperkalemia is associated with **intracellular alkalosis**.
(ii) In hyperkalemia, K^+ moves into the cell (see Figure 26-2C). In response to the K^+ uptake by the cells, H^+ and Na^+ leave the cells to maintain electroneutrality.
(iii) Hyperkalemia tends to alkalinize the cell and reduce HCO_3^- reabsorption.

b. HCO$_3^-$ regeneration. New HCO$_3^-$ can be formed by two mechanisms that provide for the excretion of H$^+$.

    (1) Nonbicarbonate buffering. Most of the H$^+$ ions secreted in excess of those required for HCO$_3^-$ reabsorption are buffered by nonbicarbonate buffers and subsequently excreted. Normally, the most important of these nonbicarbonate buffers is filtered **dibasic phosphate (HPO$_4^{2-}$)**.

        (a) When secreted H$^+$ combines with HPO$_4^{2-}$, the formation of **monobasic phosphate (H$_2$PO$_4^-$, titratable acid)** takes place in the proximal tubule.

        (b) The excretion of H$_2$PO$_4^-$ is equivalent to HCO$_3^-$ conservation and the addition of new HCO$_3^-$ to the blood (see Figure 36-3).

    (2) Glutamine catabolism. The kidneys can also add new HCO$_3^-$ to the blood via the catabolism of glutamine, which results in the formation of ammonium (NH$_4^+$) and HCO$_3^-$.

        (a) The NH$_4^+$ is excreted into the proximal lumen by the Na$^+$-NH$_4^+$ exchanger and is excreted.

        (b) The HCO$_3^-$ is transported into the peritubular capillaries and constitutes new HCO$_3^-$ formation.

3. HCO$_3^-$ secretion. All segments of the nephron normally reabsorb HCO$_3^-$, except for the **cortical collecting duct,** which, under conditions of normal acid-base balance or metabolic alkalosis, secretes HCO$_3^-$ from type B intercalated cells.

    a. The **collecting tubule** (i.e., the cortical, outer medullary, and inner medullary collecting ducts) consists of **two cell types:**

    (1) Principal cells, which reabsorb Na$^+$ and secrete K$^+$. These cells do not exhibit H$^+$ secretory activity.

    (2) Intercalated cells (Figure 26-3)

        (a) Type A intercalated cells secrete H$^+$ (i.e., reabsorb HCO$_3^-$). Because the luminal membranes of type A intercalated cells contain H$^+$-K$^+$-ATPase pumps, they can also **reabsorb K$^+$.**

        (b) Type B intercalated cells differ from type A intercalated cells in that the polarity of the membrane transporters is reversed, so that they **secrete HCO$_3^-$.**

            (i) H$^+$ ions are transported into the peritubular capillary by H$^+$-ATPase pumps located in the basolateral, rather than the luminal, membrane.

            (ii) HCO$_3^-$ ions, in comparison, are secreted into the tubular lumen by a Cl$^-$-HCO$_3^-$ exchanger located in the luminal, rather than the basolateral, membrane.

            (iii) The basolateral H$^+$-ATPase pump transports H$^+$ into the blood, where it combines with HCO$_3^-$, causing plasma [HCO$_3^-$] levels to decline and resulting in the excretion of HCO$_3^-$ in the urine.

FIGURE 26-3. There are two types of intercalated cells in the collecting tubule: H$^+$-secreting cells (type A cells) and HCO$_3^-$-secreting cells (type B cells). Both cell types contain the same transport mechanisms (i.e., H$^+$-ATPase and the HCO$_3^-$-Cl$^-$ exchanger) but they are oriented oppositely. ● = primary active transporters; ○ = secondary active transporters; → = uphill transport; ⇢ = downhill transport; *C.A.* = carbonic anhydrase.

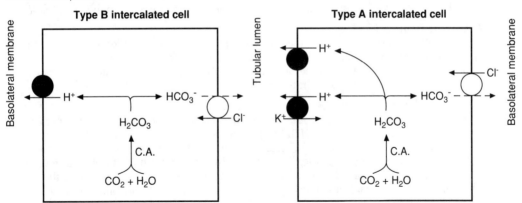

    b. The HCO_3^--secreting intercalated cell may play an important role in the excretion of HCO_3^- during metabolic alkalosis.

  4. Total (net) acid excretion (see Table 26-2).

    a. Titratable acid. Approximately 20 mEq (mmol) of H^+ per day are buffered by filtered HPO_4^{2-} and excreted as $H_2PO_4^-$. The proximal tubule is the major nephron site for titratable acid formation.

    b. Ammonia. About 40 mEq (mmol) of H^+ per day are buffered by ammonia (NH_3) and excreted as ammonium (NH_4^+).

    c. The total amount of H^+ excreted daily by an individual on a normal diet equals the sum of titratable acid and NH_4^+ excreted, or about 60 mEq (mmol) of H^+ per day: Thus, only a minute concentration of free H^+ normally exists in the final urine despite the 4300 meq (mmol) of H^+ secreted daily!

E. Cl^- transport

  1. General considerations

    a. Cl^- reabsorption occurs at a rate of 20,000 mEq/day and is inversely related to HCO_3^- reabsorption; that is, when HCO_3^- reabsorption increases, Cl^- reabsorption decreases and vice versa. Therefore, the plasma $[Cl^-]$ varies inversely with the HCO_3^- reabsorption rate.

    b. Cl^- is the chief anion to accompany Na^+ through the renal tubular epithelium. Because Na^+ reabsorption is under control of aldosterone, the plasma $[Cl^-]$ is secondarily influenced by aldosterone.

    c. Cl^- is excreted with NH_4^+ to eliminate H^+ exchange for Na^+ reabsorption.

  2. Passive transport. When Na^+ is actively reabsorbed across the luminal membrane of the proximal tubule, Cl^- is passively reabsorbed along an electrochemical gradient.

    a. Passive proximal diffusion (reabsorption) of Cl^- is aided by the increased $[Cl^-]$ within the luminal fluid of the late proximal tubule.

    b. This increased $[Cl^-]$ is due to the reabsorption of $NaHCO_3$ and, therefore, water into the early proximal tubule.

  3. Active transport. Cl^- is actively reabsorbed in the thick segment of the ascending limb of the loop of Henle via a secondary active Na^+-$2Cl^-$-K^+ symporter located in the apical membrane.

    a. Active reabsorption of Na^+ occurs in the thick segment.

    b. The reabsorption of NaCl in the thick and thin segments together with the water and impermeability of the ascending limb lead to the dilution of the tubular fluid.

F. Water reabsorption.

Water movement across membranes is determined by hydrostatic and osmotic pressure gradients. Water is reabsorbed passively by diffusing along an osmotic gradient, which primarily is established by the reabsorption of Na^+ and Cl^-.

  1. Proximal tubules

    a. In humans, about 75%–80% of the reabsorption of filtered water occurs in the proximal tubule.

    b. The proximal reabsorption of water is invariant and involves no change in osmolality.

  2. Loop of Henle

    a. Unlike the proximal tubule, the loop of Henle reabsorbs considerably more solute than water. Only about 5% of the reabsorption of filtered water occurs here.

    b. The tubular fluid entering the descending limb of the loop of Henle always is **isosmotic,** regardless of the hydration state.

  3. Distal and collecting tubules

    a. The distal and collecting tubular reabsorption of water occurs only in the presence of antidiuretic hormone (ADH). In the presence of a maximal ADH effect, 99% of the water is reabsorbed; in the absence of ADH, 88% of the water is reabsorbed.

    b. The tubular fluid entering the distal tubule always is **hyposmotic** to plasma, regardless of the hydration state. In the terminal distal tubule and the collecting ducts, the osmolality in the tubular fluid changes according to the water permeability of the tubule.

Chapter 27

Renal Concentration and Dilution of Urine

I. FUNCTIONAL CONSIDERATIONS

A. **Purpose.** The kidney can alter the composition of the urine in response to the body's daily needs, thereby maintaining the osmolality of body fluids. When it is necessary to conserve body water, the kidney excretes urine with a high solute concentration. When it is necessary to rid the body of excess water, the kidney excretes urine with a dilute solute concentration.

B. **Role of antidiuretic hormone (ADH).** The principal regulator of urine composition is the hormone ADH. In the absence of ADH, the kidney excretes a large volume of dilute urine; when ADH is present in high concentration, the kidney excretes a small volume of concentrated urine.*

C. **Components of the concentrating and diluting system.** The formation of urine that is dilute (hyposmotic to plasma) or concentrated (hyperosmotic to plasma) is achieved by the countercurrent system of the nephron (Figure 27-1).† This system consists of the:

1. Descending limb of the loop of Henle (DLH)

2. Thin and thick segments of the ascending limb of the loop of Henle (ALH)

3. Medullary interstitium

4. Distal convoluted tubule

5. Collecting duct

6. Vasa recta, which are the vascular elements of the juxtamedullary nephrons

D. **Mechanisms of dilution and concentration.** The kidney forms a dilute urine in the absence of ADH. In the presence of ADH, the kidney forms a concentrated urine via the functioning of the **countercurrent multipliers** (i.e., the loop of Henle and collecting duct) and the **countercurrent exchangers** (i.e., the vasa recta). Regardless of ADH, however, the fluid osmolality in the loop of Henle, vasa recta, and medullary interstitium always increases progressively from the corticomedullary junction to the papillary tip (see Figure 27-1D).

1. The **fundamental processes** involved in the excretion of a dilute or concentrated urine include:
   a. Variable permeability of the nephron to the passive back-diffusion (reabsorption) of water along an osmotic gradient and of urea along its concentration gradient
   b. Passive reabsorption of NaCl by the thin segment of the ALH
   c. Active reabsorption of Na^+ by the thick segment of the ALH via the Na^+-$2Cl^-$-K^+ symporter and the Na^+-K^+-ATPase pump.

2. The **formation of hyperosmotic urine** involves the following steps.
   a. The medullary interstitium becomes hyperosmotic by the reabsorption of NaCl and urea.
   b. The urine entering the medullary collecting ducts equilibrates osmotically with the hyperosmotic interstitium, resulting in the excretion of a small volume of concentrated urine in the presence of ADH.

---

*In the presence of a maximal ADH effect, 99% of the water is reabsorbed; in the absence of ADH, 88% of the water is reabsorbed.
†Only the juxtamedullary nephrons, with their long loops of Henle, contribute to the medullary hyperosmolality.

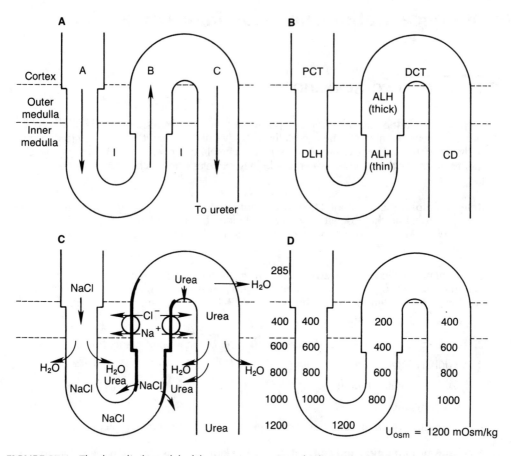

FIGURE 27-1. The three-limb model of the countercurrent multiplier system of the juxtamedullary nephron. *Model A* represents the directions of tubular flow (countercurrent). *A* = concentrating segment; *B* = diluting segment; *C* = collecting duct; and *I* = interstitium. *Model B* represents the major components of the nephron. *PCT* = proximal convoluted tubule; *DLH* = descending limb of the loop of Henle; *ALH* = ascending limb of the loop of Henle (thick and thin segments); *DCT* = distal convoluted tubule; and *CD* = collecting duct. *Model C* represents solute and solvent transfer. The *heavy line* indicates water impermeability. *Model D* represents the vertical (longitudinal) and horizontal (transverse) osmotic gradients in the presence of antidiuretic hormone (ADH). U_{osm} = urine osmolality. (After Jamison R, Maffly RH: The urinary concentrating mechanism. *N Engl J Med* 295:1059–1067, 1976.)

  3. The final osmolality of the urine is determined by the permeability of collecting ducts to water.

  4. The vascular loop (vasa recta) prevents the dissipation of the osmotic (hypertonic) "layering" in the interstitium (see III A–B).

II. COUNTERCURRENT MULTIPLIERS (see Figure 27-1)

A. General considerations

  1. The countercurrent multiplier system is analogous to a three-limb model consisting of the two limbs of the loop of Henle, which are connected by a hairpin turn, and the collecting duct. The fluid flow through the three limbs is countercurrent (i.e., in alternating opposite directions; see Figure 27-1A).

2. The countercurrent multiplier process has two important consequences (see Figure 27-1D).

  a. At equilibrium in the presence of ADH, a maximal vertical gradient of 900 mOsm/kg is established between the corticomedullary junction (300 mOsm/kg) and the renal papillae (1200 mOsm/kg). This explains the origin of the term "countercurrent multiplier."

  b. At any given level in the renal medulla, the osmolality is almost the same in all fluids except that in the ALH. The fluid in the ALH is less concentrated than that in either the DLH or the interstitium and becomes hyposmotic to plasma (see Figure 27-1D).

B. Loop of Henle

1. The **DLH** is the **concentrating segment** of the nephron (see Figures 27-1C & D, 27-2). The following characteristics of the DLH account for this.

  a. The DLH is highly permeable to water. Solute-free water leaves the DLH, causing the fluid in the DLH to become concentrated to a degree that is consistently higher than that of the fluid in the ALH.

  b. The DLH has a low permeability to NaCl and urea. The predominant solute is NaCl, with relatively small amounts of urea existing in the DLH.

  c. The fluid in the DLH nearly attains the osmolality of the adjacent medullary interstitium. [The interstitial osmolality is maintained by solvent-free solute (NaCl) that is transported out of the ALH.]

2. The **ALH**. The thick and thin segments of the ALH constitute the **diluting segments** of the nephron. Both segments are impermeable to water and both are permeable to NaCl; in addition, the thin segment of the ALH is permeable to urea.

  a. As fluid passes through the **thin segment** of the ALH, NaCl diffuses down its concentration gradient into the interstitial fluid. This contributes to the increased interstitial osmolality and renders the tubular fluid hyposmotic to the peritubular interstitium.

FIGURE 27-2. Axial profile of osmolality of the tubular fluid. Initially, the tubular fluid is isosmotic. In the descending limb of the loop of Henle (DLH), it increases in osmolality, and in the ascending limb of the loop of Henle (ALH), it decreases in osmolality. The tubular fluid enters the distal nephron hyposmotic to plasma, where it may become more hyposmotic [in the absence of antidiuretic hormone (*ADH*)], or hyperosmotic (under the influence of ADH). (Modified with permission from Guyton AC: *Textbook of Medical Physiology,* 8th ed. Philadelphia, WB Saunders, 1991, p 312.)

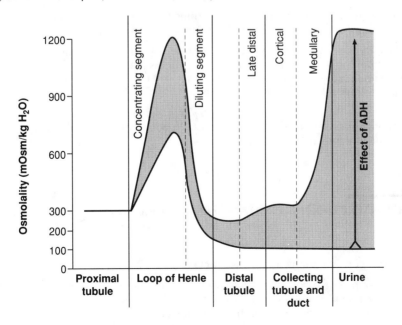

    b. The **thick segment** of the ALH actively transports Na^+ out of the lumen. The only active step in countercurrent multiplication is NaCl reabsorption in the thick ascending limb via the Na^+-$2Cl^-$-K^+ symporter.

    c. The active and passive transport of NaCl from the ALH (thick and thin segments, respectively) to the interstitium forms a horizontal osmotic gradient of up to 200 mOsm/kg between the tubular fluid of the ALH and the combined fluid of the interstitium and the DLH (see Figure 27-1D), leaving behind a smaller volume of hypotonic fluid rich in urea. This fluid flows into the distal convoluted tubule and the cortical and medullary collecting ducts.

    d. The fluid in the ALH becomes diluted by the net loss of NaCl in excess of the net gain of urea by diffusion. This efflux of NaCl causes the tubular fluid within the ALH to become hyposmotic and that in the interstitial fluid to become hyperosmotic.

C. **Simultaneous events in the interstitium**

    1. Fluid in the interstitium contains hyperosmotic concentrations of NaCl and urea.

        a. As the collecting ducts join in the inner medulla, they become increasingly permeable to urea (especially in the presence of ADH), allowing urea to flow passively along its concentration gradient into the interstitium (see Figures 27-1C, 27-2).

        b. The increase in medullary osmolality causes water to move out of the adjacent tubules, the terminal collecting ducts, and the DLH.

    2. Water is reabsorbed in the last segment of the distal convoluted tubule and in the collecting duct in the cortex and outer medulla.

    3. In the inner medulla, both water and urea are reabsorbed from the collecting duct (see Figure 27-1C).

        a. Some urea reenters the ALH but at a slower rate than the efflux of NaCl.

        b. This **medullary recycling of urea,** in addition to solute trapping by countercurrent exchange, causes urea to accumulate in large amounts in the medullary interstitium, where it osmotically abstracts water from the DLH and thereby concentrates NaCl in the DLH fluid.

D. **Collecting duct.** The cortical and upper medullary collecting ducts are relatively impermeable to water, urea, and NaCl.

    1. The relative impermeability to water occurs in the absence of ADH.

    2. The relative impermeability to NaCl permits the high interstitial concentration of NaCl to act as an effective osmotic gradient between the tubular fluid and the interstitium.

    3. When the kidney forms a concentrated urine, the collecting duct receives an isosmotic fluid from the distal tubule, the collecting duct is made permeable to water, and the urine equilibrates with the hyperosmotic medullary interstitium, resulting in the excretion of a low volume, hypertonic urine.

    4. If the collecting duct is impermeable to water (ADH absent), the dilute tubular fluid entering the collecting duct from the distal tubule remains hypotonic and is excreted as a higher volume, hypotonic urine.

III. **COUNTERCURRENT EXCHANGERS.** The vasa recta function as countercurrent diffusion exchangers, increasing the efficiency of the concentrating mechanisms.

A. **Functional considerations of the vasa recta** (Figure 27-3)

    1. The vasa recta are derived from the efferent arterioles of the juxtamedullary glomeruli and function to maintain the hyperosmolality of the medullary interstitium. The vasa recta are in juxtaposition with the loops of Henle.

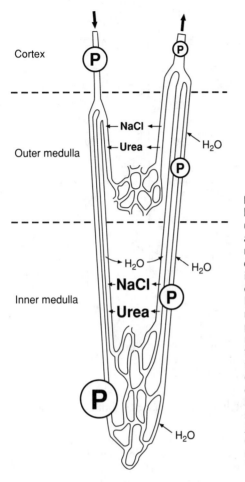

FIGURE 27-3. The vasa recta function to maintain the hyperosmolality of the medullary interstitium. NaCl and urea that have been reabsorbed from the loop of Henle and collecting tubule (respectively) accumulate in the medullary interstitium, where they are absorbed by the descending limb of the vasa recta and returned to the interstitium by the ascending limb of the vasa recta. This countercurrent exchange "traps" the solutes in the medullary interstitium. At the same time, water (H_2O) is removed from the descending vasa recta, increasing the plasma protein (P) concentration. In the ascending vasa recta, the oncotic pressure causes the capillaries to take in fluid. In this manner, water reabsorbed by the nephron is removed from the interstitium and returned to the general circulation. *Type size* indicates the relative concentration of each solute with respect to its location in the medulla, but not necessarily with respect to concentrations of other solutes. (Redrawn with permission from Jamison RL, Ghrig JJ Jr: Renal physiology VII. In *Handbook of Physiology*, edited by Windhager EE. Published for the American Psychological Society. New York, Oxford University Press, 1992, p 1268.)

2. These vessels are permeable to solute and water and reach osmotic equilibrium with the medullary interstitium.

3. The vasa recta return the NaCl and water reabsorbed in the loops of Henle and collecting ducts to the systemic circulation.

4. The concentrations of Na^+ and urea in the medullary interstitium arc kept high by the slow blood flow in the vasa recta.

B. **Countercurrent exchange effects of the vasa recta** (Figure 27-4)

1. Na^+, Cl^-, and urea are passively reabsorbed from the ALH, diffuse across the interstitial fluid into the descending limb of the vasa recta, and are returned to the interstitium by the ascending limb of the vasa recta.
   a. These solutes recirculate in the vasa recta capillary loops and increase the medullary osmolality.
   b. Na^+, Cl^-, and urea also recirculate in the loop of Henle.

2. Water diffuses from the descending limb of the vasa recta across the interstitial fluid and into the ascending limb of the vasa recta.* As a result, the cortex receives blood that is only slightly hypertonic to plasma (see Figure 27-4).

---

*Water is continually removed by the ascending limb of the vasa recta.

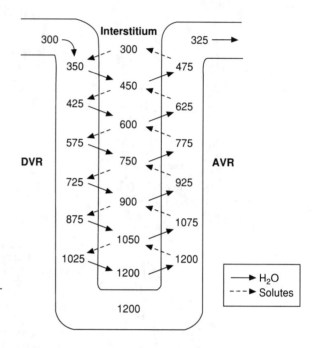

FIGURE 27-4. Countercurrent exchange in the vasa recta. Water and solute transport occur by diffusion. *AVR* = ascending vasa recta; *DVR* = descending vasa recta. (Modified and redrawn with permission from Vander AJ: *Renal Physiology,* 4th ed. New York, McGraw-Hill, 1991, p 108.)

IV. **ROLE OF UREA.** The main function of urea in the countercurrent system is to exert an osmotic effect on the DLH, promoting the abstraction of water and raising the intraluminal concentration of NaCl (see Figures 27-1, 27-3).

A. **Effect of ADH on urea concentration.** In the medulla, ADH enhances both water and urea permeabilities of the collecting duct.

1. Urea diffuses into the thin segment of the ALH via secretion but is more concentrated in the fluid entering the distal convoluted tubule (concentrated in a smaller volume) than in the filtrate entering the proximal convoluted tubule. The thin segment of the ALH is less permeable to urea than to NaCl, and the thick segment of the ALH is impermeable to urea.

2. During **antidiuresis** (i.e., under the influence of ADH), the tubular concentration of urea in the cortical collecting duct increases as a result of urea-free water reabsorption from the cortical collecting duct into the outer medullary interstitium.

3. When the tubular fluid enters the urea-permeable inner medullary collecting duct, urea, along with water, diffuses into the medullary interstitium where it is trapped by countercurrent exchange in the vasa recta, resulting in a high urea concentration in the inner medulla.

4. Urea recirculates in the loops of Henle and the vasa recta. The vasa recta are permeable to urea so that urea diffusing out of the papillary collecting duct is trapped in the medullary interstitium.
   a. Of the 1200 mOsm/kg of solute present in the renal papillary loop during **antidiuresis,** half is composed of NaCl and half of urea.
   b. During water **diuresis,** about 10% of the total medullary solute concentration is urea. The maximal urine osmolality (U_{osm}) cannot exceed that in the interstitium, and the ability to conserve water by excreting concentrated urine is reduced when the papillary concentration is reduced.

B. **Effect of urea on urine osmolality**

1. Urea increases urine osmolality via a **three-step process:**
   a. The high concentration of urea in the inner medullary collecting duct causes continued diffusion of urea out of the collecting duct and into the medullary interstitium.

b. The interstitial urea supplied by the inner medullary collecting duct removes water from the DLH and causes the tubular NaCl concentration to increase above that in the interstitium, favoring the passive reabsorption of NaCl from the thin ascending limb.

c. As water is reabsorbed from the urea-impermeable cortical and outer medullary collecting ducts, urea becomes the principle solute in tubular fluid entering the inner medullary collecting duct.

2. This process establishes the gradients for urea and NaCl in opposite directions: urea concentration in the collecting duct tubular fluid is higher than that in the interstitium, and the NaCl concentration in the collecting duct is lower than that in the interstitium (Figure 27-5).

a. These two concentration gradients for urea and NaCl are due to the higher reflection coefficient of the collecting duct for NaCl (approximately 1) than that for urea (<1).

b. Thus, despite equal osmolalities on both sides of the collecting tubules at the inner–outer medullary junction, the effective driving force for water transport favors water reabsorption.

V. MEASUREMENT OF RENAL WATER EXCRETION AND CONSERVATION (Figure 27-6)

A. **Renal water excretion.** The quantitative measure of the kidney's ability to excrete water is termed **free-water clearance**.

1. General considerations

a. Free-water clearance is not a true clearance, because no osmotically free water exists in plasma.

b. Free-water clearance denotes the volume of **pure** (i.e., solute-free) **water** that must be removed from, or added to, the flow of urine (in ml/min) to make it isosmotic with plasma.

2. Measurement. Free-water clearance (C_{H_2O}) is calculated using the equation:

$$\dot{V} = C_{osm} + C_{H_2O}$$
$$C_{H_2O} = \dot{V} - C_{osm}$$

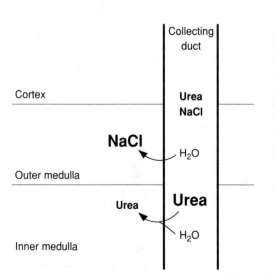

Cortex

Outer medulla

Inner medulla

Collecting duct

Urea
NaCl

NaCl ↙ H_2O

Urea **Urea** ↙ H_2O

FIGURE 27-5. The differences between the relative NaCl and urea concentrations in tubular fluid and the interstitium are represented by *type size*. The high NaCl concentration in the outer medullary interstitium [which results from NaCl reabsorption in the thick segment of the ascending limb of the loop of Henle (ALH)] and the high urea concentration in tubular fluid (as a result of water reabsorption upstream), coupled with the different reflection coefficients for NaCl and urea (approximately 1 and < 1, respectively), favors water reabsorption in the outer medullary collecting duct, despite equal osmolalities on both sides of the tubular wall. At the outer–inner medullary junction, the urea permeability of the collecting duct increases so that urea diffuses into the interstitium, increasing the osmotic gradient for additional water reabsorption.

where

$$C_{osm} = \frac{U_{osm} \cdot \dot{V}}{P_{osm}}$$

Substituting for C_{osm}, the equation becomes:

$$C_{H_2O} = \dot{V} - \frac{U_{osm} \cdot \dot{V}}{P_{osm}}$$

and by factoring out $\dot{V}$, the final equation is:

$$C_{H_2O} = \dot{V} \left(1 - \frac{U_{osm}}{P_{osm}} \right), \text{ where}$$

$\dot{V}$ = urine volume per unit time (in ml/min)
U_{osm} = urine osmolality
P_{osm} = plasma osmolality
C_{osm} = osmolal clearance, which is the volume of plasma (in ml) completely cleared of osmotically active solutes that appear in the urine each minute

a. Therefore, **during water diuresis** (i.e., in a hydrated state and in the absence of ADH), the U_{osm} is less than the P_{osm} (hyposmotic urine), and the U_{osm}/P_{osm} ratio is less than 1. Therefore, C_{H_2O} **is positive,** indicating that water is being eliminated by the excretion of a large volume of dilute urine. The maximum free-water clearance in humans is 15–20 L/day (10–15 ml/min).

b. Similarly, **during antidiuresis** (i.e., in a dehydrated state and in the presence of ADH), the U_{osm} is greater than the P_{osm} (hyperosmotic urine), and the U_{osm}/P_{osm} ratio is greater than 1. Therefore, C_{H_2O} **is negative,** indicating that water is being conserved by the excretion of a small volume of concentrated urine. (To avoid the use of the term "negative free-water clearance," the symbol $T^c_{H_2O}$ is used to denote **free-water reabsorption**. The superscript "c" signifies that net reabsorption occurs in the collecting tubule.)

B. **Renal water conservation.** A quantitative measure of the ability of the kidney to reabsorb water is termed **free-water reabsorption**.

1. **General considerations.** Free-water reabsorption denotes the volume of free water reabsorbed per unit time, or, the amount of free water that must be removed from the urine by tubular reabsorption to make it hyperosmotic with plasma. Renal water conservation is dependent on:

a. The formation and maintenance of the medullary osmotic gradient (i.e., a hyperosmotic interstitium to provide for water reabsorption by the collecting duct)

b. Equilibration of the fluid (urine) in the collecting tubules with the hyperosmotic medullary interstitium

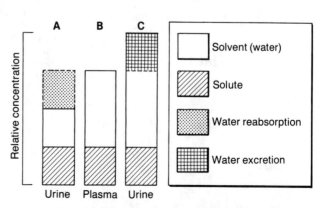

FIGURE 27-6. Renal free-water reabsorption (A) and free-water clearance (C). In column A, the *stippled area* represents the amount of free water that was removed from the urine by tubular reabsorption to **raise** the urine osmolality (U_{osm}) to the observed hypotonic value. In column C, the *cross-hatched area* denotes the amount of free water that was added to the urine by tubular excretion to **lower** the U_{osm} to the observed hypotonic value. Plasma osmolality (B) serves as the reference point for describing the osmolality of urine.

2. Measurement. Free-water reabsorption ($T^c_{H_2O}$) is calculated using the equation:

$$T^c_{H_2O} = C_{osm} - \dot{V}$$

where

$$C_{osm} = \frac{U_{osm} \cdot \dot{V}}{P_{osm}}$$

Substituting C_{osm}, the equation becomes:

$$T^c_{H_2O} = \frac{U_{osm} \cdot \dot{V}}{P_{osm}} - \dot{V}$$

and by factoring out $\dot{V}$, the final equation is:

$$T^c_{H_2O} = \left(\frac{U_{osm}}{P_{osm}} - 1 \right) \dot{V}$$

Therefore, **during antidiuresis** the U_{osm} exceeds P_{osm}, and the U_{osm}/P_{osm} ratio is greater than 1, indicating that the **$T^c_{H_2O}$ is positive**. A positive $T^c_{H_2O}$ suggests that water is being conserved by the elimination of a small volume of concentrated urine.

C. **Antidiuresis and diuresis** can be examined in terms of the volume of urine excreted per unit time ($\dot{V}$). Recall that $\dot{V}$ is the algebraic sum of osmolar clearance (C_{osm}) and either free-water clearance (C_{H_2O}) or free-water reabsorption ($T^c_{H_2O}$).

1. Antidiuresis (i.e., a decrease in $\dot{V}$) can result from:
  a. Water deprivation, which leads to an increase in plasma osmolality
  b. Reduced circulating blood volume

2. Diuresis (i.e., an increase in $\dot{V}$) can result from:
  a. Reduced osmotic reabsorption of water, leading to increased solute-free water clearance and **water diuresis**
  b. Reduced solute reabsorption (primarily Na^+ with associated anions), leading to increased osmolar clearance and **osmotic diuresis**

Chapter 28

Antidiuretic Hormone (ADH)

I. SYNTHESIS AND CHEMICAL CHARACTERISTICS

A. ADH, also known as **vasopressin,** * is a **hypothalamic hormone** synthesized in the supra-optic and paraventricular nuclei in the **hypothalamus** (ventral diencephalon). These un-myelinated neurosecretory neurons synthesize, store, and secrete ADH.

   1. The ADH-secreting neurons constitute the **supraopticohypophysial tract,** which termi-nates in the pars nervosa, or the posterior lobe of the pituitary gland (i.e., the neuro-hypophysis).

   a. ADH is stored (but not synthesized) in the pars nervosa.

   b. In the absence of a pars nervosa, the newly synthesized hormone still can be re-leased into the circulation from the hypothalamus.

   2. Because these **hypothalamoneurohypophysial neurons** produce hormones, they are known as **endocrine neurons** or **neuroendocrine cells**.

   a. Neurosecretory neurons are peptidergic neurons that conduct action potentials like all neurons, but, unlike ordinary neurons, synthesize and secrete peptide hor-mones.

   b. The hypothalamic nuclei that synthesize and secrete ADH are collectively called **magnocellular neurosecretory neurons**.

   c. ADH secretion is triggered by the depolarization of the supraopticohypophysial neu-rons, which causes Ca^{2+} influx, fusion of secretory granules with the cell mem-brane, and extrusion (exocytosis) of secretory products (ADH, oxytocin, and neuro-physin).

B. ADH is classified as a **polypeptide**. Specifically, it is an **octapeptide** or a **nonapeptide** with a molecular weight of approximately 1000.

   1. If the two cysteine residues are considered as a single cystine residue, ADH is an oc-tapeptide.

   2. ADH is a nonapeptide if the two cysteine residues are numbered individually.

C. The neurophysins are the physiologic **carrier proteins** for the intraneuronal transport of ADH and are released into the circulation with the neurosecretory products (ADH and oxytocin) without being bound to the hormone.

D. The biologic half-life of ADH is 16–20 minutes.

---

*The term vasopressin denotes an excitatory action on the blood vessels, causing vasoconstric-tion of the arterioles and an increase in systemic blood pressure. This effect is observed only when relatively large quantities of vasopressin are released from the posterior lobe (e.g., during hemorrhage) or when pharmacologic amounts are injected. Therefore, the vasopressor effect usually is not considered to be a physiologic effect. The biologically active form of ADH in hu-mans is **arginine vasopressin**.

TABLE 28-1. Factors that Influence Antidiuretic (ADH) Secretion in Humans

	Physiologic	Pathologic	Pharmacologic	Clinical
Stimuli	Hyperosmolality Upright posture (orthostatic hypotension) Exercise (epinephrine)	Decrease in effective circulating blood volume (hemorrhage, nephrotic syndrome, cirrhosis, congestive heart failure) Hypothalamic disease Pulmonary disorders (e.g., pneumonia, tuberculosis) CNS disorders (e.g., stroke, meningitis, subdural hematoma) Hypothyroidism $^+Na^+$ deficiency Vasovagal reactions (syncope) Pain Nausea Emotional stress Diabetes mellitus (glucose)	Lithium Morphine (high doses) Barbiturates Nicotine Acetylcholine Diuretics* Isoproterenol* Nitroprusside* Trimethaphan* Histamine* Bradykinin* Angiotensin II Insulin 2-deoxy-D-glucose Cholinergic drugs β-Adrenergic agonists	Positive-pressure breathing
Inhibitors	Hyposmolality Recumbent posture	Diabetes insipidus Elevation of blood pressure[‡] Hyposmolality SIADH	Norepinephrine[§] Ethanol Caffeine Anticholinergic drugs CO_2 inhalation Morphine (low doses) Phenytoin Lithium	Infusion of hypertonic or hypotonic solutions (hypovolemia) Negative-pressure breathing

CNS = central nervous system; SIADH = syndrome of inappropriate ADH secretion.
*Stimulates ADH secretion by lowering blood pressure
[†]Inhibits ADH secretion by contracting extracellular fluid (ECF) volume
[‡]ADH secretion is normal in patients with uncomplicated essential hypertension
[§]Inhibits ADH secretion by raising blood pressure

II. CONTROL OF ADH SECRETION: STIMULI AND INHIBITORS (Table 28-1)

A. **Stimuli.** The major stimuli for ADH secretion are hyperosmolality and effective circulating blood volume depletion (Figures 28-1, 28-2).

1. **Osmotic stimuli (osmoregulation)**
   a. **Hyperosmolality.** Under usual conditions, a 1%–2% increase in plasma osmolality is the prime determinant of ADH secretion, and the most common physiologic factor altering the osmolality of the blood is water depletion or water excess. (In clinical medicine, volume deficits are much more prevalent than volume excesses.)
      (1) The osmoreceptors are **located in the anterior hypothalamus** and are distinct from the cells that synthesize ADH.
         (a) The osmoreceptors have the lowest **threshold** (i.e., they are most sensitive) to changes in the **osmolal concentration of plasma**.

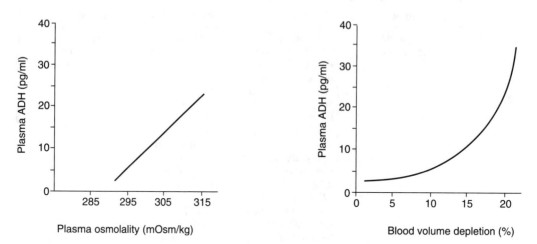

A. Osmoregulation

B. Baroregulation

FIGURE 28-1. Stimulation of antidiuretic hormone (*ADH*) secretion in response to (*A*) increases in plasma osmolality (*B*) decreases in blood volume. In general, the plasma [Na$^+$] is the primary determinant of ADH release. Note that the osmoreceptor is extraordinarily sensitive and, therefore, has the primary role in mediating the ADH response to changes in water balance. In states of hypovolemia, however, the baroreceptor stimulus becomes dominant over the chemoreceptor stimulus. The plasma [Na$^+$] remains the primary osmotic determinant of ADH release.

    (b) Stimulation of these osmoreceptors causes **reflex secretion of ADH**.
  (2) Not all solutes (e.g., urea) stimulate the osmoreceptors, despite increasing plasma osmolality. Only those solutes to which cells are relatively impermeable make up the effective osmotic pressure, in response to which ADH is secreted.
    (a) Na$^+$ and mannitol, which cross the blood–brain barrier relatively slowly, **are potent stimulators of ADH release**. Since the plasma [Na$^+$] accounts for 95% of the effective osmotic pressure, the osmoreceptors normally function as plasma Na$^+$ receptors.
    (b) Glucose. Hyperglycemia is a **less potent stimulus** for ADH production and secretion than hypernatremia for the same level of osmolality. In uncontrolled diabetes mellitus, hyperglycemia is associated with insulin deficiency. In this setting, glucose acts as an effective stimulus for ADH.
 b. Volume disturbances. The control of ADH secretion by osmolality can be overridden by volume disturbances. For example, marked hyponatremia will be tolerated in order to maintain circulating blood volume.

2. Nonosmotic stimuli
 a. Hypovolemia (i.e., decreased effective circulating blood volume) is a more dominant stimulus to ADH release than is hyperosmolality. A 10%–25% decrease in blood volume will evoke ADH release. A 10% decrease in blood volume is sufficient to cause the release of enough ADH to participate in the immediate regulation of blood pressure. Contraction of blood volume without an alteration in the tonicity of body fluids may cause ADH release.
  (1) Baroreceptors. Hemodynamic changes (i.e., changes in blood volume, pressure, or both) are mediated by autonomic (parasympathetic) afferents that arise in pressure (volume)-sensitive receptors in the atria, aortic arch, carotid sinus, great veins, and pulmonary vessels, and traject via the vagal and glossopharyngeal nerves to primary synapses in the nucleus tractus solitarius in the brain stem.
    (a) Changes in the rate of afferent discharge from these visceral afferent neurons affect the activity of the vasomotor center in the medulla. The **low-pressure (stretch) receptors** are the primary mediators of volume effects on ADH secretion.

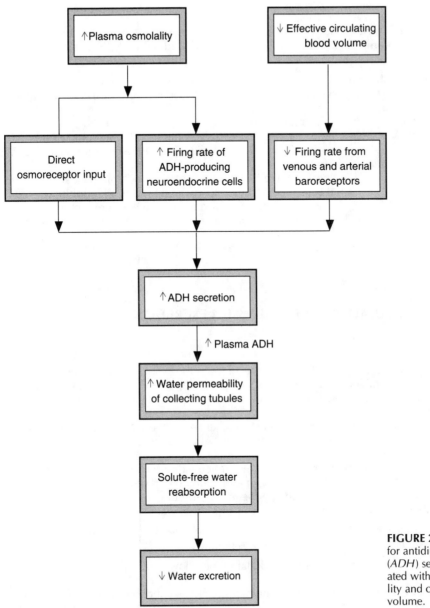

FIGURE 28-2. Pathways for antidiuretic hormone (*ADH*) secretion associated with plasma osmolality and circulating blood volume.

(b) From the synapses in the brain stem, postsynaptic fibers project to the region where the osmoreceptors are located (i.e., outside the blood–brain barrier in the circumventricular organs, primarily the organum vasculosum of the lamina terminalis).

(c) Elevation of blood pressure inhibits the ADH-producing cells. Conversely, decreased blood pressure reduces the firing frequency of the baroreceptors, causing stimulation of ADH synthesis and release. The vasopressin response to changes in blood volume is similar in response to blood pressure.

(2) Baroregulation of ADH secretion can be a dominant stimulus, but it is not the physiologic regulator of ADH secretion.

b. Other nonosmotic stimuli for ADH release are shown in Table 28-1.

B. **Inhibitors**

1. **Osmotic inhibitors.** Expansion of the intracellular volume of the osmoreceptors secondary to hyposmolality of the extracellular fluid (ECF) inhibits ADH secretion. **Water ingestion** contributes to hyposmolality of the ECF.

2. **Nonosmotic inhibitors**
   a. **Increased arterial pressure** secondary to vascular or ECF volume expansion inhibits ADH release.
      (1) Thus, ADH release is inhibited by increased tension in the left atrial wall, great veins, or great pulmonary veins secondary to increased intrathoracic blood volume due to hypervolemia, a reclining position, negative-pressure breathing, and water immersion up to the neck.
      (2) In the recumbent position, the increase in central blood volume leads to an increase in left atrial pressure and inhibition of ADH release. During sleep, the production of a concentrated urine is by and large due to a reduction of blood pressure, which offsets the effect of reduced ADH secretion due to a change in body position.
   b. **Other nonosmotic inhibitors of ADH release** are shown in Table 28-1.

III. ROLE IN REGULATION OF RENAL WATER EXCRETION

A. **Renal effects**

1. The **major site of action** of ADH is the receptor in the basolateral membrane of the principal cells of the cortical and medullary collecting ducts, where ADH increases the permeability to water. The **membrane shuttle hypothesis** may explain the action of ADH:
   a. The apical plasma membrane of the principal cell of the collecting duct has a low water permeability in the unstimulated (i.e., ADH-depleted) state, whereas the basolateral membrane is highly water-permeable.
   b. Intracellular, membrane-bound vesicles that contain proteinaceous water channels migrate and fuse with the apical plasma membrane in response to ADH.
   c. Subsequently, apical membrane permeability (and thereby transcellular water permeability) increases, leading to the reabsorption of water.

2. ADH, with its effect on increased water reabsorption, leads to the production of urine with a decreased volume and increased osmolality.

3. In addition, ADH decreases renal medullary blood flow.

B. **Extrarenal effects.** In addition, ADH stimulates the release of **adrenocorticotropic hormone (ACTH)** from the anterior lobe of the pituitary gland. ACTH plays a relatively small role in controlling aldosterone secretion (see Chapter 29 III A).

IV. ADH-RELATED DISTURBANCES

A. **Syndrome of inappropriate ADH secretion (SIADH).** SIADH leads to **water intoxication** (overhydration, or a dilution syndrome).

1. **Etiology.** SIADH is characterized by an excessive or inappropriate secretion of ADH from the posterior lobe or from an ectopic (nonhypothalamic) source, such as a malignant tumor (e.g., bronchogenic carcinoma). This syndrome is not caused by excessive water intake.

2. **Clinical characteristics.** Excessive secretion of ADH has the following effects when water is ingested:
   a. **Water retention** occurs, leading to expansion of the blood and ECF volumes.
   b. **Hypernatriuria** (i.e., increased urinary excretion of Na^+) and **hyponatremia** are present because aldosterone secretion is suppressed.
   c. **Edema.** Hyposmolality (i.e., decreased serum osmolality) results from increased water retention and urinary loss of Na^+. (It is the increased ADH secretion despite the presence of hyposmolality that is inappropriate.) The low plasma osmolality causes water to shift into the interstitial space (edema), and as the osmolality of that space decreases, there is a further shift of water into the ICF.
   d. **Urine osmolality is increased** because of the decreased urinary excretion of water and continued excretion of Na^+.
      (1) The U_{osm} exceeds the P_{osm}, and the urinary $[Na^+]$ exceeds 20 mEq/L.
      (2) **If water intake is restricted,** water retention does not occur and the excessive ADH has no effect on the plasma $[Na^+]$. Edema does not occur because the increase in volume suppresses aldosterone secretion and increases Na^+ excretion.

3. **Therapy.** SIADH is treated with demeclocycline, a tetracycline that blocks the effect of ADH on the kidney. Demeclocycline, lithium carbonate, amphotericin B, and methoxyflurane anesthesia may cause nephrogenic diabetes insipidus.

B. Diabetes insipidus

1. **Etiology.** Diabetes insipidus is characterized by a complete or partial **failure** of either **ADH secretion [central (neurogenic) diabetes insipidus]** or **the renal response to ADH (nephrogenic diabetes insipidus).**

2. **Clinical characteristics.** Regardless of the cause, diabetes insipidus is characterized by a decrease in renal water reabsorption by the collecting ducts.
   a. **Polyuria.** Decreased water reabsorption results in a diuresis of dilute urine (up to 3–20 L/day). In nephrogenic diabetes insipidus, the urine output is directly related to the volume of water delivered to the collecting ducts.
   b. **Polydipsia.** Because of the stimulation of thirst and increased water intake (polydipsia), most patients with diabetes insipidus maintain water balance with a near normal plasma $[Na^+]$.

3. **Diagnosis** of diabetes insipidus is confirmed in a polyuric patient by the demonstration of insignificant antidiuresis or by the production of hypertonic urine following water restriction of hypertonic saline infusions.
   a. A response to injected vasopressin documents the diabetes insipidus as neurogenic.
   b. Lack of a response to vasopressin is indicative of nephrogenic diabetes insipidus.

4. **Therapy**
   a. **Neurogenic diabetes insipidus**
      (1) **Hormonal therapy** involves administration of ADH as **vasopressin tannate** or **nasal lysine vasopressin.**
      (2) **Nonhormonal therapy** for neurogenic diabetes insipidus includes administration of **oral hypoglycemic agents** (e.g., chlorpropamide), **thiazide diuretics with sodium restriction, carbamazepine,** and **clofibrate**. A major side effect of chlorpropamide therapy is hypoglycemia, a common problem with hypopituitarism.
   b. **Nephrogenic diabetes insipidus. Thiazide diuretics** also are quite effective in treating nephrogenic diabetes insipidus, which does not respond to treatment with vasopressin injection or chlorpropamide. Thiazides are effective in treating diabetes insipidus because they inhibit Na^+ reabsorption in the diluting segment (i.e., the ALH). Therefore, U_{osm} does not fall below 300 mOsm/kg, and urine volume can be reduced by 50%. (Urine osmolality can be increased sixfold from a minimum of 50 mOsm/kg.)

Chapter 29

Aldosterone

I. SYNTHESIS, SECRETION, AND INACTIVATION

A. Synthesis

1. Aldosterone is a C-21 (21 carbon atoms) corticosteroid that is synthesized in the outermost area of the adrenal cortex, the **zona glomerulosa**. **Adrenocorticotropic hormone (ACTH), angiotensin II,** and **increased plasma [K$^+$]** stimulate the biosynthesis of aldosterone.

2. Aldosterone represents less than 0.5% of the corticosteroids, and as all of the corticosteroids, it is stored in very small quantities. However, aldosterone is the major **mineralocorticoid** in humans.

3. The **circulatory half-life** of aldosterone is about 30 minutes in humans during normal activity.

4. **Cholesterol (esterified)** is the precursor for steroidogenesis and is stored in the cytoplasmic lipid droplets in the adrenocortical cells. (Free plasma cholesterol appears to be the preferred source of cholesterol for corticosteroid synthesis.)

B. Secretion. Most of the secreted aldosterone is bound to **albumin,*** with a lesser amount bound to **corticosteroid-binding globulin (CBG; transcortin)**. CBG preferentially binds **cortisol,** which is a **glucocorticoid.**

C. Inactivation. Aldosterone is metabolized mainly in the liver, where greater than 90% of this corticosteroid is inactivated during a single passage.

1. Most aldosterone inactivation is by **saturation (reduction)** of the double bond in the A-ring.

2. The major metabolite is **tetrahydroaldosterone,** most of which is conjugated with glucuronic acid at the carbon-3 position of the A-ring. The resultant **glucuronides** are more polar and, therefore, more water-soluble, making them readily excreted by the kidney.

II. PHYSIOLOGIC EFFECTS. Aldosterone has the primary function of promoting Na$^+$ retention. In addition, it enhances K$^+$ and H$^+$ secretion (Figure 29-1).

A. Conservation of Na$^+$

1. Aldosterone stimulates Na$^+$ reabsorption in the connecting segment of the distal tubule and in the cortical and medullary collecting tubules.

2. Of the amount of filtered Na$^+$ (25,000 mEq/day), only 1%–2% (250–500 mEq or 6000–12,000 mg/day) are actively reabsorbed via an aldosterone-dependent mechanism in the distal nephron.†

---

*Protein-bound hormones are biologically inactive.
†Filtered load refers to the amount of Na$^+$ in the glomerular filtrate and not in the distal nephron.

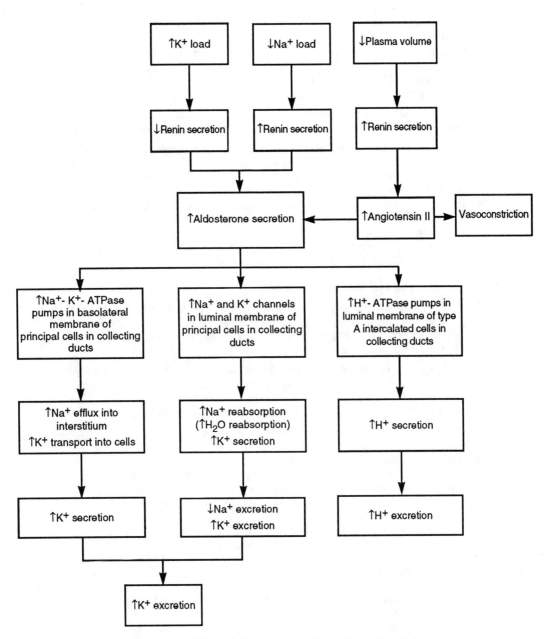

FIGURE 29-1. Effects of aldosterone on Na^+, K^+, and H^+ balance.

3. Aldosterone also promotes Na^+ reabsorption in the epithelial cells of the sweat glands, salivary glands, and the gastrointestinal mucosa (distal colon).

4. By restricting the renal excretion of Na^+, which is the main determinant of plasma osmolality, aldosterone regulates the extracellular fluid (ECF) volume. Therefore, **aldosterone regulates the total body Na^+ content,** while antidiuretic hormone (ADH) regulates the **plasma Na^+ concentration.**

B. Secretion and excretion of K^+

1. Aldosterone promotes K^+ secretion as a secondary effect of its action on Na^+ reabsorption.

    a. In the distal tubule, the linked Na^+ reabsorption–K^+ secretion is referred to as the **distal Na^+-K^+ exchange process**.

    b. Although K^+ appears to be secreted in exchange for Na^+, the distal secretion of K^+ is only indirectly related to Na^+ reabsorption. Distal Na^+ reabsorption is linked to the secretion of both K^+ and H^+.*

    c. Aldosterone increases the $[K^+]$ in sweat glands and saliva.

2. Stimulation of K^+ excretion greatly depends on dietary Na^+, as indicated by a lack of K^+ excretion after aldosterone administration in animals with Na^+-deficient diets.

3. More than 75% of the K^+ excreted in the urine is attributed to distal K^+ secretion.

    a. Excess aldosterone secretion causes a decline in the urinary Na^+/K^+ concentration ratio (from a normal value of about 2) because it decreases Na^+ excretion and increases K^+ excretion.

    b. Elevated aldosterone release increases the plasma Na^+/K^+ concentration ratio (from a normal value of 30) due to the increased excretion of K^+. Aldosterone can cause an isotonic expansion of the ECF volume with no change in the plasma $[Na^+]$ or a hypertonic expansion of the ECF volume with a rise in the plasma $[Na^+]$. In any case, the total amount of body Na^+ is increased.

C. **Water excretion and ECF volume regulation**

1. The effect of aldosterone on water excretion is not important physiologically, as aldosterone treatment fails to correct the impaired water excretion by patients with hypofunctional adrenal glands (due to Addison's disease or adrenal insufficiency) and by patients who have undergone adrenalectomy.

2. Aldosterone has no direct effect on the glomerular filtration rate (GFR), renal plasma flow, or renin production; however, by stimulating Na^+ reabsorption aldosterone causes water retention, and the resultant expansion of the ECF volume then leads to an increase in GFR and renal plasma flow and a decrease in renin production.

3. A high circulating aldosterone level is a common finding in edema and is due primarily to the increased aldosterone secretion induced by the depletion of the effective circulating blood volume.

    a. It is unlikely that the plasma $[Na^+]$ is a major regulator of aldosterone secretion because the plasma $[Na^+]$ during Na^+ depletion is normal in humans.

      (1) Hyponatremia often is a consequence of ADH secretion and is accompanied by an increase in the ECF volume, which tends to suppress rather than stimulate aldosterone secretion. When ADH and water are given to Na^+-depleted individuals, aldosterone secretion falls despite a fall in plasma $[Na^+]$.

      (2) By and large, hyponatremia in the clinical setting is due to **excess body Na^+** caused by a decreased effective blood volume. The low blood volume leads to an increase in ADH secretion and aldosterone secretion which, in turn, lead to edema. **Over 90% of the cases of hyponatremia should be treated by restriction of NaCl and water.**

    b. Therefore, it is the volume of the ECF rather than the plasma $[Na^+]$ that influences aldosterone secretion in most circumstances.

D. **Aldosterone and acid-base balance.** Aldosterone affects acid-base balance through its control of K^+ secretion. As stated above, aldosterone promotes increased distal tubular **secretion** and, therefore, **excretion** of K^+. Aldosterone also promotes the excretion of H^+ and NH_4^+.

1. **Hyperaldosteronemia** (i.e., increased aldosterone secretion) is one cause of K^+ depletion, a condition termed **hypokalemia**.

    a. Hypokalemia is characterized by an increase in intracellular $[H^+]$, which favors distal tubular secretion of H^+ over K^+ and results in a state of **metabolic alkalosis**.

---

*Na^+ transport in the proximal tubule is not associated with K^+ or H^+ exchange processes.

TABLE 29-1. Factors that Regulate Renin and Aldosterone Secretion

	Intrarenal				Extrarenal	
	-Volemia		Intraluminal [Na$^+$]		Plasma [K$^+$]	
	Hypo-	Hyper-	Depletion	Loading	Depletion	Loading
Renin	↑	↓	↑	↓	↑	↓
Aldosterone	↑	↓	↑	↓	↓	↑
	←Nonosmotic→ ←		Osmotic			→
	(Baroreceptor)		(Chemoreceptor)			

    b. Conversely, if metabolic alkalosis is the primary event, there is a decrease in intracellular [H$^+$]. This results in an increase in distal tubular intracellular [K$^+$], which favors increased urinary loss of K$^+$ over H$^+$ and results in a state of **hypokalemic metabolic alkalosis** (as occurs in hyperaldosteronemia).

  2. Hypoaldosteronemia (i.e., decreased aldosterone secretion) is one cause of excess K$^+$, a condition termed **hyperkalemia**.

    a. Hyperkalemia is characterized by an increase in intracellular [K$^+$], which favors distal tubular secretion of K$^+$ over H$^+$ and results in a state of **metabolic acidosis**.

    b. On the other hand, when metabolic acidosis is the primary event, there is an increase in intracellular [H$^+$]. This results in an increase in distal tubular intracellular [H$^+$], which favors increased secretion of H$^+$ over K$^+$ and results in a state of **hyperkalemic metabolic acidosis** (as occurs in hypoaldosteronemia).

  3. Hypokalemia causes the following conditions.

    a. Impaired renal concentrating ability leads to the formation of hyposmotic urine and polyuria that is resistant to ADH.

    b. Reduced carbohydrate tolerance leads to a decline in insulin secretion.* As a rule, fasting hyperglycemia is not present with reduced insulin secretion.

III. CONTROL OF ALDOSTERONE SECRETION. (Table 29-1). At least three well-defined mechanisms control aldosterone secretion: ACTH, plasma [K$^+$], and the renin-angiotensin system.

  A. Extrarenal control mechanisms. The following mechanisms cause the release of aldosterone by direct action on the adrenal cortex.

    1. Hypothalamic-hypophysial-adrenocortical axis. Under normal conditions, ACTH is not a major factor in the control of aldosterone synthesis or secretion. However, the pituitary gland plays an important role in the maintenance of the growth and biosynthetic capacity of the zona glomerulosa.

      a. ACTH, also known as **corticotropin,** is a 39-amino-acid polypeptide that supports steroidogenesis in the zona glomerulosa (Figure 29-2). ACTH enhances aldosterone production by stimulating the early biosynthetic pathway (i.e., the 20,22-desmolase enzyme complex that catalyzes the conversion of cholesterol to pregnenolone). ACTH also plays a minor role in mediating the diurnal rhythmic secretion of all of the corticosteroids.

      b. Corticotropin releasing hormone (CRH) is a hypothalamic **hypophysiotropic** polypeptide made up of 41 amino acid residues. CRH is secreted into the hypophysial

---

*The reduction in carbohydrate tolerance due to hypokalemia occurs in only about half of patients with elevated plasma aldosterone.

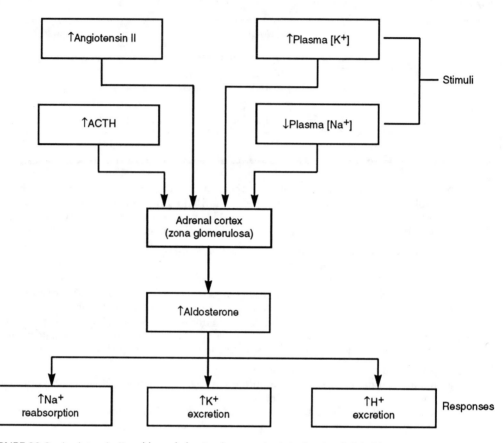

FIGURE 29-2. Angiotensin II and hyperkalemia, the two physiologic stimuli for aldosterone secretion, are more potent stimuli than adrenocorticotropic hormone (*ACTH*) and hyponatremia. Na^+ reabsorption and the secretion (and excretion) of K^+ and H^+ are the major responses to aldosterone secretion. The principal and intercalated cells of the collecting ducts are the primary target cells of aldosterone—Na^+ reabsorption and K^+ excretion take place at the principal cells, and H^+ excretion occurs in the type A intercalated cells. ACTH and hyperkalemia are extrarenal stimuli; hyponatremia and angiotensin II are intrarenal stimuli.

portal system and causes release of ACTH from the pituitary gland. Because CRH secretion is regulated by higher brain centers (e.g., the limbic system), these centers play a role in Na^+ balance.

(1) Hypophysectomized patients and those with pituitary insufficiency exhibit normal aldosterone secretion on a moderate salt intake; however, these individuals demonstrate a suboptimal aldosterone response to Na^+ restriction.

(2) Normal individuals injected chronically with ACTH show an acute rise in aldosterone secretion followed by a return to control level or below in 3–4 days, despite ACTH administration over a period of 7–8 days.

(3) When aldosterone is administered for several days to normal individuals, the kidney "escapes" from the Na^+-retaining effect but not from the K^+-excreting effect. The escape phenomenon prevents the appearance of edema in individuals treated with aldosterone for prolonged periods and in patients with primary aldosteronism.

2. **Hyperkalemia.** A 1% increase in plasma $[K^+]$ (< 0.1 mEq/L) can stimulate the synthesis and release of aldosterone by a direct action on the zona glomerulosa (see Figure 29-2). In the anephric human, K^+ appears to be the major regulator of aldosterone even though aldosterone levels are low.

a. This release probably occurs by the depolarization of the glomerulosa cell membrane by the elevated plasma $[K^+]$.

b. Stimulation of aldosterone secretion by K^+ loading is limited by the simultaneous reduction in renin release.

c. K^+ stimulates an early step in the biosynthetic pathway for aldosterone synthesis.

d. K^+ loading increases the width of the zona glomerulosa layer in experimental animals; prolonged aldosterone administration results in atrophy of this layer due to depressed renin secretion.

3. Hyponatremia. A 10% decrease in plasma $[Na^+]$ also appears to stimulate the synthesis and release of aldosterone directly at the level of the zona glomerulosa (see Figure 29-2). However, this effect usually is overridden by changes in the effective circulating volume. Thus, aldosterone secretion is increased in the hyponatremic patient who is volume-depleted but is reduced in the hyponatremic patient who is volume-repleted.

B. **Intrarenal control mechanism.** Aldosterone secretion also is regulated by the renin-angiotensin system, the major component of which is the **juxtaglomerular apparatus**. The renin-angiotensin-aldosterone system is regulated by the **sympathetic nervous system**.

1. Anatomy of the juxtaglomerular apparatus (Figure 29-3). The juxtaglomerular apparatus is a combination of specialized tubular and vascular cells located at the vascular pole where the afferent and efferent arterioles enter and leave the glomerulus. The juxtaglomerular apparatus is composed of **three cell types**.

a. Juxtaglomerular cells are specialized **myoepithelial** (modified vascular smooth muscle) cells located in the **media** of the afferent arteriole, which synthesize, store, and release a proteolytic enzyme called **renin**. Renin is stored in the granules of the juxtaglomerular cells.

(1) The juxtaglomerular cells are **baroreceptors** (tension receptors) and respond to changes in the transmural pressure gradient between the afferent arteriole and the interstitium. They are innervated by sympathetic nerve fibers.

(2) These vascular "volume" receptors monitor renal perfusion pressure and are stimulated by hypovolemia, or decreased renal perfusion pressure.

b. Macula densa cells are specialized renal tubular epithelial cells located at the transition between the thick segment of the ascending limb of the loop of Henle (ALH) and the distal convoluted tubule (see Figure 29-3).

(1) These cells are in direct contact with the mesangial cells, in close contact with the juxtaglomerular cells, and contiguous with both the afferent and efferent arterioles as the tubule passes between the arterioles supplying its glomerulus of origin.

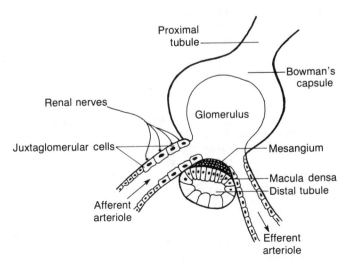

FIGURE 29-3. The anatomic components of the juxtaglomerular apparatus. (Reprinted from Yates FE, et al: The adrenal cortex. In *Medical Physiology*, 14th edition. Edited by Mountcastle VB. CV Mosby, 1980, p 1590.)

    (2) The macula densa cells are characterized by prominent nuclei in those cells on the side of the tubule that is in contact with the mesangial and vascular elements of the juxtaglomerular apparatus.

    (3) The macula densa cells function as **chemoreceptors** and are stimulated by a decreased Na$^+$ (NaCl) load. This inverse relationship between Na$^+$ load and renin release provides a reasonable explanation for the clinical problems involving a decreased filtered load of Na$^+$ and Cl$^-$ in association with increased renin release. Macula densa cells are not innervated.

  c. **Mesangial cells** also are referred to as the **polkissen** (asymmetrical cap) and are the interstitial cells of the juxtaglomerular apparatus. Mesangial cells are in contact with both the juxtaglomerular cells and the macula densa cells. A decreased intraluminal Na$^+$ load, Cl$^-$ load, or both in the region of the macula densa stimulates the juxtaglomerular cells.

2. Role of the sympathetic nervous system. The sympathetic nervous system plays an important role in the control of renin release via the **renal nerves**.

  a. The juxtaglomerular cells of the afferent arterioles are innervated directly by the sympathetic postganglionic fibers (unmyelinated). In the absence of renal nerves, the renal response to Na$^+$ depletion is attenuated.

  b. Circulating catecholamines (i.e., epinephrine and norepinephrine) and stimulation of the renal nerves produce vasoconstriction of the afferent arterioles, which causes renin release by a decrease in perfusion pressure.

    (1) This renin response caused by catecholamines and renal nerve stimulation is mediated via the β_1-adrenergic receptor and can be elicited by the synthetic sympathomimetic amine, **isoproterenol,** which is a β-agonist.

    (2) Renal denervation and β_1-adrenergic receptor blockade by **propranolol** inhibit the release of renin.

  c. Renal innervation is not a requisite for renin release because the denervated kidney can adapt to a variable salt intake.

  d. In humans, exercise for assuming an upright posture increases renal sympathetic activity, which produces renal arteriolar vasoconstriction and an increase in renin release. Thus, the sympathetic nervous system, by modulating the secretion of renin, has an **indirect effect on aldosterone secretion**.

3. Role of renin: stimuli for release

  a. Renin has a circulatory half-life of 40–120 minutes in humans. The common denominator for renin release by the intrarenal mechanism is a decrease in the effective circulating blood volume, which is induced by:

    (1) Acute hypovolemia associated with hemorrhage, diuretic administration, or salt depletion

    (2) Acute hypotension associated with ganglionic blockade or a change in posture (postural hypotension)

    (3) Chronic disorders associated with edema (e.g., cirrhosis with ascites, congestive heart failure, nephrotic syndrome)

  b. Renin release is increased by K$^+$ depletion, epinephrine, norepinephrine, isoproterenol, and standing.

4. Role of renin: inhibition of renin secretion

  a. Renin release is inhibited by angiotensin II, angiotensin III, ADH, hypernatremia, hyperkalemia, and atrial natriuretic peptide (ANP).

  b. **K$^+$ loading** leads to the inhibition of renin release and to the direct stimulation of the glomerulosa cells to secrete aldosterone. In contrast to its effect on normal individuals, aldosterone administration to patients with heart failure, cirrhosis with ascites, or nephrosis causes Na$^+$ retention without K$^+$ excretion, because of greater proximal reabsorption of Na$^+$ with less Na$^+$ available for the distal exchange with K$^+$.

5. Angiotensin synthesis

  a. Renin is secreted into the bloodstream, where it combines with the renin substrate, angiotensinogen, which is an α_2-globulin synthesized in the liver.

    (1) Renin is not saturated with its substrate in normal plasma; the same amount of renin generates more angiotensin I if the substrate concentration is increased above normal.

    (2) Oral contraceptives increase plasma angiotensinogen concentration and decrease plasma renin concentration.

b. The only physiologic effect of renin is to convert **angiotensinogen** to the biologically inactive decapeptide, angiotensin I.

c. Angiotensin I is converted primarily in the lung by pulmonary endothelial cells to the physiologically active octapeptide, angiotensin II (Figure 29-4).

    (1) The enzyme that forms angiotensin II is a peptidase (dipeptidyl carboxypeptidase) called angiotensin converting enzyme (ACE). It is found chiefly in pulmonary tissue and to a lesser degree in renal tissue and blood plasma.

    (2) The converting enzyme is identical to kininase II, which converts the nonapeptide vasodilator, bradykinin (kallidin-9), to inactive peptides, thereby diminishing the circulating levels of a vasodepressor substance and enhancing the vasoconstrictive action of angiotensin II.

d. Angiotensin II, with a circulatory half-life of 1–3 minutes, has several important physiologic actions.

    (1) It functions as the tropic hormone for the zona glomerulosa and stimulates the secretion (and synthesis) of aldosterone (see Figure 29-2). Angiotensin II is the **aldosterone-stimulating hormone**.

    (2) It is a potent **local vasoconstrictor (vasopressor)** of the renal arterioles at low plasma concentrations; at higher concentrations, angiotensin II exerts a general vasopressor effect on the smooth muscle cells of arterioles throughout the cardiovascular system, leading to an elevation of systemic mean arterial blood pressure.

        (a) Angiotensin II stimulates the secretion of ADH and ACTH.

        (b) Angiotensin II stimulates thirst, which leads to increased fluid consumption. It also stimulates the release of epinephrine and norepinephrine from the adrenal medulla.

    (3) Angiotensin II is inactivated by two angiotensinases (peptidases), and it can be converted to angiotensin III, which is a heptapeptide and a very potent stimulator of aldosterone secretion. Angiotensin III is not an effective vasoconstrictor in contrast to angiotensin II.

    (4) Renin, converting enzyme, angiotensinogen, and angiotensin II have been found in brain tissue.

6. Plasma renin activity and plasma renin concentration. Plasma renin activity is defined as the rate of angiotensin I formation when plasma renin acts on **endogenous substrate**.

FIGURE 29-4. The renin-angiotensin system and its relationship to the kinin system.

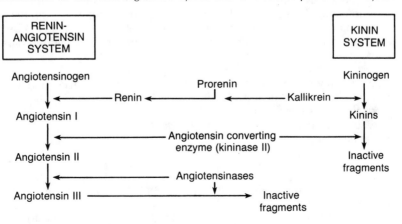

Plasma renin concentration is measured when **exogenous substrate** is added to plasma to saturate the enzyme and increase the velocity of angiotensin I formation to a maximal rate.
  a. Oral contraceptive administration is known to increase plasma renin activity and aldosterone secretion via a marked increase in renin substrate concentration, whether the woman is normotensive or hypertensive prior to the therapy.
  b. As a result of the increased angiotensin II formation, plasma renin concentration is suppressed. Thus, the increase in plasma renin activity in this situation occurs without an increase in renin concentration.
  c. Oral contraceptives are a cause of hypertension in women through this mechanism.

IV. **ALDOSTERONISM** refers to a condition of excessive aldosterone secretion.

A. **Primary hypoaldosteronism** (Conn's syndrome)

  1. **Etiology.** Primary hyperaldosteronism results from a tumor of the **zona glomerulosa**.

  2. **Characteristics** of primary hypoaldosteronism include:
     (1) Elevated plasma (and urinary) aldosterone
     (2) Hypertension due to Na^+ and water retention
     (3) Hypokalemic alkalosis with a K^+ excretion rate of greater than 40 mEq/day*
     (4) Decreased levels of angiotensin and renin
     (5) Decreased hematocrit due to the expansion of the plasma volume
     (6) Polyuria and dilute urine due to secondary nephrogenic diabetes insipidus
     (7) Absence of peripheral edema
     (8) Decreased plasma colloidal osmotic (oncotic) pressure due to ECF expansion

B. **Secondary hyperaldosteronism**

  1. **Etiology.** Secondary hyperaldosteronism is caused by the following extra-adrenal factors:
     a. Dietary therapy, which is the most common cause
     b. Extravascular loss of Na^+ and water, which is associated with edema (due to such underlying factors as nephrosis, cirrhosis, and congestive heart failure), an increase in the total ECF volume, and a loss of effective blood volume
     c. Hyperreninism caused by a tumor of the juxtaglomerular cells
     d. Renovascular disease (e.g., renal artery stenosis)

  2. **Characteristics.** Secondary hypoaldosteronism is characterized by edema and Na^+ retention. Urinary K^+ excretion is not increased because there is a reduced flow of fluid into and through the distal segments of the nephron. This low fluid flow reduces K^+ secretion and offsets the stimulating effect of aldosterone. Also, with decreased Na^+ and water delivery to the distal tubule, the quantity of K^+ (and H^+) secreted in the urine is limited. Additional characteristics of secondary hyperaldosteronism include:
     a. Increased plasma (and urinary) aldosterone
     b. Hypertension with edema (due to Na^+ retention and water accumulation in the interstitial fluid compartment) and a decrease in plasma volume
     c. Hypokalemic alkalosis
     d. Increased angiotensin and plasma renin activity†
     e. Peripheral edema

---

*Hypokalemia is not always concomitant with hypermineralocorticoidism (e.g., in 11β-hydroxylase deficiency).
†The elevated renin level is the characteristic that differentiates secondary aldosteronism from the primary form.

3. **Chronic licorice ingestion** in excessive amounts can mimic aldosteronism, because licorice contains the salt-retaining substance, glycyrrhizinic acid. Patients with this condition present with:
   a. Hypertension and hypokalemic alkalosis
   b. Suppressed plasma renin levels
   c. Reduced aldosterone secretion due to chronic volume expansion

V. ALDOSTERONE ANTAGONISM: SPIRONOLACTONE

A. **Renal effects.** Spironolactone is a steroidal aldosterone antagonist, which competitively blocks the Na^+-retaining and K^+-excreting effects that aldosterone exerts on the distal renal tubule. As a result, spironolactone leads to an increase in urinary Na^+ excretion and a decrease in K^+ excretion. This antagonist is efficacious only in the presence of aldosterone or another mineralocorticoid; spironolactone is without effect in adrenalectomized individuals.

B. **Blood pressure effect.** Because the drug enhances Na^+ diuresis, spironolactone is effective in potentiating the action of many antihypertensive drugs whose dosage should be reduced in its presence.

C. **Clinical application.** Spironolactone is useful in the differential diagnosis of primary and secondary hyperaldosteronism.

1. If both plasma $[K^+]$ and blood pressure are returned to normal with spironolactone, primary hyperaldosteronism is suspected.

2. If spironolactone causes the plasma $[K^+]$ to return to normal without the antihypertensive effect, secondary hyperaldosteronism is suspected.

Chapter 30

Atrial Natriuretic Peptide (ANP)

I. **INTRODUCTION.** Atrial natriuretic peptide (ANP) refers to a group of polypeptides produced by the atrial muscle cells. ANP exerts a hormonal influence on the kidney, which results in changes in intrarenal hemodynamics through its effects on fluid volume, electrolyte (Na$^+$) balance, and blood pressure homeostasis.

II. **SYNTHESIS**

A. In mammals, ANP is synthesized, stored, and released from **atrial cardiocytes**. The secretory activity of these cells is evidenced by the presence of membrane-bound storage granules with electron-dense cores.

B. The peptides that comprise the ANP group have molecular weights ranging from about 2500 to 13,000 daltons and lengths ranging from 21 to 73 amino acid residues. All are derived from a common 126-amino-acid precursor called **pro-ANP (atriopeptigen),** which is the predominant form of ANP in the atrium.

C. In humans, ANP is synthesized in a **prepro** (i.e., **pre-atriopeptigen**) form containing 151 amino acid residues.

D. The predominant circulating form of ANP is the 28-amino-acid peptide.

III. **SECRETION.** Stimuli for ANP release include atrial distention (hypervolemia), epinephrine, arginine vasopressin antidiuretic hormone (ADH), and acetylcholine (ACh).

IV. **PHYSIOLOGIC EFFECTS OF ANP** (Figure 30-1)

A. **Renal and adrenal responses**

1. **An increased glomerular filtration rate (GFR)** is associated with constriction of the efferent arteriole, which increases the glomerular hydrostatic pressure. It is important to note that ANP induces relaxation (dilation) of precontracted renal arteries.

2. **Dilation of the afferent arteriole** increases the hydrostatic pressure in the glomerular capillary.

3. **Natriuresis** (i.e., increased urinary Na$^+$ excretion) occurs primarily as a result of the increase in GFR, the increase in renal medullary or papillary blood flow, or both.

4. **Inhibition of aldosterone secretion.** ANP blocks aldosterone secretion that was prestimulated by Na$^+$ depletion, angiotensin II, adrenocorticotropic hormone (ACTH), K$^+$, and cyclic adenosine 3′,5′-monophosphate (cAMP).

5. **Inhibition of renin secretion** occurs via the increase in NaCl delivery to the macula densa or the increase in hydrostatic pressure at the juxtaglomerular apparatus due to afferent arteriolar vasodilation.

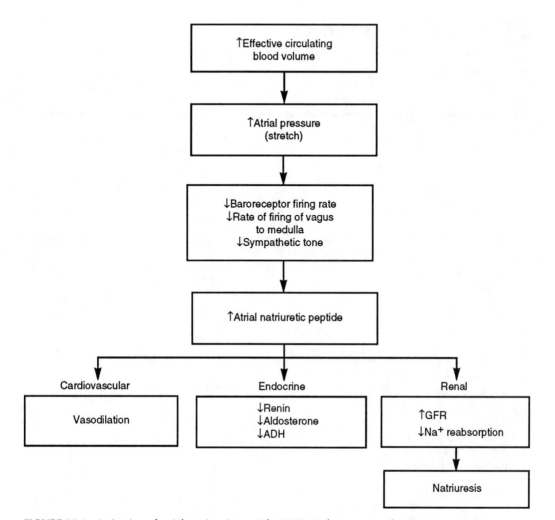

FIGURE 30-1. Activation of atrial natriuretic peptide (ANP) via low-pressure baroreceptors in the atria (especially the left atrium) leads to cardiovascular, endocrine, and renal effects. *ADH* = antidiuretic hormone; *GFR* = glomerular filtration rate.

B. **Cardiovascular effects**

    1. Decreases in mean systemic arterial blood pressure occur due to vasorelaxation or suppression of renin secretion.

    2. Reduction in cardiac output occurs primarily due to bradycardia.

STUDY QUESTIONS

DIRECTIONS: Each of the numbered items or incomplete statements in this section is followed by answers or by completions of the statement. Select the ONE lettered answer or completion that is BEST in each case.

1. The following renal function data were obtained for substance x:

Urine flow rate = 90 ml/hr
Urine concentration of substance x (U_x)
   = 480 mg/ml
Plasma concentration of substance x (P_x)
   = 6 mg/ml

What is the clearance of substance x?

(A) 12 ml/min
(B) 100 ml/min
(C) 120 ml/min
(D) 240 ml/min
(E) 480 ml/min

2. The renal transport maximum (Tm) for a substance is defined as the maximal

(A) glomerular filtration rate (GFR)
(B) urinary excretion rate
(C) tubular reabsorption or secretion rate
(D) renal clearance rate
(E) amount of a substance filtered by the glomeruli per minute

3. Of the renal systems available for the excretion of H^+, the one with the greatest activity is

(A) H^+ secretion
(B) monobasic phosphate (NaH_2PO_4) excretion
(C) sulfate (SO_4^{2-}) excretion
(D) titratable acid excretion
(E) NH_4^+ excretion

4. When the secretion of para-aminohippuric acid (PAH) reaches the renal tubular transport maximum (Tm), a further increase in plasma PAH concentration causes its clearance to

(A) increase in proportion to its plasma concentration
(B) approach glucose clearance asymptotically
(C) remain constant
(D) approach inulin clearance asymptotically
(E) increase in proportion to the glomerular filtration rate (GFR)

5. Which one of the following statements regarding Na^+ transport is correct?

(A) Active transport of Na^+ across all cells consumes most of the energy derived from cellular metabolism
(B) The Na^+ concentration is highest in the intracellular fluid (ICF)
(C) Na^+ reabsorption across proximal tubular cells is mainly active and transcellular
(D) The Na^+ concentration gradient provides energy for the cotransport of H^+
(E) The transport of Na^+ across the apical membrane of the nephron is an active transport process

6. The renal threshold for a solute denotes the

(A) maximum filtration rate
(B) maximum reabsorption rate
(C) plasma concentration at which a solute begins to appear in the urine
(D) maximum secretion rate
(E) maximum tubular secretory capacity (Ts)

7. Which one of the following conditions causes a decrease in the extracellular fluid (ECF) volume, an increase in the intracellular fluid (ICF) volume, and a decrease in the osmolar concentration of both compartments?

(A) Hyperosmotic dehydration
(B) Hyposmotic dehydration
(C) Isosmotic dehydration
(D) Hyperosmotic overhydration
(E) Hyposmotic overhydration

8. Which of the following substances has the lowest renal clearance?

(A) Glucose
(B) Urea
(C) Inulin
(D) Creatinine
(E) Para-aminohippuric acid (PAH)

9. Stimulation of renin secretion will increase the

(A) K^+ concentration in the blood
(B) volume of the extracellular fluid (ECF)
(C) hematocrit
(D) plasma colloid oncotic pressure
(E) H^+ concentration in the blood

10. Given a glomerular filtration rate (GFR) of 125 ml/min, a plasma glucose concentration of 400 mg/100 ml, a urine glucose concentration of 75 mg/ml, and a urine flow rate of 2 ml/min, what is the renal tubular transport maximum (Tm) for glucose?

(A) 300 mg/min
(B) 350 mg/min
(C) 400 mg/min
(D) 500 mg/min
(E) 550 mg/min

11. Which one of the following factors best explains an increase in the glomerular filtration rate (GFR)?

(A) Increased arterial plasma oncotic pressure
(B) Increased hydrostatic pressure in Bowman's capsule
(C) Increased glomerular capillary hydrostatic pressure
(D) Decreased net filtration pressure
(E) Vasoconstriction of the afferent arteriole

12. Measurements taken after an intravenous injection of inulin indicate that the substance appears to be distributed throughout 30%–35% of the total body water (TBW). This finding suggests that inulin most likely is

(A) excluded from the cells
(B) distributed uniformly throughout the TBW volume
(C) restricted to the plasma volume
(D) neither excreted nor metabolized by the body
(E) not freely diffusible through capillary membranes

DIRECTIONS: The numbered item or incomplete statement in this section is negatively phrased, as indicated by a capitalized word such as NOT, LEAST, or EXCEPT. Select the ONE lettered answer or completion that is BEST in each case.

13. All of the following statements concerning renin are true EXCEPT

(A) renin release from the juxtaglomerular cells is inversely related to the degree of stretch in the wall of the afferent arteriole
(B) renin is a secretory product of the juxtaglomerular cells
(C) renin substrate is a hepatic globulin
(D) the renin response to Na^+ depletion is attenuated in the absence of renal nerves
(E) renin secretion is a prerequisite for aldosterone secretion

DIRECTIONS: Each set of matching questions in this section consists of a list of four to twenty-six lettered options (some of which may be in figures) followed by several numbered items. For each numbered item, select the ONE lettered option that is most closely associated with it. To avoid spending too much time on matching sets with large numbers of options, it is generally advisable to begin each set by reading the list of options. Then, for each item in the set, try to generate the correct answer and locate it in the option list, rather than evaluating each option individually. Each lettered option may be selected once, more than once, or not at all.

Questions 14–18

Match each of the following descriptions with the most appropriate lettered region of the nephron pictured below.

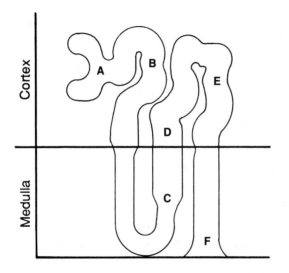

14. Tubular fluid always is hyposmotic at this site

15. The tubular fluid-to-plasma (TF/P) ratio for glucose is 1.0 at this site

16. The urea concentration is highest at this site

17. The urine osmolality (U_{osm}) can reach 1200–1400 mOsm/L at this site

18. The macula densa is closest to this site

Questions 19–21

Match each of the following substances with the appropriate lettered tubular fluid-to-plasma (TF/P) curve on the graph below.

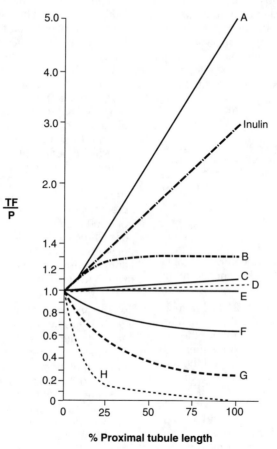

19. Inorganic phosphate (P_i)

20. Glucose

21. Glycine

ANSWERS AND EXPLANATIONS

1. The answer is C [Chapter 24 I A 1]. The renal clearance of a substance (C_x) is defined as the ratio of the renal excretion rate of the substance to its concentration in the blood plasma. Thus, it is an empiric measure of the volume of plasma that contains the same amount of the substance as is excreted in urine in 1 minute. The clearance equation is:

$$C_x = \frac{U_x \cdot \dot{V}}{P_x}, \text{where}$$

U_x and P_x are the urinary and plasma concentration of substance x in mg/ml, respectively; $\dot{V}$ is the urine flow rate in ml/min; and C_x is the clearance of substance x in ml/min. Using the data given:

$$C_x = \frac{480 \text{ mg/ml} \cdot 1.5 \text{ ml/min}}{6 \text{ mg/ml}} = 120 \text{ ml/min}$$

Notice that the units of concentration cancel out, leaving the units for clearance as ml/min.

2. The answer is C [Chapter 26 I A]. Renal tubular transport maximum, abbreviated as Tm, is defined as the upper limit for the unidirectional rate of active transport, either reabsorptive or secretory, depending on the direction of solute transport. The units for transport maximum are mg/min. Thus, there is a maximal rate of reabsorption called maximum tubular reabsorptive capacity (Tr) and a maximal rate of secretion called maximum tubular secretory capacity (Ts). Tubular maxima vary with the substance involved. Renal reabsorption of glucose and secretion of para-amino-hippuric acid (PAH) are examples of actively transported solutes exhibiting tubular transport maxima. An example of a substance that is not transport maximum–limited is the reabsorption of Na^+ along the nephron. Tubular maximum is the difference between the filtered load and the rate of excretion of a solute. The glomerular filtration rate (GFR) is defined as the volume of plasma filtered per minute (ml/min), whereas clearance is the virtual volume of plasma from which a substance is removed per minute (ml/min).

3. The answer is E [Chapter 26 III D 4]. The concentration of free H^+ at a urinary pH above 4.4 is negligible; therefore, acid must be excreted in a buffered form. The two main urinary buffers are dibasic phosphate (HPO_4^{2-}) and ammonia (NH_3). The amounts of secreted H^+ excreted bound to ammonia and phosphate are measured as ammonium (NH_4^+) and titratable acid, respectively. The sum of ammonium and titratable acid minus the amount of excreted bicarbonate (HCO_3^-) equals net acid excretion, which normally is approximately 1 mEq/kg body weight/day. Ammonia (ammonium) production occurs primarily in the proximal tubule, and it represents the major adaptive mechanism available to the kidney for the increased excretion of H^+ during states of acidosis. In contrast, the phosphate buffer enters the tubular lumen by glomerular filtration. The normal kidney excretes titratable acid (20 mEq H^+/day). Approximately 4300 mEq of H^+ must be excreted per day to accomplish the reabsorption of 4300 mEq of HCO_3^-. In this process, most of this secreted H^+ is reabsorbed in the form of water. Sulfate (SO_4^{2-}) in combination with ammonium forms a neutral salt [$(NH_4)_2SO_4$], which makes the tubular urine less acidic.

4. The answer is D [Chapter 26 II C; Figure 26-1C]. At low plasma concentrations of para-aminohippuric acid (PAH), PAH is almost completely cleared from the plasma by a combination of glomerular filtration and tubular secretion. When plasma concentrations of PAH are elevated beyond 30 mg/dl, the secretory mechanism becomes saturated, and the tubular transport maximum (Tm) is reached. As the tubular secretory mechanism becomes saturated and is exceeded by progressive increases in plasma PAH concentration, the clearance of PAH declines and becomes more a function of glomerular filtration. Because inulin clearance is essentially equal to the glomerular filtration rate (GFR), the PAH clearance asymptotically approaches the inulin clearance. Thus, the amount of PAH secreted becomes a smaller fraction of the total amount of PAH excreted. The clearance of PAH is always greater than the clearance of inulin, because some PAH is always secreted.

5. The answer is C [Chapter 23 III C; Figure 23-2]. Most of the proximal reabsorption of Na^+ occurs by active transport and is transcellular. The transcellular pathway consists of the apical and basolateral membranes. The Na^+-K^+-ATPase pump in all cell membranes is responsible for active Na^+ transport (efflux) and active K^+ transport (influx). This enzyme maintains the low intracellular Na^+ concentration and the high extracellular Na^+ concentration. Active Na^+ transport consumes 30%–50% of the energy derived from metabolism in most cells. Although the influx of Na^+ from the tubular lumen to the proximal tubular cell is in the direction favored by the electrochemical potential, this transport is mediated by specific membrane carrier proteins and not by simple diffusion. These membrane proteins couple the active movement of other solutes to the passive movement of Na^+. Examples include Na^+-glucose and Na^+-amino acid symporters and a Na^+-H^+ antiporter. In each case, the potential energy released by the downhill transport of Na^+ is used to power the uphill transport of the other substance. These transport systems are referred to as Na^+-coupled, secondary active transport processes. It is essential to understand that reabsorption includes not only Na^+ influx from the tubular lumen but also active transport of Na^+ out of the cell into the bloodstream.

6. The answer is C [Chapter 26 I A 1 c, B 1 a]. The renal threshold for a substance denotes the plasma concentration at which the solute begins to appear in the urine. It is not the plasma concentration that completely saturates the transport mechanism either for reabsorption or secretion. When the plasma concentration of a solute exceeds the renal threshold, the amount of that solute transported through the nephron exceeds the tubular transport maximum (Tm), and the solute appears in the urine in increasing amounts. The filtered load is the amount of a substance entering the tubule by filtration per unit time.

7. The answer is B [Chapter 22 III B 3 c; Figure 22-1]. Hydration states are named in terms of the extracellular fluid (ECF) compartment. Overhydration, or fluid and salt retention, results from excessive influx of water and NaCl. Dehydration, or fluid and salt depletion, usually involves both the ECF and the intracellular fluid (ICF). The volume of the ECF compartment is determined by the Na^+ content of the body, not by the Na^+ concentration of the plasma. The situation described in the question represents a state of dehydration because of the contraction of the ECF volume. Because the solute concentration of the ECF also is decreased, there is hyposmotic dehydration (e.g., as occurs with excessive salt loss in primary adrenal insufficiency). A net loss of salt in excess of water loss leads to hyposmolality

of the ECF and causes water to shift from the ECF to the ICF. Thus, the volume of the ECF is decreased, the volume of the ICF is increased, and the osmolality of both is decreased.

8. The answer is A [Chapter 24 II A 1 b (2), 2 a (2) (b)]. Renal clearance can be measured for any solute present in the plasma and excreted by the kidney; clearance best represents the rate of elimination of a substance from the plasma by the kidney. If the kidney removes a substance (e.g., inulin) from the plasma by filtration only, the tubular clearance will be zero and the excretion rate will be proportional to the plasma concentration. If the clearance involves filtration and either tubular reabsorption or secretion, then clearance depends on the plasma concentration and the tubular transport capacity for reabsorption or secretion. Thus, clearance is higher for substances that are filtered and secreted than for substances that are filtered and reabsorbed. The lowest clearance is observed with substances that are completely reabsorbed, such as glucose. Approximately 50% of the filtered urea is reabsorbed. Creatinine and para-aminohippuric acid (PAH) are filtered and secreted, which increases their clearance compared with inulin.

9. The answer is B [Chapter 29 III B 5; Figure 29-4]. The only known physiologic effect of renin is to cause the formation of angiotensin I from its plasma substrate, angiotensinogen. Angiotensin I, in turn, appears to serve only a specific substrate for angiotensin converting enzyme (ACE), a peptidase that converts angiotensin I to angiotensin II, which does have significant biologic activity. Angiotensin II exerts a potent vasoconstrictive action on the vascular smooth muscle of peripheral arterioles, which causes an increase in the mean arterial blood pressure. It also stimulates the release of aldosterone from the zona glomerulosa of the adrenal cortex, causing an increase in the extracellular fluid (ECF) volume through the increased active reabsorption of Na^+ by the collecting ducts. The increased ECF volume accounts for the decreased concentration of K^+, the decreased hematocrit, and the decreased plasma colloid oncotic pressure. Aldosterone also acts on the distal tubule and collecting duct to increase the net reabsorption of Na^+ in exchange for the secretion of K^+ and H^+. Thus, aldosterone tends to produce hypokalemia and metabolic alkalosis.

10. The answer is B [Chapter 26 I B 1–2]. The algebraic difference between filtered load and the amount excreted determines which of the two processes is used by the kidney to excrete a substance. In the case of glucose, the amount excreted is less than the amount filtered, which is consistent with reabsorption. Using the data given, the transport maximum (Tm) for glucose is found to be 350 mg/min:

$$
\begin{aligned}
Tm &= \text{amount filtered} - \text{amount excreted} \\
&= C_{in} \cdot P_G - U_G \cdot \dot{V} \\
&= (125 \text{ ml/min} \cdot 4 \text{ mg/ml}) - \\
&\quad (75 \text{ mg/ml} \cdot 2 \text{ ml/min}) \\
&= 500 \text{ mg/min} - 150 \text{ mg/min} \\
&= 350 \text{ mg/min, where}
\end{aligned}
$$

$$
\begin{aligned}
C_{in} &= \text{inulin clearance} \\
P_G &= \text{concentration of glucose in plasma} \\
U_G &= \text{concentration of glucose in urine} \\
\dot{V} &= \text{volume of urine output per minute}
\end{aligned}
$$

All substances that are reabsorbed or secreted have a transport maximum.

11. The answer is C [Chapter 25 I A 1 a, II; Figure 25-3] The major determinant of the glomerular filtration rate (GFR) is the hydrostatic pressure within the glomerulus. An increase in glomerular capillary hydrostatic pressure is a major force that drives fluid out of the glomerular capillary and into Bowman's capsule. In glomerular capillaries, the net movement of fluid is primarily out of the capillaries, whereas in systemic capillaries, the change in the balance of Starling forces is such that net movement out of the capillaries is nearly balanced by net return of fluid into the vessels. Increases in the plasma oncotic pressure, hydrostatic pressure in Bowman's space, and afferent arteriolar resistance decrease the GFR, as does a decrease in the effective filtration pressure (EFP).

12. The answer is A [Chapter 22 I A–B; II A 3, B]. The extracellular fluid (ECF) volume constitutes approximately one-third of the total body water (TBW), or about 12–19 L. This compartment includes two subcompartments separated by the capillary membrane: blood plasma (intravascular fluid) and interstitial fluid. The ECF volume is measured with a test substance that does not penetrate the cells. Therefore, it has become common to measure the volume distribution of a specific substance and refer to it as, for example, the inulin space, if the test substance is inulin. All substances used to measure the ECF volume must cross capillaries and distribute at the same concentration in plasma and interstitial fluid.

13. The answer is E [Chapter 29 III B 1–3]. Renin secretion is not always required for aldosterone secretion. Renin is a proteolytic enzyme produced by the juxtaglomerular cells, which are modified smooth muscle cells in the renal afferent arterioles. The major stimulus for the release of renin is a decrease in the perfusion pressure of blood traversing these afferent arterioles. Thus, the juxtaglomerular cells function as low-pressure baroreceptors (volume receptors) where the secretion of renin varies inversely with the degree of stretch in the wall of the afferent arteriole. The major extrinsic factor influencing renin secretion is the renal postganglionic sympathetic vasomotor nerves, which are not necessary for renin secretion but exert a stimulatory effect (via β_1-adrenergic receptors) on the magnitude of renin release in response to a given stimulus. Increased K^+ concentration directly stimulates aldosterone production by the zona glomerulosa; hyperkalemia inhibits renin secretion. Renin substrate, called angiotensinogen, is an α_2-globulin of hepatic origin containing a tetradecapeptide moiety that serves as the prohormone for the angiotensins.

14–18. The answers are: 14-C [Chapter 27 II A 2 b; Figure 27-1D], **15-A** [Chapter 23 II A 2], **16-E** [Chapter 27 IV], **17-E** [Chapter 27 IV B], **18-C** [Chapter 29 III B 1 b]. The solute concentration in the ascending limb of the loop of Henle (ALH, site C) is less than that in any segment of the descending limb. The tubular fluid leaves the ascending limb at a lower concentration than it had when it entered the descending limb. Thus, the fluid presented to the distal tubule always is hyposmotic, regardless of the body's state of hydration.

Ultrafiltration separates water and nonprotein constituents (the "crystalloids") of plasma from the blood cells and protein macromolecules (the "colloids"). Except for proteins and lipids, the concentrations of crystalloids (e.g., Na^+, glucose) in the plasma and in Bowman's capsule (site A) are nearly the same.

The wall of the ALH is relatively impermeable to water. Therefore, NaCl in this segment is reabsorbed to the virtual exclusion of water, a process that renders the medullary and papillary interstitium hyperosmotic to plasma. The medullary interstitial osmolality is higher in antidiuresis than diuresis, due to urea and

NaCl. Thus, the highest osmolality exists in the papillary interstitium. With continued reabsorption of water, urea becomes even more concentrated at the terminals of the collecting ducts (site E). During dehydration with maximal antidiuretic hormone (ADH) secretion, the urine-to-plasma osmolality ratio (U_{osm}/P_{osm}) approaches 4 to 1 (at site E) because of the increased free-water reabsorption.

The macula densa is located at the junction of the thick segment of the ALH and the distal convoluted tubule (site C).

19–21. The answers are: 19-F [Chapter 24 II A 2 b; Figure 24-2], **20-H** [Figure 24-2]; **21-H** [Figure 24-2]. Under normal conditions, inorganic phosphate (P_i) reabsorption occurs mainly in the proximal tubule. The intraluminal phosphate concentration along the proximal tubule is lower than that in the plasma filtrate, with a mean of approximately 0.7 times the plasma concentration [i.e., a tubular fluid-to-plasma (TF/P) concentration ratio close to 0.7 (curve F).

The rate of urinary glucose excretion always is less than the rate of glucose filtration at the glomerulus. Thus, there is a net reabsorption of glucose that occurs solely in the proximal tubule. The TF/P concentration ratio for glucose falls to a value of 0.1, indicating that 90% of the filtered glucose is reabsorbed in the early portion (first quarter) of the proximal tubule (curve H). The filtered glucose is reabsorbed by an active, carrier-mediated process with transport maximum–limited characteristics.

Like glucose, glycine (an amino acid) is transported from the tubular fluid into the proximal tubular cell by specific carrier molecules that also combine with Na^+. Thus, glycine transport is a Na^+-coupled, secondary active transport process. The TF/P concentration ratio for amino acids also falls to a value of 0.1, indicating that about 90% of the filtered glycine is reabsorbed in the initial 25% of the proximal tubule (curve H). The active reabsorption of amino acids also involves a transport maximum–limited process.

PART VI

ACID-BASE PHYSIOLOGY
John Bullock

Chapter 31

Acid Production and Elimination: An Overview

I. INTRODUCTION. Although the body produces large amounts of acid in two forms [i.e., carbonic (volatile) and noncarbonic (nonvolatile, or fixed) acids], body fluids are maintained in an alkaline state (pH = 7.4). Most of the hydrogen ion (H^+) is formed as an end product of metabolism. The pathways for acid removal include the kidneys, lungs, and gastrointestinal (GI) tract.

A. **Sources of H^+.** The greatest source of H^+ is the carbon dioxide (CO_2) produced as one of the end products of the oxidation of glucose and triglyceride during **anaerobic metabolism**. Unfortunately, the proton donor–acceptor terminology of Brønsted prevents the classification of CO_2 as an acid, but CO_2 functions as the single most important weak acid in the body fluids. Thus, CO_2 is an acid (H^+) generator.

B. There are **three processes** available **for maintaining the concentration of hydrogen ion within normal limits:**

1. Combination of H^+ with a blood buffer (e.g., HCO_3^-, hemoglobin) or an intracellular buffer (e.g., organic or inorganic phosphate)

2. Reduction of carbonic acid (H_2CO_3) by elimination of CO_2 via pulmonary ventilation (see III)

3. Reduction of noncarbonic acid by renal elimination of H^+

II. ACID INTAKE AND URINARY ACID. In addition to reclaiming (reabsorbing) virtually all of the filtered HCO_3^-, the kidney must excrete an amount of acid equal to that generated by endogenous acid production, which varies with the dietary intake of exogenous acid-generating foods.

A. **Western diets,** which are high in protein, generate between 50 and 100 milliequivalents (mEq) of H^+ per day. The renal tubules cannot generate a hydrogen ion concentration gradient between tubular urine and blood of more than 3 pH units (i.e., a pH lower than 4.4, which is equivalent to a urine-to-blood $[H^+]$ gradient of 1000:1). The concentration of free H^+ at a urinary pH above 4.4 is negligible and thus the acid must be excreted in buffered (combined) form.

B. The **two main urinary buffers** are HPO_4^{2-} (dibasic phosphate) and NH_3 (ammonia).

1. The amounts of secreted H^+ excreted bound to NH_3 and HPO_4^{2-} are measured as NH_4^+ (ammonium) and titratable acid ($H_2PO_4^-$), respectively.

2. The sum of NH_4^+ excretion ($U_{NH_4^+} \cdot \dot{V}$)* and titratable acidity ($U_{H_2PO_4^-} \cdot \dot{V}$) minus the amount of HCO_3^- ($U_{HCO_3^-} \cdot \dot{V}$) equals net acid secretion, which normally is approximately 1 mEq per kg body weight per 24 hours (U = urine concentration; $\dot{V}$ = rate of urine formation).

C. The great bulk of **HCO_3^- reabsorption** takes place in the proximal tubule and is Na^+-dependent, whereas the excretion of acid (i.e., the regeneration of HCO_3^-) is to a large extent a distal nephron event and mainly Na^+-independent.

III. CARBONIC ACID (H_2CO_3). Because CO_2 can be formed from H_2CO_3 and, in turn, CO_2 can be eliminated by the lungs, H_2CO_3 is called a **volatile acid**.

A. **The oxidation of most carbohydrates and triglycerides does not generate acid.** The CO_2 combines with H_2O to form H_2CO_3, and thereby acidemia is prevented by the elimination of CO_2 by pulmonary ventilation.

B. **Conditions where the oxidation of glucose and fat do lead to acid formation** include:

1. Lactic acid formation from the oxidation of carbohydrates during hypoxic states

2. Ketoacid generation from the oxidation of fat in uncontrolled diabetes mellitus

C. **Hydration–dehydration: dissociation–association reaction.** For every H_2CO_3 molecule that has its H^+ taken up by a buffer, one HCO_3^- appears in the blood. These HCO_3^- ions are distributed between the erythrocytes and the plasma. Excess fluid intake (hydration) reduces the $[HCO_3^-]$, which causes a corresponding increase in $[H^+]$ if the CO_2 tension (P_{CO_2}) is constant. This type of noncarbonic acidosis is corrected more rapidly by the renal excretion of excess water than by the renal secretion of the apparent excess of H^+. Dehydration has the opposite effect.

1. $CO_2 + H_2O \underset{\text{dehydration}}{\overset{\text{hydration}}{\rightleftharpoons}} H_2CO_3 \underset{\text{association}}{\overset{\text{dissociation}}{\rightleftharpoons}} H^+ + HCO_3^-$

2. $H^+ + HCO_3^- + Na^+ + \text{buffer} \rightleftharpoons H \cdot \text{buffer} + Na^+ + HCO_3^-$

D. **CO_2 production.** CO_2 is the **chief product of metabolism** and represents the greatest portion of acid that is continuously eliminated from the body by the lungs.

1. Most CO_2 in the body is produced from the decarboxylation reactions of the tricarboxylic (citric) acid cycle.

2. HCO_3^- is an excellent buffer with respect to noncarbonic acid added to the blood by diet, metabolism, and disease, but the HCO_3^- buffer system plays no role in the buffering of carbonic acid.

3. Comparing CO_2 and HCO_3^-, there is more CO_2 than HCO_3^- produced metabolically in the body. However, the extracellular fluid (ECF) contains a preponderance of the HCO_3^- form.

4. Under basal conditions, with a respiratory quotient (RQ) of 0.82, the average adult produces about 300 L (13 mol) of CO_2 per day (Table 31-1).

5. The lungs and kidneys are the principal routes for the elimination of protons and for maintaining normal $[HCO_3^-]/S \cdot P_{CO_2}$ ratios.

*Ammonium secretion per se does not generate HCO_3^-, but it does preserve cations and prevents urea (and H^+) from forming in the liver from the ammonium.

TABLE 31-1. CO_2 Production under Basal Conditions*

	Volume	Molarity	
Time Elapsed	**(L)**	**(mmol)**	**(mol)**
1 minute	0.2	9	9×10^{-3}
1 hour	12.0	540	0.54
1 day	300	13,500	13.5

*Data for a respiratory quotient of 1.0.

    a. In the course of a day, the equivalent of 20–40 L of 1 N acid (20,000–40,000 mEq H^+) are eliminated via the lungs, or, more correctly, the amount of CO_2 produced during a day in a normal individual is potentially capable of forming 20–40 Eq of H^+.

    b. During a 24-hour period, the equivalent of 50–150 ml of 1 N acid (50–150 mEq H^+) is excreted via the kidneys.

  6. CO_2 must also be removed from the blood by the lungs at a rate of 200 ml/min during resting conditions. With a resting cardiac output of 5 L/min, 40 ml of CO_2 must be added to each liter of blood per minute (i.e., 5 L/min • 40 ml/L = 200 ml/min).

E. **Buffering of H_2CO_3** is primarily by the intracellular **nonbicarbonate buffers** (i.e., proteins, organic and inorganic phosphate, hemoglobin). Most of the buffering occurs in the erythrocytes, where the hemoglobin buffer system is quantitatively the most important of the intracellular buffers.

IV. **NONCARBONIC ACIDS cannot be converted to CO_2** and, therefore, are called nonvolatile (or fixed) acids. Since noncarbonic acids do not form a volatile end product, they must be buffered until they are excreted by the kidneys. Noncarbonic acids are derived from three sources: diet, intermediary metabolism, and stool HCO_3^- loss (Table 31-2).

A. **Diet.** A high-protein diet accounts for the formation of more acids than bases.

  1. Foodstuffs such as glucose and triglyceride are not acids in body fluids but are converted to CO_2 during the course of their metabolism, and much of this CO_2 is hydrated to form H_2CO_3, which then dissociates into H^+ and HCO_3^-.

  2. A vegetarian diet produces an excess of alkali, which must be excreted by the kidneys as HCO_3^-.

TABLE 31-2. Sources of Noncarbonic (Nonvolatile) Acids in Humans

Source	Amount (mEq/day)	Noncarbonic Acid
Diet	30	Phosphoproteins, sulfur-containing amino acids, chloride salts
Intermediary metabolism	30	Ketoacids, lactic acid
Stool loss	30	Loss of HCO_3^-

B. **Intermediary metabolism.** The primary source of noncarbonic acid in humans is the metabolism of exogenous protein. The body produces H^+ daily from the catabolism of proteins such as phosphoproteins and methionine, which are metabolically equivalent to phosphoric and sulfuric acids, respectively (Table 31-3).

1. Most **neutral amino acids** are metabolized in the liver by their conversion into glucose or triglycerides, and energy [adenosine triphosphate (ATP)]. It is important to note that there is no net generation of acid in this setting as:

$$\text{amino acid}° \rightarrow \text{glucose}° \text{ (or triglyceride}°) + \text{urea}°$$

2. **Sulfuric acid** is produced from the catabolism of sulfur-containing neutral amino acids (i.e., methionine, cysteine, cystine). These amino acids are converted into a proton (H^+) and a SO_4^{2-} anion as:

$$\text{Methionine}° \text{ (or cysteine}°) \rightarrow \text{glucose}° \text{ (or triglyceride}°) + \text{urea}° + 2\,H^+ + SO_4^{2-}$$

Sulfuric acid is initially buffered in the ECF by HCO_3^- as:

$$H_2SO_4 + 2\,NaHCO_3 \rightarrow Na_2SO_4 + 2H_2CO_3 \rightarrow 2H_2O + CO_2$$

The excess H^+ must still be excreted by the kidneys to prevent progressive depletion of HCO_3^- and other buffers and to prevent development of metabolic acidosis.

3. **Hydrochloric acid** is formed from the catabolism of a cationic amino acid (i.e., arginine, lysine, some histidine residues) through the conversion into a proton (H^+) and a chlorine (Cl^-) anion as:
   a. $\text{Arginine}^+ \cdot Cl^- \rightarrow \text{glucose}° \text{ (or triglyceride}°) + \text{urea}° + H^+ + Cl^-$
   b. $\text{Lysine}^+ \cdot Cl^- \rightarrow CO_2° \text{ (or triglyceride}°) + \text{urea}° + H^+ + Cl^-$

4. **Phosphoric acid** is formed when the phosphoproteins (phosphoesters) are hydrolyzed as:

$$R \cdot H_2PO_4 + H_2O \rightarrow ROH + 0.8\,HPO_4^{2-}/0.2\,H_2PO_4^- + 1.8\,H^+$$

Phosphoric acid cannot be further metabolized to CO_2 and H_2O, and the elimination of this acid and the H^+ that it yields can only be accomplished by the kidneys.

5. The **metabolism of foodstuffs** is a significant source of noncarbonic acids:
   a. **Lactate** is produced from the anaerobic metabolism of glucose or glycogen.

TABLE 31-3. Metabolic Production of Noncarbonic Acids and Alkali from the Diet

Dietary Source	Production of Acid	Production of Alkali	Quantity Produced (mEq/day) Acid	Quantity Produced (mEq/day) Alkali
Carbohydrate	0	0	0	0
Fat	0	0	0	0
Amino acids				
Sulfur-containing*	H_2SO_4		70	
Cationic[†]	HCl		135	
Anionic[‡]		HCO_3^-		100
Organic anions**		HCO_3^-		60
Phosphate	$H_2PO_4^-$		30	
Net acid production			75[††]	

*Cysteine and methionine (neutral amino acids)
[†]Lysine, arginine, and histidine
[‡]Aspartate and glutamate
**Acetate, citrate, gluconate, lactate, and malate
[††]Nonvolatile acid production from 100 g protein per day is approximately 100 mmols (or mEq) per day plus 30 mmol per day of H^+ from phosphate. Since metabolism of organic anions yields 60 mmol of HCO_3^- per day, the net acid production is about 70 mmol (or mEq) per day, or about 1 mmol H^+ per day per kg of body weight.

(1) Excessive production of lactic acid during heavy exercise or hypoxia can result in a transient increase in noncarbonic acid production. This excess must be buffered until excreted or metabolized to CO_2 and H_2O.

(2) Although most organs generate lactic acid, the largest amount of lactic acid is produced by skeletal muscle, erythrocytes, and the skin. (Liver, kidney, and muscle can convert lactic acid to HCO_3^-.)

(3) Tissue hypoxia leads to **hyperlacticemia,** which reduces the plasma $[HCO_3^-]$ and increases the anion gap. The increment in the anion gap in this situation is a measure of serum lactate. Clinical conditions associated with tissue hypoxia include cardiac arrest, shock, severe cardiac failure, and severe hypoxemia.

(4) **Alkalosis** stimulates glycolysis and generates lactic acid mainly by activation of phosphofructokinase. Respiratory alkalosis is a more effective stimulus for glycolysis, because CO_2 crosses cellular membranes more readily.

b. Acetoacetic acid and β-hydroxybutyric acid are products of triglyceride metabolism.

(1) These are the ketone bodies that are noncarbonic acids produced by normal subjects during fasting. Upon eating, acetoacetic and β-hydroxybutyric acids are further catabolized to CO_2 and H_2O.

(a) Excess ketone bodies associated with insulin-dependent diabetes mellitus are excreted by the kidneys and account for the ketonuria.

(b) Clinically more important is the acidosis of uncontrolled diabetic ketoacidosis when the accumulation of acetoacetic acid and β-hydroxybutyric acids in the ECF produce coma and death.

(2) The conversion of acetoacetic acid to β-hydroxybutyric acid is catalyzed by β-hydroxybutyric dehydrogenase.

(3) The β-hydroxybutyric–acetoacetate ratio is usually about 3:1. Acetoacetic acid can be converted to acetone by nonenzymatic decarboxylation. Acetone is excreted via the lungs and kidneys with a small component being metabolized to glucose.

c. Phosphoric acid is produced from the metabolism of phosphoproteins.

d. **Uric acid** is produced from the metabolism of nucleoproteins.

e. **Acetic acid** (vinegar) functions transiently as a noncarbonic acid because humans can convert it rapidly to CO_2 and H_2O.

6. **Buffering of noncarbonic acids.** In contrast to the bicarbonate buffer system, which can buffer only noncarbonic acids, the nonbicarbonate buffer systems can buffer both noncarbonic and carbonic acids.

a. The plasma bicarbonate system is quantitatively the most important buffer in the ECF for noncarbonic acids.

b. When noncarbonic acids are being buffered in the erythrocytes, more than 60% of the buffering occurs by the hemoglobin and more than 30% by the bicarbonate system in the erythrocytes. Approximately 10% of the buffer capacity in erythrocytes is attributed to the organic phosphate esters.

c. The bodily response to mineral acids (e.g., HCl) is different in that lactic acid is distributed throughout the body water and is metabolized, whereas mineral acids are confined to the ECF and are buffered and excreted without being metabolized.

C. **Stool HCO_3^- loss**

1. The diet also contains organic cationic and anionic salts that may be metabolized to yield noncarbonic acids and bases (HCO_3^-). Organic anions that are metabolized in the body to HCO_3^- include acetate, citrate, and—in the presence of insulin—the anions of the ketoacids.

2. Digestive processes result in the loss of 20–40 mmol of alkali in the stool, and this loss is equivalent to the addition of nonvolatile acid in the body.

Chapter 32

Metabolic Origin of Alkali (Bases)

I. **AMINO ACIDS AND OTHER ORGANIC IONS** (see Table 31-3). The metabolism of foodstuffs does not always acidify the body fluids; some foodstuffs have an alkalinizing action. For example, the large amounts of organic acids ingested with fruit (e.g., lactate, isocitrate, citrate) alkalinize the fluids because these organic ions are metabolized to CO_2 and H_2O, a process that involves consumption of H^+.

A. **Daily production.** Between 40 and 60 mmols of inorganic and organic acids that are not derived from CO_2 are produced daily. About half of the metabolically produced acids are neutralized by bases in the diet, but the remainder must be neutralized by the buffer systems of the body.

B. In contrast to the production of H^+ from sulfur-containing amino acids and cationic amino acids, the metabolism of anionic amino acids (i.e., glutamate, aspartate) results in the production of HCO_3^- (see Table 31-3).*

C. The hepatic metabolism of the other dietary anions by oxidation contributes to the **formation of HCO_3^-** as:

$$Citrate^- + 4.5\ O_2 \rightarrow 5\ CO_2 + 3\ H_2O + HCO_3^-$$

The metabolism of dietary organic ions (e.g., citrate) results in the production of approximately 60 mmol/day of HCO_3^-.

II. **AMMONIUM EXCRETION AND REGULATION OF ACID-BASE BALANCE**

A. **The kidney increases the HCO_3^- content of the body through its ability to metabolize glutamine and excrete NH_4^+.**

    1. The excretion of the two ammonium ions produced by the kidney does not solely lead to net H^+ loss.

    2. Net H^+ loss occurs when NH_4^+ is excreted. H^+ also is removed when α-ketoglutarate is metabolized to the neutral end product glucose or CO_2.

    3. If glutamine catabolism is to be a HCO_3^--generating process, the NH_4^+ liberated during the conversion of glutamine to either CO_2 or glucose must be excreted (Figure 32-1, 32-2).

B. The **HCO_3^- ion** formed by the proximal tubule is transported across the basolateral membrane to the extracellular fluid (ECF), where it restores HCO_3^- that was neutralized by systemic metabolic acid production.

    1. The **NH_4^+ ions** formed by the proximal tubule are actively secreted into the luminal fluid and are excreted via the urine.

---

*The consumption of the H^+ ions by glutamate and lactate is equivalent to the production of new HCO_3^- in the body.

FIGURE 32-1. Formation and active secretion of NH_4^+ by the proximal tubules. Most of the NH_4^+ excreted in the urine is produced by the proximal tubular cells and is mainly from glutamine. The formation of urea in the liver is quantitatively the most important disposal route for ammonia, and glutamine serves as a nontoxic storage and transport form of ammonia. $\alpha\text{-}KG^{2-} =$ α-ketoglutarate. [Reprinted from Seldin DW, Giebisch G (eds): New concepts in renal ammonium excretion. In *The Regulation of Acid-Base Balance.* New York, Raven, 1989, p 170.]

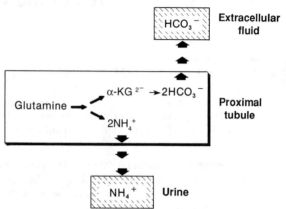

2. If the NH_4^+ is not excreted and returned to the ECF, it is incorporated into the urea by the liver. If this occurs, the two H^+ ions liberated during urea synthesis neutralize the two HCO_3^- ions produced from the oxidation of glutamine, and there is no net gain of HCO_3^- (Figure 32-2).

C. | Salient aspects of renal HCO_3^- generation

1. The "new" HCO_3^- added to the ECF in association with NH_4^+ excretion actually is produced in the proximal tubule from the metabolism of α-ketoglutarate formed from the catabolism of glutamine.

2. The addition of HCO_3^- to the renal venous blood requires the metabolism of the α-ketoglutarate anion, which removes protons and is a process equivalent to generating "new" HCO_3^- ions.

3. The NH_4^+ must be excreted in the urine; otherwise, it would return via the renal veins to the liver where it would be converted to urea plus 2 H^+ (see Figure 32-2).

FIGURE 32-2. Glutamine is metabolized by the kidneys, and NH_4^+ is excreted in the urine, while HCO_3^- is returned to the systemic circulation to replenish the HCO_3^- that is lost in the titration of nonvolatile acids. It is important to recognize that the formation of new HCO_3^- by this reaction depends upon the ability of the kidney to excrete NH_4^+ in the urine. If NH_4^+ is not excreted in the urine but instead enters the systemic circulation, it will titrate plasma HCO_3^-, thus negating the process of new HCO_3^- generation.

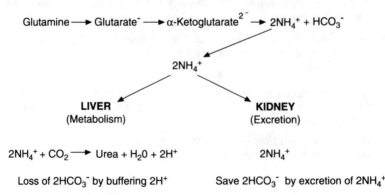

Chapter 33

The Hydrogen Ion and pH

I. **FUNDAMENTAL CHEMISTRY** (Table 33-1). H^+ is a proton (i.e., a hydrogen atom without its orbital electron); H^+ in aqueous solution exists as a hydrated proton called the hydronium ion, or H_3O^+. pH refers to the negative Briggsian logarithm of the H^+ concentration. The gain and loss of protons constitutes acid-base chemistry. The currently accepted model of acid-base relationships is that proposed by Brønsted.

A. An **acid** is a substance that acts as a proton donor.

B. A **base** is a substance that accepts protons (i.e., H^+) in solution. Thus, bicarbonate ion (HCO_3^-), phosphate ion (HPO_4^{2-}), ammonia (NH_3), and acetate ion (CH_3COO^-) all are bases.

C. Some substances are nearly equally divided between the acidic and basic forms at the normal H^+ concentration of the body. For example, the imidazole side groups of hemoglobin undergo the following reaction:

$$HHb \rightleftharpoons H^+ + Hb^-$$

The acid, deoxyhemoglobin (HHb), dissociates to form H^+ and the conjugate base, Hb^-. HHb and Hb^- occur in about equal concentrations in blood cells.

II. **CONCEPT OF pH AND H^+ CONCENTRATION** (Table 33-2)

A. **H^+ concentration is expressed in two different ways,** either directly as $[H^+]$ or indirectly as **pH**. (The symbol, $[H^+]$, refers to H^+ concentration in mol/L or Eq/L.) The relationship between $[H^+]$ and pH can be expressed as:

1. $pH = \log_{10} \dfrac{1}{[H^+]}$

2. $pH = -\log_{10} [H^+]$

3. $[H^+] = 10^{-pH}$

TABLE 33-1. Important Buffer Acids at Physiologic $[H^+]$

Proton Donor (Conjugate Acid)*		Proton (H^+)		Proton Acceptor (Conjugate Base)*
$(CO_2)H_2CO_3$	⇌	H^+	+	HCO_3^-
$H_2PO_4^-$	⇌	H^+	+	HPO_4^{2-}
H • Protein	⇌	H^+	+	Proteinate$^-$
$HHbO_2$	⇌	H^+	+	HbO_2^-
HHb	⇌	H^+	+	Hb^-

*A conjugate acid can be an anion or a cation; a conjugate base usually is an anion; HCO_3^-, a conjugate base, also can be an acid.

TABLE 33-2. Relationship between pH and [H$^+$]

pH	[H$^+$] (nEq/L)
7.70	20
7.40 (plasma)	40
7.30 (CSF)	50
7.10 (ICF)	80
7.00	100
6.90	126

CSF = cerebrospinal fluid; ICF = intracellular fluid.

B. **pH is a dimensionless number** and should be treated as such; it should not be referred to in concentration units or as "pH concentration." In fact, the quantity whose logarithm determines the pH is a volume per equivalent, which is the inverse of concentration. Thus, pH could be correctly conceptualized as a logarithmic expression of the volume required to contain 1 equivalent of H$^+$. In human plasma at pH 7.4 [i.e., [H$^+$] of 40 × 10^{-9} mol (Eq)/L or 40 nmol (nEq)/L], that volume is 25 million L!

C. Because pH is the logarithmic expression of [H$^+$], it permits a graphic representation of a wide range of [H$^+$] values. (It is important to note that **pH and [H$^+$] are inversely related**.) Another advantage of the pH concept is that when the pK′ of a buffer system is known, it is immediately possible to determine the effective pH range of the buffer.*

D. **A disadvantage of the pH system** is that it both inverts and uses the logarithmic scale to express [H$^+$]. For example, it is not immediately apparent that a decrease in pH from 7.4 to 7.1 represents a doubling of the [H$^+$] from 40 nmol/L to 80 nmol/L.

III. H$^+$ CONCENTRATION OF BODY FLUIDS (Figure 33-1; Table 33-3)

A. **Blood and plasma.** With regard to [H$^+$], the body fluid compartment most studied is arterial blood plasma. The term **blood pH** always refers to **plasma pH** (7.4), which is higher than the intracellular pH of the erythrocyte (7.2).

1. In normal individuals, the [H$^+$] is approximately 40 nmol (nEq)/L, which is equivalent to pH 7.4. The range of [H$^+$] that is compatible with life is 20–126 nEq/L, which is equivalent to a pH range of 7.7–6.9.

2. In normal individuals at rest, the pH of mixed venous blood is 7.38 compared to 7.41 for arterial blood because of the uptake of CO$_2$ by blood as it perfuses the tissues.

3. The [H$^+$] of plasma is very small compared to that of other ions in plasma (e.g., the plasma Na$^+$ concentration is about 142 million nmol/L, and the K$^+$ concentration is about 4 million nmol/L).

B. **Cerebrospinal fluid (CSF)** is essentially a bicarbonate buffer with a negligible concentration of protein (between 2 × 10^{-2} g/dl and 4 × 10^{-2} g/dl) or other nonbicarbonate buffers.

---

*K = the ionization or dissociation constant; pK = the negative logarithm of K (−log K) and is equal to the pH at which half of the acid molecules are dissociated and half are undissociated; pK′ = the apparent pK (see IV B 1).

TABLE 33-3. H^+ Concentration ($[H^+]$) and pH of Biologic Fluids

Fluid	pH	$[H^+]$ (nEq/ or nmol/L)*	$[H^+]$ (Eq/L or mol/L)
Pure water	7.0	100	1×10^{-7}
Blood			
Normal mean	7.40	40	3.98×10^{-8}
Normal range	7.36–7.44	44–36	$4.36 \times 10^{-8} - 3.63 \times 10^{-8}$
Acidosis (severe)	6.9	126	1.26×10^{-7}
Alkalosis (severe)	7.7	20	2.00×10^{-8}
CSF (normal range)	7.36–7.44	44–36	$4.36 \times 10^{-8} - 3.63 \times 10^{-8}$
Pure gastric juice (normal)	1.0	100,000,000	1×10^{-1}
Urine			
Normal average	6.0	1000	1×10^{-6}
Maximum acidity	4.5	31,600	3.16×10^{-5}
Maximum alkalinity	8.0	10	1×10^{-8}
ICF (muscle)	6.8	158	1.58×10^{-7}

CSF = cerebrospinal fluid; ICF = intracellular fluid. (Adapted from Brobeck JR (ed): Regulation of hydrogen ion concentration in body fluids. In *Best and Taylor's Physiological Basis of Medical Practice,* 10th ed. Baltimore, Williams & Wilkins, 1979, pp 5–13.)
*n = nano- = 10^{-9}. Thus, 100 nEq/L = 100×10^{-9} Eq/L, where nEq = nmol of a monovalent ion.

1. The arterial pH (7.4) is higher than that of CSF (7.32), because the CO_2 tension (P_{CO_2}) is about 48 mm Hg in the CSF and about 40 mm Hg in the arterial blood (see Figure 33-1).

2. Generally, it is possible to characterize acid-base disturbances precisely in terms of the blood data; however, the extracellular changes do not always reflect the intracellular changes. Also, acid-base alterations in arterial blood may produce similar or opposite changes in the CSF depending on whether the blood $[H^+]$ changes are due to respiratory or metabolic abnormalities.

   a. An **increase in arterial CO_2 tension (respiratory acidosis)** leads to a parallel rise in the CSF CO_2 tension, with a resulting increase in the $[H^+]$ of both arterial blood and CSF.

   b. A **decrease in arterial CO_2 tension (respiratory alkalosis)** leads to a parallel decline in CSF CO_2 tension, with a resulting decrease in the $[H^+]$ of both arterial blood and CSF.

FIGURE 33-1. Comparison of the relationship between pH, CO_2 tension (P_{CO_2}), and $[HCO_3^-]$ in arterial blood and cerebrospinal fluid (*CSF*) in normal adult humans. Each vertical bar represents ± 1 SD. (Reprinted from Seldin DW, Giebisch G (eds): Acid-base balance in specialized tissues: central nervous system. In *The Regulation of Acid-Base Balance.* New York, Raven, 1989, p 109.)

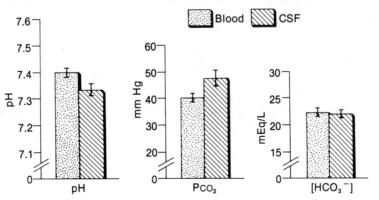

    c. An **increase in arterial [H$^+$] (metabolic acidosis)** stimulates ventilation, lowering the CO_2 tension of the arterial blood and CSF, resulting in CSF alkalosis and blood acidosis.

    d. A **decrease in arterial [H$^+$] (metabolic alkalosis)** decreases ventilation, raising the CO_2 tension of the arterial blood and CSF, resulting in CSF acidosis and blood alkalosis.

IV. HENDERSON-HASSELBALCH EQUATION

A. **Significance.** Acid-base balance is maintained primarily through the control of two organ systems. The lungs control the CO_2 tension through the regulation of alveolar ventilation, and the kidneys control the HCO_3^- concentration ([HCO_3^-]). The classic description of the acid-base state is based on the **Henderson-Hasselbalch equation,** which is an expression of three variables (pH, P_{CO_2}, and [HCO_3^-] and two constants (pK' and S).

B. **Definition of parameters.** The Henderson-Hasselbalch equation could functionally be written as:

$$pH = pK' + \log \frac{kidneys}{lungs} \tag{1)*}$$

However, the equation is expressed more usefully as:

$$pH = pK' + \log \frac{[HCO_3^-]}{S \cdot P_{CO_2}} \tag{2}$$

or, using specific values for pK' and S (defined in IV B 1 and 2), as:

$$pH = 6.1 + \log \frac{[HCO_3^-]}{0.03 \cdot P_{CO_2}} \tag{3}$$

From equation (3) it is clear that the value of arterial pH depends on the ratio of [HCO_3^-] to $S \cdot P_{CO_2}$, not on the individual value of each variable. In clinical medicine, pH, CO_2 tension, and [HCO_3^-] can be measured directly. With equation (3), however, any one of the variables can be calculated if the other two are known.

1. **pK** is defined as the negative logarithm of the [H$^+$] at which half the acid molecules are undissociated and half are dissociated. When equimolar concentrations of weak acid and conjugate base exist, the pH value equals the pK (i.e., the log of 1 is 0).

    a. The **actual dissociation constant (K)** for carbonic acid (H_2CO_3) in dilute aqueous solution at 38°C is 1.6×10^{-4} mol/L (pK = 3.8). Thus, H_2CO_3 is almost completely dissociated in the body where [H$^+$] = 4×10^{-8} mol/L, and it exists in quantities that are too small to be analyzed (i.e., 2.4×10^{-4} mEq/L). The formation and dissociation of H_2CO_3 is expressed as:

$$\underset{\substack{alveolar \\ gas}}{CO_2} \rightleftharpoons \underset{plasma}{CO_2} + H_2O \underset{500:1}{\overset{carbonic\ anhydrase}{\rightleftharpoons}} H_2CO_3 \underset{4000:1}{\rightleftharpoons} H^+ + HCO_3^- \tag{4}$$

At equilibrium there are approximately 500 mmol of CO_2 for every 1 mmol of H_2CO_3 and approximately 4000 mmol of H_2CO_3 for every 1 mmol of H$^+$. Because

---

*The kidneys primarily regulate [HCO_3^-], the numerator; the lungs mainly regulate CO_2 tension, the denominator.

of the presence of carbonic anhydrase, equilibrium between CO_2 and H_2CO_3 is rapid and constant.*

b. Because the denominator of equation (3) is increased by a factor of 500, the **apparent dissociation constant (K′)** for the CO_2/HCO_3^- buffer system in plasma at 38°C is correspondingly smaller (8×10^{-7} mol/L or 800 nmol/L), and the **apparent pK (pK′)** for this same buffer pair is correspondingly larger (6.1). As a rule, the optimal buffer region of a buffer pair system is within a range of ± 1 pH units of its pK value.

c. It is important to note that, because CO_2 increases $[H^+]$, as shown in equation (4), it is considered an acid even though it is an acid **anhydride**. Thus, dissolved CO_2 is present as a potential H^+ donor, and its concentration is proportionate to the true donor, H_2CO_3. For these reasons, it is more meaningful to characterize acid-base disturbances in terms of the CO_2/HCO_3^- buffer system instead of the H_2CO_3/HCO_3^- system. For all practical purposes, the H_2CO_3/HCO_3^- buffer pair can be considered to be composed of HCO_3^- (conjugate base) and dissolved CO_2 (conjugate "acid").

2. **S** is defined as the solubility constant for CO_2 in plasma at 38° C and is equal to 0.03 mmol/L/mm Hg. S represents the proportionality constant between CO_2 and CO_2 tension. For blood plasma at 38°C, the amount of dissolved CO_2 is expressed as:

$$\text{dissolved } CO_2 = 0.03 \cdot P_{CO_2}$$

where CO_2 = the millimoles of dissolved CO_2 per liter of plasma. At an arterial P_{CO_2} of 40 mm Hg, the CO_2 concentration ($[CO_2]$) is expressed more usefully as:

$$0.03 \text{ mmol/L/mm Hg} \cdot 40 \text{ mm Hg} = 1.2 \text{ mmol/L}$$

Multiplying the proportionality constant by milliliters of CO_2 per millimole (22.3 ml/mmol) yields a solubility constant of 0.67 ml/L/mm Hg for CO_2 in plasma. At this solubility constant and at a CO_2 tension of 40 mm Hg, the concentration of dissolved CO_2 is expressed as:

$$0.67 \text{ ml/L/mm Hg} \cdot 40 \text{ mm Hg} = 26.8 \text{ ml/L}^\dagger$$

3. **$[HCO_3^-]$** denotes the bicarbonate ion concentration, which is expressed in millimolarity rather than molarity. The normal value of $[HCO_3^-]$ in plasma is 24 mmol (mEq)/L.

C. Calculations with the Henderson-Hasselbalch equation

1. **Traditional calculation of pH.** pH in normal arterial plasma = 7.4; normal CO_2 tension = 40 mm Hg, and normal $[HCO_3^-]$ = 24 mmol/L (or 24 mEq/L). Applying the Henderson-Hasselbalch equation, pH is calculated as:

$$pH = 6.1 + \log \frac{[HCO_3^-]}{0.03 \cdot P_{CO_2}}$$

$$= 6.1 + \log \frac{24 \text{ mmol/L}}{0.03 \cdot 40 \text{ mm Hg}}$$

$$= 6.1 + \log \frac{24 \text{ mmol/L}}{1.2 \text{ mmol/L}}$$

$$= 6.1 + \log 20$$

$$= 7.4$$

---

*Carbonic anhydrase is found in erythrocytes, gastric parietal cells, renal tubular cells, pancreatic and pulmonary tissue, bone, and the eye; it is not found in muscle, peripheral nerves, or skin.
†This amounts to about 5% of the total amount of CO_2 carried in the arterial blood.

2. **Simple calculation of [H$^+$] and pH**
   a. **Conversion of pH to [H$^+$]: a close approximation**
      (1) **Equation.** The [H$^+$] in plasma is expressed in nmol/L. The equation for [H$^+$] now becomes:

$$\log [H^+] - 9 = -pH \text{ or } [H^+] = \text{antilog } (9 - pH)$$

      (2) **Examples** (see Table 33-2)
         (a) At pH 7.4:

$$
\begin{aligned}
[H^+] &= \text{antilog } (9 - 7.4) \\
&= \text{antilog } 1.6 \\
&= 39.6 \text{ nmol/L}
\end{aligned}
$$

         (b) At pH 6.9:

$$
\begin{aligned}
[H^+] &= \text{antilog } (9 - 6.9) \\
&= \text{antilog } 2.1 \\
&= 126 \text{ nmol/L}
\end{aligned}
$$

   b. **Conversion of [H$^+$] to pH**
      (1) **Equation.** The [H$^+$] in plasma is expressed in nmol/L. The equation for pH is:

$$pH = 9 - \log [H^+]$$

      (2) **Examples** (see Table 33-2)
         (a) At [H$^+$] = 40 nmol/L:

$$
\begin{aligned}
pH &= 9 - \log 40 \\
&= 9 - 1.6 \\
&= 7.4
\end{aligned}
$$

         (b) At [H$^+$] = 126 nmol/L:

$$
\begin{aligned}
pH &= 9 - \log 126 \\
&= 9 - 2.1 \\
&= 6.9
\end{aligned}
$$

V. HENDERSON EQUATION

A. **Use of the Henderson equation.** This equation provides a simple means for converting pH to [H$^+$], because CO_2 tension and [HCO_3^-] will have been provided. Additionally, this equation provides a simple **arithmetic estimate** of [HCO_3^-], because the value of [H$^+$] often will have been approximated from the pH value reported by the laboratory.

B. **Calculations with the Henderson equation.** The Henderson equation represents the non-logarithmic (arithmetic) method of determining [H$^+$] or [HCO_3^-] as:

$$[H^+] = K' \frac{P_{CO_2}}{[HCO_3^-]} \text{ or } [HCO_3^-] = K' \frac{P_{CO_2}}{[H^+]} \tag{5}$$

1. A value for K' is derived by converting the apparent pK for carbonic acid (6.1) into a dissociation constant (expressed in nmol/L) and multiplying that value by the solubility constant for CO_2 in plasma at 38°C (i.e., 0.03). K' has a value of 800 nmol/L [see IV B 1 (b)], which, when multiplied by 0.03, equals 24.

2. Using normal values for CO_2 tension, [H$^+$], and [HCO_3^-], the Henderson equation is applied as:

$$[H^+] = 24 \frac{P_{CO_2} \text{ (mm Hg)}}{[HCO_3^-] \text{ (mmol/L)}}$$

$$= 24 \frac{40}{24} = 40 \text{ nmol/L}$$

Similarly,

$$[HCO_3^-] = 24 \ \frac{P_{CO_2} \ (mm \ Hg)}{[H^+] \ (nmol/L)}$$

$$= 24 \ \frac{40}{40} \ = 24 \ mmol/L$$

Additionally,

$$P_{CO_2} = \frac{[H^+] \cdot [HCO_3^-]}{24}$$

$$= \frac{40 \cdot 24}{24} \ = 40 \ mm \ Hg$$

Chapter 34

Body Buffer Systems

I. **BUFFER SYSTEMS.** A buffer is a solution consisting of a weak acid and its conjugate base. Buffering is the primary means by which large changes in [H$^+$] are minimized.

A. Types of body buffer systems

1. **Blood buffers** (Table 34-1)
   a. It is important to note that the blood buffers are not solely plasma buffers, but that hemoglobin, HCO$_3$$^-$, and phosphate are found in erythrocytes and act as important blood buffers.
      (1) In blood, the chief H$^+$ acceptor is HCO$_3$$^-$, which exists in a concentration of 20–30 mEq/L.
      (2) H$^+$ acceptor is available in the hemoglobin system, where reduced hemoglobin (deoxyhemoglobin) is a stronger base than oxyhemoglobin (i.e., deoxyhemoglobin has a stronger capacity to combine with H$^+$).
      (3) The hemoglobin buffer system is quantitatively as important as the bicarbonate buffer system.
   b. The major buffer anions of whole blood—HCO$_3$$^-$, protein, and hemoglobin—have a total concentration of approximately 48 mEq/L (see Chapter 37 II A 1 b).
   c. With the bicarbonate and hemoglobin systems taken together, 5 L of blood of a normal adult have sufficient buffer capacity to combine with almost 150 mmol of protons (150 ml of 1 N HCl) before the pH of body fluids becomes dangerously acidic.

2. **Tissue buffers.** The major buffer capacity of the body is not in the blood but in the H$^+$ acceptors found in other tissues, principally in the muscle and in bone. These tissues can neutralize about five times as much acid as the blood buffers.
   a. **Muscle**
      (1) Since skeletal muscle represents about half of the cellular mass, most intracellular buffering presumably occurs in muscle.
      (2) For any individual, the body HCO$_3$$^-$ concentration averages 13 mEq/kg body weight. Muscle cells contain HCO$_3$$^-$ at a concentration of about 12 mEq/L, and

TABLE 34-1. Whole Blood Buffers

Buffer Type	Buffering Capacity of Whole Blood (%)
Bicarbonate	
Plasma	35
Erythrocyte	18
Total bicarbonate	53
Nonbicarbonate	
Hemoglobin and oxyhemoglobin	35
Plasma proteins	7
Organic phosphate	3
Inorganic phosphate	2
Total nonbicarbonate	47

Adapted from Brobeck JR (ed): Regulation of hydrogen ion concentration in body fluids. In *Best and Taylor's Physiological Basis of Medical Practice,* 10th ed. Baltimore, Williams & Wilkins, 1979, pp 5–14.

most other cells contain it at higher concentrations. The intracellular fluid (ICF) and extracellular fluid (ECF) compartments each contain about 50% of the total body HCO_3^-.

b. Bone plays an important role in buffering H^+.

(1) The total body store of bone carbonate is about 50 times the amount of HCO_3^- in the ICF and ECF compartments together. Bone carbonate appears to be the predominant source of base for neutralizing excess noncarbonic acid in the ECF. Indeed, it has long been recognized that chronic noncarbonic acidosis causes bone dissolution (resorption) through the loss of calcium carbonate ($CaCO_3$). The early carbonate release is in the form of sodium carbonate (Na_2CO_3).

(2) Bone contains about 80% of the total CO_2 (including CO_3^{2-}, HCO_3^-, and CO_2) in the body. About two-thirds of this CO_2 is in the form of CO_3^{2-} complexed with Ca^{2+}, Na^+, and other cations located in the lattice of the bone crystals. The other third consists of HCO_3^- and is located in the hydration shell of the hydroxyapatite crystal, an inorganic compound found in the matrix of bone and teeth.

B. **Distribution of body buffer systems in major compartments**

1. **Blood,** in regard to its buffering activity, usually is considered as a whole rather than in terms of its components. Blood buffers are described here in terms of separate compartments (i.e., plasma and erythrocytes) for didactic purposes only.

   a. Important concepts

   (1) Whole blood is an excellent buffering system for noncarbonic acids because of its nonbicarbonate buffers as well as its bicarbonate buffer system.

   (2) More than 90% of the blood's capacity to buffer carbonic acid is attributed to the hemoglobin buffer system. Thus, the nonbicarbonate buffer in the erythrocyte is quantitatively more important than the bicarbonate buffer in that compartment. The bicarbonate buffer system remains quantitatively important, however, in the erythrocyte (see Table 34-1).

      (a) The bicarbonate buffer system does not function as a buffer for carbonic acid.

      (b) The nonbicarbonate buffer systems can buffer both noncarbonic and carbonic acids.

   b. Buffer capacity of blood components

   (1) **Plasma,** which contains three buffer systems, has a considerable capacity for buffering noncarbonic acids but a much smaller capacity for buffering carbonic acid.

      (a) **Bicarbonate buffer system** (HCO_3^-/H_2CO_3 or HCO_3^-/CO_2). This buffer exists in a concentration of 24 mmol/L of plasma. When **noncarbonic acid** is added to normal plasma, more than 75% of the buffering capacity of plasma is due to the HCO_3^-/CO_2 system. Most of the remaining buffering of noncarbonic acid involves the plasma protein buffers and, to a small degree, the phosphate buffer system. Again, the HCO_3^-/CO_2 system plays no role in the buffering of carbonic acid.

      (b) **Nonbicarbonate buffer systems**

         (i) **Plasma protein** (Protein$^-$/H • Protein). Plasma is a salt solution containing 7% protein, which exists as polyanions at the pH of plasma. Plasma protein H^+ acceptors exist in a concentration of 1.5 mmol/L of plasma and account for less than one-sixth of the total buffering capacity of whole blood.

         (ii) **Inorganic orthophosphate** ($HPO_4^{2-}/H_2PO_4^-$). Because this buffer system exists in a concentration of only 0.66 mmol/L of plasma, it contributes little to the total buffering activity of plasma.* At a plasma

---

*The pK' of the acid form (i.e., $H_2PO_4^-$) is 6.8. With this pK', the $HPO_4^{2-}/H_2PO_4^-$ system would be a more effective buffer than the HCO_3^-/CO_2 system (pK' = 6.1) if it were present in an appreciable concentration. Many of the organic phosphate compounds found in the body have pK' values within half of a pH unit from 7.0.

pH of 7.4, the concentration ratio of $HPO_4^{2-}/H_2PO_4^-$ is 4:1. Therefore, 80% of the inorganic phosphate exists as disodium phosphate,
and 20% is in the form of monosodium phosphate. The $HPO_4^{2-}/$
$H_2PO_4^-$ system is a major elimination route for H^+ via the urine,
which has a relatively high phosphate content.

  (2) Erythrocytes. Although the blood contains other cells, the erythrocyte is the
only important cellular component of the blood buffer system. The erythrocyte
contains four buffer systems.

    (a) Bicarbonate buffer system. The HCO_3^-/CO_2 system exists in a concentra-
tion of 15 mmol/L of erythrocytes, compared to a concentration of 21
mmol/L and 25 mmol/L of whole blood and plasma, respectively.

    (b) Nonbicarbonate buffer systems

      (i) Hemoglobin buffers (Hb^-/HHb and $HbO_2^-/HHbO_2$) [see II]. One L
of **erythrocytes** contains 334 g (5.1 mmol) of hemoglobin. One L of
whole blood contains 150 g (2.3 mmol) of hemoglobin.

      (ii) Organic phosphate. Although the erythrocyte contains a significant
amount of organic phosphate buffer, this amount is quantitatively
small compared to the bicarbonate and hemoglobin buffer concentrations in the erythrocyte.

      (iii) Inorganic orthophosphate. The $HPO_4^{2-}/H_2PO_4^-$ system exists in a
concentration of 2 mmol/L of erythrocytes.

2. Interstitial fluid (including lymph)

  a. Bicarbonate buffer system. The HCO_3^-/CO_2 system is quantitatively the most important buffer of noncarbonic acid. The $[HCO_3^-]$ of the interstitial fluid is 27
mmol/L, which is similar to, or about 5% higher than, the $[HCO_3^-]$ of plasma. It is
important to consider that, in humans, the interstitial fluid volume is about three
times that of plasma; therefore, the total capacity of the interstitial fluid to buffer
noncarbonic acid is considerably greater than that of the total blood volume to
buffer these acids. **On a per unit of volume basis,** however, **the interstitial fluid has
nearly the capacity of plasma to buffer noncarbonic acids.**

  b. Nonbicarbonate buffer system. The $HPO_4^{2-}/H_2PO_4^-$ system exists in a concentration of 0.7 mmol/L of interstitial fluid; therefore, this compartment has little capacity
to buffer carbonic acid. The interstitial fluid is essentially free of protein.

3. Intracellular fluid (ICF; excluding erythrocytes)

  a. Bicarbonate buffer system. The ICF contains only about 12 mmol of HCO_3^-/L in
skeletal and cardiac muscle.

  b. Nonbicarbonate buffer systems. Protein and organic phosphate compounds exist in
quantitatively significant amounts in the ICF, giving this compartment the capacity
to effectively buffer both noncarbonic and carbonic acids as well as alkali. The ICF
concentrations of these major buffer anions are:

    (1) $HPO_4^{2-}/H_2PO_4^-$ (skeletal muscle) = 6 mmol/L

    (2) $Protein^-/H \cdot Protein$ (skeletal muscle) = 6 mmol/L

    (3) Organic anions (skeletal muscle) = 84 mmol/L

II. HEMOGLOBIN: AN "EXTRACELLULAR" BUFFER

A. **Important concepts.** Although hemoglobin is found intracellularly, it is more conventionally
regarded as extracellular and, therefore, part of the extracellular buffer system because:

  1. Hemoglobin is confined to the erythrocyte, which is a cellular component of the ECF.

  2. Hemoglobin is readily available for the buffering of extracellular acids.

  3. Hemoglobin is the primary nonbicarbonate buffer of the blood.

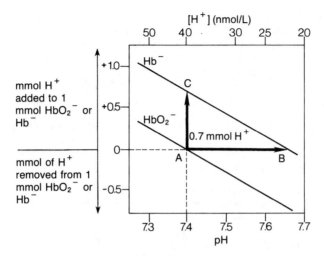

FIGURE 34-1. Titration curves of oxyhemoglobin (HbO_2^-) and deoxyhemoglobin (Hb^-), illustrating the importance of hemoglobin as a buffer. The complete deoxygenation of 1 mmol of HbO_2^- to liberate 1 mmol of O_2 results in the neutralization of 0.7 mmol of H^+ without a change in pH. *Arrow AC* indicates the amount of H^+ that can be added during the reduction of hemoglobin without causing a pH change. *Arrow AB* represents the pH change that would occur if the oxyhemoglobin at pH 7.4 was completely reduced. The reduction of HbO_2^- to Hb^- would cause a large increase in pH if CO_2 and, hence, H^+ were not added simultaneously to the system. Reduction of hemoglobin denotes the O_2-free state without a change in the valence of iron. (Adapted from White A, et al: Hemoglobin and the chemistry of respiration. In *Principles of Biochemistry*, 5th ed. New York, McGraw-Hill, 1973, p 843.)

B. **Hemoglobin as a buffer.** Like all proteins, hemoglobin is a buffer. At pH 7.2 (i.e., the pH of normal arterial erythrocytes), the buffering action of hemoglobin is due mainly to the imidazole groups of the histidine residues.

1. The titration curves of deoxyhemoglobin and oxyhemoglobin in Figure 34-1 illustrate the basis for the ability of hemoglobin to neutralize H^+ formed subsequent to the diffusion of CO_2 into the erythrocyte. Most of the $H+$ is buffered by hemoglobin and most of the HCO_3^- diffuses into the plasma.*

2. Oxyhemoglobin dissociates more completely than does deoxyhemoglobin, and, as a result, deoxyhemoglobin produces less H^+ at a given pH than does oxyhemoglobin, which is a stronger acid. Thus, hemoglobin becomes a more effective buffer when CO_2 and, hence, H^+ are added from the tissues. This is important because the diffusion of CO_2 from the tissues to the capillary blood is accompanied by the simultaneous reduction of oxyhemoglobin.

3. As the uptake of CO_2 depends on H^+ acceptors, this increase in H^+ acceptors in the form of deoxyhemoglobin facilitates the uptake and buffering of the H^+ generated by the hydration of CO_2 and the dissociation of H_2CO_3. As a result of these reactions, CO_2 is converted into HCO_3^- within the erythrocyte.
   a. For each mmol of oxyhemoglobin that is reduced, about 0.7 mmol of H^+ can be taken up and, consequently, 0.7 mmol of CO_2 can enter the blood without a change in pH (see Figure 34-1, *point A* to *point C*).
   b. A reaction that causes no change in $[H^+]$ (or pH) is called **isohydric buffering**.

4. **Respiratory quotient (RQ).** When the body is at rest, the rate of O_2 consumption ($\dot{V}O_2$) under standard conditions is 250–350 ml/min, and the rate of CO_2 production ($\dot{V}CO_2$) is

---

*The buffer capacity of nonbicarbonate buffers is dependent primarily on the hemoglobin concentration.

200–250 ml/min. (The dot over the symbol V denotes **volume per unit time**.) RQ represents the ratio of CO_2 production to O_2 consumption. Normally, on a mixed diet, the O_2 consumption exceeds the CO_2 production, and the RQ is less than 1.

    a. The metabolic RQ equals the **molar ratio** of CO_2 production rate to the corresponding O_2 consumption rate by metabolizing tissues.

    b. If the RQ rate is 0.7, then, for 1 mmol of O_2 consumed, 0.7 mmol of CO_2 is produced, which, when converted to H_2CO_3, yields 0.7 mmol of H^+ upon dissociation.

    c. The complete deoxygenation of 1 mmol of oxyhemoglobin to liberate 1 mmol of O_2 results in the neutralization of 0.7 mmol of H^+ without a change in pH. Thus, all of the H^+ produced when the RQ is 0.7 can be buffered by deoxyhemoglobin with no change in pH (see Figure 34-1).

5. At pH 7.2, about 84% of the deoxyhemoglobin is in the form of HHb, whereas only about 23% of the oxyhemoglobin is in the form of $HHbO_2$. Of the total oxyhemoglobin that causes O_2 to form deoxyhemoglobin:

    a. 23% was already combined with H^+,

    b. 16% will not combine with H^+, and

    c. 61% will take up H^+ before a pH decrease occurs.

6. CO_2 entering the erythrocytes rapidly undergoes two reactions.

    a. CO_2 is hydrated to form H_2CO_3, a reaction that is catalyzed by carbonic anhydrase in the erythrocyte. The H_2CO_3 dissociates to form HCO_3^- and H^+, which is buffered primarily by the hemoglobin buffers. Much of the HCO_3^- formed within the erythrocyte diffuses into the plasma in exhange for Cl^-.

    b. CO_2 combines with the amino groups of deoxyhemoglobin to form carbaminohemoglobin. The carbaminohemoglobin dissociates to a carboxylate anion and H^+, which is buffered primarily by the hemoglobin buffers. In summary:

$$Hb \cdot NH_2 + CO_2 \rightleftharpoons Hb \cdot NHCOO^- + H^+$$

7. The unloading of O_2 from oxyhemoglobin to the tissues causes the formation of deoxyhemoglobin that is better able to tie up the H^+ produced by the simultaneous uptake of CO_2. The loss of O_2 from hemoglobin facilitates the uptake of CO_2 in the form of carbaminohemoglobin by the erythrocytes. Oxyhemoglobin contains about 0.1 mmol and deoxyhemoglobin about 0.3 mmol of carbaminohemoglobin per mmol.

Chapter 35

Respiratory Regulation of Acid-Base Balance

I. **INTRODUCTION.** It is not immediately apparent that the actual quantities of CO_2 and O_2 transported in the blood are far greater than the amounts of these gases in physical solution, because these gases are transported mainly in the form of chemical derivatives. Further, it is even less apparent that there is far more total CO_2 than O_2 in every liter of blood. At the level of the lung, the HCO_3^- combines with the potential protons (i.e., the protons associated with $HHbO_2$) to produce HbO_2^- and H_2CO_3 resulting in the elimination of the protons as CO_2 and water.* Thus, free H^+ never appear to any appreciable extent but are in the form of HHb or H_2CO_3. **The entire respiratory cycle can be regarded as an exchange of a HCO_3^- for a HbO_2^-,** with the HCO_3^- transported from the tissues to the lungs and the HbO_2^- transported to the tissues from the lungs.

II. **BLOOD FORMS OF CO_2**

A. **CO_2 is transported in three forms.** However, H_2CO_3 ultimately must be converted to CO_2 in order to be eliminated via the lungs.

  1. HCO_3^- accounts for 90% of the total CO_2 in plasma.

  2. Carbamino compounds (i.e., carbamates of hemoglobin and protein) represent the combination of CO_2 with free NH_2 groups of blood proteins according to the equation:

  $$R \cdot NH_2 + CO_2 \rightleftharpoons R \cdot NHCOO^- + H^+$$

  About 5% of the CO_2 normally carried in arterial blood is in the form of carbamino compounds (carbamates).

  3. A small amount of CO_2 gas (5%) is physically dissolved in plasma and also hydrated as H_2CO_3. Also, approximately 500 mmol of CO_2 exist for every 1 mmol of H_2CO_3 (see Chapter 33 IV B 1 a).

B. It is important to recognize that the term "bicarbonate" is used interchangeably with the term "total CO_2 combining power" or "total CO_2." This must never be confused with the term "partial pressure of CO_2," which is the Pco_2 of the arterial blood gas measurement.

C. In blood with an arterial O_2 tension of 100 mm Hg and a mixed venous O_2 tension of 40 mm Hg, there are 200 ml O_2/L and 150 ml O_2/L of arterial and mixed venous blood, respectively. However, the arterial CO_2 tension and mixed venous CO_2 tension are associated with 480 ml CO_2/L and 520 ml CO_2/L of arterial and mixed venous blood, respectively.

---

*$HHbO_2$ and HbO_2^- are forms of oxygenated hemoglobin and constitute a buffer pair where the protonated form, $HHbO_2$, is the conjugate acid and the unprotonated form, HbO_2^-, is the conjugate base.

III. H_2CO_3 AND OTHER CO_2-FORMING ACIDS

A. Source

1. The oxidation of glucose and triglyceride leads to the formation of CO_2, much of which is hydrated to H_2CO_3. H_2CO_3 in turn dissociates into H^+ and HCO_3^-.

2. CO_2 and water are the most abundant end products of metabolism.
   a. Lactic acid, a noncarbonic acid, normally is metabolized to CO_2 and water.
   b. The ketone bodies, β-hydroxybutyric acid and acetoacetic acid, also are noncarbonic acids, which are produced primarily by the liver during fasting. Upon reingestion, these acids that have accumulated are further catabolized to CO_2 and water by the extrahepatic tissues.

B. Elimination

1. Respiration accounts for the greatest portion of acid eliminated continuously from the body.

2. The lungs eliminate H_2CO_3 in the dehydrated (anhydrous) form of CO_2. Indeed, the unique property of the CO_2/HCO_3^- buffer system lies in the ability of the lungs to eliminate undissociated H_2CO_3 as nonionizable CO_2.

C. Buffering of H_2CO_3. Because CO_2 readily penetrates cellular membranes, H_2CO_3 is buffered by the entire body.

1. Most of the buffering of H_2CO_3 occurs via the nonbicarbonate buffer systems within erythrocytes. The H^+ derived from the dissociation of H_2CO_3 is buffered primarily by the hemoglobin buffer system.

2. The erythrocyte bicarbonate buffer and the plasma bicarbonate buffer play no role in the buffering of H_2CO_3.

Chapter 36

Renal Regulation of Acid-Base Balance

A. Normal acid-base conditions met by the kidneys

1. The kidneys are responsible for ridding the body of metabolically produced noncarbonic acids.
   a. Among these are sulfuric and phosphoric acids and smaller amounts of hydrochloric, lactic, uric, β-hydroxybutyric, and acetoacetic acids. The most abundant of the weak acid waste products of metabolism is acid phosphate ($H_2PO_4^-$).
   b. About half of these metabolically produced acids are neutralized by base in the diet. The other half must be neutralized by buffer anion systems of the body. Of those noncarbonic acids buffered in the extracellular fluid (ECF), 97%–98% are buffered by reacting with HCO_3^-.

2. Type of diet is a major determinant of the daily acid-base conditions that must be regulated by the kidneys.
   a. High-protein diets contain large amounts of sulfur in the form of sulfhydryl groups. The sulfur is oxidized to sulfate ion (SO_4^{2-}), a process that tends to lead to metabolic acidosis.
   b. Vegetarian diets are associated with large intakes of lactate and acetate, and the metabolites of these anions tend to lead to metabolic alkalosis.

B. Kidney function

1. The primary role of the kidneys in acid-base regulation is to conserve major cations and anions in the body fluids. To maintain the total quantities and concentrations of the major electrolytes within normal limits, the kidneys perform two major functions.
   a. The kidneys stabilize the standard HCO_3^- pool by obligatory reabsorption (mainly by the proximal tubule) and by controlled reabsorption of filtered HCO_3^- (by the distal and collecting tubules).
   b. The kidneys excrete a daily load of 50–100 mEq of metabolically produced noncarbonic acid.* This represents a H^+ excretion of 1 mEq/kg of body weight/day.
      (1) In most cases, 25% of this noncarbonic acid (10–30 mEq/day) is excreted in the form of **titratable acid** (see IV).
      (2) About 75% (30–50 mEq/day) is excreted in the form of acid combined with **ammonium** (i.e., NH_4^+; see V).
      (3) The normal urinary ratio of NH_4^+ to titratable acid is between 1 and 2.5.

2. The major sites of urine acidification are the distal and collecting tubules.
   a. Essentially all of the H^+ within the tubular lumen is from the tubular secretion of H^+ generated by metabolism. There is no significant contribution of H^+ from the glomerular filtrate, which accounts for less than 0.1 mmol of H^+ per day.
   b. Since the lowest pH attainable in urine is 4.4 (i.e., a $[H^+]$ of 40×10^{-6} Eq/L) and the plasma $[H^+]$ is 40×10^{-9} Eq/L, the kidney can cause a 1000-fold $[H^+]$ gradient between plasma and urine.

---

*In normal individuals in Western countries, a net of 40–60 mEq of noncarbonic acid is excreted daily.

3. An important difference between the acidification of urine and free H^+ excretion is that the ability to reduce urinary pH (acidification) does not reveal a great deal about the amount of free H^+ excreted. This is because most of the H^+ that is excreted occurs in association with an anion (mostly as $H_2PO_4^-$) or in combination with ammonia (as NH_4^+).

4. The majority of secreted H^+ is used to bring about HCO_3^- reabsorption and, therefore, is not excreted.

II. BASIC ION EXCHANGE MECHANISMS (Figures 36-1; 36-2)

A. **Transport of Na^+.** Entry of Na^+ into the luminal (adluminal) membrane of the tubular cell is operationally linked to the secretion of H^+ into the tubular lumen. (It is a cation-exchange process that maintains intracellular electroneutrality.)

1. Although the movement of Na^+ across the luminal membrane is favored by the electrochemical gradient, the transport is mediated by specific membrane transport proteins (i.e., it is carrier-mediated) and is *not* by simple diffusion.

2. Secretion of H^+ is active in that it occurs against an electrochemical gradient.

3. The process of H^+ secretion and HCO_3^- reabsorption occurs throughout the nephron, with the exception of the descending limb of the loop of Henle.

4. The Na^+-H^+ antiporter is the major mechanism for H^+ secretion in the proximal tubule and thick ascending limb of the loop of Henle. This antiporter system is termed "secondary active" transport because the energy is derived from the Na^+ concentration gradient and not directly from the hydrolysis of adenosine triphosphate (ATP).

FIGURE 36-1. Major cellular and luminal processes in HCO_3^- reabsorption in the proximal tubule. The proximal nephron accounts for the largest fraction of HCO_3^- reabsorption and it occurs by secondary active transport (antiport) via the Na^+-H^+ exchanger rather than by direct reabsorption. The secreted H^+ reacts with filtered HCO_3^- to form carbonic acid, which is dehydrated to CO_2 and H_2O, both of which diffuse into the cell. Note that the secreted H^+, which combined with HCO_3^- in the lumen, is not excreted. Whether H^+ ions are secreted by a Na^+-H^+ exchanger or by a H^+ pump at the luminal membrane, the filtered HCO_3^- will be converted to H_2CO_3 and subsequently to CO_2 and H_2O. C.A. = carbonic anhydrase; ● = adenosine triphosphatase (ATPase) pump; ○ = protein carrier; F = filtered; S = secreted.

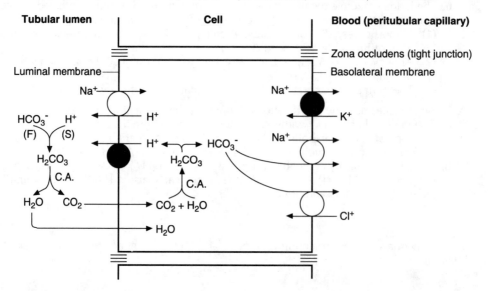

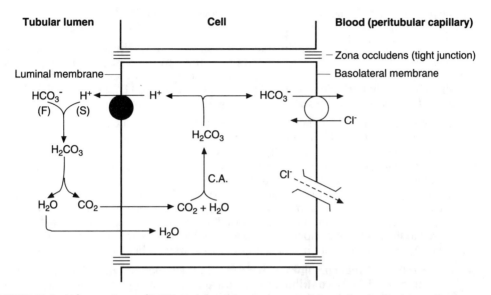

Tubular lumen   **Cell**   **Blood (peritubular capillary)**

FIGURE 36-2. H$^+$ secretion and HCO$_3^-$ reabsorption in the type A intercalated cell of the collecting tubules. Unlike the proximal tubule, the collecting duct lacks luminal carbonic anhydrase (*C.A.*). ● = adenosine triphosphatase (ATPase) pump; ○ = protein carrier; ⇒⇐ = channel; *F* = filtered; *S* = secreted.

    5. The Na$^+$-H$^+$ antiporter exchanges one Na$^+$ ion for one H$^+$ ion and, therefore, is electroneutral.

B. **Active transport of Na$^+$.** Subsequent to the downhill transport of Na$^+$ across the luminal membrane, Na$^+$ is actively reabsorbed across the basolateral (abluminal) membrane into the peritubular capillary in association with HCO$_3^-$ that was formed within the cell. The secretion of 1 mol of H$^+$ leads to the reabsorption of 1 mol of Na$^+$ and 1 mol of HCO$_3^-$ in the peritubular blood.

    1. H$^+$ for secretion originates from the hydration-dissociation reaction within the tubular cell [see Chapter 33 IV B 1 a, equation (4)] and from the splitting of water, which liberates hydroxyl ion (OH$^-$) within the cell; the OH$^-$ then reacts with CO$_2$ to form HCO$_3^-$.

    2. H$^+$ secretion exceeds H$^+$ excretion; the amount of H$^+$ excreted in the form of free H$^+$ is negligible.

    3. The net effect of H$^+$ secretion is to facilitate the transfer of HCO$_3^-$ from tubular lumen to tubular cell.

    4. It is probable that H$^+$ secretion along the entire tubule is mediated by an active transport process.

III. **REABSORPTION OF HCO$_3^-$** (see Figures 36-1; 36-2). The filtered HCO$_3^-$ is not directly reabsorbed by the renal tubules; instead, it must first be converted to H$_2$CO$_3$, with the H$^+$ secreted into the tubular lumen in exchange for the Na$^+$ that is transported from the lumen. The HCO$_3^-$ is reabsorbed indirectly by conversion to CO$_2$ within the tubular lumen, and most of the H$^+$ is reabsorbed following its conversion to water within the lumen.

A. **Proximal reabsorption.** Approximately 90% of the filtered HCO$_3^-$ is reabsorbed into the proximal tubules as a result of the Na$^+$-H$^+$ exchange and represents more than 4000 mEq/day.

1. The presence of carbonic anhydrase on the microvilli of the luminal border (brush border) of the proximal tubules catalyzes the rapid dehydration of H_2CO_3 to form CO_2 and water.

2. CO_2 diffuses back into the proximal tubular cell, where it is rehydrated by intracellular carbonic anhydrase into H_2CO_3. The HCO_3^- formed by the dissociation of H_2CO_3 is passively reabsorbed into the peritubular blood along with equimolar amounts of Na^+, which is actively transported into the peritubular blood. The H^+ formed by the dissociation of H_2CO_3 serves as a source for another H^+ to be secreted.

3. The entry of Na^+ into the tubular cell is balanced electrochemically by two mechanisms.
   a. Cl^-, which is the only quantitatively important reabsorbable anion in the filtrate, is transported from lumen to cell by passive diffusion.
   b. In the presence of carbonic anhydrase and with OH^- as its substrate, HCO_3^- is regenerated from the cellular CO_2 and water.
   c. The net result of either the Cl^- diffusion into the cell or the regeneration of HCO_3^- is a net reabsorption of sodium chloride (NaCl) or sodium bicarbonate ($NaHCO_3$), respectively, into the peritubular capillaries. It is important to note that the Na^+ delivered to the peritubular capillaries comes from the filtrate in the lumen, but the HCO_3^- transported into the capillaries is synthesized within the proximal tubular cell.

4. **Any secreted H^+ that combines with HCO_3^- in the lumen to bring about HCO_3^- reabsorption DOES NOT contribute to the urinary excretion of acid.** Thus, the majority of the secreted H^+ is used to accomplish HCO_3^- reabsorption.

5. Referring to this process as HCO_3^- reabsorption, then, seems inaccurate, since the HCO_3^- that appears in the peritubular capillaries is not the same HCO_3^- that was filtered. Moreover, most of the H^+ that was secreted into the lumen is not excreted in the urine but is incorporated into water and reabsorbed.

6. There is no absolute maximal tubular transport capacity (Tm) for HCO_3^- since the reabsorptive capacity for HCO_3^- varies directly with the fractional reabsorption of Na^+.

B. Reabsorption from the distal and collecting tubules

1. The remaining 10%–15% of the filtered HCO_3^- is reabsorbed by the distal and collecting tubules via a mechanism that involves the exchange of Na^+ for K^+ or H^+. As in the proximal tubules, every H^+ secreted into the distal tubular lumen leaves a HCO_3^- within the tubular cell.

2. Except for the Na^+-H^+ antiporter or Na^+-K^+ exchange pump at the luminal border of the distal cellular membranes and the absence of carbonic anhydrase on the distal luminal border, the reabsorption pathways for HCO_3^- are analogous to those described for the proximal tubule.*

C. Addition of new HCO_3^-. In addition to the renal conservation of HCO_3^-, the kidneys add newly synthesized HCO_3^- to the plasma so that the quantity of HCO_3^- in the renal vein exceeds the amount that entered the kidneys.

1. The addition of new HCO_3^- to the plasma does not involve the HCO_3^- reabsorbed into the tubule but the HCO_3^- generated within the tubular cell via the hydration of CO_2 and dissociation of H_2CO_3. This process is similar to the scheme for the reabsorption of filtered HCO_3^-; however, **the HCO_3^- that is generated within the tubular cell does not represent filtered HCO_3^-.**

2. The renal contribution of new HCO_3^- is accompanied by the excretion of an equivalent amount of acid in the urine in the form of titratable acid, NH_4^+, or both.

3. The amount of new HCO_3^- formed per day (approximately 70–100 mEq) is much less than the quantity of filtered HCO_3^- reabsorbed per day (more than 4300 mEq).

---

*Intracellular carbonic anhydrase is found throughout the renal tubule, including the distal convoluted tubule and collecting duct. In the proximal convoluted tubule, carbonic anhydrase also is located in the luminal cell membranes.

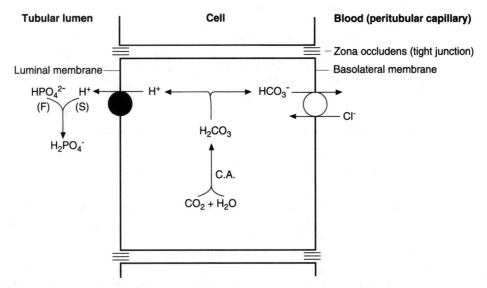

FIGURE 36-3. Formation of titratable acid in the type A intercalated cell of the collecting tubule. Formation of titratable acid is due primarily to buffering of secreted H^+ by filtered phosphate (HPO_4^{2-}). *C.A.* = carbonic anhydrase; ● = adenosine triphosphatase (ATPase) pump; ○ = protein carrier; *F* = filtered; *S* = secreted.

IV. EXCRETION OF TITRATABLE ACID (Figures 36-3 and 36-4; Table 36-1)

A. **Filtration of HPO_4^{2-}.** Titratable acid denotes that portion of H^+ bound to filtered buffers and equals the amount of alkali, in the form of sodium hydroxide (NaOH), required to titrate urine back to the normal pH of blood (i.e., the number of milliequivalents of H^+ added to the tubular fluid that combined with phosphate or organic buffers).

FIGURE 36-4. Formation of titratable acid in the proximal tubule. *C.A.* = carbonic anhydrase; ● = adenosine triphosphatase (ATPase) pump; ○ = protein carrier; *F* = filtered; *S* = secreted.

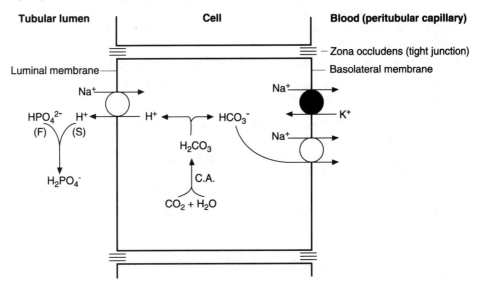

TABLE 36-1. Urinary Acid Excretion in Health and Disease (in mEq H^+/day)

Urinary Acid	Normal Excretion	Excretion in Diabetic Ketoacidosis
Titratable acid	10–30	75–250
Ammonium	30–50	300–500
Total	40–80	375–750

1. Titratable acid is largely attributed to the conversion of HPO_4^{2-} to $H_2PO_4^-$. Additional buffer species that may contribute to titratable acid are creatinine, uric acid, β-hydroxybutyrate, and SO_4^{2-}.

2. Titratable acid is a poor measure of the total amount of H^+ secreted by the tubules, since most of the acid produced by this secretion is H_2CO_3, which disappears from the urine as CO_2; however, the H^+ that is trapped by anions of noncarbonic acids remains in the final urine in the form of titratable acid.

3. Titratable acid is a measure of the content of weak acids. It is not a measure of the H^+ that combines with ammonia (NH_3) to yield NH_4^+, because the pK' of the NH_3/NH_4^+ buffer system is high (9.2).

4. The proximal tubule is the major nephron site where titratable acid is formed. Additional titratable acid is generated along the collecting duct by a H^+-ATPase pump.

B. **Excretion of $H_2PO_4^-$.** The exchange of H^+ for Na^+ converts dibasic sodium phosphate (Na_2HPO_4) in the glomerular filtrate into dihydrogen phosphate (NaH_2PO_4), which is excreted in the urine as titratable acid. The H^+ secreted into the tubules, therefore, can react with filtered HPO_4^{2-} rather than the filtered HCO_3^-.

1. Besides HCO_3^-, HPO_4^{2-} represents a major filtered conjugate base.* Furthermore, the $HPO_4^{2-}/H_2PO_4^-$ buffer pair provides an excellent buffer system because it has a pK' of 6.8.

2. The blood $HPO_4^{2-}/H_2PO_4^-$ molar ratio is 4:1. The H^+ secretory system converts much of the HPO_4^{2-} to $H_2PO_4^-$ in the tubular fluid, resulting in a urinary $HPO_4^{2-}/H_2PO_4^-$ molar ratio of 1:4. The acidification of the phosphate buffer system occurs significantly in the proximal tubule, and much of the H^+ generated by the formation of H_2SO_4 and H_3PO_4 during protein and phospholipid metabolism is excreted in this way.

3. H^+ secreted into the lumen that reacts with HCO_3^- is not excreted, whereas H^+ secreted into the lumen that reacts with a nonbicarbonate buffer remains in the tubular fluid and is excreted.

4. Urinary phosphate is the major anion contributing to urine acid excretion.

5. Virtually all of the filtered phosphate is reabsorbed by the Na^+-phosphate symporter of the proximal tubular brush border.

C. **Role of aldosterone in H^+ excretion.** Aldosterone exerts its major renal effect on the collecting duct (type A intercalated cells).

1. Aldosterone plays an important role in regulating collecting duct HCO_3^- reabsorption via its stimulatory effect on H^+ secretion.

2. Aldosterone stimulates H^+ secretion directly on the tubular cell.

3. Aldosterone stimulates H^+ secretion indirectly by increasing Na^+ reabsorption.

---

*Since about 75% of the filtered HPO_4^{2-} is reabsorbed, only about 25% of the filtered HPO_4^{2-} is available for buffering.

V. **EXCRETION OF AMMONIUM (NH$_4$$^+$)** [Figures 36-4 and 36-5; see Table 36-1]. Unlike phosphate, NH$_3$ enters the tubular lumen not by filtration but by tubular synthesis and secretion, which normally are confined to the distal and collecting tubules.

A. **Secretion of NH$_3$.** NH$_3$ is synthesized mainly by the deamidization and deamination of glutamine in the presence of glutaminase. Most of the NH$_4$$^+$ excreted in the urine is produced in the proximal tubular cells from amino acids, primarily glutamine.

1. The NH$_4$$^+$ that are formed are actively secreted into the lumen and are excreted in the final urine at a rate equal to their secretion rate.

2. With regard to proximal NH$_4$$^+$ secretion, it must be emphasized that the traditional view that states that NH$_3$ combines with the H$^+$ derived from the dissociation of H$_2$CO$_3$ is incorrect. Instead, the H$^+$ produced in the metabolism of glutamine results in the formation of NH$_4$$^+$ within the proximal tubular cell (see Figure 36-4).

3. The mechanism of distal nephron NH$_4$$^+$ excretion is different from that in the proximal tubule (see Figures 32-1; 36-5).

B. **Formation of NH$_4$$^+$.** The NH$_3$/NH$_4$$^+$ system has a very high pK' (about 9.2), which means that, at the usual urine pH, virtually all the nonpolar NH$_3$ that enters the distal tubular lumen immediately combines with H$^+$ to form NH$_4$$^+$, which is nondiffusible because it is lipid insoluble. The renal excretion of NH$_4$$^+$ causes the net addition of HCO$_3$$^-$ to the plasma.

1. The important physiologic characteristic of the NH$_3$/NH$_4$$^+$ buffer pair is that, as H$^+$ is combined with intraluminal buffer (NH$_3$), H$^+$ is excreted in the urine as NH$_4$$^+$, a substance that does not cause the pH of urine to fall.

2. Under most conditions, the excretion of NH$_4$$^+$ is quantitatively more important to the acid-base balance of the body than is the excretion of titratable acid. The normal kidney excretes almost twice as much acid combined with NH$_3$ (30–50 mEq/day) as titratable acid (10–30 mEq/day) [see Table 36-1].
   a. In chronic severe metabolic acidosis, NH$_3$ serves as the major urinary buffer, and NH$_4$$^+$ excretion can increase from a normal value of 30 mEq/day to 500 mEq/day. However, in chronic acidosis, the excretion of H$_2$PO$_4$$^-$ may increase by only 20–40 mEq/day.
   b. In diabetic ketoacidosis, the excretion of titratable acid may reach 75–250 mEq/day, while the excretion of acid combined with NH$_3$ may reach amounts of 300–500 mEq/day. However, the ratio of NH$_4$$^+$ to titratable acid remains within the normal range of 1 to 2.5.

C. **Excretion of H$^+$.** The sum of titratable acid and urinary NH$_4$$^+$ excretion represents the net gain of HCO$_3$$^-$ for the body fluids.

1. The net amount of fixed acid excreted (in mEq/day) equals the sum of titratable acid and NH$_4$$^+$ minus the urinary [HCO$_3$$^-$].

2. Normally, the urine is free of HCO$_3$$^-$ because all the HCO$_3$$^-$ remaining in the distal and collecting tubules has combined with secreted H$^+$ and been reabsorbed.

3. Aldosterone enhances NH$_3$ production via an effect on cellular metabolism.

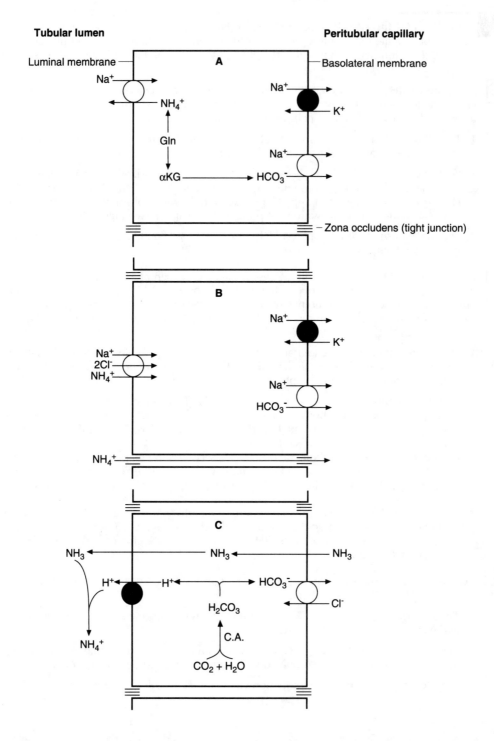

FIGURE 36-5. Mechanism of NH_4^+ transport in the proximal tubule (*A*), the thick ascending limb of the loop of Henle (*B*), and the collecting duct (*C*). In *A*, glutamine (*Gln*) is taken up by the cells and metabolized into NH_4^+ and α-ketoglutarate (*αKG*). In *B*, NH_4^+ is transported by the Na^+-$2Cl^-$-NH_4^+ symporter and by paracellular transport. In *C*, NH_3 diffuses from the interstitial fluid into the lumen, where it combines with secreted H^+ to form NH_4^+. *C.A.* = carbonic anhydrase; ● = adenosine triphosphatase (ATPase) pump; ○ = protein carrier.

Chapter 37

Primary and Secondary Acid-Base Abnormalities

I. DEFINITIONS AND BASIC CONCEPTS (Tables 37-1; 37-2)

A. Acidosis and alkalosis

1. **Acidosis** or **acidemia** is an abnormal clinical condition or process caused by the bodily accumulation of acid (or the loss of base) sufficient to decrease pH below 7.36 or to increase the $[H^+]$ above 43.6 nEq/L of blood in the absence of compensatory (secondary) changes.

2. **Alkalosis** or **alkalemia** is an abnormal condition or process caused by the accumulation of base (or the loss of acid) sufficient to raise pH above 7.44 or to decrease the $[H^+]$ below 36.3 nEq/L of blood in the absence of compensatory changes.

B. Respiratory and metabolic

1. The adjective **respiratory** denotes that the primary abnormality involves impairment in alveolar ventilation, which results in an abnormally high or low [total CO_2] of the extracellular fluid (ECF).*
   a. **Respiratory acidosis** refers to a condition of abnormally high arterial CO_2 tension (Pco_2), which is termed **hypercapnia** (hypercarbia).
   b. **Respiratory alkalosis** refers to a condition of abnormally low arterial CO_2 tension, which is called **hypocapnia** (hypocarbia).

2. The adjective **metabolic** denotes that the primary abnormality involves an abnormal gain or loss of noncarbonic acid by the ECF, which affects $[HCO_3^-]$.
   a. **Metabolic acidosis** refers to a disturbance that leads to the accumulation of noncarbonic acid in the ECF or to the loss of HCO_3^- from the ECF.
   b. **Metabolic alkalosis** refers to an imbalance characterized by a loss of noncarbonic acid or a gain of HCO_3^- by the ECF.

C. Primary and secondary factors

1. The factor in the $[HCO_3^-]/S \cdot Pco_2$ ratio (i.e., $[HCO_3^-]$ or Pco_2) that undergoes the greater degree of displacement (i.e., the larger proportional change) indicates the **primary abnormality** (see Tables 37-1; 37-2).
   a. If the $[HCO_3^-]/S \cdot Pco_2$ ratio becomes less than 20:1, by either decreasing the numerator or increasing the denominator, pH falls ($\uparrow [H^+]$), and acidosis occurs.
   b. If the $[HCO_3^-]/S \cdot Pco_2$ ratio becomes more than 20:1, by either increasing the numerator or decreasing the denominator, pH rises ($\downarrow [H^+]$), and alkalosis occurs.

2. A **secondary response** in the alternate variable, which occurs in the same direction as the primary abnormality, counteracts the effect of the primary abnormality on the $[HCO_3^-]/S \cdot Pco_2$ ratio. The secondary change represents **compensation** and acts to minimize the pH alteration produced by the primary disorder.

D. Simple and mixed acid-base disturbances

1. **Simple acid-base imbalances** are caused by one primary factor.

---

*[Total CO_2] refers to the dissolved CO_2 ($S \cdot Pco_2$) plus the $[HCO_3^-]$.

TABLE 37-1. Characteristics of the Uncompensated Acid-Base Abnormalities

Acid-Base Abnormality	Primary Disturbance	Effect on:			Compensatory Response
		$[HCO_3^-]/S \cdot P_{CO_2}$	$[H^+]$	pH	
Acidosis					
Respiratory	↑ P_{CO_2}	< 20	↑	↓	↑ $[HCO_3^-]$
Metabolic	↓ $[HCO_3^-]$	< 20	↑	↓	↓ P_{CO_2}
Alkalosis					
Respiratory	↓ P_{CO_2}	> 20	↓	↑	↓ $[HCO_3^-]$
Metabolic	↑ $[HCO_3^-]$	> 20	↓	↑	↑ P_{CO_2}

Note—These conditions are simple disturbances (i.e., they represent the effects of one primary etiologic factor) and, thus, are termed uncompensated or pure acid-base abnormalities. The compensatory response always occurs in the same direction as the primary disturbance.

2. **Mixed acid-base imbalances** are caused by more than the primary factor.
   a. Mixed-type disturbances are not uncommon. Sometimes a primary abnormality of one type is superimposed on a primary abnormality of another type.
      (1) A patient with respiratory acidosis from pulmonary emphysema may develop metabolic acidosis from uncontrolled diabetes or metabolic alkalosis from large doses of corticosteroids used in the treatment of an attack of status asthmaticus.
      (2) A patient with a metabolic acid-base disturbance, in turn, may develop a respiratory acid-base abnormality.
   b. The changes in the variables may be difficult to interpret in mixed acid-base disturbances, because manifestations of one primary abnormality may be either cancelled out or augmented by those of the other primary abnormality. Therefore, **a normal pH may not necessarily mean a normal compensation in mixed acid-base imbalances**.

E. **Plasma $[HCO_3^-]$, CO_2 tension, and [total CO_2].** For an adequate analysis of any acid-base disorder, it is necessary to measure pH, CO_2 tension, and the total CO_2 content of blood.

1. **An acid-base disorder cannot be diagnosed with certainty from the plasma $[HCO_3^-]$ alone.** Although a reduction in the plasma $[HCO_3^-]$ may be due to metabolic acidosis,

TABLE 37-2. Primary and Secondary* Changes in the CO_2/HCO_3^- System

	Acidosis		Alkalosis	
Respiratory	$\dfrac{[HCO_3^-]}{S \cdot P_{CO_2}}$	< 20:1	$\dfrac{[HCO_3^-]}{S \cdot P_{CO_2}}$	> 20:1
	[total CO_2]† ↑		[total CO_2] ↓	
Metabolic	$\dfrac{[HCO_3^-]}{S \cdot P_{CO_2}}$	< 20:1	$\dfrac{[HCO_3^-]}{S \cdot P_{CO_2}}$	> 20:1
	[total CO_2]		[total CO_2]	

Note—*Dashed arrows* denote direction of compensatory responses. *Thin solid arrows* show changes in [total CO_2] during respiratory imbalances. *Wide arrows* depict direction of change of primary acid-base imbalance and direction of change in [total CO_2]. (Adapted from Christensen HN: *Diagnostic Biochemistry.* New York, Oxford University Press, 1959, p 106.)

*A compensatory (secondary) response in the alternate variable occurs in the same direction and counteracts the effect of the primary disturbance by returning the $[HCO_3^-]/S \cdot P_{CO_2}$ ratio toward normal (20:1).

†[Total CO_2] does not always increase in alkalosis or decrease in acidosis. Dissolved CO_2 contributes little to the [total CO_2]. [Total CO_2] provides little information about pulmonary function.

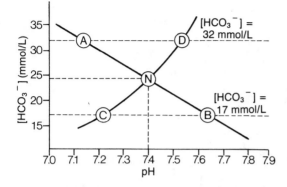

FIGURE 37-1. A pH-[HCO_3^-] diagram illustrating that acidosis (*points A and C*) and alkalosis (*points B and D*) cannot be differentiated solely on the basis of [HCO_3^-] or [total CO_2]. The [HCO_3^-] pertains to plasma. A = respiratory acidosis; B = respiratory alkalosis; C = metabolic acidosis; D = metabolic alkalosis; and N = normal point. The slope of *line ANB* is a measure of the nonbicarbonate buffer capacity of whole blood. (After Davenport HW: *The ABC of Acid-Base Chemistry,* 5th edition. Chicago, University of Chicago Press, 1969, p 65.)

it also can indicate a renal compensation for respiratory alkalosis. Similarly, an elevated [HCO_3^-] can result from a metabolic alkalosis or the secondary response to respiratory acidosis. Since the aim of therapy is to bring about acid-base balance and not to normalize [HCO_3^-], it is necessary to measure the pH (i.e., [H^+]) when acid-base imbalance is suspected.

2. **Low CO_2 tension** may be due to respiratory alkalosis or a respiratory compensation to metabolic acidosis. Similarly, a **high CO_2 tension** may be caused by respiratory acidosis or a respiratory compensatory response to metabolic alkalosis.

3. **[Total CO_2]** does not always increase in alkalosis or decrease in acidosis (Figure 37-1; see Table 37-2). About 90% of the [total CO_2] of plasma is contributed by [HCO_3^-], and 5% is contributed by dissolved CO_2 and H_2CO_3; therefore, dissolved CO_2 contributes little to the [total CO_2]. The [total CO_2] gives little information about the functional status of the lungs.

4. The physician accepts [total CO_2] and [HCO_3^-] as essentially interchangeable terms. The [total CO_2] normally exceeds the [HCO_3^-] by 1.2 mmol. Most laboratories measure the [total CO_2] and not the [HCO_3^-].

II. CLINICAL EXPRESSIONS FOR EVALUATION OF ACID-BASE STATUS

A. **Base excess (BE)** [Figure 37-2; Table 37-3]

1. **Definition.** BE refers to the change in the concentration of buffer base [BB] from its normal value, or, the **observed [BB] minus the normal [BB]** in mEq/L of whole blood. The range of normality for BE is -2 to $+2$ mEq/L in arterial whole blood.*
   a. The [BB] in normal whole blood is 48 mEq/L, which is arbitrarily set to 0, and deviations (i.e., ± BE) are measured from this reference point.
   b. The normal [BB] of 48 mEq/L equals the sum of all the conjugate bases in 1 L of arterial whole blood. These bases and their concentrations are:
   (1) [HCO_3^-] = 24 mEq/L
   (2) [$Protein^-$] = 15 mEq/L
   (3) [Hb^-/HbO_2^-] = 9 mEq/L

2. **Clinical application.** BE refers principally to the [HCO_3^-] but also to other bases in the blood (mainly plasma protein and hemoglobin). However, the [HCO_3^-] or [total CO_2] is used in acid-base problems as the indicator of BE.

---

*Because BE is a negative or positive value, the term "base deficit" is avoided.

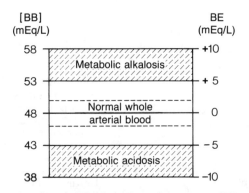

FIGURE 37-2. Diagram illustrating the relationship between base excess *[BE]* and the acid-base status of normal whole arterial blood. BE is the base concentration (in mEq/L), as measured by the titration with strong acid to pH 7.4, at CO_2 tension of 40 mm Hg, and at 37° C. For negative values of BE, the titration must be carried out with strong base. BE measures the change in the concentration of the buffer base (*[BB]*) from its normal value, which is 48 mEq/L in normal whole arterial blood. [BB] is the sum of the buffer anions of blood or plasma. When observed [BB] is less than normal [BB], BE is a negative value; when observed [BB] is greater than normal [BB], BE is a positive value; and when observed [BB] equals normal [BB], BE equals zero. (After Winters RW, et al: *Acid-Base Physiology in Medicine.* Westlake, Ohio, London Co, 1967, p 45.)

    a. Because $[HCO_3^-]$ and BE are influenced only by metabolic processes, there are only two acid-base conditions associated with abnormalities in $[HCO_3^-]$ and BE.

        (1) Metabolic acidosis. When a metabolic process leads to the accumulation of noncarbonic acid in the body or the loss of HCO_3^-, the $[HCO_3^-]$ falls, and the BE value becomes negative. Metabolic acidosis is associated with a BE below -5 mEq/L.

        (2) Metabolic alkalosis. When a metabolic process causes a loss of acid or an accumulation of HCO_3^-, the $[HCO_3^-]$ rises above normal, and the BE value becomes positive. Metabolic alkalosis is associated with a BE above $+5$ mEq/L.

    b. The BE value serves only as a rough guide for alkalosis and acidosis therapy.

B. | **Anion gap**

    1. Definition

        a. The ionic profile of normal serum is depicted in Figure 37-3. The **law of electroneutrality** states that the number of positive charges in any solution must equal the number of negative charges. If every ion present in serum were measured, the concentration of cations would equal the concentration of anions if these concentrations were expressed in mEq/L. Routine serum electrolyte determinations measure essentially all cations but only a fraction of the anions. **This apparent disparity between the total cation concentration and the total anion concentration is termed the anion gap.** It is a virtual measurement and does not represent any specific ionic constituent.

        b. The anion gap, which has a normal value of 12 ± 4 mEq/L, reflects the concentrations of those anions actually present but routinely undetermined (i.e., other than $[Cl^-]$ and $[HCO_3^-]$), such as:

        (1) Polyanionic plasma proteins (primarily albumin)

TABLE 37-3. Base Excess and Metabolic Acid-Base Abnormalities

$[HCO_3^-]$	Base Excess	Metabolic Abnormality	Characteristics
↑	+	Metabolic alkalosis	Noncarbonic acid is lost; HCO_3^- is gained
↓	−	Metabolic acidosis	Noncarbonic acid is gained; HCO_3^- is lost

Adapted from Broughton JO: *Understanding Blood Gases.* Madison, Wisconsin, Ohio Medical Products, form no. 456, 1979, p 8.

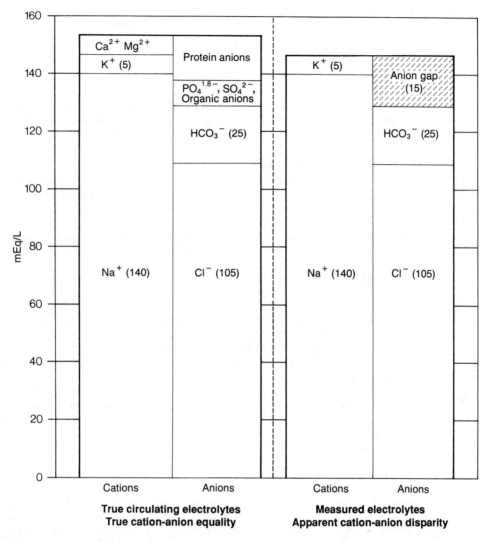

FIGURE 37-3. Normal plasma concentrations of anions and cations in mEq/L show that circulating cations always counterbalance anions, thereby maintaining electroneutrality (*left*). Routine electrolyte assays only measure a portion of circulating anions, and an apparent cation-anion disparity exists (*right*). This difference, in mEq/L, is termed the anion gap. *Parentheses* indicate concentrations in mEq/L. (Reprinted from Narins RG, et al: Lactic acidosis and elevated anion gap (I). *Hosp Pract* 15:125–136, 1980. Original drawing by Albert Miller.)

     (2) Inorganic phosphates
     (3) Sulfate
     (4) Ions of organic acids (e.g., lactic, β-hydroxybutyric, and acetoacetic acids)
   c. When the anion gap is increased, unmeasured anions fill this gap. (Most of these anions usually are the products of metabolic processes that generate H^+.)

2. Determination of anion gap
   a. The anion gap (AG) equals 16 mEq/L of anions other than HCO_3^- and Cl^- when determined by the following equation

$$AG = ([Na^+] + [K^+]) - ([HCO_3^-] + [Cl^-])$$
$$= (142 \text{ mEq/L} + 4 \text{ mEq/L}) - (25 \text{ mEq/L} + 105 \text{ mEq/L})$$
$$= 146 \text{ mEq/L} - 130 \text{ mEq/L}$$
$$= 16 \text{ mEq/L}$$

b. Often the anion gap is calculated with Na^+ as the major cation, because K^+ has a relatively minor quantitative contribution. The equation, then, becomes:

$$AG = [Na^+] - ([HCO_3^-] + [Cl^-])$$
$$= 142\ mEq/L - (25\ mEq/L + 105\ mEq/L)$$
$$= 142\ mEq/L - 130\ mEq/L$$
$$= 12\ mEq/L$$

c. The plasma proteins contribute a significant amount of negative charges and, thus, anionic equivalence (16 mEq/L). The plasma proteins account, almost stoichiometrically, for the difference between the $[Na^+]$ and the sum of $[HCO_3^-]$ and $[Cl^-]$.

3. Clinical application

a. Acidification with acids other than HCl

(1) Organic acids increase the anion gap. In this case, the lost (neutralized) HCO_3^- is *not* replaced by the routinely unmeasured anions of noncarbonic acids such as lactic acid and the ketoacids. Thus, the anion gap is increased by all metabolic acidoses except the **hyperchloremic acidoses**. With the accumulation of acid, there is rapid extracellular buffering by HCO_3^-. If the acid is HCl, then

$$HCl + NaHCO_3 \rightleftharpoons NaCl + H_2CO_3$$

The net effect is the milliequivalent-for-milliequivalent replacement of extracellular HCO_3^- by Cl^-. Since the sum of $[Cl^-]$ and $[HCO_3^-]$ remains constant, the anion gap is unchanged in hyperchloremic acidosis.

(2) It is apparent that the decrease in plasma $[HCO_3^-]$ equals the increase in the anion gap (see Figure 37-3). The presence of an increased anion gap usually indicates an excess of H^+ derived from noncarbonic acid. Whatever the size of the anion gap, **it is the retention of H^+—not the particular anion—that is responsible for the acidosis**.

(a) An increased anion gap due to excess H^+ derived from noncarbonic acid occurs in **metabolic acidosis;** this also may result from a compensatory increase in lactic acid production in response to respiratory alkalosis via activation of the phosphofructokinase step in glycolysis.

(b) In **respiratory acidosis** the anion gap is not increased, because excess H^+ is derived from the H_2CO_3 pool, not the noncarbonic acid pool.

b. Conditions that increase the anion gap include diabetic and alcoholic ketoacidosis, intoxicant and lactic acidoses, and renal failure.

c. Conditions that cause metabolic acidosis without an increase in the anion gap are associated with a high serum $[Cl^-]$ and include diarrhea; pancreatic drainage; ureterosigmoidostomy; ileal loop conduit; treatment with acetazolamide, ammonium chloride, or arginine-HCl; renal tubular acidosis; and, rarely, intravenous hyperalimentation.

C. **Osmolar gap**

1. Definition. Osmolar gap refers to the disparity between the measured serum (plasma) osmolality and the calculated serum osmolality.

2. Clinical application. Osmolar gap measurement provides a reasonably good screening procedure for toxins. Other causes of metabolic acidosis do not affect the osmolar gap, since the metabolic acid simply replaces the HCO_3^- with another anion and HCO_3^- is lost as CO_2.

a. Serum osmolality is estimated by dividing blood urea nitrogen (BUN) and glucose concentrations (in mg/L) by their molecular weights,* thereby converting them to milliosmoles (mOsm)/L. Doubling the serum $[Na^+]$ (in mEq/L) provides an estimate

---

*The molecular weight of urea (60) is not used because it is BUN that is measured, and urea contains 2 nitrogen atoms $(2 \cdot 14 = 28)$.

of the serum ion concentration. With a BUN of 14 mg/dl, plasma glucose of 90 mg/dl, and a plasma [Na^+] of 142 mEq/L, serum (plasma) osmolality (P_{osm}) is estimated as

$$P_{osm} = 2 [Na^+] + [BUN (mg/L)/28] + [glucose (mg/L)/180]$$
$$= 2 (142) + 140/28 + 900/180$$
$$= 286 + 5 + 5 = 296 \text{ mOsm/kg}$$

b. Circulating intoxicants increase measured serum osmolality without altering serum [Na^+]. Therefore, the measured serum osmolality exceeds the calculated serum osmolality, and the difference closely reflects the osmolar concentration of the circulating toxins.

III. GRAPHIC EVALUATION OF ACID-BASE STATUS: pH-[HCO_3^-] DIAGRAM
(Figures 37-4; 37-5)

A. **Value of diagram.** When acid-base data obtained from blood are plotted on the pH-[HCO_3^-] diagram, the physician has a tool that can be used to:

1. Diagnose the primary cause of an acid-base abnormality

2. Determine the degree of compensatory response of the kidneys and of the lungs by monitoring the [HCO_3^-] and CO_2 tension, respectively

3. Estimate the concentration of noncarbonic and carbonic acids in the ECF

4. Aid in the choice of therapy to correct an acid-base imbalance

5. Measure the nonbicarbonate buffer power of whole blood by the slope of the buffer line

6. Indicate not only the total buffer activity but also the distribution of the buffering activity between the bicarbonate and nonbicarbonate buffer systems

FIGURE 37-4. The pH-[HCO_3^-] diagram showing the buffer curves of separated plasma (*S*) and true oxygenated plasma (*T*). *Point N* denotes the intercept for a normal plasma [HCO_3^-] of 24 mmol/L and a normal pH of 7.4. The slope of the normal in vitro buffer line of true plasma is a function of the hemoglobin content of blood. *Lines S* and *T* are called nonbicarbonate buffer curves, and the steepness of the slopes of these lines is a quantitative assessment of the amount of nonbicarbonate buffer present in the system. These same two lines represent CO_2-titration curves. The steeper the buffer line, the smaller the pH change resulting from a given increase or decrease in CO_2 tension (Pco_2). *Vertical arrows* indicate increases and decreases in CO_2 tension, fixed acid, or fixed base. (Adapted from Davenport HW: *The ABC of Acid-Base Chemistry,* 5th edition. Chicago, University of Chicago Press, 1969, p 48.)

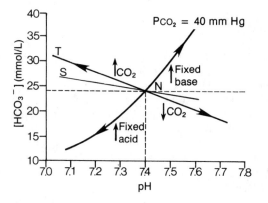

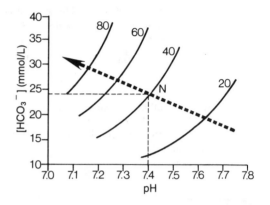

FIGURE 37-5. The pH-[HCO_3^-] diagram with arterial CO_2 tension (P_{CO_2}) isobars for 20, 40, 60, and 80 mm Hg. Each arterial P_{CO_2} isobar is the titration curve of a HCO_3^--H_2CO_3 solution with arterial P_{CO_2} held constant. *Point N* denotes the intercept for a normal plasma [HCO_3^-] of 24 mmol/L and a normal pH of 7.4. The *dashed arrow* represents the buffer line. (After Davenport HW: *The ABC of Acid-Base Chemistry,* 5th edition. Chicago, University of Chicago Press, 1969, p 45.)

B. **Construction of diagram: effects of H^+ and CO_2 on pH.** The basis for plotting pH and [HCO_3^-] on cartesian coordinates is best understood from a consideration of the CO_2/HCO_3^- buffer system:

$$CO_2 + H_2O \rightleftharpoons H_2CO_3 \rightleftharpoons H^+ + HCO_3^-$$

1. **Effect of fixed acid or base on pH**
   a. **Addition of H^+ at a constant CO_2 tension**
      (1) When H^+ ions, in the form of fixed (noncarbonic) acid, are added to the CO_2/HCO_3^- system, most of them combine with HCO_3^- to form H_2CO_3, which dehydrates to form CO_2 and water. Thus, the series of reactions in the equation is driven to the left.
      (2) The decrease in [HCO_3^-] closely approximates the amount of H^+ added only if the solution has no other nonbicarbonate buffer substances. Otherwise, the amount of acid added would be greater than the decrease in [HCO_3^-].
   b. **Addition of base at a constant CO_2 tension**
      (1) When a base is added to the system depicted in the equation, some of the base combines with H^+, causing more H_2CO_3 to dissociate into H^+ and HCO_3^-. Here the series of reactions proceeds to the right.
      (2) The increase in [HCO_3^-] determines the amount of base added.
   c. **Principles** (see Figure 37-4)
      (1) When fixed acid or base is added under the conditions described, there is an inverse relationship between [H^+] and [HCO_3^-], and the primary acid-base disturbance is metabolic. Thus, when fixed acid or base is added, the [H^+] and [HCO_3^-] change in opposite directions.
      (2) Or, if the direction of change in [HCO_3^-] is the same as that for pH, the primary acid-base abnormality is metabolic.

2. **Effect of CO_2 on pH**
   a. **Addition of CO_2** (see Figure 37-5)
      (1) An increase in arterial CO_2 tension titrates the blood (and ECF) in the acid direction. This is clear from the equation in III B, which shows that an increase in CO_2 tension causes more CO_2 to hydrate to form H_2CO_3; the H_2CO_3 in turn dissociates into H^+ and HCO_3^-, moving the reactions to the right.
      (2) In the CO_2/HCO_3^- system, every molecule of CO_2 that hydrates and dissociates forms one H^+ and one HCO_3^-; therefore, the changes in [H^+] and [HCO_3^-] are exactly equal.
      (3) The final [H^+] depends not only on the change in CO_2 tension, but also on the buffers in the system. For every H_2CO_3 molecule that has its H^+ taken up by a buffer ion, one HCO_3^- appears in the solution. For example, the reaction with hemoglobin is:

$$H_2CO_3 + Hb^- \rightleftharpoons HHb + HCO_3^-$$

(4) If CO_2 is added to whole blood (which contains many buffer systems), the H^+, rather than remaining in solution, are mostly bound to protein buffers (Pr^-) as:

$$CO_2 + H_2O \rightleftharpoons H_2CO_3 \rightleftharpoons HCO_3^- + H^+$$
$$+$$
$$Pr^-$$
$$\updownarrow$$
$$H \cdot Pr$$

Therefore, **the CO_2/HCO_3^- system alone is not effective in buffering the changes in pH that are induced by the addition of CO_2.**

b. Removal of CO_2

   (1) When CO_2 is removed from the CO_2/HCO_3^- system, fewer CO_2 molecules are available for combination with water to form HCO_3^-.

   (2) Thus, not only is CO_2 decreased, but $[H^+]$ and $[HCO_3^-]$ also are decreased and the reactions shift to the left.

c. Principles. The concentration of acid added to a buffer solution by a change in CO_2 tension is equal to the change in $[HCO_3^-]$.*

   (1) If CO_2 is added or removed, the $[H^+]$ and $[HCO_3^-]$ change in the same direction, and the primary acid-base disturbance is respiratory.

   (2) Or, if the direction of change in $[HCO_3^-]$ is opposite to that for pH, the primary acid-base disturbance is respiratory.

   (3) As CO_2 tension is increased or decreased, the values for pH and $[HCO_3^-]$ form coordinates that define a nearly straight line termed the buffer line or the CO_2 titration curve (see Figure 37-5).

d. Summary

   (1) When H^+ is added or removed, the $[H^+]$ and $[HCO_3^-]$ change in opposite directions and the acid-base imbalance is metabolic in origin.

   (2) When CO_2 is added or removed, the $[H^+]$ and $[HCO_3^-]$ change in the same direction and the acid-base imbalance is respiratory in origin.

C. **Properties of CO_2 tension (Pco_2) isobars**

1. At constant $[HCO_3^-]$, CO_2 tension is proportional to $[H^+]$. This property can be illustrated using the following form of the Henderson equation to solve for Pco_2 (see Chapter 33 V):

$$[H^+] = 24 \; \frac{Pco_2}{[HCO_3^-]}$$

$$Pco_2 = \frac{[H^+] \, [HCO_3^-]}{24}$$

2. At constant pH (i.e., along any vertical line in Figures 37-4 and 37-5), CO_2 tension is proportional to $[HCO_3^-]$.

IV. **INTERPRETATION OF ACID-BASE ABNORMALITIES USING THE pH-$[HCO_3^-]$ DIAGRAM** (Figure 37-6). Acid-base imbalances are determined graphically with reference to the intercept of the Pco_2 isobar of 40 mm Hg and the normal buffer line. The intercept of these two curves marks the point of normality (*N*), which is associated with a pH of 7.4 (the *abscissa*) and a $[HCO_3^-]$ of 24 mmol/L (the *ordinate*). **Point N is the triple intercept that defines the pH, $[HCO_3^-]$, and CO_2 tension of true arterial plasma of a normal individual.**

---

*In humans, the serum $[HCO_3^-]$ is 24×10^{-3} Eq/L (24×10^{-3} mol/L), whereas the serum $[H^+]$ is 40×10^{-9} Eq/L (40×10^{-9} mol/L). Therefore, the $[HCO_3^-]$ is 6×10^5 greater than $[H^+]$, and any shift (left or right) in equilibrium has a much greater proportionate effect on $[H^+]$ than on $[HCO_3^-]$.

FIGURE 37-6. Pathways of acid-base imbalance. The *[HCO$_3^-$]* pertains to plasma. *Line ANB* represents the normal blood buffer line, and *line CND* represents the normal CO$_2$ tension (*PCO_2*) isobar (40 mm Hg). *N* = normal point; *A* = respiratory acidosis (uncompensated); *B* = respiratory alkalosis (uncompensated); *C* = metabolic acidosis (uncompensated); *D* = metabolic alkalosis (uncompensated); *E* = respiratory acidosis + metabolic acidosis; *F* = respiratory acidosis + metabolic alkalosis; *G* = respiratory alkalosis + metabolic acidosis; and *H* = respiratory alkalosis + metabolic alkalosis. (After Davenport HW: *The ABC of Acid-Base Chemistry*, 5th edition. Chicago, University of Chicago Press, 1969, p 65.)

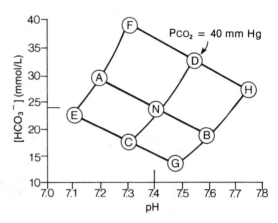

A. **Points to the left** of *point N* (*points A, C, E,* and *F*) indicate acidosis. **Points to the right** of *Point N* (*points B, D, G,* and *H*) indicate alkalosis.

    1. *Line NA* represents the direction of an individual's response to an increase in PCO_2. Points to the left of the normal PCO_2 isobar (*points A, E,* and *F*) indicate **respiratory acidosis**.

        a. *Point A* denotes a condition of uncompensated respiratory acidosis, which is characterized by a low pH, high [HCO$_3^-$], and high CO$_2$ tension.

        b. *Point E* denotes respiratory acidosis (↑ PCO_2) with metabolic acidosis (↓ [HCO$_3^-$]).

        c. *Point F* denotes respiratory acidosis (↑ PCO_2) with metabolic alkalosis (↑ [HCO$_3^-$]).

    2. *Line NB* represents the direction of an individual's response to a decrease in CO$_2$ tension. Points to the right of the normal PCO_2 isobar (*points B, G,* and *H*) indicate **respiratory alkalosis**.

        a. *Point B* denotes a condition of uncompensated respiratory alkalosis, which is characterized by high pH, low [HCO$_3^-$], and low CO$_2$ tension.

        b. *Point G* denotes respiratory alkalosis (↓ PCO_2) with metabolic acidosis (↓ [HCO$_3^-$]).

        c. *Point H* denotes respiratory alkalosis (↓ PCO_2) with metabolic alkalosis (↑ [HCO$_3^-$]).

B. **Points below** the normal buffer line (*points C, E,* and *G*) indicate conditions with a component of metabolic acidosis (↓ [HCO$_3^-$]). **Points above** the normal buffer lines (*points D, F,* and *H*) indicate conditions with a component of metabolic alkalosis (↑ [HCO$_3^-$]).

    1. *Line NC* represents the direction of the development of metabolic acidosis in an individual; the [HCO$_3^-$] of this individual moves down the normal PCO_2 isobar.

        a. *Point C* denotes a condition of uncompensated metabolic acidosis, which is characterized by low pH, low [HCO$_3^-$], and normal CO$_2$ tension.

        b. *Point E* denotes metabolic acidosis (↓ [HCO$_3^-$]) with respiratory acidosis (↑ PCO_2).

        c. *Point G* denotes metabolic acidosis (↓ [HCO$_3^-$]) with respiratory alkalosis (↓ PCO_2).

    2. *Line ND* represents the direction of the development of metabolic alkalosis in an individual; the [HCO$_3^-$] of this individual moves up the normal PCO_2 isobar.

        a. *Point D* denotes a condition of uncompensated metabolic alkalosis, which is characterized by high pH, high [HCO$_3^-$], and normal CO$_2$ tension.

        b. *Point F* denotes metabolic alkalosis (↑ [HCO$_3^-$]) with respiratory acidosis (↑ PCO_2).

        c. *Point H* denotes metabolic alkalosis (↑ [HCO$_3^-$]) with respiratory alkalosis (↓ PCO_2).

C. **Mixed acid-base disturbances,** in which the primary state of acidosis or alkalosis has an additional abnormality with respect to CO$_2$ tension and [HCO$_3^-$], are denoted by *points E, F, G,* and *H*.

Chapter 38

Compensatory Mechanisms For Primary Acid-Base Abnormalities

I. TERMINOLOGY

A. **Compensation** is the secondary physiologic process occurring in response to a primary acid-base disturbance by which the deviation of blood pH is ameliorated. Thus, abnormal pH is returned toward normal by **altering the component that is not primarily affected**.

    1. The terms secondary and compensatory may be used to describe a change in the composition of the blood or to describe a process.

    2. The terms secondary and compensatory may be used to describe a process, but they should not be used to describe acidosis or alkalosis.

B. **Correction** denotes the therapeutic course of action that is directed toward **counteracting or altering the factor that is primarily affected**. Correction involves the amelioration of the primary underlying abnormality in $[H^+]$, $[HCO_3^-]$, or CO_2 tension. In correction, as in compensation, the pH is returned toward normal.

II. BASIC MECHANISMS OF COMPENSATORY RESPONSES

A. **Buffers of the extracellular fluid (ECF)** represent the body's first defense mechanism for neutralizing noncarbonic acid. H^+ is transferred across tissue cell membranes in a direction that normalizes blood pH.

B. A deviation in plasma pH acts on the respiratory center to change alveolar ventilation, which, in turn, alters the alveolar CO_2 tension and arterial CO_2 tension so as to counteract the change in plasma pH. This is called **respiratory compensation** of a primary noncarbonic acid excess or deficit; respiratory compensation usually is not complete (i.e., does not restore pH to normal).

C. **Carbonate ion is released from bone,** which is the predominant source of alkali for neutralizing excess noncarbonic acid added to the ECF.

D. Renal excretion of noncarbonic acid or base increases in order to eliminate an excess of noncarbonic acid or base or to compensate for the pH changes due to an abnormal arterial CO_2 tension. This is called **renal compensation** of a primary hyper- or hypocapnia.

E. The responses to respiratory acidosis and alkalosis differ from responses to metabolic acid-base disorders in that there is virtually no extracellular buffering of H_2CO_3 in respiratory acidosis because HCO_3^- is not an effective buffer for H_2CO_3.

III. PRINCIPLES OF RESPIRATORY AND RENAL COMPENSATORY RESPONSES

A. **General considerations**

    1. Physiologic compensation for major acid-base abnormalities rarely is complete. Therefore, the singular concern of clinical diagnosis is to differentiate the primary cause of the imbalance from the secondary (compensatory) response.

2. Most data indicate that the pH ($[H^+]$) is the most important factor in determining the bodily response to acid-base imbalances. (In both acid and base disturbances, the abnormal pH is returned toward normal.) However, in spite of its critical biologic significance, **a change in pH does not provide all the information needed for quantitative assessment and, thus, for planning therapy.**
   a. The key determinants of the cause and compensation of acid-base abnormalities are CO_2 tension and $[HCO_3^-]$, not pH.
   b. The most important factor in determining the efficacy of the bodily responses to deviations from the normal acid-base status is pH or $[H^+]$.

B. **Respiratory and renal processes** (see Tables 37-1; 37-2)

1. Both respiratory and renal compensatory responses tend to restore the abnormal $[HCO_3^-]/S \cdot P_{CO_2}$ ratio toward its normal value of 20:1.
   a. Primary acid-base disturbances of **metabolic origin** lead to secondary adjustment of the CO_2 tension by changes in the rate of alveolar ventilation.
   b. Primary acid-base disturbances of **respiratory origin** lead to secondary changes in blood $[HCO_3^-]$ by appropriate adjustment of the rate of H^+ secretion/excretion from the renal tubular cell into the tubular lumen.

2. These two processes interact to control the $[H^+]$ of the ECF, and this integration is clearly understood using the mathematical relationship of the Henderson equation (see Chapter 33 V):

$$[H^+] = 24 \frac{P_{CO_2}}{[HCO_3^-]}$$

From this equation it is apparent that:
   a. An increase in the $[H^+]$, regardless of cause, can be reduced toward normal by a decrease in CO_2 tension, an increase in plasma $[HCO_3^-]$, or by both changes.
   b. A decrease in the $[H^+]$, regardless of cause, can be increased toward normal by an increase in CO_2 tension, a decrease in plasma $[HCO_3^-]$, or by both changes.

IV. ETIOLOGY OF ACID-BASE DISORDERS AND THEIR COMPENSATORY RESPONSES

A. **Metabolic acidosis** is a disorder characterized by a low arterial pH ($\uparrow [H^+]$) or a reduced plasma $[HCO_3^-]$.

1. **Causes** of metabolic acidosis include:
   a. Gain of noncarbonic (fixed) acid by the ECF, such as excess quantities of the keto-acids (β-hydroxybutyric acid and acetoacetic acid) or lactic acid or the ingestion of alcohol, salicylates, or NH_4Cl
   b. Loss of HCO_3^- and other conjugate bases, such as occurs with severe diarrhea and fistulas (which cause loss of bowel fluids containing HCO_3^-)

2. **Compensation** for metabolic acidosis includes:
   a. Extracellular buffering primarily by HCO_3^-
   b. Intracellular buffering primarily by proteins and phosphates
   c. Respiratory compensation by an increase in alveolar ventilation resulting in a decline in the CO_2 tension
   d. Renal compensation by an increase in H^+ excretion and an increase in HCO_3^- reabsorption

B. **Metabolic alkalosis** is a disorder characterized by an elevated arterial pH ($\downarrow [H^+]$) or an increased plasma $[HCO_3^-]$.

1. **Causes** of metabolic alkalosis include:
   a. Decreased production of noncarbonic acid or a loss of noncarbonic acid via the kidneys or the gastrointestinal (GI) system (vomiting)
   b. Excess HCO_3^- or other conjugate base by ingestion or infusion
   c. Excessive renal reabsorption of HCO_3^-
   d. Overtreatment with HCO_3^- or lactate

2. **Compensation** for metabolic alkalosis includes:
   a. Respiratory compensation by hypoventilation resulting in an increase in plasma P_{CO_2}
   b. Renal compensation by an increase in HCO_3^- excretion

C. **Respiratory acidosis** is a disorder characterized by a reduced arterial pH (↑ $[H^+]$), an elevated CO_2 tension (hypercapnia), and a variable increase in the plasma $[HCO_3^-]$.

1. **Causes**
   a. The common denominator in respiratory acidosis is a reduction in alveolar ventilation.
   b. The most common causes are chronic obstructive pulmonary disease (COPD) and the overuse of respiratory depressant drugs.

2. **Compensation** for respiratory alkalosis includes:
   a. Increased renal reabsorption of HCO_3^-
   b. Increased renal H^+ secretion and excretion

D. **Respiratory alkalosis** is a disorder characterized by an elevated arterial pH (↓ $[H^+]$), a low CO_2 tension (hypocapnia), and a variable decrease in the plasma $[HCO_3^-]$.

1. **Causes** of respiratory alkalosis include:
   a. Hyperventilation (hyperpnea)
   b. Anxiety and hysteria
   c. Salicylate overdosage

2. **Compensation** for respiratory alkalosis includes:
   a. Decreased urinary H^+ excretion
   b. Increased urinary HCO_3^- excretion

V. **PATHWAYS FOR COMPENSATION OF PRIMARY ACID-BASE DISTURBANCES: THE pH-$[HCO_3^-]$ RELATIONSHIP** (Figure 38-1)

A. **Respiratory acidosis** (*point A*). The system at fault in this condition is the respiratory system, and compensation occurs through metabolic processes.

1. The kidneys excrete more acid and reabsorb more HCO_3^-, returning the $[HCO_3^-]/S \cdot P_{CO_2}$ ratio toward 20:1 and, therefore, returning pH toward normal (*point A_1*).

2. If the CO_2 tension is elevated but the pH is normal, the kidneys had time to retain HCO_3^- to compensate for the elevated CO_2 tension and the process is not acute; that is, the condition has existed at least a few days to give the kidneys time to compensate (*point A_2*).

3. Usually, the body does not fully compensate for respiratory acidosis.

B. **Respiratory alkalosis** (*point B*). In this condition, the acid-base abnormality again is respiratory, and compensation again occurs through metabolic means.

1. The kidneys compensate by excreting HCO_3^-, thus returning the $[HCO_3^-]/S \cdot P_{CO_2}$ ratio toward 20:1; this compensation takes 2–3 days (*point B_1*).

2. Of the four acid-base abnormalities, only in respiratory alkalosis is the body able to compensate fully; the $[HCO_3^-]/S \cdot P_{CO_2}$ ratio and pH, therefore, return entirely to normal (*point B_2*).

FIGURE 38-1. Effects of renal and respiratory compensation on the pH and plasma [HCO_3^-]. N = normal point; A = respiratory acidosis; A_1 = respiratory acidosis with partial renal compensation; A_2 = respiratory acidosis with complete renal compensation; B = respiratory alkalosis; B_1 = respiratory alkalosis with partial renal compensation; B_2 = respiratory alkalosis with complete renal compensation; C = metabolic acidosis; C_1 = metabolic acidosis with partial respiratory compensation; D = metabolic alkalosis; and D_1 = metabolic alkalosis with partial renal compensation. (After Davenport HW: *The ABC of Acid-Base Chemistry,* 5th edition. Chicago, University of Chicago Press, 1969, p 65.)

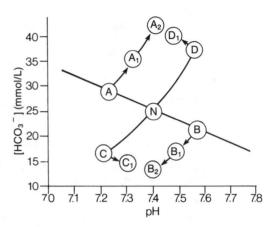

C. **Metabolic acidosis** (*point C*). In this condition, the major abnormality is low [HCO_3^-] or negative base excess (BE), and the compensation is a respiratory process.

　　1. By hyperventilation, the CO_2 tension is lowered so that the [HCO_3^-]/S • PCO_2 ratio is returned toward 20:1 (*point C_1*).

　　2. Since the compensatory system is the lungs, compensation can occur rapidly. If the metabolic acidosis is severe, however, the lungs may not be able to expel sufficient CO_2 to compensate fully.

D. **Metabolic alkalosis** (*point D*). In this condition, the major abnormality is a high [HCO_3^-], and the compensation is a respiratory mechanism.

　　1. By hypoventilation, the CO_2 tension is elevated so that the [HCO_3^-]/S • PCO_2 ratio is increased toward normal (*point D_1*).

　　2. The body usually cannot compensate fully for metabolic alkalosis.

VI. **SUMMARY: CHARACTERISTICS OF UNCOMPENSATED ACID-BASE DISTURBANCES AND THEIR COMPENSATORY RESPONSES** (see Table 37-1)

A. **Metabolic acidosis**

　　1. An uncompensated metabolic acidosis is characterized by a low pH, a low CO_2 content, and a normal CO_2 tension.

　　2. When a metabolic acidosis is compensated by a respiratory alkalosis, the pH tends to return to normal but the CO_2 content and tension decrease.

B. **Metabolic alkalosis**

　　1. An uncompensated metabolic alkalosis is characterized by high pH, a high CO_2 content, and a normal CO_2 tension.

　　2. When a metabolic alkalosis is compensated by a respiratory acidosis, the pH tends to return to normal but the CO_2 content and tension increase.

C. **Respiratory acidosis**

　　1. An uncompensated respiratory acidosis is characterized by a low pH, a high CO_2 content, and a high CO_2 tension.

　　2. When a respiratory acidosis is compensated by a metabolic alkalosis, the pH tends to return to normal but the CO_2 content increases; the CO_2 tension remains unchanged.

D. Respiratory alkalosis

1. An uncompensated respiratory alkalosis is characterized by a high pH, a low CO_2 content, and a low CO_2 tension.

2. When a respiratory alkalosis is compensated by a metabolic acidosis, the pH tends to become normal but the CO_2 content decreases; the CO_2 tension remains unchanged.

STUDY QUESTIONS

Directions: Each of the numbered items or incomplete statements in this section is followed by answers or by completions of the statement. Select the **one** lettered answer or completion that is **best** in each case.

1. Most of the body's total daily acid production is derived from

(A) protein catabolism
(B) triglyceride catabolism
(C) phospholipid catabolism
(D) oxidative metabolism

2. Acid-base disturbances associated with an elevated arterial CO_2 content include which one of the following conditions?

(A) Diabetic ketoacidosis
(B) Respiratory alkalosis
(C) Metabolic acidosis
(D) Metabolic alkalosis

3. CO_2, which is not an acid, can lead to an increase in the $[H^+]$ of body fluids through the formation of

(A) HCO_3^-
(B) H_2CO_3
(C) lactic acid
(D) acetic acid
(E) phosphoric acid

4. In metabolic acidosis, the fall in arterial pH is associated with which of the following arterial blood conditions?

	$[HCO_3^-]$	P_{CO_2}
(A)	↑	normal
(B)	↑	↑
(C)	↓	↓
(D)	↑	↓
(E)	↓	normal

5. Respiratory alkalosis is characterized by which of the following arterial blood conditions?

	pH	P_{CO_2}
(A)	↑	↓
(B)	↓	↑
(C)	↓	↓
(D)	↑	↑
(E)	↑	normal

6. Which of the following sets of values is indicative of compensated metabolic alkalosis?

	$[HCO_3^-]$ (mEq/L)	P_{CO_2} (mm Hg)	pH
(A)	20	25	7.50
(B)	40	46	7.56
(C)	17	30	7.30
(D)	34	10	7.70
(E)	17	19	7.90

7. Compared to normal, the urinary excretion of acid in metabolic acidosis is best characterized by which of the following patterns?

	Urinary excretion of		
	$[HCO_3^-]$	NH_4^+	Titratable acid
(A)	↑	↑	↑
(B)	↓	↑	↓
(C)	↓	↑	↑
(D)	↓	↓	↑
(E)	↓	↓	↓

8. The following arterial blood data are obtained from a hospitalized 55-year-old man:

$$[Na^+] = 140 \text{ mEq/L}$$
$$[HCO_3^-] = 15 \text{ mEq/L}$$
$$[Cl^-] = 113 \text{ mEq/L}$$
$$pH = 7.2$$

From these data, which of the following statements about this patient is true?

(A) The patient has an elevated anion gap
(B) The patient has respiratory acidosis
(C) The $[H^+]$ of this patient is higher than 100 nmol/L
(D) The arterial CO_2 tension of this patient is approximately normal
(E) The pH of this patient's cerebrospinal fluid (CSF) is decreased markedly

9. An hysterical, 35-year-old woman is admitted to the hospital and the following blood data are collected: $[HCO_3^-]$ = 22.2 mmol/L, P_{CO_2} = 30 mm Hg, and P_{O_2} = 98 mm Hg. From these data, the blood $[H^+]$ of this patient would be expected to be

(A) 17.4 nmol/L
(B) 22.5 nmol/L
(C) 28.1 nmol/L
(D) 32.4 nmol/L
(E) 36.7 nmol/L

10. A semi-comatose, 63-year-old male cigarette smoker presents with labored breathing, chronic coughing, and drowsiness. He has had a "smoker's cough" for 15 years. Physical examination reveals wheezing and slowing of forced expiration as well as cyanosis and bilateral leg edema. Initial laboratory data include the following:

hemoglobin concentration = 18.5 mg/dl
serum pH = 7.32
P_{CO_2} = 68 mm Hg
P_{O_2} = 33 mm Hg

From the above findings on history, physical examination, and laboratory testing, the most likely diagnosis is

(A) metabolic acidosis
(B) metabolic alkalosis
(C) respiratory acidosis
(D) respiratory alkalosis

11. The following blood data are collected from a 58-year-old man: $[H^+]$ = 49 nEq/L, P_{CO_2} = 30 mm Hg, and P_{O_2} = 95 mm Hg. From these data, this patient's blood pH would be expected to be

(A) 4.23
(B) 7.31
(C) 8.24
(D) 3.56
(E) 6.17

Questions 12–13

The blood pressure falls during a surgical procedure on an anesthetized animal. The arterial blood becomes cyanotic despite the maintenance of normal ventilation. The arterial pH is 7.25, and the arterial CO_2 tension is 40 mm Hg.

12. The acid-base status of the animal is most likely to be

(A) metabolic acidosis with respiratory compensation
(B) respiratory acidosis with metabolic compensation
(C) metabolic acidosis
(D) respiratory acidosis
(E) respiratory alkalosis with renal compensation

13. To increase pH toward normal, in which direction would the ventilation rate be changed and what would be the corresponding change in arterial CO_2 tension?

	Ventilation Rate	Arterial P_{CO_2}
(A)	↑	↑
(B)	↑	↓
(C)	↓	↑
(D)	↓	↓
(E)	↑	no change

Questions 14–17

The following data were obtained from an arterial blood sample drawn from a hospitalized patient: pH = 7.55, P_{CO_2} = 25 mm Hg, and $[HCO_3^-]$ = 22.5 mEq/L.

14. This patient's arterial blood findings are consistent with a diagnosis of

(A) metabolic alkalosis
(B) respiratory alkalosis
(C) metabolic acidosis
(D) respiratory acidosis

15. These findings indicate that the ratio of $[HCO_3^-]$ to dissolved CO_2 is

(A) 5:1
(B) 10:1
(C) 20:1
(D) 30:1

16. The data indicate that the CO_2 content is approximately

(A) 22 mmol/L
(B) 23 mmol/L
(C) 24 mmol/L
(D) 25 mmol/L
(E) 26 mmol/L

17. The major compensatory response for this patient's acid-base disorder is

(A) hyperventilation
(B) hypoventilation
(C) increased renal HCO_3^- excretion
(D) increased H^+ excretion

Questions 18–21

A 25-year-old man suffers from severe diarrhea for about 1.5 hours and reports to the emergency room, where several tests are performed.

18. Based on his history, this patient's arterial pH would be expected to be approximately

(A) 7.32
(B) 7.40
(C) 7.42
(D) 7.45
(E) 7.50

19. The pH of this patient's cerebrospinal fluid (CSF) would be expected to be approximately

(A) 7.12
(B) 7.22
(C) 7.30
(D) 7.32
(E) 7.40

20. During the compensatory response to his acid-base disorder, this patient's arterial CO_2 tension would most likely be

(A) 32 mm Hg
(B) 40 mm Hg
(C) 46 mm Hg
(D) 50 mm Hg
(E) 60 mm Hg

21. Following the compensatory response, a urinalysis of this patient would most likely indicate

(A) high pH
(B) decreased HCO_3^- excretion
(C) decreased titratable acid excretion
(D) increased Na^+ excretion

DIRECTIONS: Each of the numbered items or incomplete statements in this section is negatively phrased, as indicated by a capitalized word such as NOT, LEAST, or EXCEPT. Select the ONE lettered answer or completion that is BEST in each case.

22. All of the following are intracellular buffers EXCEPT

(A) hemoglobin
(B) protein
(C) organic phosphate
(D) HCO_3^-
(E) carbonate

23. Extracellular HCO_3^- is an effective buffer for all the following acids EXCEPT

(A) phosphoric acid
(B) lactic acid
(C) sulfuric acid
(D) carbonic acid
(E) β-hydroxybutyric acid

24. Which of the following statements about the pH of cerebrospinal fluid (CSF) is FALSE?

(A) It is the same as plasma pH during normal acid-base conditions
(B) It decreases during metabolic alkalosis
(C) It increases during metabolic acidosis
(D) It increases during hyperventilation
(E) It changes in the opposite direction as plasma pH during metabolic acidosis

25. All of the following statements correctly describe noncarbonic acids EXCEPT

(A) they are called fixed acids
(B) they are classified as nonvolatile acids
(C) they are excreted by the kidneys
(D) They are buffered by both HCO_3^- and non-HCO_3^- buffers
(E) they are primarily derived from the metabolism of fats

26. All of the following acids can be transiently classified as a nonvolatile acid when present in high amounts EXCEPT

(A) lactic acid
(B) acetic acid
(C) citric acid
(D) acetoacetic acid
(E) carbonic acid

Directions: Each group of items in this section consists of lettered options followed by a set of numbered items. For each item, select the **one** lettered option that is most closely associated with it. Each lettered option may be selected once, more than once, or not at all.

Questions 27–30

For each acid-base disorder listed below, select the appropriate pair of $[HCO_3^-]$ and CO_2 tension values.

	$[HCO_3^-]$ (mmol/L)	P_{CO_2} (mm Hg)
(A)	34	65
(B)	10	25
(C)	20	20
(D)	40	45

27. Metabolic acidosis

28. Metabolic alkalosis

29. Respiratory acidosis

30. Respiratory alkalosis

Questions 31–35

For each patient described below, select the set of arterial blood values that coincides with that patient's acid-base condition.

Patient	pH	P_{CO_2} (mm Hg)	$[HCO_3^-]$ (mmol/L)
(A)	7.00	70	16
(B)	7.10	27	8
(C)	7.34	70	39
(D)	7.50	48	36
(E)	7.56	26	23

31. Patient with metabolic alkalosis and appropriate respiratory compensation

32. Patient with metabolic acidosis and respiratory acidosis

33. Patient with the highest CO_2 content

34. Patient with the highest $[H^+]$

35. Patient with the highest $[HCO_3^-]/S \cdot P_{CO_2}$ ratio

Questions 36–39

Points A–D on the pH-[HCO_3^-] diagram below indicate states of acid-base imbalance; point N indicates a normal acid-base state. Match each of the following conditions with the appropriate lettered point on the diagram.

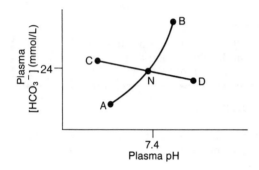

36. Hypocapnia

37. Hypercapnia

38. Ketoacidosis

39. $NaHCO_3$ ingestion

Questions 40–45

For each patient described below, select the set of arterial blood values that coincides with that patient's acid-base condition.

Patient	pH	[HCO_3^-] (mEq/L)	[Pco_2] (mm/Hg)	Po_2 (mm/Hg)
(A)	7.2	20	50	62
(B)	7.3	33	60	45
(C)	7.3	16	30	105
(D)	7.5	30	38	95
(E)	7.6	20	20	120

40. Patient with chronic respiratory acidosis and partial renal compensation

41. Patient with metabolic acidosis and partial respiratory compensation

42. Patient with uncompensated metabolic alkalosis

43. Patient with respiratory alkalosis and partial renal compensation

44. Patient with highest dissolved CO_2

45. Patient with highest CO_2 content

ANSWERS AND EXPLANATIONS

1. The answer is D [Chapter 31 I A; III D]. Most of the body's daily CO_2 production occurs from chemical reactions of the tricarboxylic acid cycle. From this metabolic activity, humans produce about 13,000 mmol of CO_2 daily, or, in acid-base terms, about 13,000 mEq of H^+ per day. The mammalian body produces large amounts of acids from two major sources. The volatile acid H_2CO_3 is produced from CO_2, the end product of oxidative metabolism. A variety of nonvolatile acids (e.g., H_2SO_4, H_3PO_4) are produced from dietary substances.

2. The answer is D [Chapter 37 I E 2; Figure 37-5; Table 37-2]. The total CO_2 content is mainly a function of HCO_3^- and dissolved CO_2. Therefore, CO_2 content is elevated in metabolic alkalosis and respiratory acidosis. Conversely, CO_2 content declines with decreases in $[HCO_3^-]$ and $[CO_2]$, conditions that are associated with metabolic acidosis and respiratory alkalosis, respectively. Note that $[CO_2]$ denotes the concentration of dissolved CO_2 and is calculated as:

$$[CO_2] = S \cdot P_{CO_2}$$
$$= 0.03 \cdot 40 \text{ mm Hg}$$
$$= 1.2 \text{ mmol/L (normal)}$$

3. The answer is B [Chapter 31 I A]. Carbon dioxide (CO_2) is a hydrogen ion (H^+) generator. It is evident that the total reservoir of carbonic acid (H_2CO_3) [i.e., both dissolved CO_2 and H_2CO_3 termed the total "carbonic acid pool"] should be considered the acid component of the HCO_3^- buffer system, and HCO_3^- the conjugate base. The HCO_3^-/CO_2 buffer system is the most important extracellular fluid (ECF) buffer pair. CO_2, which is not an acid, increases the acidity of a solution through the formation and dissociation of H_2CO_3. Students often are confused by the following series of reactions

$$CO_2 + H_2O \rightarrow H_2CO_3 \rightarrow H^+ + HCO_3^-$$

because both H^+ and HCO_3^- are formed simultaneously, and there should be no change in pH. However, the HCO_3^- concentration is several orders of magnitude (6×10^5) greater than the H^+ concentration and is correspondingly less affected by any change in the concentration of CO_2 or of H_2CO_3.

4. The answer is C [Chapter 38 IV A; V C; Table 37-1]. Metabolic acidosis is an acid-base disturbance characterized by a decreased arterial pH (or increased $[H^+]$), a decreased plasma $[HCO_3^-]$, and a compensatory hyperventilation resulting in a decreased arterial CO_2 tension.

5. The answer is A [Chapter 38 IV D; V B; Table 37-1]. Respiratory alkalosis is an acid-base disturbance characterized by an increased arterial pH (or decreased $[H^+]$), a decreased CO_2 tension (hypocapnia), and a variable reduction in arterial $[HCO_3^-]$ due to renal compensation.

6. The answer is B [Chapter 37 IV B 2; Chapter 38 IV B 2; Figure 37-6]. The hallmarks of metabolic alkalosis are elevated $[HCO_3^-]$, which is the primary event, and elevated CO_2 tension, which is the compensatory event. Also, the $[H^+]$ decreases and the $[HCO_3^-]$ increases. The only other patient set of values showing an increased $[HCO_3^-]$—*set D*—does not exhibit an increase in CO_2 tension.

7. The answer is C [Chapter 36 IV B; V B 2, C; Chapter 38 IV A 2; Table 36-1]. The kidney responds to an increased acid load by augmenting renal ammonia (NH_3) production and, consequently, ammonium (NH_4^+) excretion. There is a smaller increase in the excretion of titratable acid (primarily in the form of $H_2PO_4^-$). Thus, the decreased pH and decreased HCO_3^- excretion and increased H^+ excretion (in the form of NH_4^+ and $H_2PO_4^-$) are consistent with metabolic acidosis.

8. The answer is D [Chapter 33 IV C; V B; Figure 37-6]. From the data given, this patient's arterial CO_2 tension is determined to be 38 mmol/L, which is close to normal (40 mmol/L). The Henderson equation is used to calculate CO_2 tension. The equation typically is expressed as:

$$[H^+] = 24 \frac{P_{CO_2}}{[HCO_3^-]}$$

Note that when $[HCO_3^-]$ is low, as in this patient, a change in CO_2 tension will have a greater effect on $[H^+]$ than when $[HCO_3^-]$ is normal or high.

To determine CO_2 tension, the Henderson equation is rearranged as:

$$Pco_2 = \frac{[H^+]\,[HCO_3^-]}{24}$$

First, the $[H^+]$ is calculated from the pH value as:

$$
\begin{aligned}
[H^+] &= \text{antilog } (9 - pH) \\
&= \text{antilog } (9 - 7.22) \\
&= \text{antilog } (1.78) \\
&= 60 \text{ nmol/L}
\end{aligned}
$$

Then, CO_2 tension can be determined as:

$$
\begin{aligned}
Pco_2 &= \frac{[H^+]\,[HCO_3^-]}{24} \\[2mm]
&= \frac{(60)\,(15)}{24} = 38 \text{ mm Hg}
\end{aligned}
$$

Thus, the CO_2 tension of this patient is within normal limits and, therefore, respiratory compensation has not occurred. The acid-base disturbance of this patient is attributable to a hyperchloremic metabolic acidosis.

Notice that, in this type of acidosis, the anion gap (AG) is normal, as

$$
\begin{aligned}
AG &= [Na^+] - [HCO_3^- + Cl^-] \\
&= 140 \text{ mEq/L} - (15 \text{ mEq/L} + 113 \text{ mEq/L}) \\
&= 140 \text{ mEq/L} - 128 \text{ mEq/L} = 12 \text{ mEq/L}
\end{aligned}
$$

The pH of the cerebrospinal fluid (CSF) in metabolic acidosis usually is elevated because of the compensatory hyperventilatory response to the metabolic acidosis. In this patient, there is no evidence of significant hyperventilation and, therefore, the pH of the CSF would not increase significantly. The axiom to remember is that metabolic acid-base alterations evoke opposite changes in the pH of the CSF and blood, while respiratory acid-base disturbances bring about parallel changes in the pH of CSF and blood.

9. The answer is D [Chapter 33 V B 2]. From the data given, this patient's arterial $[H^+]$ is determined to be 32.4 nmol/L. The Henderson equation is used in this case, which is:

$$[H^+] = 24 \frac{Pco_2}{[HCO_3^-]}$$

It is necessary to keep in mind that the units for these factors are: $[H^+]$ (nmol/L), $[HCO_3^-]$ (mmol/L), and Pco_2 (mm Hg). Substituting,

$$
\begin{aligned}
[H^+] &= 24 \frac{30}{22.2} \\
&= 32.4 \text{ nmol/L}
\end{aligned}
$$

This patient has a partially compensated respiratory alkalosis. Note that her arterial $[H^+]$ and $[HCO_3^-]$ exhibit a parallel decrease.

10. The answer is C [Chapter 33 IV C; V B; Chapter 38 IV C]. This patient has an acidosis caused by CO_2 retention (i.e., respiratory acidosis). The condition is partially compensated by renal retention of HCO_3^-. The $[HCO_3^-]$ can be determined using either the Henderson-Hasselbalch equation or the Henderson equation.

Using the Henderson-Hasselbalch equation, the $[HCO_3^-]$ is determined as

$$pH = pK' + \log \frac{[HCO_3^-]}{S \cdot Pco_2}$$

$$7.32 = 6.1 + \log \frac{[HCO_3^-]}{0.03 \cdot 68 \text{ mm Hg}}$$

$$7.32 = 6.1 + \log\,[HCO_3^-] - \log 2.04$$

$$
\begin{aligned}
\log\,[HCO_3^-] &= 7.32 - 6.1 + 0.31 \\
&= 1.53
\end{aligned}
$$

$$\text{antilog } 1.53 = 33.9 \text{ mmol/L}$$

Using the Henderson equation, the $[HCO_3^-]$ is determined as

$$
\begin{aligned}
[HCO_3^-] &= 24 \frac{Pco_2 \text{ (mm Hg)}}{[H^+] \text{ (nmol/L)}} \\[2mm]
&= 24 \frac{68 \text{ mm Hg}}{47.9 \text{ nmol/L}} \\[2mm]
&= 34.1 \text{ mmol/L}
\end{aligned}
$$

The patient's hypoxemia could be due to a pulmonary diffusion barrier. The cyanosis is caused by a greater than normal concentration of reduced hemoglobin (greater than 5 g/dl of arterial blood in the deoxy state). The leg edema could be due to the increased retention of HCO_3^- and Na^+ resulting in an increase in plasma volume. The hypoxemia causes polycythemia and constriction of vascular smooth muscle of the pulmonary circulation, leading to pulmonary hypertension, elevated right atrial pressure, and increased capillary pressure. The increase in capillary pressure causes fluid movement into the interstitial space and, ultimately, edema.

11. The answer is B [Chapter 33 IV C 1; V B 2]. From the data given, this patient's blood pH is determined to be 7.31. To determine the arterial pH, it is first necessary to calculate the arterial [HCO_3^-] using the Henderson equation:

$$[HCO_3^-] = 24\,\frac{P_{CO_2}}{[H^+]}$$

$$= 24\,\frac{30}{49}$$

$$= 14.7\ mmol/L$$

Then, applying the Henderson-Hasselbalch equation, the arterial pH is determined as

$$pH = pK' + \log\frac{[HCO_3^-]}{S\cdot P_{CO_2}}$$

$$= 6.1 + \log\frac{14.7\ mmol/L}{0.9\ mmol/L}$$

$$= 6.1 + \log 16.3$$

$$= 6.1 + 1.21$$

$$= 7.31$$

This patient has a partially compensated metabolic acidosis. It is important to note that the [HCO_3^-]/S • P_{CO_2} ratio is equal to 16.3, indi-

cating a significant respiratory response (hyperventilation) to the metabolic problem. Thus, this metabolic acidosis is alleviated by a response that involves the respiratory system.

12–13. The answers are: 12-C, 13-B [Chapter 37 IV B 1; Chapter 38 IV A 2]. The acid-base abnormality is metabolic acidosis without respiratory compensation. Since the respiratory response to metabolic acidosis is hyperpnea, hypocapnia would be expected, which is not observed in this case. Moreover, the imbalance cannot be attributed to respiratory acidosis, which would be characterized by an elevated arterial CO_2 tension. It is apparent that a mechanical respirator maintained the animal's CO_2 tension at a constant level.

The pH of arterial blood is a function of two variables: [HCO_3^-] and CO_2 tension. A compensatory response occurs in the alternate variable and in the same direction as the primary abnormality. In this case, the decline in [HCO_3^-] (i.e., metabolic acidosis) would be associated with a parallel decline in CO_2 tension, the alternate variable. To bring this about, there would be a hyperventilatory response and a resultant decline in CO_2 tension.

14–17. The answers are: 14-B, 15-D, 16-B, 17-C [Chapter 37 I E 3; IV A 2; Chapter 38 IV D; V B; Figure 37-1; Table 37-2]. The blood findings indicate that this patient has a respiratory alkalosis, an acid-base disturbance characterized by increased arterial pH (or decreased [H^+]), decreased arterial CO_2 tension (hypocapnia), and decreased plasma [HCO_3^-]. It should be noted that both [H^+] and [HCO_3^-] are decreased in this patient, which is consistent with the axiom that [H^+] and [HCO_3^-] change in the same direction in respiratory acid-base imbalances. The decline in [HCO_3^-] indicates that renal compensation has begun.

In alkalotic states, the [HCO_3^-]/S • P_{CO_2} ratio exceeds the normal 20:1, due to either an increase in [HCO_3^-] (metabolic alkalosis) or a decrease in CO_2 tension (respiratory alkalosis). The normal ratio of 20:1 is derived as:

$$\frac{[HCO_3^-]}{S\cdot P_{CO_2}} = \frac{24\ mmol/L}{0.03\cdot 40\ mm\ Hg} = \frac{24\ mmol/L}{1.2\ mmol/L} = \frac{20}{1}$$

In this alkalotic patient, the [HCO_3^-]/S • P_{CO_2} is 30:1. This ratio can be determined by substituting the patient's blood data into the above equation, as:

$$\frac{[HCO_3^-]}{S\cdot P_{CO_2}} = \frac{22.5\ mmol/L}{0.03\cdot 25\ mm\ Hg} = \frac{22.5\ mmol/L}{0.75\ mmol/L} = \frac{30}{1}$$

The total CO_2 content for this patient is approximately 23 mmol/L. Total CO_2 equals the sum of all forms of CO_2 in the blood (i.e., HCO_3^-, H_2CO_3, dissolved CO_2, and CO_2 bound to proteins). Dissolved CO_2 content ([CO_2]) is calculated as

$$[CO_2] = 0.03\cdot P_{CO_2}$$
$$= 0.03\cdot 40\ mm\ Hg$$
$$= 1.2\ mmol/L$$

Since $[H_2CO_3]$ is negligible, the total CO_2 content normally exceeds the $[HCO_3^-]$ by 1.2 mmol/L and, therefore, equals 25.2 mmol/L when normal plasma $[HCO_3^-]$ and CO_2 tension values exist, as:

$$\text{total } CO_2 \text{ content} = [HCO_3^-] + (S \cdot P_{CO_2})$$
$$= 24 \text{ mmol/L} + 1.2 \text{ mmol/L}$$
$$= 25.2 \text{ mmol/L}$$

Total CO_2 content is decreased in respiratory alkalosis. From this patient's blood data, the CO_2 content is calculated as

$$\text{total } CO_2 \text{ content} = [HCO_3^-] + (S \cdot P_{CO_2})$$
$$= 22.5 \text{ mmol/L} + 0.75 \text{ mmol/L}$$
$$= 23.25 \text{ mmol/L}$$

Respiratory alkalosis decreases the renal reabsorption of HCO_3^-, causing a transient HCO_3^- diuresis and a decline in net acid secretion. Since the major change in respiratory alkalosis is a decrease in arterial CO_2 tension, the compensation will be in the alternate variable (kidney) and in the same direction as the primary event. Thus, there is a decline in plasma $[HCO_3^-]$, which is indicative of a partial renal response (i.e., increased HCO_3^- excretion).

18–21. The answers are: 18-A, 19-E, 20-A, 21-B [Chapter 33 III B; Chapter 38 IV A; V C; VI A; Figures 33-1, 37-5, 37-6]. The fluid in the small and large intestine is relatively high in HCO_3^-. Therefore, severe diarrhea is a cause of metabolic acidosis, which is the likely diagnosis in this patient. Only one of the given arterial pH values—7.32—is consistent with acidosis.

Increased arterial $[H^+]$ stimulates ventilation, lowering the CO_2 tension of the blood and cerebrospinal fluid (CSF), resulting in blood acidosis and CSF alkalosis. Thus, in metabolic acid-base alterations the pH of the blood and CSF change in opposite directions, whereas in respiratory acid-base abnormalities the pH of the blood and CSF change in the same direction. Since the normal pH of CSF is 7.32, the pH would increase in metabolic acidosis. The only given pH value indicating a CSF alkalosis is 7.40.

Metabolic acidosis is compensated by hyperventilation, which reduces the CO_2 tension and thereby attenuates the reduction in pH. This hyperventilatory response to metabolic acidosis is due to direct stimulation of the medullary respiratory center and of the peripheral chemoreceptors in the carotid and aortic bodies. Only one of the given CO_2 tension values—32 mm Hg—is associated with hypocarbia.

In the absence of therapy with $NaHCO_3$, the renal compensation for a metabolic acidosis requires an increased HCO_3^- reabsorption together with an increased H^+ secretion and excretion. This adaptive response is accomplished by an increased NH_4^+ excretion with a limited ability to enhance titratable acidity via $H_2PO_4^-$ excretion. Thus, the urine will exhibit a low pH.

22. The answer is E [Chapter 34 I A 2 b, B 3; II A]. Carbonate is a major buffer system in bone and, therefore, is not an intracellular buffer; bone contains approximately 35,000 mEq of carbonate that contributes to the buffering of acid and base loads. The intracellular fluid (ICF) does not have a high concentration of the bicarbonate (HCO_3^-) buffer system but does contain significant amounts of the non-HCO_3^- buffer systems. Intracellular protein, with its histidine residues, and organic phosphate are the quantitatively significant non-HCO_3^- buffer systems. Thus, the ICF functions to buffer both noncarbonic and carbonic acids well. The red blood cell compartment, by virtue of its hemoglobin content, is regarded in the physiologic context of body buffers as a part of the extracellular buffer system, although it is clearly intracellular. Hemoglobin is quantitatively more important than the erythrocyte HCO_3^- buffer.

23. The answer is D [Chapter 31 IV B 6; Chapter 35 III C]. Extracellular HCO_3^- is not an effective buffer for H_2CO_3. The HCO_3^- buffer system plays no role in the buffering of H_2CO_3, but it is an effective buffer for noncarbonic acids. HCO_3^- cannot buffer H_2CO_3, because the combination of H^+ with HCO_3^- results in regeneration of H_2CO_3 as:

$$H_2CO_3 + HCO_3^- \leftrightarrow HCO_3^- + H_2CO_3$$

Most buffering of H_2CO_3 occurs within the erythrocytes by hemoglobin. In contrast, the non-HCO_3^- buffer systems (hemoglobin, protein, phosphate) can buffer both noncarbonic acids and H_2CO_3. The HCO_3^-/CO_2 buffer system accounts for 97%–98% of the buffering in the extracellular fluid (ECF), which includes the interstitial fluid, lymph, and cerebrospinal fluid (CSF). A stipulation here is that HCO_3^- is a major buffer for noncarbonic acids.

24. The answer is A [Chapter 33 III B]. The $[HCO_3^-]$ of cerebrospinal fluid (CSF) and arterial blood are similar (about 24 mmol/L); however, the arterial pH (7.4) is higher than the pH of CSF (about 7.32), because the CO_2 tension of CSF is about 48 mm Hg compared to 40 mm Hg in arterial blood. Acute metabolic alkalosis depresses ventilation, increasing the CO_2 tension of the blood and CSF, resulting in CSF acidosis and blood alkalosis. Thus, in metabolic acid-base disturbances, the $[H^+]$ of CSF and blood change in opposite directions, and in respiratory acid-base disturbances, the $[H^+]$ of CSF and blood change in the same direction.

25. The answer is E [Chapter 31 IV; Chapter 34 I B 1–3]. Noncarbonic acids include all acids other than H_2CO_3, such as lactic, acetoacetic, β-hydroxybutyric, hydrochloric, and sulfuric acids. Some noncarbonic acids (lactic, acetoacetic, and β-hydroxybutyric acids) can be converted to CO_2 and eliminated by ventilation (the fate of "volatile" acids); however, when present in large amounts, these acids are eliminated by the kidney (the fate of "nonvolatile" acids). Sulfuric, hydrochloric, and phosphoric acids cannot be catabolized to CO_2 and H_2O and, thus, must be eliminated by renal excretion. Noncarbonic acids can be buffered by both HCO_3^- and non-HCO_3^- buffer systems. Fixed acid is another name for nonvolatile acid. Metabolites of amino acids constitute the major portion of nonvolatile acids.

26. The answer is E [Chapter 31 IV B]. Acids that can be converted to carbonic acid are classified as volatile acids. Carbonic acid is called a volatile acid because it can be dehydrated into CO_2 and H_2O and excreted from the body by pulmonary ventilation. In high amounts, many of these volatile acids are excreted by the kidneys without being converted to CO_2 and H_2O and, thus, are classified as nonvolatile acids. Examples of such acids are citric, isocitric, acetic, and lactic acids and the ketoacids. Acetic acid can increase following the ingestion of even small amounts of vinegar. All of these acids can be either eliminated by pulmonary ventilation or excreted by the kidneys. Thus, it is important to remember that nonvolatile (fixed) acids are dissociated at

body pH, and the H^+ can be excreted via the urine.

27–30. The answers are: 27-B, 28-D, 29-A, 30-C [Chapter 37 I A–C; IV A–B; Chapter 38 VI; Figure 37-6; Table 37-1]. In analyzing these four acid-base disorders, it is necessary to consider both the primary abnormality (i.e., the variable that undergoes the greater degree of change) and the compensatory response (i.e., the alternate variable, which undergoes a lesser degree of change in the same direction).

In metabolic acidosis, there is a primary decrease in $[HCO_3^-]$ and a compensatory decrease in CO_2 tension. Only one pair of values (*B*) shows this pattern. In metabolic alkalosis, there is a primary increase in $[HCO_3^-]$ and a compensatory increase in CO_2 tension. Only one pair of values (*D*) shows this pattern. In respiratory acidosis, there is a primary increase in CO_2 tension and a compensatory increase in $[HCO_3^-]$. Only one pair of values (*A*) shows this pattern. In respiratory alkalosis, there is a primary decrease in CO_2 tension and a compensatory decrease in $[HCO_3^-]$. Only one pair of values (*C*) shows this pattern.

31–35. The answers are: 31-D, 32-B, 33-C, 34-A, 35-E [Chapter 33 IV C 2; V B 2; Chapter 37 I C 1 b, E 3; IV B; Chapter 38 IV A, B; VI]. *Patient D* has a metabolic alkalosis with partial respiratory compensation. Metabolic alkalosis can result from ingestion of alkaline substances (e.g., antacids) or from loss of acid (as occurs in vomiting). The total CO_2 content also is increased in metabolic alkalosis. The compensation for this condition is hypoventilation or excretion of an alkaline urine. Note that *patient D* shows partial compensation, as evidenced by the increased CO_2 tension. The $[HCO_3^-]/S \cdot P_{CO_2}$ ratio and pH are increased.

Patient B has a metabolic acidosis with partial respiratory compensation. Metabolic acidosis can result from increased fixed (noncarbonic) acid production (as occurs in diabetic ketoacidosis), increased acid retention (as occurs in renal failure with subsequent accumulation of $H_2PO_4^-$), or loss of base (as occurs in diarrhea). The total CO_2 content also is decreased in metabolic acidosis. The compensation for this condition is hyperventilation or excretion of an acid urine. Note that *patient B* shows partial compensation, as evidenced by the decrease in CO_2 tension. The $[HCO_3^-]/S \cdot P_{CO_2}$ ratio and pH are decreased.

Patient C has the highest plasma CO_2 content (i.e., $[HCO_3^-] + S \cdot P_{CO_2}$). This patient has respiratory acidosis, a condition caused by CO_2 retention resulting from hypoventilation, which may be due to pulmonary disease. The compensation for this condition is an increase in both HCO_3^- reabsorption and in H^+ excretion in the form of $H_2PO_4^-$ and NH_4^+. Note that *patient C* shows nearly complete compensation, as evidenced by the increased $[HCO_3^-]$ and return of pH toward normal. The $[HCO_3^-]/S \cdot P_{CO_2}$ ratio is only slightly decreased (i.e., 18:1 versus the normal 20:1).

Patient A has the lowest pH and, thus, the highest $[H^+]$. This patient has metabolic acidosis without compensation. $[H^+]$ can be calculated as

$$[H^+] = \text{antilog } (9 - \text{pH})$$
$$= \text{antilog } (9 - 7)$$
$$= \text{antilog } 2$$
$$= 100 \text{ nmol/L}$$

or estimated as

$$[H^+] = 24 \; \frac{P_{CO_2}}{[HCO_3^-]}$$

$$= 24 \; \frac{70}{16}$$

$$= 24 \, (4.38)$$

$$= 105 \text{ nmol/L}$$

Patient E has the highest $[HCO_3^-]/S \cdot P_{CO_2}$ ratio (i.e., 29.5:1). This patient has respiratory alkalosis, a condition caused by loss of CO_2 due to anxiety or hysteria leading to hyperventilation. The compensation for respiratory alkalosis is excretion of an alkaline urine. This patient shows partial renal compensation, as evidenced by the slight decline $[HCO_3^-]$. *Patient E* has a slightly lower than normal CO_2 content.

36–39. The answers are: 36-D, 37-C, 38-A, 39-B [Chapter 37 IV A–B; Chapter 38 IV; Figure 37-6; Table 37-1]. In analyzing these four uncompensated acid-base disturbances, it is important to consider the CO_2 tension and $[HCO_3^-]$ rather than the pH, because CO_2 tension and $[HCO_3^-]$ are the key determinants of the cause of, and compensation for, these disturbances.

Point D represents a patient with increased pH ($\downarrow [H^+]$) brought about by respiratory alkalosis. This condition is characterized by a primary decrease in CO_2 tension (hypocapnia) and a variable secondary decrease in plasma $[HCO_3^-]$. Metabolic acidosis also is characterized by declines in these two variables, but

the pH is decreased ($\uparrow [H^+]$) as well. Respiratory alkalosis is defined as alveolar ventilation in excess of the existing need of the body to eliminate CO_2. This excess in alveolar ventilation, called hyperventilation, results in a reduced arterial CO_2 tension. Hyperpnea is the general term used to describe any increase in ventilatory effort. With respiratory alkalosis there is a decline in the [total CO_2].

Point C represents a patient with decreased pH ($\uparrow [H^+]$) caused by respiratory acidosis. This clinical disorder is characterized by a primary increase in CO_2 tension (hypercapnia) and a variable secondary increase in plasma $[HCO_3^-]$. The common denominator in respiratory acidosis is hypoventilation, which is defined as alveolar ventilation insufficient to excrete CO_2 rapidly enough to meet the existing needs of the body. With respiratory acidosis there is a relatively small increment in the [total CO_2], because the major fraction of the CO_2 content is comprised of HCO_3^-.

Point A represents a patient with diabetes mellitus, which is the most common cause of ketoacidosis. This overproduction of ketoacids is caused by a deficiency of insulin, which leads to: (1) increased lipolysis and an increased delivery of free fatty acids to the liver and (2) the preferential conversion of free fatty acids to ketoacids rather than to triglycerides. Thus, metabolic acidosis is characterized by a low arterial pH ($\uparrow [H^+]$), a reduced $[HCO_3^-]$, and a compensatory hyperventilation resulting in hypocapnia. The renal compensatory response for respiratory alkalosis also diminishes the plasma $[HCO_3^-]$, but the pH in that disorder is elevated ($\downarrow [H^+]$). Overproduction of ketoacids causes acidosis by two mechanisms: (1) a decrease in plamsa $[HCO_3^-]$ with an increase in the anion gap and (2) overloading of the renal capacity to excrete H^+ resulting in a loss of Na^+ and a failure to recover $NaHCO_3$. In metabolic acidosis, there is a decline in the [total CO_2]. Furthermore, ketoacidosis, like lactic acidosis, differs from other forms of metabolic acidosis in that the anion associated with H^+ can be metabolized back to HCO_3^-, as

$$\beta\text{-hydroxybutyrate}^- + O_2 \rightarrow$$
$$CO_2 + H_2O + HCO_3^-$$

β-hydroxybutyrate represents about 75% of the circulating ketoacids in diabetic ketoacidosis. It can be seen from the above chemical equation that the metabolism of the β-hydroxybutyrate anion results in the regeneration of the HCO_3^- that was neutralized in buffering the H^+. Since the HCO_3^- is replaced by an anion that is metabolized back to HCO_3^-, there is no actual loss of HCO_3^- from the body in ketoacidosis (or lactic acidosis). Insulin administration decreases the accumulation of β-hydroxy-butyric acid and allows the metabolism of the acid ions back to HCO_3^-.

Point B represents a patient with metabolic alkalosis. Excessive ingestion of $NaHCO_3$ can result in metabolic alkalosis and an increase in pH ($\downarrow [H^+]$). Metabolic alkalosis is characterized by an increase in the plasma $[HCO_3^-]$ and a compensatory increase in the CO_2 tension produced by a decline in alveolar ventilation. Since elevation of plasma $[HCO_3^-]$ can be due to the renal compensation for chronic respiratory acidosis, the diagnosis of metabolic alkalosis cannot be made without measuring the pH. Metabolic alkalosis is associated with a large increase in the [total CO_2].

40–45. The answers are: 40-B, 41-C, 42-D, 43-E, 44-B, 45-B [Chapter 33 IV B 2, 3; Chapter 37 I E 3; IV A–B; Chapter 38 IV–VI; Figures 37-5, 37-6]. *Patient B* has respiratory acidosis with partial renal compensation. Respiratory acidosis is caused by increased CO_2 tension, which reduces the $[HCO_3^-]/S \cdot P_{CO_2}$ ratio and, thus, the pH. Whenever CO_2 tension rises, $[HCO_3^-]$ also must increase somewhat because of the dissociation of H_2CO_3 formed by the hydration of CO_2. The kidney responds by conserving $[HCO_3^-]$ via increased H^+ secretion, with acid excreted as $H_2PO_4^-$ and NH_4^+.

Patient C has metabolic acidosis with partial respiratory compensation. Metabolic acidosis

means a primary decrease in $[HCO_3^-]$, which decreases the $[HCO_3^-]/S \cdot P_{CO_2}$ ratio and, thus, the pH. The $[HCO_3^-]$ may be lowered by the addition of H^+ or by the loss of HCO_3^-. Respiratory compensation occurs by increased ventilation via the action of H^+ in the peripheral and central chemoreceptors.

Patient D has uncompensated metabolic alkalosis. In this condition, the increase in $[HCO_3^-]$ causes an increase in the $[HCO_3^-]/S \cdot P_{CO_2}$ ratio and, thus, the pH. Respiratory compensation occurs by a reduction in alveolar ventilation, which tends to raise CO_2 tension (not evident in *patient D*).

Patient E has respiratory alkalosis with partial renal compensation. Respiratory alkalosis is caused by a decrease in CO_2 tension, which increases the $[HCO_3^-]/S \cdot P_{CO_2}$ ratio and, thus, the pH. Renal compensation occurs by increased HCO_3^- excretion.

Patient B has both the highest amount of dissolved CO_2 and the highest total CO_2 content. Dissolved CO_2 is determined by multiplying the CO_2 tension (in mm Hg) by the CO_2 tension solubility constant, S, as

$$\begin{aligned} \text{dissolved } CO_2 &= S \cdot P_{CO_2} \\ &= 0.03 \cdot 60 \\ &= 1.8 \text{ mmol/L} \end{aligned}$$

Total CO_2 content is the sum of $[HCO_3^-]$ and dissolved CO_2. It is calculated in *patient B* as

$$\begin{aligned} \text{total } CO_2 \text{ content} &= [HCO_3^-] + S \cdot P_{CO_2} \\ &= 33 + 0.03 \cdot 60 \\ &= 33 + 1.8 \\ &= 34.8 \text{ mmol/L} \end{aligned}$$

PART VII

GASTROINTESTINAL PHYSIOLOGY
Michael B. Wang

Chapter 39

Structure and Function of the Gastrointestinal (GI) Tract

I. **STRUCTURE.** The digestive system is composed of a long muscular tube—the gastrointestinal (GI) tract, or alimentary canal—and a set of accessory organs (Figure 39-1).

FIGURE 39-1. The gastrointestinal (GI) tract.

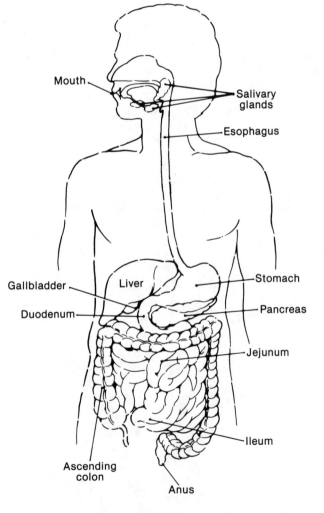

A. The **GI tract** consists of the oral cavity, pharynx, esophagus, stomach, small intestine, large intestine, rectum, and anal canal.

B. The **accessory organs** include the tongue, teeth, salivary glands, pancreas, liver, and gall-bladder.

II. **FUNCTION.** The digestive system is responsible for breaking down food and supplying the body with water, nutrients, and electrolytes needed to sustain life. Before food can be used by the body, it must be ingested, digested, and absorbed—processes that involve coordinated movement of muscle and secretion of various substances.

A. **Ingestion** involves:

1. Placing food into the mouth

2. Chewing the food into smaller pieces (**mastication**)

3. Moistening the food with salivary secretions

4. Swallowing the food (**deglutition**)

B. **Digestion.** During **digestion,** food is broken down into small particles by the grinding action of the GI tract and then degraded by digestive enzymes into usable nutrients.

1. Starches are degraded by amylases into monosaccharides.

2. Proteins are degraded by a variety of enzymes (e.g., pepsin, trypsin) into dipeptides and amino acids.

3. Fats are degraded by lipases and esterases into monoglycerides and free fatty acids.

C. **Absorption.** During **absorption,** nutrients, water, and electrolytes are transported from the GI tract (principally from the small intestine) to the circulation.

Chapter 40

Ingestion of Food

I. MASTICATION. After food is placed into the mouth, it is cut and ground into smaller pieces by chewing (mastication).

A. **The chewing reflex.** Although chewing is a voluntary act, it is coordinated by reflex centers in the brain stem that facilitate the opening and closing of the jaw.

1. When the mouth opens, stretch receptors in the jaw muscles initiate a reflex contraction of the masseter, medial pterygoid, and temporalis muscles, causing the mouth to close.

2. When the mouth closes, food comes into contact with buccal receptors eliciting a reflex contraction of the digastric and lateral pterygoid muscles, causing the mouth to open.

3. When the jaw drops, the stretch reflex causes the entire cycle to be repeated.

4. The tongue contributes to the grinding process by positioning the food between the upper and lower teeth.

B. **Function of mastication**

1. Chewing **breaks food into smaller pieces,** which:
   a. Makes it easier for the food to be swallowed
   b. Breaks off the undigestible cellulose coatings of fruits and vegetables
   c. Increases the surface area of the food particles, making it easier for them to be digested by the digestive enzymes

2. Chewing **mixes the food with salivary gland secretions,** which:
   a. Initiates the process of starch digestion by salivary amylase
   b. Initiates the process of lipid digestion by lingual lipase
   c. Lubricates and softens the bolus of food, making it easier to swallow

3. Chewing **brings food into contact with taste receptors** and **releases odors** that stimulate the olfactory receptors. The sensations generated by these receptors increase the pleasure of eating and initiate gastric secretions.

II. SALIVA helps ready the food to be moved from the mouth to the esophagus.

A. **Salivary glands.** Saliva is secreted primarily by three pairs of glands.

1. The **parotid glands,** located near the angle of the jaw, are the largest glands. They secrete a watery fluid.

2. The **submandibular** and **sublingual glands** secrete a fluid that contains a higher concentration of proteins and so is more viscous.

3. Smaller glands are located throughout the oral cavity. Those in the tongue secrete lingual lipase.

B. **Composition of saliva.** The salivary glands secrete a relatively high volume of fluid (0.5–1 L/day) containing electrolytes and proteins.

1. **Electrolytes.** Electrolyte concentration and osmolality both vary with secretory flow rates, but generally, in comparison to plasma, saliva is hypotonic and contains higher concentrations of K^+ and bicarbonate (HCO_3^-) and lower concentrations of Na^+ and Cl^-.

    2. Proteins. Two types of proteins are found in saliva.
      a. The **enzymes,** α-amylase (ptyalin) and lingual lipase, begin the process of starch and fat digestion.
      b. Mucin is a glycoprotein that lubricates the food.

C. **Control of salivary secretion**

    1. Salivary secretion is controlled entirely by **autonomic nervous system (ANS) reflexes**.
      a. Parasympathetic nerve stimulation causes the salivary gland cells to secrete a large volume of watery fluid that is high in electrolytes but low in proteins.
      b. Sympathetic nerve stimulation causes the salivary glands to secrete a small volume of fluid containing a high concentration of mucus.

    2. Salivary reflexes are elicited by the thought, aroma, or taste of food or by the presence of food within the alimentary canal.

    3. Salivary gland metabolism and growth are both stimulated by increased ANS activity.

D. **Functions of saliva.** Salivary secretions perform a number of important functions.

    1. Protection. Salivary secretions protect the mouth by:
      a. Cooling hot foods
      b. Diluting any hydrochloric acid (HCl) or bile regurgitated into the mouth
      c. Washing food away from the teeth and destroying harmful bacteria within the mouth

    2. Digestion. Salivary secretions begin the process of starch and fat digestion.
      a. α-Amylase can digest most of the ingested starches into disaccharides before they reach the small intestine. α-Amylase is ultimately inactivated by the low pH of the stomach.
      b. Lingual lipase begins to break down ingested fats while they are in the mouth, stomach, and upper portions of the small intestine.

    3. Lubrication. Salivary secretions lubricate the food, making swallowing easier, and moisten the mouth, facilitating speech.

III. PERISTALSIS

A. **Definition.** Peristalsis is a coordinated pattern of smooth muscle contraction and relaxation.

B. **Function.** Peristalsis helps move food through the pharynx and esophagus and within the stomach. Peristalsis plays a minor role in propelling food through the intestine.

C. **Mechanics.** During peristalsis, contraction of a small section of proximal muscle is followed immediately by relaxation of the muscle just distal to it. The resulting wavelike motion moves food along the gastrointestinal (GI) tract in a proximal (orad) to distal (caudad) direction (see also IV A 3; Chapter 41 II C; Chapter 42 III B 2; Chapter 43 II B 2).

IV. DEGLUTITION (SWALLOWING)

A. **Phases of swallowing**

    1. Oral (voluntary) phase. During the voluntary phase, the tongue forms a bolus of food and forces it into the oropharynx by pushing up and back against the hard palate.

2. Pharyngeal phase

a. The pharyngeal phase is coordinated by a swallowing center in the medulla and lower pons. It begins when the food reaches the oropharynx and progresses as follows.

   (1) The nasopharynx is closed by the soft palate, preventing regurgitation of food into the nasal cavities.

   (2) The palatopharyngeal folds are pulled medially, forming a passageway for the food to move into the pharynx.

   (3) The glottis and vocal cords are closed and the epiglottis swings down over the larynx, guiding the food toward the esophagus and away from the airways.

   (4) The bolus of food is pushed into the esophagus by the peristaltic contractions of the pharynx and the opening of the upper esophageal sphincter.

b. Respiration is inhibited for the duration of the pharyngeal phase of swallowing (1–2 seconds).

3. Esophageal phase. After reaching the esophagus, food is propelled into the stomach by peristalsis. The strength of peristaltic contractions is proportional to the size of the bolus entering the esophagus.

a. Esophageal structure. The esophagus is about 20 cm in length and is isolated from the oral cavity by the **upper esophageal sphincter (UES)** and from the stomach by the **lower esophageal sphincter (LES).**

   (1) The **UES** is a true anatomical sphincter formed by the cricopharyngeal muscle. The UES is a **striated muscle** and is completely under the control of the vagal fibers that innervate the esophagus. UES tone is maintained by the continual firing of the vagal fibers originating from the **nucleus ambiguus**. The neurotransmitter released by these fibers is **acetylcholine (ACh)**.

   (2) The **upper third of the esophagus,** like the UES, is striated muscle that is under the control of vagal fibers emerging from the nucleus ambiguus.

   (3) The **lower two-thirds of the esophagus** is composed of **smooth muscle**. Its activity is regulated by vagal fibers originating within the **dorsal motor nucleus**. These fibers innervate intrinsic neurons within the muscle layers of the esophagus. These interneurons release an inhibitory neurotransmitter that is either **vasoactive intestinal peptide (VIP) or nitric oxide**.

   (4) The distal 2 cm of the esophagus forms the **LES**. Although the LES is not a separately identifiable muscle, its contractile characteristics are quite different from the rest of the esophageal smooth muscle. During quiescent (nonperistaltic) periods, tonic activity of the vagal fibers innervating the esophagus causes the smooth muscle of the LES to contract. In contrast, the remainder of the smooth muscle within the esophagus is flaccid when not undergoing peristaltic contractions.

b. Types of esophageal peristalsis. There are two types of esophageal peristalsis, primary and secondary.

   (1) Primary esophageal peristalsis is initiated by swallowing. It begins when food passes into the esophagus from the pharyngeal cavity.

   (a) As soon as the food enters the esophagus, the UES contracts to prevent regurgitation of food into the mouth, and the LES relaxes so that food can pass from the esophagus into the stomach.

   (b) The peristaltic wave travels rather slowly (3–4 cm/sec), taking about 9 seconds to push the food from mouth to stomach (a distance of about 30 cm). The force of gravity causes liquids to pass through the esophagus at a much faster rate.

   (c) After food enters the stomach, the LES contracts to prevent regurgitation of food into the esophagus.

   (d) If swallowing is not accompanied by the passing of food into the esophagus, the ensuing peristaltic wave will be very weak or may not occur at all.

   (2) Secondary peristalsis is initiated by the presence of food within the esophagus. Any material remaining in the esophagus stimulates mechanical or irritant receptors.

        (a) After primary peristalsis is completed, any food remaining in the esophagus stretches mechanical receptors, initiating another peristaltic wave.

        (b) Secondary peristaltic waves continue until all of the swallowed food is removed from the esophagus.

    c. Coordination of esophageal peristalsis

        (1) Primary esophageal peristalsis is coordinated by vagal fibers emerging from the swallowing center within the medulla that are activated as part of the swallowing reflex. **Vagotomy,** which eliminates the efferent fibers emerging from the swallowing center, would prevent the initiation of primary esophageal peristalsis.

        (2) Secondary esophageal peristalsis is coordinated by the intrinsic nervous system of the esophagus. Afferent fibers innervate stretch receptors within the wall of the esophagus and thereby activate the appropriate fibers in the intrinsic nervous system. Since intrinsic rather than vagal nerves are involved, **vagotomy** would have little or no effect on secondary esophageal peristalsis.

B. Disorders of swallowing

    1. Esophageal reflux may occur if the intragastric pressure rises high enough to force the LES open, if the LES is unable to maintain its normal tone, or if the LES is forced through the diaphragm and into the thoracic cavity (i.e., hiatal hernia).

        a. During pregnancy, the growing fetus may push the top of the stomach into the thorax. The low intrathoracic pressure (compared to the higher intra-abdominal pressure) causes the LES to expand, allowing reflux to occur.

        b. Reflux of stomach acid causes esophageal pain (heartburn) and may lead to esophagitis.

    2. Belching (eructation). Following a heavy meal or the ingestion of large amounts of gas (e.g., from carbonated beverages), the gas bubble that is usually in the fundus of the stomach is displaced to the cardia (see Figure 41-1). When the LES relaxes during the swallowing process, gas enters the esophagus and is regurgitated.

    3. Achalasia is a neuromuscular disorder of the lower two-thirds of the esophagus that leads to absence of peristalsis and failure of the LES to relax. Food accumulates above this sphincter, taking hours to enter the stomach and dilating the esophagus.

Chapter 41

Stomach

I. ANATOMY (Figure 41-1)

A. Functional components

1. The three functional parts of the stomach are the **fundus, corpus** (body), and **antrum.** Gastric contents are isolated from other parts of the gastrointestinal (GI) tract by the lower esophageal sphincter (LES) proximally and by the pylorus (pyloric sphincter) distally.

2. The antrum and pylorus are anatomically continuous and respond to nervous control as a unit.

B. Musculature

1. As elsewhere in the gut, each muscle layer in the stomach forms a functional syncytium and therefore acts as a unit. In the fundus, where the layers are relatively thin, strength of contraction is weak; in the antrum, where the muscle layers are thick, strength of contraction is greater.

2. The stomach and duodenum (the uppermost part of the small intestine) are divided by a thickened muscle layer called the pyloric sphincter.

C. Innervation

1. **Intrinsic.** The interconnected **myenteric (Auerbach's) plexus** and **submucosal (Meissner's) plexus** within the stomach wall comprise the intrinsic innervation of the stomach, as they do elsewhere in the gut. They are **directly responsible for peristalsis** and other contractions. Because this system is continuous between the stomach and duodenum (see Figure 41-3), peristalsis in the antrum influences the duodenal bulb.
   a. The **myenteric plexus** is located between the layers of the circular and longitudinal muscles of the stomach.

FIGURE 41-1. The stomach.

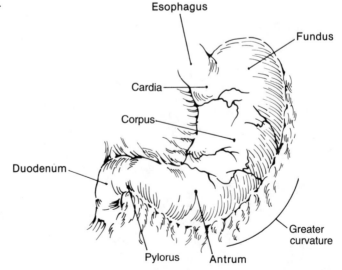

422

b. The **submucosal plexus** is located between the layers of the circular muscle and mucosa on the luminal surface of the stomach.

2. Extrinsic. Autonomic innervation is dual—both **sympathetic,** via the celiac plexus, and **parasympathetic,** via the vagus nerve. Sympathetic innervation inhibits motility, and parasympathetic innervation stimulates motility. Together, the two systems modify the coordinated motor activity that arises independently in the intrinsic system.

II. **MOTILITY.** Complex patterns of motility move food through the stomach, where it can be broken down further by gastric secretions and then propelled into the small intestine. Only small amounts of food are digested or absorbed in the stomach.

A. **Function.** Gastric motility serves three basic functions.

1. Storage. When food enters the stomach, the orad region—primarily the fundus—enlarges to accommodate the food by a process referred to as **receptive relaxation**.

2. Mixing. The presence of food in the caudad stomach—primarily the corpus and antrum—increases the contractile activity of the stomach.
   a. The enhanced contractile activity (a combination of **peristalsis** and **retropulsion**) mixes the food with stomach acid and enzymes, breaking it into smaller and smaller pieces.
   b. When the food is mixed into a pasty consistency, it is called **chyme**.

3. Emptying. When the chyme is broken down into small enough particles, it is propelled through the pyloric sphincter into the intestine.

B. **Receptive relaxation.** When food is passed from the esophagus to the stomach, the contractile activity of the fundus is inhibited, enabling it to easily accommodate 1–2 L of food.

1. Innervation. Vagal nerve fibers innervate the **intrinsic (enteric) nervous system** of the orad stomach to cause receptive relaxation. During this process, postganglionic fibers within the enteric nervous system release a transmitter whose identity is unknown but may be vasoactive intestinal peptide (VIP) or nitric oxide.

2. Initiation. Receptive relaxation may be initiated as part of the peristaltic process causing swallowing and esophageal motility or in response to a bolus of food entering the stomach. Stretch receptors in the orad stomach detect the presence of food and initiate a vagovagal reflex producing receptive relaxation.

3. Effects of vagotomy. Sectioning the vagus nerve will prevent or greatly diminish receptive relaxation since the process is controlled by vagal fibers.

C. **Peristalsis.** Peristaltic contractions are initiated near the fundal-corpus border and proceed caudally, producing a peristaltic wave that propels the food towards the pylorus.

1. Mechanics of peristalsis. Peristaltic contractions are produced by periodic changes in membrane potential, called **slow waves,** or the **basic electrical rhythm (BER)** [Figure 41-2]. These waves are responsible for the rhythm and force of gastric contractions.
   a. Gastric slow waves originate in a pacemaker within the longitudinal muscle high on the greater curvature of the orad corpus.
   b. Slow waves consist of an **upstroke** and **plateau** phase and occur at a rate of approximately 3–4/min.
   c. The **velocity** of the waves is 1 cm/sec when they sweep over the corpus and increases to 3–4 cm/sec in the antrum.
   d. Although the electrophysiological basis for the slow waves is not entirely known, it is assumed that the upstroke is due to the flow of Na^+ and Ca^{2+} into the cell and that the plateau is dependent primarily on the flow of Ca^{2+} into the cell.

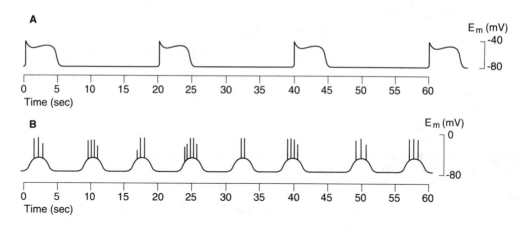

FIGURE 41-2. Basic electrical rhythm (BER), or slow waves, as recorded from smooth muscle cells of (A) the stomach and (B) the middle of the intestine. The slow waves of the stomach are 5–7 seconds in duration and occur at a rate of 3–5 waves/min, while the slow waves of the intestine are more frequent (12 waves/min in the duodenum and 8 waves/min in the ileum) and have spikes superimposed on their plateaus. Ca^{2+} entering the smooth muscle cell during the slow wave produces mechanical activity. In the stomach, acetylcholine (ACh) increases contractile activity by increasing the amplitude and duration of the plateau phase of the slow wave. In the intestine, ACh increases the strength of contraction by increasing the frequency of spikes appearing on top of the slow wave.

> **2. Force of peristalsis.** The force of peristaltic contractions is regulated by gastrin and acetylcholine (ACh). These hormones:
>> **a. Increase the size of the slow wave plateau potential,** which increases the amount of Ca^{2+} entering the cell from the extracellular fluid (ECF)
>> **b. Activate second messengers** that release Ca^{2+} from the sarcoplasmic reticulum (SR)

D. **Retropulsion.** Retropulsion is the back and forth movement of the chyme caused by the forceful propulsion of food against the closed pyloric sphincter (Figure 41-3).

> **1.** The wave of peristaltic contraction reaches the pyloric sphincter before the chyme does. Thus, when the chyme reaches the sphincter, it is pushed back into the body of the stomach.

FIGURE 41-3. Peristaltic contractions begin in the midstomach (A) and proceed caudally, pushing the food toward the pylorus (B). When the food reaches the pylorus (where small enough pieces of food flow into the duodenum), a mass contraction of the terminal antrum pushes the food back toward the corpus through a narrow antral ring (C). The backward movement of the food is called retropulsion.

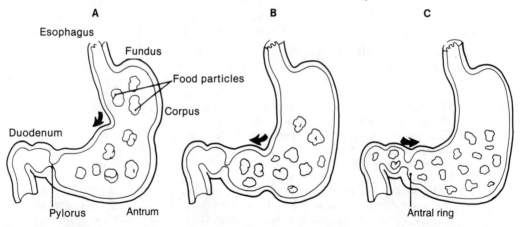

2. The forward and backward movement of the chyme (caused by peristalsis and retropulsion) breaks the chyme into smaller and smaller pieces and mixes it with the gastric secretions present within the stomach.

E. **Gastric emptying** occurs when the chyme is decomposed into small enough pieces to fit through the pyloric sphincter.

1. Each time the chyme is pushed against the pyloric sphincter, a small amount (2–7 ml) may escape into the duodenum.

2. The amount of chyme passing through the pylorus depends on the size of the particles. If the particles are too large, none of the chyme will enter the duodenum.

3. Thus, the rate of gastric emptying of solids depends on the rate at which the chyme is broken down into small particles.

4. Liquids empty much faster than solids. The rate at which liquids empty is proportional to pressure within the orad stomach, which increases slowly during the digestive period.

F. **Regulation of gastric emptying**

1. Local reflexes
   a. Excitatory reflexes, initiated by expansion of the antrum, are responsible for increasing gastric motility. Although these reflexes do not require the vagus nerve, vagotomy decreases the magnitude and coordination of stomach contractions.
   b. Inhibitory reflexes. A variety of stimuli act on the duodenum to initiate **enterogastric reflexes** that slow the rate of gastric emptying.
      (1) Purpose. Enterogastric reflexes prevent the flow of chyme from exceeding the ability of the intestine to handle it.
      (2) Causes. High osmolarity, low pH, fat and protein digestion products, low osmolality, the caloric content of food, and distention of the duodenal wall all elicit an enterogastric reflex.

2. Hormones released from the stomach and intestine also influence gastric motility.
   a. Excitatory effects. Gastrin, released into the circulation in response to antral distention or food breakdown products, enhances gastric contractions.
   b. Inhibitory effects. A variety of intestinal hormones, collectively called **enterogastrones,** inhibit gastric contractions. **Cholecystokinin (CCK)** and **secretin** are two known enterogastrones. The identity and mode of action of other enterogastrones remain to be discovered.
      (1) CCK is released from the duodenum in response to fat or protein digestion products. CCK probably acts by blocking the excitatory effects of gastrin on gastric smooth muscle (see Chapter 42 IV C 3).
      (2) Secretin is released from the duodenum in response to the presence of acid. Secretin most likely has a direct inhibitory effect on smooth muscle (see Chapter 42 IV C 3).

3. Migrating motor complex (MMC). During the interdigestive period, any food left in the stomach is removed by the MMC.
   a. The MMC is a peristaltic wave that begins within the esophagus and travels through the entire GI tract (see Chapter 42 III B 3).
   b. The peristaltic wave occurs every 60–90 minutes during the interdigestive period.
   c. The hormone **motilin,** which is released from endocrine cells within the epithelium of the small intestine, increases the strength of the MMC.

G. **Vomiting** (or **emesis**) is the forceful expulsion of the food from the stomach and intestine.

1. Initiation. Vomiting may be initiated by direct activation of the **vomiting center** in the medulla or by activation of the **chemoreceptor trigger zone** within the area postrema of the brain stem.
   a. The **vomiting center** may be directly activated by afferent fibers or by irritation due to injury or increases in intracranial pressure. When the vomiting center is directly activated, it causes **projectile vomiting**—a rapid, forceful emesis not accompanied by nausea.

    b. The **chemoreceptor trigger zone** may be activated by afferent nerves originating within the GI tract or by circulating emetic agents such as apomorphine or copper sulfate. Vomiting caused by activation of the chemoreceptor trigger zone is accompanied by nausea.

2. Mechanical sequence of vomiting
    a. Vomiting begins with a deep inspiration followed by the closing of the glottis.
    b. Next, a pressure wave originating in the intestine propels chyme into the orad stomach.
    c. Finally, an increase in abdominal pressure forces the chyme into the esophagus and out of the mouth.
    d. Retching may proceed vomiting. Retching involves all of the involuntary motions of vomiting but without the production of vomitus. The chyme is not ejected because the abdominal and thoracic pressures are not sufficient to overcome the resistance of the upper esophageal sphincter (UES).

III. GASTRIC SECRETION

A. General considerations

1. Function. Gastric secretions aid in the breakdown of food into small particles and continue the process of digestion begun by salivary enzymes. About 2 L/day of gastric secretions are produced.

2. Phases of gastric secretion
    a. The cephalic phase of gastric secretion is initiated by the thought, sight, taste, or smell of food. It is dependent on the integrity of the vagal fibers innervating the stomach.
        (1) Secretion of hydrochloric acid (HCl) from parietal cells, gastrin from G cells, and pepsinogen from peptic (chief) cells is stimulated by vagal efferent fibers.
        (2) Almost half of the gastric secretions released during a meal occur as a result of cephalically induced vagal stimulation.
    b. The gastric phase of secretion is initiated by the entry of food into the stomach. Food entering the stomach buffers acid, raises pH, and allows other stimuli (e.g., vagal fibers, gastrin) to release acid.
        (1) Distention of the corpus, acting through local and vagovagal reflexes, results in an increase in HCl secretion.
        (2) Distention of the antrum initiates vagally mediated and local reflexes that result in gastrin release from antral G cells. Gastrin release is inhibited at low pH (< 3).
        (3) Low pH activates local reflexes, which enhance pepsinogen secretion.
        (4) Although the rate of gastric secretion during the gastric phase is less than during the cephalic phase, it continues for a longer time. Thus, the two phases contribute about the same amount of secretion.
    c. The intestinal phase of secretion begins as the chyme begins to empty from the stomach into the duodenum. Overall, little gastric secretion occurs during the intestinal phase.

3. Gastric secretory cells are located on the surface of the stomach and in glands that are buried within the mucosa.
    a. Oxyntic glands are located in the fundus and corpus of the stomach. They contain three types of secretory cells.
        (1) The **parietal (oxyntic) cells** secrete **HCl**. These cells are also responsible for the secretion of **intrinsic factor,** which is necessary for the absorption of vitamin B_{12} by the ileum of the small intestine (see III E; Chapter 42 VII E 3).
        (2) Peptic (chief) cells secrete **pepsinogen,** the precursor for the proteolytic enzyme **pepsin.**
        (3) Mucous cells secrete mucus.

   b. Pyloric glands are located in the antrum and pyloric regions of the stomach. They contain **G cells** and some mucous cells. **G cells** are responsible for the release of the hormone **gastrin**.

     (1) There are two forms of gastrin, G-17 (little gastrin, a 17-amino-acid peptide) and G-34 (big gastrin, a 34-amino-acid peptide). Although G-17 is more potent than G-34, the larger gastrin is found in higher concentrations within the circulation.

     (2) Gastrin is released from the basolateral surface of the G cells, enters the circulation, and travels to the orad stomach, where it stimulates parietal-cell HCl secretion.

B. HCl secretion

   1. Functions of HCl

     a. HCl participates in the breakdown of protein.

     b. It provides an optimal pH for the action of pepsin.

     c. It hinders the growth of pathogenic bacteria.

   2. Mechanism of HCl secretion (Figure 41-4)

     a. HCl is secreted into the parietal cell **canaliculi** by a three-step process.

     (1) The active transport process is begun by the transport of K^+ and Cl^- into the canaliculi. Cl^- is transported either by a pump or through a channel. The flow of Cl^- creates a negative potential inside the canaliculi, causing K^+ to flow passively into the canaliculi.

FIGURE 41-4. Hydrochloric acid (HCl) being formed by the parietal cell. Cl^- and K^+ are secreted into canaliculi by separate transporters, which may be channels or carriers. H^+ is exchanged for K^+ by an ATPase active transport system, allowing HCl to be secreted into the stomach. Numerous mitochondria provide energy for the active transport process. (Adapted from Guyton AC: *Textbook of Medical Physiology,* 8th edition. Philadelphia, WB Saunders, 1991, p 64.)

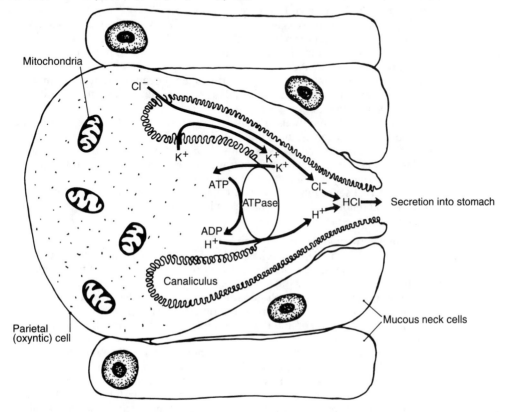

      (2) H^+ is then exchanged for K^+ by the H^+-K^+-ATPase pump.
      (3) Water enters the canaliculi down the osmotic gradient created by the movement of HCl into the canaliculi.
  b. The H^+ entering the canaliculi is supplied by the dissociation of carbonic acid (H_2CO_3) into H^+ and bicarbonate (HCO_3^-) within the parietal cell.
      (1) H_2CO_3 is formed from the reaction:

$$CO_2 + H_2O \rightarrow H_2CO_3 \rightarrow H^+ + HCO_3^-$$

      (2) The formation of H_2CO_3 from CO_2 is catalyzed by the enzyme **carbonic anhydrase**. Acetazolamide, a carbonic anhydrase inhibitor, blocks the formation of HCl by the parietal cell.
      (3) The HCO_3^- diffuses back into the plasma (creating the alkaline tide associated with gastric secretion) in exchange for Cl^-, thus providing Cl^- for the initial step in the secretory process [see III B 2 a (1)].
  c. Most of the HCl that is secreted into the stomach is neutralized and reabsorbed within the small intestine. However, if the gastric contents are lost before they enter the small intestine (e.g., by vomiting), a severe alkalosis may ensue.
  d. The active transport processes involved in the generation of HCl require a large amount of adenosine triphosphate (ATP). The ATP is generated by mitochondria found in very high concentration (40% of cell volume) within the parietal cell.
  e. The pH of the parietal cell secretion can be as low as 0.8 (i.e., a H^+ concentration of approximately 150 mmol, or almost 4 million times as great as the H^+ concentration of plasma).
  f. The **H^+-K^+-ATPase pump** can be irreversibly inhibited by the drug **omeprazole**, which is now used for the treatment of duodenal and gastric ulcers.

3. Substances affecting HCl secretion
  a. Stimulation of HCl secretion. ACh, histamine, and gastrin act directly on the parietal cell to stimulate HCl secretion (Figure 41-5). In addition, ACh and gastrin may directly stimulate the mast cell to secrete histamine.
      (1) ACh, a neurotransmitter, is released from nerve cells innervating the parietal cell.
      (2) Histamine is released from mast cells located within the corpus.
         (a) Histamine can stimulate HCl secretion directly or can potentiate the secretion produced by ACh or gastrin.
         (b) Histamine is classified as a **paracrine agent** because it diffuses from its release site to the parietal cells (rather than traveling within the circulation as does a hormone).
         (c) The most commonly used anti-ulcer drugs (i.e., cimetidine and ranitidine) are histamine antagonists that block the H_2 receptor on the parietal cell.
      (3) Gastrin is released from G cells in the distal stomach [see III A 3 b]. Gastrin is classified as a hormone because it travels to its target cell through the circulation. A variety of substances affect gastrin secretion (see III C).
  b. Inhibition of HCl secretion. Somatostatin inhibits HCl secretion by parietal cells and gastrin secretion by G cells. Somatostatin is released from interneurons within the enteric nervous system.

4. Regulation of gastric acid (HCl) secretion
  a. Stimulation during the cephalic phase. The vagus nerve stimulates the release of ACh and inhibits the release of somatostatin from interneurons within the enteric nervous system, thus enhancing the secretion of HCl (see Figure 41-5).
  b. Stimulation during the gastric phase. The **amount of ingested protein** is the most important determinant of acid secretion during the gastric phase.
      (1) Protein is a good buffer and thus keeps the pH of the stomach at an optimal level for acid secretion.
      (2) Amino acids and peptides directly stimulate parietal cells to secrete acid.
      (3) Alcohol and caffeine also cause the release of HCl and gastrin.
  c. Inhibition during the gastric phase. The most potent inhibitor of HCl secretion during the gastric phase is the presence of acid in the stomach. If the pH of the stomach falls below 2, acid secretion stops. Acid secretion is inhibited by two mechanisms.

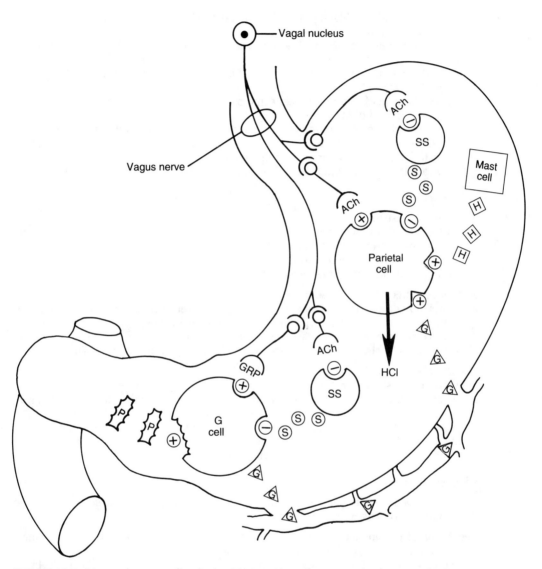

FIGURE 41-5. Many substances affect hydrochloric acid (*HCl*) secretion by the parietal cell. Acetylcholine (*ACh*) from the vagus nerve, histamine (*H*) from mast cells, and gastrin (*G*) from G cells all stimulate the parietal cell directly to secrete HCl. The release of gastrin into the circulation, in turn, is stimulated by gastrin-releasing peptide (*GRP*) and protein digestion products (*P*). Somatostatin (*S*), released from somatostatin cells (*SS*), inhibits the release of both gastrin and HCl. Thus, the stimulation of vagal fibers, which causes the release of ACh and GRP but inhibits the release of somatostatin, has an amplified positive effect on parietal-cell HCl secretion. *Plus signs* = stimulation; *Minus signs* = inhibition. (Adapted from Johnson LR [ed]: *Gastrointestinal Physiology,* 3rd edition. St. Louis, CV Mosby, 1985, p 72.)

        (1) A low pH directly inhibits HCl and gastrin secretion.
        (2) Lowering the pH also releases somatostatin, which inhibits the secretion of gastrin by the G cells and HCl by the parietal cells (see Figure 41-5).
    d. Stimulation during the intestinal phase. The presence of protein digestion products within the duodenum causes an increase in HCl secretion.
        (1) Although G cells have been identified within the duodenum, gastrin is not thought to cause the increase in acid secretion.
        (2) An as yet unidentified hormone, called **entero-oxyntin,** is postulated to be responsible for the increase in acid secretion.

(3) Amino acids circulating in the blood after being absorbed from the intestine may also stimulate HCl secretion.

e. Inhibition during the intestinal phase. Acid secretion is inhibited when food enters the duodenum. Because the major effect of food entering the duodenum is inhibition of acid secretion, loss of the proximal intestine increases gastric secretion (and motility).

   (1) H^+, fatty acids, and increased osmolarity stimulate the release of enterogastrones from the duodenum.

   (2) The most important of the enterogastrones may be **gastric inhibitory peptide (GIP),** which inhibits both gastrin release and parietal cell secretion of HCl. GIP is thought to act by stimulating the release of somatostatin, which, in turn, inhibits the parietal and G cells. GIP is also involved in the release of insulin.

C. **Gastrin secretion**

1. **Functions of gastrin**
   a. Gastrin stimulates HCl secretion.
   b. It increases gastric and intestinal motility.
   c. It increases pancreatic secretions.
   d. It is necessary for the proper growth of GI mucosa.

2. **Substances affecting gastrin secretion**
   a. Stimulation of gastrin secretion. Bombesin [gastrin-releasing peptide (GRP)] is most likely the neurotransmitter responsible for stimulating G cells to secrete gastrin. The vagus nerve increases the release of GRP during the cephalic phase (see Figure 41-5).
   b. Inhibition of gastrin secretion. Somatostatin inhibits gastrin secretion. The vagus nerve inhibits the release of somatostatin during the cephalic phase.

3. **Regulation of gastrin secretion**
   a. In general, gastrin secretion is regulated by the same mechanisms that regulate HCl secretion (i.e., vagal stimulation, pH, enterogastrones).
   b. In addition, several foods and food breakdown products (**secretagogues**) directly stimulate the release of gastrin. These include protein digestion products, alcohol, and coffee (both caffeinated and decaffeinated).

D. **Pepsinogen secretion**

1. **Function of pepsinogen. Pepsin,** the active form of pepsinogen, is a proteolytic enzyme that begins the process of protein digestion [see Chapter 42 VII B 2 a].

2. **Regulation of pepsinogen secretion.** Pepsinogen is released from the chief cells of the oxyntic glands during all three phases of digestion.
   a. Cephalic phase. During the cephalic phase of digestion, vagally stimulated cholinergic neurons within the enteric nervous system directly stimulate chief cells to release pepsinogen.
   b. Gastric phase. During the gastric phase of digestion, low pH activates local reflexes that enhance pepsinogen secretion. The low pH of the stomach is also responsible for converting pepsinogen into pepsin. Again, ACh is the transmitter that stimulates the chief cells.
   c. Intestinal phase. Secretin enhances pepsinogen release. Thus, the presence of H^+ within the duodenum during the intestinal phase of digestion may contribute to pepsinogen secretion.

E. **Intrinsic factor**

1. **Definition.** Intrinsic factor is a glycoprotein secreted by the parietal cells of the gastric mucosa, chiefly by those in the fundus.

2. **Function.** Intrinsic factor is required for the **absorption of vitamin B_{12}.**
   a. Intrinsic factor forms a complex with vitamin B_{12}.
   b. The intrinsic factor–B_{12} complex is carried to the terminal ileum, where the vitamin is absorbed (see Chapter 42 VII E 3).

IV. **GASTRIC MUCOSAL BARRIER.** The gastric mucosal barrier protects the gastric lining cells from damage by intraluminal HCl, or **autodigestion**. Its chief component is a thick viscous alkaline mucous layer that measures over 1 mm thick and is secreted by the mucous cells. The mucous cells cover the surface between the various gastric glands and outnumber all other cell types found in the gastric mucosa.

A. The turnover rate of the gastric mucosa is extremely high; 5×10^5 mucosal cells are shed each minute, and the entire mucosa is replaced in 1–3 days.

B. **Mild injury** results in increased mucus secretion and surface desquamation followed by regeneration.

C. **More serious injury** denudes the mucosal surface, forming an ulcer, and produces bleeding. Ulceration results when damage to the mucosa (e.g., due to a highly concentrated HCl, 10% ethanol, salicylic acid, or acetylsalicylic acid) allows acid to penetrate the mucosal barrier and destroy mucosal cells. This liberates histamine, which increases acid secretion and produces increased capillary permeability and vasodilation. The latter two effects lead to edema. It is the exposure of mucosal capillaries to the digestive process that leads to bleeding.

D. The **rate of repair** of mucosal injury depends on the extent of injury and varies from as little as 48 hours for restricted desquamation to up to 3–5 months if damage has left only the deepest portions of the gastric pits intact.

E. **Ulcer therapy.** Ulcer repair is aided by drugs that neutralize gastric acid (e.g., nonprescription antacids) or prevent acid release, such as proton pump inhibitors (see III B 2 f) and H_2 blockers [see III B 3 a (2) (c)]. Recent evidence implicates the *Helicobacter pylori* bacillus in the production of gastric acid and duodenal ulcers; some ulcers are being treated with antibiotics.

V. **GASTRIC DIGESTION AND ABSORPTION**

A. **Digestion**

1. **Carbohydrate digestion** in the stomach depends on the action of salivary amylase, which remains active until halted by the low pH in the stomach.

2. **Protein digestion.** About 10% of ingested protein is broken down completely in the stomach. Gastric pepsin facilitates later digestion of protein by breaking apart meat particles.

3. **Fat digestion** is minimal due to the restriction of gastric lipase activity to triglycerides containing short-chain (< 10 carbons) fatty acids. Acid and pepsin break emulsions so that fats coalesce into droplets, which float and empty last.

B. **Absorption**

1. **Nutrients.** Very little absorption of nutrients takes place in the stomach. The only substances absorbed to any appreciable extent are highly lipid-soluble substances (e.g., the un-ionized triglycerides of acetic, propionic, and butyric acids). Aspirin at gastric pH is un-ionized and fat soluble; after absorption, it ionizes intracellularly, damaging mucosal cells and ultimately producing bleeding. Ethanol is rapidly absorbed in proportion to its concentration.

2. **Water** moves in both directions across the mucosa. It does not, however, follow osmotic gradients. Water-soluble substances, including Na^+, K^+, glucose, and amino acids, are absorbed in insignificant amounts.

Chapter 42

Small Intestine

I. **OVERVIEW.** The small intestine is the major site of **digestion** and **absorption** of carbohydrates, proteins, and fats in the gastrointestinal (GI) tract. The action and secretions of several **accessory organs** (see II B) are essential to the digestive and absorptive functions of the small intestine. Nutrients and fluids that are not absorbed in the small intestine are passed on to the colon.

II. **ANATOMY**

A. **Small intestine**

1. The small intestine has three parts: the **duodenum,** the **jejunum,** and the **ileum** (see Figure 39-1).

2. Although the small intestine is approximately 5 m long, it has an **absorptive area** of over 250 m².
   a. Its large surface area is created by numerous folds of the intestinal mucosa (**valvulae conniventes**); by densely packed **villi,** which line the entire mucosal surface; and by **microvilli,** which protrude from the surface of the intestinal cells.
      (1) The epithelial cells from which the microvilli protrude are called **enterocytes.**
      (2) The microvilli (about 1 μm long and 0.1 μm in diameter) give the intestinal mucosa its characteristic **brush border** appearance.
   b. The **blood supply** of the villus is ideally organized to collect the nutrients after they are absorbed across the brush border membrane.
      (1) Each villus is supplied by an arteriole, which gives rise to a capillary tuft at the tip of the villus. The capillaries coalesce into venules, which drain into the portal vein. The portal vein carries the absorbed nutrients to the liver.
      (2) Branches of the lymphatics, called **lacteals,** also extend to the tip of the villus. These vessels carry absorbed fats to the thoracic duct from which they enter the general circulation.

B. The **accessory organs** involved in intestinal digestion and absorption are the **pancreas,** the **liver,** and the **gallbladder.**

1. The **pancreas** secretes various substances that aid in intestinal digestion, including HCO_3^-, which neutralizes the acidic content of chyme entering the small intestine.

2. The **liver** secretes bile, which is necessary for fat digestion and nutrient absorption.
   a. Bile travels from the liver through **bile ducts** to reach the duodenum of the small intestine.
   b. The **sphincter of Oddi,** located at the distal end of the duodenum, forms the opening that connects the small intestine to the **common bile duct.**
      (1) This sphincter controls the flow of bile into the small intestine [see V A 4, E 2 a].
      (2) When the sphincter is closed, bile cannot enter the small intestine and must be stored in the gallbladder.
   c. The **portal circulation** carries bile that has been absorbed from the terminal ileum back to the liver (see V C; Figure 42-1).

3. The **gallbladder** stores bile during the interdigestive period.

III. MOTILITY

A. Contractile activity

1. **Function.** Contractile activity of the smooth muscles lining the small intestine serves two major functions:
   a. **Mixing the chyme** with the digestive juices and bile to facilitate digestion and absorption
   b. **Propelling the chyme** from the duodenum to the colon

2. **Transit time.** It usually takes about 2–4 hours for the chyme to move from one end of the small intestine to the other.

B. Types of movements

1. **Segmentation** is the most common type of intestinal contraction.
   a. During segmentation, about 2 cm of the intestinal wall contracts, forcing the chyme back toward the stomach (oradly) and toward the colon (aborally).
   b. When the muscle relaxes, the chyme returns to the area from which it was displaced.
   c. This back-and-forth movement enables the chyme to become thoroughly mixed with the digestive juices and to make contact with the absorptive surface of the intestinal mucosa.
   d. Segmentation contractions occur about 12 times/min in the duodenum and 8 times/min in the ileum. The contractions last for 5–6 seconds.
   e. Segmentation occurs throughout the digestive period.

2. **Peristaltic contractions** also occur in the small intestine.
   a. Although peristaltic contractions occasionally propel food along the entire length of the intestine, they rarely involve more than a short segment of the intestine.
   b. Peristalsis is not considered to be an important component of intestinal transit.

3. The **migrating motor complex (MMC;** see Chapter 41 II F 3) spreads over the intestine during the interdigestive period.
   a. The MMCs sweep out the chyme remaining in the small intestine during the interdigestive period.
   b. MMCs occur every 60–90 minutes and last for about 10 minutes.

C. Propulsion of chyme. During the digestive period, the higher frequency of segmentation in the proximal intestine (duodenum) than in the distal intestine (ileum) propels the chyme slowly toward the colon.

1. Thus, when the chyme is pushed aborally, it is less likely to be forced back by a segmentation contraction.

2. In contrast, when the chyme is pushed oradly, it will be quickly pushed toward the colon again by a segmentation contraction in the more proximal region of the small intestine.

D. Control of intestinal motility. The frequency and strength of segmentation contractions in the intestine are controlled by the **slow waves** (see Figure 41-2).

1. **Generation.** Segmentation contractions can occur only if the slow waves produce **spikes,** or **action potentials.** Spikes appear on the slow waves when the membrane potential is sufficiently depolarized.

2. **Frequency**
   a. The frequency of segmentation contractions is directly related to the frequency of the slow waves.
   b. Slow wave frequency is controlled by **pacemaker cells** within the wall of the intestine and is not influenced by neural activity or circulating hormones.

3. Strength

a. The strength of a segmentation contraction is proportional to the frequency of the spikes generated by the slow wave. This frequency is controlled by the **amplitude of the slow wave**. Thus, the greater the slow wave amplitude, the greater the frequency of spikes generated and the greater the strength of the contraction.

b. Slow wave amplitude is controlled by the hormones released during digestion.

(1) Gastrin, cholecystokinin (CCK), motilin, and insulin increase the slow wave amplitude.

(2) Secretin and glucagon reduce the slow wave amplitude.

IV. PANCREATIC SECRETIONS

A. **Pancreatic cell types and their functions.** The pancreas contains endocrine, exocrine, and ductal cells.

1. The **endocrine cells,** arranged in small islets within the pancreas, secrete **insulin, glucagon, somatostatin,** and **pancreatic polypeptide** directly into the circulation.

2. The **exocrine cells** are organized into acini that produce four types of digestive enzymes: **peptidases, lipases, amylases,** and **nucleases,** which are responsible for digesting proteins, fats, carbohydrates, and nucleic acids, respectively. In their absence, malabsorption syndromes develop.

3. Each day, the **ductal cells** secrete about 1200–1500 ml of pancreatic juice containing a high concentration of **bicarbonate (HCO_3^-)**. The HCO_3^- neutralizes gastric acid and regulates the pH of the upper intestine. Failure to neutralize the chyme as it enters the intestine will result in duodenal ulcers.

B. **Composition of pancreatic secretions**

1. Electrolytes

a. Na^+ and K^+ concentrations in pancreatic juice are the same as those in plasma water (i.e., 142 mEq/L and 4.8 mEq/L, respectively).

b. HCO_3^- concentration in pancreatic juice is much higher than it is in plasma water (100 mEq/L as opposed to about 24 mEq/L).

c. Pancreatic juice also contains small amounts of other ions such as Ca^{2+}, magnesium (Mg^{2+}), zinc (Zn^{2+}), phosphate (HPO_4^{2-}), and sulfate (SO_4^{2-}).

d. Secretion of HCO_3^- by the ductal cells requires at least one active transport process.

(1) HCO_3^- and H^+ are formed from the dissociation of carbonic acid (H_2CO_3) via the same reaction that occurs in the parietal cells of the stomach [see Chapter 41 III B 2 b].

(2) H^+ is actively transported out of the cell across its basal membrane by an Na^+-H^+ antiporter.

(3) HCO_3^- is transported across the apical membrane of the ductal cells. The transport process responsible for HCO_3^- secretion is not yet known.

(4) Na^+ follows the HCO_3^- into the pancreatic ducts. Most of the Na^+ flows passively between the ductal cells. However, some may be actively transported across the apical membranes.

(5) Water flows into the ducts down the osmotic gradient established by the secretion of sodium bicarbonate ($NaHCO_3$). Flow rates as high as 1 ml/min can be established.

(6) As the pancreatic juice flows along the ducts, Cl^- is exchanged for HCO_3^-. The higher the flow rate, the smaller the exchange. Thus, HCO_3^- concentration is highest when pancreatic secretion is greatest.

2. Enzymes. Three major types of pancreatic enzymes are secreted by the pancreas: **amylases, lipases,** and **proteases**.

a. **Pancreatic α-amylase** is secreted in its active form. It hydrolyzes glycogen, starch, and most other complex carbohydrates, except cellulose, to form disaccharides.

b. Pancreatic lipases (lipase, cholesterol lipase, and phospholipase) [see VII C 1 a] are secreted in their active forms. The enzymes that hydrolyze water-insoluble esters require bile salts to work. Water-soluble esters can be hydrolyzed without the action of bile salts.

c. Pancreatic proteases (**trypsin** and **the chymotrypsins**) are secreted in their inactive zymogen form (trypsinogen and the chymotrypsinogens, respectively) [see VII B 2 b].
  (1) Trypsinogen is converted to trypsin by enterokinase (also called enteropeptidase) or by trypsin itself (autocatalysis).
  (2) The chymotrypsinogens are converted to their active form by trypsin.

d. Trypsin inhibitor is secreted by the same cells and at the same time as the pancreatic proenzymes. Trypsin inhibitor protects the pancreas from autodigestion.

C. **Control of pancreatic secretion.** Like gastric secretion, pancreatic secretion is divided into the following three phases.

1. Cephalic phase. The thought, sight, smell, or taste of food produces the cephalic phase of pancreatic secretion. Both acinar and ductal cell secretions are enhanced by vagal stimulation.
  a. Enzyme secretion by the acinar cells is stimulated directly by vagal fibers that release acetylcholine (ACh) or by cholinergic interneurons that are stimulated by the vagal preganglionic fibers.
  b. Although the vagus nerve can also cause HCO_3^- secretion by ductal cells, its ability to stimulate HCO_3^- secretion is not nearly as great as its ability to stimulate the release of enzymes.

2. Gastric phase. Pancreatic secretion is enhanced during the gastric phase by distension and food breakdown products.
  a. Distention of the antrum and corpus initiates a vagovagal reflex resulting in a low volume of pancreatic secretion containing both HCO_3^- and enzymes. ACh is the transmitter.
  b. Food breakdown products (primarily amino acids and peptides) can stimulate pancreatic secretions because of their ability to cause the G cells of the antrum to release gastrin. Gastrin produces a low-volume, high-enzyme pancreatic secretion.

3. Intestinal phase. The major stimulants for pancreatic secretion are the hormones cholecystokinin (CCK) and secretin. They are released from endocrine cells in the duodenum and jejunum during the intestinal phase of pancreatic secretion.
  a. CCK, in addition to its effect on the gallbladder [see V E 2 a], is a potent stimulant of pancreatic enzyme secretion.
    (1) Like gastrin, CCK is found in two physiologically active forms, an octapeptide called CCK-8 and a 33-chain polypeptide, CCK-33.
    (2) The actions of CCK are potentiated by secretin. By itself, secretin has no effect on enzyme secretion.
  b. Secretin was the first hormone ever discovered. Its primary effect is to increase HCO_3^- secretion by the pancreas.
    (1) The actions of secretin are potentiated by CCK. By itself, CCK has no effect on HCO_3^- secretion.
    (2) Because they are potentiators of each other's action, small concentrations of CCK and secretin together can produce significant amounts of pancreatic HCO_3^- and enzyme secretions, while either one alone would have little or no effect.
  c. Control of CCK and secretin release. CKK and secretin are secreted from endocrine cells in response to the entrance of chyme into the small intestine.
    (1) Amino acids (primarily **phenylalanine**), **fatty acids,** and **monoglycerides** are the major stimuli for CCK secretion.
    (2) Low pH (< 4.5), caused by the presence of gastric acid (HCl) in the intestine, is a potent stimulus for the release of secretin.
  d. A vagovagal reflex, which greatly potentiates the effects of secretin and CCK, is activated during the intestinal phase of digestion.

e. ACh potentiates the effects of both CCK and secretion. Thus, vagal stimulation is much more potent in stimulating pancreatic secretions when CCK and secretin are present in the plasma.

V. BILIARY SECRETIONS

A. General features of bile

1. **Function.** Bile is required for the digestion and absorption of fats and for the excretion of water-insoluble substances such as cholesterol and bilirubin.

2. **Formation.** Bile is formed by liver epithelial cells, called **hepatocytes,** and by epithelial cells lining the bile ducts, called **ductal cells**. Between 250 and 1100 ml of bile are secreted daily.

3. **Storage.** Although it is secreted continuously, bile is stored in the gallbladder during the interdigestive period.

4. **Release.** Bile is released into the duodenum during the digestive period only after chyme has triggered the release of CCK, which then produces contraction of the gallbladder and relaxation of the sphincter of Oddi.

B. Composition of bile

1. **Bile acids**
   a. Primary bile acids (trihydroxycholic acid and **dihydroxychenodeoxycholic acid)** are synthesized from cholesterol and converted into bile salts by the hepatocytes as follows.
      (1) Cholesterol is absorbed through microvilli lining the serosal (antiluminal) border of the hepatic epithelial cells.
      (2) The bile acids are conjugated with either taurine or glycine to form bile salts.
      (3) The bile salts are actively secreted into a canaliculus on the lateral (luminal) surface of the hepatocyte, from which they then drain into the bile duct.
      (4) Because bile salts are not lipid soluble, they remain within the intestine until reaching the ileum, where they are actively absorbed (see VII C 2).
   b. Secondary bile acids are formed by deconjugation and dehydroxylation of the primary bile salts by intestinal bacteria, forming **deoxycholic acid** and **lithocholic acid**.

2. **Bile pigments**
   a. Bilirubin and **biliverdin,** the two principal bile pigments, are metabolites of hemoglobin formed in the liver and conjugated as glucuronides for excretion. They are responsible for the golden yellow color of bile.
   b. Intestinal bacteria metabolize bilirubin further to **urobilin,** which is responsible for the brown color of stool.
   c. If bilirubin is not secreted by the liver, it builds up in the blood and tissues, producing **jaundice** (see V F 2).

3. **Phospholipids** (primarily **lecithins**) are, after bile salts, the most abundant organic compound in bile.
   a. Although the phospholipids are normally insoluble in water, they are solubilized by the bile salt micelles.
   b. The **micelles** are able to solubilize other lipids more effectively when they are composed of bile salts and phospholipids than when they are composed of bile salts alone.

4. **Cholesterol,** although present only in small amounts, is an important component of bile.
   a. Cholesterol is essentially insoluble in water and thus must be solubilized by bile salt micelles before it can be secreted in the bile (see VII C 2).
   b. Biliary secretion of cholesterol is important because it is one of the few ways in which cholesterol stores can be regulated.

5. Electrolytes. The electrolyte composition of bile is similar to that of pancreatic juice and plasma (see IV B 1).

C. **Enterohepatic circulation** is the recirculation of bile salts from the liver to the small intestine and back again. This circulation is necessary because of the limited pool of bile salts available to help break down and absorb fat (Figure 42-1).

1. **Path of circulation.** Bile salts travel from the liver to the duodenum via the common bile duct. When the bile salts reach the terminal ileum, they are reabsorbed into the portal circulation. The liver then extracts them from the portal blood and secretes them once again into the bile.
   a. Bile salts are reabsorbed only in the terminal ileum. No reabsorption of bile salts occurs in the duodenum or jejunum.
   b. From 90%–95% of the bile salts that enter the small intestine are actively reabsorbed from the lower ileum back into the portal circulation.
   c. The remaining bile salts are excreted into the feces.

2. **Circulating pool.** The total circulating pool of bile salts (consisting of primary and secondary bile acids) is approximately 3.6 g. Because 4–8 g of bile salts are required to digest and absorb a meal (more if the meal is high in fat), the total pool of salts must circulate twice during the digestion of each meal. Consequently, the bile salts usually circulate 6–8 times daily.

3. **Bile salt synthesis and replacement.** The rate of bile salt synthesis is determined by the rate of return to the liver. The usual rate is 0.2–0.4 g/day, which replaces normal fecal losses. The maximal rate is 3–6 g/day. If fecal losses exceed this rate, the total pool size decreases.

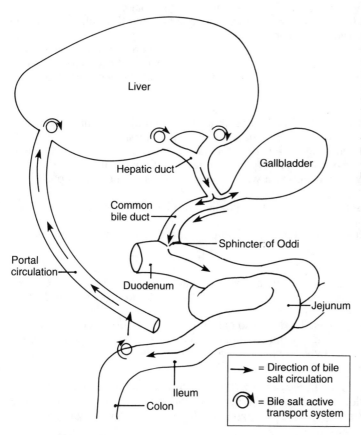

FIGURE 42-1. The enterohepatic circulation. Bile acids are absorbed from the terminal ileum by a Na^+-dependent active transport system and then rapidly sequestered by hepatocytes in the liver and returned to the gallbladder or duodenum. Each day, about 20% of the bile acid pool escapes the enterohepatic circulation (thus becoming lost to excretion) and must be resynthesized by the liver.

4. Clinical implications. Because bile salts are required for proper digestion and absorption of fats, any condition that disrupts the enterohepatic circulation (e.g., ileal resection or small intestinal diseases such as sprue or Crohn's disease) leads to a decreased bile acid pool and malabsorption of fat and fat-soluble vitamins. The clinical manifestations of such conditions are steatorrhea and nutritional deficiency. An increase in fecal losses of bile salts results in watery diarrhea, since bile salts inhibit water and Na^+ absorption in the colon.

D. **Control of biliary secretion.** The volume of biliary secretion and the amount of bile in that secretion are regulated separately.

1. **The bile-independent fraction of biliary secretion** refers to the amount of **fluid,** composed of electrolytes and water, that is secreted each day by the liver. Although this fluid, by definition, is secreted with the bile, its secretion is controlled separately from bile secretion.
   a. Secretion of this fluid is controlled by the hormone **secretin**.
   b. The fluid resembles the secretion of the pancreatic ductal cells in the following ways (see IV B 1, C).
      (1) The fluid is secreted by ductal cells.
      (2) Its secretion is controlled by secretin.
      (3) It has a high concentration of HCO_3^-.

2. **The bile-dependent fraction of biliary secretion** refers to the quantity of **bile salts** secreted by the liver.
   a. The amount of bile salts secreted is directly related to the amount of bile reabsorbed by the hepatocytes (i.e., the more bile reabsorbed from the portal circulation, the more bile secreted by the liver).
      (1) The total amount of bile is relatively constant. Because the liver has limited synthetic capacity, there is a limit to the amount of bile that can be secreted.
      (2) Substances that enhance bile secretion are called **choleretics**. Bile salts and bile acids are the major choleretics.
   b. Unlike the bile-independent secretion, the **synthesis and secretion of bile** by the liver is not under any direct hormonal or nervous control. However, CCK increases bile flow indirectly by increasing the release of bile from the gallbladder [see V E 2 a].

E. **Gallbladder**

1. **Functions.** The gallbladder stores and concentrates the bile during the interdigestive period and empties its contents into the duodenum during digestion.
   a. **Storage.** During the interdigestive period, the bile secreted by the liver is collected in the gallbladder. The gallbladder typically stores 20–50 ml of bile.
      (1) The bile is highly **concentrated** within the gallbladder by the reabsorption of water.
      (2) Water is reabsorbed by the osmotic gradient produced by the active reabsorption of Na^+ and HCO_3^-.
   b. **Contraction.** During digestion, the gallbladder contracts, emptying its contents into the duodenum.

2. **Control**
   a. **CCK is the major stimulus** for gallbladder contraction and sphincter of Oddi relaxation. When chyme enters the small intestine, **fat** and **protein digestion products** directly stimulate the secretion of CCK [see IV C 3 c (1)].
   b. **Vagal stimulation** of the gallbladder also causes gallbladder contraction and sphincter of Oddi relaxation. Vagal stimulation occurs directly during the cephalic phase of digestion and indirectly via a vagovagal reflex during the gastric phase of digestion.

3. **Effects of cholecystectomy.** Bile, not the gallbladder, is essential to digestion. After removal of the gallbladder, bile empties slowly but continuously into the intestine, allowing digestion of fats sufficient to maintain good health and nutrition. Only high-fat meals need to be avoided.

4. Gallstones form in an estimated 10%–30% of the population, although only a fraction, perhaps 20% of these, ever produce symptoms. In Western societies, about 85% of gallstones are composed chiefly of cholesterol; the remainder are pigment stones, composed chiefly of calcium bilirubinate.

    a. Cholesterol and lecithin, which are both insoluble in water, are kept in solution in bile through the formation of micelles [see VII C 2 a]. When the proportions of lecithin, cholesterol, and bile salts are altered, cholesterol crystallizes, leading to stone formation. Cholesterol stones are radiolucent.

    b. Calcium bilirubinate stones can form when infection of the biliary tree leads to bacterial deconjugation of conjugated bilirubin. Unconjugated bilirubin, which is insoluble in bile, then precipitates to begin the stone-forming process. Calcium bilirubinate stones are radiopaque.

F. **Bilirubin metabolism**

1. Formation of bilirubin. Bilirubin is a yellowish pigment formed as an end product of hemoglobin catabolism.

    a. Hemoglobin is released from red blood cells when their membranes rupture at the end of their life span.

    b. Hemoglobin is taken up by the cells of the reticuloendothelial system where the porphyrin ring and iron of the heme moiety are separated.

    c. The porphyrin ring is then converted to bilirubin and gradually released into the plasma.

    d. The bilirubin combines with plasma albumin and circulates to the liver where it is absorbed by the hepatocytes, conjugated with glucuronic acid, and secreted into the gallbladder along with the bile salts.

2. Jaundice is a yellowing of the skin due to the accumulation of bilirubin within the tissues. Jaundice may result from:

    a. Excess production of bilirubin caused by excessive red blood cell destruction (e.g., in hemolytic anemia)

    b. Obstruction of the bile ducts or liver cells preventing the secretion of bilirubin

VI. INTESTINAL SECRETIONS

A. **Mucus** most likely serves a protective role, preventing HCl and chyme from damaging the intestinal wall. Mucus is secreted by:

1. Brunner's glands, which are located within the duodenum

2. Goblet cells located along the length of the intestinal epithelium and in the intestinal crypts, called the crypts of Lieberkühn

B. **Enzymes** capable of breaking down small peptides and disaccharides are associated with the microvilli of the epithelial cells lining the intestine. Although these enzymes are not secreted into the intestine, they are able to digest small peptides and disaccharides during the absorptive process.

C. **Water and electrolytes** are secreted by all the epithelial cells of the intestine.

1. The watery secretion provides a solvent into which the products of digestion are dissolved.

2. If excessive amounts of fluid are produced (as happens when the enterotoxin responsible for cholera stimulates massive fluid secretion), potentially life-threatening watery diarrhea can result.

VII. DIGESTION AND ABSORPTION

A. **Carbohydrates.** The **three major carbohydrates** in the human diet are the disaccharides, **sucrose** (cane sugar) and **lactose** (milk sugar), as well as the polysaccharide starches (which may be in either the straight chain form, **amylose,** or the branched chain form, **amylopectin**). **Cellulose,** another plant polysaccharide, is present in the diet in large amounts, but no enzymes in the human digestive tract can digest it, so it is excreted unused. Dietary intake of carbohydrates is 250–800 g/day, which represents 50%–60% of the diet.

1. **Digestion.** Carbohydrates must be digested into monosaccharides before being absorbed from the GI tract.
   a. Although starch digestion, by **salivary α-amylase,** begins in the mouth, almost all carbohydrate digestion occurs within the small intestine.
   b. **Pancreatic** α-amylase digests carbohydrates into a variety of oligosaccharides.
   c. The oligosaccharides are digested into monosaccharides by brush border enzymes such as **maltase, lactase,** and **sucrase**.
   d. The end products of carbohydrates are **fructose, glucose,** and **galactose**.

2. **Mechanisms of absorption**
   a. **Glucose** and **galactose** are absorbed by a common **Na^+-dependent active transport system**.
      (1) The carrier has two binding sites for Na^+ and one to which either one molecule of glucose or galactose can bind.
      (2) Because two Na^+ are transported down their electrochemical gradient, a large amount of energy is available for transport; thus, almost all of the glucose and galactose present in the intestine can be absorbed.
   b. **Fructose** is absorbed by **facilitated transport**. Fructose absorption occurs readily because most of the fructose is rapidly converted into glucose and lactic acid within the intestinal epithelial cells, thus maintaining a high concentration gradient for diffusion.
   c. After being absorbed into the enterocytes, the monosaccharides are transported across the basolateral membrane by facilitated diffusion. They then diffuse from the intestinal interstitium into the capillaries of the villus.
   d. Absorption of monosaccharides is not regulated. The intestine can absorb over 5 kg of sucrose each day.
   e. Failure to absorb carbohydrates results in diarrhea and intestinal gas.
      (1) The unabsorbed carbohydrates act as osmotic particles and draw excessive fluids into the intestine, which results in diarrhea.
      (2) The flora of the intestine and colon metabolize the unabsorbed carbohydrates, producing a variety of gases [hydrogen (H_2), methane (CH_4), and CO_2], as well as a variety of intestinal irritants.
      (3) **Lactose intolerance** is the most common cause of carbohydrate malabsorption. It results from the inability of the goblet cells to produce lactase.
         (a) Avoidance of milk or milk products prevents the symptoms from developing.
         (b) Lactose intolerance in adults, where it is most common, is not usually a problem. However, in infants, the diarrhea-produced dehydration can be life threatening.

B. **Proteins.** The daily dietary protein requirement for adults is 0.5–0.7 g/kg of body weight. For children 1–3 years old, it is 4 g/kg.

1. **Sources.** The protein that is found in the intestines comes from two sources.
   a. **Endogenous proteins,** totaling 30–40 g/day, are secretory proteins as well as the protein components of desquamated cells.
   b. **Exogenous proteins** are dietary proteins, which total at least 75–100 g daily in the average American diet.

2. **Digestion.** Proteins must be digested into small polypeptides and amino acids before being absorbed.
   a. About 10%–15% of the protein entering the GI tract is digested by **gastric pepsin** secreted by chief cells. Protein digestion within the stomach is important primarily because the protein digestion products act as secretagogues, stimulating the secretion of proteases by the pancreas.
   b. **Pancreatic proteases** play a major role in protein digestion. The proteases, such as **trypsin,** are secreted in an inactive form and must be converted into an active form within the intestine [see IV B 2 c].
      (1) **Enterokinase,** an enzyme secreted by the epithelial cells of the duodenum and jejunum, converts the inactive trypsinogen into trypsin.
      (2) Trypsin then autocatalyzes the conversion of trypsinogen to trypsin as well as activating the other proteases.
   c. **Peptidases,** secreted by the intestinal epithelial cells, continue the digestive process begun by the pancreatic proteases, eventually converting the ingested proteins to small polypeptides and amino acids.

3. **Mechanisms of absorption**
   a. A variety of **Na^+-dependent active transport systems** have been identified for the transport of tripeptides, dipeptides, and amino acids. Polypeptides with more than three peptides are poorly absorbed.
      (1) Separate transporters are present for the absorption of basic, acidic, and neutral amino acids. At least two different polypeptide transporters exist.
      (2) Tripeptides and dipeptides are absorbed in greater quantities than amino acids.
   b. Once inside the enterocytes, intercellular peptidases digest some of the polypeptides to amino acids.
   c. Amino acids and the remaining polypeptides are transported across the basolateral membrane of the enterocytes by facilitated or simple diffusion. They then enter the capillaries of the villus by simple diffusion.
   d. Almost all of the ingested protein is absorbed by the intestine. Any protein that appears in the stool derives from the bacteria within the colon or from cellular debris.
   e. **Malabsorption** of amino acids due to lack of adequate transporters (e.g., the malabsorption of neutral amino acids that occurs in **Hartnup disease**) is relatively rare. Inadequate absorption of proteins due to lack of trypsin is a common consequence of pancreatic diseases.

C. Fats. Daily dietary fat intake varies widely from 25–160 g.

1. **Digestion.** Although the serous glands of the tongue secrete lingual lipase, very little, if any, lipid digestion occurs in the mouth or stomach. Unlike carbohydrates and proteins, lipids are absorbed from the GI tract by **passive diffusion.** However, before the lipids can be absorbed, they must first be made soluble in water. **Bile salts** are required for the solubilization of lipids.
   a. **Pancreatic lipases.** The pancreas secretes three different lipases [see IV B 2 b].
      (1) **Pancreatic lipase** is a fairly specific lipase that cleaves fatty acids from the 1 and 1' positions of triglycerides, leaving a 2-monoglyceride.
      (2) **Cholesterol esterase** cleaves the fatty acid from cholesterol esters, leaving free cholesterol.
      (3) **Phospholipase A_2** cleaves the fatty acids from phospholipids such as phosphatidylcholine.
   b. **Emulsification of lipids.** Lipids must be broken down into small droplets (less than 1 μm in diameter) or **emulsified** into fat globules by bile acids and lecithin (a component of bile) before being digested.
   c. Fat digestion by the pancreatic lipases occurs very rapidly after emulsification because of the large surface-to-volume ratio of the small globules.

2. **Mechanism of absorption**
   a. **Micelle formation.** The emulsified products of lipid digestion (e.g., monoglycerides, cholesterol) must form **micelles** with bile salts before they can be absorbed.

(1) **Micelles** are small (about 5 nm in diameter) spherical aggregates containing some 20–30 molecules of lipids and bile salts.

(2) The **bile salts** are on the outside of the micelle. The 2-monoglycerides and lysophosphatides have their hydrophobic chains facing the interior of the micelle and their polar ends facing the surrounding water phase. The cholesterol and fat-soluble vitamins are located within the fat-soluble interior of the micelle.

b. **Absorption of lipids and bile salts from micelles**

(1) The micelles move along the microvilli surface allowing their lipids to diffuse across the microvilli membrane and into the enterocytes.

(2) Lipids, cholesterol, and the fat-soluble vitamins are removed rapidly from the micelles once the micelles make contact with the microvilli.

(3) The rate-limiting step in lipid absorption is the migration of the micelles from the intestinal chyme to the microvilli surface.

(4) The bile salts, freed of their associated lipids, are absorbed in the terminal ileum by a Na^+-dependent active transport process.

(5) Normally, all of the ingested lipid is absorbed. Fat present in the stool is derived from the intestinal flora.

c. **Formation of chylomicrons by enterocytes**

(1) Once inside the enterocytes, the digested lipids enter the **smooth endoplasmic reticulum** (ER), where they are reconstituted.

(a) 2-Monoglycerides are combined with fatty acids to produce triglycerides.

(b) Lysophosphatides are combined with fatty acids to form phospholipids.

(c) Cholesterol is re-esterified.

(2) The re-formed lipids coalesce into **chylomicrons** (small lipid droplets about 1 nm in diameter) within the smooth ER.

(3) The chylomicrons are transported out of the cell by exocytosis. **β-lipoprotein,** which is synthesized by the enterocytes, covers the surface of the chylomicrons. In the absence of β-lipoprotein, exocytosis will not occur, and the enterocytes become engorged with lipids.

d. **Transport of lipids into circulation**

(1) After exiting the cell, the chylomicrons merge into larger droplets that vary in size from 50–500 nm, depending on the amount of lipids being absorbed.

(2) The large lipid droplets then diffuse into the lacteals, from which they enter the lymphatic circulation.

(3) Almost all digested lipids are totally reabsorbed by the time the chyme reaches the midjejunum, with most of the absorption occurring in the duodenum.

e. **Lipid malabsorption** is much more common than carbohydrate or protein malabsorption. It usually results from one of the following two conditions.

(1) The pancreas does not secrete sufficient quantities of lipase.

(2) The liver does not secrete sufficient quantities of bile.

D. | **Water and electrolytes**

1. **Water**

a. The small intestine, in addition to absorbing most of the dietary Na^+ and water, must also absorb the 7–8 L of water and 20–30 g of Na^+ that are contained in salivary, gastric, biliary, and pancreatic secretions. Failure to reabsorb water from the intestine can lead to rapid dehydration and circulatory collapse.

b. Water undergoes **passive, isosmotic reabsorption** in the small intestine.

(1) Active reabsorption of electrolytes and nutrients creates an osmotic gradient favoring the reabsorption of water.

(2) Because osmotic equilibrium is rapidly achieved, the fluid in the intestine is always isotonic to plasma.

c. In the duodenum, the osmotic pressure created by the entering chyme causes water to flow into the intestine.

d. In the jejunum and ileum, the reabsorption of sodium chloride (NaCl) creates an osmotic gradient favoring the reabsorption of water.

2. NaCl
   a. Na^+ reabsorption is a two-step process.
      (1) First, Na^+ and Cl^- are transported from the lumen into the enterocyte.
      (2) Then, they are transported across the basolateral membrane into the intestinal interstitium.
   b. Na^+ enters the enterocyte in three ways.
      (1) About 30% is transported into the cell by a Na^+-glucose, Na^+-amino acid, or Na^+–(di- or tri-) peptide cotransport system.
      (2) About 30% is transported into the cell by a neutral Na^+-Cl^- cotransport system.
      (3) The remainder enters the cell passively down an electrochemical gradient.
   c. Once inside the enterocyte, Na^+ is transported across the basolateral membrane by a Na^+-K^+-ATPase active transport system.
   d. For the most part, Cl^- flows passively through the enterocyte down the electrochemical gradient established by the active transport of Na^+.

E. | **Vitamins and minerals**

1. Fat-soluble vitamins (A, D, E, and **K)** become part of the micelles formed by bile salts and are absorbed along with other lipids in the proximal intestine.

2. Water-soluble vitamins (C, and the **B vitamins biotin, folic acid, nicotinic acid, B_6 or pyridoxine, B_2 or riboflavin,** and **B_1 or thiamine)** are absorbed by facilitated transport or a Na^+-dependent active transport system in the proximal small intestine.

3. Vitamin B_{12} absorption is more complex than that of other vitamins.
   a. In the stomach, vitamin B_{12} is bound to an **R protein,** which is a specific binding protein.
   b. The gastric parietal cells secrete another vitamin B_{12}–binding protein called **intrinsic factor**. However, the affinity of intrinsic factor for vitamin B_{12} is less than that of R protein, so most of the B_{12} is bound to R protein in the stomach.
   c. In the intestine, pancreatic proteases cleave vitamin B_{12} from the R protein, allowing it to bind to intrinsic factor.
   d. The intrinsic factor–B_{12} complex binds to a receptor on ileal enterocytes.
      (1) Absorption of the vitamin B_{12} from the intrinsic factor–B_{12} complex can occur only after the complex binds to the receptor.
      (2) In the absence of intrinsic factor, minimal amounts of vitamin B_{12} can be absorbed by diffusion. Thus, if large amounts of the vitamin are ingested, enough B_{12} can be absorbed to prevent **pernicious anemia**.

4. Ca^{2+} absorption within the small intestine is regulated to maintain Ca^{2+} balance. Normally, about 25%–80% of the daily intake of Ca^{2+} (1000 mg) is absorbed.
   a. Ca^{2+} absorption occurs via a membrane-bound carrier that is activated by **vitamin D**.
      (1) Vitamin D_3 is converted to **25-hydroxyvitamin D_3** by the liver.
      (2) The kidney converts the 25-hydroxyvitamin D_3 to **1,25-dihydroxyvitamin D_3** by a process that is regulated by parathyroid hormone.
      (3) 1,25-dihydroxyvitamin D_3 then enters the enterocyte where it induces the formation of a Ca^{2+} carrier that inserts on the luminal surface of the enterocyte.
   b. Ca^{2+} is transported out of the cell by a Ca^{2+}-ATPase active transport system and by a Na^+-Ca^{2+} exchange system.

5. Iron absorption is necessary to maintain normal iron balance. However, very little (0.75 mg for men and 1.5 mg for women) of the 15–25 mg of iron ingested each day is actually absorbed.
   a. Iron is absorbed primarily within the **duodenum** and **jejunum**.
   b. Iron can be absorbed either as **heme** (derived from meat) or as a **free ion**.
   c. **The ferrous ion (Fe^{2+})** is absorbed more efficiently than the **ferric ion (Fe^{3+})**.
   d. **Ascorbic acid (vitamin C)** promotes iron absorption by reducing Fe^{3+} to Fe^{2+} and by preventing iron from forming insoluble complexes within the chyme.
   e. **Stomach acid** tends to break insoluble iron complexes apart and thus facilitates iron absorption.

   f. Four separate steps are involved in the transport of iron from the intestine to the plasma.

     (1) First, the iron is transported across the apical membrane of the enterocyte by a specific iron carrier system.

     (2) Second, the iron binds to **apoferritin,** an iron-binding protein, to form **ferritin**.

     (3) In order to leave the enterocyte, the iron must dissociate from ferritin and bind to an intracellular carrier protein that shuttles it to the basolateral membrane, where it is transported out of the cell.

     (4) Upon entering the intestinal interstitium, the iron is transported to the plasma by **transferrin,** a β**-globulin**.

   g. The **amount of iron absorbed** depends on the amount of intracellular and extracellular transport protein (transferrin) compared to the amount of ferritin.

     (1) If a large amount of transferrin is available, iron can be transported rapidly from the enterocyte to the plasma.

     (2) If little transferrin is available, most of the iron remains trapped in the enterocyte and is eventually excreted when the cells are desquamated.

     (3) When iron stores are depleted, such as after a hemorrhage, transferrin synthesis increases.

Chapter 43

Large Intestine

I. **OVERVIEW.** The **colon,** or large intestine (Figure 43-1), absorbs some of the nutrients and most of the fluids passed into it from the small intestine. Under normal circumstances, all but 50–100 ml of the 1500 ml received from the small intestine is absorbed. Any nutrients or fluids that cannot be absorbed are passed into the feces.

II. **MOTILITY**

A. **Function.** The **contractile activity** of the large intestine serves two functions.

   1. It enhances the efficiency of water and electrolyte absorption.

   2. It promotes the excretion of the fecal material remaining in the colon.

B. **Types of movements**

   1. Haustral shuttling
      a. Bands of muscle divide the large intestine into sac-like segments called **haustrations**. Although the haustrations are present when the colon is empty, the entry of food into the colon causes an increase in colonic contractile activity.
      b. The dynamic formation and disappearance of haustrations squeeze the chyme, moving it back and forth along the colon in a manner similar to that described for the segmentation contractions in the small intestine (see Chapter 42 III A 1).

   2. Peristalsis, here, as elsewhere in the gut, is a progressive contractile wave preceded by a wave of relaxation. Peristaltic-like segmentation contractions move the chyme very slowly (5 cm/hr) along the colon. It can take up to 48 hours for chyme to traverse the colon.

   3. Mass movements. Occasionally (three to four times daily) the chyme is swept rapidly along the colon by a peristaltic wave called a mass movement. The mass movement forces fecal material into the rectum.

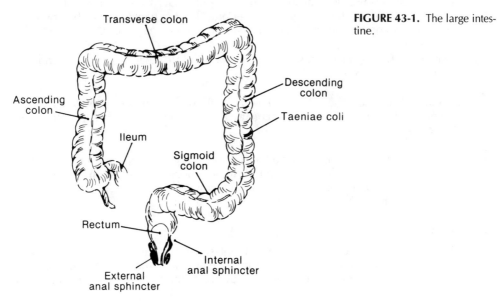

FIGURE 43-1. The large intestine.

4. The **frequency** of contractions is greater in the rectum than in the sigmoid colon, causing retrograde movement of fecal material. Because of this orad movement of fecal material, the rectum is usually empty and material placed into it, such as a suppository, will be pushed up into the colon.

5. The overall effect of the **neural input to the colon is inhibitory**. Thus, elimination of the enteric nervous system, as occurs in **Hirschsprung's disease,** leads to a large increase in colonic tone.

C. Defecation

1. Fecal material entering the rectum is evacuated by defecation, during which:
  a. The smooth muscles of the distal colon and rectum contract, propelling the fecal material into the anal canal
  b. The **internal and external anal sphincters** both relax
  c. The abdominal and diaphragmatic muscles contract, increasing the intra-abdominal pressure and forcing the feces through the anal canal

2. Defecation involves both voluntary and reflex activity.
  a. When fecal material expands the rectum, a **rectosphincteric reflex** relaxes the internal anal sphincter, contracts the external anal sphincter, and generates the urge to defecate.
  b. Voluntary control mechanisms then either maintain contraction of the external anal sphincter (which is composed of skeletal muscle innervated by the pudendal nerves) to prevent defecation or allow it to relax so that defecation can occur.
  c. If defecation does not occur, the internal anal sphincter closes, and the rectum relaxes to accommodate the fecal material within it.
  d. Individuals lacking α-motoneuronal control over the external anal canal will defecate whenever the rectum is filled with fecal material.

III. ABSORPTION, SECRETION, AND GAS PRODUCTION

A. **Water.** The colon is unable to absorb more than 2–3 L/day. Thus, if most of the 8–10 L entering the intestine (either as ingested water or as gastric, pancreatic, or biliary secretions) is not absorbed in the small intestine, severe diarrhea can occur.

B. **Na^+ and Cl^-.** The colon absorbs most of the Na^+ and Cl^- that escapes absorption in the small intestine.

C. **K^+,** on the other hand, is secreted by the colon. Its concentration typically rises from its ileal concentration of 9 mEq/L to 75 mEq/L by the time the fluid reaches the end of the large intestine.

D. **Aldosterone.** While the small intestine has no way to regulate Na^+ or K^+ absorption, in the colon, the hormone aldosterone controls these processes. **Aldosterone** enables the colon to absorb all of the Na^+ in the fecal fluid. However, in doing so, it causes significant amounts of K^+ to be lost from the body.

E. Intestinal gas

1. There are three sources of gas in the gastrointestinal (GI) tract.
  a. **Swallowed air,** including air released from food and carbonated beverages, enters the stomach, from which it is removed by eructation or passed into the intestines with chyme.
  b. Gas is formed by **bacterial action** in the ileum and large intestine.
  c. Some gases diffuse into the GI tract from the **bloodstream.**

2. Gas in the colon differs in volume and source from gas in the small intestine.

   a. Small intestine. The small amount of gas present is usually the result of swallowed air. This gas most likely will be passed on to the colon.

   b. Colon

      (1) Colonic gas, or flatus, is produced in large volumes—up to 7–10 L/day.

      (2) The gas is produced chiefly through the breakdown of undigested nutrients that reach the colon.

      (3) The main components of flatus are CO_2, CH_4, H_2, and nitrogen gas (N_2). Since all of these gases except N_2 diffuse readily through the intestinal mucosa, the volume of flatus expelled is reduced to about 600 ml/day.

STUDY QUESTIONS

Directions: Each of the numbered items or incomplete statements in this section is followed by answers or by completions of the statement. Select the **one** lettered answer or completion that is **best** in each case.

1. Which of the following secretions is most dependent on vagal stimulation?

(A) Saliva
(B) Hydrochloric acid (HCl)
(C) Pepsin
(D) Pancreatic juice
(E) Bile

2. The major stimulus for primary peristalsis in the esophagus is

(A) presence of food in the esophagus
(B) swallowing
(C) regurgitation of food from the stomach
(D) closing of the upper esophageal sphincter (UES)
(E) opening of the lower esophageal sphincter (LES)

3. Which of the following will inhibit stomach contractions?

(A) Acetylcholine (ACh)
(B) Motilin
(C) Gastrin
(D) Secretin
(E) Histamine

4. Gastric acid secretion increases when food enters the stomach because

(A) protein digestion products directly stimulate the parietal cells to release hydrochloric acid (HCl)
(B) food raises the pH of the stomach, allowing more acid to be released
(C) both
(D) neither

5. Gastric parietal cells secrete

(A) gastrin
(B) motilin
(C) cholecystokinin (CCK)
(D) intrinsic factor
(E) secretin

6. Which of the following can occur without brain stem coordination?

(A) Chewing
(B) Swallowing
(C) Primary esophageal peristalsis
(D) Vomiting
(E) Gastric emptying

7. The major stimulus for gastric acid (HCl) secretion during the cephalic phase is

(A) histamine
(B) gastrin
(C) secretin
(D) somatostatin
(E) acetylcholine (ACh)

8. The major stimulus for the release of secretin is

(A) protein digestion products
(B) histamine
(C) somatostatin
(D) hydrochloric acid (HCl)
(E) cholecystokinin (CCK)

9. Fats are transported from intestinal cells to blood plasma primarily in the form of

(A) micelles
(B) chylomicrons
(C) triglycerides
(D) fatty acids
(E) monoglycerides

10. Acetylcholine (ACh) is required for the contraction of

(A) the lower esophageal sphincter (LES)
(B) the upper esophageal sphincter (UES)
(C) both the LES and UES
(D) the antrum

11. The major stimulus for receptive relaxation of the stomach is

(A) food in the stomach
(B) food in the intestine
(C) secretin
(D) cholecystokinin (CCK)
(E) motilin

12. The motility pattern primarily responsible for the propulsion of chyme along the small intestine is

(A) the migrating motor complex (MMC)
(B) peristaltic waves
(C) myogenic contractions
(D) haustrations
(E) segmentation

13. Gastric acid (HCl) secretion is inhibited by

(A) somatostatin
(B) entero-oxyntin
(C) high pH
(D) amino acids
(E) acetylcholine (ACh)

14. The major factor controlling the secretion of bile salts from the liver is the amount of

(A) secretin released during a meal
(B) fat entering the small intestine
(C) bile acids produced by the liver
(D) bile reabsorbed from the intestine
(E) cholecystokinin (CCK) released during a meal

15. Micelle formation is necessary for absorption of

(A) bile salts
(B) iron
(C) cholesterol
(D) alcohol
(E) B vitamins

16. Secondary bile acids are formed

(A) in the liver from cholesterol
(B) by the conjugation of bile acids with taurine or glycine
(C) both
(D) neither

DIRECTIONS: Each of the numbered items or incomplete statements in this section is negatively phrased, as indicated by a capitalized word such as NOT, LEAST, or EXCEPT. Select the ONE lettered answer or completion that is BEST in each case.

17. Na$^+$-dependent transport is responsible for the absorption of all of the following EXCEPT

(A) vitamin E
(B) amino acids
(C) glucose
(D) bile salts

18. Intestinal motility is increased by all of the following EXCEPT

(A) cholecystokinin (CCK)
(B) secretin
(C) gastrin
(D) insulin
(E) motilin

19. All of the following stimulate cholecystokinin (CCK) secretion EXCEPT

(A) amino acids
(B) fatty acids
(C) hydrochloric acid (HCl)
(D) bile acids

20. All of the following effects are caused by secretin EXCEPT

(A) stimulation of pancreatic bicarbonate (HCO$_3{}^-$) secretion
(B) enhancement of bile acid secretion
(C) potentiation of pancreatic enzyme secretion by cholecystokinin (CCK)
(D) inhibition of gastric muscle contraction

21. A Na$^+$-dependent active transport system is necessary for the intestinal absorption of all of the following EXCEPT

(A) dipeptides
(B) bile salts
(C) fructose
(D) vitamin C
(E) glucose

Directions: Each group of items in this section consists of lettered options followed by a set of numbered items. For each item, select the **one** lettered option that is most closely associated with it. Each lettered option may be selected once, more than once, or not at all.

Questions 22–26

For each of the following gastrointestinal (GI) processes, choose the stimulus that is most important for its regulation.

(A) Secretin
(B) Histamine
(C) Cholecystokinin (CCK)
(D) Bombesin
(E) Motilin

22. Gastric acid (HCl) secretion

23. Gastrin secretion

24. Pancreatic enzyme secretion

25. Gallbladder emptying

26. Emptying of the intestine during the interdigestive period

Questions 27–31

For each of the following functions of the gastrointestinal (GI) tract, choose the area where it occurs.

(A) Fundus of the stomach
(B) Antrum of the stomach
(C) Duodenum of the intestine
(D) Ileum of the intestine
(E) Colon

27. Absorption of bile acids

28. Secretion of gastrin

29. Secretion of intrinsic factor

30. Secretion of K^+

31. Absorption of iron

ANSWERS AND EXPLANATIONS

1. The answer is A [Chapter 40 II C 1; Chapter 41 III B 3–4, D 2 a–b; Chapter 42 IV C, V D 2]. Salivary flow is entirely dependent on the autonomic nervous system (ANS). Vagal stimulation produces a large volume of watery fluid, while sympathetic stimulation causes the secretion of proteins (mucus and some enzymes). Secretion of hydrochloric acid (HCl), pepsin, pancreatic juice, and bile is influenced by vagal stimulation but can occur without it.

2. The answer is B [Chapter 40 IV A 3 b (1)]. Primary esophageal peristalsis is part of the swallowing response and occurs whether or not food enters the esophagus. The intensity of the peristalsis, however, increases if food is present. If the esophagus is not emptied by primary peristalsis, the presence of food in the esophagus will initiate another peristaltic reflex, called scondary peristalsis.

3. The answer is D [Chapter 41 II F 2 b (2)]. Secretin has a direct inhibitory effect on the smooth muscle fibers forming the stomach wall. Acetylcholine (ACh) and motilin, and possibly gastrin, increase the force of stomach contractions. Histamine has no direct effect on stomach contractions.

4. The answer is C [Chapter 41 III B 4 b]. Gastric acid secretion is increased directly by protein digestion products and inhibited when the pH of the stomach is reduced. The buffering action of food promotes gastric acid secretion by keeping the pH from falling too low.

5. The answer is D [Chapter 41 III A 3 a (1)]. Parietal cells, located in the oxyntic glands of the orad stomach (fundus and corpus), secrete hydrochloric acid (HCl) and intrinsic factor. Gastrin is secreted by G cells located in the pyloric glands of the distal stomach (antrum). Secretin and cholecystokinin (CCK) are secreted by endocrine cells in the proximal intestine. Motilin is secreted from endocrine cells within the epithelium of the small intestine.

6. The answer is E [Chapter 41 II E, F 1 a]. Gastric emptying of solids occurs when the contractile activity of the stomach reduces the size of the particles within the food sufficiently for them to pass through the pyloric sphincter. Stomach contractions are elicited by reflexes initiated by antral distention and by gastrin. Although the strength of the contractions is reduced by vagotomy, contractions can still occur. Chewing, swallowing, primary (but not secondary) peristalsis, and vomiting all are coordinated by specific regions of the brain stem.

7. The answer is E [Chapter 41 III B 4]. During the cephalic phase of gastric acid (HCl) secretion, the sight, smell, or thought of food activates cholinergic [acetylcholine (ACh)-releasing] vagal fibers, which stimulate the release of HCl from antral parietal cells.

8. The answer is D [Chapter 42 IV C 3 c (2)]. Secretin and cholecystokinin (CCK) are hormones released from endocrine cells located in the proximal intestine. Although the release of both hormones is stimulated by the presence of chyme in the small intestine, it is the low pH resulting from hydrochloric acid (HCl) in the chyme that is the major stimulus for the release of secretin. Protein digestion products (e.g., amino acids, particularly phenylalanine) stimulate both secretin and CCK secretion, but are the major stimuli for CCK secretion. CCK potentiates the effects of secretin but does not affect its release. Somatostatin and histamine have no effect on CCK or secretin secretion.

9. The answer is B [Chapter 42 VII C 2 c]. Chylomicrons are small lipid droplets within the enterocytes that are reconstituted from the lipid digestion products formed in the intestine. The chylomicrons are transported from the enterocytes by exocytosis by the intestinal fluid surrounding the intestine. The chylomicrons are then absorbed into the intestinal lacteals, from which they enter the circulation.

10. The answer is B [Chapter 40 IV A 3 a]. The upper esophageal sphincter (UES) is composed of striated muscle and is stimulated by cholinergic [acetylcholine (ACh)-releasing] vagal fibers. The lower esophageal sphincter (LES) is composed of smooth muscle, which normally is maintained in a contracted state by a myogenic process. ACh regulates the force of peristaltic contractions in the antrum but is not required to stimulate them.

11. The answer is A [Chapter 41 II B 2]. When food distends the orad stomach (fundus and corpus), it produces a vagovagal reflex by which noncholinergic, nonadrenergic fibers relax the stomach. About 2 L of food can be accommodated in the stomach when receptive relaxation is at its maximum.

12. The answer is E [Chapter 42 III B 1]. Although the major functions of segmentation are the mixing of chyme with digestive juices and exposing the products of digestion to the intestinal wall, segmentation is also responsible for pushing the chyme along the intestine. Propulsion occurs because the frequency of segmentation is higher in the more proximal intestine than it is in the distal intestine. Thus, it is more likely for the chyme to move toward the colon than it is to move toward the stomach. Peristaltic waves will move chyme along the intestine, but these are not frequent enough to propel the food into the colon. The migrating motor complex (MMC) empties the intestine of the small amount of chyme remaining in the intestine during the interdigestive period.

13. The answer is A [Chapter 41 III B 3 b]. Gastric acid (HCl) secretion is inhibited by low pH in the stomach and by somatostatin released from interneurons within the enteric nervous system. Acetylcholine (ACh), the hormone known as entero-oxyntin, and amino acids all stimulate gastric acid secretion.

14. The answer is D [Chapter 42 V D 2 a]. The amount of bile synthesized each day is not sufficient to absorb all of the fat digested. However, because the bile salts are absorbed by the intestine and returned to the liver, they can be used over and over again. The amount of bile secreted each day is thus proportional to the amount absorbed from the intestine.

15. The answer is C [Chapter 42 VII C 2 a, b (2)]. Micelles are necessary for absorption of dietary lipids such as cholesterol. Bile salts, iron, and B vitamins are absorbed by membrane-bound active transport systems. Alcohol is both water- and fat-soluble and so can diffuse directly across the membranes.

16. The answer is D [Chapter 42 V B 1]. Secondary bile acids are formed in the intestine by the deconjugation and dehydroxylation of primary bile acids. Primary bile acids are syn-

thesized from cholesterol in the liver and then conjugated with taurine or glycine to form primary bile salts.

17. The answer is A [Chapter 42 VII A 2 a, B 3 a, C 2 b (4), E 1]. Amino acids and glucose are absorbed in the proximal intestine by Na^+-dependent active transport systems. Bile salts are reabsorbed by a Na^+-dependent active transport system located within the terminal ileum. Vitamin E is a fat-soluble vitamin and is absorbed by passive diffusion along with other lipids in the proximal intestine.

18. The answer is B [Chapter 42 III D 3 b]. Secretin decreases intestinal and gastric contractile force. Cholecystokinin (CCK), gastrin, insulin, and motilin all increase intestinal contractions.

19. The answer is D [Chapter 42 IV C 3 c]. Amino acids, particularly phenylalanine, fatty acids, and monoglycerides all directly stimulate the release of cholecystokinin (CCK) from intestinal endocrine cells. Hydrochloric acid (HCl), which lowers the pH of chyme entering the intestine, also causes CCK secretion. Bile acids do not influence CCK secretion.

20. The answer is B [Chapter 41 II F 2 b; Chapter 42 IV C 3 b, V D 1 a]. Although secretin regulates the volume of biliary secretion, it neither enhances nor inhibits the secretion of bile salts. Secretin's major effect is to stimulate bicarbonate (HCO_3^-) secretion by the pancreas and the liver. In addition, it potentiates the effect of cholecystokinin (CCK) on pancreatic enzyme secretion. Secretin also diminishes gastric emptying, probably by directly reducing gastric smooth muscle contractility.

21. The answer is C [Chapter 42 VII A 2 a–b, B 3 a, C 2 b (4), E 2]. Unlike glucose, which depends on a Na^+-dependent active transport system for absorption, fructose is absorbed by facilitated diffusion. Active transport for fructose is not required because the intracellular fructose concentration is maintained at a low value by intracellular enzymes that rapidly convert fructose to glucose and lactose. Bile salts, vitamin C, and dipeptides all rely on Na^+-dependent active transport systems to be absorbed.

22–26. The answers are: 22-B, 23-D, 24-C, 25-C, 26-E [Chapter 41 II F 3 c, III B 3 a, C 2 a; Chapter 42 IV C 3 a, V E 2 a]. Gastric acid (HCl) secretion is stimulated by acetylcholine (ACh), histamine, and gastrin. It is inhibited by somatostatin.

Gastrin secretion is stimulated by bombesin [also called gastrin-releasing peptide (GRP)] and inhibited by somatostatin.

Cholecystokinin (CCK) and secretin both stimulate the pancreas. CCK, however, is responsible for the secretion of pancreatic enzymes. CCK is also responsible for stimulating the gallbladder to contract.

Motilin, a hormone released from the small intestine, increases the strength of the migrating motor complex (MMC) and may be responsible for initiating it. The MMC, which occurs every 60–90 minutes during the interdigestive period, starts in the stomach and sweeps along the entire gastrointestinal (GI) tract. It is thought to be responsible for emptying the small intestine of any chyme remaining after the completion of a meal.

27–31. The answers are: 27-D, 28-B, 29-A, 30-E, 31-C [Chapter 41 III A 3 a (1), b; Chapter 42 VII C 2 b (4), E 5 a; Chapter 43 III C]. Bile acids are absorbed in the terminal ileum by a Na^+-dependent cotransport process.

Gastrin is secreted by G cells contained in the pyloric glands of the distal stomach (antrum and pylorus).

Intrinsic factor, necessary for the absorption of vitamin B_{12}, is secreted by parietal cells located in the proximal stomach (fundus).

The colon secretes K^+ into the chyme. About 10% of the daily K^+ load is excreted by the colon; the remainder is excreted by the kidney.

Iron is absorbed primarily within the duodenum and jejunum.

PART VIII

ENDOCRINE PHYSIOLOGY
John Bullock

Chapter 44

Physical and Chemical Characteristics of Hormones

I. **DEFINITION.** In the classic definition, hormones are secretory products of the ductless glands, which are released in catalytic amounts into the bloodstream and transported to specific target cells (or organs), where they elicit physiologic, morphologic, and biochemical responses. In reality, the requirement that hormones be secreted into the bloodstream is too restrictive, because they also can act locally (Figure 44-1). For example:

A. **Paracrine hormones** can be conveyed over short distances by diffusion through the interstitial space, to act on neighboring cells as regulatory substances

FIGURE 44-1. The three methods of hormone information transfer. Cell-to-cell signaling can occur over long distances via hormone secretion into the bloodstream (*A*) or over short distances via hormone infusion through the interstitium (*B*). A hormone also may act directly on the cell that produces it (*C*). (Adapted from Darnell J, Lodish H, Baltimore D (eds): *Molecular Cell Biology*, 2nd edition. New York, WH Freeman, 1990, p 710.)

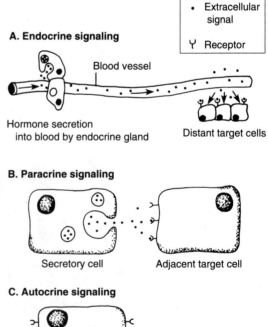

- Extracellular signal

Ⴘ Receptor

A. Endocrine signaling

Blood vessel

Hormone secretion into blood by endocrine gland

Distant target cells

B. Paracrine signaling

Secretory cell   Adjacent target cell

C. Autocrine signaling

Target sites on same cell

B. | **Autocrine hormones** can regulate the activity of the same cells that produce them

I. | **HORMONE-SECRETING TISSUES.** Virtually all organs in the body exhibit endocrine function.

A. | The most-studied endocrine organs and examples of the hormones they produce are listed in Table 44-1.

B. | Other organs with endocrine function and the hormones they produce are:

1. Heart: atrial natriuretic peptide (ANP)

2. Kidney: 1,25-dihydroxycholecalciferol (calcitriol)

3. Liver: 25-hydroxycholecalciferol (calcidiol), somatomedin

4. Pineal gland: melatonin

5. Skin: calciferol (vitamin D_3)

6. Gastrointestinal (GI) tract: gastrin, cholecystokinin (CCK), secretin, vasoactive intestinal peptide (VIP)

II. | **FUNCTIONS.** Hormones regulate existing fundamental bodily processes but do not initiate cellular reactions de novo. In contrast to vitamins, hormones serve no nutritive role in responsive tissues and are not incorporated as a structural moiety into another molecule.

A. | **Regulation of biochemical reactions.** As regulators, hormones stimulate or inhibit the rate and magnitude of biochemical reactions by their control of enzymes and, thereby, cause morphologic, biochemical, and functional changes in target tissues. Although they are not used as energy sources in biochemical reactions, hormones modulate energy-producing processes and regulate the circulating levels of energy-yielding substrates (e.g., glucose, fatty acids).

TABLE 44-1. Principal Endocrine Organs and Hormones They Produce

Organ	Examples of Hormones
Pituitary gland	Tropic hormones (e.g., adrenocorticotropic hormone, growth hormone, prolactin)
Hypothalamus	Hypophysiotropic hormones [e.g., releasing hormones (e.g., thyrotropin releasing hormone), antidiuretic hormone, oxytocin]
Thyroid gland	Thyroxine, 3,5,3'-triiodothyronine
Adrenal glands	Mineralocorticoids (e.g., aldosterone), glucocorticoids (e.g., cortisol), catecholamines (e.g., epinephrine, norepinephrine)
Parathyroid glands	Parathyroid hormone
Gonads	Testosterone, estradiol, progesterone
Pancreatic islets	Insulin, glucagon, somatostatin

B. **Regulation of bodily processes.** Hormones regulate growth, maturation, differentiation, regeneration, reproduction, pigmentation, behavior, metabolism, and chemical homeostasis. Slower processes (e.g., growth, reproduction, metabolism) require longer periods of continual hormone stimulation in contrast to rapid coordination of the body (e.g., reflex contraction of a somatic muscle), which is regulated by the nervous system.

IV. PHYSICAL CHARACTERISTICS

A. **Chemical composition.** The three major classes of hormones are: steroids, proteins and polypeptides, and amino acid derivatives (i.e., catecholamines and thyroid hormones). No polysaccharides or nucleic acids are known to function as hormones.

B. **Plasma concentration.** Hormones usually are secreted into the circulation in extremely low concentrations.

1. Peptide hormone concentration is between 10^{-12} mol/L and 10^{-10} mol/L.

2. Epinephrine and norepinephrine concentrations are 2×10^{-10} mol/L and 13×10^{-10} mol/L, respectively.

3. Steroid and thyroid hormone concentrations are 10^{-9} mol/L and 10^{-6} mol/L, respectively.

C. **Latent period** is the time interval between the application of a stimulus and a response. In contrast to a latent period of 8 msec between a neural stimulus and the contraction of a muscle, the latent period associated with hormones can be as long as seconds, minutes, hours, or days.

1. Following the administration of **oxytocin,** milk ejection occurs in a few seconds.

2. The metabolic response to **thyroxine** can take as long as 3 days.

D. **Postsecretory modification** of hormones occurs by the proteolytic cleavage of peptide hormones or by enzymatic conversion of steroids and thyroid hormones at sites beyond the site of secretion. This peripheral conversion to more active hormonal forms occurs in the liver, kidney, fat, or bloodstream as well as in the target tissues themselves.

E. **Circulating forms.** The binding of serum proteins (e.g., globulin) to hormones protects the hormones against clearance by the kidneys, slows the rate of degradation by the liver, and provides a circulating reserve of hormones. Only unbound hormones pass through capillaries to produce their effects or to be degraded.

F. **Hormone receptors** are unique molecular groups in or on target cells that interact with hormones to initiate a characteristic response. The specificity of a hormone depends on the formation of a strong noncovalent bond with its hormone receptor (see Chapter 45 I).

G. **Half-life.** Most hormones are metabolized rapidly after secretion. In general, peptide hormones are short-lived in the circulation, whereas steroids have a significantly longer half-life.

H. **Degradation.** The interaction of hormones with their target cells is followed by intracellular degradation.

1. Degradation of protein hormones and amines occurs after binding to membrane receptors and internalization of the hormone-receptor complex.

2. Degradation of steroids and thyroid hormones occurs after binding of the hormone-receptor complex to the chromatin.

I. **Inactivation and excretion.** Only a small fraction of the circulating hormone is removed by most target tissues. Hormone inactivation occurs in the liver and kidney.

1. Hormone degradation uses many enzymatic mechanisms such as hydrolysis, oxidation, hydroxylation, methylation, decarboxylation, sulfation, and glucuronidation.

2. Only a small fraction ($< 1\%$) of any hormone is excreted intact in the urine or feces.

V. **CHEMISTRY** (Table 44-2)

A. **Proteins and polypeptides** generally are water-soluble and circulate unbound in plasma.

1. **Structure**
   a. The peptide and protein hormones vary greatly in size. For example, **thyrotropin releasing hormone (TRH)** is a tripeptide, whereas **human chorionic gonadotropin (HCG)** consists of 243 amino acid residues.
   b. The molecular weights of the pituitary tropic hormones, which consist of about 200 amino acid residues, vary from 23,000 to 25,000 daltons. (The average molecular weight of an amino acid residue is 120; thus, multiplying the number of residues by 120 is a good estimate of the molecular weight of a peptide or protein.)

2. **Synthesis.** Many of the protein-type hormones are synthesized on the rough endoplasmic reticulum (ER) as **prohormones** or **preprohormones**. These precursor hormones undergo posttranslational cleavage by an endopeptidase within the Golgi complex prior to secretion of the biologically active hormone.
   a. **Growth hormone (GH)** and **prolactin** are synthesized as prohormones.
   b. **Proopiomelanocortin (POMC),** synthesized in the pituitary and hypothalamus, is a prohormone complex that contains peptide hormone moieties including adrenocorticotropic hormone (ACTH; corticotropin), melanotropin, lipotropin, and endorphins.
   c. **Insulin** and **parathyroid hormone (PTH)** are synthesized as preprohormones, which are hydrolyzed to prohormones and then further hydrolyzed to the hormone that is secreted.

3. **Storage and secretion.** Protein and polypeptide hormones are secreted by endocrine organs derived from ectoderm (pituitary gland*; tuberoinfundibular, supraoptic, and paraventricular nuclei) as well as organs derived from endoderm (pancreatic islets of Langerhans, thyroid gland, parathyroid glands). The primordial cells that give rise to the parafollicular cells (C cells) of the thyroid gland are derived from neural crest precursors.
   a. Protein and polypeptide hormones probably are stored exclusively in subcellular membrane-bound secretory granules within the cytoplasm of endocrine cells.
   b. These hormones are released into the blood by exocytosis, which involves fusion of the secretory granule and cell membrane followed by extrusion of the granular contents into the bloodstream.

4. **Half-life** of various peptide/protein hormones is as follows:
   a. Antidiuretic hormone (ADH) and oxytocin: < 1 minute
   b. Insulin: 7 minutes
   c. Prolactin: 12 minutes
   d. ACTH: 15–25 minutes
   e. Luteinizing hormone (LH): 15–45 minutes
   f. Follicle-stimulating hormone (FSH): 180 minutes

---

*The anterior and posterior lobes of the pituitary gland are derivatives of buccal ectoderm and neural ectoderm, respectively.

TABLE 44-1. Characteristics of the Principal Classes of Hormones

Characteristic	Peptides	Steroids and Calcitriol*	Amines	
			Catecholamines	Thyroid Hormone
Solubility				
In aqueous solvents	Excellent	Limited	Good	Limited
In nonaqueous solvents	Poor	Excellent	Limited	Good
Biosynthetic pathway	Single peptide, prohormone, or preprohormone	Multiple enzymes	Multiple enzymes	Multiple enzymes
Postsecretory modifications	Very rare**	Common	None	Common
Storage of preformed hormone	Often substantial	Minimal	Substantial	Substantial
Degradation products	Irreversibly inactive	Sometimes retain or regain activity	Inactive	Inactive
Plasma binding proteins	Very rare	Yes	Limited	Yes
Half-life	Short (minutes)	Long (hours)	Very short (seconds)	Very long (hours to days)
Receptors	Cell surface	Nucleus	Cell surface	Nucleus
Site of action	Plasma membrane	Nucleus	Plasma membrane	Nucleus
Mechanism of action	Stimulates production of second messenger	Stimulates production of specific mRNAs	Stimulates production of second messenger	Stimulates production of specific mRNAs

*Calcitriol = 1,25-dihydroxyvitamin D_3.
**Although most translational processing is presecretory (i.e., it occurs within the cell), in some instances, postsecretory modifications occur (i.e., additional proteolytic modifications take place outside of the cell).

B. **Amino acid derivatives.** Catecholamines, which are water-soluble, and thyroid hormones, which are lipid-soluble, circulate in the plasma bound mainly to binding globulins.

1. **Structure**
   a. Catecholamines are derived from the amino acid tyrosine. Thyroid hormones are derived from two iodinated tyrosine residues. Hormones derived from tyrosine also are called **phenolic derivatives**.
   b. Catecholamines and thyroid hormones both retain the aliphatic α-amino group. Introduction of a second hydroxyl group in the ortho- position on the benzene ring is characteristic of the catecholamines, whereas iodination of the benzene ring distinguishes the thyroid hormones.
   c. Thyroid hormones are the only substances in the body that contain iodine.

2. **Synthesis.** Epinephrine and norepinephrine are synthesized in the chromaffin cells, which are modified postganglionic neurons (see Chapter 49 I B–D). Thyroid hormones are synthesized in thyroid follicular cells (see Chapter 52 I; IV).

3. **Storage and secretion.** The amine hormones are secreted by endocrine tissues derived from the neural crest (adrenal medulla) and from endoderm (thyroid gland).
   a. Catecholamines are stored in secretory granules. Secretion occurs when the membrane of the chromaffin granules fuses with the plasma membrane, causing the granular contents to be extruded into the circulation.
   b. Thyroid gland secretions are called **iodothyronines**—compounds resulting from the coupling of two iodinated tyrosine molecules. Thyroid hormones are stored outside follicular cells in the form of **thyroglobulin,** a glycoprotein precursor found in the lumen of the cells. Several weeks' supply of thyroid hormone is stored in this form. Following endocytosis and proteolysis of thyroglobulin, thyroid hormone is secreted into the bloodstream by simple diffusion.

4. **Circulation and half-life**
   a. Epinephrine and norepinephrine exist in plasma either in the free form or in conjugation with sulfate or glucuronide. Most circulating epinephrine is bound to blood proteins (mainly albumin); norepinephrine does not bind to blood proteins to any appreciable degree.
   b. Most thyroid hormones are bound to thyroxine-binding globulin.
   c. The half-life of various amine hormones is as follows:
      (1) Epinephrine: 10 seconds
      (2) Norepinephrine: 15 seconds
      (3) Triiodothyronine: 1 day
      (4) Thyroxine: 7 days

C. **Steroid hormones** are lipid soluble and circulate in the plasma bound to carrier proteins called **steroid-binding globulins**.

1. **Structure.** Steroids are a group of biologically active substances, including androgens, estrogens, progesterone, glucocorticoids, and mineralocorticoids. 25-Hydroxyvitamin D_3 and 1,25-dihydroxyvitamin D_3 are modified steroids called **secosteroids**.
   a. Steroids consist of three cyclohexyl rings and one cyclopentyl ring combined into a single structure. They are derived from the cyclopentanoperhydrophenanthrene nucleus consisting of a fully hydrogenated phenanthrene (rings A, B, and C), to which is attached a hydrogenated cyclopentane ring (D-ring). This fully saturated, four-ring structure consisting of 17 carbon atoms is the hypothetical parent compound, gonane or sterane.
   b. The secosteroids, which are vitamin hormones, lack a B-ring and, therefore, consist of two cyclohexyl rings (rings A and C) and one cyclopentyl ring (D-ring).

2. **Synthesis and secretion.** Steroids are synthesized and secreted by the endocrine organs derived from mesoderm (adrenal cortex, testis, ovary). The placenta also synthesizes and secretes steroids.

    a. Steroids are derived from cholesterol. The first, and rate-limiting, step in steroid synthesis is conversion of cholesterol to pregnenolone. Depending on the product, hydroxylations of the steroid nucleus may occur at carbons 11, 17, 18, or 21.

    b. There is little storage of steroids. Instead, steroid-producing cells store esterified cholesterol in the form of lipid droplets, which serve as precursors. Steroids are released in the circulation by simple diffusion.

3. Half-life of various steroid hormones is as follows:

    a. Aldosterone: 30 minutes

    b. Cortisol: 90–100 minutes

    c. 1,25-Dihydroxyvitamin D_3: 15 hours

    d. 25-Hydroxyvitamin D_3: 15 days

Chapter 45

Mechanisms of Hormone Action

I. **HORMONE-RECEPTOR INTERACTION.** Hormones produce their effects by first combining with specific cell components called receptors. Only cells with receptors for a specific hormone respond. Hormones may be found on the target cell membrane (**external receptors**) or within the cytoplasm or nucleus (**internal receptors**).

A. **Mechanisms** (Figure 45-1). Hormones and receptors interact in the following ways to affect intracellular metabolism.

1. Polypeptide hormones and catecholamines bind irreversibly to **fixed receptors** on the outer surface of target cells, which initiates the transduction of the hormone signal across the membrane. Transduction occurs after activation of adenyl cyclase (see II A).
   a. Approximately 10^4–10^5 receptors exist on the surface of a polypeptide hormone target cell.
   b. Polypeptide hormones that are known to enter cells include insulin, prolactin, parathyroid hormone (PTH), and gonadotropins.
      (1) Human chorionic gonadotropin (HCG) remains bound to its receptor when it enters ovarian cells.

FIGURE 45-1. Mechanisms of hormone-receptor interaction. (*A*) Some hormones (e.g., polypeptide hormones, catecholamines) bind to specific receptors on the surface of target cells, triggering an increase or decrease in the concentration of cyclic adenosine 3′,5′-monophosphate (cAMP) or some other second messenger (e.g., calcium, 1,2-diacylglycerol). (*B*) Other hormones (e.g., steroids, thyroid hormones) are transported by carrier proteins in the blood and, once dissociated from the carrier, enter the target cell and bind to specific receptors in the cytoplasm or nucleus. These hormone-receptor complexes then act on nuclear DNA to alter transcription of specific genes. (Adapted from Darnell J, Lodish H, Baltimore D (eds): *Molecular Cell Biology,* 2nd edition. New York, WH Freeman, 1990, p 711.)

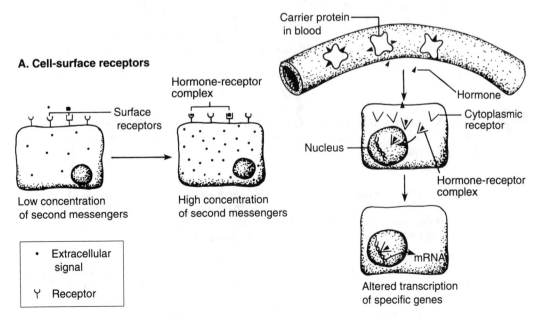

B. Cytoplasmic or nuclear receptors

Carrier protein in blood

A. Cell-surface receptors

Hormone-receptor complex

Surface receptors

Hormone

Cytoplasmic receptor

Nucleus

Hormone-receptor complex

Low concentration of second messengers

High concentration of second messengers

• Extracellular signal

Y Receptor

mRNA

Altered transcription of specific genes

(2) Because receptors may enter cells, it is possible that receptors (not hormones) are the biologically active component. For example, insulin antibody causes the insulin receptor to change its configuration, allowing the receptors to exert insulin-like effects on cells.

2. Steroids and thyroid hormones enter target cells and bind to specific **nuclear receptors**. The hormone-receptor complex then binds to specific DNA sequences to initiate transcription and translation (see II B). Approximately 3000 to 10^4 intracellular receptors exist per steroid target cell.

B. **Effects**

1. **Increased membrane permeability.** Hormonal control of membrane permeability is well documented for ions and for metabolites such as glucose and amino acids.
   a. The increased entry of glucose into cells, which is mediated by a special transport system, is the major method by which insulin controls glucose utilization by muscle.
   b. Other peptide hormones that change membrane permeability include growth hormone (GH), adrenocorticotropic hormone (ACTH), calcitonin, thyroid-stimulating hormone (TSH), and thyroxine.
   c. Steroids that alter the membrane permeability include mineralocorticoids, glucocorticoids, estrogens, and androgens.

2. **Changes in receptor number (down-regulation, up-regulation)**
   a. Hormone-sensitive cells respond to high concentrations of certain hormones by reducing the number of cell surface receptors, thus decreasing sensitivity to the circulating hormone. For example, elevated ambient insulin concentrations cause a loss or inactivation of insulin receptors in liver cells, fat cells, and white blood cells.
   b. Catecholamines exert their effects via plasma membrane receptors, and thyroid hormones have receptors in the target cell nucleus. Excess thyroid hormone leads to an increased number of catecholamine receptors in the myocardium of experimental animals. This may explain the catecholamine-like effects noted in hyperthyroid patients who produce normal amounts of catecholamines.
      (1) Hyperthyroid patients have tachycardia and palpitations, effects observed with excessive catecholamines (pheochromocytoma). These cardiac symptoms can be ameliorated by the administration of a β-blocker such as propranolol.
      (2) The increase in cardiac β-adrenergic receptors in hyperthyroidism makes the heart more responsive to catecholamines.
   c. Angiotensin II decreases the number of its receptors on adrenocortical cells and increases the number of its receptors in vascular smooth muscle.

II. **MECHANISMS OF INFORMATION TRANSFER: DISTANT SIGNALS.** For hormones that communicate via the bloodstream, two mechanisms for hormone-receptor coupling operate.

A. **Cyclic AMP-mediated hormone activity** (see Figure 45-1) involves cyclic adenosine 3',5'-monophosphate (cAMP) as a second messenger that changes enzyme activities. cAMP is the intracellular nucleotide messenger that mediates the effects of certain hormones on subcellular processes, leading to a variety of physiologic responses.

1. Most polypeptide hormones, many biogenic amines, and some prostaglandins activate their target cells by stimulating the synthesis of cAMP, which causes enzyme phosphorylation, a reaction usually associated with enzyme activation. The first messenger is the hormone that binds to the membrane receptor and leads to **activation of membrane-bound adenyl cyclase**. This enzyme converts adenosine triphosphate (ATP) to cAMP in the presence of magnesium ion (Mg^{2+}).

   2. cAMP exerts biologic activity via the **phosphorylation of cAMP-dependent protein kinases** (e.g., protein kinase A).

     a. Kinases are a family of enzymes that phosphorylate their substrate. Protein kinases, a subgroup of the kinases, can be soluble or membrane-bound. Protein kinases transfer a phosphate group from ATP to the hydroxyl group of the substrate. Only the protein kinases that are regulated by cAMP are called cAMP-dependent protein kinases.

     b. The phosphorylation of a cAMP-dependent protein kinase can lead to the activation (e.g., via phosphorylase kinase, phosphorylase, and triglyceride lipase) or in the inactivation (e.g., via glycogen synthetase and pyruvate dehydrogenase) of the substrate.

     c. The effects of the cAMP-dependent protein kinases on their substrates are reversed by a group of enzymes called **phosphoprotein phosphatases,** which remove the phosphate group from the protein enzyme by hydrolysis.

     d. The enzyme that inactivates cAMP is called **cyclic nucleotide phosphodiesterase**. Most of the phosphodiesterases are cytosolic enzymes.

   3. An **example of cAMP-mediated hormone activity** is the hormone regulation of glycogen metabolism. Glucagon and epinephrine stimulate glycogenolysis and also inhibit glycogen synthesis.* Insulin has the opposite effect on both processes.

     a. The enzymes that promote glycogenolysis are active in their phospho- form and inactive in their dephospho- form. The reverse is true for glycogen synthetase, which is the major enzyme in glycogen synthesis.

     b. In muscle, a cAMP-dependent protein kinase is activated either by catecholamines (via their β-adrenergic receptors) or by glucagon (via its receptor in the liver).

     c. This activated protein kinase phosphorylates another protein kinase, phosphorylase kinase, thereby activating it.

     d. The activated phosphorylase kinase phosphorylates the enzyme, phosphorylase, which initiates glycogen breakdown. Phosphorylase kinase also phosphorylates glycogen synthetase, thereby inactivating it.

B. **Transcription and translation effects** (see Figure 45-1B). Steroids and thyroid hormones modulate transcription in specific areas of the nuclear chromatin by interacting with DNA molecules in the chromatin to cause **enzyme induction**. Steroid-receptor complexes are translocated through the cytosol and enter the nucleus. Thyroid hormones enter the nucleus in the free state and then combine with nuclear receptors.

   1. Chromatin consists of DNA, histone proteins, and nonhistone (acidic) proteins. The specificity of nuclear binding is a property of a particular acidic protein.

   2. As a result of the interaction of steroid and thyroid hormones with the chromatin, transcription is stimulated and specific messenger RNA (mRNA) synthesis increases.

   3. The specific mRNAs enter the cytoplasm, where they direct the synthesis (translation) of specific proteins. These proteins may be enzymes, structural proteins, receptor proteins, or secretory proteins.

III. **MECHANISMS OF INFORMATION TRANSFER: LOCAL SIGNALS.** Some cell-to-cell signaling occurs over very small distances.

A. **Paracrine communication** (see Figure 44-1B) involves local diffusion of a peptide or other regulatory molecule to its target cell through the interstitium. Target cells are in the vicinity of the paracrine cell that releases the messenger. Although paracrine substances may diffuse into the blood, it is not necessary that they do.

---

*The glycogenolytic effect of epinephrine in human liver probably occurs via activation of a cAMP-independent phosphorylase.

1. Somatostatin is a paracrine substance whose secretion is stimulated by glucose, glucagon, and gut hormones. Somatostatin is enzymatically degraded in the blood, a process that protects distant cells from the paracrine substance, should it enter the blood.

2. Signal transmission also occurs through the release of peptides into the lumen of the gastrointestinal (GI) tract, where these peptides interact with endocrine cells to cause the release of a second endocrine or paracrine messenger. Gastrin, somatostatin, and substance P are released into the gut lumen following nerve stimulation. Possibly, these substances are released as precursor molecules that are activated by digestive enzymes. Gastrin also is secreted into the bloodstream.

B. **Neurocrine communication** involves the release of chemical messengers from nerve terminals. Neurocrine substances may reach their target cells via one of three routes.

1. The **neurotransmitter can be released directly** into the intercellular space, cross the synaptic junction, and inhibit or activate the postsynaptic cell.
   a. Neurocrine substances are inactivated by degrading enzymes and by reuptake of the substances by neurons.
   b. Examples of neurocrine substances secreted by this route are acetylcholine (ACh) and norepinephrine.

2. A **neural signal also can be transferred via a gap junction,** which is a membrane specialization between nerve cells, between nerve terminals and endocrine cells, and between endocrine cells. Gap junctions allow the movement of small molecules and electric signals from one cell to another, creating a functional **syncytium**.

3. The third potential route for the transmission of a neural signal is identical to the classic neurosecretory mechanism, which involves the **release of a peptide or neurohormone from a neurosecretory neuron** into the blood followed by the interaction of this neurohormone with specific receptors on distant target cells (see Chapter 46 III; IV). Examples of such neurocrine substances are oxytocin and antidiuretic hormone (ADH). The effector sites of neurohormones are not always endocrine cells.

C. **Autocrine communication** (see Figure 44-1C). **Autacoid** is a term used to designate a compound that is synthesized at, or close to, its site of action. This is in contrast to the circulating hormones, which act on tissues distant from their site of synthesis. Prostaglandins are autacoids.

Chapter 46

Hormonal Control Systems

I. **HOMEOSTASIS AND STEADY-STATE.** A major function of the endocrine system is to maintain the homeostasis of the internal environment. This condition of relative constancy in the concentration of dissolved substances, in temperature, and in pH is a basic requirement for the normal function of cells.

A. The concept of **homeostasis** as a constancy of physiologic variables must be modified, because many regulated organismic processes are not constant but conform to a persistent endogenous or exogenous **rhythm**. For example, humans demonstrate a **circadian pattern** in the levels of plasma 17-hydroxycorticosteroids. Such 24-hour cycles are not solely a response to fluctuating environmental stimuli but also are a result of internal endogenous oscillators whose phases are influenced by environmental stimuli.

1. In humans, certain corticosteroids (e.g., cortisol) have a rhythmic pattern of secretion, with secretory rates highest early in the morning and lowest late at night. Accordingly, plasma cortisol concentration is at a peak between 6 A.M. and 8 A.M. and at a nadir between midnight and 2 A.M.
   a. This circadian rhythm persists but shifts to correspond with a change in sleeping habit (e.g., during illness, night work, changes in longitude, and total bed rest or confinement).
   b. For this reason, treatment of patients with exogenous corticosteroids is on an alternate-day dosage regimen, whereby the entire dose is given in the morning of every other day. This dosage schedule simulates the normal adrenocortical secretory rhythm.

2. The rhythmic pattern of corticosteroid secretion occurs in isolated adrenal glands and even in single adrenocortical cells.

B. The term **steady-state** indicates that a function or a system is unvarying with time, but that the system is not in true equilibrium. The system is said to be in a **dynamic equilibrium,** because matter and energy flow into the system at a rate equal to that at which matter and energy flow out of the system.

II. **HYPOTHALAMIC-HYPOPHYSIAL AXES AND FEEDBACK CONTROL.** The hypothalamus has **neural control** over hormone secretion by the posterior lobe of the pituitary gland. The secretory activity of the anterior lobe is controlled by **hypothalamic hormones,** which are secreted into the **hypothalamic-hypophysial portal system** (the hypophysial portal system). Only those hypothalamic hormones that regulate the anterior pituitary are hypophysiotropic hormones.

A. **Hypophysial portal system** (Figure 46-1). The median eminence has a poorly developed blood–brain barrier, and there is relatively little arterial blood perfusing the cells of the anterior lobe. The blood supply of the anterior lobe is derived from branches of the internal carotid arteries (mainly the superior hypophysial artery).

1. The posterior lobe derives its blood from a capillary plexus emanating from the inferior hypophysial artery. This capillary plexus drains into the dural sinus. The neural tissue of the upper infundibular stem (neural stalk) and of the median eminence is supplied largely by branches of the superior hypophysial artery. The median eminence is the specialized area of the hypothalamus located beneath the inferior portion of the third ventricle. It is a release center for hypophysiotropic hormones.

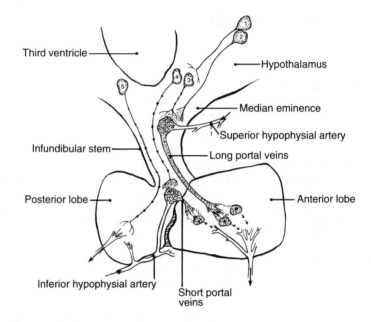

FIGURE 46-1. Anatomic relationship of the hypothalamus and pituitary gland, showing the hypophysial portal system and neurons involved in control of the pituitary gland. Neuron five (5) represents the peptidergic neurons of the supraopticohypophysial and paraventriculohypophysial tracts. Neurons four (4) and three (3) are the peptidergic neurons of the tuberohypophysial tract. Neuron one (1) and neuron two (2) are monoaminergic neurons. (Adapted from Gay VL: The hypothalamus: physiology and clinical use of releasing factors. *Fertil Steril* 23:51, 1972.)

Third ventricle

Hypothalamus

Median eminence

Superior hypophysial artery

Infundibular stem

Long portal veins

Posterior lobe

Anterior lobe

Inferior hypophysial artery

Short portal veins

2. The **primary capillary plexus,** which emanates from the superior hypophysial artery, forms a set of long portal veins that carry blood downward into the anterior lobe.
   a. The portal veins, which give rise to the **secondary capillary plexus,** constitute about 90% of the blood supply to the cells of the anterior lobe. The secondary capillary plexus drains into the dural sinus.
   b. The anterior lobe receives its remaining blood from the short portal veins, which originate in the capillary plexus of the inferior hypophysial artery at the base of the infundibular stem.

B. **Feedback control** is an important mechanism regulating hormone synthesis and secretion (Figure 46-2).

1. **Hypothalamic-pituitary-target gland model.** The paradigm for feedback control is the interaction of the pituitary gland with target endocrine tissues (thyroid gland, adrenal cortex, gonads). Unbound circulating hormones produced by target endocrine organs inhibit the hypothalamic-pituitary system, causing a decrease in the secretion of pituitary tropic hormones, which, in turn, control the secretion by the endocrine target glands. Virtually all hormone secretions are controlled by some type of feedback control.

2. **Negative feedback control** occurs on three levels.
   a. **Long-loop feedback.** Peripheral gland hormones and substrates arising from tissue metabolism can exert long-loop feedback control on both the hypothalamus and the anterior lobe of the pituitary gland. Long-loop feedback usually is negative but occasionally can be positive and is particularly important in the control of thyroidal, adrenocortical, and gonadal secretions.
   b. **Short-loop feedback.** Negative feedback also can be exerted by the anterior pituitary tropic hormones on the synthesis or release of the hypothalamic releasing or inhibiting hormones, which collectively are called **hypophysiotropic hormones**.
   c. **Ultrashort-loop feedback.** Evidence suggests that the hypophysiotropic hormones may inhibit their own synthesis and secretion via a control system referred to as ultrashort-loop feedback.

III. **NEUROSECRETORY NEURONS** (see Figure 46-1). Neural control of the pituitary gland is exerted through neurohumoral secretions that arise from specialized neurosecretory neurons (peptidergic neurons) and are carried by the bloodstream to a target site.

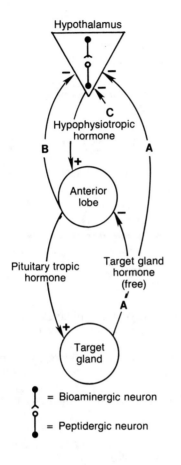

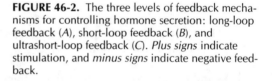

FIGURE 46-2. The three levels of feedback mechanisms for controlling hormone secretion: long-loop feedback (*A*), short-loop feedback (*B*), and ultrashort-loop feedback (*C*). *Plus signs* indicate stimulation, and *minus signs* indicate negative feedback.

A. | Structure and function

1. Neurosecretory neurons are glandular, unmyelinated secretory cells with two functions.
   a. They function as typical neurons, in that they conduct action potentials.
   b. They also function as endocrine glands, in that they synthesize and release neurohormones either directly into the general circulation (as in the case of the neurosecretory neurons of the pars nervosa) or into a portal system (as in the case of the hypophysiotropic neurons, which release their neurohormones into the primary plexus of the hypophysial portal system).

2. A neurosecretory cell system consists of axons that terminate directly on or near blood vessels. This differentiates these cells from typical neurons, which release neurotransmitters at localized synaptic regions. The functional complex of a neurosecretory neuron together with a blood vessel (hemocoele) is called a **neurohemal organ**.

B. | Classification. Neurosecretory neurons in humans are restricted to the hypothalamus, where they occur as two distinct populations of cells that secrete neurohormones (Tables 46-1, 46-2).

1. The **magnocellular neurosecretory system** refers to the neurosecretory neurons of the supraoptic and paraventricular nuclei, which together form the supraopticohypophysial tract. Magnocellular neurons synthesize and secrete the neurohormones antidiuretic hormone (ADH) and oxytocin.

2. The **parvicellular** (also called parvocellular) **neurosecretory system** refers to the neurosecretory neurons of the tuberoinfundibular tract. These neurons of the medial basal

TABLE 46-1. Neuroendocrine Transducer Systems

Neuroendocrine System	Hormone
Magnocellular neurosecretory neurons Posterior lobe	
Supraoptic	Antidiuretic hormone (ADH)
Paraventricular	Oxytocin
Parvicellular neurosecretory neurons Median eminence	Hypophysiotropic hormones
Preganglionic fibers Adrenal medulla	Epinephrine*
Postganglionic fibers Pineal gland	Melatonin[†]
Juxtaglomerular apparatus	Renin[†]

After Martin JB, et al: Neuroendocrine transducers and neurosecretion. In *Clinical Endocrinology.*
Philadelphia, FA Davis, 1977, p 4.
*Acetylcholine (ACh) is the neurotransmitter preceding epinephrine release.
[†]Norepinephrine precedes the release of both melatonin and renin.

hypothalamus have axons that terminate directly on the capillaries of the portal vessels in the median eminence, and they form a final common pathway for neuroendocrine function.

  a. The parvicellular neurosecretory neurons mainly are peptidergic. An important exception, however, is the dopaminergic neurosecretory neurons that form and secrete **prolactin-inhibiting factor (PIF)**.

  b. The secretory products of the parvicellular neurosecretory neurons are called hypophysiotropic hormones.

C. **Neural control.** Neural information is transmitted to the parvicellular and magnocellular neurosecretory cells by **monoaminergic neurons**. Most of the cell bodies of the monoaminergic neurons are located in the mesencephalon and lower brain stem.

  1. The monoaminergic neurons that innervate the parvicellular neurons produce and secrete biogenic amines, which modulate the hypothalamic release of the hypophysiotropic hormones.

TABLE 46-2. Characteristic Features of the Neurosecretory Control Systems

Feature	Parvicellular System	Magnocellular System
Neural input	Norepinephrine, dopamine, serotonin	Acetylcholine
Nuclei	Arcuate nucleus	Supraoptic and paraventricular nuclei
Tract	Tuberoinfundibular	Supraopticohypophysial
Terminus	Median eminence, upper infundibular stem	Pars nervosa (infundibular process)
Neurohormones	Neuropeptides (hypophysiotropic hormones), polypeptides	Neuropeptides (arginine-ADH, oxytocin), nonapeptides
Type of endocrine neuron	Peptidergic	Peptidergic
Stimuli	Monoamines	Acetylcholine
Vascular elements	Hypophysial portal system	Capillary bed (systemic)

2. The function of the magnocellular neurosecretory neurons is controlled by cholinergic and noradrenergic neurotransmitters.

  a. Acetylcholine (ACh) stimulates the release of ADH and oxytocin.

  b. Norepinephrine inhibits the secretion of ADH and oxytocin.

3. Since the secretion of the parvicellular and magnocellular peptidergic neurons is regulated by biogenic amines, the neurosecretory neurons can correctly be viewed as **neuroeffector** cells.

IV. **NEUROENDOCRINE TRANSDUCERS** are endocrine glands that convert neural signals into hormonal signals. Neural control of endocrine tissues occurs in three ways (Figure 46-3).

A. **Direct innervation of autonomic secretomotor neurons**

  1. Pancreatic islets of Langerhans

   a. The islets of Langerhans have a postganglionic parasympathetic innervation. Increased vagal activity to the beta cells stimulates insulin release only during periods of elevated blood sugar.

FIGURE 46-3. The three types of neuroendocrine transducers. (*A*) Secretomotor neurons control endocrine glands by direct innervation via autonomic fibers; the adrenal medulla is the only autonomic neuroeffector that is innervated by preganglionic neurons. *A1* = the autonomic nervous system (ANS); *A2* = the sympathoadrenomedullary axis; *ACh* = acetylcholine. (*B*) Magnocellular neurosecretory neurons control the posterior lobe of the pituitary gland, and (*C*) parvicellular neurosecretory neurons control the anterior lobe. (Reprinted from Martin JB, et al: Neuroendocrine transducers and neurosecretion. In *Neuroendocrinology.* Philadelphia, FA Davis, 1977, p 5.)

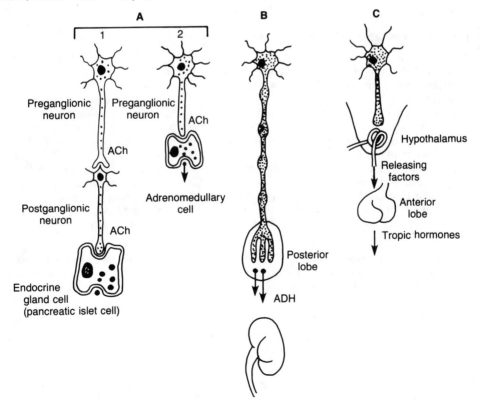

b. The islets of Langerhans also have a postganglionic sympathetic innervation. When the sympathetic nerves to the beta cells are stimulated or when norepinephrine or epinephrine is infused, the predominant effect is inhibition of insulin secretion.

2. Pineal gland. This endocrine structure of the diencephalon is classified as a periventricular organ because it borders on the third ventricle.
   a. The pinealocytes are innervated by the postganglionic (adrenergic) sympathetic fibers, which originate in the superior cervical ganglia of the sympathetic chain.
   b. When the neurotransmitter norepinephrine is released by the autonomic fibers, it stimulates the synthesis and release of melatonin and other indoleamine hormones.
   c. The pineal gland, like other periventricular organs (e.g., the median eminence), has a poorly developed blood–brain barrier.

3. Juxtaglomerular cells. These granular cells of the juxtaglomerular apparatus receive a postganglionic input, which, when stimulated, leads to the release of the proteolytic enzyme renin. Renin can be classified, according to the neuroendocrine transduction concept, as a hormone.

4. Adrenal medulla. This endocrine structure is composed of chromaffin cells and is innervated by preganglionic (cholinergic) sympathetic fibers, which, when stimulated, cause the release of the adrenomedullary hormones epinephrine and norepinephrine. The release of ACh at the synapses causes the secretion of these catecholamines.

B. **Magnocellular neurosecretory regulation of the posterior lobe** (see Table 46-2)

1. Depolarization of the magnocellular neurosecretory cells by ACh released at synapses on the cell bodies of these neurons causes the release of ADH and oxytocin. The axons of these neurons terminate directly on the blood vessels of the posterior lobe, but they do not innervate the vessels.

2. The neural input to the cell bodies of the magnocellular neurons is cholinergic, and the hormonal output consists of peptide hormones.

C. **Parvicellular neurosecretory regulation of the anterior lobe** (see Table 46-2)

1. The anterior lobe lacks a direct nerve supply, but the pituitary gland does possess an innervation. The neurons in the anterior lobe are exclusively postganglionic sympathetic, which are **vasomotor fibers** and not secretomotor fibers.
   a. The hypothalamic regulation of the anterior lobe is achieved through the tuberohypophysial neurons of the medial basal hypothalamus. These peptidergic neurons synthesize and secrete specific hypophysiotropic hormones, which enter the hypophysial portal system and stimulate or inhibit the secretion of anterior pituitary hormones.
   b. The arcuate nucleus (nucleus infundibularis) is the main site of origin of the fine unmyelinated axons of the tuberoinfundibular pathway; however, tuberohypophysial neurons exist throughout the hypophysiotropic area, including the ventromedial nuclei and the periventricular and preoptic areas. The arcuate nucleus serves as the final neural link in the neurovascular connection between the hypothalamus and the anterior lobe.

2. The cells of the intermediate lobe, unlike those of the anterior lobe, receive a direct bioaminergic secretomotor supply from the hypothalamus and are not perfused directly by the hypophysial portal system.

Chapter 47

Endogenous Opioid Peptides

I. TERMINOLOGY

A. **Opioid** denotes opiate-like in terms of a functional similarity with a chemical dissimilarity.

1. **Opioid peptides** refer to endogenous or synthetic compounds that have a spectrum of pharmacologic activity similar to that of morphine.

2. Because the opioid peptides bind with morphine receptors in the central nervous system (CNS), they are called **morphinomimetic peptides**.

B. The two chemical groups of opioid peptides are called **endorphins** and **enkephalins**.

1. The term endorphin originally was used to designate all opiate-like compounds occurring in the brain. This term now is restricted to the specific endogenous morphine-like substances, α-, β-, and γ-endorphins.

2. The enkephalins include met-enkephalin, leu-enkephalin, dynorphin, and α-neo-endorphin.

II. ENDORPHINS

A. **Chemistry.** Endorphins are structural derivatives of β-lipotropic hormone (β-LPH). The lipotropin-related peptides (β-LPH and β-endorphin) and adrenocorticotropic hormone (ACTH) have a common glycoprotein precursor molecule (molecular weight of about 31,000 daltons), which is **proopiomelanocortin (POMC)**.

1. POMC is activated by proteolytic cleavage to yield 10 individual peptides that have been **classified into four groups:**
   a. ACTH and corticotropin-like intermediate lobe peptide (CLIP)
   b. Lipotropic hormones (lipotropins), represented by β- and γ-LPH
   c. Melanocyte-stimulating hormones (MSHs), represented by α-, β-, and γ-MSH
   d. Opioid peptides or endorphins, represented by α- , β-, and γ-endorphins

2. POMC consists of **three major chemical moieties**.
   a. The N-terminal fragment (16 K) is cleaved to form γ-MSH.
   b. ACTH consists of the fragment containing amino acid residues 1–39.
      (1) ACTH is cleaved to produce α-MSH, which consists of the fragment containing amino acid residues 1–13. In humans, α-MSH normally is not synthesized in significant quantities.
      (2) ACTH also is cleaved into CLIP, which consists of the fragment containing amino acid residues 18–39.
   c. β-LPH consists of the fragment containing amino acid residues 1–91. β-LPH is the precursor of the endogenous opioids, γ-LPH and β-endorphin.
      (1) γ-LPH consists of the β-LPH fragment containing amino acid residues 1–58.
         (a) In nonhuman species, γ-LPH is converted by proteolytic cleavage to β-MSH, which is the fragment containing amino acid residues 41–58.

          (b) β-MSH and α-MSH are not formed in humans in significant quantities because the intermediate lobe is vestigial in adult humans. β-MSH does not exist as such in normal human plasma.*

      (2) β-Endorphin consists of the β-LPH fragment containing amino acid residues 61–91.

          (a) β-Endorphin is cleaved into γ-endorphin, which is the fragment containing amino acid residues 61–77.

          (b) γ-Endorphin forms α-endorphin, which is the fragment containing amino acid residues 61–76.

B. **Distribution.** The endogenous opioids (endorphins) do not cross the blood–brain barrier.

  1. **POMC** is synthesized in the anterior and intermediate lobes of the pituitary gland, in the hypothalamus and other areas of the brain, and in several peripheral tissues including the placenta, gastrointestinal (GI) tract, and lung.

    a. In the human **pituitary gland,** POMC exists principally in the anterior lobe, where it is synthesized by basophils. POMC is the precursor of ACTH and β-LPH, which are secreted together from the basophils.

      (1) ACTH and β-LPH occur within the same pituitary cell and possibly within the same secretory granule.

      (2) The intermediate lobe does not produce ACTH and β-LPH as final secretory products.

    b. In the **brain,** the concentrations of ACTH and β-LPH are much lower than in the anterior lobe, and all neural cells or fibers that contain ACTH also contain β-LPH. The highest extrapituitary concentrations of ACTH, α-MSH, β-LPH, γ-LPH, and β-endorphin exist in the hypothalamus followed by the limbic system.

  2. **β-Endorphin**

    a. β-Endorphin is the principal opioid peptide in the **pituitary gland,** and it exists in the highest concentration in the intermediate lobe of experimental animals. In the human pituitary, β-endorphin generally is confined to the cells of the anterior lobe. In response to acute stress, the pituitary gland secretes concomitantly ACTH and β-endorphin.

    b. **Other sites**

      (1) β-Endorphin has been found in the human pancreas, placenta, semen, and in the male reproductive tract.

      (2) Immunoassayable β-endorphin also exists in the plasma and cerebrospinal fluid (CSF).

  3. **CNS endorphins.** The endorphin system exists in the CNS and is characterized by long fiber projection systems from the arcuate region of the hypothalamus to the periventricular region.

    a. Only in the arcuate area do ACTH, β-LPH, and the endorphins occur within **cell bodies** of neurons. Therefore, endorphin cell bodies exist only in the ventral hypothalamus.

    b. Outside the arcuate area, the endorphins are found within the **fibers** of neurons, which project mainly to the mesencephalic periaqueductal gray area. Other areas of projection include the periventricular thalamus, medial amygdala, locus ceruleus, and the zona incerta.

    c. Brain β-endorphin content is unaltered by hypophysectomy, suggesting that the pituitary gland is not the source of CNS endorphins.

C. **Physiologic effects**

  1. **Opiate-receptor binding.** Endorphins bind to opiate receptors in the brain to cause analgesia, sedation, respiratory depression, and miosis (pupillary contraction).

---

*β-MSH has been isolated from the human pituitary gland, but the MSH activity in human plasma probably is caused largely by the MSH-related peptides, β-LPH and γ-LPH.

    a. The intraventricular administration of β-endorphin in humans relieves intractable pain.

        (1) It is believed that this analgesic effect involves the pituitary release of endorphins because the effect is abolished with hypophysectomy.

        (2) The characteristic effect of opiates in humans is less a specific blunting of pain than it is an induced state of indifference or emotional detachment from the experience of suffering.

    b. The intracerebral administration of endorphins produces a profound sedation and immobilization (catatonia).

    c. Endorphins play a central role in controlling affective states and may be involved in controlling the drive for food, water, and sex, all of which are associated with the limbic nervous system.

2. Stimulation of the hypothalamus. The intracisternal or intracerebral injection of β-endorphin elicits a hypophysiotropic effect via the stimulation of the hypothalamus.

    a. This hypothalamic effect on pituitary function causes the secretion of growth hormone (GH), prolactin, ACTH, and antidiuretic hormone (ADH).

    b. In addition, there is a diminished pituitary secretion of thyroid-stimulating hormone (TSH) and the gonadotropic hormones, luteinizing hormone (LH) and follicle-stimulating hormone (FSH).

    c. β-Endorphin in the GI tract stimulates the secretion of glucagon and insulin but inhibits the secretion of somatostatin.

3. β-LPH is a more potent stimulator of **aldosterone synthesis** than is angiotensin II, and β-LPH is thought to be the stimulator of the zona glomerulosa in primary aldosteronism.

4. β-LPH, γ-LPH, and MSH have a lipolytic effect resulting in **fat mobilization;** however, these compounds do not have significant regulatory effects on fat metabolism in humans.

    (1) β-MSH has been isolated from the human pituitary gland.

    (2) MSH activity in human plasma is caused largely by the MSH-related peptides, β-LPH and γ-LPH.

5. β-LPH, γ-LPH, and ACTH have a **weak MSH activity,** which explains the hyperpigmentation associated with increased ACTH secretion.

III. ENKEPHALINS

A. Chemistry

1. Enkephalins, represented by met-enkephalin and leu-enkephalin, are pentapeptide opioids, which are not breakdown products of POMC. The precursor substance for met- and leu-enkephalin is **proenkephalin A.**

2. Proenkephalin B contains the sequences of α-neo-endorphin, dynorphins A and B, and leu-enkephalin. Although α-neo-endorphin and dynorphin contain the leu-enkephalin sequence, neither is considered a precursor for leu-enkephalin.

B. Distribution

1. The pituitary gland is virtually devoid of enkephalins.

    a. Of the pituitary enkephalins, including dynorphin (amino acid residues 1–13), the greatest concentrations exist in the pars nervosa.

    b. Brain enkephalins are not depleted following hypophysectomy, suggesting a neural origin for these peptides.

2. CNS. Enkephalins usually are localized in neuronal processes and terminals. The enkephalin-containing neurons have the widest distribution throughout the CNS and are characterized by short axon projections. Dynorphin also is distributed widely in the CNS, with the highest concentrations in the hypothalamus, medulla-pons, midbrain, and spinal cord.

   a. The highest concentrations of met-enkephalin are in the **basal ganglia** (globus pallidus) and substantia nigra.*

   b. In the **spinal cord,** enkephalins exist in highest concentration in the dorsal gray matter, which corresponds to the synapses of primary sensory nerve endings. Enkephalins also are found in other areas of the spinal cord, notably the substantia gelatinosa, which is known to be involved in the transmission of pain impulses.

   c. Many **limbic structures** have relatively high enkephalin levels, including the lateral septal nucleus, the interstitial nucleus of the striae terminalis, the prefornical area, and the central amygdala.

3. **Other sites**
   a. **Enkephalin-like material** has been demonstrated in the CSF and has been isolated from the adrenal medulla, where it exists in high concentrations both in axon terminals of the splanchnic nerve and in adrenomedullary chromaffin cells.
      (1) **Met-enkephalin** in the circulation originates from the adrenal medulla.
      (2) High plasma concentrations of met-enkephalin are found in patients with **pheochromocytoma**.

   b. **Enkephalin immunoreactivity** has been reported in the human GI tract (myenteric plexus and mucosa), gallbladder, and pancreas.

   c. **Ectopic production of enkephalins** has been reported in carcinoid tumors of the lung and thymus.

C. **Physiologic effects.** Enkephalins are most appropriately classified as neurotransmitters or neuromodulators. A **neuromodulator** is a substance that modifies (positively or negatively) the action of a neurotransmitter at a presynaptic site (by modulating the release of a neurotransmitter) or a postsynaptic site (by modulating the neurotransmitter action).

1. **Spinal cord.** The enkephalins in the dorsal gray matter of the spinal cord function to suppress substance P–containing nerve endings and provide an analgesic (antinociceptive) role. Electric stimulation of the periaqueductal gray area of the spinal cord of humans, which causes analgesia, is associated with an increase in the concentration of enkephalins in the CSF.
   a. Opioids are believed to modulate pain impulses by acting to reduce sensitivity to pain peripherally as well as centrally by activating opiate receptors in the periaqueductal gray area.
   b. Acupuncture is associated with the release of enkephalins into the CSF.

2. **Medulla.** The vagal nuclear localization of enkephalins corresponds to the emetic actions and antitussive properties of morphine.

3. **Brain stem.** Morphine and enkephalins inhibit the firing of the noradrenergic neurons of the locus ceruleus, which is the origin of the ascending noradrenergic fibers.
   a. The localization of enkephalins in the locus cereleus may account for the euphoria-producing actions of morphine.
   b. The amygdala is considered the prime site for morphine-generated euphoria.
   c. The enkephalin tracts and opiate receptors in the limbic system may explain the euphoric effects of opiates.
   d. Respiratory depression, which accounts for the lethal effects of opiates, may involve receptors in the nucleus solitarius of the brain stem, which regulates visceral reflexes including respiration.

---

*The substantia nigra is classified by some neuroanatomists as a component of the basal ganglia.

Chapter 48

Pituitary Gland

I. **EMBRYOLOGY.** The pituitary gland is in close anatomic relation to the hypothalamus. This relationship has both embryologic and functional significance.

A. The anterior lobe of the pituitary gland, the **adenohypophysis,** is derived from the primitive gut by an upward extension (Rathke's pouch) of the epithelium of the primitive mouth cavity (stomodeum). The adenohypophysis is a derivative of buccal ectoderm.

B. The neural or posterior lobe, the **neurohypophysis,** develops as a downward evagination of the neural tube at the base of the hypothalamus (infundibulum) and, therefore, represents a true extension of the brain. Neuroregulation of this structure is achieved by direct neural connections. The neurohypophysis is a derivative of neural ectoderm.

II. **MORPHOLOGY**

A. **Gross anatomy** (see Figure 46-1)

1. **General structure**
   a. The pituitary gland lies in a bony walled cavity, the **sella turcica,** in the sphenoid bone at the base of the skull.
   b. The **dura mater** completely lines the sella turcica and nearly surrounds the gland.
   c. The **pituitary (hypophysial) stalk** and its blood vessels reach the main body of the gland through the diaphragma sellae. The pituitary stalk consists of the infundibular stem and the adenohypophysial tissue that is continuous with the infundibular stem.

2. **Adenohypophysial structure.** The anterior lobe has three components.
   a. The **pars distalis** represents the bulk of the anterior lobe in humans and receives most of its blood supply from the superior (anterior) hypophysial artery, which gives rise to the hypophysial portal system. The pars distalis is the source of the pituitary tropic hormones.
   b. The **pars intermedia** lies between the pars distalis and the neural lobe and is a vestigial structure in humans. It is relatively avascular and is considered almost nonexistent in humans.
   c. The **pars tuberalis** is an elongated collection of secretory cells, which superficially envelops the infundibular stem and extends upward as far as the basal hypothalamus. It is the most vascular portion of the anterior lobe.

3. **Neurohypophysial structure**
   a. **Components**
      (1) The **median eminence,** located beneath the third ventricle, is a small, highly vascular protrusion of the dome-shaped base of the hypothalamus, which is designated grossly as the tuber cinereum. The floor of the third ventricle is designated as the infundibulum because it resembles a funnel.
      (2) The **infundibular stem** (neural stalk) of the posterior lobe arises in the median eminence.
      (3) The **pars nervosa** retains its neural connection with the ventral diencephalon.
   b. **Dominant features** of the posterior lobe are the **neurosecretory neurons** that form the **magnocellular neurosecretory system.** These unmyelinated nerve tracts arise from the supraoptic and paraventricular nuclei within the ventral diencephalon and descend through the infundibulum and neural stalk to terminate in the posterior lobe. The posterior lobe is a storage and secretory site for hormones and, therefore, is not correctly termed an endocrine gland.

B. Histology

1. **Cells.** Two major types of cells are found in equal numbers in the anterior lobe.
   a. **Chromophils** (granular secretory cells) exist in two forms.
      (1) **Acidophils** (eosinophils) account for about 80% of the chromophils and are the cellular source of prolactin and growth hormone (GH).
      (2) **Basophils** comprise about 20% of the chromophils and are the source of thyroid-stimulating hormone (TSH), adrenocorticotropic hormone (ACTH), luteinizing hormone (LH), follicle-stimulating hormone (FSH), and β-lipotropic hormone (β-LPH).
   b. **Chromophobes** (agranular cells) are not precursors of the chromophils and are now known to have an active secretory function. Most of these cells likely are degranulated secretory cells.

2. **Neurons** in the anterior lobe are almost exclusively postganglionic sympathetic fibers that innervate blood vessels.

3. **Nerve fibers** of the neurohypophysial system terminate mostly in the pars nervosa. Interspersed between these neurosecretory fibers are numerous glial cells called **pituicytes,** whose function, other than structural support, remains unknown.

C. Vascular supply (see also Chapter 46 II A and Figure 46-1). A basic tenet of the neurovascular hypothesis is that the concentration of the hypophysiotropic hormones is greater in hypophysial portal blood than at any other site in the vasculature.

1. In humans, the capillaries at the base of the hypothalamus are formed directly from branches of the superior hypophysial arteries, which arise from the internal carotid arteries. There are few vascular anastomoses between the hypothalamic artery and the superior hypophysial artery. The crucial regulatory connection between the hypothalamus and the anterior lobe is via the hypophysial portal vessels.

2. The intermediate lobe is not perfused directly by the hypophysial portal system but is regulated by bioaminergic secretomotor fibers originating in the hypothalamus.

3. The blood supply to the posterior lobe is largely separate from that of the anterior lobe. The blood supply to the median eminence is greater than that to the entire pituitary gland.

III. HORMONES OF THE POSTERIOR LOBE: ANTIDIURETIC HORMONE (ADH) AND OXYTOCIN (see Table 46-1). The physiologic aspects of ADH are described in Chapter 28. The physiologic aspects of oxytocin are described below.

A. Synthesis and storage

1. Like ADH, oxytocin is a nonapeptide* that is synthesized within the cell bodies of the peptidergic neurons of the magnocellular neurosecretory system.

2. This polypeptide is synthesized mainly in the paraventricular nuclei of the hypothalamus and, like ADH, is stored in the posterior lobe of the pituitary gland.

B. Stimuli for release. Oxytocin secretion is brought about by stimulation of cholinergic nerve fibers.

1. Stimulation of the tactile receptors in the areolar region of the female breast during suckling activates somesthetic neural pathways, which transmit this signal to the

*If the two cysteine residues are counted together as a single cystine residue, ADH and oxytocin are correctly classified as octapeptides.

hypothalamus. This leads to the reflex secretion of oxytocin into the bloodstream and to milk release following a latent period of 30–60 seconds. This reflex is called the **milk let-down** or **milk ejection reflex**.

    a. Oxytocin causes milk release in lactating women by contraction of the myoepithelial cells, which cover the stromal surface of the epithelium of the alveoli, ducts, and cisternae of the mammary gland.

    b. Oxytocin secretion can be conditioned so that the physical stimulation of the nipple no longer is required. Thus, lactating women can experience milk release in response to the sight and sound of a baby.

    c. While oxytocin aids in the process, its presence is not absolutely required for successful nursing in humans.

2. Oxytocin secretion also can occur in response to genital tract stimulation, such as that which occurs during coitus and parturition.

3. Oxytocin is produced in men and also is released during genital tract stimulation. The role of this neurohypophysial hormone in men is unknown.

C. Inhibition of release

1. Milk let-down can be inhibited by emotional stress and psychic factors such as fright.

2. Excitation of adrenergic fibers to the hypothalamus inhibits peptide release. Activation of the sympathetic neurons with the concomitant release of norepinephrine and epinephrine inhibits oxytocin secretion.

3. Ethanol inhibits endogenous oxytocin release, resulting in reduced myometrial contractility.

4. Enkephalins also inhibit oxytocin release.

D. Physiologic effects

1. Oxytocin stimulates contraction of the smooth muscle (myoepithelium) of the lactating mammary gland.

2. It also stimulates contraction of the smooth muscle of the uterus (myometrium).

    a. The sensitivity of the myometrium to exogenous oxytocin during pregnancy increases as pregnancy advances.

    b. Oxytocin plays a role in labor and has been shown to be a useful therapeutic agent in the induction of labor.

IV. HORMONES OF THE ANTERIOR LOBE.

The principal hormones of the anterior lobe of the pituitary gland can be classified conceptually into two groups: hormones that stimulate other endocrine glands to secrete hormones and hormones that have a direct effect on nonendocrine target tissues. Only the latter group is described here, using GH and prolactin as examples.

A. GH also is known as human growth hormone (HGH), somatotropic hormone (STH), and somatotropin.

1. Synthesis, chemistry, and general characteristics

    a. GH is synthesized by the acidophils of the anterior lobe and is stored in very large amounts in the human pituitary gland. GH represents approximately 4%–10% of the wet weight of the pituitary gland, which is equivalent to 5–15 mg.

    b. GH is a single, unbranched polypeptide chain containing 191 amino acid residues. It has been synthesized in bacteria using recombinant DNA techniques.

    c. Humans exhibit a **species specificity** for GH, and only human and monkey GH preparations have biologic activity in humans.

d. Like all other pituitary hormones, GH is secreted episodically in periods of 20–30 minutes. The large diurnal fluctuations represent integrations of many small secretory episodes. A regular nocturnal peak in GH secretion occurs 1–2 hours after the onset of deep sleep, which correlates with stage 3 or stage 4 slow-wave sleep.

e. The plasma GH concentration in the growing child is not significantly higher than that in the adult whose growth has ceased.

2. Control of secretion. The release of GH is primarily under the control of two hypophysiotropic hormones.

a. Stimuli for release. Somatotropin releasing hormone (SRH) is the releasing hormone for GH that has been identified as a 44-amino acid peptide. However, several pharmacologic, physiologic, and psychic agents are known to stimulate GH release.

(1) Release of GH is mediated by monoaminergic and serotoninergic pathways; thus, α-adrenergic, dopaminergic, and seritoninergic agonists as well as β-adrenergic antagonists all stimulate GH release in humans.

(2) Bromocriptine (a dopamine agonist), enkephalins, endorphins (β-endorphin), and opiates stimulate GH secretion.

(3) Insulin-induced hypoglycemia is a potent stimulus of GH secretion as are pharmacologic doses of glucagon and vasopressin.

(4) Physiologic stimuli include hypoglycemia, increased plasma concentrations of amino acids (arginine, leucine, lysine, tryptophan, and 5-hydroxytryptophan), and decreased free fatty acid concentrations. In addition, estrogens stimulate GH synthesis and secretion.

(5) GH secretion is stimulated by moderate to vigorous exercise, emotional stress, and stress owing to fever, surgery, anesthesia, trauma, pyrogen administration, and repeated venipuncture. Fasting leads to elevated GH secretion after 2 or 3 days.

b. Inhibitors of secretion

(1) GH can inhibit its own secretion via a short-feedback loop mechanism that operates between the anterior lobe and the median eminence. Somatotropin-inhibiting hormone (SIH; somatostatin) inhibits the synthesis and release of GH.

(a) Somatostatin is a tetradecapeptide (14 amino acid residues) that has been chemically synthesized.

(b) It is a product of the parvicellular neurosecretory neurons that terminate in the median eminence and produce hypophysiotropic hormones.

(c) Somatostatin also is found in other parts of the brain, in the gastrointestinal (GI) tract, and in the delta cells of the pancreatic islets.

(d) In addition to inhibiting GH secretion, somatostatin blocks the secretion of insulin, glucagon, and gastrin and inhibits the intestinal absorption of glucose. These effects of somatostatin produce a state of hypoglycemia.

(2) The secretion of GH in response to the aforementioned stimuli often is blunted in obese individuals.

(3) Glucocorticoids decrease GH secretion, but their predominant effect is the interference with the metabolic actions of GH.

(4) A decline in GH secretion is observed in late pregnancy, despite the presence of high estrogen levels.

(a) Impaired glucose tolerance is common, and clinical diabetes occurs frequently, despite the above normal insulin secretion in response to a glucose load during pregnancy.

(b) Pregnancy regularly antagonizes the action of insulin and increases the pancreatic secretory capacity of both normal and diabetic individuals.

(c) The development of gestational diabetes probably is because of a greater degree of insulin antagonism caused by normal plasma concentrations of human placental lactogen (HPL).

3. Physiologic and metabolic effects

a. Stimulation of growth of bone, cartilage, and connective tissue

(1) The effects of GH on skeletal growth are mediated by a family of polypeptides called **somatomedins** [also termed insulin-like growth factors (IGF)], which are synthesized mainly in the liver. (Thyroid hormone and insulin also are necessary for normal osteogenesis.)

(2) Somatomedin may be produced in nonhepatic tissue as well, in that somatomedin activity has been found in the serum, kidney, and muscle tissue.

    (a) Receptors for somatomedin exist in chondrocytes, hepatocytes, adipocytes, and muscle cells.

    (b) Somatomedin has insulin-like effects on tissues, including lipogenesis in adipose tissue, increased glucose oxidation in fat, and increased glucose and amino acid transport by muscle.

(3) GH, through somatomedin, stimulates proliferation of chondrocytes and the appearance of osteoblasts. The increase in the thickness of the epiphysial (cartilaginous) end-plate accounts for the increase in linear skeletal growth.

(4) After epiphysial fusion, bone length can no longer be increased by GH, but bone thickening can occur through periosteal growth. It is this growth that accounts for the changes seen in hypersomatotropism (acromegaly).

(5) These reactions are the biochemical correlates of protein synthesis in general body growth and also account for the hyperplasia and hypertrophy associated with increased tissue mass.

b. Protein metabolism

(1) GH has predominantly anabolic effects on skeletal and cardiac muscle, where it stimulates the synthesis of protein, RNA, and DNA.

(2) GH reduces circulating levels of amino acids and urea (i.e., it promotes nitrogen retention), which accounts for the term **positive nitrogen balance**. Urinary urea concentration also is decreased.

(3) GH promotes amino acid transport and incorporation into proteins.

c. Fat metabolism

(1) GH has an overall catabolic effect on adipose tissue. It stimulates the mobilization of fatty acids from adipose tissue, leading to a decreased triglyceride content of fatty tissue and increased plasma levels of free fatty acids and glycerol.

(2) GH increases hepatic oxidation of fatty acids to the ketone bodies, acetoacetate and β-hydroxybutyrate.

d. Carbohydrate metabolism. GH is a diabetogenic hormone. Because of its anti-insulin effect, GH has the tendency to cause hyperglycemia.

(1) GH can produce an insulin-resistant diabetes mellitus primarily because of its lipolytic effect.

    (a) Free fatty acids can antagonize the effect of insulin to promote glucose uptake by skeletal muscle and adipose tissue.

    (b) Free fatty acids can stimulate gluconeogenesis.

    (c) Excess acetyl coenzyme A (acetyl-CoA) production favors gluconeogenesis, because pyruvate carboxylase requires acetyl-CoA to form oxaloacetate from pyruvate. Oxaloacetate is the rate-limiting factor in gluconeogenesis.

    (d) Acetyl-CoA inhibits glycolysis by inhibition of pyruvate kinase.

    (e) Free fatty acids stimulate hepatic glucose synthesis mainly via the stimulation of fructose biphosphatase. At the same time, pyruvate kinase and phosphofructokinase both are inhibited by free fatty acids, which block glycolysis and favor gluconeogenesis. Citrate also blocks glycolysis at the phosphofructokinase step.

(2) GH induces an elevation in basal plasma insulin levels.

(3) Because of its anti-insulin effect, GH inhibits glucose transport in adipose tissue. Because adipose tissue requires glucose for triglyceride synthesis, GH antagonizes insulin-stimulated lipogenesis.

e. Mineral metabolism. GH promotes renal reabsorption of Ca^{2+}, phosphate, and Na^+.

4. Endocrinopathies

a. Disorders associated with increased GH

(1) **Growth retardation** can occur when GH levels are increased and somatomedin levels are depressed (e.g., in kwashiorkor). (In the African pygmy, who is resistant to the action of GH, both GH and somatomedin levels are normal. This condition is caused by a decrease in GH receptors.)

(2) Overproduction of GH during adolescence results in **giantism,** which is characterized by excessive growth of the long bones. Patients may grow to heights of as much as 8 feet.

(3) Excessive GH secretion during adulthood, after the epiphysial plates of long bones have fused, causes growth in those areas where cartilage persists. This leads to **acromegaly,** a condition characterized by coarse facial features, underbite (prognathism), prominent brow, enlarged hands and feet, and soft tissue hypertrophy (e.g., cardiomegaly, hepatosplenomegaly, and renomegaly).

b. Disorders associated with decreased GH

(1) Decreased GH secretion in immature persons leads to stunted growth, or **dwarfism,** which is accompanied by sexual immaturity, hypothyroidism, and adrenal insufficiency.

(2) GH deficiency may be part of an overall lack of anterior pituitary hormones (**panhypopituitarism**) or from an isolated genetic deficiency. Selective GH deficiency is rare in adults; clinical manifestations may include impaired hair growth and a tendency toward fasting hypoglycemia.

c. Treatment

(1) The treatment of choice for hypersomatotropism is selective surgical extirpation of the pituitary adenoma without damage to other pituitary functions. Bromocriptine is effective in suppressing, but not normalizing, GH levels in most acromegalic patients. This substance tends to stimulate GH secretion in normal individuals.

(2) Disorders associated with GH deficiency can be treated with HGH.

B. **Prolactin** also is known as lactogenic hormone, mammotropic hormone, and galactopoietic hormone.

1. Synthesis, chemistry, and general characteristics
   a. Prolactin is synthesized in the pituitary acidophils.
   b. Human prolactin is a single peptide chain containing 198 amino acid residues.
   c. It does not regulate the function of a secondary endocrine gland in humans.

2. Control of secretion. Two hypothalamic neurosecretory substances have been implicated in prolactin secretion.
   a. Stimuli for release. Prolactin releasing factor (PRF) is the putative releasing hormone for prolactin. PRF has not been identified or chemically synthesized; however, one of these releasing factors is thyrotropin releasing hormone (TRH), which causes the release of TSH.

   (1) Prolactin secretion increases about 1 hour after the onset of sleep, and this increase continues throughout the sleep period. The nocturnal peak occurs later than that for GH.

   (2) Prolactin secretion is enhanced by exercise and by stresses such as surgery under general anesthesia, myocardial infarction, and repeated venipuncture.

   (3) Plasma prolactin levels begin to increase by the eighth week of pregnancy and usually reach peak concentrations by the thirty-eighth week.

   (4) Nursing and breast stimulation are known to stimulate prolactin release. Oxytocin has been shown to act directly on lactotrophs to stimulate prolactin release, which suggests a physiologic role for oxytocin as a prolactin releasing factor. This role is most evident around the time of ovulation when estrogens are also high, which stimulates the release of oxytocin.

   (5) Serum prolactin levels are elevated in those patients with primary hypothyroidism who are believed to have high TRH levels in the hypophysial portal circulation.

   (6) Dopamine antagonists (phenothiazine and tranquilizers), adrenergic blockers, and serotonin agonists stimulate prolactin secretion.

   (7) Pituitary stalk section and lesions that interfere with the portal circulation to the pituitary gland also cause prolactin release.

b. Inhibitors of release. Normally, the control of prolactin secretion is under constant inhibition via prolactin inhibition factor (PIF).

  (1) Dopamine, physiologically the most important PIF, is secreted into the hypophysial portal vessels.

  (2) Serotonin antagonists and dopamine agonists (bromocriptine) block the secretion of prolactin. Bromocriptine administered during the postpartum period reduces prolactin secretion to nonlactating levels and terminates lactation.

3. Physiologic effects

  a. Prolactin does not have an important role in maintaining the secretory function of the corpus luteum and, therefore, is not a gonadotropic hormone in women.

  b. Prolactin plays an important role in the development of the mammary glands and in milk synthesis.

  (1) During pregnancy, the mammary duct gives rise to lobules of alveoli, which are the secretory structures of this tissue. This differentiation requires prolactin, estrogens, and progestogens. Once the lobuloalveolar system is developed, the role of prolactin and corticosteroids in milk production, although essential, becomes minimal. GH and thyroid hormone enhance milk secretion.

  (2) Immediately following pregnancy, prolactin stimulates galactosyltransferase activity, leading to the synthesis of lactose.

  (3) In women, high serum levels of prolactin are associated with suppressed LH secretion and anovulation, which account for an absence of menses (amenorrhea) during postpartum lactation.

  (a) With continued nursing, FSH levels rise, but LH levels remain low.

  (b) In the early postpartum period, both the FSH and LH levels are low.

4. Endocrinopathies

  a. Hyperprolactinemia is not a rare condition but frequently is undiagnosed because galactorrhea occurs in only about 30% of cases.

  (1) In women, elevated serum prolactin manifests as infertility and amenorrhea.

  (2) In men, hyperprolactinemia is a cause of impotence and decreased libido.

  b. Treatment of prolactin hypersecretion includes administration of bromocriptine, a dopamine-agonist that lowers prolactin levels and usually restores normal gonadal function.

Chapter 49

Adrenal Gland

I. ADRENAL MEDULLA

A. Embryology

  1. The neural crest gives rise to neuroblasts, which eventually give rise to the autonomic postganglionic neurons, the adrenal medulla, and the spinal ganglia.

  2. The adrenal medulla consists of chromaffin cells (pheochromocytes), which are neuro-ectodermal derivatives and the functional analogs of the sympathetic postganglionic fibers of the autonomic nervous system (ANS).

  3. In early fetal life, the adrenal medulla contains only norepinephrine.

B. Morphology

  1. Gross anatomy
  a. The adrenal medulla represents essentially an enlarged and specialized sympathetic ganglion and is called a **neuroendocrine transducer** because a neural signal to this organ evokes hormonal secretion.
  b. The adrenal medulla is the only autonomic neuroeffector organ without a two-neuron motor innervation. It is innervated by long sympathetic preganglionic, cholinergic neurons that form synaptic connections with the chromaffin cells.
  c. Small clumps of chromaffin cells also can be found outside the adrenal medulla, along the aorta and the chain of sympathetic ganglia.

  2. Histology
  a. **Cells.** There are two types of adrenomedullary chromaffin cells. Individual cells contain either **norepinephrine** or **epinephrine,** which is stored largely in subcellular particles called chromaffin granules. These granules are osmiophilic, electron-dense, membrane-bound secretory vesicles.
    (1) Approximately 80% of the chromaffin granules in the human adrenal medulla synthesize epinephrine (adrenalin). The remaining 20% synthesize norepineph-rine (noradrenalin).
    (2) The chromaffin granules contain catecholamines, protein, lipids, and adenine nucleotides [mainly adenosine triphosphate (ATP)].
      (a) One of the proteins localized in the particulate fraction is the enzyme, dopamine-β-hydroxylase.
      (b) Soluble acidic proteins found in the granules are called **chromagranins**.
  b. **Neurons**
    (1) The preganglionic sympathetic fibers that innervate the adrenal medulla traverse the splanchnic nerve, which contains myelinated (type B) secretomotor fibers emanating mainly from lower thoracic segments (T5 and T9) of the ipsilateral intermediolateral gray column of the spinal cord.
    (2) The chromaffin cells do not have axons; therefore, they are functional ana-logues of the postganglionic neurons.

  3. Vascular supply (Figure 49-1)
  a. **Arterial blood** to the adrenal gland reaches the outer capsule from branches of the renal and phrenic arteries, with a less important arterial input directly from the aorta. The adrenal medulla is perfused by blood vessels in two ways.
    (1) A type of **portal circulation** exists in the adrenal gland where the cortex and medulla are in juxtaposition. From the capillary plexus on the outer adrenal capsule most of the blood enters venous sinuses, which drain into and supply the medullary tissue. Thus, most of the blood perfusing the adrenal medulla is derived from the portal system and is partly deoxygenated.

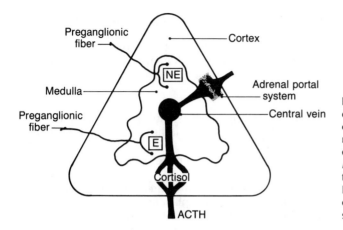

FIGURE 49-1. The adrenal portal vascular system constitutes a functional connection between the cortex and medulla and has a high cortisol concentration. *NE* = norepinephrine; *E* = epinephrine; *ACTH* = adrenocorticotropic hormone. (Adapted from Pohorecky LA, Wurtman RJ: Adrenocortical control of epinephrine synthesis. *Pharmacol Rev* 23(1):1–35, 1971.)

    (2) There also exists a direct arterial blood supply to the medulla via the **medullary arteries,** which traverse the cortex.
    b. Venous blood drains via a single central vein, composed almost entirely of bundles of longitudinal smooth muscle fibers, which passes along the longitudinal axis of the gland.

C. **Adrenomedullary hormones: monoamines**

    1. The adrenal medulla synthesizes and secretes biogenic amines. These dihydroxylated phenolic amines, or **catecholamines,** are epinephrine and norepinephrine. Most of the met-enkephalin in the circulation also originates in the adrenal medulla. Enkephalins are pentapeptides that function as neurotransmitters or neuromodulators, which normally are localized in neuronal processes and terminals (see Chapter 47 III).
      a. Epinephrine is produced almost exclusively in the adrenal medulla, with smaller amounts synthesized in the brain. Essentially all circulating epinephrine is derived from the adrenal medulla.
      b. Norepinephrine is widely distributed in neural tissues, including the adrenal medulla, sympathetic postganglionic fibers, and central nervous system (CNS). In the brain, the concentration of norepinephrine is the highest in the hypothalamus. The norepinephrine content of a tissue reflects the extent of density of its sympathetic innervation. Norepinephrine has been demonstrated in almost all tissues except the placenta, which is devoid of nerve fibers.

    2. Bilaterally adrenalectomized human patients excrete practically no epinephrine in the urine. Urinary levels of norepinephrine remain within normal limits, however, indicating that the norepinephrine originates from extra-adrenal sources (i.e., the terminals of the postganglionic sympathetic fibers and the brain).

D. **Control of catecholamine synthesis**

    1. The biosynthetic pathway originates with **L-tyrosine,** which is derived from the diet or from the hepatic hydroxylation of L-phenylalanine by phenylalanine hydroxylase. Tyrosine is hydroxylated in the cytoplasm by tyrosine hydroxylase to L-dopa (3,4-dihydroxyphenylalanine).

    2. Dopa is converted in the cytosol to **dopamine** (3,4-dihydroxyphenylethylamine) by a nonspecific aromatic L-amino acid decarboxylase (dopa decarboxylase).

    3. Dopamine enters the chromaffin granule, where it is converted to **L-norepinephrine** by dopamine-β-hydroxylase, which exists exclusively in the granule.
      a. Norepinephrine is the end product in approximately 20% of chromaffin cells.
      b. In about 80% of chromaffin cells, norepinephrine diffuses back into the chromaffin cytoplasm. There, it is N-methylated by phenylethanolamine-N-methyltansferase (PNMT) using S-adenosylmethionine as a methyl donor to form L-epinephrine.

(1) PNMT is selectively localized in the adrenal medulla, the only site where it exists in significant concentrations.

(2) PNMT activity is induced by very high local concentrations of glucocorticoids, which are found only in the adrenal portal blood draining the adrenal cortex.

E. **Control of catecholamine secretion**

1. General considerations

a. Acetylcholine (ACh) provides the major physiologic stimulus for the secretion of the adrenomedullary hormones. In addition, angiotensin II, histamine, and bradykinin stimulate catecholamine secretion.

(1) Catecholamine release is stimulated by ACh from the preganglionic sympathetic nerve endings innervating the chromaffin cells.

(2) The final common effector pathway activating the adrenal medulla is the cholinergic preganglionic fibers in the greater splanchnic nerve.

(3) ACh causes the depolarization of the chromaffin cells followed by the release of catecholamines by **exocytosis**. Ca^{2+} influx secondary to membrane depolarization is the central event in **stimulus-secretion coupling**.

b. Cortisol and adrenocorticotropic hormone (ACTH). Because catecholamine synthesis is dependent on cortisol, the functional integrity of the adrenal medulla indirectly depends on a functional pituitary gland for ACTH secretion and a functional median eminence for corticotropin releasing hormone (CRH) secretion. CRH is a hypophysiotropic hormone produced by the parvicellular nuclei of the ventral diencephalon.

2. Physiologic and psychological stimuli for release. The adrenal medulla constitutes the neuroeffector of the sympathoadrenomedullary axis that is activated during states of emergency. This response to stress is called the **fight-or-flight reaction**. Among the conditions in which the sympathetic nervous system is activated are fear, anxiety, pain, trauma, hemorrhage and fluid loss, asphyxia and hypoxia, changes in blood pH, extreme cold or heat, severe exercise, hypoglycemia, and hypotension. During hypoglycemia, the adrenal medulla is activated selectively. In humans, epinephrine and norepinephrine appear to be released independently by specific stimuli.

a. Anger and active aggressive states or situations are associated with increased norepinephrine secretion.

b. States of anxiety are associated with increased epinephrine secretion. In addition, epinephrine release is increased by tense but passive emotional displays or threatening situations of an unpredictable nature.

c. Angiotensin II potentiates the release of catecholamines.

d. Plasma concentrations of epinephrine vary according to physiologic or pathologic state as follows:

(1) Basal level: 25–50 pg/ml (6×10^{-10} mol/L)

(2) Hypoglycemia: 230 pg/ml

(3) Diabetic ketoacidosis: 500 pg/ml

(4) Severe hypoglycemia: 1500 pg/ml

3. Regulation of adrenergic receptors. A reciprocal relationship exists between catecholamine concentration and the number and function of adrenergic receptors.

a. A sustained decrease in catecholamine secretion is associated with an increased number of adrenergic receptors in target cells and an increased responsiveness to catecholamines. Conversely, a chronic increase in catecholamine secretion is associated with a decreased number of adrenergic receptors in target cells and a decreased responsiveness to catecholamines.

b. This relationship may account for the phenomenon of **denervation hypersensitivity,** which is observed in sympathetic neuroeffectors following autonomic fiber denervation.

F. **Metabolism and inactivation of circulating catecholamines**

1. General considerations. The plasma half-life of epinephrine and norepinephrine is 10 seconds and 15 seconds, respectively. The biologic effects of circulating catecholamines are terminated rapidly by both nonenzymatic and enzymatic mechanisms.

    a. Neuronal uptake. Sympathetic nerve endings have the capacity to take up amines actively from the circulation. This uptake of circulating catecholamines leads to nonenzymatic inactivation by intraneuronal storage and to enzymatic inactivation by a mitochondrial enzyme called **monoamine oxidase (MAO).**

    b. Extraneuronal uptake. The formation of catecholamine metabolites locally in innervated tissues and systemically in the liver, kidney, lung, and gut implies catecholamine uptake by a variety of cells.

    c. Inactivation. Circulating epinephrine and norepinephrine are metabolized predominantly in the liver and kidney.

2. Metabolic pathways for catecholamine inactivation

    a. MAO is found in very high concentrations in the mitochondria of the liver, kidney, stomach, and intestine. MAO catalyzes the oxidative deamination of a number of biogenic amines, including the intraneuronal and circulating catecholamines.

       (1) The combined actions of MAO and **aldehyde oxidase** on epinephrine and norepinephrine produce 3,4-dihydroxymandelic acid by oxidative deamination.

       (2) The combined actions of MAO and aldehyde oxidase on the meta-O-methylated metabolites of epinephrine and norepinephrine (metanephrine and normetanephrine, respectively) produce 3-methoxy-4-hydroxymandelic acid (vanillylmandelic acid or VMA) by oxidative deamination.

    b. Catechol-O-methyltransferase (COMT) is found in the soluble fraction of tissue homogenates with the highest levels in liver and kidney. COMT is considered mainly as an extraneuronal enzyme, but it is also found in postsynaptic membranes. COMT metabolizes circulating catecholamines in the kidney and liver and metabolizes locally released norepinephrine in the effector tissue.

       (1) COMT requires S-adenosylmethionine as a methyl donor.

       (2) The action of COMT produces normetanephrine from norepinephrine, metanephrine from epinephrine, and VMA from 3,4-dihydroxymandelic acid by 3-O-methylation.

3. Significance of catecholamine metabolites

    a. Only 2%–3% of the catecholamines are excreted directly into the urine, mostly in conjugation with sulfuric or glucuronic acid. Most of the catecholamines produced daily are excreted as the deaminated metabolites, VMA and 3-methoxy-4-hydroxyphenylglycol (MOPG). Only a small fraction is excreted unchanged or as metanephrines.

    b. Under normal circumstances, epinephrine accounts for a very small proportion of urinary VMA and MOPG. Because the majority is derived from norepinephrine, urinary VMA and MOPG reflect the activity of the nerve terminals of the sympathetic nervous system rather than that of the adrenal medulla.

    c. The excretion of unchanged epinephrine or plasma epinephrine provides a better index of the physiologic activity of the sympathoadrenomedullary system than does the excretion of catecholamine metabolites, because the latter reflects, to a considerable extent, norepinephrine that is metabolized within nerve endings and the brain and never released at adrenergic synapses in the active form.

G. **Physiologic actions of catecholamines** (Table 49-1). The effects of adrenomedullary stimulation and sympathetic nerve stimulation generally are similar. In some tissues, however, epinephrine and norepinephrine produce different effects owing to the existence of two types of adrenergic receptors, **alpha** (α) and **beta** (β) receptors, which have different sensitivities for the various catecholamines and, therefore, produce different responses. **Epinephrine** is the single most active endogenous amine on both α and β receptors.

1. The **α-adrenergic receptors** are sensitive to both epinephrine and norepinephrine. These receptors are associated with most of the excitatory functions of the body and with at least one inhibitory function (i.e., inhibition of intestinal motility).

2. The **β-adrenergic receptors** respond to epinephrine and, in general, are relatively insensitive to norepinephrine. These receptors are associated with most of the inhibitory functions of the body and with one important excitatory function (i.e., excitation of the myocardium).

TABLE 49-1. Some Physiologic Effects of Catecholamines and Types of Adrenergic Receptors

Effector Organ	Receptor Type	Response
Eye		
Radial muscle	α	Contraction (mydriasis)
Ciliary muscle	β	Relaxation for far vision
Heart		
Sinoatrial node	β	Increase in heart rate (increase in rate of diastolic depolarization and decrease in duration of phase 4 of sinoatrial nodal action potential)
Atrioventricular node	β	Increase in conduction velocity and shortening of functional refractory period
Atria	β	Increase in contractility
Ventricles	β	Increase in contractility and irritability
Blood vessels	α	Constriction (arterioles and veins)
	β	Dilation (predominates in skeletal muscle)*
Bronchial muscle	β	Relaxation (bronchodilation)
Gastrointestinal tract		
Stomach	β	Decrease in motility
Intestine	α, β	Decrease in motility
Sphincters	α	Contraction
Urinary bladder		
Detrusor muscle	β	Relaxation
Trigone and sphincter	α	Contraction
Skin		
Pilomotor muscles	α	Piloerection
Sweat glands	α	Selective stimulation (adrenergic sweating)*
Uterus	α	Contraction
	β	Relaxation
Liver	α	Glycogenolysis
Muscle	β	Glycogenolysis
Pancreatic islets	α	Inhibition of insulin secretion
	β	Stimulation of insulin secretion

Adapted from Morgan HE: Function of the adrenal glands. In *Best and Taylor's Physiological Basis of Medical Practice,* 10th edition. Edited by Brobeck JR. Baltimore, Williams & Wilkins, 1979.
*Mediated via sympathetic cholinergic fibers.

H. **Biochemical effects of catecholamines.** Norepinephrine has little direct effect on carbohydrate metabolism; however, both norepinephrine and epinephrine can inhibit glucose-induced secretion of insulin from the beta cells of the pancreatic islets of Langerhans.

    1. Carbohydrate metabolism. Because hepatic stores of glycogen are limited (about 100 g) and decrease only transiently after epinephrine activation, lactate derived from muscle glycogen (300 g) is the major precursor for hepatic **gluconeogenesis,** the process that sustains hepatic glucose formation and secretion. Gluconeogenesis mainly accounts for the hyperglycemic action of epinephrine in normal physiologic states. In pathologic states (e.g., pheochromocytoma), the diabetogenic action of catecholamines is caused by the inhibition of insulin secretion and the gluconeogenic effect of these hormones (usually norepinephrine), which are secreted in excessive amounts. Because propranolol attenuates hyperlactacidemia and hyperglycemia, this implies that epinephrine-induced glucogenolysis in muscle and in the liver is mediated by the β and α receptors, respectively (see Table 49-1).

a. Glycogenolysis in the liver
  (1) Epinephrine stimulates glycogenolysis in the liver via the Ca^{2+}-activated glycogen phosphorylase and the inhibition of glycogen synthetase. Glucose-6-phosphatase, found mainly in the liver and in lesser amounts in the kidney, forms free glucose, which increases blood glucose.
  (2) Because glucagon stimulates and insulin suppresses hepatic glycogenolysis, the effects of epinephrine on insulin secretion (suppression) and glucagon secretion (stimulation) reinforce the breakdown of glycogen and the increase in hepatic glucose secretion.
  (3) Epinephrine also increases the hepatic production of glucose from lactate, amino acids, and glycerol, all of which are gluconeogenic substances.
b. Glycogenolysis in muscle
  (1) Epinephrine stimulates glycogenolysis in muscle by a β-adrenergic receptor mechanism involving the stimulation of adenyl cyclase and cyclic adenosine 3′,5′-monophosphate (cAMP)-induced stimulation of glycogen phosphorylase. Concomitantly, glycogen synthetase activity is reduced.
  (2) Muscle lacks glucose-6-phosphatase, and epinephrine-induced glycogenolysis in muscle **does not directly increase blood glucose**. The glucose-6-phosphate is metabolized to lactate or pyruvate, which is converted to glucose by the liver.
  (3) The ultimate physiologic effect of epinephrine-stimulated glycogenolysis in muscle is increased hepatic glucose secretion (hyperglycemia) via the hepatic conversion of muscle lactate to glucose.
c. Hyperglycemic effects of epinephrine on the liver are important only in conjunction with the effects of epinephrine on glucagon and insulin secretion together with its glycogenolytic effect on muscle in acute emergency situations. Epinephrine in physiologic concentrations does not have a direct glycogenolytic effect in the liver.
  (1) Much higher amounts of epinephrine than glucagon are required to cause hyperglycemia. However, epinephrine has a more pronounced hyperglycemic effect than glucagon for the following important reasons.
    (a) Epinephrine inhibits insulin secretion, while glucagon stimulates insulin secretion; therefore, the hyperglycemic effect of glucagon is attenuated by insulin.
    (b) Epinephrine stimulates glycogenolysis in muscle, thereby providing lactate for hepatic gluconeogenesis.
    (c) Epinephrine stimulates glucagon secretion, which amplifies its hyperglycemic effect.
    (d) Epinephrine stimulates ACTH secretion, which then stimulates cortisol secretion. Cortisol also is a potent gluconeogenic hormone via the hepatic conversion of alanine to glucose.
    (e) Circulating catecholamines inhibit muscle glucose uptake, which is in contrast to the effect of glucagon.
  (2) Epinephrine indirectly inhibits insulin-mediated facilitated diffusion of glucose by muscle and adipose tissue via its blockade of insulin secretion.
  (3) Catecholamines also directly inhibit glucose uptake by the suppression of glucose transporter proteins in the cell membranes of skeletal and cardiac muscle cells and adipocytes.

2. **Fat metabolism.** In humans, the major site of lipogenesis from glucose is the liver.
  a. A man of average size has fat stores that contain about 15 kg of triglyceride, some of which can be metabolized to free fatty acids.
  b. Epinephrine stimulates lipolysis by activating triglyceride lipase, which is called the intracellular **hormone-sensitive lipase**. The activation of this enzyme is via the β-adrenergic receptor (i.e., cAMP).
  c. Mobilization of free fatty acids from stores in adipose tissue supplies a substrate for ketogenesis in the liver. Acetoacetate and β-hydroxybutyrate are transported from the liver to the peripheral tissues, where they are quantitatively important as energy sources.

    (1) Cardiac muscle and the renal cortex use fatty acids and acetoacetate in preference to glucose, whereas resting skeletal muscle uses fatty acids as the major source of energy.

    (2) During extreme conditions, such as starvation and diabetes, the brain adapts to the use of ketoacids. Ketoacids also are oxidized by skeletal muscle during starvation.

3. Gluconeogenesis refers to the formation of glucose from noncarbohydrate sources. Gluconeogenesis occurs in the liver and the kidney.

    a. Gluconeogenic substances include pyruvate, lactate, glycerol, odd-chain fatty acids, and amino acids. However, the major source of endogenous glucose production is protein, with a small fraction available from the glycerol contained in fat. All of the constituent amino acids in protein tissue, with the exception of leucine, can be converted to glucose.

    b. The conversion of even-chain fatty acids is not possible in the mammalian liver because of the absence of the enzymes necessary for the de novo synthesis of the four-carbon dicarboxylic acids from acetyl coenzyme A (acetyl-CoA).

I. Endocrinopathies

1. Hyposecretion of catecholamines, such as occurs during tuberculosis and malignant destruction of the adrenal glands or following adrenalectomy, probably produces no symptoms or other clinical features.

    a. Catecholamine production from the sympathetic nerve endings appears to satisfy the normal biologic requirements, because the adrenal medulla is not necessary for life.

    b. The functional integrity of the adrenal medulla can be determined experimentally by the administration of 2-deoxy-D-glucose. This nonmetabolizable carbohydrate induces intracellular glycopenia and extracellular hyperglycemia.

    c. Spontaneous deficiency of epinephrine is unknown as a disease state, and adrenalectomized patients do not require epinephrine replacement therapy.

2. Hypersecretion of catecholamines from chromaffin cell tumors (pheochromocytomas) produces demonstrable clinical features.

    a. Pheochromocytoma patients have sustained or paroxysmal hypertension. Unlike patients with essential hypertension, pheochromocytoma patients exhibit orthostatic hypotension.

    b. The hypersecretion of catecholamines is associated with severe headache, sweating (cold or **adrenergic sweating**), palpitations, chest pain, extreme anxiety with a sense of impending death, pallor of the skin caused by vasoconstriction, and blurred vision.

    c. Most pheochromocytomas contain predominantly norepinephrine, and most affected patients secrete predominantly norepinephrine into the bloodstream. Evidence of epinephrine hypersecretion increases the likelihood that the tumor origin is in the adrenal medulla. However, an extra-adrenal site should not be excluded.

        (1) If epinephrine is secreted primarily, the heart rate is increased.

        (2) If norepinephrine is the predominant hormone, the pulse rate decreases reflexly in response to marked hypertension.

    d. Urinary excretion of catecholamines, metanephrines, and VMA is increased.

3. Clinical tests

    a. The **adrenolytic test** involves the administration of an α-blocker (phenoxybenzamine) to observe the effect of systemic blood pressure. A dramatic fall in blood pressure is pathognomonic of pheochromocytoma because this procedure is without significant effect in normotensive subjects.

    b. The **provocative test** involves the administration of histamine, glucagon, or tyramine, which will cause a rise in blood pressure.

4. Treatment requires surgical removal of the tumor.

II. **ADRENAL CORTEX.** This section describes the physiologic and biochemical aspects of **glucocorticoids,** the most important of which is **cortisol** (also known as **hydrocortisone**).

A. **Embryology.** Morphologically and physiologically, the fetal adrenal gland differs strikingly from that of the adult. However, through all stages of life, the function of the adrenal cortex is dependent on ACTH.

1. The adrenal cortex is a mesodermal derivative. It is axiomatic that all endocrine glands derived from mesoderm synthesize and secrete steroid hormones.

2. The outer **neocortex,** which is the progenitor of the adult cortex, comprises about 15% of the total volume of this organ, and the inner **fetal zone** (inner zone or **fetal cortex**) constitutes about 85%.

3. The adrenal gland is larger at birth than it is during adulthood. This is because the fetal zone undergoes rapid involution during the first few months of extrauterine life and completely disappears by 3–12 months postpartum.

4. Near term, the fetal cortices of the fetal adrenal glands secrete 100–200 mg of steroids daily in the form of sulfoconjugates, the principal one of which is the biologically weak androgen, **dehydroepiandrosterone (DHEA) sulfate.** This 17-ketosteroid is an androgen containing 19 carbon atoms.

B. **Morphology**

1. **Gross anatomy**
   a. The adrenal glands are paired structures situated above the kidneys.
   b. Normally, each gland weighs about 5 g, of which the cortex constitutes approximately 80%.

2. **Histology and function.** The adrenal cortex consists of three distinct layers or **zones** of cells.
   a. The outermost layer, the **zona glomerulosa,** is the site of **aldosterone** and **corticosterone** synthesis. Aldosterone is the principal mineralocorticoid of the human adrenal cortex.
   b. The wider, middle zone is the **zona fasciculata,** and the innermost layer is the **zona reticularis**. The two inner zones of the adrenal cortex should be considered a functional unit, where mainly **cortisol** (and some corticosterone) and **DHEA** are synthesized.

C. **Adrenocortical hormones: corticosteroids** (Tables 49-2, 49-3)

1. **Secretion**
   a. The human adrenal cortex secretes two glucocorticoids (cortisol and corticosterone*), one mineralocorticoid (aldosterone), biosynthetic precursors of three end products (progesterone, 11-deoxycorticosterone, and 11-deoxycortisol), and androgenic substances (DHEA and its sulfate ester).
   b. The normal human adrenal cortex does not secrete physiologically effective amounts of testosterone or estrogenic substances.

2. **Transport**
   a. Under physiologic circumstances, about 75% of the plasma cortisol is bound to **cortisol-binding globulin** (CBG; transcortin), which is an α-globulin. (In addition to binding cortisol, transcortin has a high binding affinity for progesterone, deoxycorticosterone, corticosterone, and some synthetic analogs.) About 15% of the plasma cortisol is bound to plasma **albumin,** and about 10% is **unbound** and represents the physiologically active steroid.

---

*At physiologic concentrations, corticosterone has glucocorticoid activity.

TABLE 49-2. Average 8:00 A.M. Plasma Concentration and Secretion Rate of Corticosteroids in Adults

Corticosteroid	Plasma Concentration (μg/dl)	Secretion Rate (mg/day)
Cortisol	13	15
Corticosterone	1	3
11-Deoxycortisol	0.16	0.40
Deoxycorticosterone	0.07	0.20
Aldosterone	0.009	0.15
18-Hydroxycorticosterone	0.009	0.10
Dehydroepiandrosterone (DHEA)	0.5	15
DHEA sulfate	115	15

Adapted with permission from Genuth SM: The adrenal glands. In *Physiology, 3rd ed.* Edited by Berne RM, Levy MN. St. Louis, CV Mosby, 1993, p 958.

    b. The 90% of the plasma cortisol bound to plasma protein represents the metabolically inactive pool, which serves as a reservoir for free hormone.

3. Corticosteroidogenesis (Figure 49-2)
    a. Uptake of cholesterol. Free cholesterol is the preferred precursor of the corticosteroids, although the adrenal cortex can form cholesterol from acetyl-CoA. Most stored adrenal cholesterol is esterified with fatty acids; it is the cholesterol ester content that is reduced following ACTH stimulation of the fasciculata–reticularis complex.
    b. Side-chain cleavage of cholesterol. The rate-limiting step in corticosteroidogenesis is the mitochondrial conversion of cholesterol to pregnenolone by 20,22-desmolase.
    c. Pregnenolone: the common precursor of all steroid hormones. Conversion of pregnenolone to progesterone requires two enzymes—3β-hydroxysteroid dehydrogenase and Δ^5-isomerase—which are found in the endoplasmic reticulum (microsomes).
    d. Hydroxylation reactions follow sequentially after the formation of pregnenolone and progesterone and require $NADPH_2$ and molecular O_2.
        (1) The hydroxylation reactions in the biosynthesis of aldosterone occur sequentially at the C-21, C-11, and C-18 positions.
        (2) The hydroxylation reactions in the biosynthesis of cortisol occur sequentially at the C-17, C-21, and C-11 positions.
    e. Sex steroid pathways. Both pregnenolone and progesterone are substrates for the microsomal enzyme, 17α-hydroxylase, which is found not only in the adrenal cortex but also in the testis and ovary.

TABLE 49-3. Blood Production Rates of Adrenal Androgens

Androgenic Steroid	Plasma Concentration (μg/dl)	Blood Production Rate (mg/day)
Testosterone		
Men	0.8	7.0
Women*	0.034	0.34
Androstenedione		
Men	0.06	1.4
Women	0.14	3.4

Reprinted from Mulrow PJ: The adrenals. In *Physiology and Biophysics.* Edited by Ruch TC, Patton HD. Philadelphia, WB Saunders, 1973, p 229.
*In women, about 50% of the blood testosterone is derived from androstenedione.

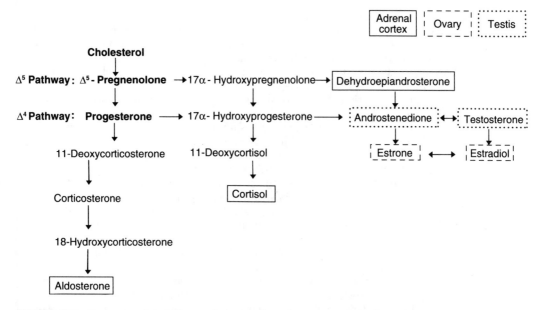

FIGURE 49-2. Summary of steroidogenesis in the adrenal cortex, testis, and ovary. Notice that there is no interconversion between mineralocorticoids and glucocorticoids.

   (1) Pregnenolone is converted to 17α-hydroxypregnenolone, which can either continue along the Δ^5-pathway to the synthesis of androgens or enter the glucocorticoid pathway.
   (2) Progesterone is converted to 17α-hydroxyprogesterone, which can either proceed along the Δ^4-pathway to the synthesis of androgens or enter the mineralocorticoid pathway.
   (3) The zona glomerulosa lacks 17α-hydroxylase and does not have the capacity to synthesize 17α7-hydroxyprogesterone. For this reason, the zona glomerulosa cannot synthesize cortisol.
      (a) The conversion of pregnenolone to DHEA is in the Δ^5-pathway.
      (b) The conversion of progesterone to androstenedione is in the Δ^4-pathway.
   f. Glucocorticoid/mineralocorticoid pathways. Both progesterone and 17α-hydroxyprogesterone are substrates for the microsomal enzyme, 21-hydroxylase.
      (1) Progesterone is converted to 11-deoxycorticosterone, which is in the mineralocorticoid pathway.
      (2) 17α-Hydroxyprogesterone is converted to 11-deoxycortisol, which is in the glucocorticoid pathway.
      (3) 11-Deoxycorticosterone and 11-deoxycortisol are not precursors for the synthesis of each other.
   g. Diverging pathways. Both 11-deoxycorticosterone and 11-deoxycortisol are acted upon by mitochondrial 11β-hydroxylase.
      (1) 11-Deoxycorticosterone forms corticosterone in the mineralocorticoid pathway.
      (2) 11-Deoxycortisol is converted to cortisol, the principal glucocorticoid of the human adrenal cortex.
      (3) Corticosterone differs from cortisol only in that the former lacks a 17α-hydroxyl group. Although this structural difference is small, it accounts for a very large difference in the biologic activity of these two hormones.
   h. Aldosterone synthesis. Some of the corticosterone serves as the substrate for another mitochondrial enzyme, 18-hydroxylase.
      (1) This reaction leads to the formation of 18-hydroxycorticosterone.
      (2) 18-Hydroxycorticosterone is converted to aldosterone by the mitochondrial enzyme, 18-hydroxysteroid dehydrogenase, which is found only in the zona glomerulosa.

i. Synthesis and secretion of sex steroids

(1) Androgens. The adrenal cortex secretes four androgenic hormones: androstene-dione, testosterone, DHEA, and DHEA sulfate. Quantitatively, the most important sex steroids produced by the human adrenal cortex are DHEA and DHEA sulfate. Except for testosterone, adrenocortical androgens are relatively weak and serve as precursors for hepatic conversion to testosterone.

(a) DHEA is derived from 17α-hydroxypregnenolone by the action of 17,20-desmolase.

(b) DHEA is mainly conjugated with sulfuric acid and, as such, is bound to plasma protein. While in this bound form, DHEA is not readily excreted but circulates in higher concentrations than any other steroid.

(c) DHEA is the principal precursor of urinary 17-ketosteroids; however, the most abundant urinary 17-ketosteroids are androsterone and etiocholanolone.

(d) DHEA and DHEA sulfate have androgenic activity by virtue of their peripheral conversion to testosterone. DHEA sulfate is active as a minor precursor of other 19-carbon steroids formed in the gonads and placenta, which are sites of sulfatase activity.

(i) Normal excretion rates for 17-ketosteroids are 5–14 mg/day in women and 8–20 mg/day in men.

(ii) Normally, adrenocortical precursors represent the bulk of the urinary 17-ketosteroid pool, with a smaller contribution from the gonads.

(iii) The main androgen secreted by premenopausal women is androstene-dione, about 60% of which is of adrenocortical origin.

(2) Estrogens. The human adrenal cortex can synthesize minute amounts of estrogen. The adrenal cortex makes its major contribution to the body's estrogen (estrone) pool indirectly by supplying androstenedione together with DHEA and its sulfate as substrates for conversion to estrogens by subcutaneous fat, hair follicles, mammary adipose tissue, and other tissues.

4. Metabolism of corticosteroids

a. General considerations

(1) Corticosteroid inactivation occurs by:

(a) Enzymatic reduction of the Δ^{4-5} double bond in the A-ring to form **dihydro-cortisol**

(b) Enzymatic reduction of the ketonic oxygen substituent at the C-3 position to form **tetrahydrocortisol**

(c) Conjugation with glucuronic acid to form a water-soluble metabolite that is readily excreted by the kidney

(2) The major urinary metabolite of cortisol is **tetrahydrocortisol glucuronide**.

b. Hepatic conversion. The liver is the major extra-adrenal site of corticosteroid metabolism.

(1) Cortisol can be enzymatically converted to cortisone by 11β-hydroxysteroid dehydrogenase and excreted as **tetrahydrocortisone glucuronide**.

(2) The ketonic oxygen substituent on the C-20 position of cortisol and other steroids can be enzymatically converted to a hydroxyl group. These steroids can be subjected to A-ring reduction, conjugated, and excreted as glucuronides.

(a) Cortisol is converted to **cortol glucuronide**.

(b) Cortisone is converted to **cortolone glucuronide**.

(c) Progesterone forms **pregnanediol glucuronide**.

(d) 17α-Hydroxyprogesterone forms **pregnanetriol glucuronide**.

(3) Steroids that contain a 17α-hydroxyl group are ketogenic. Cortisol can be enzymatically converted to a 17-ketosteroid by 17,20-desmolase. About 5% of cortisol appears in the urine as a 17-ketosteroid.

(4) Aldosterone also can undergo A-ring reduction and conjugation to form **tetrahydroaldosterone glucuronide**.

c. Conversion in other extra-adrenal tissues. Muscle, skin, fibroblasts, intestine, and lymphocytes also can carry out oxidation-reduction reactions at the C-3, C-11, C-17, and C-20 positions of the corticosteroid molecule.

D. **Control of adrenocortical function** (see Figure 46-2). Physiologic control of the rate of cortisol secretion occurs via a double negative feedback loop, a mechanism that is characteristic of most neuroendocrine control systems.

1. **Regulation of ACTH secretion**
   a. The parvicellular peptidergic neurons of the hypothalamus release **CRH,** which stimulates the secretion of ACTH from the anterior lobe of the pituitary gland via the hypophysial portal system. CRH is a polypeptide consisting of 41 amino acid residues.
   b. CRH secretion from the tuberoinfundibular neurons is regulated by neurotransmitters secreted by the monoaminergic neurons that innervate the peptidergic neurons.
      (1) Hypothalamic secretion of CRH is stimulated by cholinergic neurons. Serotonin also is a stimulatory signal to CRH neurons.
      (2) Adrenergic neuron activity inhibits release of CRH. **Gamma-aminobutyric acid (GABA)** also is a known inhibitor of CRH secretion.

2. **Negative feedback action of corticosteroids**
   a. Of the endogenous corticosteroids, only cortisol has ACTH-suppressing activity. The synthetic glucocorticoid, **dexamethasone,** is a potent inhibitor of ACTH secretion and, therefore, of endogenous glucocorticoid secretion.
   b. The negative feedback of cortisol is exerted at the level of the pituitary gland and the ventral diencephalon.
      (1) If free cortisol levels are supraphysiologic, ACTH secretion is suppressed and the adrenal cortex ceases its excretory activity and undergoes **disuse atrophy**.
      (2) Conversely, if plasma free cortisol levels are subnormal, the anterior lobe is released from inhibition by cortisol, ACTH secretion rises, and the adrenal cortex secretes more cortisol and becomes hypertrophic.

3. **Hypophysial-adrenocortical rhythm.** Normally, blood ACTH levels are higher in the morning than in the evening. This accounts for the diurnal rhythms in cortisol secretion, plasma cortisol concentration, and 17-hydroxycorticosteroid excretion (see Chapter 46 I A).

4. **Hypophysial-adrenocortical response to stress.** The normal hypothalamic-hypophysial-adrenocortical control system can be overridden by a variety of challenges, which collectively are referred to as **stress**.
   a. Among the stresses shown to induce increased activity in this system are severe trauma, pyrogens, hypoglycemia, histamine injection, electroconvulsive shock, acute anxiety, burns, hemorrhage, exercise, infections, chemical intoxication, pain, surgery, psychological stress, and cold exposure.
   b. In stress conditions, ACTH secretion is stimulated despite the fact that systemic levels of cortisol are much higher than those required to inhibit ACTH secretion completely in unstressed conditions.

E. **Physiologic effects of glucocorticoids.** Of the naturally occurring steroids, only cortisol, cortisone, corticosterone, and 11-dehydrocorticosterone have appreciable glucocorticoid activity. Full recovery from hypothalamic-hypophysial-adrenocortical suppression may require as long as 1 year following cessation of all steroid therapy.

1. **Anti-inflammatory effects.** Glucocorticoids inhibit inflammatory and allergic reactions in several ways.
   a. They stabilize lysosomal membranes, thereby inhibiting the release of proteolytic enzymes.
   b. They decrease capillary permeability, thereby inhibiting diapedesis of leukocytes.
   c. Glucocorticoids reduce the number of circulating lymphocytes, monocytes, eosinophils, and basophils.
      (1) The decreased number of these formed elements in blood is caused primarily by a redistribution of the cells from the vascular compartment into the lymphoid tissue (e.g., spleen, lymph nodes, bone marrow). Cellular lysis is not a major mechanism for decreasing the number of these cells in the human circulation.

(2) The decrease in circulating basophils accounts for the fall in blood histamine levels and the abrogation of the allergic response.

    (a) The migration of inflammatory cells from capillaries is decreased.

    (b) Glucocorticoids lessen the formation of edema and, thereby, reduce the swelling of inflammatory tissue.

(3) Glucocorticoids cause an increase in the number of circulating neutrophils owing to the accelerated release from bone marrow and a reduced migration from the circulation. Steroids also inhibit the ability of neutrophils to marginate to the vessel wall.

d. Glucocorticoids cause involution of the lymph nodes, thymus, and spleen, which leads to decreased antibody production.

    (1) This lymphocytopenic effect aids in the prevention or reduction of the immune response by an organ transplant recipient.

    (2) Because recipients pretreated with large doses of glucocorticoids are susceptible to intercurrent infections, antibiotics are a necessary adjunct to the steroid therapy.

    (3) Antibody production is not suppressed in humans at conventional steroid doses; however, chronic administration of high doses of glucocorticoids leads to an impairment of host defense mechanisms.

e. Glucocorticoids lead to an increase in the total blood count because of the increased numbers of neutrophils, erythrocytes, and platelets. The increase in circulating erythrocytes (polycythemia) is from the stimulation of hematopoiesis.

2. Renal effects

a. Glucocorticoids restore glomerular filtration rate (GFR) and renal plasma flow to normal following adrenalectomy. Mineralocorticoids do not have these effects.

b. Glucocorticoids facilitate free-water excretion (clearance) and uric acid excretion.

3. Gastric effects. Cortisol increases gastric flow and gastric acid secretion, while it decreases gastric mucosal cell proliferation. The latter two effects lead to peptic ulceration following chronic cortisol treatment.

4. Psychoneural effects of glucocorticoids have been noted following chronic high-dose therapy with these steroids. Patients may become initially euphoric and then psychotic, paranoid, and depressed.

5. Antigrowth effects

a. Large doses of cortisol have been shown to antagonize the effect of active vitamin D metabolites on the absorption of Ca^{2+} from the gut, to inhibit mitosis of fibroblasts, and to cause degradation of collagen. All of these effects lead to osteoporosis, which is a reduction in bone mass per unit volume with a normal ratio of mineral-to-organic matrix.

b. Glucocorticoids delay wound healing because of the reduction of fibroblast proliferation. Connective tissue is reduced in quantity and strength.

c. Chronic supraphysiologic doses of glucocorticoids suppress growth hormone (GH) secretion and inhibit somatic growth.

d. Although glucocorticoids increase the ability of muscle to perform work, large doses lead to muscle atrophy and muscular weakness.

6. Vascular effect. Cortisol in pharmacologic doses enhances the vasopressor effect of norepinephrine. In the absence of cortisol, the vasopressor action of catecholamines is diminished, and hypotension ensues. Thus, corticosteroids have a role in the maintenance of normal arterial systemic blood pressure and volume through their support of vascular responsiveness to vasoactive substances. (Cortisol enhances catecholamine synthesis via its activation of the epinephrine-forming enzyme, PNMT.)

7. Stress adaptation. Glucocorticoids allow mammals to adapt to various stresses in order to maintain homeostasis.

a. Resistance to stress is not increased by the administration of glucocorticoids.

b. Stress is associated with the activation of the hypothalamic-hypophysial-adrenal axis.

F. **Metabolic effects of glucocorticoids**

1. **Carbohydrate metabolism.** Cortisol is a carbohydrate-sparing hormone and, therefore, exerts an anti-insulin effect, which leads to hyperglycemia and insulin-resistance.

   a. Glucocorticoids maintain blood glucose and the glycogen content of the liver by promoting the conversion of amino acids to carbohydrates and the storage of carbohydrate as hepatic glycogen.

   b. Cortisol is hyperglycemic principally because of its gluconeogenic activity, which is related to its protein catabolic effect on extrahepatic tissues, especially muscle.

      (1) The proteolytic effect of glucocorticoids results in the mobilization of amino acids from muscle and in an increase in plasma amino acid concentration.

      (2) Alanine is quantitatively the major gluconeogenic amino acid precursor in the liver. Like acetyl-CoA, alanine inhibits pyruvate kinase activity.

   c. Cortisol also exerts an anti-insulin effect by blocking glucose transport in muscle and adipose tissue. This effect accounts for the phenomenon of glucose intolerance or an eventual "steroid diabetes."

   d. Glucocorticoids augment the activity of key gluconeogenic enzymes by the induction of hepatic enzyme synthesis.

      (1) The gluconeogenic pathway has three steps that differ from those in the glycolytic pathways as a result of their thermodynamic irreversibility. These enzymes and their substrates are as follows.

         (a) Glucose-6-phosphate → (glucose-6-phosphatase)

         (b) Fructose 1,6-bisphosphate → fructose-6-phosphate (fructose 1,6-bisphosphatase)

         (c) Conversion of pyruvate to phosphoenolpyruvate requires two steps:

            (i) Pyruvate → oxaloacetate (pyruvate carboxylase)

            (ii) Oxaloacetate → phosphoenolpyruvate (phosphoenolpyruvate carboxykinase)

      (2) Glucocorticoids are associated with activation of glycogen synthetase by glucose-6-phosphate and of pyruvate carboxylase by acetyl-CoA. They also are associated with indirect inhibition of glycolysis by free fatty acids, resulting in increased glucogenesis and glycogenesis.

   e. Cortisol also indirectly inhibits the activities of glycolytic enzymes, which accounts for its anti-insulin effect. Enzymes that are blocked by the effect of glucocorticoids include:

      (1) Glucokinase

      (2) Phosphofructokinase

      (3) Pyruvate kinase

   f. Glucocorticoids mobilize fatty acids from adipose tissue to the liver, where metabolism of free fatty acids may lead to products that inhibit glycolytic enzymes and favor gluconeogenesis.

      (1) The glycerol released from the fat cell with the fatty acids also serves as a secondary substrate for gluconeogenesis.

      (2) The cortisol-inhibited glycolysis in peripheral tissue probably is indirectly blocked via the inhibition of the key glycolytic enzyme, phosphofructokinase, by the elevated concentration of plasma free fatty acids.

   g. Glucocorticoids are associated with compensatory hyperinsulinemia following hyperglycemia.

2. **Protein metabolism.** The most important gluconeogenic substrates are amino acids derived from proteolysis in skeletal muscle.

   a. Cortisol enhances the release of amino acids from proteins in skeletal muscle and other extrahepatic tissues, including the protein matrix of bone.

      (1) The amino acids released, especially the glucogenic amino acid **alanine,** are transported to the liver and converted to glucose.

      (2) Increased glucose production by cortisol via gluconeogenesis is associated with increased urea production via the conversion of amino nitrogen to urea. This effect accounts for the increased urinary nitrogen excretion.

      (3) The proteolysis in skeletal muscle brings about a negative nitrogen balance.

b. The amino acids taken up by the liver are used not only to form glucose or glycogen but also to build new protein. This protein anabolic effect at the level of the liver is an important exception to the overall protein catabolic effect of cortisol.

c. Equally important is the ability of glucocorticoids to inhibit the de novo synthesis of protein, probably at the translational level. This is called an **antianabolic effect** of cortisol.

3. Fat metabolism. Glucocorticoids are lipolytic hormones. Their lipolytic effect is in part from the potentiation of the lipolytic actions of other hormones, such as GH, catecholamines, glucagon, and thyroid hormone.

a. Glucocorticoids favor the mobilization of fatty acids from adipose tissue to the liver, where the metabolism of fatty acids inhibits glycolytic enzymes and promotes gluconeogenesis. As a result of increased fatty acid oxidation, glucocorticoids may lead to ketosis, especially in the context of diabetes mellitus.

(1) The major site of stimulation is the gluconeogenic enzyme, fructose 1,6-bisphosphatase, which is activated by fatty acids.

(2) At the same time, pyruvate kinase and phosphofructokinase are inhibited by fatty acids. Thus, glycolysis is inhibited while gluconeogenesis proceeds.

b. Glucocorticoids also indirectly stimulate lipolysis by blocking peripheral glucose uptake and utilization. They inhibit re-esterification of fatty acids within adipocytes by inhibiting the use of glucose.

c. Fatty acid synthesis is inhibited in the liver by cortisol, an effect not observed in adipose tissue. The overall effect of glucocorticoids on fat is to induce a redistribution of fat together with an increase in total body fat (i.e., truncal obesity). The increase in body weight is not a result of a growth effect that is the accretion of protein. (Recall that cortisol is an antigrowth hormone.)

(1) There is characteristic centripetal distribution of fat (i.e., an accumulation of fat in the central axis of the body).

(a) The deposition of fat in the facies is called "moonface."

(b) The deposition of fat in the suprascapular region is referred to as "buffalo hump."

(c) Excessive fat distribution leads to a pendulous abdomen.

(2) Glucocorticoid-induced obesity reflects increased food intake rather than a change in the rate of lipid metabolism.

d. Chronic excessive amounts of cortisol lead to hyperlipidemia and hypercholesterolemia. Glucocorticoids increase appetite and, thereby, play a role in obesity.

Chapter 50

Reproduction

I. **TESTIS**

A. **Embryology**

1. **Internal genitalia** (Figure 50-1). The internal reproductive tract in males and females is derived from one of two pairs of genital ducts. The **wolffian (mesonephric) ducts** give rise to the male internal genitalia, and the **müllerian ducts** give rise to the female internal genitalia.

   a. In the male, the internal genitalia (also called **accessory sex organs**) consist of the following ducts and glands for the conveyance of sperm:
   - **(1)** Seminiferous tubules
   - **(2)** Rete testis*
   - **(3)** Ductuli efferentes
   - **(4)** Epididymis
   - **(5)** Vas deferens
   - **(6)** Ejaculatory duct
   - **(7)** Seminal vesicles
   - **(8)** Prostate gland
   - **(9)** Bulbourethral glands (Cowper's glands)

   b. The wolffian ducts are excretory ducts of the mesonephric kidney and are attached to the primordial gonad; the müllerian duct is formed parallel to the wolffian duct and is not contiguous with the primitive gonad.

   c. At 9–10 weeks gestation, following the appearance of the **Leydig (interstitial) cells** in the fetal testis, the two wolffian ducts begin to differentiate and give rise to the following male internal genitalia:
   - **(1)** Epididymis
   - **(2)** Vas deferens
   - **(3)** Seminal vesicles
   - **(4)** Ejaculatory duct

   d. It is the primary function of **androgen** to induce the formation of the male accessory sex organs during fetal life. Specifically, testicular androgen is required for the differentiation of the male genital (wolffian) duct system.

   e. Genetic sex determines gonadal sex, and gonadal sex determines phenotypic sex. The control over the formation of the male phenotype requires the action of the following three hormones.
   - **(1)** **Müllerian duct inhibiting factor** is produced by the Sertoli (sustentacular) cells of the fetal testis and induces the regression of the müllerian duct system. Müllerian duct inhibiting factor functions as a paracrine secretion that diffuses to the paired müllerian ducts.
   - **(2)** **Testosterone** is secreted by the fetal testis and promotes growth and differentiation of the wolffian duct system and the accessory sex organs.
   - **(3)** Testosterone also is the precursor for **dihydrotestosterone,** which is necessary for the growth and differentiation of the male external genitalia and prostate gland (see I A 2 b).

2. **External genitalia.** The external genitalia also begin to differentiate between 9 and 10 weeks gestation.

   a. In contrast to the internal genitalia, the external genitalia and urethra in both sexes develop from common anlagen, which are the urogenital sinus, the genital sinus,

---

*The duct system distal to the rete testis is known as the excurrent (excretory) duct system.

INDIFFERENT STAGE

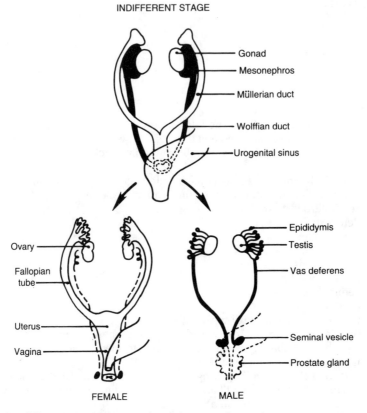

- Gonad
- Mesonephros
- Müllerian duct
- Wolffian duct
- Urogenital sinus

Ovary
Fallopian tube
Uterus
Vagina

FEMALE

Epididymis
Testis
Vas deferens
Seminal vesicle
Prostate gland

MALE

FIGURE 50-1. Sex differentiation of the gonad and the internal genitalia. Up to 6 weeks gestation, the gonad is a bipotential structure (*indifferent stage*), and the urogenital tract in both sexes consists of two pairs of genital ducts (i.e., the wolffian and müllerian ducts) and a mesonephros, all of which terminate in the urogenital sinus. In the female, the gonad develops into an ovary and the müllerian ducts become organized into the fallopian tubes, uterus, and upper vagina, while the wolffian ducts remain vestigial. In the male, a testis develops and the wolffian ducts differentiate into the epididymis, vas deferens, and seminal vesicle, while the müllerian ducts regress. (Reprinted from Wilson JD: Embryology of the genital tract. In *Campbell's Urology*, 4th edition. Edited by Harrison JH, et al. Philadelphia, WB Saunders, 1977, p 1473.)

the genital tubercle, the genital swelling, and the genital (urethral) folds. The anlagen of the male external genitalia give rise to the following structures.

(1) The urogenital sinus forms the male urethra and prostate gland.

(2) The genital tubercle grows into the glans penis.

(3) The genital swelling forms the scrotum.

(4) The urethral folds enlarge to form the outer two-thirds of the penile urethra and corpora spongiosa.

b. The growth and development of the male external genitalia require dihydrotestosterone, which is formed from the conversion of the fetal testicular testosterone within the urogenital sinus and lower urogenital tract.

3. Testis (see Figure 50-1). The primitive gonads develop midabdominally in association with the mesonephric ridges.

a. Through 5 weeks gestation, the gonads are undifferentiated and consist of a medulla, a cortex, and primordial germ cells. These germ cells are embedded in a layer of cortical epithelium surrounding a core of medullary mesenchymal tissue.

b. At 6 weeks, the seminiferous tubules begin to form from the medulla. The cortical region of the primitive gonad undergoes regression. (The female gonad develops from the cortical region.) The primordial germ cells (gonocytes) arise outside the gonads and migrate from the endodermal yolk sac epithelium to the urogenital ridge. During migration, the gonocytes undergo continuous mitosis.

c. At 7 weeks gestation, the **Sertoli cells** begin to form and secrete the **H-Y (histocompatibility-Y) antigen,** which is under control of the Y chromosome. The Sertoli cells are derivatives of the mesenchymal tissue of the urogenital ridge.

d. At 8–9 weeks of fetal life, the Leydig cells are formed and secrete testosterone in response to **chorionic gonadotropin,** which is secreted by the placenta. The Leydig cells also are mesenchymal derivatives, which appear in the connective tissue surrounding the seminiferous cords.

e. At 9 weeks gestation, definitive testes are present.

f. At 7–9 months gestation, the testes normally descend through the inguinal rings into the scrotum.

B. **Morphology**

1. Gross anatomy

a. The adult testis is an ovoid gland, which is approximately 4–5 cm long and 2.5–3.0 cm wide and ranges in weight from 10–45 g.

b. The testes normally are situated in the scrotum, where they are maintained at a temperature that is about 2° C lower than normal body temperature. The lower temperature is necessary for normal spermatogenesis.

c. Each gland is surrounded by a fibrous connective tissue membrane called the **tunica albuginea,** which, in turn, is surrounded by a serous membrane called the **tunica vaginalis**.

d. By puberty, the testes usually have developed sufficiently to perform the functions of spermatogenesis and steroidogenesis. Generally, puberty begins between the ages of 12 and 14 years. In the United States, 95% of normal boys show signs of puberty by the age of 16 years.

2. Functional histology of testicular parenchyma. The testicular parenchyma consists mainly of the seminiferous tubules and Leydig cells, which comprise about 90% and 10% of the testicular volume, respectively. These two compartments are separated by boundary tissue, which includes a basement membrane.

a. Tubular tissue. The seminiferous tubules have a total collective length of approximately 800 m (a half mile)!

(1) The spermatogenic tubules are organized into coiled loops, each of which begins and terminates in a single duct called the **tubulus rectus**.

(2) The tubuli recti join to form the **rete testis** and eventually drain via the **ductuli efferentes** into the **epididymis**. The epididymis is the primary storage and final maturation site for spermatozoa.

(3) From the epididymis, the spermatozoa are transmitted via the **vas deferens** and **ejaculatory duct** into the penile urethra.

b. Tubular cells. The basic cellular components of the tubules are the germinal cells and the nongerminal (Sertoli) cells. The spermatogonia and the Sertoli cells are the only tubular cells that lie on the tubular membrane called the **basal lamina**. The other germinal cells (spermatocytes) are found between the Sertoli cells.

(1) **Spermatogonia** are nonmotile stem cells that divide by mitosis to form two cellular pools: a pool of additional stem cells, which undergo continual renewal by mitosis, and a pool of **type A spermatogonia,** which enter the maturation process called **spermatogenesis**. Each spermatogonial cell (type A) gives rise to 64 sperm cells.

(a) Spermatogenesis requires about 64 days in man, and the transport of sperm cells through the epididymis to the ejaculatory duct requires an additional 12–21 days. There are three phases of spermatogenesis.

(i) The type A spermatogonia divide by mitosis to form **primary spermatocytes,** which are **diploid** cells.

(ii) The euploid primary spermatocytes continue the gametogenic process, which leads to the formation of haploid **secondary spermatocytes** and then to the formation of **spermatids haploid,** which also are haploid cells.

(iii) The third phase of spermatogenesis produces mature **spermatozoa** via the metamorphosis of spermatids. This process, called **spermiogenesis,** is characterized by the absence of cell division.

(b) Because testosterone and follicle-stimulating hormone (FSH) act directly on the seminiferous epithelium, these hormones are required for spermatogenesis.

(2) Sertoli cells are nonmotile and, in the mature testis, nonproliferating tubular cells that lie on the basal lamina.

(a) Structural features. The Sertoli cells extend through the entire thickness of the germinal epithelium from the basement membrane to the lumen. The tight junctions between the bases of the Sertoli cells serve two functions.

(i) They divide the seminiferous tubular epithelium into two functional pools: a basal compartment containing the spermatogonia and an adluminal compartment containing the spermatogonia and spermatids.

(ii) They form an effective permeability barrier within the seminiferous epithelium, which is defined in man as the blood–testis barrier that limits the transport of many substances from the blood to the seminiferous tubular lumen. This barrier maintains germ cells in an immunologically privileged location, because mature sperm cells are very immunogenic when introduced into the systemic circulation.

(b) Functions

(i) Sertoli cells provide mechanical support for the maturing gametes.

(ii) They also have a role in spermatogenesis, which may be attributable to their high glycogen concentration being a potential energy source.

(iii) In the fetal testis, the Sertoli cells secrete H-Y antigen, which is a product of the testis-organizing genes and is the cell surface glycoprotein responsible for the induction of testicular organogenesis. H-Y antigen has been found in all cell membranes from normal XY males except the cell membranes of immature germ cells.

(iv) The Sertoli cells secrete müllerian duct inhibiting factor, the glycoprotein that causes regression of the müllerian duct system.

(v) With the onset of fetal testicular differentiation induced by H-Y antigen and the incorporation of primitive germ cells into the seminiferous tubules, the Sertoli cells secrete a meiosis-inhibiting factor, which suppresses germ cell proliferation and differentiation beyond the spermatogonial stage.

(vi) Sertoli cells participate in the release of spermatids by enveloping the residual lobules of spermatid cytoplasm. Thus, the residual bodies are not cast off into the lumen but are retained within the epithelium throughout the spermiation process.

(vii) Sertoli cells secrete a watery, solute-rich (K^+ and HCO_3^-) fluid into the seminiferous lumen. These cells actively pump ions into the intercellular spaces to create a standing osmotic gradient that moves water from the Sertoli cell base to the free surface of the lumen. This fluid movement provides a driving force for conveying sperm from the testis to the epididymis, where most of this isosmotic fluid is reabsorbed.

(viii) Sertoli cells synthesize estradiol from androgenic precursors.

(ix) Sertoli cells also produce and secrete an **androgen-binding protein** and **inhibin** (see I D 1 a).

c. Leydig cells are located between the seminiferous tubules and produce androgenic steroids. These cells also are called **interstitial cells**.

(1) Leydig cells are extensive at birth but virtually disappear within the first 6 months of postnatal life. Their reappearance marks the onset of puberty.

(2) At puberty, the fibroblast-like cells of the testis serve as stem cells that differentiate into Leydig cells.

C. Hormones of the testis: steroids

1. Secretion and transport

a. Testosterone is the major hormone produced by the Leydig cells of the testis. Like all naturally occurring androgens, testosterone consists of 19 carbon atoms.

(1) Testosterone is not stored in the testis. Cholesterol esters, the major precursor for testosterone biosynthesis, are stored in the lipid droplets in the Leydig cells.

(2) A normal man secretes 4–9 mg of testosterone daily. More than 97% of secreted testosterone is bound to plasma proteins; 68% is bound to albumin, and 30% is bound to testosterone-binding globulin (also called sex hormone–binding globulin, because it binds estradiol as well). A very small percentage of the plasma testosterone is unbound.

b. Androstenedione also is secreted by the testis at a rate of about 2.5 mg/day and is an important steroid precursor for blood estrogens in men.

(1) Many nonendocrine tissues (e.g., brain, skin, fat, liver) have the cytochrome P-450-dependent aromatase, which converts androgens to estrogens.

(2) Major portions of blood estradiol and estrone in normal men are derived from blood testosterone and androstenedione, respectively.

(3) In addition, the Sertoli and Leydig cells of the testis secrete small amounts of estradiol.

c. Dihydrotestosterone is synthesized by the testis, probably because of the action of **5α-reductase** from the Sertoli cells on testosterone secreted by the Leydig cells.

(1) Only 20% of plasma dihydrotestosterone is synthesized in the testis. The remainder is derived from the peripheral conversion of testosterone, which serves as a prohormone in the skin and male reproductive tract (prostate gland and seminal vesicles).

(2) Dihydrotestosterone has more than twice the biologic activity of testosterone.

2. Testicular steroidogenesis (see Figure 49-2)

a. The key step in steroidogenesis is the conversion of cholesterol to pregnenolone. This reaction is catalyzed by 20,22-desmolase, the rate-limiting enzyme for steroid biosynthesis in all steroid-producing tissues.

(1) Luteinizing hormone (LH)* activates 20,22-desmolase and, therefore, is the pituitary gonadotropic hormone that regulates testosterone synthesis by the Leydig cells.

(2) In the developing male fetus, the stimulus for testosterone synthesis is chorionic gonadotropin, which is the placental hormone secreted in the highest amounts during the first trimester.

b. The Leydig cells contain 17α-hydroxylase, which hydroxylates pregnenolone at position 17.

(1) Some pregnenolone may be converted via 3-hydroxysteroid dehydrogenase/Δ^5-isomerase to progesterone, which can be hydroxylated at the C-17 position prior to its conversion to androstenedione along the Δ^4-pathway.

(2) In the human testis, the Δ^5-pathway is the preferential pathway for the synthesis of testosterone, which is a Δ^4-steroid.

c. Androstenedione is the common final precursor in the synthesis of testosterone.

3. Metabolism

a. Dihydrotestosterone formation. Testosterone can serve as a prohormone and be metabolized by 5α-reductase to the more active androgen, dihydrotestosterone. The activity of 5α-reductase is high in the skin, prostate gland, seminal vesicles, epididymis, and liver. This enzyme also is present in testicular tissue.

(1) Androgen target tissues are thought to be the principal sites of dihydrotestosterone formation. The nuclear membrane and microsomes of androgen-sensitive tissues are the physiologically important sites for conversion of testosterone to dihydrotestosterone. Therefore, dihydrotestosterone not only is secreted by the testis into the circulation but also is synthesized from testosterone that has entered the cells of androgen-dependent tissues.

(2) The formation of dihydrotestosterone is an irreversible reduction reaction; therefore, dihydrotestosterone cannot serve as a proestrogen.

---

*In the male, LH is known as **interstitial cell–stimulating hormone (ICSH)**.

    (3) The anlagen of the prostate gland and external genitalia can form dihydrotest-
osterone prior to the onset of virilization. Dihydrotestosterone is formed from
testosterone prior to the secretion of significant amounts of testosterone from the
testis. The wolffian duct derivatives can form dihydrotestosterone after the onset
of androgen secretion and after differentiation of the male genital system is far
advanced.

    (4) Dihydrotestosterone can be metabolized further to 17-ketosteroids and polar
derivatives found in the urine.

  b. Estradiol and estrone formation. Testosterone and androstenedione can be converted to estradiol and estrone, respectively, by the action of aromatase. Thus, the
estrogens in the male are derived from direct secretion by the testis and from peripheral conversion of circulating androstenedione and testosterone.

    (1) Aromatization of circulating androgens is the major pathway for estrogen forma-
tion in the male.

    (2) Aromatases are membrane-bound enzymes found in the brain, skin, liver, mam-
mary tissues, and most significantly, the adipose tissue.

  c. 17-Ketosteroid formation. Testosterone can be metabolized to less active metabolites that are conjugated in the liver and excreted into the urine as 17-ketosteroids.

    (1) Androsterone and **etiocholanolone** are the major urinary metabolites of test-
osterone. **Testosterone glucuronide** and **5α-androstanediole glucuronides** are
among the other androgenic metabolites measured in urine.

      (a) Testosterone glucuronide originates mainly in the liver from testosterone,
androstenedione, and dihydrotestosterone. The measurement of plasma testosterone is the mainstay for assessing Leydig cell function.

      (b) 5α-Androstanediol glucuronides arise from the testosterone metabolites in
both the liver and the skin. Because an increased 5α-reductase activity in
the skin and other extrahepatic tissues is produced by increased androgen
secretion, the measurement of urinary 5α-androstanediol glucuronides has
been recommended as an index of clinical androgenicity.

    (2) The **excretory rate** for urinary 17-ketosteroids in normal men is 15–20 mg/day.

      (a) Of this amount, 20%–40% are of testicular origin. The remainder are
adrenocortical secretions, the major one of which is **dehydroepiandrosterone (DHEA).**

      (b) Because the urinary 17-ketosteroid pool reflects mainly adrenocortical ac-
tivity, a measurement of the 17-ketosteroid secretion is not a good index of
testicular function.

D. **Hormonal control of testicular function** (see Figure 45-1)

  1. Hypothalamic-hypophysial-seminiferous tubular axis

    a. LH and FSH are required for spermatogenesis. Because the effects of LH are medi-
ated by testosterone, testosterone and FSH are two hormones that act directly on
Sertoli cells to promote gametogenesis.

      (1) Exogenous testosterone alone does not promote spermatogenesis in men lacking
Leydig cells. Spermatogenesis requires that a high concentration of testosterone
be produced locally by LH action on Leydig cells.

      (2) Sertoli cells synthesize **androgen-binding protein** by an FSH-dependent process.
This protein binds testosterone and dihydrotestosterone and functions as a local
androgenic pool to support spermatogenesis.

      (3) Sertoli cells also synthesize **inhibin** in response to FSH secretion. This protein
inhibits FSH secretion by direct negative feedback on the pituitary gland. Inhibin is not known to suppress secretion of FSH releasing hormone (FSH-RH), a
decapeptide produced by the parvicellular peptidergic neurons. The selective
rise in plasma FSH levels in individuals with damaged seminiferous tubules is
from a reduced secretion of inhibin.

    b. Normal spermatogenesis occurs in men with a 5α-reductase deficiency. Dihydrotes-
tosterone is not required for normal sperm development.

    c. Plasma physiologic levels of androgens have little effect on the inhibition of FSH
secretion.

d. Testosterone administration has little effect on FSH secretion, and very large doses are required to suppress FSH in the male.

2. **Hypothalamic-hypophysial-Leydig cell axis** (see Figure 46-2)

   a. The rate of testosterone synthesis and secretion by Leydig cells is stimulated primarily by LH. The secretion of testosterone, in turn, inhibits LH secretion. It is unbound testosterone that suppresses LH secretion. The likely major target of negative feedback is the hypothalamus.

   b. Both testosterone and estradiol can inhibit LH secretion; however, since dihydrotestosterone can also suppress LH, androgen conversion to estrogen is not a prerequisite for this inhibitory action on the hypothalamus and pituitary gland.

   c. Some neurosecretory neurons of the hypothalamus secrete a releasing hormone into the hypophysial portal system, which then conveys it to the anterior lobe of the pituitary gland.

   (1) This decapeptide, called **gonadotropin-releasing hormone (Gn-RH),** stimulates the pituitary basophils.

   (2) Gn-RH also is called LH releasing hormone (LH-RH) and FSH-RH because it elicits the secretion of both LH and FSH.

E. **Physiologic effects of androgens** (Table 50-1)

1. **Reproductive function.** Androgens are essential for the control of spermatogenesis, the maintenance of the secondary sex characteristics, and the functional competence of the accessory sex organs.

   a. The accessory sex organs consist of excretory ducts and glands that transmit spermatozoa and that secrete seminal fluid necessary for the survival and motility of spermatozoa after ejaculation. The major accessory sex organs are the prostate gland and seminal vesicles.

   b. The secondary sex characteristics are the physiologic characteristics of masculinity (e.g., growth of facial hair, recession of hair at the temples, enlargement of the larynx, thickening of the vocal cords).

   c. Seminal plasma, the fluid in which spermatozoa normally are ejaculated, originates almost entirely from the prostate gland and seminal vesicles.

   (1) The volume of the human ejaculate (semen) is 2–5 ml, most of which is contributed by the seminal vesicles.

   (a) The prostate gland is the origin of citric acid, acid phosphatase, zinc, and spermine.

   (b) The seminal vesicles are the source of prostaglandins, fructose, ascorbic acid, and phosphorylcholine. Metabolism of fructose provides energy for sperm motility.

   (2) Sperm represents less than 10% of the ejaculate volume. The ejaculate contains 100–300 million spermatozoa.

   (a) Euspermia is defined as greater than 20 million sperm/ml of ejaculate.

   (b) Oligospermia is defined as 5–20 million sperm/ml of ejaculate.

   (c) Azoospermia is defined as 5 million sperm/ml of ejaculate.

TABLE 50-1. Major Actions of Androgenic Hormones

Life Stage	Testosterone	Dihydrotestosterone
Fetal period	Development of epididymis, vas deferens, and seminal vesicles	Development of penis, penile urethra, scrotum, and prostate gland
Puberty	Growth of penis, seminal vesicles, musculature, skeleton, and larynx	Growth of scrotum, prostate gland, pubic hair, and sebaceous glands
Adulthood	Spermatogenesis	Prostatic secretions

Reprinted from Genuth SM: The reproductive glands. In *Physiology*, 2nd edition. Edited by Berne RM, Levy MN. St. Louis, CV Mosby, 1988, p 999.

2. **Biologic effects.** Androgens stimulate cell division as well as tissue growth and maturation and are classified as protein anabolic hormones. (This anabolic effect in muscle is referred to as the **myotropic effect** of androgens.) Only testosterone and dihydrotestosterone have significant biologic activity. The 17-ketosteroids, androstenedione, DHEA, and etiocholanolone, are weak androgens but important metabolites of testosterone.
   a. In the adolescent, androgens produce linear growth, muscular development, and retention of nitrogen, potassium, and phosphorus. Testosterone also accelerates epiphysial fusion of the long bones. The skeletal development during puberty, particularly of the shoulder girdle, is pronounced. Dihydrotestosterone is the active androgen in all adult tissues containing 5α-reductase, with the notable exception of muscle.
   b. Testosterone stimulates differentiation of the wolffian duct system into the epididymis, vas deferens, and seminal vesicles; dihydrotestosterone stimulates organogenesis of the urogenital sinus and tubercle into the prostate gland, penis, urethra, and scrotum.
   c. Androgens produce a low-pitched voice; stimulate growth of chest, axillary, and facial hair; and cause temporal hair recession.
   d. Androgens are responsible for libido.
   e. Testosterone is a requisite for normal spermatogenesis.

II. OVARY

A. Embryology

1. **Internal genitalia** (see Figure 50-1). The primordia of both male and female genital ducts, which are derived from the mesonephros, are present in the fetus at 7 weeks gestation. The paired **müllerian ducts** form parallel to the paired **wolffian ducts**. In the female fetus, the upper ends ot the müllerian ducts are the anlagen of the fallopian tubes (oviducts), whereas the lower ends join to form the uterus, cervix, and upper end of the vagina.
   a. The uterus and fallopian tubes, which develop from the müllerian ducts, do not require the presence of an ovary.
   b. In the absence of a fetal testis, müllerian duct inhibiting factor and testosterone are not secreted by the fetal Sertoli cells and Leydig cells, respectively. Moreover, dihydrotestosterone is not formed from testosterone. Without the presence of these three hormones, at 10–11 weeks gestation, the müllerian ducts begin to differentiate and the wolffian ducts undergo regression.
   c. This process of female genital duct development is completed at 18–20 weeks gestation.

2. **External genitalia**
   a. The external genitalia of both sexes begin to differentiate at 9–10 weeks gestation. In the female, this process proceeds without any known hormonal influence.
   b. The external genitalia and urethra in both sexes develop from common anlagen, which are the urogenital sinus and the genital tubercle, folds, and swelling.
      (1) The urogenital sinus gives rise to the lower portion of the vagina and to the urethra.
      (2) The genital tubercle is the origin of the clitoris.
      (3) The genital swelling is the primordium of the labia majora.
      (4) The genital folds develop into the labia minora.

3. **Ovary.** In the absence of the H-Y antigen, the gonadal primordium develops into an ovary, provided that germ cells are present. Ovarian development occurs several weeks later than does testicular differentiation.
   a. At 8 weeks, when the testicular secretion of testosterone begins and before the ovarian differentiation is completed, the fetal "ovary" has the capacity to synthesize **estradiol**.

(1) It is unlikely that the fetal ovary contributes significantly to the circulating estrogens in the fetus. The fetus is exposed to estrogen (**estriol**) of placental origin. The site of estradiol synthesis by the primordial ovary is not known.

(2) The ovary has no role in sex differentiation of the female genital tract.

b. Also at about 8 weeks gestation, the cortex of the primitive gonad undergoes active mitosis, and epithelial cells infiltrate the gonad as a syncytium of tubules and cords. Primordial germ cells are carried along with this inward migration.

(1) Proliferation of the cortex ceases at about 6 months.

(2) The **rete ovarii** secretes a meiosis-inducing factor.

(3) The germ cells begin to proliferate to form **oogonia**. This mitotic process is maximal between 8 and 20 weeks, after which it diminishes, ceases, and never is resumed.

c. Unlike the fetal testis, the fetal ovary begins gametogenesis. **Oogenesis,** the formation of primary oocytes from oogonia, begins at 15 weeks and reaches a peak between 20 and 28 weeks gestation.

(1) The **primary oocytes** enter a prolonged prophase (diplotene stage) of the first meiotic division and remain in this state until ovulation occurs between 10 and 45 years later!

(2) The **diploid primary oocytes** become enveloped by a single layer of flat granulosa cells and in this form are called **primordial follicles**. The formation of primordial follicles reaches a peak between 20 and 25 weeks.

d. Also between 20 and 25 weeks, the gonad has acquired the morphologic appearance of an ovary. During this period, the plasma concentration of pituitary FSH reaches a peak and the first **primary follicles** appear [see II E 2 a (1)].

B. | **Oogenesis**

1. Fetal oogenesis. The period of oogonial proliferation results in a peak population of about 6–7 million germ cells in the two ovaries at 5 months gestation. Included in this group of cells are oogonia, oocytes in various stages of prophase, and degenerating germ cells. This total number of germ cells decreases to 2 million at term. The number of primordial follicles present in the ovary at birth rapidly diminishes thereafter.

2. Postnatal oogenesis. By 6 months postpartum, all of the oogonia have been converted to primary oocytes. By the onset of puberty, the number of primary oocytes has decreased to about 400,000.

3. Prepubertal oogenesis. Between birth and puberty, the primary oocyte is surrounded by the zona pellucida and six to nine layers of granulosa cells. These follicles are in varying stages of development.

4. Pubertal oogenesis. In contrast to the male who produces spermatogonia and primary spermatocytes continuously throughout life, the female cannot form oogonia beyond 28 weeks gestation and must function with a declining pool of oocytes.

a. Meiosis in the female results in the formation of one viable oocyte (**ootid**). In contrast, each primary spermatogonium in the male ultimately gives rise to 64 spermatozoa.

b. Oogenesis in the female begins in utero in response to meiosis-stimulating factor, whereas in the male, spermatogenesis is arrested at the spermatogonial stage in response to meiosis-inhibiting factor.

c. Just prior to ovulation, the first polar body is extruded from the primary oocyte, which completes the first meiotic division, and forms a secondary oocyte.

(1) This haploid cell immediately begins the second meiotic division but remains in metaphase.

(2) Extrusion of the second polar body (**polocyte**) does not occur until the mature ovum (ootid) is fertilized by a sperm cell. Fertilization normally occurs in the ampulla of the fallopian tube.

C. **Morphology**

1. **Gross anatomy**
   a. The ovaries are ovoid glands with a combined weight of 10–20 g during the reproductive years.
   b. The ovaries are anchored to the **broad ligament** by the **mesovarium**.

2. **Functional histology of the ovary and uterus**
   a. **Ovary**
      (1) **Structural divisions** include the **cortex** (which is lined by the germinal epithelium and contains all of the oocytes), the **inner medulla,** and the **hilus** (i.e., the point where the ovary attaches to the mesentery).
      (2) **Functional subunits** include the **follicle** and **oocyte** (each consisting of **theca cells** and **granulosa cells**) and the **corpus luteum.**
      (3) **Gametogenesis** in the female denotes **folliculogenesis,** which leads to the formation of a mature ovum.
         (a) During the preovulatory phase, the functional unit of the ovary is the follicle.
         (b) During the postovulatory phase, the functional unit of the ovary is the corpus luteum.
      (4) **Steroidogenesis** in the ovary is the synthesis and secretion of **estradiol** and **progesterone**. Although steroidogenesis occurs in three morphologic units (i.e., the follicle, corpus luteum, and stroma), only the follicle and corpus luteum are major steroid-producing units.
         (a) Theca interna cells produce androstenedione and testosterone, which diffuse into the granulosa cells.
         (b) Granulosa cells synthesize estradiol and estrone from androgenic precursors produced by the theca interna cells.
   b. **Uterus**
      (1) **Layers.** The uterus consists of two major tissue layers.
         (a) The outer layer, the **myometrium,** is a thick layer of smooth muscle.
         (b) The inner layer of the uterus is the **endometrium**. At the height of its development, during the luteal phase, the endometrium is approximately 5 mm thick. On the basis of blood supply, the endometrium can be divided into two major layers.
            (i) The **stratum basale** is the abluminal (deeper) layer of the endometrium. This layer functions as the regenerative layer in the growth and differentiation of endometrial tissue that is sloughed during menses.
            (ii) The **stratum functionale** is the adluminal (superficial) layer of the endometrium, which is shed during menses.
      (2) **Endometrial blood supply**
         (a) The stratum basale receives its vascular supply from the **basal (straight) arteries,** which arise from the uterine radial arteries.
         (b) The stratum functionale is perfused by the **spiral (coiled) arteries,** which also emanate from the radial arteries.
         (c) Thus, the arcuate arteries give off radial arteries, and the radial arteries bifurcate to form the basal and coiled arteries.

D. **Hormones of the ovary: steroids.** Ovarian hormones include two phenolic steroids—estradiol (C-18) and estrone (C-18)—and the progestogen, **progesterone** (C-21).

1. **Secretion and transport** (Tables 50-2, 50-3)
   a. **Estrogens.** Over 70% of circulating estrogens are bound to sex steroid–binding globulin, and 25% are bound to plasma albumin.
      (1) **Estradiol** is the principal and biologically most active estrogen secreted by the ovary. Ovarian estradiol accounts for more than 90% of the circulating estradiol.
      (2) **Estrone,** a weak ovarian estrogen, also is formed by the peripheral conversion of **androstenedione**.
         (a) In premenopausal women, most of the circulating estrone is derived from estradiol by conversion via 17-hydroxysteroid dehydrogenase.

TABLE 50-2. Types of Steroids and their Systemic Concentrations and Rates of Synthesis in Women

Steroid	Plasma Concentration (ng/dl)	Production Rate (μg/day)
Estrogens (C-18)		
Estradiol		
Early follicular phase	6	80
Late follicular phase	50	700
Middle luteal phase	20	300
Estrone		
Early follicular phase	5	100
Late follicular phase	20	500
Middle luteal phase	10	250
Progestogens (C-21)		
17-Hydroxyprogesterone		
Early follicular phase	30	600
Late follicular phase	200	4000
Middle luteal phase	200	4000
Progesterone		
Follicular phase	100	2000
Luteal phase	1000	25,000
Testosterone	40	250
Dihydrotestosterone	20	50
Androstenedione	150	3000
Dehydroepiandrosterone	500	8000

Reprinted from Lipsett MB: Steroid hormones. In *Reproductive Endocrinology: Physiology, Pathophysiology and Management.* Edited by Yen SSC, Jaffe RB. Philadelphia, WB Saunders, 1978, p 84.

   (b) In postmenopausal women, estrone is the dominant plasma estrogen and is derived via the prohormone pathway. Specifically, estrone is derived from the conversion of adrenocortical androstenedione in peripheral tissues (mainly liver). In obese women, there is a significant peripheral conversion of androstenedione to estrone by adipose tissue. This extraovarian synthesis of estrogen is implicated in the higher incidence of endometrial carcinoma in obese women.
   (3) Estriol, the weakest of all the naturally occurring estrogens, is synthesized by the placenta and the liver but not the ovary. In nongravid women, estriol is formed in the liver as a conversion product of estradiol and estrone.
  b. Progesterone is not bound to sex hormone–binding globulin. The progesterones are bound primarily to CBG (transcortin) and albumin.

 2. **Ovarian steroidogenesis** (see Figure 49-2). The two pathways for biosynthesis of ovarian steroids have in common the conversion of cholesterol to pregnenolone, a reaction stimulated by LH and FSH via 20,22-desmolase.
  a. One pathway proceeds by way of the Δ^5-pathway, which involves the synthesis of 17α-hydroxypregnenolone and DHEA via 17α-hydroxylase and 17,20-lyase, respectively.
   (1) DHEA,* a 17-ketosteroid, is converted to another androgenic 17-ketosteroid, androstenedione,* by 3β-hydroxysteroid dehydrogenase and Δ^5-reductase.
   (2) Androstenedione can be reduced by 17-hydroxysteroid dehydrogenase to testosterone, which is a reversible reaction.
   (3) Testosterone is a precursor of estradiol via an aromatase reaction.

---

*These 17-ketosteroids are steroids that consist of 19 carbon atoms.

TABLE 50-3. Serum FSH and LH Concentrations During the Life Cycle of the Normal Female*

Stage of Life	FSH[†]	LH[†]
Prepubertal period (5–11 years)	4.5	3.9
Puberty (11–13 years)	6.8	8.2
Reproductive period		
Follicular phase	8.3	12.8
Midcycle	19.3	83.5
Luteal phase	6.9	11.6
Postmenopausal period	96.0	66.0

Adapted from Ontjes DA, Walton J, Ney RL: The anterior pituitary gland. In *Metabolic Control of Disease*, 8th edition. Edited by Bondy PK, Rosenberg LE. Philadelphia, WB Saunders, 1980, p 1192.
*Mean values without standard deviations.
[†]Concentrations expressed in milli-International units.

  b. The other pathway proceeds via the conversion of pregnenolone to progesterone by 3β-hydroxysteroid dehydrogenase and Δ^5-isomerase. Progesterone is the initial compound in the Δ^4-pathway.
    (1) Progesterone is converted to 17α-hydroxyprogesterone by 17α-hydroxylase.
    (2) 17α-Hydroxyprogesterone is another precursor for androstenedione via another 17, 20-lyase step.
    (3) Androstenedione and testosterone are interconvertible with the enzyme 17-hydroxysteroid dehydrogenase.
    (4) Androstenedione and testosterone are converted to estrone and estradiol, respectively, by the action of aromatase. These two 18-carbon steroids (estrogens) also are interconvertible with 17-hydroxysteroid dehydrogenase.

3. Metabolism of ovarian steroids. The liver is the major site of steroid metabolism.
  a. Catabolism of progestogens
    (1) Progesterone is converted to **pregnanediol**.
    (2) 17α-Hydroxyprogesterone is catabolized to **pregnanetriol**.
  b. Catabolism of estrogens
    (1) Large quantities of both estradiol and estrone are hydroxylated (primarily in the liver) at the C-16 position to form **estriol**.
    (2) Another major catabolic route for estrogens is hydroxylation at the C-2 and C-4 positions, which yields the **catecholestrogens**.
  c. Estrogens are excreted in the urine in the form of soluble conjugates.
    (1) Estriol, catecholestradiol, and catecholestrone are excreted primarily as glucuronidates.
    (2) Estrone is excreted primarily as a sulfate conjugate.

E. **Ovarian function**

1. Menstruation
  a. Menarche refers to the onset of menstruation, which normally occurs between the ages of 12 and 14 years. Prior to menarche, minimal amounts of estrogen are produced by the peripheral conversion of androgens.
  b. Menstrual cycle
    (1) Duration. Although a duration of 25–30 days is considered typical, a cycle length of 28 days is the exception rather than the rule in adult women. In early adolescence, the cycle is characterized by irregular menses and anovulation.
    (2) Temporal reference points (Figure 50-2). The menstrual cycle conventionally begins with the first day of menstruation, when the endometrial lining is shed

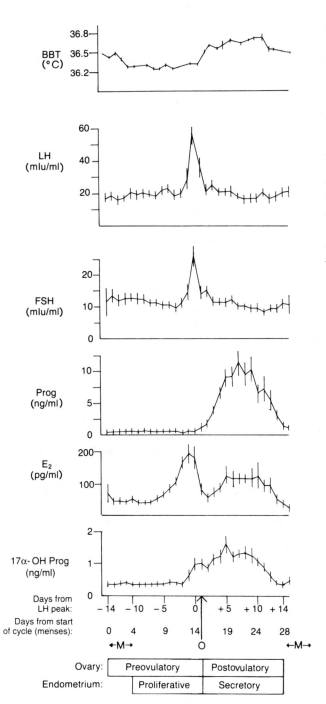

FIGURE 50-2. Hormonal (ovarian and pituitary), uterine (endometrial), and basal body temperature (*BBT*) correlates of the normal menstrual cycle. Mean plasma concentrations (± SEM) of luteinizing hormone (*LH*), follicle-stimulating hormone (*FSH*), progesterone (*Prog*), estradiol (*E$_2$*) (day +1) and 17α-hydroxyprogesterone (*17α-OHProg*) are shown as a function of time. Ovulation occurs on day 15 (day +1) following the LH surge, which occurs at midcycle on day 14 (day 0). *M* = menses, *O* = ovulation. (Adapted from Thorneycroft IA, et al: The relation of serum 17-hydroxyprogesterone and estradiol 17-β levels during the human menstrual cycle. *Am J Obstet Gynecol* 111:947–951, 1971.)

along with blood and uterine secretions. Days of the menstrual cycle are measured using two different reference points.

    (a) One system designates the first day of menses as **day 0** and the last day as **day 28**.

    (b) The other system designates the day of the LH peak (ovulation) as **day 0,** with preovulatory days indicated with a **minus sign** and postovulatory days indicated with a **plus sign**.

  c. Menopause refers to the cessation of menses, which typically occurs at about age 50 following a gradual decrease in frequency. Menopause is a result of primary hypogonadism (i.e., cessation of ovarian steroid secretion) and is associated with an increased gonadotropin (predominantly FSH) secretion.

2. Ovarian cycle

  a. Preovulatory phase. This phase is marked by follicular growth and maturation and by endometrial proliferation. The preovulatory phase generally lasts 8–9 days but can be quite variable (10–25 days). During this phase, the stratum basalis regenerates a stratum functionalis, and, by the end of this phase, one follicle (rarely more) has reached the final stage of growth.

    (1) A **primary follicle** begins as an oocyte surrounded by a single layer of cuboidal epithelial cells called granulosa cells. The primary follicle becomes multilaminar by the mitosis of the granulosa cells, which occurs with maturation.

    (2) Upon cavitation of the granulosa, an **antrum** is formed, which is filled with **liquor folliculi** secreted by the granulosa cells. The developing follicle now is called the **secondary follicle** (vesicular follicle or, more commonly, graafian follicle).

      (a) The granulosa cells secrete a protective shell, the **zona pellucida,** which surrounds the oocyte.

      (b) The stroma gives rise to a bilaminar **theca,** which surrounds the granulosa cells but is separated from them by a basal lamina (**lamina propria**).

        (i) The **theca interna** is a well-vascularized layer consisting of steroid-secreting cells that lie on the basal lamina. The blood vessels do not penetrate this membrane and, therefore, the granulosa cells are avascular until after ovulation.

        (ii) The **theca externa** is peripheral to the theca and is composed mainly of fibrous connective tissue. It is less vascular than the theca interna.

    (3) Just prior to ovulation, the primary oocyte of the secondary follicle completes the first meiotic division, which began prior to birth, and forms a secondary oocyte with a haploid nucleus and the first polar body. The second meiotic division takes place in the ampulla of the oviduct and occurs only if fertilization occurs.

  b. Ovulation. The secondary oocyte is released from the secondary follicle by a process called **ovulation,** which usually occurs on **day 15** of the average cycle. (In reference to the LH peak, this midcycle event occurs on **day 1,** that is, one day following the LH surge.)

  c. Postovulatory phase. The next 13–14 days constitute the postovulatory phase, during which the endometrium is prepared for the possible implantation of the fertilized ovum, which is a blastocyst when it arrives in the uterine cavity. This phase is relatively constant in duration. As a result, the day of ovulation can be estimated by subtracting 14 days from the total length of the menstrual cycle.

    (1) If fertilization does not occur, implantation also does not occur because the hormonal maintenance of the endometrial growth and differentiation is withdrawn.

    (2) Without conception, ischemia and necrosis of the luminal endometrium result after 14 days, and the ensuing menses marks the beginning of another menstrual cycle.

    (3) If conception occurs, the functional lifespan of the corpus luteum is extended, and it continues to secrete estradiol and progesterone at increasing rates during the first 6–8 weeks gestation.

F. **Neuroendocrine control of ovarian function** (see Figure 46-2). The ovarian cycle is associated with the secretion of ovarian steroids (estradiol and progesterone), which, in turn, are regulated by FSH and LH from the pituitary gland. A single hypothalamic hormone, Gn-RH, differentially regulates the secretion of FSH and LH.

1. Ovarian cycle

a. Preovulatory phase. Under the influence of FSH and LH, the primary follicle begins to develop. The combined effects of LH on the theca* cells to produce androgenic proestrogens and of FSH on the granulosa cells to aromatize these androgens to estrogens (estradiol) result in a slowly increasing blood estradiol concentration. During the preovulatory phase, the dominant gonadotropic hormone is FSH, and the dominant steroid is estradiol. Therefore, the preovulatory phase of the ovarian cycle also is referred to as the **follicular phase** and the **estrogenic phase**.

(1) Plasma estradiol concentration reaches a peak about 24 hours prior to the surge in LH secretion, or about 48 hours prior to ovulation. The peak in plasma estradiol concentration occurs on day 13 (day −1).[†]

  (a) As the plasma estradiol concentration increases, it exerts a negative feedback on the hypothalamic-hypophysial complex, resulting in a gradual decline in FSH. LH levels rise slightly through the follicular phase.

  (b) Inhibin secretion by the granulosa cells also exerts a negative feedback effect on FSH secretion.

(2) The peak in plasma estradiol concentration exerts a positive feedback on the hypothalamic-hypophysial axis, causing a reflex release of Gn-RH and a concomitant surge in pituitary LH secretion 24 hours later, on day 14 (day 0).

  (a) A lesser increase in plasma FSH secretion occurs on day 14 (day 0).

  (b) Increased plasma estradiol levels inhibit FSH secretion both directly and indirectly at the level of the pituitary gland and ventral diencephalon, respectively.

  (c) The midcycle surge of LH requires estradiol in plasma concentrations of approximately 150 pg/ml for about 50 hours.

(3) The follicular phase usually lasts 14 days, but any variability in the duration of the menstrual cycle usually is attributable to variability in the length of the follicular phase. It should be noted that the follicular phase begins with the first day of menses, while the proliferative phase of the endometrium begins with the last day of menses.

(4) There is a fall in the plasma estradiol level following the estradiol peak. Note in Figure 50-2 that this fall in plasma estradiol precedes ovulation.

(5) During the preovulatory phase, the plasma progesterone concentration remains low.

  (a) The bulk of this progesterone is derived from the peripheral conversion of adrenal progestogens; however, large amounts of progesterone exist within the follicular antrum.

  (b) The principal progestin secreted by the granulosa cells during the late follicular phase is 17α-hydroxyprogesterone.

b. Ovulatory phase. Ovulation occurs on day 15 (day +1) in response to the surge of LH secretion that occurred 24 hours earlier.

(1) Ovulation refers to the extrusion of a haploid secondary oocyte into the peritoneal cavity. The oocyte enters the oviduct (fallopian tube) where fertilization occurs.

(2) The rupture of the secondary follicle by the midcycle surge of LH leads to the formation of a new endocrine tissue (i.e., the corpus luteum), which involves proliferation, vascularization, and luteinization of the theca and granulosa cells. LH, therefore, is called the **luteotropic hormone of the menstrual cycle**.

---

*For the remainder of this chapter, the term theca denotes theca interna.
[†]Also throughout this chapter, the cycle days are numbered with reference to the onset of menses and with reference to the LH peak (in parentheses).

c. **Postovulatory phase.** LH also maintains the functional status of the corpus luteum during the postovulatory phase. The hormone secreted in the greatest amounts by the corpus luteum during this phase is progesterone. For these reasons, the postovulatory phase also is called the **luteal phase** and the **progestational phase**.

    (1) The decline in estrogen secretion prior to ovulation and at the time of ovulation removes the positive feedback effect of estradiol on gonadotropin secretion.

    (2) Both FSH and LH levels fall after their midcycle peaks but remain sufficiently high to stimulate the newly formed lutein-theca cells and lutein-granulosa cells to secrete estradiol, estrone, and progesterone.

    (3) About 6 days after ovulation, on day 21 (day +7), the plasma concentrations of progesterone and 17α-hydroxyprogesterone peak coincidently with the second peak of plasma estradiol concentration. Note in Figure 50-2 that this second peak is lower and broader than the estradiol peak that occurs during the preovulatory phase.

    (4) The plasma concentrations of progesterone during the entire ovarian cycle are higher than those of estradiol. It is imperative to note that the units of concentration for progesterone (ng/ml) are 1000 times greater than for estradiol (pg/ml).

    (5) The effect of the raised plasma estradiol and progesterone levels is a negative feedback on FSH and LH, respectively. Progesterone acts as an antiestrogen at this time because it inhibits LH secretion when the second estradiol peak occurs.

    (6) LH levels continue to decline during the luteal phase, while FSH levels begin to rise progressively during the late luteal phase.

    (7) The total amount of estradiol secreted during the follicular phase is comparable to that secreted during the luteal phase. This can be appreciated by comparing the areas under the estradiol curve during these two phases, using day 15 (day +1) as the dividing line between the follicular and luteal phases.

    (8) Unless conception occurs and is followed by implantation of the blastocyst, the corpus luteum undergoes involution following the reduction in gonadotropin secretion.

        (a) The declines in estradiol and progesterone secretion remove the negative feedback effect on the hypothalamic-hypophysial complex.

        (b) The corpus luteum regresses after about 12 days of steroid hormone secretion.

    (9) Progesterone is associated with a 0.2° C–0.5° C rise in basal body temperature, which occurs immediately following ovulation and which persists during most of the luteal phase (see Figure 50-2).

        (a) The basal body temperature dips during the follicular phase.

        (b) This temperature increment is used clinically as an index of ovulation.

        (c) Progesterone halts endometrial mitosis but causes maturation and differentiation of the endometrium.

2. **Endometrial cycle**

    a. **Hormonal effects on the myometrium.** Estradiol and progesterone are antagonistic with respect to their effects on the myometrium: estrogens promote uterine motility, and progestogens inhibit myometrial contractility.

    b. **Hormonal effects on the endometrium.** Estradiol and progesterone are synergistic with respect to their effects on the endometrium during the proliferative and secretory phases.

        (1) The **proliferative (preovulatory) phase** of the menstrual cycle refers to the endometrial changes that occur in response to estradiol.

            (a) Estrogens stimulate mitosis of the stratum basale, which regenerates the stratum functionale.

            (b) Estrogens stimulate angiogenesis (neovascularization) in the stratum functionale as well as stimulate the growth of secretory glands. The blood vessels become the spiral arteries that perfuse the stratum functionale. The glands contain glycogen but are nonsecretory at this time.

> **(c)** The cervical epithelium secretes a watery mucus in response to estrogen stimulation.

(2) The **secretory (postovulatory) phase** is characterized by secretion of large amounts of both progesterone and estradiol by the corpus luteum. The endometrium during this secretory phase is hyperemic and has a "lace curtain" or "Swiss cheese" appearance.

> **(a)** Progesterone promotes differentiation of the endometrium, including elongation and coiling of the mucous glands (which secrete a thick viscous fluid containing glycogen) and spiraling of the blood vessels.
> **(b)** Unless fertilzation occurs, hormone secretion by the hypothalamic-hypophysial complex and ovarian steroid secretion decline on about day 25 (day +11).
>> **(i)** Menses, beginning on day 0 (day − 14) of the following cycle, starts with vasoconstriction of the spiral arteries, which causes ischemia and necrosis.
>> **(ii)** The necrotic tissue releases vasodilator substances, causing vasodilation. The necrotic walls of the spiral arteries rupture, causing hemorrhage and shedding of cells over a period of 4–5 days.

G. Physiologic effects of ovarian steroids

1. Estrogens have important protein anabolic effects. Estrogens are responsible for the growth and development of the fallopian tubes, uterus, vagina, and external genitalia as well as the maintenance of these organs in adulthood. These steroids also promote cellular proliferation in the musocal linings of these structures.

a. Endometrium. Estrogens stimulate the regeneration of the stratum functionalis during the proliferative phase of the endometrial cycle.

(1) The water content and blood flow to the endometrium are increased markedly.

(2) The spiral arteries of the stratum functionalis are especially sensitive to estrogens and grow rapidly under their influence.

b. Myometrium. Estrogens increase the amount of contractile proteins (i.e., actin and myosin) in the myometrium and, thereby, increase spontaneous muscular contractions. Estrogens also sensitize the myometrium to the action of oxytocin, which promotes uterine contractility.

c. Cervix. Under the influence of estrogens, the uterine cervix secretes an abundance of thin, watery mucus.

(1) This fluid can be drawn into very long threads when placed between two glass slides. This is a clinical index of estrogen activity called **spinnbarkheit**.

(2) Cervical mucus also demonstrates the phenomenon of crystallization when it is dried on a glass slide. The characteristic **ferning pattern** is from the accumulation of sodium chloride. This phenomenon also is used diagnostically as an index of endogenous estrogen secretion.

d. Breast. Estrogens promote the development of the tubular duct system of the mammmary gland. Estrogens are synergistic with progesterone in stimulating the growth of the lobuloalveolar portions of this gland.

e. Bone. Estrogens, like androgens, exert a dual effect on skeletal growth in that they cause an increase in osteoblastic activity, which results in a growth spurt at puberty.

(1) Estrogens hasten bone maturation and promote the closure of the epiphysial plates in the long bones more effectively than does testosterone. Therefore, the female skeleton usually is shorter than the male skeleton.

(2) Estrogens are responsible for the oval or roundish shape of the female pelvic inlet. This inlet in the male is spade-shaped.

(3) Estrogens, to a lesser degree than testosterone, promote the deposition of bone matrix by causing Ca^{2+} and HPO_4^{2-} retention. In large amounts, estrogens also promote retention of Na^+ and water.

f. Liver. Estrogens stimulate the hepatic synthesis of the transport globulins, including thyroxine-binding globulin and transcortin.

(1) This results in increased plasma concentrations of thyroxine and cortisol but unchanged amounts of free thyroxine.

(2) Pregnant women often are in a state of mild hyperadrenocorticism because the elevated placental progesterone competes with cortisol for binding sites on transcortin, thus increasing plasma free cortisol.

2. Progesterone

    a. Endometrium. The endometrium, which proliferates under the influence of estrogens, becomes a secretory structure under the influence of progesterone.

        (1) The endometrial glands become elongated and coiled and secrete a glycogen-rich fluid.

        (2) Progesterone accounts for the differentiation of the stratum functionalis.

    b. Cervix. Under the influence of progesterone, the mucus secreted by the cervical glands is reduced in volume and becomes thick and viscid. This consistency of cervical mucus together with the absence of "ferning" provide presumptive evidence that ovulation and luteinization have occurred.

    c. Myometrium. Progesterone decreases the frequency and amplitude of myometrial contractions.

    d. Breast. This steroid also promotes lobuloalveolar growth in the mammary gland.

    e. Kidney. Progesterone promotes renal excretion of Na^+.

III. ENDOCRINE PLACENTA

A. Placenta formation

    1. Timetable of early placental function (days measured from ovulation). The gestational period, measured from the time of conception (ovulation) to parturition, is 38 weeks (266 days).

        a. Day 0: Fertilization occurs in the distal portion of the oviduct, or **ampulla**.

            (1) Fertilization triggers the final stage of the second meiotic division of the oocyte. The second polar body is extruded from the oocyte, and a haploid number of chromosomes are present in the female pronucleus.

            (2) The lifespan of an unfertilized ovum is less than 20 hours following ovulation; sperm cells are viable for about 24 hours after ejaculation.

        b. Day +3 or +4: The morula enters the uterine cavity.

        c. Day +5 or +6: The morula forms a cavity, the **blastocoele,** which is transformed into a **blastocyst.**

        d. Day +7: The blastocyst is implanted into the endometrium. By the end of 1 week, a primitive uteroplacental circulation begins to develop.

        e. Day +21: The placenta is fully functional.

    2. Early placental formation

        a. At the time of implantation, or **nidation,** the blastocyst consists of two cellular masses (Figure 50-3).

            (1) The inner cell mass, the **embryoblast,** will form the embryo and, eventually, the fetus.

            (2) The outer rim of cells, the **trophoblast,** forms the attachment to the endometrium and gives rise to the fetal membranes.

        b. The endometrium, under the influence of progesterone secreted by the corpus luteum, is transformed into a **decidua,** which is the maternal portion of the placenta that surrounds the conceptus.

    3. Placental development. The trophoblast, which is entirely fetal in origin, develops into the placenta. The trophoblast forms two cell layers (see Figure 50-3).

        a. The inner layer forms the **cytotrophoblast,** which is on the fetal side of the blastocyst and is the progenitor of the syncytiotrophoblast.

        b. The outer layer forms the **syncytiotrophoblast.**

B. Hormones of the placenta (Figures 50-4, 50-5, 50-6). The fetus, placenta, and mother are interdependent and constitute a functional unit called the **feto-placento-maternal unit.**

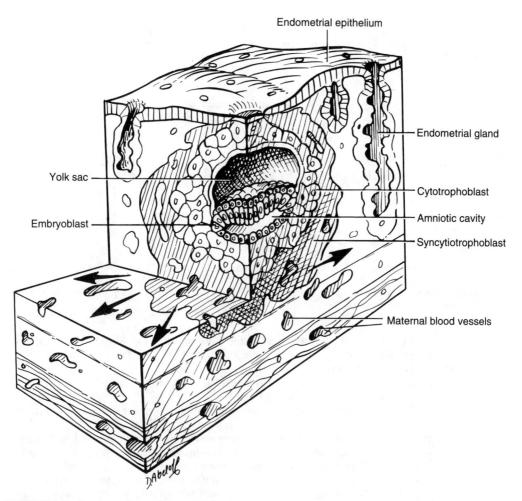

FIGURE 50-3. Structures formed by an implanting conceptus within the uterine endometrium. (Reprinted from Johnson KE: *Human Developmental Anatomy.* Baltimore, Williams & Wilkins, 1988, p 52.)

The pregnant woman at or near term produces 15–20 mg/day of estradiol, 50–100 mg/day of estriol, 250–300 mg/day of progesterone, 1–2 mg/day of aldosterone, and 3–8 mg/day of deoxycorticosterone (DOC). By itself, the placenta is an incomplete steroid-producing organ.

1. **Human chorionic gonadotropin (HCG)** is a polypeptide containing 236 amino acid residues, making it the largest active peptide hormone produced in humans.
   a. Synthesis and secretion. The syncytiotrophoblast is the source of HCG, which is secreted soon after fertilization.
   b. Plasma concentration
      (1) HCG reaches a plasma peak between 60 and 90 days gestation. This peak is 200 times greater than the LH peak at the height of the ovulatory surge.
      (2) HCG is detectable in maternal blood as early as 6–8 days after conception, which forms the basis for the immunologic pregnancy test.
      (3) HCG measurement in maternal blood is a useful index of the functional status of the trophoblast.

2. **Progesterone** (see Figures 50-4, 50-5, 50-6)
   a. Synthesis and secretion
      (1) Placental progesterone is derived mainly from maternal cholesterol; the fetus does not contribute significantly to placental progesterone formation.

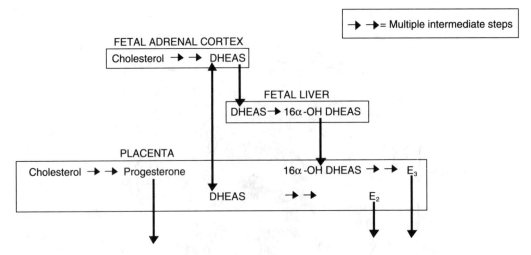

FIGURE 50-4. Pathways of steroid hormone biosynthesis in the feto-placental unit. *DHEAS* = dehydroepiandrosterone sulfate; *16α-OH DHEAS* = 16α-hydroxydehydroepiandrosterone sulfate; E_2 = estradiol; E_3 = estriol. (Adapted from Wilson JD, Foster DW: *Williams Textbook of Endocrinology,* 7th edition. Philadelphia, WB Saunders, 1985, p 423.)

    (2) This C-21 steroid is synthesized by the trophoblast. Most (85%) of the progesterone formed in the trophoblast is secreted into the maternal compartment.
  b. Plasma concentration of placental progesterone rises steadily throughout gestation, reaching a maximal plateau at 36–40 weeks. There is no significant drop in plasma progesterone concentration prior to labor.
    (1) Both 17α-hydroxyprogesterone and progesterone from the corpus luteum reach a peak 3–4 weeks postconception in response to HCG secretion.
    (2) At 6–8 weeks postconception, progesterone reaches a nadir, while 17α-hydroxyprogesterone continues to decline. The 17α-hydroxyprogesterone levels reflect corpus luteal secretion. The secondary rise in plasma progesterone reflects placental (trophoblast) secretion and is referred to as the luteal-placental shift.
    (3) An intact materno-placental circulation will produce essentially normal progesterone levels, even in the event of fetal death.
  c. Metabolism. The principal urinary metabolite of progesterone is pregnanediol, which is an estimate of placental function.
  d. Conversion to fetal corticoids
    (1) The placenta produces pregnenolone from maternal cholesterol, but it lacks the enzymes necessary for androgen synthesis (i.e., 17α-hydroxylase and 17,20-desmolase).
    (2) Pregnenolone synthesized by the placenta is oxidized to progesterone by 3β-hydroxysteroid dehydrogenase/Δ^5-isomerase.
    (3) Placental progesterone circulates to the fetal adrenal cortex, where it is hydroxylated at positions C-17, C-21, and C-11 to form aldosterone and cortisol. Thus, in early pregnancy, the fetus requires placental progesterone to synthesize corticoids, because the fetal zone of the adrenal cortex has a relative block in the 3β-hydroxysteroid dehydrogenase/Δ^5-isomerase system.
    (4) Beyond 10 weeks gestation, the fetal adrenal cortex no longer depends on placental progesterone for synthesis of aldosterone and cortisol.
  3. Estrogens (see Figures 50-4, 50-5, 50-6). Quantitatively, estriol is the major estrogen of human pregnancy, with smaller amounts of estradiol and estrone produced.
  a. Synthesis and secretion
    (1) These C-18 steroids are synthesized in the placental trophoblasts.
    (2) Estriol is produced primarily from androgenic precursors formed in the fetal zone of the adrenal cortex and the liver. The principal adrenal steroid is **DHEA,**

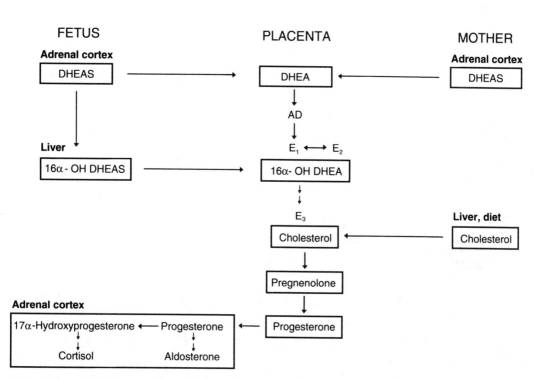

```
                                                        ┆↓ = Multiple intermediate steps
```

FIGURE 50-5. The primary pathways of estrogen and progesterone synthesis in the human feto-materno-placental unit. *DHEA(S)* = dehydroepiandrosterone (sulfate); *16α-OH DHEA(S)* = 16α-hydroxydehydro-epiandrosterone (sulfate); *AD* = androstenedione; E_1 = estrone; E_2 = estradiol; E_3 = estriol.

which is a ketosteroid. DHEA is sulfoconjugated (sulfurylated) by sulfokinase* in the fetal adrenal to **DHEAS**.

(a) DHEAS is the predominant precursor for estradiol and estrone synthesis after DHEAS is deconjugated by placental sulfatase. These two estrogens contribute to the estriol pool by conversion in the maternal liver.

(b) DHEAS is converted in the fetal liver to 16α-hydroxydehydroepiandrosterone sulfate by 16α-hydroxylase. This enzyme is not found in the placenta.

b. **Plasma concentration.** Like progesterone, the plasma estriol concentration rises steadily throughout gestation, reaching a maximal plateau at 36–40 weeks. The secretory curve for estriol parallels that for progesterone and correlates well with the fetal growth curve.

(1) Plasma estriol concentrations reflect the functional status of the feto-placental unit. Falling levels indicate impending fetal death.

(2) Even with fetal death, plasma progesterone levels can remain within normal limits.

C. **Physiologic effects of placental hormones**

1. **HCG** is classified as an anterior pituitary-like hormone with biologic actions that mimic those of LH (i.e., it can be used to induce ovulation).

---

*Fetal sulfokinase conjugates metabolites of pregnenolone, progesterone derivatives, and the androgens (DHEA) found in the high levels of the feto-placental unit.

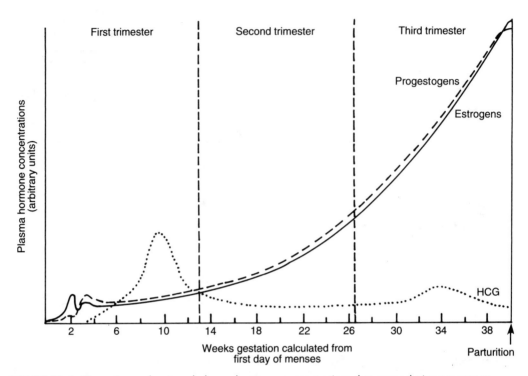

FIGURE 50-6. The patterns of maternal plasma hormone concentrations that occur during pregnancy. *HCG* = human chorionic gonadotropin. (Adapted from Laycock J, Wise P: *Essentials of Endocrinology,* 2nd edition. New York, Oxford University Press, 1983, p 155.)

   a. HCG is a second luteotropic hormone, in that it maintains the function of the corpus luteum until the feto-placental unit is autonomous in terms of hormone synthesis (about 6–7 weeks postconception).

   b. HCG converts the corpus luteum of menstruation into the corpus luteum of pregnancy, thereby extending the functional lifespan of the corpus luteum.

   c. HCG stimulates the corpus luteum of early pregnancy to secrete 17α-hydroxyprogesterone and lesser amounts of progesterone, which reach a peak 3–4 weeks postconception. Blood 17α-hydroxyprogesterone level is an excellent indicator of corpus luteal function during early pregnancy, because the placenta lacks significant 17α-hydroxylase activity.

   d. HCG stimulates the fetal testis to secrete testosterone at a time prior to fetal pituitary LH secretion.

   e. HCG may serve as a tropic agent for the fetal zone of the adrenal cortex, which secretes DHEA.

  2. Progesterone

   a. Progesterone inhibits uterine motility by hyperpolarization of the uterine myometrium.

   b. It converts the secretory endometrium of the luteal phase of the menstrual cycle to the decidua during pregnancy. Progesterone maintains the decidua.

   c. Synergistic action of progesterone and estrogen is required to induce development of the lubuloalveolar compartment, which prepares the breasts for lactation. Progesterone acts primarily on the lubuloalveolar compartment.

   d. Progesterone has an immunosuppressive role in protecting the fetus.

   e. Progesterone contributes to the growth and development of the fetus (e.g., by acting as a precursor for corticoid synthesis by the fetal adrenal cortex).

   f. Progesterone promotes renal excretion of Na^+, which antagonizes the effects of increased aldosterone levels found in pregnancy.

3. **Estrogens.** The estrogenic effects of pregnancy are primarily caused by estradiol, the most potent of the estrogens.

   a. Estrogens mediate the growth and development of the maternal reproductive organs.

      (1) The gravid uterus increases 18-fold (about 1700%) in weight, beginning as a 60 g organ in the nongravid state.

      (2) During the gestational period, the uterus lengthens from 7 cm to 30 cm.

      (3) Uterine volume at term is 500–1000 times greater than that before pregnancy.

      (4) The increase in uterine size during pregnancy occurs by stretching and hypertrophy of the myometrium.

   b. Estrogens stimulate hepatic synthesis of thyroxine-binding globulin, steroid hormone–binding globulin, and angiotensinogen as well as renal renin secretion. The latter two effects lead to increased angiotensin II synthesis.

   c. Estrogens stimulate development of the lactiferous ductal system in the mammary gland.

   d. Just before term, the estrogen-to-progesterone ratio increases and the uterus is dominated by estrogen.

Chapter 51

Pancreas

I. ENDOCRINE PANCREAS

A. **Histology and function of the islets of Langerhans** (Figure 51-1). The endocrine pancreas consists of **islet of Langerhans,** which form less than 2% of the pancreatic tissue.

1. **Cells.** Four types of cells have been identified.
   a. **Alpha cells** make up about 25% of the islet cells and are the source of glucagon, which consists of 29 amino acid residues.
   b. **Beta cells** constitute about 60% of the islet cells and are associated with insulin synthesis. This polypeptide consists of 51 amino acid residues.
   c. **Delta cells** comprise about 10% of the islet cells and are the source of somatostatin, which is a tetradecapeptide.
   d. **Pancreatic polypeptide cells** comprise approximately 5% of the islet cells and synthesize a polypeptide that contains 36 amino acid residues.

2. **Neurotransmitters and epinephrine.** Unmyelinated postganglionic sympathetic and parasympathetic nerve fibers terminate close to the three cell types (alpha, beta, and delta cells) and modulate pancreatic endocrine function via the secretion of neurotransmitters.
   a. **Acetylcholine (ACh)** secretion causes insulin release only when glucose levels are elevated. ACh appears to inhibit somatostatin release.
   b. **Norepinephrine** secretion from sympathetic stimulation via activation of the α-receptor leads to inhibition of insulin release. Norepinephrine stimulates somatostatin release.
   c. **Epinephrine.** Despite the dual α- and β-adrenergic receptor system in beta cells, the α-adrenergic action of epinephrine predominates, so that insulin secretion is inhibited. (Insulin release is mediated by a β-adrenergic receptor.)

3. **Control of secretions.** The alpha, beta, and delta cells constitute a functional syncytium, which forms a paracrine control system for the coordinated secretion of pancreatic polypeptides (see Figure 51-1).
   a. Insulin inhibits alpha cell (glucagon) secretion, which increases peripheral glucose uptake and opposes glucagon-mediated glucose production.
   b. Glucagon stimulates beta cell (insulin) secretion and delta cell (somatostatin) secretion, which increases hepatic glucose production and opposes hepatic glucose storage.

FIGURE 51-1. The paracrine system of the pancreatic islet cells. The pattern of islet cell hormone secretion represents an integrated response by all of the islet cells to humoral, neural, and paracrine regulation. *Plus signs* indicate stimulation, and *minus signs* indicate inhibition. (Reprinted from Tepperman J: Endocrine function of the pancreas. In *Metabolic and Endocrine Physiology,* 4th edition. Chicago, Year Book, 1980, p 233.)

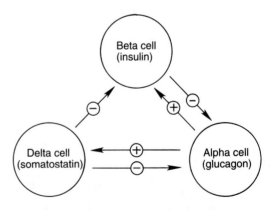

    c. Somatostatin inhibits alpha cell (glucagon) and beta cell (insulin) secretion, which produces hypoglycemia and inhibition of intestinal glucose absorption. The lowering of blood glucose levels by somatostatin in diabetic patients probably is from both inhibition of glucagon secretion and reduced intestinal absorption of glucose.

    d. Pancreatic polypeptide inhibits insulin and somatostatin secretion via a direct pancreatic effect.

B. **Biosynthetic organization of the beta cell**

1. Human proinsulin is a single-chain polypeptide of 86 amino acid residues, with a molecular weight of approximately 9000 daltons.

    a. Intracellular proteolytic cleavage of proinsulin forms insulin and C-peptide.

    b. The conversion of proinsulin is not fully completed, and about 5% of the secretory product of the beta cell is proinsulin.

    c. The biologic activity of proinsulin is about 5%–10% that of insulin.

    d. The plasma half-life of proinsulin is 15 minutes.

2. C-peptide is the connecting peptide remaining after cleavage of proinsulin to insulin. In humans, it consists of 31 amino acid residues.

    a. Beta cell secretory products consist of equimolar amounts of insulin and C-peptide; therefore, circulating C-peptide concentrations reflect beta cell activity.

    b. The normal fasting concentration of C-peptide in peripheral blood is approximately 1.0–3.5 ng/ml.

    c. C-peptide has no detectable biologic activity.

    d. The plasma half-life of C-peptide is 30 minutes.

3. Insulin is stored in the beta cell granules as a crystalline hexamer complex with two atoms of zinc per hexamer. In plasma, insulin is transported as a monomer.

    a. Insulin has a molecular weight of about 6000 daltons and contains 51 amino acid residues.

    b. Insulin secretion requires the presence of extracellular Ca^{2+}. Inside the beta cell, Ca^{2+} binds to a Ca^{2+}-binding protein called **calmodulin**.

    c. The plasma half-life of insulin is 5 minutes.

C. **Control of insulin secretion**

1. Carbohydrates

    a. Monosaccharides that can be metabolized (e.g., hexose, triose) are more potent stimuli of insulin secretion than carbohydrates that cannot be metabolized (e.g., mannose, 2-deoxy-D-glucose).

    b. The principal stimulus for insulin release is glucose. As the blood glucose level rises above 4.5 mmol/L (80 mg/dl), it stimulates the release and synthesis of insulin.

    c. Substances that inhibit glucose metabolism (e.g., 2-deoxy-D-glucose, D-mannoheptulose) interfere with insulin secretion.

    d. The reduction of glucose to sorbitol may contribute to insulin secretion.

    e. Glucose also stimulates somatostatin release.

2. Gastrointestinal (GI) hormones

    a. The plasma concentration of insulin is higher after oral administration of glucose than after it has been administered intravenously, even though the arterial blood glucose concentration remains lower. This augmented release of insulin following an oral glucose dose is a result of the secretion of GI hormones, including:

        (1) Gastric inhibitory peptide (GIP), which appears to be the principal GI potentiator of insulin release

        (2) Gastrin

        (3) Secretin

        (4) Cholecystokinin (CCK)

    b. GI hormones also augment somatostatin release.

3. Amino acids

    a. Amino acids vary in their ability to stimulate beta cells. Among the essential amino acids, in decreasing order of effectiveness, are arginine, lysine, and phenylalanine.

    b. The stimulation of insulin secretion by oral administration of amino acids exceeds that of intravenously administered amino acids. Protein-stimulated secretion of CCK, gastrin, or both may mediate this effect.

    c. The analogs of leucine and arginine that cannot be metabolized also stimulate insulin secretion.

4. Fatty acids and ketone bodies are not known to have an important role in the regulation of insulin secretion in humans. The ingestion of medium-chain triglycerides causes a small increment in insulin levels.

5. Islet hormones. Glucagon stimulates insulin secretion and somatostatin inhibits insulin secretion. Somatostatin inhibits gastrin and secretin secretion, glucose absorption, and GI motility (Table 51-1).

6. Other hormones

    a. Growth hormone (GH) induces an elevation in basal insulin levels, which precedes a change in blood glucose levels, suggesting a direct beta-cytotropic effect.

    b. Hyperinsulinemia also has been observed with exogenous and endogenous increments of corticosteroids, estrogens, progestogens, and parathyroid hormone (PTH). Since blood glucose concentrations are not reduced with these hormones, it is inferred that these hormones have an anti-insulin effect.

7. Obesity. Hyperinsulinemia is observed in obese patients. An increase in body weight in the absence of a disproportionate increase in body fat does not affect insulin levels.

8. Ions. Both K^+ and Ca^{2+} are necessary for normal insulin and glucagon responses to glucose. Therefore, hypokalemia leads to glucose intolerance.

9. Cyclic nucleotides. Cyclic adenosine $3',5'$-monophosphate (cAMP) is a releaser of insulin.

D. **Physiologic actions of insulin** (Figure 51-2). Target cells with insulin receptors have been demonstrated in the liver, muscle, adipose tissue, lymphocytes, monocytes, and granulocytes.

1. Carbohydrate metabolism

    a. Liver

        (1) The liver is freely permeable to glucose, and glucose transport can occur without insulin by simple diffusion.

            (a) Insulin acts on the liver to promote glucose uptake and to inhibit enzymatic processes involved in glucose production and release (glycogenolysis).

            (b) Because the hepatocyte is permeable to glucose, uptake of glucose in the liver is not rate-limiting.

        (2) A control point in glucose metabolism occurs when metabolism is initiated by the phosphorylation of glucose to glucose-6-phosphate, which is catalyzed by hexokinase and glucokinase.

TABLE 51-1. Physiologic Effects of Somatostatin

Site	Action
Anterior pituitary	Inhibits secretion of growth hormone and thyrotropin
Pancreas	Inhibits secretion of insulin, glucagon, and pancreatic polypeptide
Gastrointestinal tract	Inhibits secretion of gut hormones (gastrin, secretin, VIP, cholecystokinin), gastric acid, and pepsin; decreases blood flow, motility, and carbohydrate absorption; increases water and electrolyte absorption

VIP = vasoactive intestinal peptide.

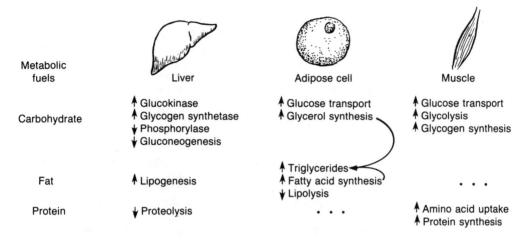

Metabolic fuels	Liver	Adipose cell	Muscle
Carbohydrate	↑ Glucokinase ↑ Glycogen synthetase ↓ Phosphorylase ↓ Gluconeogenesis	↑ Glucose transport ↑ Glycerol synthesis	↑ Glucose transport ↑ Glycolysis ↑ Glycogen synthesis
Fat	↑ Lipogenesis	↑ Triglycerides ↑ Fatty acid synthesis ↓ Lipolysis	. . .
Protein	↓ Proteolysis	. . .	↑ Amino acid uptake ↑ Protein synthesis

FIGURE 51-2. The major target sites and metabolic actions of insulin. Insulin is primarily involved in the regulation of metabolic processes, the principal manifestation of which is the control of plasma glucose concentration. (Reprinted from Felig P: Pathophysiology of diabetes mellitus. *Med Clin North Am* 55:821–834, 1971.)

- **(a) Hexokinase** is saturated at normal plasma glucose concentrations and is not regulated by insulin.
- **(b) Glucokinase** is only half-saturated at blood glucose concentrations between 90 and 100 mg/dl (5–5.5 mmol/L). Thus, the activity of this enzyme is insulin- and glucose-dependent.
- **(3)** The phosphorylation of fructose-6-phosphate by phosphofructokinase is enhanced by insulin. A decrease in phosphofructokinase activity favors the reversal of glycolysis.
- **(4)** Insulin diminishes hepatic glucose output by activating glycogen synthetase and by inhibiting gluconeogenesis. The key intermediary reaction in gluconeogenesis is between pyruvate and phosphoenolpyruvate, which requires the enzymes pyruvate carboxylase and phosphoenolpyruvate carboxykinase. The latter enzyme is inhibited in the presence of glucose and insulin.
- **b. Muscle.** The **insulin-dependent facilitated diffusion mechanism** for glucose is found in skeletal and cardiac muscle.
  - **(1)** Glucose transport across muscle cell membranes requires insulin.
  - **(2)** Insulin activates glycogen synthetase and phosphofructokinase, which cause glycogen synthesis and glucose utilization, respectively.
  - **(3)** It should be emphasized that glucose uptake in exercising muscle is not dependent on *increased* insulin secretion. In resting muscle, glucose is a relatively unimportant fuel, with the oxidation of fatty acids supplying most of the energy.
- **c. Adipose tissue.** The **insulin-dependent facilitated diffusion mechanism** for glucose is found also in adipose tissue.
  - **(1)** Insulin acts primarily to stimulate glucose transport.
  - **(2)** It activates glycogen synthetase and phosphofructokinase.
  - **(3)** The major end products of glucose metabolism in fat cells are fatty acids and α-glycerophosphate. The fat cell depends on glucose as a precursor of α-glycerophosphate, which is important in fat storage because it esterifies with fatty acids to form triglycerides.

- **2. Fat metabolism.** Insulin is a lipogenic as well as an antilipolytic hormone.
  - **a. Liver**
    - **(1)** When insulin and carbohydrate are available, the human liver is quantitatively a more important site of fat synthesis than is adipose tissue.
    - **(2)** In the absence of insulin, the liver does not actively synthesize fatty acids, but it is capable of esterifying fatty acids with **glycerol,** which is phosphorylated by glycerokinase.

    (a) Glycerol must be phosphorylated before it can be used in the synthesis of fat.

    (b) In the absence of glycolytic breakdown of glucose to α-glycerophosphate, glycerokinase permits the esterification of fatty acids.

  (3) In the absence of insulin, there is an increase in fat oxidation and the production of ketone bodies. Insulin exerts a potent antiketogenic effect.

  (4) Insulin promotes the synthesis and release of **lipoprotein lipase,** which is an extracellular enzyme that hydrolyzes both chylomicron and very-low-density lipoprotein (VLDL) triglyceride.

    (a) Lipoprotein lipase catalyzes the hydrolysis of circulating lipoprotein triglyceride to fatty acids and glycerol.

    (b) Lipoprotein lipase is the key enzyme in the removal of lipoprotein triglyceride and thereby is important in the formation of both light and heavy lipoprotein (LDL and HDL).

    (c) Lipoprotein lipase is active at the luminal surface of the capillary endothelial cell and under normal conditions is completely absent from the circulation.

    (d) The highest lipoprotein lipase activity is found in the heart, but its distribution includes adipose tissue, lactating mammary gland (and milk), lung, skeletal muscle, aorta, corpus luteum, brain, and placenta.

  b. Adipose tissue

    (1) Insulin deficiency also decreases the formation of fatty acids in adipose tissue.

    (2) The major effect of insulin-stimulated glucose uptake in human fat cells is to provide α-glycerophosphate for esterification of free fatty acids. The absence of α-glycerophosphate formation from glycolysis during insulin deficiency prevents the esterification of free fatty acids, which are constantly released from triglycerides in the adipocytes.

    (3) The lipolytic effect in the absence of insulin is caused by an increase in the hormone-sensitive lipase known as triglyceride lipase, the activity of which is normally inhibited by insulin.

3. Amino acid and protein metabolism. Insulin is an important protein anabolic hormone, and it is necessary for the assimilation of a protein meal. The protein anabolic effect of insulin is not dependent on increased glucose transport.

  a. In diabetic patients, the muscle uptake of amino acids is reduced and elevated postprandial blood levels are observed.

  b. During severe insulin deficiency, **hyperaminoacidemia** involving branched-chain amino acids (i.e., valine, leucine, and isoleucine) is present.

  c. Insulin increases uptake of most amino acids into muscle and increases the incorporation of amino acids into protein.

  d. Insulin increases body protein stores by four mechanisms:

    (1) Increased tissue uptake of amino acids

    (2) Increased protein synthesis

    (3) Decreased protein catabolism

    (4) Decreased oxidation of amino acids

4. Electrolyte metabolism

  a. Insulin lowers serum K^+ concentration. This hyperkalemic action of insulin is caused by stimulation of K^+ uptake by muscle and hepatic tissue.

  b. Diabetic patients have a proclivity toward developing hyperkalemia in the absence of acidosis.

  c. Insulin has an antinatriuretic effect.

5. Membrane polarization

  a. Insulin decreases membrane permeability to both Na^+ and K^+, but it decreases Na^+ permeability to a greater extent, causing hyperpolarization of mammalian muscle.

  b. The membrane hyperpolarization produced by insulin is the cause of the net shift of K^+ from the extracellular to the intracellular space, and not the result of the shift.

6. Integration of insulin action: a summary (Figure 51-3)

  a. Insulin is a very effective hypoglycemic hormone for two major reasons.

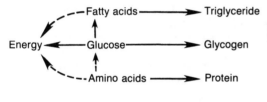

FIGURE 51-3. Insulin exerts integrated and synergistic actions in the promotion of the storage of body fuels. It enhances the storage of fat and protein, and it promotes both the storage and utilization of carbohydrates. *Solid arrows* denote stimulation, and *dashed arrows* indicate inhibition. (Reprinted from Felig P: Disorders of carbohydrate metabolism. In *Metabolic Control and Disease,* 8th edition. Edited by Bondy PK and Rosenberg LE. Philadelphia, WB Saunders, 1980, p 294.)

    (1) It promotes both hepatic and muscle glycogen deposition.
    (2) It enhances glucose utilization (glycolysis).
  b. In terms of glucose transport, the major insulin-independent tissues are brain, erythrocytes, liver, and epithelial cells of the kidney and intestine.
  c. Insulin is the primary anabolic hormone in the body for the following reasons.
    (1) Inhibition of hepatic gluconeogenesis decreases the hepatic requirement for amino acids.
    (2) The protein anabolic effect of insulin reduces the output of amino acids from muscle, thereby decreasing the availability of glucogenic amino acids for gluconeogenesis.
    (3) Glucose uptake by muscle is stimulated, providing an energy source to spare fatty acids, the release of which is inhibited by the antilipolytic action of insulin.
    (4) Fat accumulation is enhanced by increased hepatic lipogenesis.
    (5) The antilipolytic action of insulin (inhibition of hepatic oxidation of fatty acids) is a result of the formation of α-glycerophosphate from glucose in the fat cell.
    (6) The antilipolytic action of insulin at the level of the adipose cell reinforces the insulin-mediated inhibition of hepatic ketogenesis and gluconeogenesis by depriving the liver of precursor substrates for ketogenesis and the energy source (fatty acids) and cofactors (acetyl-CoA) necessary for gluconeogenesis.

E. **Control of glucagon secretion**

1. Metabolic fuels
  a. Hypoglycemia stimulates and hyperglycemia inhibits glucagon secretion.
  b. Amino acids (e.g., arginine, alanine) also are stimuli for glucagon release.
  c. Decreasing circulatory levels of fatty acids are associated with glucagon release.

2. GI hormones
  a. CCK, gastrin, secretin, and GIP stimulate glucagon secretion.
  b. The potentiation of glucagon secretion by the ingestion of a protein meal is probably mediated via CCK secretion.

3. Fatty acids inhibit glucagon release.

F. **Physiologic actions of glucagon.** The major site of action of glucagon is the liver.

1. Carbohydrate metabolism
  a. Glucagon has a hyperglycemic action, resulting primarily from stimulation of hepatic glycogenolysis. It should be emphasized, however, that the hyperglycemic effect of glucagon does not involve inhibition of the peripheral utilization of glucose.
  b. Glucagon is an important gluconeogenic hormone.
  c. The hyperglycemic action of epinephrine is amplified by its stimulation of glucagon secretion and its inhibition of insulin secretion.
  d. Suppression of glucagon secretion by glucose is not essential for normal glucose tolerance as long as insulin is available.

2. Fat metabolism
  a. Glucagon is a lipolytic hormone *because* of its activation of glucagon-sensitive lipase (triglyceride lipase) in adipose tissue by cAMP.

b. Glucagon causes an elevation in the plasma level of fatty acids and glycerol.
  (1) Glycerol is utilized as a glyconeogenic substrate in the liver.
  (2) The oxidation of fatty acids as an energy substrate accounts for the glucose-sparing effect of glucagon.
  (3) Glucagon is essential for the ketogenesis brought about by the oxidation of fatty acids. In the absence of insulin, glucagon can accelerate ketogenesis, which leads to metabolic acidosis.

3. Protein metabolism
  a. Glucagon has a net proteolytic effect in the liver.
  b. This peptide is gluconeogenic, an effect that leads to increased amino acid oxidation and urea formation.
  c. In addition to its protein catabolic effect, glucagon has an antianabolic effect—inhibition of protein synthesis.

II. ENDOCRINE CELLS OF THE GASTROENTEROPANCREATIC (GEP) SYSTEM

A. **Neuropeptides of the GEP system**

  1. The endocrine cells of the gut produce peptides that also are found in a variety of other tissues, including the pancreas, pituitary gland, and central and peripheral nerves. The same peptide might function as both a gut hormone and a neurotransmitter. The pancreatic hormones insulin, glucagon, and somatostatin, are considered to be part of the GEP system.

  2. The cells of the GEP system share with certain neural tissues the capacity of **a**mine **p**recursor **u**ptake and **d**ecarboxylation and, therefore, are called **APUD cells**. APUD cells are located in the GI tract and pancreatic islets.
  a. The endocrine cells of the GI tract are found singly or in small groups and are dispersed among the other epithelial cells of the mucosa.
  b. The GEP cells usually are situated on or near the base of the intestinal glands.

B. **APUD cells** represent a group of neurons and endocrine cells, which take up amino acids and modify them into amines and peptides. APUD cells synthesize and secrete all of the hormones of the body except the steroids.

  1. Most APUD cells are of ectodermal origin. Those in the gut and pancreas, however, originate from endoderm. Some APUD cells, such as those of the adrenal medulla and neurons of sympathetic ganglia, are derived from the neuroectodermal component called the **neural crest**.

  2. Many APUD endocrine cells contain dopamine, serotonin, and histamine as well as polypeptide hormones, and these substances tend to be released together.
  a. In some cases, enough hormone is secreted to enter the plasma and be transported to other tissues in the body and, therefore, an **endocrine** function is performed.
  b. In other instances, little hormone is bound to distant receptors (because of too few receptors, too little available hormone to bind them, or both) and, therefore, a **paracrine** action results.

Chapter 52

Thyroid Gland

I. HISTOLOGY AND FUNCTION

A. **Thyroid components.** The functional unit of the thyroid gland is the **follicle (acinus)** surrounded by a rich capillary plexus.

1. The follicular (acinar) **epithelium** consists of a single layer of cuboidal cells.
   a. The cell height of the follicular epithelium varies with the degree of stimulation of thyroid-stimulating hormone (TSH).
   b. The glandular epithelium varies with the degree of stimulation, becoming columnar when active and flat when inactive.

2. The **lumen** of the follicle is filled with a clear, amber proteinaceous fluid called **colloid,** which is the major constituent of the thyroid mass.

3. **Microvilli** extend into the colloid from the apical (adluminal) border, which is the site of the iodination reaction. The initial phase of thyroid hormone secretion (i.e., resorption of the colloid by endocytosis) also occurs in the apical border.

4. The **parafollicular (C) cells,** which secrete **calcitonin,** do not border on the follicular lumen.

B. **Thyroid functions**

1. **Hormone secretion.** The thyroid gland secretes two hormones, which are **iodothyronines** and, therefore, derivatives of the amino acid **tyrosine.**
   a. The major secretory product of the thyroid gland is **3,5,3′,5′-tetraiodothyronine (thyroxine),** which is abbreviated as T_4 to denote the four iodide atoms. The other thyroid hormone is **3,5,3′-triiodothyronine,** which is abbreviated as T_3. T_3 is secreted in small amounts.
   b. Only these two thyronines have biologic activity.
      (1) The **molar activity** ratio of T_3 to T_4 is 3–5:1
      (2) The **secretory ratio** of T_4 to T_3 is 10–20:1
      (3) The **plasma concentration ratio** of free T_4 to free T_3 is 2:1. Most of the T_3 in the plasma is derived from monodeiodination of T_4 by the action of monodeiodinase (5′-deiodinase) found in peripheral tissue.
   c. Reverse **3,3′,5′-triiodothyronine (rT_3)** is a biologically inactive thyronine formed by peripheral conversion catalyzed by 5-deiodinase.

2. **Related functions.** The thyroid cell performs two parallel functions in the synthesis of thyroid hormone.*
   a. It synthesizes a protein substrate called **thyroglobulin.**
      (1) This glycoprotein serves as a matrix in which thyroid hormone is formed.
      (2) Thyroglobulin also is the storage form of thyroid hormone.
      (3) Each thyroglobulin molecule contains approximately 120 tyrosyl residues.
   b. The thyroid cell accumulates inorganic iodide from the plasma.

II. DISTRIBUTION OF THYROID IODIDE

A. **Iodide intake**

1. In the United States, the daily dietary iodine intake is about 500 μg.

---

*The term "thyroid hormone" denotes thyroxine (T_4) and triiodothyronine (T_3).

2. About 1 mg of iodide is required per week (or 150 μg/day) to maintain euthyroidism.

3. The thyroid gland stores enough thyroid hormone to maintain a euthyroid state for 3 months without hormone synthesis.

B. Thyroid iodide

1. The thyroid gland contains 5–7 mg of iodide.
   a. Of the total iodide, 95% is in the extracellular space (i.e., stored in the colloid as thyroglobulin).
      (1) Two-thirds of the total iodide content in the colloid is in the form of biologically inactive **iodotyrosines**.
      (2) One-third of the colloid iodide content is in the form of biologically active **thyronines** (i.e., T_4 and T_3).
      (3) The molar storage ratio of T_4 to T_3 is 9:1, and the molar storage ratio of iodotyrosines to iodothyronines is 2:1.
   b. The remaining 5% of the total thyroid iodide is in the intracellular space of the follicular epithelium.

2. The thyroid gland contains the body's largest iodide pool.

III. **HORMONE TRANSPORT: EXTRACELLULAR BINDING PROTEINS.** Extracellular binding proteins for thyroid hormone are in the plasma, while the storage form of thyroid hormone (thyroglobulin) is in the follicular lumen (colloid).

A. Thyroxine-binding proteins. Virtually all (99.95%) of T_4 is bound to plasma proteins, leaving about 0.05% unbound (free). This portion of unbound thyroxine represents the biologically active hormone. Thyroxine is mainly associated with two of the three binding proteins.

1. Thyroxine-binding globulin (TBG) binds about 75% of the plasma T_4. In normal individuals, less than half of the available binding sites on TBG are saturated with T_4.

2. Thyroxine-binding prealbumin (TBPA) binds about 15%–20% of the circulating T_4.

3. About 9% of the T_4 is bound to albumin.

B. Triiodothyronine-binding proteins

1. Almost all (99.5%) of T_3 is transported bound to TBG.

2. Very little T_3 is bound to albumin, and practically none is bound to TBPA.

3. About 0.5% of the T_3 is unbound. The lower affinity of T_3 for the plasma binding proteins (and, thus, the higher concentration of unbound T_3) contributes to the greater biologic activity of T_3.

IV. **BIOSYNTHESIS AND RELEASE OF THYROID HORMONE.** The thyroid gland accumulates or "traps" iodide by an active transport mechanism that operates against a concentration and an electric gradient. The normal thyroid iodide-to-plasma iodide concentration ratio is 25–40:1.

A. Synthesis. All of the biosynthetic steps are stimulated by TSH.

1. **Iodide uptake** (Figure 52-1)
   a. Active iodide uptake occurs at the basal membrane of the follicular cell and is not an essential step in thyroid hormone synthesis.

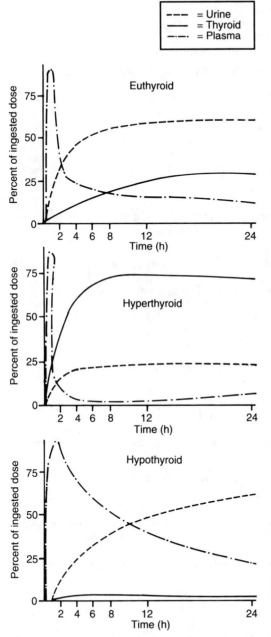

FIGURE 52-1. Radioactive iodine uptake by individuals on a relatively low-iodine diet. Percentages are plotted against time after an oral does of radioactive iodine for plasma, urine, and the thyroid gland. In hyperthyroidism, plasma radioactivity falls rapidly, then rises again as a result of release of labeled T_4 and T_3 from the thyroid gland. (Adapted from Ingbar SH, Woeber KA: *Textbook of Endocrinology,* 4th edition. Edited by Williams RH. Philadelphia, WB Saunders, 1968, p 144.)

    b. Iodide diffuses along an electric gradient into the lumen, where the luminal iodide-to-follicular cell iodide concentration ratio is 5:1.

    c. Radioactive iodide uptake by the thyroid gland is a useful therapeutic index of the functional status of the thyroid gland. A 24-hour uptake normally ranges between 10% and 35% of the administered dose.

  2. Oxidation of iodide is mediated by a peroxidase and forms active iodide, which may be in the form of iodinum ion (I^+), a free radical of iodine (IO_3^-), or iodine (I_2).

  3. Iodination of active iodide denotes the addition of iodide to the tyrosyl residues of thyroglobulin. The substrate for iodination is thyroglobulin.

    a. Iodination leads to the formation of the iodotyrosines within the preformed thyro-globulin molecule rather than in free amino acids that are then incorporated into protein. The hormonally inactive substrates formed by the iodination of thyroglobulin are called **mono-** and **diiodotyrosine**.

    b. Iodination occurs by the catalytic action of thyroid peroxidase.

4. Coupling (condensation) of iodotyrosines occurs and forms biologically active thyronines (T_3 and T_4).

    a. The iodothyronines are formed at the apical border of the follicular cell and are held in peptide linkage with thyroglobulin (as are the tyrosines).

    b. T_4 synthesis requires the fusion of two diiodotyrosine molecules, and T_3 synthesis requires the condensation of a monoiodotyrosine molecule with a diiodotyrosine molecule.

    c. Thyroid peroxidase also mediates the coupling reaction.

B. **Release**

1. Secretion begins with endocytosis of the colloid at the apical border. This brings colloid "droplets" into contact with protease-containing lysosomes.

2. The release of hormones involves the following reactions:

    a. Hydrolysis of thyroglobulin by the thyroid protease and by peptidases, which liberate free amino acids

    b. Secretion of iodothyronines into the blood and deiodination of iodotyrosines, which form a second iodide pool that can be recycled into hormone synthesis

V. METABOLISM AND EXCRETION OF THYROID HORMONE

A. T_4 is the iodothyronine found in highest concentration in plasma and is the only one that arises solely by direct secretion from the thyroid gland.

B. Most of the T_3 present in plasma is derived from the peripheral conversion of T_4 by mono-deiodination via 5'-deiodinase.

1. The extrathyroid deiodination of T_4 accounts for over 80% of the circulating T_3.

2. The liver and kidney deiodinate T_4 to form T_3.

C. Thyroid hormone is metabolized by deiodination, deamination, and by conjugation with glucuronic acid. The conjugate then is secreted via the bile duct into the intestine.

D. In normal individuals, T_4 and T_3 are excreted mainly in the feces, with a small amount appearing in the urine.

VI. CONTROL OF THYROID FUNCTION (see Figure 46-2)

A. **Hypothalamic-hypophysial-thyroid axis.** TSH secretion is influenced by four factors: thyrotropin releasing hormone (TRH) secretion from the median eminence, the blood level of unbound T_4, the blood level of unbound T_3 generated by the peripheral conversion of T_4 to T_3, and the peripheral conversion of T_4 to T_3 within the pituitary gland.

1. **TRH** is a tripeptide synthesized by the parvicellular peptidergic neurons in the hypothalamus.

    a. TRH is transported to the median eminence, where it is stored. From there, TRH is released into the hypophysial portal system and is carried to the anterior lobe of the pituitary gland.

b. TRH stimulates some of the basophils (**thyrotrophs**) to secrete TSH.

2. **TSH** stimulates the thyroid follicle to secrete thyroid hormone, most of which is bound to plasma protein carriers.
   a. TSH stimulates the series of chemical reactions that lead to the synthesis of the iodothyronines.
   b. It is the circulating free T_3 and T_4 concentrations that influence (regulate) TSH release by exerting a negative feedback effect at the level of the anterior lobe and probably at the level of the hypothalamus. Both free T_3 and free T_4 are effective inhibitors of TSH secretion when their plasma concentrations are increased.
   c. The pituitary gland also converts T_4 to T_3 by monodeiodinase, and this intrapituitary T_3 plays a major role in the negative feedback that occurs at the pituitary level.

3. **Other regulators**
   a. Estrogens enhance TSH secretion.
   b. Large doses of iodide inhibit thyroid hormone release and, thereby, cause decreases in serum T_4 and T_3 concentrations and an increase in TSH secretion.
   c. Somatostatin inhibits TSH secretion and the response to TRH.
   d. Dihydroxyphenylethylamine (dopamine), dopa, and bromocriptine decrease the basal secretion of TSH.

B. **Thyroid autoregulation**

1. Thyroid function also is regulated by an intrinsic control system that maintains the constancy of thyroid hormone stores.
   a. The high concentrations of intrathyroidal **inorganic** iodide lead to the inhibition of thyroid release.
   b. High concentrations of **organic** iodide (thyroid hormone) lead to a decrease in iodide uptake.

2. Both of these effects reduce the fluctuation in thyroid hormone secretion when an acute change occurs in the availability of a requisite substrate (e.g., iodide).

C. **Goiter**

1. Any enlargement of the thyroid gland is called a **goiter,** and antithyroid substances that cause thyroid enlargement are called **goitrogens**.
   a. Goitrogens are substances that block the synthesis of thyroid hormone.
   b. A goiter does not define the functional state of the thyroid gland.

2. If the goitrogen reduces thyroid hormone synthesis to subnormal levels, TSH secretion is enhanced.

3. Goitrogens lead to the increased synthesis of endogenous TSH, which is responsible for the formation of a hypertropic thyroid gland (goiter).

4. Goitrogenic agents include:
   a. Perchlorate, thiocyanate, and pertechnetate, which are monovalent anions that block iodide trapping
   b. Thionamides (propylthiouracil and methimazole), which block the coupling of iodotyrosines
   c. Iodide deficiency
   d. Excess iodide

VII. **PHYSIOLOGIC EFFECTS OF THYROID HORMONE**

A. Thyroid hormone increases the **basal metabolic rate (BMR)** of most cells in the body. The normal BMR for adult euthyroid males is 35–40 kcal/m^2 body surface/hr. Normal BMR is 6%–10% lower in euthyroid females.

1. Exceptions to this effect occur in the gonads, brain, lymph nodes, thymus, lung, spleen, dermis, and some accessory sex organs.

2. A correlate of the increase in BMR is an increase in the size and the number of mitochondria together with an increase in the enzymes that regulate oxidative phosphorylation.

3. The increase in BMR also is associated with an increase in Na^+-K^+-ATPase (Na^+-K^+ pump) activity. The fluxes of Na^+ (efflux) and K^+ (influx) are estimated to require 10%–30% of the total energy consumed by cells.

4. The increase in BMR accounts for the thermogenic effect of thyroid hormone.

B. Thyroid hormone is essential for normal **bone growth and maturation** as well as for the maturation of neurologic tissue, especially the brain.

1. In hypothyroidism, there is a marked decrease in the myelination and arborization of neurons in the brain.

2. If hypothyroidism is untreated, mental retardation occurs.

C. Thyroid hormone is necessary for normal **lactation**.

VIII. METABOLIC EFFECTS OF THYROID HORMONE

A. Carbohydrate metabolism

1. In physiologic amounts, thyroid hormone potentiates the action of insulin and promotes glycogenesis and glucose utilization.

2. In pharmacologic amounts, thyroid hormone is a hyperglycemic agent.
   a. Thyroid hormone potentiates the glycogenolytic effect of epinephrine, causing glycogen depletion.
   b. Thyroid hormone is gluconeogenic in that it increases the availability of precursors (lactate and glycerol).
   c. In large doses, thyroid hormone promotes intestinal glucose absorption.

B. Protein metabolism

1. In physiologic amounts, thyroid hormone has a potent protein anabolic effect.

2. In large doses, thyroid hormone has a protein catabolic effect.

C. Fat metabolism

1. Thyroid hormone stimulates all aspects of lipid metabolism, including synthesis, mobilization, and utilization. On a net basis, the lipolytic effect is greater than the lipogenic effect.

2. There is a general inverse relationship between thyroid hormone levels and plasma lipids.
   a. Elevated thyroid hormone levels are associated with decreases in blood triglycerides, phospholipids, and cholesterol.
   b. High levels of thyroid hormone also are associated with increases in plasma free fatty acids and glycerol.

D. **Vitamin metabolism.** The metabolism of fat-soluble vitamins is affected by thyroid hormone. For example, thyroid hormone is required for the synthesis of vitamin A from carotene and the conversion of vitamin A to retinene.

1. In hypothyroid states, the serum carotene is elevated, and the skin becomes yellow.

2. This skin condition differs from that observed in jaundice in that the sclera of the eye is not yellow.

Chapter 53

Parathyroid Hormone, Calcitonin, and Vitamin D

I. ROLE OF Ca^{2+} IN PHYSIOLOGIC PROCESSES

A. **Hemostasis.** Ca^{2+} is necessary for the activation of clotting enzymes in plasma.

B. Ca^{2+} controls **membrane excitation,** and Ca^{2+} influx occurs during the excitatory process of nerve and muscle.

1. Excitable membranes contain specific Ca^{2+} channels.

2. Ca^{2+} entry does not require an active transport process, because the concentration gradient across the membrane is larger for Ca^{2+} than for any other ion.
   a. Ca^{2+} concentration ($[Ca^{2+}]$) in the intracellular fluid (ICF) is about 10^{-7} mol/L.
   b. $[Ca^{2+}]$ in the extracellular fluid (ECF) is about 10^{-3} mol/L (the actual value is 2.5×10^{-3} mol/L).
   c. The $[Ca^{2+}]$ gradient from outside to inside the cell is on the order of 10,000 to 1!

C. Ca^{2+} is bound to cell surfaces and has a role in the **stabilization of the membrane** and **intercellular adhesion**.

D. Ca^{2+} is necessary for **muscle contraction** [excitation–contraction (EC) coupling].

E. Ca^{2+} is essential in all **excitation–secretion processes,** such as the release of hormone by endocrine cells and the release of other products by exocrine cells. It is also essential for **neurotransmitter release**.

F. Ca^{2+} is necessary for the production of **milk** and the formation of **bone** and **teeth**.

II. Ca^{2+} DISTRIBUTION

A. **Skeletal storage.** More than 99% of the total body Ca^{2+} is stored in the skeleton.

1. The skeleton of a 70-kg individual contains about 1000 g of Ca^{2+} compared to about 1 g in the extracellular pool.

2. The skeleton also serves as a storage depot for phosphorus and contains about 80% of the total body phosphorus.

3. Bone serves as a third-line defense in acid-base regulation by virtue of its CO_3^{2-}, HCO_3^-, and PO_4^{3-} content.

B. **Plasma.** The plasma concentration of total (ionized and nonionized) Ca^{2+} is about 10 mg/dl, which is equivalent to 5 mEq/L or 2.5 mmol/L.

1. Ca^{2+} is present in the plasma as:
   a. Ionized or free (45%)
   b. Complexed with HPO_4^{2-}, HCO_3^-, or citrate ion (10%)
   c. Bound to protein [primarily to albumin] (45%)

2. The sum of the ionized and complexed Ca^{2+} constitutes the diffusible fraction (55%) of Ca^{2+}. The protein-bound form constitutes the nondiffusible fraction (45%).

3. Parathyroid hormone (PTH), calcitonin, and vitamin D regulate the serum-ionized Ca^{2+} concentration.

C. **Ca^{2+} pools.** Total body Ca^{2+} can be conceptualized as two major "pools."

1. The larger Ca^{2+} pool, which contains over 99% of the total Ca^{2+}, consists of stable (mature) bone. This represents the Ca^{2+} pool that is not readily exchangeable, and it is not available for rapid mobilization.

2. The smaller Ca^{2+} pool, which contains less than 1% of the total body Ca^{2+}, consists of labile (young) bone. This Ca^{2+} pool is readily exchangeable because it is in physicochemical equilibrium with the ECF. The pool consists of calcium phosphate salts and provides an immediate reserve for sudden decreases in blood $[Ca^{2+}]$.

III. BONE CHEMISTRY

A. Bone Ca^{2+} is found in the form of **hydroxyapatite crystals**. The empirical chemical formula for this substance is $Ca_{10}(PO_4)_6(OH)_2$ or $[(Ca_3PO_4)_2]_3 \cdot Ca(OH)_2$. Fluoride ion can replace the OH^- group and form **fluoroapatite,** or $[(Ca_3PO_4)_2]_3 \cdot CaF_2$.

1. The calcium:phosphorus ratio in bone is about 1.7:1.

2. A large surface area is provided by the microcrystalline structure of bone; it is estimated to be 100 acres in humans!

B. Dry, fat-free bone consists of two-thirds mineral (inorganic) and one-third organic matrix.

1. Over 90% of the organic matrix is collagen.

2. The inorganic crystalline structure of bone imparts to it an elastic modulus similar to that of concrete.

IV. BONE DEVELOPMENT. Bone is both a tissue and an organ. It consists of cells and an extracellular matrix containing organic and inorganic components. In its early development, bone exists as **osteoid,** an organic, unmineralized matrix surrounding the bone cells that deposited it.

A. **Mesoderm** is the embryonic germ layer that gives rise to cartilage, bone, and muscle.

B. **Neural crest** cells also can differentiate as bone cells, including:

1. Cells that deposit dentin in teeth

2. Cartilage and bone of the head

V. Ca^{2+} REGULATION: OVERVIEW. Ca^{2+} regulation involves three tissues (bone, intestine, and kidney), three hormones (PTH, calcitonin, and activated vitamin D_3), and three cell types (osteoblasts, osteocytes, and osteoclasts).

A. **Hormonal control of Ca^{2+} metabolism.** The pituitary gland does not play a major role in regulating the cells that produce PTH, calcitonin, and activated vitamin D_3.

1. **PTH** is a polypeptide containing 84 amino acid residues. PTH is secreted by chief cells of the four parathyroid glands.

    a. PTH is the **hypercalcemic hormone** of the body; it exerts its effects on the bone, intestine, and kidney.

    b. PTH regulates only the plasma $[Ca^{2+}]$. An inverse linear relationship exists between plasma $[Ca^{2+}]$ and PTH secretion (Figure 53-1).

        (a) When plasma $[Ca^{2+}]$ falls, PTH secretion increases.

        (b) As plasma $[Ca^{2+}]$ increases, PTH secretion decreases.

2. Calcitonin is a 32 amino acid residue polypeptide secreted by the parafollicular (C) cells of the thyroid gland.

    a. Calcitonin is the **hypocalcemic hormone** of the body; it exerts a biologic effect on the bone, intestine, and kidney.

    b. A positive linear relationship exists between plasma $[Ca^{2+}]$ and calcitonin secretion (see Figure 53-1).

        (1) As plasma $[Ca^{2+}]$ increases, calcitonin secretion increases.

        (2) When plasma $[Ca^{2+}]$ decreases, calcitonin secretion decreases.

    c. Calcitonin release is stimulated by pentagastrin.

3. Vitamin D_3 (cholecalciferol) is a secosteroid containing 27 carbon atoms, which makes it the largest steroid hormone. The term "vitamin D" denotes both vitamins D_2 and D_3. In humans, the storage, transport, metabolism, and potency of vitamin D_2 and vitamin D_3 are identical.

    a. Active metabolites. Only the active metabolites of vitamin D exert biologic activity.

        (1) Calcidiol (25-hydroxyvitamin D_3; 25-hydroxycholecalciferol) is the major blood form of vitamin D. This active metabolite is two to five times more effective than vitamin D_3 in preventing rickets.

        (2) Calcitriol (1,25-dihydroxyvitamin D_3; 1,25-dihydroxycholecalciferol) is another active metabolite of vitamin D. On a molar basis, it is 100 times more potent than calcidiol.

    b. Synthesis of active vitamin D_3

        (1) In the epidermis, the previtamin 7-dehydrocholesterol is transformed into the lipid-soluble vitamin D_3 by nonenzymatic photoactivation upon exposure to the sun's ultraviolet rays. Photoactivation converts the 4-ring sterol into a 3-ring sterol.

            (a) Exposure of the skin to sunlight for 15–20 min/day or the irradiation of food has an antirachitic effect.

            (b) The plant sterol, **ergosterol,** is transformed into vitamin D_2 (ergocalciferol) by irradiation and has been the main source of vitamin D that is added to foods (e.g., milk).

        (2) In the liver, vitamin D_3 is converted to calcidiol by 25-hydroxylase.

        (3) In cells of the proximal convoluted tubule, calcidiol is converted to calcitriol by the action of 1α-hydroxylase. The activity of this enzyme is enhanced by PTH.

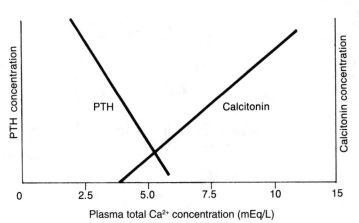

FIGURE 53-1. Relationship between plasma total Ca^{2+} concentration and blood levels of parathyroid hormone (*PTH*) and calcitonin. (Reprinted from Tepperman J: Hormonal regulation of calcium homeostasis. In *Metabolic and Endocrine Physiology,* 4th edition. Chicago, Year Book, 1980, p 297.)

B. **Cellular aspects of bone metabolism** (Figure 53-2)

1. **Osteoblasts** are highly differentiated cells that are nonmitotic in their differentiated state. They are the **bone-forming cells** and are located on the bone-forming surface.
   a. Osteoblasts **synthesize and secrete collagen**.
   b. They contain abundant alkaline phosphatase activity.
   c. They are derived from bone marrow mesenchyme.

2. **Osteocytes** are osteoblasts that have become **buried in bone matrix** and are the most numerous of the bone cells in mature bone.
   a. Each cell is surrounded by its own lacuna, but an extensive canalicular system connects osteocytes and surface osteoblasts, forming a functional syncytium.
   b. In the osteocytic form, these cells no longer synthesize collagen.
   c. Osteocytes have an **osteolytic activity,** which is stimulated by PTH. The "osteocytic osteolysis" in the bone matrix provides for the rapid movement of Ca^{2+} from bone into the ECF space.

3. **Osteoclasts** are large, multinucleated cells containing numerous lysosomes. They mediate **bone resorption** at bone surfaces.
   a. These cells contain acid phosphatase.
   b. Osteoclasts are stimulated by PTH and form significant amounts of lactic and hyaluronic acids.
   c. Osteoclasts might cause bone dissolution via an increased local concentration of H^+, which solubilizes bone mineral and increases the activity of enzymes that degrade matrix.
   d. They are derived from circulating monocytes.

VI. PHYSIOLOGIC ACTIONS OF PTH (Figure 53-3)

A. **Osseous tissue**

1. PTH action on bone is increased mobilization of Ca^{2+} and phosphate (i.e., bone dissolution) from the **nonreadily exchangeable Ca^{2+}** pool.
   a. The long-term effects of PTH on bone (i.e., the release of Ca^{2+} from bone) may be related to its effects on bone remodeling, which involves bone resorption and accretion.
   b. PTH also is known to stimulate bone synthesis.

FIGURE 53-2. Anatomic relationships among the three types of bone cells. The canalicular system provides the structure for a functional syncytium between the osteocytes and osteoblasts. These intercellular connections between these two cell types are disrupted at sites where osteoblasts are found. (Reprinted from Avioli LV, Raisz LG: Bone metabolism and disease. In *Metabolic Control and Disease,* 8th edition. Philadelphia, WB Saunders, 1980, p 1715.)

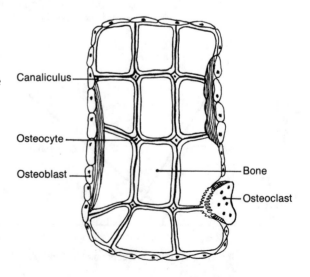

Canaliculus

Osteocyte

Osteoblast

Bone

Osteoclast

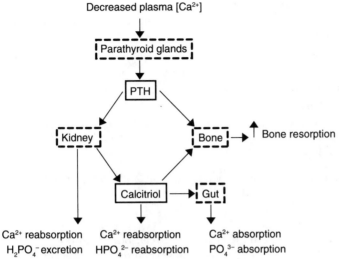

Decreased plasma [Ca^{2+}]

Parathyroid glands

PTH

Kidney   Bone → ↑ Bone resorption

Calcitriol → Gut

| Ca^{2+} reabsorption | Ca^{2+} reabsorption | Ca^{2+} absorption |
| $H_2PO_4^-$ excretion | HPO_4^{2-} reabsorption | PO_4^{3-} absorption |

FIGURE 53-3. Physiologic actions of parathyroid hormone (*PTH*) and calcitriol on major target organs.

    c. The effects of PTH on osteogenesis can be both anabolic and catabolic in terms of collagen metabolism.

  2. PTH has three important effects on bone that account for its overall osteolytic activity.
    a. It stimulates osteoclastic and osteocytic activity.
    b. It stimulates the fusion of progenitor cells to form the multinucleated osteoclastic cells.
    c. It causes a transient suppression of osteoblastic activity.

  3. Bone forms from cartilage, which serves as a template for cortical bone by periosteal apposition and for trabecular bone by endochondral ossification.
    a. Some bones, particularly those in the skull, are formed without a cartilage anlage by intramembranous bone formation.
    b. In adults, hematopoietic tissue is more abundant in trabecular bone.

  4. Cyclic adenosine 3′,5′-monophosphate (cAMP) is a mediator of bone resorption because PTH stimulates adenyl cyclase in bone cells.

  5. With the dissolution of stable bone, **hydroxyproline** is excreted in the urine. This forms the basis for assessing collagen metabolism and thereby the relative rate of bone resorption.

B. **Intestinal tissue**

  1. Ca^{2+} and phosphate are absorbed in the intestine by both active and passive transport, but most of the intestinal absorption of Ca^{2+} occurs via facilitated diffusion.

  2. PTH alone does not directly affect the intestinal absorption of Ca^{2+}. PTH and calcitriol act synergistically to absorb Ca^{2+} and phosphate.
    a. Intestinal absorption of Ca^{2+} does reflect parathyroid status, in that hypoparathyroid states are associated with low absorption and hyperparathyroid states are associated with high absorption.
    b. The increased intestinal absorption of Ca^{2+} promoted by PTH is mediated indirectly through the increased synthesis of calcitriol.
    c. Calcitriol acts on the intestine to promote the transport of Ca^{2+} and phosphate.

C. **Renal tissue**

  1. PTH increases the renal threshold for Ca^{2+} by promoting the active reabsorption of Ca^{2+} by the distal nephron, including the distal tubule, the cortical thick ascending limb of the loop of Henle, and the connecting segment. PTH inhibits the proximal tubular reabsorption of Ca^{2+}.

2. PTH inhibits phosphate reabsorption in the proximal tubules [i.e., it lowers the renal threshold for HPO_4^-, which leads to a phosphate diuresis (phosphaturia)]. Thus, PTH decreases the tubular transport maximum for phosphate.

3. Both increased PTH secretion and phosphate depletion stimulate the formation of calcitriol via the activation of 1α-hydroxylase.

4. The phosphaturic effect of PTH may be mediated by cAMP, because PTH activates adenyl cyclase in the renal cortex.

5. PTH also increases the urinary excretion of Na^+, K^+, and HCO_3^- and decreases the excretion of NH_4^+ and H^+. These effects account for the metabolic acidosis that occurs in hyperparathyroid states.

D. **Summary.** The unopposed effects of PTH on bone, intestine, and kidney include:

1. Hypercalcemia

2. Hypophosphatemia

3. Hypocalciuria, initially resulting from increased Ca^{2+} reabsorption (however, in chronic hyperparathyroid states, the hypercalcemia exceeds the renal threshold for Ca^{2+}, and **hypercalciuria** is observed)

4. Hyperphosphaturia

VII. PHYSIOLOGIC ACTIONS OF CALCITONIN

A. **Osseous tissue**

1. Calcitonin **inhibits osteoclastic activity,** and the antihypercalcemic effect of calcitonin is caused principally by the direct inhibition of bone resorption. This effect is not dependent on a functioning kidney, intestine, or parathyroid gland.

2. Calcitonin also diminishes the osteolytic activity of osteoclasts and osteocytes.

3. Calcitonin activity is associated with an increase in alkaline phosphatase synthesis from the osteoblasts.

B. **Intestinal tissue**

1. Calcitonin inhibits gastric motility and gastrin secretion; however, it stimulates intestinal secretion.

2. Calcitonin inhibits the intestinal (jejunal) absorption of Ca^{2+} and PO_4^{3-}.

C. **Renal tissue**

1. Calcitonin promotes the urinary excretion of phosphate, Ca^{2+}, and Na^+.

2. It also inhibits renal 1α-hydroxylase activity, which leads to a decrease in the synthesis of calcitriol.

VIII. PHYSIOLOGIC ACTIONS OF THE BIOLOGICALLY ACTIVE CALCIFEROLS.
The actions of calcitriol raise the plasma Ca^{2+} in concert with PTH. This vitamin hormone acts directly on the bone, small intestine, and kidney.

A. **Osseous tissue**

1. Calcitriol, together with PTH, increases the mobilization of Ca^{2+} and phosphate from the bone.

    a. Paradoxically, by raising serum Ca^{2+} and HPO_4^{2-} levels, it also fosters bone deposition.

    b. This effect of calcitriol on Ca^{2+} and HPO_4^{2-} mobilization is not observed with vitamin D_3 (cholecalciferol) or vitamin D_2 (ergocalciferol).

2. Ca^{2+}-binding protein in bone can be activated by PTH and calcitriol.

3. The antirachitic action of vitamin D has traditionally been measured by an increase in bone formation following vitamin D administration to vitamin D–depleted rats.

    a. It appears that the direct effect of the calciferols on bone is resorption.

    b. The antirachitic influence of the calciferols appears to be caused by an indirect effect on bone through the direct stimulating effect of the calciferols on the intestinal absorption of Ca^{2+} and phosphate.

4. In summary, calcitriol acts synergistically with PTH to cause bone dissolution through the proliferation of osteoclasts. In short, increased osteoclastic activity by PTH requires calcitriol.

B. Intestinal tissue

1. Calcitriol is the principal factor in increased Ca^{2+} absorption in intestinal tissue.

    a. Calcitriol is the mediator hormone for the intestinal actions of PTH and calcitonin.

    b. In turn, the action of calcitriol is mediated by the induction of a Ca^{2+}-binding protein.

2. Calcitriol causes a lesser increase in intestinal HPO_4^{2-} absorption.

C. Renal tissue

1. Calcitriol promotes the distal tubular reabsorption of Ca^{2+}.

2. It promotes proximal tubular reabsorption of HPO_4^{2-}.

3. The renal effects of PTH and calcitriol are similar in that both promote Ca^{2+} reabsorption; however, the renal effects of these two hormones on phosphate reabsorption are different.

    a. PTH promotes phosphate diuresis.

    b. Calcitriol promotes phosphate reabsorption; however, pharmacologic doses of calcitriol do have phosphaturic effects.

STUDY QUESTIONS

Directions: Each of the numbered items or incomplete statements in this section is followed by answers or completions of the statement. Select the **one** lettered answer or completion that is **best** in each case.

1. Which of the following peptides is synthesized by neurosecretory neurons?

(A) Epinephrine
(B) Norepinephrine
(C) Somatomedin
(D) Somatostatin
(E) Somatotropin

2. Which of the following adrenomedullary enzymes is correctly paired with its substrate?

(A) Phenylethanolamine-N-methyltransferase (PNMT)/epinephrine
(B) Phenylalanine hydroxylase/tyrosine
(C) Dopa decarboxylase/phenylalanine
(D) Dopamine β-hydroxylase/ dihydroxyphenylethylamine
(E) Tyrosine hydroxylase/norepinephrine

3. When is the second meiotic division of the developing ovarian follicle completed?

(A) At puberty
(B) Just prior to ovulation
(C) During the follicular phase of the menstrual cycle
(D) Just after conception
(E) During fetal development

4. In terms of serum concentration, the major postmenopausal steroid hormone and pituitary tropic hormone are

(A) estradiol and follicle-stimulating hormone (FSH)
(B) estradiol and luteinizing hormone (LH)
(C) estrone and FSH
(D) estrone and LH
(E) estriol and LH

5. A decrease in cortisol secretion would lead to

(A) increased storage of glycogen in the liver
(B) decreased adrenocorticotropic hormone (ACTH) secretion
(C) decreased adrenomedullary synthesis of epinephrine
(D) increased plasma glucose concentration
(E) increased hepatic protein synthesis

6. Glucose transport occurs by insulin-dependent facilitated diffusion in which of the following tissues?

(A) Cardiac muscle
(B) Intestinal epithelium
(C) Renal epithelium
(D) Brain

7. Because hepatic glycogen stores are limited and decrease only temporarily after epinephrine secretion, muscle glycogenolysis is the major mechanism for providing gluconeogenic precursors for hepatic glucogenesis. This gluconeogenic substance derived from muscle is

(A) lactate
(B) acetyl coenzyme A (acetyl-CoA)
(C) glucose
(D) glycerol
(E) alanine

8. Progesterone secretion during the second and third trimesters of pregnancy is a measure of the functional status of the

(A) materno-placental unit
(B) corpus luteum
(C) fetal liver
(D) fetal adrenal gland
(E) maternal ovary

9. The most biologically active iodothyronine secreted by the thyroid follicles is

(A) triiodothyronine (T_3)
(B) tetraiodothyronine (T_4)
(C) reverse triiodothyronine (rT_3)
(D) thyroglobulin
(E) triiodothyroacetic acid

10. The primary site of 1,25-dihydroxycholecalciferol formation from its immediate precursor is the

(A) bone
(B) liver
(C) skin
(D) nephron
(E) bloodstream

11. Which of the following endocrine organs is larger at birth than in adulthood?

(A) Hypophysis
(B) Thyroid gland
(C) Adrenal gland
(D) Parathyroid glands
(E) Endocrine pancreas

12. The major steroid hormone secreted by the inner zone of the fetal adrenal cortex is

(A) cortisol
(B) dehydroepiandrosterone (DHEA)
(C) progesterone
(D) estriol
(E) corticosterone

13. Prostaglandins found in the seminal fluid are secretory products of the

(A) prostate gland
(B) Sertoli cells
(C) seminal vesicles
(D) Leydig cells
(E) epididymis

14. Which of the following is a secretory product of the lutein-granulosa cells in non-gravid women?

(A) Androstenedione
(B) Pregnenolone
(C) Pregnanediol
(D) Estriol
(E) Estrone

15. The cells that contribute most to testicular volume are the

(A) interstitial cells
(B) tubular cells
(C) spermatocytes
(D) connective tissue cells

16. An increase in plasma parathyroid hormone (PTH) level would lead to an increase in which of the following?

(A) The number of active osteoblasts
(B) Plasma inorganic phosphate concentration
(C) Renal synthesis of calcitriol
(D) Collagen synthesis
(E) Renal proximal tubular reabsorption of Ca^{2+}

17. Active vitamin D_3 (calcitriol) and parathyroid hormone (PTH) have many similar effects. Which of the following physiologic effects is specific only for calcitriol?

(A) Increased renal phosphate reabsorption
(B) Increased renal Ca^{2+} reabsorption
(C) Increased intestinal Ca^{2+} absorption
(D) Increased plasma $[Ca^{2+}]$
(E) Decreased plasma $[HPO_4^{2-}]$

18. A 24-year-old woman has regular menstrual cycles of 21–23 days. Ovulation can be expected to occur between cycle days

(A) 7 and 9
(B) 10 and 12
(C) 13 and 15
(D) 16 and 18
(E) 19 and 21

19. A 17-year-old patient with a normal female phenotype is referred to an endocrinology clinic because of sparse pubic and axillary hair and amenorrhea. Chromosomal analysis reveals a male genotype. Further evaluation reveals intra-abdominal testes and circulating testosterone and estrogen concentrations that are characteristic of a normal man. The physician concludes that the patient has testicular feminization syndrome, a genetic end-organ insensitivity to androgen, caused by the absence of androgen receptors. Which of the following findings would be consistent with this syndrome?

(A) Normal wolffian duct development
(B) Normal müllerian duct development
(C) Regression of the internal genitalia
(D) Beard growth following androgen treatment
(E) Normal fertility

Questions 20–23

The graph below shows plasma steroid hormone levels as a function of time during a normal ovarian cycle in a 22-year-old woman. The woman becomes pregnant during this cycle.

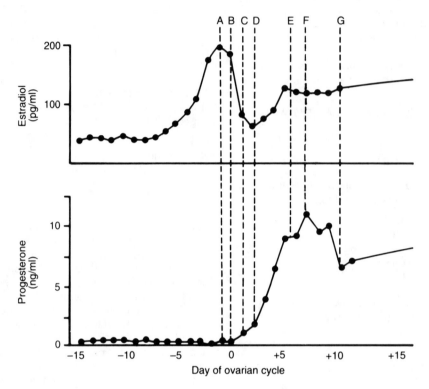

Adapted from Goodman HM: *Basic Medical Endocrinology*. New York, Raven, 1988, p 306.

20. Ovulation is indicated by which of the following lettered points on the graph?

(A) A
(B) B
(C) C
(D) D
(E) E

21. The LH peak is indicated by which of the following lettered points on the graph?

(A) A
(B) B
(C) C
(D) D
(E) E

22. Implantation of the blastocyst corresponds to which of the following points on the graph?

(A) C
(B) D
(C) E
(D) F
(E) G

23. The gradual increase in plasma concentrations of estradiol and progesterone beginning at point G indicate steroid secretion by the

(A) placenta
(B) corpus luteum
(C) adrenal cortex
(D) theca interna
(E) ovarian follicle

DIRECTIONS: Each of the numbered items or incomplete statements in this section is negatively phrased, as indicated by a capitalized word such as NOT, LEAST, or EXCEPT. Select the ONE lettered answer or completion that is BEST in each case.

24. All of the following statements regarding testosterone are true EXCEPT

(A) it is produced by the fetal testis
(B) it inhibits luteinizing hormone (LH) secretion from the pituitary gland
(C) it is a proestrogen
(D) it is inactivated after conversion to dihydrotestosterone
(E) it accelerates epiphysial closure of the long bones

25. All of the following are biochemical effects of cortisol EXCEPT

(A) hepatic lipogenesis
(B) hepatic gluconeogenesis
(C) muscle proteolysis
(D) hepatic protein anabolism
(E) hepatic glycogenesis

26. Abnormally high glucocorticoid levels would be associated with an increase in all of the following activities in the liver EXCEPT

(A) gluconeogenesis
(B) glycogenesis
(C) glycogenolysis
(D) glucose production
(E) protein synthesis

27. Characteristics of estriol include all of the following EXCEPT

(A) production by the liver
(B) production by the placenta
(C) secretion by theca interna cells of the ovary
(D) quantitatively the major urinary metabolite of the estrogens
(E) the least biologically active of the endogenous estrogens

28. Thyroid peroxidase is required for all of the following steps in thyroid hormone synthesis EXCEPT

(A) iodide uptake
(B) oxidation of iodide
(C) iodination of active iodide
(D) coupling of iodotyrosines
(E) synthesis of iodothyronines

29. Oxytocin is a neurosecretory hormone released from the pars nervosa. Oxytocin secretion promotes all of the following actions EXCEPT

(A) myometrial contraction
(B) lactogenesis
(C) milk ejection
(D) myoepithelial cell contraction

30. Activation of the sympathetic nervous system would lead to all of the following responses EXCEPT

(A) inhibition of peristalsis
(B) contraction of the radial ocular muscle
(C) renin secretion
(D) insulin secretion
(E) vasodilation in skeletal muscle

31. Insulin exerts all of the following effects EXCEPT

(A) hyperpolarization of skeletal muscle cells
(B) promotion of lipogenesis
(C) stimulation of glycogen synthase activity
(D) increase in secondary active transport of glucose into muscle cells
(E) increase in glucose transport in adipocytes

32. Progesterone serves several important functions during pregnancy. All of the following physiologic or biochemical effects require progesterone EXCEPT

(A) stimulation of myometrial contraction
(B) promotion of differentiation and growth of the lactiferous ducts
(C) inhibition of renal Na^+ excretion
(D) formation of cortisol by the fetal adrenal cortex in early pregnancy
(E) formation of aldosterone by the fetal adrenal cortex in early pregnancy

33. All of the following are substrates for monoamine oxidase (MAO) EXCEPT

(A) norepinephrine
(B) vanillylmandelic acid (VMA)
(C) epinephrine
(D) metanephrine
(E) normetanephrine

34. Estriol synthesis during gestation requires all of the following organs EXCEPT the

(A) fetal pituitary gland
(B) fetal liver
(C) neocortex of the fetal adrenal gland
(D) placenta
(E) trophoblast

35. All of the following are neuropeptide hormones EXCEPT

(A) antidiuretic hormone (ADH)
(B) β-endorphin
(C) oxytocin
(D) somatomedin
(E) thyrotropin releasing hormone (TRH)

36. All of the following are stimuli for growth hormone (GH) release EXCEPT

(A) bromocriptine
(B) hypoglycemia
(C) stress
(D) obesity
(E) vigorous exercise

37. True statements about β-endorphin include all of the following EXCEPT

(A) it reacts with the same receptors that bind morphine
(B) it is synthesized by pituitary basophils
(C) it is synthesized from corticotropin
(D) it is a proteolytic cleavage product of proopiomelanocortin (POMC)

38. Hormones that may cause negative nitrogen balance include all of the following EXCEPT

(A) glucagon
(B) thyroid hormone (excess)
(C) cortisol
(D) growth hormone (GH)

39. Epinephrine is a potent hyperglycemic agent because of its ability to do all of the following EXCEPT

(A) stimulate glucagon secretion
(B) stimulate adrenocorticotropic hormone (ACTH) secretion
(C) stimulate hepatic and muscle glycogenolysis
(D) inhibit insulin secretion
(E) inhibit cortisol secretion

40. Administration of exogenous thyroid hormone would likely lead to all of the following EXCEPT

(A) negative feedback inhibition of thyroid-stimulating hormone (TSH) secretion
(B) decreased secretion of triiodothyronine (T_3)
(C) decreased iodide uptake by the thyroid gland
(D) increased O_2 consumption by the brain

Directions: Each group of items in this section consists of lettered options followed by a set of numbered items. For each item, select the **one** lettered option that is most closely associated with it. Each lettered option may be selected once, more than once, or not at all.

Questions 41–44

Match each statement concerning plasma catecholamines with the specific catecholamine(s).

(A) Epinephrine
(B) Norepinephrine
(C) Both
(D) Neither

41. Decreases markedly following bilateral adrenalectomy

42. Mainly derived from tissues other than the adrenal medulla

43. Enzymatically inactivated by catechol-O-methyltransferase (COMT)

44. Elevated in most pheochromocytoma patients

Questions 45–48

Match each hormone classification with the hormone(s) to which it applies.

(A) Somatostatin
(B) Insulin
(C) Both
(D) Neither

45. Hypoglycemic hormone

46. Polypeptide hormone

47. Neurosecretory hormone

48. Hypophysiotropic hormone

Questions 49–54

Match each physiologic effect with the appropriate adrenergic receptor(s).

(A) Alpha receptor
(B) Beta receptor
(C) Both
(D) Neither

49. Increased insulin secretion

50. Increased pupillary diameter

51. Bronchodilation

52. Increased renin secretion

53. Decreased peristalsis

54. Relaxation of the detrusor muscle

ANSWERS AND EXPLANATIONS

1. The answer is D [Chapter 48 IV A 2 b (1) (b)]. Somatostatin is a hypophysiotropic hormone produced by the parvicellular neurosecretory neurons that terminate in the median eminence. Somatostatin inhibits secretion of growth hormone (GH; somatotropin), a pituitary tropic hormone. The catecholamines, norepinephrine and epinephrine, are mainly products of the sympathetic postganglionic neurons and the adrenal medulla, respectively. Somatomedin is a hepatic hormone.

2. The answer is D [Chapter 49 I D]. The substrate for dopamine β-hydroxylase is dihydroxyphenylethylamine (dopamine), which is converted to norepinephrine within the cytoplasmic granules of the adrenomedullary chromaffin cell or within the vesicle of the postganglionic sympathetic neuron. The other substrates and their enzymes are: epinephrine or norepinephrine/catechol-O-methyltransferase (COMT) or monoamine oxidase (MAO), tyrosine/tyrosine hydroxylase, phenylalanine/phenylalanine hydroxylase, and dihydroxyphenylalanine (dopa)/dopa decarboxylase.

3. The answer is D [Chapter 50 II B 4 c (2)]. The second meiotic division of the oocyte is not completed until just after fertilization. From the third month of embryonic life, the primary oocytes are arrested in the diplotene phase of the early prophase of meiosis. Completion of this first division occurs just prior to ovulation by the extrusion of the first polar body, resulting in the formation of a secondary oocyte with a haploid number of chromosomes (22 autosomes and 1 sex chromosome). The onset of ovulatory menstrual cycles may lag several months behind menses, which usually occurs between the ages of 12 and 14. The second meiotic division occurs just after conception, yielding a fertilized ovum (zygote) and a second polar body.

4. The answer is C [Chapter 50 II D 1 a (2) (b); Table 50-3]. In terms of serum concentration, the major postmenopausal steroid and pituitary tropic hormones are estrone and follicle-stimulating hormone (FSH). Menopause, defined as the physiologic cessation of menses, results from a loss in the cyclic ovarian function caused by the failure of the ovary to respond to gonadotropins. The decline in ovarian function causes an increase in pituitary tropic hormones, with a striking increase in plasma FSH concentration relative to the increase in plasma luteinizing hormone (LH) level. Some estrogens continue to be produced in postmenopausal women by the extraovarian conversion of androstenedione of adrenal origin to estrone.

5. The answer is C [Chapter 49 I D 3 b (2)]. Because cortisol activates the epinephrine-forming enzyme phenylethanolamine-N-methyltransferase (PNMT), a decrease in cortisol secretion would lead to decreased adrenomedullary synthesis of epinephrine. Another effect of cortisol is to elevate blood glucose through gluconeogenesis and inhibition of peripheral glucose transport and utilization. Cortisol's proteolytic effect in extrahepatic tissues leads to increased transport of amino acids to the liver, where they are used for gluconeogenesis, glycogenesis, and protein synthesis. Hypocortisolism, therefore, would result in decrements in plasma glucose concentration, hepatic glycogen, and hepatic protein. A decline in free blood cortisol levels would also lead to an increase in adrenocorticotropic hormone (ACTH; corticotropin) secretion due to the reduced negative feedback effect of cortisol on the hypothalamic-pituitary complex.

6. The answer is A [Chapter 51 I D 1 b]. Glucose is transported by facilitated diffusion in erythrocytes, skeletal and cardiac muscle, and adipocytes. Facilitated diffusion for glucose transport requires a protein carrier molecule in the cell membrane. The insulin-dependent facilitated diffusion mechanism for glucose is found in skeletal and cardiac muscle and in adipose tissue. Some tissues obtain their glucose requirements by insulin-independent mechanisms. For example, central nervous system (CNS) cells and hepatocytes transport glucose by simple diffusion, whereas erythrocytes transport glucose by insulin-independent facilitated diffusion. In renal and gastrointestinal (GI) epithelium, glucose transport occurs by a secondary active transport system that is not insulin-dependent but is Na^+-dependent. Since Na^+ and glucose are transported in the same directions, this active transport system is called a symport system.

7. The answer is A [Chapter 49 I H 1 b (2)]. After an overnight fast, about 75% of the hepatic glucose output is derived from glycogen and about 25% from gluconeogenesis. If fasting is prolonged or combined with exercise, the hepatic glycogen stores are depleted much more rapidly and the percentage contribution from gluconeogenesis increases. The gluconeogenic substrates are lactate, glycerol, and amino acids. The lactate is derived from incomplete oxidation of glucose by the action of epinephrine on muscle, which comprises about 45% of the body mass. Thus, net hepatic glucose synthesis and secretion requires lactate prodcution from muscle glycogenolysis. The major gluconeogenic source of endogenous glucose production by the action of cortisol is alanine, with a smaller fraction available from the glycerol released from triglycerides hydrolyzed in adipose tissue. Acetyl coenzyme A (acetyl-CoA) is not a gluconeogenic substance in mammalian liver. Thus, epinephrine causes hyperglycemia because it stimulates hepatic glycogenolysis and gluconeogenesis and muscle glycogenolysis while it inhibits insulin secretion.

8. The answer is A [Chapter 50 III B 2 (b) (3)]. The major source of progesterone during the second and third trimesters of pregnancy is the trophoblast of the placenta. Maternal cholesterol is the principal source of precursor substrate for the biosynthesis of progesterone. Thus, progesterone levels during pregnancy are an index of the functional status of the materno-placental unit. Plasma progesterone levels during pregnancy are of limited value as an indicator of placental function alone because of the variability of normal levels.

9. The answer is A [Chapter 52 III B]. Triiodothyronine (T_3) is the most biologically active iodothyronine secreted by the thyroid follicles. In the secretory process, thyroglobulin (the major storage form of thyroid hormone) is degraded to free amino acids, including tetraiodothyronine (T_4), T_3, monoiodotyrosine, and diiodotyrosine. Of these, only T_4 and T_3 are released into the bloodstream, in a ratio of about 20:1. T_3 has three to five times the biologic activity of T_4 and is considered the most biologically active form of thyroid hormone. T_3 also can be formed from T_4 by the action of 5′-deiodinase in peripheral tissues, especially the liver and kidney. The other circulating iodothyronines, such as reverse triiodothyronine (rT_3), tetraiodothyroacetic acid, and triiodothyroacetic acid have much less biologic activity than T_3 and T_4.

10. The answer is D [Chapter 53 V A 3 b (3)]. Active vitamin D_3 (1,25-dihydroxycholecalciferol; calcitriol) is formed in the proximal convoluted tubule by conversion (via the action of 1α-hydroxylase) from its immediate precursor, 25-hydroxycholecalciferol (calcidiol). Vitamin D activation begins in the skin, where the previtamin, 7-dehydrocholesterol, is photoactivated by sunlight to lipid-soluble vitamin D_3. Vitamin D_3 then is converted in the liver to calcidiol, the major circulating form of vitamin D_3.

11. The answer is C [Chapter 49 II A 3]. In the fetus, the adrenal glands are much larger relative to body size than in the adult. In absolute size, they are almost as large at term as the fetal kidneys and are as large as the adult adrenal glands. The fetal zone, or inner zone, of the fetal adrenal cortex undergoes complete involution 4–12 weeks postpartum. The outer zone, or neocortex, undergoes further differentiation into the adult adrenal cortex.

12. The answer is B [Chapter 50 III B 3 a (2)]. The chief product of the fetal zone of the adrenal cortex is the weak androgen dehydroepiandrosterone (DHEA), which is secreted as the inactive sulfate ester. DHEA sulfate is converted in the fetal liver to 16α-hydroxydehydroepiandrosterone sulfate, which, upon aromatization in the placenta, becomes estriol (estriol is synthesized only by the placenta and adult liver). After the first trimester, the inner zone of the fetal adrenal cortex also can secrete cortisol and aldosterone. Progesterone and corticosterone are synthesized, but not secreted, by the fetal adrenal cortex.

13. The answer is C [Chapter 50 I E 1 c (1) (b)]. Prostaglandins found in seminal fluid are secretory products of the seminal vesicles. Chemical analysis of seminal fluid provides an indirect measure of testicular function, because the male accessory sex organs (i.e., the seminal vesicles and prostate gland) are androgen-dependent. The seminal vesicles secrete fructose, prostaglandins, and ascorbate into the seminal plasma; the prostate gland secretes acid phosphatase and citrate into the seminal fluid. Originally, the prostate was believed to be the source of the prostaglandins.

14. The answer is E [Chapter 50 II F 1 c (2)]. The secretory products of the ovary are estradiol and estrone, which are produced by the granulosa cells of the unruptured follicle and the lutein-granulosa cells of the corpus luteum. Luteinization of the granulosa cells, which depends on LH, involves the appearance of lipid droplets in the cytoplasm. The estrogenic precursors produced by the theca interna (and lutein-theca interna) are androstenedione and testosterone, which are converted by the granulosa (and lutein-granulosa) cells into estrone and estradiol, respectively. The lutein-granulosa cells also produce and secrete progesterone. Thus, androstenedione and pregnenolone are produced, but not secreted, by the ovary. Estriol is not an ovarian product but is synthesized in the liver from estradiol and estrone and in the placenta by the conversion of imported androgens. Pregnanediol is the urinary metabolite of progesterone and is formed in the liver.

15. The answer is B [Chapter 50 I B 2]. Ninety percent of the volume of the testis is composed of tubular tissue, with the remaining 10% consisting mainly of nontubular, or interstitial, cells (also called Leydig cells). The seminiferous tubular epithelium contains three cell types: spermatogonia, spermatocytes, and Sertoli cells. Spermatogonia and spermatocytes are germinal cells, and the Sertoli cells are nongerminal cells.

16. The answer is C [Chapter 53 VI C 3]. The major renal effect of parathyroid hormone (PTH) is stimulation of proximal tubular 1α-hydroxylase, an enzyme that converts calcidiol to calcitriol. PTH affects renal Ca^{2+} reabsorption in two ways: it reduces Ca^{2+} reabsorption in the proximal tubule and increases Ca^{2+} reabsorption in the distal tubule. The net effect is an increase in tubular Ca^{2+} reabsorption. This renal effect of PTH can be lessened by PTH's effect on bone. PTH increases plasma $[Ca^{2+}]$ by promoting bone resorption, which causes an increase in the renal filtered load of Ca^{2+} and, thus, hypercalciuria. Also, PTH inhibits the proximal reabsorption of phosphate, thereby favoring phosphate excretion. PTH acts on bone by dissolving the nonreadily exchangeable calcium phosphate "pool" known as stable bone. PTH activates the osteoclasts—cells that cause osteolysis by their high content of lysosomal enzymes. PTH also stimulates osteocytes, which are bone-bound osteoblasts that mediate osteocytic osteolysis.

17. The answer is A [Chapter 53 VIII C 3]. Parathyroid hormone (PTH) and active vitamin D_3 (calcitriol) have several similar effects, particularly in Ca^{2+} regulation. However, these two hormones exert different effects on renal phosphate handling: calcitriol promotes phosphate reabsorption, whereas PTH promotes phosphate diuresis. PTH and calcitriol both promote bone resorption, which leads to an increase in serum $[Ca^{2+}]$. Calcitriol promotes intestinal absorption of Ca^{2+} and phosphate. Although PTH has no direct intestinal effect on Ca^{2+} or phosphate absorption, it does stimulate renal 1α-hydroxylase activity and subsequent calcitriol formation; thus, PTH indirectly promotes intestinal absorption of Ca^{2+} and phosphate. In the kidney, PTH promotes distal tubular Ca^{2+} reabsorption, inhibits proximal phosphate reabsorption, and activates proximal 1α-hydroxylase. Calcitriol enhances renal tubular reabsorption of both Ca^{2+} and phosphate. At pharmacologic doses, calcitriol mimics the effects of PTH (i.e., it promotes phosphate excretion).

18. The answer is A [Chapter 50 II E 2]. In this woman with a menstrual cycle of 21–23 days, ovulation would be expected to occur between day 7 and day 9, counting from the first day of menses. The normal menstrual cycle has an average duration of 28 days. The luteal, or postovulatory, phase corresponds to the secretory phase of the endometrium or the progestational phase of the corpus luteum. This phase of the menstrual cycle usually is very constant, lasting about 14 days, while the duration of the follicular phase can be highly variable. Because of the constancy of the postovulatory phase, the time of ovulation can be estimated by subtracting 14 days from the cycle length.

19. The answer is C [Chapter 50 I A 1 e (1), (2)]. Patients with testicular feminization have a female phenotype but a male genotype. They have intra-abdominal testes that produce testosterone and estrogen concentrations that are characteristic of normal men, but their tissues are totally unresponsive to androgens because of a lack of androgen receptors and they do not produce sperm. The external genitalia are female because the primordial tissue develops in the female pattern unless stimulated by androgen. The androgen resistance syndromes are disorders in which müllerian duct regression and testosterone synthesis are normal. Male development of the embryo requires müllerian duct inhibiting factor (to prevent the development of female internal genitalia), testosterone (to mediate wolffian duct development), and dihydrotestosterone (to stimulate development of the prostate and male external genitalia). This patient lacks female internal genitalia (because the müllerian duct inhibiting factor exerts its normal action) and lacks male wolffian duct development (because testosterone does not exert its normal stimulatory action). Since both duct systems regress, neither male nor female internal genitalia develop. Secondary sex characteristics, including breast development, appear at puberty in response to the unopposed action of estrogen formed extragonadally from testosterone. This patient has male pseudohermaphroditism, in that the female phenotype in this genetic male is not a result of the presence of both testes and ovaries.

20–23. The answers are: 20-C, 21-B, 22-D, 23-B [Chapter 50 II E 2 b, c (3), F 1 b; III A 1 d; Figure 50-2]. In a normal ovarian cycle, the plasma concentration of estradiol peaks (*point A*), triggering a sudden discharge of pituitary gonadotropins [luteinizing hormone (LH) and follicle-stimulating hormone (FSH)] 12–24 hours later. This preovulatory gonadotropin surge (*point B*) is more pronounced for LH than for FSH. Ovulation (*point C*) occurs 16–20 hours after the LH peak and marks the formation of the corpus luteum. LH maintains the functional and morphologic integrity of the corpus luteum for 14 days.

The ovum can be fertilized between 6 and 20 hours following ovulation (*point D*). If fertilization occurs, a morula is formed, which enters the uterine cavity on the third to fourth day postovulation (*point E*). The morula is converted to the blastocyst during the fifth to sixth day postovulation, which is implanted in the endometrial wall of the uterus on the seventh day (*point F*). At *point G*, human chorionic gonadotropin (HCG) rescues the corpus luteum.

The corpus luteum is the principal source of estrogen (estradiol) and progesterone during the first 6–8 weeks gestation. The principal gonadotropic hormone during this period is HCG, which is secreted by the syncytiotrophoblast of the placenta. This anterior pituitary-like hormone has mainly LH activity and is the luteotropic hormone of pregnancy.

The trophoblast takes over as the major source of progesterone and estrogen (estriol) secretion by 8 weeks gestation. HCG also stimulates the fetal testis to produce testosterone and the fetal adrenal gland to produce corticoids in early pregnancy.

24. The answer is D [Chapter 50 I C 1 c]. In skin and target cells of the male reproductive tract, testosterone may be reduced to the more potent androgen, dihydrotestosterone, in a one-way reaction by the enzyme 5α-reductase. Testosterone also is metabolized to estradiol by the action of aromatase in various tissues, including brain, breast, and adipose tissue. During fetal organization, the testis is stimulated by placental human chorionic gonadotropin (HCG) to produce testosterone. The characteristic adolescent "growth spurt" in the male results from an interplay of testosterone and growth hormone (GH), which promote growth of the vertebrae, shoulder girdle, and long bones. This growth is self-limited, as androgens also accelerate epiphysial closure. In adulthood, testosterone is secreted from the testis in response to LH secretion from the pituitary gland; testosterone acts as a negative feedback regulator of LH secretion and, thus, its own secretion.

25. The answer is A [Chapter 49 II F 3 a–c]. The overall metabolic effects of cortisol are the release of amino acids from muscle and both the storage and release of glucose and fatty acids. Cortisol inhibits fatty acid synthesis in the liver, and it increases blood glycerol and fatty acid concentrations in concert with other hormones (norepinephrine, epinephrine, glucagon) to increase lipolysis in adipose tissues. Glycerol is an excellent index of lipolysis, because, unlike free fatty acids, it is not reused by adipocytes in the resynthesis of triglycerides. Rather, glycerol is used by the liver as a gluconeogenic substrate. Cortisol also promotes synthesis of gluconeogenic enzymes needed to convert the amino acids released from muscle to carbohydrate synthesis in the liver. Glucocorticoids promote proteolysis and inhibit protein synthesis in most tissues except the liver.

26. The answer is C [Chapter 49 II F 1 a]. Glucocorticoids are potent hypoglycemic hormones that, at high levels, lead to glucose intolerance. Glucocorticoids stimulate the liver to produce glucose and glycogen owing, in part, to the increased synthesis of key gluconeogenic enzymes. They also promote muscle proteolysis, resulting in the release of amino acids, which (aside from leucine) are gluconeogenic. The synthesis of hepatic gluconeogenic enzymes accounts for the protein anabolic effect of glucocorticoids in liver. Hepatic glycogenesis occurs via the activation of glycogen synthase by glucocorticoids. The increased hepatic glucose production and secretion involves gluconeogenesis, not glycogen degradation (glycogenolysis).

27. The answer is C [Chapter 50 II D 1 a (3); III B 3]. Estriol, the least biologically active natural estrogen, is produced only in the liver and placenta. It is not an ovarian product but represents the predominant urinary end product of estrogen metabolism. In pregnancy, the levels of urinary estriol increase 1000-fold, because estriol is formed by the trophoblast of the placenta. The liver converts estrone and estradiol of ovarian origin into estriol, whereas the trophoblast converts 16α-hydroxydehydroepiandrosterone (16α-OH DHEA) sulfate of fetal adrenal origin into estriol.

28. The answer is A [Chapter 52 IV A]. The biosynthetic pathway for thyroid hormone has four steps: active uptake of inorganic iodide, oxidation of iodide to active iodide, iodination of the tyrosyl residues on the thyroglobulin molecule, and coupling of iodotyrosines to form iodothyronines. Only the first step does not require thyroid peroxidase. Secretion depends on a lysosomal protease.

29. The answer is B [Chapter 48 III B, D; IV B 3]. Prolactin, not oxytocin, causes milk synthesis (lactogenesis); oxytocin evokes milk secretion (ejection). Oxytocin is an octapeptide produced mainly in the paraventricular nucleus of the ventral diencephalon. Because it is synthesized in the magnocellular neurosecretory neurons, oxytocin is classified as a neuropeptide, or neurosecretory hormone. The unmyelinated axons of the paraventricular nucleus contribute to the supraopticohypophysial tract, which terminates in the pars nervosa. The pars nervosa is a release center and not the site of oxytocin synthesis. Oxytocin has two important smooth muscle effects: It causes milk ejection via contraction of the myoepithelial cells of the mammary gland, and it promotes myometrial (uterine) contraction.

30. The answer is D [Chapter 49 I H 1 c (1) (a)]. Activation of the sympathetic nervous system involves stimulation of both α- and β-adrenergic receptors by both norepinephrine and epinephrine. The α-adrenergic response of the pancreatic islet cells is dominant during stimulation; thus, insulin secretion (a response mediated by β-adrenergic receptors) is suppressed. Other responses that are mediated by β-adrenergic receptors include renin secretion, vasodilation, and inhibition of intestinal motility. Responses mediated by α-adrenergic receptors include inhibition of intestinal motility, contraction of the radial eye muscle (mydriasis), and vasoconstriction. Vasodilation in skeletal muscle is caused by epinephrine and by stimulation of sympathetic cholinergic nerves.

31. The answer is D [Chapter 51 I D 1 b]. Insulin decreases cell membrane permeability to both Na^+ and K^+, but it decreases Na^+ permeability more, causing hyperpolarization of skeletal and cardiac muscle cells and adipocytes but not liver or pancreatic cells. Insulin also promotes lipogenesis, protein anabolism, glycogenesis, and glycolysis. Insulin promotes glucose uptake by muscle and adipose tissue but not by the liver, and it promotes storage and utilization of glucose in all three tissues. Glucose storage by muscle and liver is achieved by formation of glycogen via an increase in glycogen synthase activity. Glucose transport across skeletal and cardiac muscle cells and adipocytes occurs by insulin-dependent facilitated diffusion.

32. The answer is A [Chapter 50 III B 2 d, C 2]. By 8 weeks gestation, maternal cholesterol can be converted to progesterone by the placental trophoblast, which becomes the major producer of progesterone. Progesterone is necessary for maintenance of the decidual cells of the endometrium, inhibition of myometrial contraction by the hyperpolarization of uterine smooth muscle cells, and formation of a dense, viscous mucus that seals off the uterine cavity. Progesterone, in synergy with estrogen, promotes the growth and branching of the lobuloalveolar ductal system of the mammary gland. In early pregnancy, placental progesterone serves as a precursor for the synthesis of cortisol and aldosterone by the fetal zone of the adrenal cortex under the stimulation of human chorionic gonadotropin (HCG). Beyond 10 weeks gestation, the outer zone of the fetal adrenal cortex (neocortex) can produce these two corticoids under the stimulation of adrenocorticotropic hormone (ACTH). Progesterone also exhibits a natriuretic effect by antagonizing the action of aldosterone on the renal tubule.

33. The answer is B [Chapter 49 I F 2 a (1), (2)]. Vanillylmandelic acid (VMA) is the major urinary end product of catecholamine metabolism. Monoamine oxidase (MAO) is a mitochondrial enzyme that catalyzes the oxidative deamination of the catecholamines—dopamine, norepinephrine, and epinephrine. MAO also inactivates the indolamine, serotonin (5-hydroxytryptamine). Substrates for MAO also include normetanephrine and metanephrine.

34. The answer is C [Chapter 50 III B 3 a]. The neocortex of the fetal adrenal gland does not participate in estriol synthesis during gestation. Estriol is a product of the combined activities of the fetal pituitary gland, the fetal zone of the adrenal cortex, the fetal liver, and the trophoblast of the placenta. The fetal zone of the adrenal gland is the inner zone that persists only during gestation; the neocortex is the outer zone of the fetal adrenal gland, which is the progenitor of the adult adrenal cortex. In the first trimester, the fetal zone is primarily stimulated by human chorionic gonadotropin (HCG), with adrenocorticotropic hormone (ACTH) [from the fetal pituitary] gaining prominence thereafter. The fetal zone synthesizes dehydroepiandrosterone (DHEA) and its sulfate (DHEAS), which are transported to the placenta after 16α-hydroxylation by the fetal liver. It is 16α-OH DHEAS that is the precursor of placental estriol.

35. The answer is D [Chapter 48 IV A 3 a (1)]. The protein anabolic effects of growth hormone (GH) are mediated through a GH-dependent peptide known as somatomedin, which is synthesized in the liver. Important biologic effects of somatomedin include mitogenesis of chondrocytes and bone cells, stimulation of lipogenesis and muscle glycogenesis, and GH-like actions in cartilage (e.g., sulfation of chondroitin). Antidiuretic hormone (ADH) and oxytocin are neuropeptides synthesized mainly in the neurosecretory neurons of the supraoptic and paraventricular nuclei, respectively. Thyrotropin releasing hormone (TRH) is produced in the neurosecretory neurons of the arcuate nucleus. Endorphins are secretory products of basophils in the adenohypophysis

and of brain neurons. Thus, endorphins also can be classified as neuropeptides.

36. The answer is D [Chapter 48 IV A 2 a (2)]. Secretion of growth hormone (GH) in response to various stimuli often is blunted in obesity. Stimuli for GH secretion include norepinephrine, bromocriptine, hypoglycemia, stress, vigorous exercise, insulin, and arginine. The long-term metabolic effects of GH include hyperglycemia owing to gluconeogenesis, glycogenolysis, and lipolysis, all of which explain its anti-insulin effects. GH has a fat catabolic effect, which increases plasma free fatty acid levels. It should be noted that these free fatty acids are not substrates for gluconeogenesis.

37. The answer is C [Chapter 47 II A, B]. The polypeptide known as β-endorphin is formed from β-lipotropin [not from adrenocorticotropic hormone (ACTH)], which is a cleavage product of the prohormone proopiomelanocortin (POMC). POMC and its derivatives are synthesized in the basophils of the pars distalis of the anterior lobe of the pituitary gland. Endorphins, β-lipotropin, and ACTH (corticotropin) also are found in the brain. ACTH can be cleaved to form α-melanocyte-stimulating hormone (α-MSH) and a corticotropin-like peptide called CLIP. β-Endorphin binds with morphine receptors and produces morphine-like responses, such as respiratory depression, analgesia, and miosis.

38. The answer is D [Chapter 48 IV A 3 b (2)]. Growth hormone (GH), like insulin and normal levels of thyroid hormone, is a potent protein anabolic hormone (i.e., it promotes nitrogen retention and, thus, a positive nitrogen balance). Negative nitrogen balance results when nitrogen losses exceed nitrogen intake, as occurs with deficient dietary protein or in stressful conditions (e.g., tissue trauma, disease states, burns, surgery). Utilization of amino acids in hepatic gluconeogenesis results in negative nitrogen balance. Glucocorticoids (e.g., cortisol) promote muscle proteolysis and inhibit protein synthesis, leading to increased BUN and enhanced nitrogen excretion. Alanine is the predominant amino acid released by muscle. Glucagon is a catabolic hormone for protein in muscle, and it also is an important gluconeogenic hormone. At normal levels, thyroid hormone stimulates protein synthesis and degradation. The positive effect of thyroid hormone on body growth is derived largely from stimulation of protein synthesis.

In excess amounts, thyroid hormone causes accelerated protein catabolism, leading to increased nitrogen excretion.

39. The answer is E [Chapter 49 I H 1 c]. Epinephrine is a potent hyperglycemic hormone owing mainly to its effects on the liver and pancreas. In the liver, it promotes glycogenolysis and gluconeogenesis, and the glucose-6-phosphate formed by glycogenolysis is hydrolyzed to glucose. Epinephrine-induced glycogenolysis in muscle leads to formation of lactic acid, which is converted to glucose in the liver, further elevating blood glucose. In the pancreas, epinephrine stimulates glucagon secretion and inhibits insulin secretion, both hyperglycemic effects. Epinephrine also stimulates adrenocorticotropic hormone (ACTH) secretion, which, in turn, leads to secretion of cortisol, which is a major hyperglycemic hormone.

40. The answer is D [Chapter 52 IV A; VI A 2 b; VII A]. Normally, thyroid hormone (T_3 and T_4) is regulated by thyroid-stimulating hormone (TSH), which, in turn, is controlled by thyrotropin releasing hormone (TRH). The circulating levels of T_3 and T_4 also influence TSH release by exerting negative feedback control at the adenohypophysial and, to a lesser extent, hypothalamic levels. TSH influences the structure and function of thyroid follicular cells and regulates all phases of thyroid hormone synthesis, storage, and secretion. Administration of exogenous thyroid hormone leads to disuse atrophy of the thyroid gland via TSH suppression. Thus, iodide trapping and endogenous thyroid hormone secretion are reduced when TSH is suppressed by exogenous thyroid hormone. In most cells of the body, thyroid hormone stimulates the cell membrane enzyme Na^+-K^+-ATPase, thereby increasing O_2 consumption. This effect is not exerted on cells of the brain, lymph nodes, gonads, lungs, spleen, and dermis.

41–44. The answers are: 41-A, 42-B, 43-C, 44-B. [Chapter 49 I C 2, F 2 b, I 2 c]. The adrenal medulla is a modified sympathetic ganglion consisting of chromaffin cells (pheochromocytes) that are the functional analogs of the postganglionic neurons. About 80% of the adrenomedullary secretion is epinephrine, and 20% is norepinephrine. The plasma concentration ratio of epinephrine to norepinephrine normally is 1:4, the inverse of the secretory ratio. All circulating epinephrine originates in the adrenal medulla; therefore,

bilaterally adrenalectomized patients have very low plasma epinephrine levels. In contrast, plasma norepinephrine levels in these patients remain within normal limits, which demonstrates that the major source of circulating norepinephrine is not the adrenal medulla.

Most circulating norepinephrine is derived from the postganglionic sympathetic neurons. Norepinephrine has been demonstrated in almost all tissues except the placenta, which lacks nerve fibers. The norepinephrine content of a tissue or organ serves as an index of the density of its sympathetic autonomic innervation.

Catecholamines are catabolized by two enzymes: catechol-O-methyltransferase (COMT) and monoamine oxidase (MAO). Norepinephrine and epinephrine are metabolized extracellularly by COMT to normetanephrine and metanephrine, respectively. COMT is localized within the cytosol of sympathetic effector cells.

Tumors may arise within the sympathoadrenomedullary system. Adrenal chromaffin cell tumors, known as pheochromocytomas, usually arise within the adrenal medulla, are benign, and secrete primarily excessive amounts of norepinephrine. However, pheochromocytoma patients may have signs of both norepinephrine and epinephrine hypersecretion. Most patients have paroxysms of hypertension, tachycardia, sweating, tremor, palpitations, and nervousness. Most lose weight and are hyperglycemic because of the catecholamine-induced inhibition of insulin secretion.

45–48. The answers are: 45-C, 46-C, 47-A, 48-A [Chapter 48 IV A 2 b (1); Chapter 51 I A 1 b, c, 3 c, C 6 a]. Somatostatin and insulin both are hypoglycemic hormones. Somatostatin decreases blood glucose levels by decreasing hepatic production. It also inhibits the secretion of growth hormone (GH), glucagon, and insulin and decreases intestinal glucose absorption. Insulin exerts its hypoglycemic action by promoting glucose uptake by muscle and adipose tissue, by increasing storage and utilization of glucose (mainly by liver, muscle, and adipose tissue), and by suppressing hepatic production and release of glucose.

Somatostatin and insulin are polypeptides, consisting of 14 and 51 amino acid residues, respectively. Insulin is synthesized in the pancreatic beta cells; somatostatin is synthesized in the pancreatic delta cells as well as many other sites, including the hypothalamus, cerebrum, thymus, thyroid, gastric and intestinal epithelium, skin, and heart. Insulin inhibits glucagon secretion, and somatostatin inhibits secretion of GH, thyroid-stimulating hormone (TSH), insulin, glucagon, pancreatic polypeptide, gut hormones, gastric acid, and pepsin.

Somatostatin is synthesized in the arcuate nucleus of the tuberoinfundibular neurosecretory neurons that terminate in the median eminence. Somatostatin of hypothalamic origin functions as a neurosecretory hormone. It also is a hypophysial portal system. Insulin is neither a neurosecretory hormone nor a hyophysiotropic hormone.

49–54. The answers are: 49-B, 50-A, 51-B, 52-B, 53-C, 54-B [Chapter 49 I G; Table 49-1]. Insulin secretion is elicited by stimulation of the β-adrenergic receptor on the surface of the pancreatic beta cell.

In order to increase pupillary diameter, the radial eye muscle must contract. Contraction of this intrinsic ocular muscle occurs by stimulation of the α-adrenergic receptor. The radial ocular muscle is innervated only by postganglionic sympathetic nerves. A decrease in pupillary diameter occurs by stimulation of the parasympathetic innervation of the pupillary sphincter.

Bronchodilation is caused by stimulation of β-adrenergic receptors. There are no α-adrenergic receptors in bronchial smooth muscle. Bronchoconstriction occurs by stimulation of the parasympathetic innervation of the bronchi.

Renin is synthesized by the juxtaglomerular cells of the afferent arteriole, which receive postganglionic sympathetic innervation. Thus, these autonomic nerves secrete norepinephrine, which stimulates β-adrenergic receptors on the surface of the juxtaglomerular cells.

Intestinal smooth muscle contains both α- and β-adrenergic receptors. Intestinal motility is inhibited by stimulation of either receptor type. Sphincters in the gastrointestinal (GI) tract contain only α-adrenergic receptors, which, when stimulated, lead to sphincter contraction.

The detrusor muscle of the urinary bladder, in contrast to the trigone and sphincter, contains only β-adrenergic receptors, which, when stimulated, evoke relaxation of this smooth muscle. Stimulation of the α-adrenergic receptors of the bladder produces contraction of these smooth muscles. Thus, sympathetic stimulation of the bladder blocks micturition.

CASE STUDIES IN CLINICAL DECISION MAKING

Case 1

A 57-year-old white man enters his physician's office complaining of tiredness, shortness of breath, and swelling of the ankles. His vital signs are blood pressure, 128/80; heart rate, 112 beats/min; respiratory rate, 18/min.

1. What organ system is most likely involved if the patient has these symptoms?

DISCUSSION

Whereas tiredness could be related to depression or severe anemia, and shortness of breath could be related to respiratory dysfunction, and swelling of the ankles could be caused by renal disease with proteinuria, it makes more sense to connect all of these symptoms. Dysfunction of the cardiovascular system could explain all of these symptoms. Depressed cardiac function could cause an inadequate supply of nutrients and oxygen, which may result in tiredness. A decrease in myocardial contractility could also produce shortness of breath by several mechanisms. The most direct cause is an increase in ventricular end diastolic volume (VEDV) because of a reduction in stroke volume (SV), which in turn increases the ventricular end diastolic pressure (VEDP). This backs up blood into the left atrium, pulmonary veins, and pulmonary capillaries. Engorgement of the pulmonary vasculature makes it more difficult to breathe and may also interfere with gas exchange.

Upon further questioning, the patient states that these symptoms have been gradually getting worse over the last 6 months. He says that he occasionally feels some palpitations in his chest and that he has a "heavy feeling" in his chest when he climbs stairs or walks up hills. An electrocardiogram (EKG) is performed. A portion of lead II is shown below.

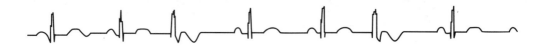

2. What does the EKG reveal, and how is this related to the patient's symptoms?

DISCUSSION

The palpitations are the result of ventricular extrasystoles. The patient usually feels the beat following the extrasystole because it is stronger due to the added stroke volume. Lead II shows several ventricular extrasystoles (beats 3 and 6), which can be caused by many different cardiac diseases. The "heavy feeling" in the patient's chest probably indicates early angina pectoris or myocardial ischemia.

3. What could be the cause and consequences of this condition?

DISCUSSION

The most likely cause of this condition is atherosclerosis of the coronary arteries. The formation of atherosclerotic plaques causes narrowing of the coronary arteries and limits coronary blood flow. The decrease in blood flow results in an inadequate delivery of oxygen and nutrients to the heart, which causes hypoxia because oxygen utilization exceeds the oxygen supply. During hypoxia the ventricles cannot generate normal systolic pressures so that myocardial fiber shortening is reduced; that is, there is a decrease in myocardial contractility.

4. What is the consequence of a decreased myocardial contractility?

DISCUSSION

The decreased contractility results in a reduced stroke volume by the ventricle. Because the amount of fiber shortening is reduced, there is an increase in the end systolic volume of the ventricles. This increased volume coupled with a normal venous return leads to an increase in the end diastolic volume and pressure (preload) of the ventricle. The increase in preload causes the heart to utilize the Frank-Starling mechanism in order to maintain stroke volume. While stroke volume may return almost to normal, the heart is at a disadvantage because of its larger size and the need to generate increased tension due to the Laplace effect. In addition, the ventricular wall becomes thinner, since a larger ventricle has a larger surface area over which the muscle must be distributed. This, again, means that each myocardial fiber must generate more tension. With time, the need for increased tension generates an increase in the thickness of the myocardial cells (hypertrophy), which can compensate somewhat for the need for increased tension. However, the increase in myocardial cell size means that oxygen must diffuse further to reach the center of the cells, which may exacerbate the hypoxia. As the disease progresses, the heart increasingly relies on an increase in heart rate to maintain cardiac output. This is one reason why the patient's heart rate is elevated. He may also be nervous about seeing a physician or about finding out what is wrong, which could also contribute to an increased heart rate.

5. What is the next logical step in taking care of this patient so that he can carry out his normal activities?

DISCUSSION

The physician prescribes skin patches containing nitrates (a vasodilator), and sends the patient for a cardiac stress test. The vasodilators may cause only minimal changes to the resistance of the coronary arteries because the atherosclerotic plaques are a rigid segment of the vessels that cannot dilate. Dilation of the more distal coronary vessels can provide some benefit from vasodilators. However, vasodilators may affect other systemic vascular beds and decrease the afterload on the ventricles. By decreasing afterload, the ventricles are able to eject more of the end diastolic volume, reduce end systolic volume, and finally lower the radius of the ventricles so that wall tension is reduced by means of the Laplace effect. This reduces ventricular work and provides additional reserve for the ventricles.

The patient underwent a stress test that included measurements of cardiac output and pulmonary wedge pressure.

6. What is pulmonary wedge pressure and what does it measure?

DISCUSSION

Pulmonary wedge pressure is obtained by means of a Swan-Ganz catheter. The catheter is advanced into a branch of the pulmonary artery and wedged in place. A small balloon on the distal end of the catheter is inflated to ensure complete occlusion of that arterial branch. When blood flow in the vessel ceases, there is a reflection backward of the pressure in the left atrium. Therefore, the pulmonary artery wedge pressure measures the filling pressure of the left ventricle and can provide a measure of the left ventricular end diastolic pressure (VEDP). Changes in left VEDP are indicative of changes in left VEDV.

During the patient's cardiac stress test, the following information was obtained.

Parameter	Predicted	Rest	Maximal Exercise
Heart rate (beats/min)	75–80	115	175
Cardiac output (L/min)	5.5	4.25	7.0
Pulmonary wedge pressure (mm Hg)	< 10	17	25
Maximal oxygen consumption (L/min)	2.0		0.75

7. What is the significance of the data?

DISCUSSION

These are very abnormal results. Even while at rest, the patient shows abnormal values for heart rate and pulmonary wedge pressure. The high pulmonary wedge pressure probably indicates a large left VEDV yet the stroke volume is approximately half the normal value. This means that the ejection fraction (SV/VEDV), which normally is 0.5–0.6, is very small. During exercise, the patient was able to increase cardiac output less than twice the normal resting value, although his heart rate increased to a near maximal rate. This indicates that the patient was cooperating fully and exercising to his maximal capacity but his heart was limiting his ability to exercise. The large rise in the pulmonary wedge pressure again indicates that the ventricle is unable to eject sufficient blood. This indicates that the heart is dilating markedly in response to exercise, which is abnormal.

8. What are the consequences of an increase in ventricular end diastolic pressure?

DISCUSSION

An increase in VEDP causes a back up of pressure through the venous and the capillary systems. In order to have blood flow continue, the pressure must be higher at the source; that is, pulmonary artery pressure > pulmonary capillary pressure > pulmonary venous pressure > left atrial pressure > left VEDP. The increased pulmonary capillary pressure can lead to additional filtration of fluid through the endothelium (Starling hypothesis) and the onset of pulmonary edema. Edema in the lungs interferes with gas exchange and also makes breathing more difficult and causes shortness of breath (dyspnea).

The patient is referred for coronary angiography in order to visualize the coronary arteries. The angiography reveals that the patient has several branches of the left coronary artery that are markedly narrowed by atherosclerotic plaques. The patient is then scheduled for coronary angioplasty via a balloon catheter. (In this procedure, a thin catheter is inserted into a peripheral artery and is advanced into the involved branch of the coronary artery. A small balloon at the catheter tip is inflated to dilate the coronary artery in order to increase coronary blood flow.) The coronary angioplasty was successful, and the patient returns in 1 month feeling much better and is able to climb two flights of stairs without any discomfort or shortness of breath.

9. What is the next step in treating this patient?

DISCUSSION

The patient is put on a low-fat, low-cholesterol diet and a supervised, graded exercise program to increase his aerobic capacity and to minimize the chances of recurring coronary insufficiency.

Case 2

A 48-year-old government employee presents complaining of breathlessness on exertion, a cough, and yellowish sputum production, especially in the morning. He has had these symptoms for years but has noted a worsening of his shortness of breath in the last year. Now he can climb only one flight of stairs before having to stop to catch his breath.

---

1. What is the significance of the shortness of breath, cough, and sputum production?

DISCUSSION

The cough and sputum production indicate chronic bronchitis; that is, chronic infection and inflammation of the airways. The presence of inflammatory changes, with edema of the airway walls, infiltration by inflammatory cells, and production of mucus, can narrow the airway lumen. The narrowing leads to increased airway resistance, increased work of breathing, and marked unevenness of the distribution of the tidal volume in the lungs. All of these changes in the airways alter the ventilation–perfusion ratio at the acinar level and can interfere with gas exchange. Thus, the patient has to work harder to breathe and probably suffers from inadequate oxygen uptake into his arterial blood due to ventilation–perfusion imbalance.

For the past several months, the patient has noted swelling of his ankles after he has been awake for several hours. The edema disappears while he is sleeping.

2. What is the connection between the respiratory symptoms and the ankle swelling?

DISCUSSION

The ankle swelling is most likely due to an increased pressure in the systemic venous system, which causes an increased filtration of fluid in dependent capillaries. The increased venous pressure is caused by an elevation of the right ventricular diastolic pressure secondary to an increased afterload. The increased afterload indicates an increased pulmonary vascular resistance and elevation of the pulmonary artery pressure. The increased resistance could be caused by either destruction of significant portions of the pulmonary vascular bed secondary to the pulmonary disease or to hypoxic pulmonary vasoconstriction (HPV). Administration of supplemental oxygen to this patient while measuring his pulmonary artery pressure would differentiate between these causes. If the elevated pulmonary artery pressure was due to HPV, administration of oxygen would relieve the vasoconstriction, and the pulmonary artery pressure would decrease significantly.

The patient served in the Air Force and was honorably discharged with no disability at the age of 38 years. He is a moderate smoker, smoking about one pack per day for 30 years (30 pack years). The patient denies exposure to hazardous dusts or chemicals related to any of his jobs. His mother and father both died in their sixties from heart disease. He has a 54-year-old sister who is alive and well. None of the patient's aunts or uncles died at an early age from severe lung disease.

3. What is the significance of this part of the patient's history?

DISCUSSION

The patient has had no apparent exposure to any industrial hazards. There is no evidence of any genetic defect, such as α_1-antitrypsin deficiency, which can lead to severe emphysema at an early age, given the negative family history. Therefore, the patient's symptoms are most likely related to his history of smoking.

During the physical examination, the patient's height was recorded as 72 inches and his weight was 120 pounds. His vital signs were: pulse, 104 beats/min; temperature, 37°C; blood pressure, 135/90; and respiratory rate, 10/min.

4. What is the significance of the patient's vital signs?

DISCUSSION

The elevation of the heart rate and blood pressure may be secondary to hypoxia, which stimulates the peripheral chemoreceptors. Presumably, there is an increased cardiac output as a compensatory means of increasing the oxygen delivery to the tissues. The relatively slow respiratory rate is more indicative of an obstructive lung disease. The slow rate increases the duration of the respiratory cycle, which provides additional time for lung deflation.

Physical examination revealed a thin male who appeared older than his stated age and was using accessory muscles of inspiration. His chest was hyperresonant to percussion, and there was retraction of the supraclavicular area during inspiration and bulging during expiration. Auscultation of the chest revealed distant breath sounds throughout all lung fields. Expiration was prolonged with wheezes at both lung bases posteriorly. The liver was not enlarged by percussion but was palpable two fingers below the right costal margin.

5. What is the significance of the observations from the patient's physical examination?

DISCUSSION

The use of accessory inspiratory muscles indicates an increased respiratory drive, which may be secondary to hypoxia. The retraction and bulging of the supraclavicular area indicates the interpleural pressure has an increased negativity during inspiration and becomes positive during expiration. Thus, expiration is not passive in this patient as it is normally. Both of these findings indicate an increased airway resistance, which requires much higher pressure gradients between the atmosphere and the alveoli to generate adequate gas flow. The patient is working much harder than normal in order to breathe. The hyperresonance and distant breath sounds indicate an increased gas volume in the thorax, which is characteristic of obstructive airway disease. Prolongation of expiration is another indication of a high airway resistance. Because of the high resistance, there is a reduction of expiratory flow rate, so it requires a longer time to expel the tidal volume. The expiratory wheezes indicate turbulent air flow caused by the narrowed airways. The palpable but not enlarged liver is caused by a depression of the respiratory diaphragm secondary to the enlarged thoracic gas volume. When the diaphragm is depressed, it assumes a flatter configuration, which makes it much less effective in increasing the volume of the thorax during inspiration.

The patient's jugular veins were distended 2–3 cm above the clavicle when he sat at a 45-degree angle. There was moderate ankle edema bilaterally. The second (pulmonary) component of the second heart sound was accentuated at the second left sternal border.

6. What conclusions can be drawn regarding the right ventricular filling pressure and afterload?

DISCUSSION

The jugular veins are normally collapsed because of the low diastolic pressure in the right ventricle. The distention of the jugular veins and the ankle edema are additional evidence of an increased right ventricular diastolic pressure. Thus, the right ventricle is utilizing a Frank-Starling mechanism (an increased end-diastolic volume) to maintain an adequate cardiac output. The accentuated pulmonary component of the second heart sound typically occurs with an elevated pulmonary artery pressure as discussed previously. This patient shows evidence of mild right ventricular decompensation secondary to the chronic lung disease.

The patient had cyanosis of his nail beds and mucous membranes.

7. Based on the observation of cyanosis in this patient, what might be the patient's hemoglobin level and arterial oxygenation?

DISCUSSION

The cyanosis indicates that there is at least 5 g/dl of deoxygenated hemoglobin in the capillaries and could indicate hypoxia as well as an increased hemoglobin concentration. Both of these conditions are likely because of the chronic airway disease causing ventilation–perfusion imbalance with poor gas exchange, which, in turn, leads to hypoxia with stimulation of erythropoietin release and an increased hemoglobin concentration.

A complete blood count (CBC) with differential and arterial blood gas analysis were performed on the patient. The CBC with differential was normal with the exception of slightly elevated RBCs and hemoglobin. The patient's pH was 7.37. Results of the blood gas analysis were: HCO_3^-, 26 mEq/L; $PaCO_2$, 46 mm Hg; PaO_2 (rest), 58 mm Hg; PaO_2 (exercise), 45 mm Hg.

8. Which of the laboratory results are abnormal and what is the significance of the abnormal results?

DISCUSSION

The RBC count and the hemoglobin concentration are both elevated. The arterial carbon dioxide tension ($PaCO_2$) is minimally elevated indicating that this patient's ventilation may suddenly deteriorate if the work of breathing increases further. The arterial oxygen tension (PaO_2) is well below normal, and the further decrease with exercise indicates a diffusion limitation. The decrease also may be due to the effect of venous desaturation and a shunt effect during exercise. This patient certainly has enough evidence of pulmonary disease and exercise limitation.

The patient's pulmonary function test results are shown in the following table.

Test	Units	Predicted	Patient
Total lung capacity	(L)	6.8	8.2
Vital capacity	(L)	4.6	3.1
Functional residual capacity	(L)	3.7	5.1
FEV_1	(L)	3.1	1.2
DL_{co}	(ml/min/mm Hg)	32	18
C_{stat}	(L/cm H_2O)	0.23	0.34
C_{dyn}	(L/cm H_2O)	0.23	0.06

C_{dyn} = dynamic compliance; C_{stat} = static compliance; DL_{co} = diffusing capacity for carbon monoxide; FEV_1 = forced expiratory volume in 1 second.

9. *Which of the pulmonary function tests are consistent with obstructive lung disease?*

DISCUSSION

All of the patient's pulmonary function test results are consistent with obstructive lung disease. The patient has increases in total lung capacity (TLC), functional residual capacity (FRC), and residual volume (RV) [RV = TLC − vital capacity (VC) = 5.1 L]. The FEV_1 is severely reduced and the $FEV_1\%$ = 40% (normal is 80%). The diffusing capacity of the lung is reduced, which may be due to a loss of alveolar surface area for gas exchange. The decreased diffusing capacity is more consistent with emphysema or an emphysema component than it is with pure chronic bronchitis. Pulmonary gas exchange may be worse than the results indicate because the patient has an increased hemoglobin concentration, which increases the diffusing capacity results. The patient has increased static compliance, which again is consistent with emphysematous changes caused by destruction of alveolar septa. The decreased dynamic lung compliance does not indicate increased elastance but is caused by the high airway resistance, which prevents the lungs from being fully expanded by the negative interpleural pressure change during inspiration. The patient has frequency-dependent compliance, which is another indication of a high airway resistance.

The patient was treated with bronchodilators, antibiotics, and chest physiotherapy. This reduced his mucous production and coughing somewhat. The patient continued to smoke four to five cigarettes per day. During the next 3 years, there was little change in the results of the patient's pulmonary function tests, except that the diffusing capacity progressively decreased. During this time, there was a gradual increase in breathlessness and a loss of exercise tolerance until the patient could walk only about 30 feet before having to stop to catch his breath. The patient was placed on low-flow, portable oxygen therapy, which relieved some of his breathlessness and allowed him to walk several blocks before having to rest.

10. *What else could have been done to help this patient?*

DISCUSSION

This patient certainly should have been urged to end his smoking habit, which has continued to impair his lung function. He could have been referred to smoking cessation clinics for support and help in this effort. Presumably, the patient's chronic airway infection has been controlled with the use of antibiotics, physiotherapy, and bronchodilators. However, now the patient is a respiratory cripple whose activities are severely limited. As a last resort, the patient should be considered for a lung transplant in order to improve his gas exchange and allow him a greater degree of activity.

Case 3

A 32-year-old woman consults her physician because of intermittent swelling of her face, hands, and legs during the past 4 weeks. Though she felt well, she became increasingly aware of periorbital puffiness in the morning, tightening of her rings, and ankle swelling toward the end of the day. More recently, her legs were swollen on awakening. Despite no changes in her normally good appetite or in her usual food intake, she noted a gradual 10-pound weight gain.

1. *What clinical conditions are associated with the development of edema, and what are the basic mechanisms of edema formation?*

DISCUSSION

Edema is defined as a palpable swelling resulting from the expansion of the interstitial fluid component of the extracellular fluid volume. It is clear that this woman has a generalized edema because of her weight gain and swelling of her face, hands, and legs.

There are three possible causes of edema in this patient: congestive heart failure (CHF), cirrhosis of the liver, and nephrotic syndrome. Less likely considerations are cyclic edema, which is often characterized by abrupt changes in weight correlating with the menstrual cycle, and hypothyroidism, where swelling of subcutaneous tissue (myxedema) can be mistaken for pitting edema. In patients with hypothyroidism, extravasated proteins bind to excess interstitial mucopolysaccharides, thereby preventing removal of the proteins by the lymphatics and resulting in myxedema.

The development of edema requires an alteration in one or more of the Starling forces in a direction that favors an increase in net capillary filtration (Figure C3-1). Edema can be produced by elevations in capillary hydrostatic (hydraulic) pressure, capillary permeability, or interstitial oncotic (colloid osmotic) pressure, or by a reduction in plasma oncotic pressure. Edema also can be caused by lymphatic obstruction, where some of the fluid filtered at the arteriolar end of the capillaries is not returned to the systemic venous circulation or the lymphatic circulation. Venous and lymphatic obstruction are ruled out in this patient because of edema in the hands and face as well as the legs.

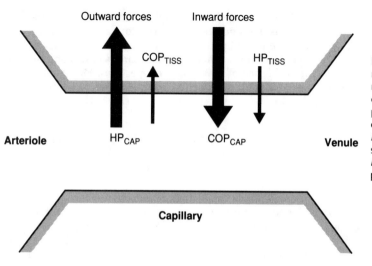

FIGURE C3-1. Capillary hemodynamics and fluid movement. COP_{CAP} = capillary colloid osmotic (oncotic) pressure; COP_{TISS} = tissue colloid oncotic pressure; HP_{CAP} = capillary hydrostatic (hydraulic) pressure; HP_{TISS} = tissue hydrostatic pressure.

2. What is the role of sodium in edema?

DISCUSSION

In any type of edema, sodium retention is always the necessary antecedent factor, although it can result from one of two causes. Accordingly, there are primary and secondary edema. In primary edema, there is a diminution in the ability of the kidneys to excrete the ingested sodium and water, which leads to an initial expansion of the plasma and extracellular fluid (ECF) volumes. In secondary edema, the kidneys function as if they were inadequately perfused. The major abnormality in secondary edema is a deficit—apparent or real—in plasma or whole blood volume. As a consequence of reduced plasma volume, renal hypoperfusion occurs, which initiates the process of renal sodium conservation via the activation of the renin-angiotensin-aldosterone system.

Further history reveals that the patient previously had excellent health and two uneventful full-term pregnancies. She takes no medication, does not smoke, and averages one alcoholic drink per week. There is no history of hepatitis or jaundice. She tolerates activity and exercise quite well, without experiencing dyspnea or fatigue; she is comfortable in cold weather when suitably dressed; she has normal bowel movements; and her menstrual cycle is normal and regular. Recently, the patient experienced nocturia once or twice nightly, and has noticed some foaminess in the toilet bowl following urination.

3. In reviewing the etiology of edema, which organ systems can be ruled out and which systems should be evaluated in more detail?

DISCUSSION

Good exercise tolerance and the absence of dyspnea and fatigue on exertion exclude CHF. Moreover, there is no history of congenital heart disease, acute rheumatic fever, hypertension, or murmurs, which would be expected in a young woman with CHF. Absence of drug or alcohol abuse together with no history of jaundice rule out cirrhosis of the liver. Tolerance to cold, normal bowel habits, and good exercise tolerance exclude thyroid hypofunction. Foamy urine is characteristic in patients with massive proteinuria. At this point, the evidence strongly suggests that the kidney is the cause of the illness.

Physical examination of this patient reveals a well-nourished patient. Her vital signs are: blood pressure, 110/70; pulse, 80 beats/min and regular; respiratory rate, 12/min; and body temperature, 36.6°C. No jaundice or pallor is noted. The neck veins are not distended and examination of the heart and lungs is normal. There are no rales and breath sounds are normal (i.e., there is no evidence of pulmonary congestion). Heart sounds are normal. The spleen and liver are not palpable and ascites is not evident. No definite facial edema is noted at this time, but her legs show pitting edema (2+). The neurologic exam is within normal limits.

4. Based on the physical examination, what possible diagnoses can be ruled out at this point?

DISCUSSION

The physical examination confirms that cirrhosis and CHF can be ruled out. Myxedema, which was a remote possibility, is eliminated in the absence of facial edema and the presence of normal reflexes. The only abnormal finding is edema of the legs and a clue to its cause. The nephrotic syndrome is strongly suspected in the context of the clinical triad of edema, hypoalbuminemia, and severe proteinuria.

5. What additional tests would be useful to support the diagnosis of nephrotic syndrome?

DISCUSSION

Estimation of the glomerular filtration rate (GFR) is an essential part of the evaluation of patients with renal disease. It is important to determine whether the GFR is changing, which can usually be determined from the plasma creatinine concentration alone. Thus, elevation in the blood urea nitrogen (BUN) and plasma creatinine concentration will usually distinguish underlying renal disease from heart failure. A 24-hour creatinine clearance can also be performed.

6. *What are the characteristics of edema in nephrotic syndrome?*

DISCUSSION

The nephrotic syndrome is characterized by an increase in the permeability of the glomerular capillary membranes to proteins, which leads to an increase in protein excretion. Primary renal sodium retention and consequent expansion of the interstitial volume component of the ECF appears to be the major factor in nephrotic edema. In this disorder, the glomeruli are diseased and the GFR is reduced. In the early stages of the disease, tubular function is normal.

7. *What volume changes occur in the ECF, interstitial fluid, and plasma volume of a patient with nephrotic syndrome?*

DISCUSSION

The urinary loss of proteins (proteinuria, which is caused mainly by the excretion of albumin) occurs because of an abnormal glomerular permeability, resulting in hypoalbuminemia and a decreased plasma oncotic pressure. Consequently, tissue capillary filtration is increased, which causes plasma water and electrolytes to move into the interstitial space. This, in turn, reduces the plasma volume. The fall in plasma protein concentration leads to a parallel decline in the interstitial oncotic pressure due to less entry of albumin into the interstitium. A positive sodium balance (i.e., increase in sodium content) follows, causing a dependent edema with expansion of the ECF (which exacerbates the hypoalbuminemia). These changes lead to a reduction in the effective circulating blood volume. It is important to note that generalized edema will not develop if sodium retention is prevented by eliminating sodium from the diet. In this setting, the initial movement of fluid into the interstitium will significantly reduce the plasma volume. This, in turn, decreases both the arterial and venous pressures and, consequently, the capillary hydrostatic pressure, thereby diminishing further fluid entry into the interstitium.

8. *What is meant by the term "effective circulating blood volume?"*

DISCUSSION

The term "effective circulating blood volume" refers to the fluid that is actually perfusing the tissues. It is an unmeasured entity and refers to the part of the ECF that is in the arterial system (normally about 700 ml in a 70-kg man). A better physiologic definition is the perfusion pressure in the region of the arterial baroreceptors in the aortic arch and carotid sinus and also in the glomerular afferent arterioles, since it is a change in pressure (or stretch) rather than volume or flow that is generally sensed at these sites.

The effective circulating blood volume varies directly with the ECF volume (Figure C3-2). The ECF volume is proportional to the total body sodium content, since sodium salts are the primary extracellular solutes that act to hold water within the ECF space. As a result, the regulation of sodium balance and the maintenance of the effective circulating blood volume are closely related functions.

9. *If the effective circulating blood volume is reduced, what changes in renal function would be expected?*

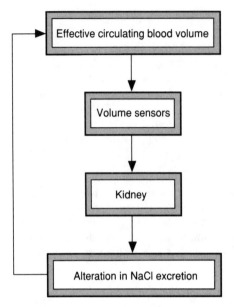

FIGURE C3-2. General scheme for monitoring and controlling effective circulating blood volume. The kidneys alter the excretion of NaCl in response to changes in the effective circulating blood volume.

DISCUSSION

A reduction in the effective circulating blood volume will lead sequentially to a reduction in the GFR, an increased proximal reabsorption of sodium and water, and an enhanced passive proximal urea reabsorption with a rise in blood urea nitrogen (BUN).

The patient's laboratory tests revealed a total plasma protein concentration of 5.1 g/dl (albumin was 2.1 g/dl and globulin was 3.0 g/dl) and a normal hemoglobin, hematocrit, and white blood cell count. Urinalysis showed a pH of 6.1, a trace of glucose, a protein of 4+, and one white blood cell and one red blood cell per high-power field. Her 24-hour urine protein excretion was 6.4 g.

DISCUSSION

The heavy proteinuria supports the diagnosis of nephrotic syndrome. Advanced renal insufficiency is probably not present because there is a normal hemoglobin. The few red and white blood cells on urine examination suggests the absence of a typical glomerular inflammatory process. A trace of glucose in the urine may suggest diabetes (a possible cause of nephrotic syndrome), but mild glycosuria is often present in massive albuminuria as a consequence of impaired proximal tubular function. In patients with diabetes mellitus, the increased capillary permeability may be mediated in part by the accumulation of glycosylation end-products derived from the nonenzymatic linkage of glucose with circulating plasma proteins. This can exacerbate the edema of nephrotic syndrome.

10. *If the total body sodium content is increased in nephrotic syndrome, why is the plasma sodium concentration low?*

DISCUSSION

The decrease in the effective circulating blood volume triggers the secretion of antidiuretic hormone (ADH) via a baroreceptor mechanism. This response is appropriate because the water retention induced by ADH both lowers the plasma osmolality and raises the ECF volume. The other major stimulus for ADH secretion is hyperosmolality.

In this patient, the effective circulating blood volume depletion is the dominant stimulus that overrides the negative signal for ADH secretion (i.e., the decrease in the osmolar concentration of plasma). Sodium concentration is a better measure of water metabolism than sodium metabolism. Thus, ADH regulates the sodium concentration by its effect on water reabsorption.

11. *What happens to aldosterone secretion in patients with nephrotic syndrome?*

DISCUSSION

Recent evidence indicates that edematous patients with the nephrotic syndrome (or with CHF) may not always have hyperaldosteronism. They can exhibit either increased or suppressed renin and aldosterone secretion. The high renin (high aldosterone) form of this syndrome is usually associated with minimal lesion renal disease. This form of the disease would be associated with markedly reduced sodium excretion. Patients with nephrotic syndrome who have suppressed activity of the renin-angiotensin-aldosterone system generally have more advanced renal disease.

12. *What mechanism explains the retention of sodium in nephrotic syndrome?*

DISCUSSION

In this disorder, the glomeruli are diseased and the filtration rate is reduced, which favors sodium retention. The net effect is that the kidney functions as though it were underperfused and avidly reabsorbs sodium, particularly in the collecting tubules. Sodium reabsorption in the proximal tubule may also occur in some patients with nephrotic syndrome, depending on the nature of the underlying disease.

With volume depletion, a decrease in sodium excretion results. It is mostly due to enhanced collecting tubule reabsorption, which is mediated at least in part by an increase in aldosterone secretion and probably by a reduction in atrial natriuretic peptide (ANP). In hypovolemic states, proximal sodium reabsorption is also enhanced, and it is associated with an increased activity of the luminal sodium–hydrogen ion antiporter. This exchanger is responsible for both sodium bicarbonate and sodium chloride reabsorption. Increased secretion of norepinephrine and angiotensin II in response to volume depletion also play an important role in the stimulation of proximal sodium reabsorption.

13. *What is the treatment for nephrotic syndrome?*

DISCUSSION

The main aim of therapy is, if possible, to reverse the glomerular disease. A loop diuretic and dietary sodium restriction can be used to control edema. Resistance to a loop diuretic may occur in patients with marked hypoalbuminemia due, in part, to the binding of free drug in the tubular lumen by filtered albumin, thereby rendering the diuretic ineffective. In addition, the marked hypoalbuminemia can lead to diminished plasma protein binding of the diuretic, with a resultant increased diffusion of the unbound drug from the vascular space into the interstitium.

14. *Summarize the major defects associated with nephrotic syndrome.*

DISCUSSION

There is a triad of clinical findings in nephrotic syndrome: edema, proteinuria, and hypoalbuminemia.

Hypoalbuminemia occurs secondary to albuminuria, which leads to a loss of fluid from the ECF and a decline in plasma volume with an increase in interstitial volume.

Patients with nephrotic syndrome have a decreased effective circulating blood volume, which stimulates ADH secretion and reduces delivery of tubular fluid to the diluting segment of the nephron (the ascending limb of the loop of Henle). As a result, renal water excretion is diminished and edema ensues.

Primary renal salt retention occurs secondary to glomerular injury. Nausea and malaise may occur when the sodium concentration falls below 120 mEq/L. The main clinical manifestation of hypotonic states are due to brain swelling and increased intracranial pressure as the brain cells take up water from the hypotonic ECF. Headache and lethargy follow. Seizures and coma are often seen with serum sodium concentrations below 110 mEq/L.

Case 4

A 22-year-old man with insulin dependent diabetes mellitus was admitted to the hospital with fever and abdominal pain. His temperature was 102.6°F, his blood pressure was 130/84 and his pulse rate was 86 beats/min. His diabetes was controlled with human recombinant DNA insulin, 15 units in the morning and 10 units at night. A chest x-ray showed a left lobe pneumonia.

Laboratory data

Serum chemistries	Arterial blood gases
[Na$^+$]: 132 mEq/L	pH: 7.32
[K$^+$]: 5.2 mEq/L	Pco$_2$: 20 mm Hg
[Cl$^-$]: 100 mEq/L	Po$_2$: 95 mm Hg
[HCO$_3{}^-$]: 10 mEq/L	Hemoglobin saturation: 95%
[Glucose]: 600 mg/dl	
BUN: 60 mg/dl	
[Creatinine]: 2.0 mg/dl	
Serum ketones: mildly positive	

1. What steps are taken in the evaluation of this acid-base disorder?

2. What is the diagnosis?

DISCUSSION

The first step is to verify that the pH is correct. This can be quickly and easily done with the Henderson equation, which determines the [H$^+$] as:

$$[H^+] = 24 \cdot \frac{Pco_2}{[HCO_3{}^-]}$$
$$= 24 \cdot \frac{20}{10} = 48 \text{ nmol/L}$$

Calculation of pH with the Henderson-Hasselbalch equation is:

$$pH = 9 - \log [H^+]$$
$$= 9 - \log 48$$
$$= 9 - 1.68$$
$$= 7.32$$

The arterial blood gases and the pH measurements are correct, and the patient is diagnosed as having acidosis since the pH is less than 7.36.

The second step is to determine the primary acid-base imbalance by examining the serum [HCO$_3{}^-$] and the Pco$_2$. Acidosis could result from a decrease in [HCO$_3{}^-$] (metabolic) or an increase in Pco$_2$ (respiratory). This patient has metabolic acidosis. If the hyperventilation that this patient manifested was the primary cause of the acid-base disturbance, the arterial blood pH would have indicated alkalosis with a pH more than 7.36. Thus, at this point, the patient has metabolic acidosis with respiratory alkalosis.

The third step is to determine the anion gap.

$$
\begin{aligned}
AG &= [Na+] - ([HCO_3^-] + [Cl^-]) \\
&= 132 - (10 + 100) \\
&= 132 - 110 \\
&= 22 \text{ mEq/L (normal is } 12 \pm 2 \text{ mEq/L)}
\end{aligned}
$$

The increase in the anion gap represents those acid anions (e.g., lactate, ketoacid anions, sulfate, phosphate) associated with an equimolar reduction in $[HCO_3^-]$. Thus, in this patient, the increase in the anion gap of 10 mEq/L is equivalent to the 10 mEq/L reduction in $[HCO_3^-]$ ($22 - 12 = 10$ mEq/L). Note that the addition of acid produced a fall in $[HCO_3^-]$, whereas the $[Cl^-]$ remained relatively stable. In patients with an anion gap acidosis, the reciprocity between the anion gap and the $[HCO_3^-]$ decline should always be identified. If this reciprocal relationship is not seen, a complicating acid-base disturbance may coexist.

3. *How would the cause of the elevated anion gap be determined in this patient?*

DISCUSSION

There are four major causes of high anion gap metabolic acidosis: renal failure, lactic acidosis, ketoacidosis (e.g., starvation, diabetes, chronic alcoholism), and toxins (e.g., methyl alcohol, ethylene glycol, paraldehyde, salicylate poisoning). In this patient, the history and physical did not reveal toxin ingestion or alcoholism. The absence of hypoxia (95% hemoglobin saturation with O_2; PO_2 of 95 mmHg) rules out lactic acidosis. The detection of ketones in the blood and a history of insulin-dependent diabetes mellitus indicates that ketoacidosis caused the increased anion gap.

4. *What was the cause of ketoacidosis in a diabetic controlled with exogenous insulin?*

DISCUSSION

The fundamental cause of diabetic ketoacidosis is a relative or absolute deficiency of insulin. The **major causes of insulin deficiency** are:
  a. **Infections** may cause patients to develop anorexia or reduce food intake. Because of the anorexia or fear of developing insulin-induced hypoglycemia, they may stop taking insulin entirely. Nausea and vomiting are common complaints in diabetic patients, occurring during acute ketoacidosis. Frequently, these patients have symptoms of anorexia, early satiety, and postprandial abdominal fullness due to gastroparesis. These symptoms can limit oral nutrition. In patients with severe hyperglycemia, failure to take insulin for 1 or 2 days, particularly when the patient does not eat, may lead to rapid development of ketoacidosis. The well-trained patient with diabetes often takes additional insulin during an obvious infectious process. In this patient, pneumonia, which was confirmed by chest x-ray, was the precipitating factor for ketoacidosis. The pneumonia was also the cause of the fever in this patient.
  b. **Surgical emergency, trauma, myocardial infarction, and severe emotional stress** represent frequent precipitating causes of uncontrolled diabetes. Abdominal pain in association with ketoacidosis is often misdiagnosed as an acute surgical emergency; however, the abdominal pain in this patient is secondary to the diabetic ketoacidosis.
  c. **Chronic, severe overinsulinization** with repeated hypoglycemic reactions may lead to hepatic glycogen depletion and severe ketosis that is disproportionate to the extent of glucosuria and hyperglycemia.
  d. **Associated endocrine disorders** (e.g., acromegaly, Cushing's syndrome, thyrotoxicosis, pheochromocytoma) can antagonize the metabolic effects of insulin.
  e. In patients with latent or subclinical diabetes, acute deficiency may be produced by **overeating,** particularly of carbohydrates.

5. *Describe the pathogenesis of the hyperglycemia, hyperosmolality, shifts of body fluids, and hypovolemia that develop in diabetic ketoacidosis.*

DISCUSSION

The relative or absolute lack of insulin impairs glucose utilization by most tissues (excluding brain tissue), and insulin deficiency combined with excessive glucagon increases hepatic production of glucose from noncarbohydrate sources (i.e., gluconeogenesis is stimulated). Both processes—underutilization of glucose and excessive glucose production—cause progressive hyperglycemia. It is the increased filtered load of glucose, which is due to hyperglycemia, that exceeds the renal reabsorptive capacity and causes glucosuria.

With the relative or absolute lack of insulin, the insulin-sensitive cells (muscle, adipose) are relatively impermeable to glucose; thus, the increasing hyperglycemia progressively raises the osmolality of the extracellular fluid (ECF). In this patient, the osmolar concentration is calculated as:

$$C_{osm} = 2\,[Na^+] + \frac{BUN\ (in\ mg/dl)}{2.8} + \frac{Blood\ glucose\ (in\ mg/dl)}{18}$$

$$= 2(132) + \frac{60}{2.8} + \frac{600}{18}$$

$$= 264 + 21 + 33$$

$$= 314\ mOsm/kg\ H_2O$$

As the osmolality of the ECF increases, water diffuses from the intracellular fluid (ICF) to the ECF, thereby reestablishing transcellular osmotic equivalence and leading to the dilution of the remaining solute, the most abundant of which are the sodium salts. This process progressively dehydrates cells while expanding the ECF. This expansion is transient because of the resulting osmotic diuresis caused by the glucosuria and obligatory salt loss.

During the osmotic diuresis, urinary water losses are disproportionately greater than electrolyte loss, which contributes further to the progressive rise in serum osmolality. Thus, the water loss in patients with diabetic acidosis represents the most serious form of hyperosmotic dehydration and leads to extreme intracellular depletion of water. The water losses through the skin, lungs, and gastrointestinal tract are aggravated by the limited ability of the patient to replace water because of anorexia, vomiting, or coma. This depletion of body fluids leads to contraction of the vascular volume, hypotension, and, often, shock. The depletion of vascular volume and increased blood viscosity leads to a reduction in renal blood flow and glomerular filtration rates (GFR) which further impairs renal compensatory mechanisms (acid excretion). This results in a more rapidly developing and severe acidosis.

6. *How does protein catabolism contribute to hyperglycemia in insulin-dependent diabetics?*

DISCUSSION

Insulin deficiency (and excess glucagon) increases protein catabolism, which leads to nitrogen loss and increased circulating levels of amino acids, which are gluconeogenic substances (except leucine) and can exacerbate hyperglycemia. The increased endogenous production of urea secondary to tissue breakdown causes an increase in blood urea nitrogen (BUN) due to a decrease in GFR. It also causes a marked increase in renal urea excretion, which contributes a further osmotic effect in addition to that of glucosuria. The destruction of tissue protein is associated with the liberation of intracellular potassium, phosphate, and magnesium, which are lost by urinary excretion as diabetic acidosis proceeds.

7. *What does the presence of ketoacids indicate?*

DISCUSSION

The generation and maintenance of ketoacid production involves three tissues: adipose tissues, the liver, and extrahepatic tissue (muscle, brain). Insulin deficiency and glucagon cause uncontrolled lipolysis, which increases plasma concentrations of both glycerol and free fatty acids (FFA). Glycerol also serves as a gluconeogenic substrate potentiating the hyperglycemia. The ability of liver to rapidly metabolize FFA limits the plasma concentration of these substances and prevents a significant fall in pH. The increased delivery of FFA to the liver increases ketone body production (ketogenesis), which soon leads to increased blood levels of ketone bodies (ketonemia). The liver, unlike other peripheral tissues, is unable to metabolize the ketoacids it synthesizes. Acetoacetate and β-hydroxybutyrate are taken up from the circulation by such tissues as muscle and kidney and are oxidized to CO_2 and H_2O. However, a deficiency of insulin increases fat mobilization and stimulates hepatic ketogenesis. The increased concentration of ketones spills over into the urine (ketonuria). The excretion of ketoacid anions exacerbates the electrolyte (Na^+ and K^+) depletion.

8. *What is the best treatment for this patient who has diabetic ketoacidosis?*

DISCUSSION

Ketoacidosis may safely, simply, and effectively be treated with cumulative doses of regular insulin under 100 units. Low doses of insulin have been shown to inhibit hepatic glucose production and stimulate peripheral utilization of glucose. This not only effectively reverses hyperglycemia, but it also reverses the ketosis, thereby negating the notion that ketosis or acidemia induces insulin resistance.

The addition of ketoacids to the ECF causes a loss of HCO_3^-, fall in systemic pH, and a compensatory decrease in P_{CO_2}. With therapy, β-hydroxybutyrate and acetoacetate are metabolized to HCO_3^-, thereby improving the ketoacidosis. Thus, it has been argued that since insulin therapy reverses the biochemical abnormalities (including the HCO_3^- deficit) $NaHCO_3$ therapy is not only unnecessary but may be detrimental. It remains necessary to establish volume repletion with normal (0.9%) saline followed by supplements of K^+ and HPO_4^{2-} to allow for the increased uptake of these two ions by insulin treatment as well as their loss in the urine with improvement in GFR.

Case 5

A 33-year-old woman presented to the emergency department with lower back pain that appeared suddenly as she was carrying groceries to her car. Further history revealed a 40-pound weight gain over the past 2.5 years. The woman also complained of emotional changes ranging from euphoria to depression, which disturbed her sleep cycles. Her menstrual cycles were irregular, occurring three or four times a year over the past 3 years.

Physical examination showed an obese, rosy-cheeked woman with excess fat deposits in the central axis, including her face, shoulders, and abdomen. Her extremities were thin and exhibited muscle atrophy. There were several ulcers on her left leg that did not heal. Her skin was thin and dry with large, purple bruises on her abdomen. She also had increased facial hair, especially on her upper lip.

Laboratory findings included a white blood cell count that showed 8% lymphocytes (low), 92% neutrophils (high), and fewer than 100 eosinophils/mm^3 (low). The patient's blood pressure was 155/100 mm Hg; her pulse was 80 beats/min and regular; her fasting blood glucose was 150 mg/dl; and her serum electrolytes were as follows: [HCO_3^-], 28 mEq/L; [Na^+], 140 mEq/L; [K^+], 3.6 mEq/L. The woman's serum cortisol level was 22 µg/dl in the morning (normal is 7–18 µg/dl) and 18 µg/dl at night (normal is 2–9 µg/dl). Her neurologic examination showed muscle weakness with normal deep tendon reflexes.

---

1. What is the most likely cause of this endocrinopathy?

DISCUSSION

This is a classic case of Cushing's syndrome. This patient has centripetal obesity, muscle weakness, abdominal striae, hirsutism, purpura, insomnia, and backache.

2. How can the major clinical manifestations of Cushing's syndrome be described in more detail?

DISCUSSION

In patients with Cushing's syndrome, the muscles become weak (paresis), bones become thin (osteoporosis), and the skin is unable to sustain the stress of normal activity, which leads to stria and nonhealing ulcers. The vasculature breaks down, and subcutaneous hemorrhages (ecchymoses) result. The immune system is less efficient, which leads to opportunistic infections. Glucose intolerance results from enhanced gluconeogenesis and the insulin-antagonistic action of cortisol.

Most patients with Cushing's syndrome gain weight and show a characteristic centripetal (truncal) obesity with a characteristic accentuation of fat deposits in the face, around the base of the neck (supraclavicular or suprascapular), and in the abdomen.

To the extent that sex steroid precursors are produced, women will manifest some degree of masculinization (including hirsutism), and men will manifest some degree of feminization. Amenorrhea, oligomenorrhea, or menstrual irregularity occurs in the majority of premenopausal women. Males typically have reduced libido, and they may have testicular softening and true or pseudogynecomastia.

Because excess cortisol has an aldosterone-like effect, patients will manifest arterial hypertension and hypokalemic alkalosis. The characteristic laboratory findings associated with hypercortisolism are confined mainly to elevated plasma cortisol levels; loss of cyclical secretion of cortisol, which normally shows a peak at 8:00 A.M. and a nadir around midnight; granulocytosis (neutrophilia); and lymphocytopenia.

3. *What are some possible mechanisms of action to explain the hypertension of excessive cortisol secretion?*

DISCUSSION

High plasma levels of glucocorticoids, like mineralocorticoids, may increase blood pressure. Excessive cortisol, while mainly a glucocorticoid, may exert enough sodium-retaining activity to cause significant volume expansion and hypertension. Beyond their mineralocorticoid properties, glucocorticoids can also increase blood pressure through other mechanims. First, glucocorticoids increase the production of angiotensinogen (renin substrate) by the liver. This may promote an increase in angiotensin I and II generation. Second, glucocorticoids increase cardiac output via a direct effect on the heart. Third, cortisol also increases peripheral vascular sensitivity to norepinephrine, a potent vasoconstrictor.

4. *Why is the patient's fasting blood glucose level elevated?*

DISCUSSION

The most important overall action of cortisol is to promote muscle proteolysis, which provides amino acid substrates for hepatic glucose formation (gluconeogenesis). This effect leads to hyperaminoacidemia together with the storage and release of glucose by the liver. Cortisol also decreases the use of glucose by cells, which involves direct inhibition of glucose transport into the cells. The resultant hyperglycemia stimulates chronic insulin secretion and insulin resistance.

5. *Why does this patient exhibit abnormalities in plasma concentrations of potassium and bicarbonate?*

DISCUSSION

Approximately 30%–40% of patients with Cushing's syndrome develop hypokalemic alkalosis and hypertension. The metabolic alkalosis usually results principally from the mineralocorticoid properties of cortisol.

Glucocorticoids produce a profound kaliuresis that is always associated with a rise in urine flow rate as well as increase in flow rate through the distal tubule and collecting duct. The rise in the tubular fluid and urinary flow rates has been attributed to the glucocorticoid-induced increase in glomerular filtration rate (GFR) and to a decrease in the water permeability of the distal tubule, which reduces water reabsorption.

6. *How would the determination be made as to whether this patient had Cushing's syndrome caused by an adrenal lesion or by an extra-adrenal lesion?*

DISCUSSION

The disorders causing cortisol excess may originate within the adrenal cortex or, more commonly, from oversecretion of adrenocorticotropic hormone (ACTH) either by the pituitary gland or by ectopic ACTH-producing hormone tumors (Figure C5-1).

If the neoplasm arises in the adrenal gland, pituitary ACTH secretion will be suppressed by negative feedback from the high circulating levels of free cortisol. With hypersecretion of cortisol due to a pituitary or an ectopic tumor, plasma levels of ACTH are elevated. These two Cushing's syndromes associated with elevated plasma ACTH levels can be further differentiated by the suppression test with dexamethasone. Corticotropin secretion caused by a pituitary neoplasm is suppressible with dexamethasone, whereas corticotropin of an extrapituitary origin is not suppressible.

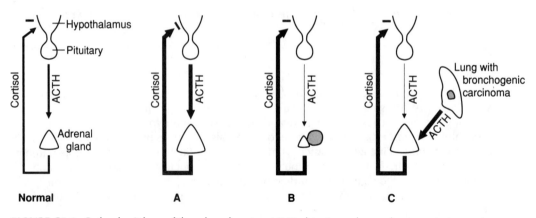

FIGURE C5-1. Pathophysiology of the adrenal cortex. (*A*) Cushing's syndrome due to a pituitary adenoma (Cushing's disease). (*B*) Cushing's syndrome due to an adrenal adenoma. (*C*) Ectopic adrenocorticotropic hormone (*ACTH*) syndrome. Panels *A* and *C* depict types of secondary hypercortisolemia. Panel *B* depicts primary hypercortisolemia. The *thickness of the arrows* indicates relative plasma concentration.

It is necessary to measure plasma cortisol to establish Cushing's syndrome of any cause. If the plasma cortisol level is elevated while the plasma ACTH is low, the patient is classified as having adrenal Cushing's syndrome (primary hypercortisolemia). If the plasma levels of cortisol and ACTH are both elevated, the patient is classified as having either pituitary Cushing's syndrome (secondary hypercortisolemia) or ectopic ACTH syndrome (also secondary hypercortisolemia). These two types of Cushing's syndrome patients with elevated ACTH can be differentially diagnosed by the pituitary ACTH suppression test. In addition, pituitary Cushing's syndrome patients respond to exogenous corticotropin-releasing hormone (CRH). Thus, if a patient with Cushing's syndrome has an elevated plasma ACTH and a positive response to CRH, the patient has Cushing's disease (hypercortisolism due to the elevated pituitary ACTH).

7. How can the metabolic actions of cortisol be summarized?

DISCUSSION

Cortisol enhances gluconeogenesis, glycogenesis, proteolysis (muscle) and lipolysis. The net effect is hyperglycemia, hyperproteinemia, and elevated plasma free fatty acids. Cortisol promotes hepatic glycogen formation and hepatic protein synthesis. The net effect of cortisol is an increase in body fat. It is important to note that cortisol promotes both the hepatic storage of glucose (glycogenesis) and the hepatic synthesis and release of glucose via gluconeogenesis (hyperglycemia).

COMPREHENSIVE EXAM

QUESTIONS

DIRECTIONS: Each of the numbered items or incomplete statements in this section is followed by answers or by completions of the statement. Select the ONE lettered answer or completion that is BEST in each case.

Questions 1–3

The following data were obtained from intracellular recordings of cardiac cells.

	Cell A	Cell B	Cell C
Resting membrane potential (mV)	−50	−60	−80
Capacity for diastolic depolarization	Yes	Yes	No
Intrinsic rate of depolarization (beats/min)	90	45	None

1. Cell A would most likely be found in the

(A) sinoatrial (SA) node
(B) atrial muscle
(C) atrioventricular (AV) node
(D) Purkinje fibers
(E) ventricular muscle

2. Pacemaker activity is exhibited by

(A) cell A
(B) cell B
(C) cell C
(D) cells A and B
(E) cells A, B, and C

3. Cells lacking fast Na^+ current during phase 0 of the cardiac action potential include

(A) cell A
(B) cell B
(C) cell C
(D) cells A and B
(E) cells A, B, and C

Questions 4–5

Lung compliance in a 32-year-old woman is studied. Data collected under control and experimental conditions are listed in the following table.

	Respiratory Rate (breaths/min)	Tidal Volume (ml)	Change in Interpleural Pressure during Inspiration (cm H_2O)
Control	15	600	4
Experimental	25	600	10

4. This patient's lung compliance during control and experimental conditions was

(A) unchanged
(B) 40 ml/breath and 24 ml/breath, respectively
(C) 150 ml/cm H_2O and 60 ml/cm H_2O, respectively
(D) 150 cm H_2O/ml and 60 cm H_2O/ml, respectively

5. This patient can be characterized as having frequency-dependent compliance, which indicates

(A) abnormal surfactant function
(B) obstructive lung disease
(C) restrictive lung disease
(D) pulmonary vascular disease

6. The curves below represent the clearances of various substances as a function of their plasma concentrations. Curve E represents the clearance curve for glucose. Following the administration of a substance (phlorizin) that blocks epithelial transport of glucose, the clearance curve for glucose would resemble which of the following curves?

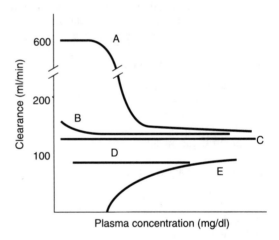

Adapted from Bauman JW, Chinard FP: *Renal Function: Physiological and Medical Aspects.* St. Louis, Mosby, 1975, p 43.

(A) A
(B) B
(C) C
(D) D
(E) E

7. Which of the following reflexes is most dependent on a vagovagal reflex?

(A) Chewing
(B) Swallowing
(C) Receptive relaxation
(D) Gastric emptying
(E) Intestinal segmentation

8. O_2 delivery to the tissues would be cut in half by a 50% decrease in the normal value of

(A) arterial O_2 tension
(B) minute ventilation
(C) hemoglobin concentration
(D) inspired O_2 tension

9. The bones of the middle ear are primarily responsible for

(A) amplifying the sound waves reaching the ear
(B) detecting the presence of a sound stimulus
(C) locating the source of a sound
(D) discriminating among different frequencies of sound
(E) adapting to a prolonged monotonous sound

10. The hexaxial reference system consists of reference lines generated from

(A) leads V_1–V_6
(B) standard limb leads
(C) augmented limb leads
(D) standard and augmented limb leads
(E) bipolar limb leads

11. Which of the following factors best explains an increase in the filtration fraction?

(A) Increased ureteral pressure
(B) Increased efferent arteriolar resistance
(C) Increased plasma protein concentration
(D) Decreased glomerular capillary hydrostatic pressure
(E) Decreased glomerular filtration area

12. Cortisol affects the biosynthesis of epinephrine in the adrenal medulla by

(A) augmenting the conversion of dopamine to epinephrine
(B) activating the epinephrine-forming enzyme
(C) inhibiting the methylation of norepinephrine
(D) activating catechol-O-methyltransferase
(E) increasing the release of acetylcholine (ACh)

13. A 32-year-old woman is admitted to the hospital with suspected partially compensated respiratory acidosis. Which of the following sets of laboratory data would confirm this suspicion?

	$[HCO_3^-]$ (mEq/L)	P_{CO_2} (mm Hg)	pH
(A)	17	19	7.9
(B)	31	80	7.22
(C)	9.8	30	7.14
(D)	24	45	7.5
(E)	20	25	7.5

Questions 14–16

A lightly anesthetized man who is breathing spontaneously has a cardiac output of 6 L/min, a heart rate of 75 beats/min, an O_2 consumption of 240 ml/min, and a mixed venous O_2 tension of 40 mm Hg. He is ventilated for 10 minutes at his normal tidal volume but at twice the normal frequency with a gas mixture that is 20% O_2 and 80% N_2. The airway pressure at the end of inspiration is 5 cm H_2O. On cessation of artificial ventilation, the patient does not breathe for 1 minute.

14. The most important factor responsible for the temporary apnea in this patient is reduced activity of the

(A) peripheral chemoreceptors, because of the high O_2 tension
(B) peripheral chemoreceptors, because of the low CO_2 tension
(C) pulmonary stretch receptors that inhibit inspiration
(D) medullary chemoreceptors, because of the low CO_2 tension
(E) medullary chemoreceptors, because of the high O_2 tension

15. During the artificial ventilation, the mixed venous O_2 tension in this patient would be

(A) equal to 40 mm Hg because the metabolic rate was unchanged
(B) greater than 40 mm Hg because the hyperventilation would cause vasoconstriction and less gas exchange with the tissues
(C) greater than 40 mm Hg because of the significant increase in P_{50} of hemoglobin
(D) less than 40 mm Hg because of the hypoxia that would occur from breathing this gas mixture
(E) less than 40 mm Hg because of a decrease in cardiac output

16. During the period of spontanous breathing, this patient had an arteriovenous O_2 difference of

(A) 4 ml/dl
(B) 5 ml/dl
(C) 6 ml/dl
(D) 40 ml/dl
(E) 80 ml/dl

17. Which of the following conditions is most likely to cause acidosis with marked dehydration?

(A) Severe diarrhea
(B) Severe, persistent vomiting
(C) Excessive sweating
(D) Drinking sodium lactate solution
(E) Complete water deprivation for 24 hours

18. Which of the following hormones is correctly paired with its effect on renal electrolyte reabsorption?

(A) Calcitriol/increased HPO_4^{2-} reabsorption
(B) Calcitonin/increased Ca^{2+} reabsorption
(C) Aldosterone/increased K^+ reabsorption
(D) Progesterone/increased Na^+ reabsorption
(E) Calcitonin/increased HPO_4^{2-} reabsorption

19. A red blood cell is placed in a solution. The cell initially shrinks and then returns to its original volume. The red cell's reaction indicates that the solution is

(A) hyperosmotic and hypertonic
(B) hyposmotic and hypertonic
(C) hyperosmotic and isotonic
(D) hyperosmotic and hypotonic
(E) hyposmotic and isotonic

20. Leads V_1 through V_6 measure the electrical activity in the

(A) frontal plane and are bipolar
(B) horizontal plane and are bipolar
(C) frontal plane and are unipolar
(D) frontal plane and are part of a standard 12-lead electrocardiogram (EKG)
(E) horizontal plane and are part of a standard 12-lead EKG

21. Spironolactone, which is an aldosterone antagonist, is injected into the renal artery of a laboratory animal. What are the effects on Na^+ and K^+ excretion, assuming that this drug does not affect the glomerular filtration rate (GFR) or renal blood flow?

	Na^+	K^+
(A)	↑	↑
(B)	↓	↓
(C)	↑	↓
(D)	↓	↑
(E)	↑	unchanged

22. Fatigue-resistant muscle fibers are characterized by high

(A) mitochondria concentrations
(B) myosin–adenosine triphosphatase (ATPase) activity
(C) velocity of shortening
(D) strength-generating capability
(E) glycolytic enzyme concentration

23. Two patients are studied and the following data are collected:

Patient	Respiratory Rate (breaths/min)	Tidal Volume (ml)	Dead Space (ml)
A	20	200	150
B	10	400	150

Which of the following statements about these two patients is true?

(A) The alveolar ventilation is greater in patient A than in patient B
(B) The alveolar ventilation is greater in patient B than in patient A
(C) The alveolar ventilation in both patients is equal
(D) The dead space ventilation in both patients is equal

24. A delirious 5-year-old boy is brought to the emergency room. His parents report that their son was in good health until 2 hours beforehand, when his mental state suddenly began to deteriorate. They suspect he may have swallowed a bottle of aspirin, because they discovered an empty bottle before leaving for the hospital. Laboratory evaluation reveals the following arterial blood data: $[H^+] = 18$ nmol/L, $[HCO_3^-] = 13$ mmol/L, $P_{CO_2} = 10$ mm Hg. This patient's history and blood data are most likely associated with

(A) respiratory alkalosis with partial renal compensation
(B) metabolic alkalosis with partial respiratory compensation
(C) a reduced $[HCO_3^-]$/dissolved CO_2 ratio
(D) a greater than normal CO_2 content

25. The rate of gastric emptying is controlled primarily by reflexes that occur

(A) during chewing
(B) during swallowing
(C) when chyme enters the stomach
(D) when chyme enters the intestine
(E) during the interdigestive period

26. The regression of the corpus luteum at the end of the postovulatory phase is caused by

(A) a decrease in follicle-stimulating hormone (FSH) secretion
(B) a decrease in luteinizing hormone (LH) secretion
(C) an increase in human chorionic gonadotropin (HCG) secretion
(D) a reduced capacity of the corpus luteum to synthesize steroids
(E) ovarian failure

27. A positive QRS complex in leads aV$_R$ and aV$_F$ indicates

(A) a normal axis
(B) right axis deviation (RAD)
(C) left axis deviation (LAD)
(D) indeterminate axis
(E) a mean electrical axis (MEA) between 0° and 90°

28. Following the intravenous administration of 1 L of a 150 mmol NaCl solution into a patient with a blood loss, there will be

(A) a decrease in the plasma Na$^+$ concentration
(B) an increase in the osmolarity of the intracellular fluid (ICF) compartment
(C) an increase in the volume of the ICF compartment
(D) edema
(E) a decrease in the colloid oncotic pressure of the plasma

29. When the acetylcholine (ACh) receptors on the pacemaker cells of the heart are activated, there is an increase in the membrane conductance to

(A) K$^+$
(B) Na$^+$
(C) Ca^{2+}
(D) Cl$^-$
(E) K$^+$ and Na$^+$

30. Epinephrine inhibits glucose uptake by muscle and adipose tissue. This inhibitory effect is attributed to

(A) glucagon secretion
(B) thyroid hormone secretion
(C) inhibition of insulin secretion
(D) inhibition of growth hormone (GH) secretion
(E) inhibition of cortisol secretion

31. At the end of a normal expiration, a young woman has an interpleural pressure of -5 cm H_2O. Without expiring further, she closes off her nose and mouth and performs a Valsalva maneuver against the closed airway. If the airway pressure is $+20$ cm H_2O during the expiratory maneuver, the interpleural pressure would be

(A) -5 cm H_2O
(B) 5 cm H_2O
(C) 10 cm H_2O
(D) 15 cm H_2O
(E) 20 cm H_2O

Questions 32–34

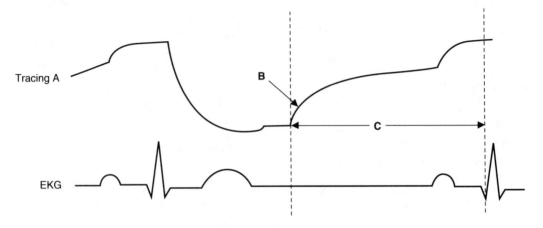

Tracing A

B

C

EKG

32. Tracing A is a

(A) ventricular pressure pulse
(B) ventricular volume curve
(C) atrial pressure pulse
(D) atrial volume pulse
(E) aortic pressure pulse

33. The point labeled B occurs in what phase of the cardiac cycle?

(A) Atrial contraction
(B) Isovolumic contraction
(C) Rapid ejection
(D) Reduced ejection
(E) Rapid filling

34. The duration of the interval labeled C is decreased by

(A) vagal stimulation
(B) baroreceptor stimulation
(C) arteriolar constriction
(D) an increased heart rate

35. Lesions within the flocculonodular lobe of the cerebellum prevent an individual from

(A) making rapid, alternating movements
(B) moving smoothly when reaching toward a target
(C) keeping the limbs still when resting
(D) grasping an object tightly
(E) maintaining balance when walking

36. Assuming a total body fluid volume of 40 L, 15 L of which are extracellular fluid (ECF) with an osmolarity of 300 mOsm/L, what will be the equilibrium intracellular fluid (ICF) osmolarity if 500 ml of a 0.15 mol/L NaCl solution are infused intravenously?

(A) 164 mOsm/L
(B) 277 mOsm/L
(C) 286 mOsm/L
(D) 300 mOsm/L
(E) 324 mOsm/L

37. A 45-year-old woman with severe vomiting caused by pyloric obstruction would be expected to show

(A) hyperchloremia
(B) an increase in plasma bicarbonate concentration [HCO_3^-]
(C) an increase in alveolar ventilation
(D) acid urine
(E) a decrease in arterial CO_2 tension (Pa_{CO_2})

38. In the formation of HCl by the parietal cells, the transport of H^+ across the parietal cell is coupled with the transport of

(A) K^+
(B) Cl^-
(C) HCO_3^-
(D) H^+
(E) Na^+

39. A newborn genotypic male is found to have an adrenogenital syndrome due to a 17α-hydroxylase defect. Which of the following biochemical reactions in the biosynthesis of gonadal hormones is decelerated in this case of congenital adrenal hyperplasia?

(A) Pregnenolone → 17α-hydroxypregnenolone
(B) Cholesterol → pregnenolone
(C) Progesterone → corticosterone
(D) Deoxycorticosterone → corticosterone
(E) 17α-Hydroxypregnenolone → 17α-hydroxyprogesterone

40. Gastric digestion is most important for which of the following substances?

(A) Fats
(B) Carbohydrates
(C) Proteins
(D) Vitamins
(E) Minerals

41. The following blood data are collected from a 27-year-old male patient: pH = 7.50, $[HCO_3^-]$ = 38 mmol/L, P_{O_2} = 80 mm Hg. Given these findings, what is the expected CO_2 tension (P_{CO_2}) for this patient?

(A) 30 mm Hg
(B) 40 mm Hg
(C) 50 mm Hg
(D) 60 mm Hg
(F) 70 mm Hg

42. The mean electrical axis (MEA) of a ventricular depolarization is −30°. Analysis of the resultant electrocardiogram (EKG) should reveal the largest positive QRS complex in lead

(A) I
(B) II
(C) aV$_R$
(D) aV$_L$
(E) aV$_F$

Questions 43–46

The data below represent the volume of distribution of tritiated water, inulin, and Evans blue dye in a 60-kg man after allowing time for equilibration.

Space	Volume (L)
Tritiated water	35
Inulin	8
Evans blue	3

43. With a hematocrit ratio of 0.40, the man's blood volume is

(A) 2 L
(B) 3 L
(C) 4 L
(D) 5 L
(E) 6 L

44. The interstitial fluid volume is

(A) 3 L
(B) 5 L
(C) 8 L
(D) 11 L
(E) 24 L

45. The extracellullar fluid (ECF) volume is

(A) 2 L
(B) 4 L
(C) 6 L
(D) 8 L
(E) 10 L

46. Assuming that the total body water (TBW) constitutes 70% of the lean body mass (LBM), the amount of body fat in this man is approximately

(A) 5 kg
(B) 10 kg
(C) 15 kg
(D) 20 kg
(E) 25 kg

47. A 57-year-old man who is breathing air at sea level has a respiratory exchange ratio of 1. Arterial blood gas analysis of this patient reveals the following: P_{O_2} = 85 mm Hg, P_{CO_2} = 30 mm Hg, and pH = 7.52. This patient's blood data indicate which of the following?

(A) His alveolar-to-arterial O_2 tension difference exceeds 20 mm Hg
(B) His plasma bicarbonate levels are increased
(C) He has been hypoventilating
(D) He has metabolic alkalosis
(E) He has chronic obstructive pulmonary disease (COPD)

48. A person with normal vision has a total converging power (without accommodation) of 60 diopters (D). In order to focus an object placed 25 cm from the eye, the converging power of the lens must increase by approximately

(A) 1 D
(B) 2 D
(C) 4 D
(D) 5 D
(E) 10 D

49. Primary hyperkalemic periodic paralysis is an inherited disease characterized by alterations in K^+ homeostasis. The K^+ defect leads to intermittent periods of muscle weakness due to failure of action potential propagation. This failure results from

(A) inactivation of Na^+ channels
(B) membrane hyperpolarization
(C) both
(D) neither

Questions 50–55

The following questions refer to the pressure–volume relationships depicted below. Loop ABEH represents a normal pressure–volume loop.

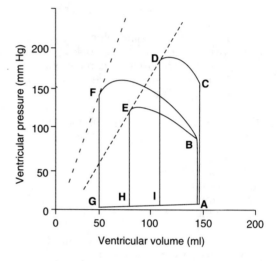

50. Loop ACDI shows the effect of

(A) increased stroke volume
(B) increased contractility
(C) increased ventricular end-diastolic volume (VEDV)
(D) increased afterload
(E) decreased preload

51. Loop ABFG shows the effect of

(A) increased heart rate
(B) increased contractility
(C) increased ventricular end-diastolic volume (VEDV)
(D) increased afterload
(E) decreased preload

52. In loop ABEH, mitral valve closure occurs

(A) at point A
(B) at point B
(C) at point E
(D) at point H
(E) between points E and H

53. In loop ABEH, ventricular filling occurs during the interval between points

(A) A and B
(B) B and E
(C) E and H
(D) H and A

54. The slope of the line from point G to point A represents

(A) diastolic compliance
(B) systolic compliance
(C) diastolic elastance
(D) systolic elastance

55. The volume of blood in the ventricles at points D, E, or F is referred to as the ventricular

(A) stroke volume
(B) end-diastolic volume (VEDV)
(C) end-systolic volume (VESV)
(D) diastolic reserve volume
(E) residual volume

56. Acetazolamide is administered to a glaucoma patient. Given that this drug inhibits carbonic anhydrase in the renal proximal tubule, which of the following substances will be excreted at a lower rate?

(A) Na^+
(B) H_2O
(C) HCO_3^-
(D) NH_4^+
(E) K^+

57. Reciprocal innervation is most accurately described as

(A) inhibition of flexor muscles during an extension
(B) activation of contralateral extensors during a flexion
(C) reduction of Ia fiber activity during a contraction
(D) simultaneous stimulation of alpha and gamma motoneurons
(E) inhibition of alpha motoneurons during a contraction

58. The graph below shows the diurnal variation in the plasma concentration of which of the following hormones?

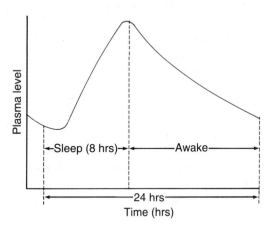

(A) Thyroxine
(B) Insulin
(C) Testosterone
(D) Cortisol
(E) Estradiol

59. During moderate exercise, a patient has a cardiac index of 6.5 L/min/m², a hemoglobin concentration of 12 g/dl, a venous O_2 tension of 30 mm Hg, and a venous O_2 saturation of 50%. Assuming 100% hemoglobin saturation in arterial blood, what is this patient's O_2 consumption?

(A) 150 ml/min/m²
(B) 275 ml/min/m²
(C) 520 ml/min/m²
(D) 790 ml/min/m²
(E) 1030 ml/min/m²

60. The normal sequence of phases of the menstrual cycle is

(A) menses, preovulatory, ovulatory, estrogenic
(B) preovulatory, ovulatory, progestational, menses
(C) ovulatory, progestational, menses, luteal
(D) progestational, menses, follicular, preovulatory
(E) menses, follicular, ovulatory, estrogenic

61. The following arterial blood data are collected from a 42-year-old female patient: $[H^+] = 49$ nEq/L, $P_{CO_2} = 30$ mm Hg, $P_{O_2} = 95$ mm Hg. Given these findings, what is the expected arterial bicarbonate concentration $[HCO_3^-]$ for this patient?

(A) 13.2 mEq/L
(B) 14.7 mEq/L
(C) 15.8 mEq/L
(D) 16.5 mEq/L
(E) 17.1 mEq/L

62. If acidosis and hypokalemia result from loss of fluid from the gastrointestinal (GI) tract, the fluid was most likely drained from the

(A) stomach
(B) intestine
(C) gallbladder
(D) pancreas
(E) colon

Questions 63–64

The left ventricular and aortic pressure tracings below were recorded during cardiac catheterization of a 62-year-old patient who complains of chest pain and dizziness on exertion.

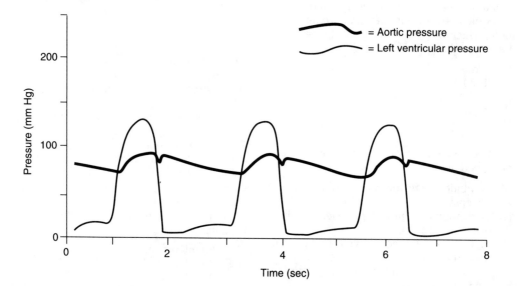

63. The left ventricular and aortic tracings indicate that this patient has

(A) pulmonary valve stenosis
(B) aortic valve stenosis
(C) mitral valve stenosis
(D) aortic valve insufficiency
(E) mitral valve insufficiency

64. Physical examination of this patient would most likely reveal

(A) a systolic murmur
(B) a diastolic murmur
(C) a presystolic murmur
(D) a middiastolic murmur
(E) no first heart sound (S_1)

Questions 65–67

The following arterial blood data are obtained from a patient who is cyanotic at rest: P_{O_2} = 60 mm Hg, hemoglobin saturation = 85%, P_{CO_2} = 40 mm Hg, pH = 7.39, and hemoglobin concentration = 18 g/dl. The arterial O_2 tension rises to 295 mm Hg after the patient breathes 100% O_2 at sea level for 20 minutes. Cardiac catheterization reveals normal pressures and an O_2 tension of 40 mm Hg in the right atrium, right ventricle, and pulmonary artery while the patient breathes 100% O_2.

65. The most likely cause of hypoxia in this patient is

(A) a ventilation–perfusion abnormality (physiologic shunt)
(B) a right-to-left anatomic shunt
(C) a left-to-right anatomic shunt
(D) a diffusion defect
(E) hypoventilation

66. The fraction of the cardiac output that represents shunted blood is

(A) 0.2
(B) 0.3
(C) 0.4
(D) 0.5
(E) 0.6

67. The most useful data for locating the site of the shunt would be

(A) O_2 tension in the left ventricle and left atrium
(B) airway resistance
(C) right ventricular and pulmonary arterial pressures
(D) left ventricular pressures
(E) compliance of the lungs

68. In contrast to motor units that fire later during a movement, motor units that fire at the beginning of a movement

(A) can be tetanized at a lower frequency of stimulation
(B) generate a greater amount of force
(C) have a greater amount of glycogen stored within them
(D) fatigue more rapidly
(E) are innervated by larger alpha motoneurons

69. A 45-year-old man is studied and found to have a respiratory rate of 15 breaths/min, a tidal volume of 0.5 L, and a dead space of 200 ml. The patient is asked to increase his respiratory rate to 30 breaths/min, and his tidal volume is measured at 350 ml. Assuming no change in dead space, which of the following is true regarding alveolar CO_2 tension?

(A) The alveolar CO_2 tension will increase because of the decreased ventilation
(B) The alveolar CO_2 tension will decrease because of the increased ventilation
(C) The alveolar CO_2 tension will not change because it is not affected by respiration
(D) The alveolar CO_2 tension will not change because alveolar ventilation remains constant
(E) The arterial blood pH will decrease because of the increased ventilation

70. A decrease in the total osmotic pressure of arterial blood would lead to an increase in urine volume by

(A) increasing the hydrostatic pressure inside the glomerulus
(B) increasing the permeability of the glomerular capillaries to water
(C) inhibiting antidiuretic hormone (ADH) secretion
(D) stimulating the secretion of aldosterone
(E) directly inhibiting the reabsorption of water by the collecting ducts

71. A 23-year-old woman with diabetes mellitus is admitted to the hospital. She is dehydrated and hyperpneic. Laboratory examination reveals high urinary concentrations of acetoacetic acid and glucose, blood pH of 7.39, plasma [$HCO_3{}^-$] of 19.0 mmol/L, and plasma P_{CO_2} of 33 mm Hg. These data are most suggestive of

(A) metabolic alkalosis with complete renal compensation
(B) metabolic acidosis with complete respiratory compensation
(C) respiratory acidosis with partial renal compensation
(D) respiratory alkalosis with partial renal compensation

Questions 72–74

A man undergoes lung volume studies using the helium dilution method. The test begins at the end of a normal expiration. The initial fraction of helium in the spirometer is 0.05, and the helium fraction after equilibration with the lungs is 0.03. The volume of gas in the spirometer is kept constant at 4 L during the procedure by the addition of O_2. According to a spirogram, this patient's vital capacity (VC) is 5 L and his expiratory reserve volume (ERV) is 2 L.

72. What is this patient's functional residual capacity (FRC)?

(A) 1.0 L
(B) 1.7 L
(C) 2.7 L
(D) 3.0 L
(E) 5.0 L

73. What is this patient's residual volume (RV)?

(A) 0.7 L
(B) 1.0 L
(C) 1.7 L
(D) 2.7 L
(E) 3.0 L

74. What is this patient's total lung capacity (TLC)?

(A) 1.7 L
(B) 2.7 L
(C) 3.0 L
(D) 5.0 L
(E) 5.7 L

75. A 32-year-old male electrician consults an internist and complains of episodes of palpitations and sweating that occur when he climbs a ladder at work. He feels "washed out" after these attacks, which he ascribes to "nerves." Subsequent examination by the physician leads to a diagnosis of pheochromocytoma. Body fluid analysis of this patient is most likely to reveal a low plasma concentration of

(A) free fatty acids
(B) insulin
(C) fasting glucose
(D) lactate
(E) pyruvate

76. Unloading of muscle spindles can be prevented by

(A) alpha motoneurons
(B) gamma motoneurons
(C) Ia afferent fibers
(D) Ib afferent fibers
(E) C fibers

77. A 32-year-old man can generate an inspiratory pressure of −50 mm Hg intermittently for several minutes. How deep can this man lie underwater while breathing through a tube, if the tube offers no significant resistance to air flow?

(A) 37 mm
(B) 37 cm
(C) 68 mm
(D) 68 cm
(E) 74 cm

78. A 27-year-old, anxious man is examined in the emergency room, after which the following arterial blood data are obtained:

Blood Chemistry	Blood Gas
[Na^+] = 140 mEq/L	pH = 7.6
[K^+] = 4 mEq/L	Pco_2 = 20 mm Hg
[HCO_3^-] = 19 mEq/L	Po_2 = 98 mm Hg
[Cl^-] = 109 mEq/L	

From these data, the most likely diagnosis is

(A) hypoxia
(B) metabolic alkalosis
(C) metabolic acidosis with respiratory compensation
(D) obstructive lung disease
(E) respiratory alkalosis

79. Electrically excitable channels are required for

(A) the propagation of the action potential in nerve fibers
(B) the release of synaptic transmitter from presynaptic nerve terminals
(C) both
(D) neither

80. Elimination of the terminal ileum will result in malabsorption of

(A) vitamin B_{12}
(B) fats
(C) both
(D) neither

81. Resistance to blood flow through the kidney can be determined by

(A) measuring the clearance of para-aminohippuric acid (PAH)
(B) measuring the hydrostatic pressure difference between the renal artery and renal vein
(C) measuring the renal blood flow
(D) dividing the arteriovenous hydrostatic pressure difference by the renal blood flow
(E) dividing the renal blood flow by the arteriovenous hydrostatic pressure difference

82. The variation in auditory threshold as a function of frequency (the minimum audibility curve) is most related to the properties of the

(A) outer ear
(B) auditory canal
(C) middle ear
(D) tympanic membrane
(E) basilar membrane

Questions 83–84

The following two questions refer to the pressure–volume loop below.

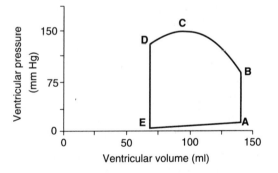

83. Which point on the loop represents the opening of the mitral valve?

(A) Point A
(B) Point B
(C) Point C
(D) Point D
(E) Point E

84. Isovolumic ventricular contraction occurs during the interval between points

(A) A and E
(B) A and B
(C) B and C
(D) C and D
(E) D and E

85. Normally, most of the H^+ is excreted by the kidneys in the form of

(A) HCO_3^-
(B) phosphate ion
(C) NH_4^+
(D) titratable acid
(E) β-hydroxybutyrate ion

86. Following a massive hemorrhage during delivery, a 34-year-old woman experiences a failure to lactate and to menstruate. Which of the following is most likely to be associated with this clinical picture?

(A) Elevated prolactin secretion
(B) Excessive urinary Na^+ excretion
(C) Excessive water excretion
(D) Increased sensitivity to insulin
(E) Elevated gonadotropin secretion

87. A semicomatose 19-year-old woman is brought to the emergency room with dry skin, hyperventilation, hypotension, and a rapid pulse rate. The following blood data are obtained:

pH = 7.14
$[Na^+]$ = 140 mEq/L
$[K^+]$ = 4.5 mEq/L
$[Cl^-]$ = 82 mEq/L
$[HCO_3^-]$ = 11 mEq/L
Pco_2 = 30 mm Hg
[glucose] = 180 mg/dl

From the above history and laboratory data, the most likely diagnosis is

(A) metabolic alkalosis
(B) metabolic acidosis
(C) respiratory alkalosis
(D) respiratory acidosis
(E) hypoglycemia

88. Which of the following statements best characterizes the transpulmonary pressure at the base of the lung of a person who is standing?

(A) It is independent of lung volume
(B) It is equal to the transpulmonary pressure at the apex of the lung
(C) It may be negative if the lung is at residual volume (RV)
(D) It causes the basal alveoli to be more dilated than the apical alveoli
(E) It causes the bronchioles at the lung base to be more dilated than those at the apex

Questions 89–92

A patient presents with crushing chest pain, shortness of breath, and marked anxiety. A preliminary diagnosis of acute myocardial infarction is made. Physical examination reveals evidence of pulmonary edema, cardiomegaly, peripheral edema, and pulmonary hypertension. Arterial blood gas analysis, on room air, reveals the following: $P_{CO_2} = 40$ mm Hg, $P_{O_2} = 60$ mm Hg, pH = 7.32, and $[HCO_3^-] = 20$ mEq/L.

89. This patient's blood data are most indicative of

(A) a diffusion abnormality
(B) hypoventilation
(C) a ventilation–perfusion abnormality
(D) left-to-right cardiac shunt

90. The patient is admitted to the CCU, sedated, and given 40% O_2 by respirator, which is set to deliver a tidal volume of 0.6 L at a rate of 16 breaths/min. An inspiratory pressure of 20 mm Hg is required to deliver the tidal volume. If the patient's predicted dead space is 150 ml, then his calculated alveolar ventilation is approximately

(A) 5 L/min
(B) 7 L/min
(C) 9 L/min
(D) 12 L/min
(E) 16 L/min

91. While ventilation continues with 40% O_2, a Swan-Ganz catheter is inserted into the patient's pulmonary artery. A repeat arterial blood analysis reveals a P_{O_2} of 120 mm Hg, a P_{CO_2} of 43 mm Hg, and pH of 7.31. These findings indicate that

(A) the pulmonary edema has cleared
(B) gas exchange is completely normal
(C) hemoglobin concentration has increased
(D) the alveolar-to-arterial P_{O_2} difference is increased

92. This patient's respiratory compliance is

(A) 0.03 L/mm Hg
(B) 0.1 L/mm Hg
(C) 9.0 L/mm Hg
(D) 12.0 mm Hg/L
(E) 26.7 mm Hg/L

93. The greatest amount of fat absorption occurs in the

(A) stomach
(B) duodenum
(C) ileum
(D) colon
(E) rectum

94. The following data were obtained from a 55-year-old man during cardiac catheterization: O_2 consumption = 210 ml/min, O_2 content of right ventricular blood = 11 ml/dl, O_2 content of brachial artery blood = 18 ml/dl, heart rate = 75 beats/min. These data are compatible with which one of the following statements?

(A) The tissues receive 29 ml of O_2/dl of blood
(B) Cardiac output is approximately 1470 ml/min
(C) Pulmonary venous O_2 content is approximately 145 ml/dl of blood
(D) Right ventricular stroke volume averages approximately 40 ml
(E) Cardiac output is extremely high

95. During the process of vitamin B_{12} absorption, almost all of the ingested vitamin B_{12}

(A) binds to intrinsic factor in the stomach
(B) is absorbed in the stomach
(C) both
(D) neither

96. During the first 6–8 weeks of pregnancy, progesterone is secreted mainly by the

(A) maternal adrenal glands
(B) maternal theca interna
(C) corpus luteum
(D) fetal adrenal gland
(E) decidua

Questions 97–99

A 50-year-old man who has smoked two packs of cigarettes a day for 35 years complains of shortness of breath, chronic cough, and production of yellowish, foul-smelling sputum. He has clubbing of his finger nails, and the nail beds and lips are noted to be cyanotic. Arterial blood gas analysis of this patient reveals the following: P_{O_2} = 55 mm Hg, P_{CO_2} = 56 mm Hg, HCO_3^- = 35 mEq/L, and pH = 7.4.

97. These blood gas data are most consistent with

(A) acute respiratory failure
(B) inadequate alveolar ventilation
(C) anatomic shunt
(D) anemia
(E) carbon monoxide (CO) poisoning

98. Which of the following pathophysiologic phenomena most likely initiated this patient's syndrome?

(A) Decreased alveolar ventilation followed by increased work of breathing
(B) Bronchial narrowing due to inflammation and edema followed by pulmonary hypertension
(C) Bronchial narrowing due to inflammation and edema followed by increased work of breathing
(D) Hypercapnia followed by pulmonary hypertension
(E) Pulmonary hypertension followed by hypercapnia

99. The acid-base status of this patient's blood is best categorized as

(A) respiratory acidosis, uncompensated
(B) metabolic acidosis, compensated
(C) respiratory acidosis, compensated
(D) lactic acidosis secondary to hypoxia
(E) respiratory alkalosis

Questions 100–103

Questions 100–103 refer to the following diagram, which is an experimental record obtained from an anesthetized dog.

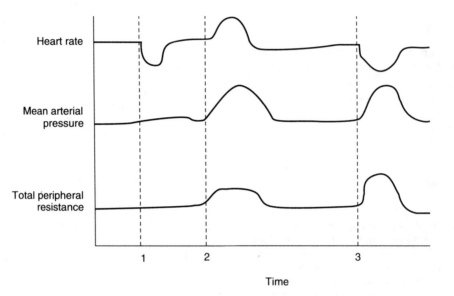

Time

100. The experimental intervention at time 1 most likely represents

(A) electrical stimulation of the lumbar sympathetic nerve roots
(B) electrical stimulation of the superior cervical ganglion (cardiac sympathetic nerves)
(C) administration of a β-adrenergic blocking drug
(D) stimulation of the right vagus nerve
(E) administration of a cholinergic blocking drug

101. The experimental intervention at time 2 most likely represents

(A) electrical stimulation of the sacral sympathetic nerve roots
(B) electrical stimulation of the superior cervical ganglion (cardiac sympathetic nerves)
(C) administration of a sympathetic blocking drug
(D) stimulation of the right vagus nerve
(E) administration of a cholinergic blocking drug

102. A drug was administered at time 3. This drug is most likely classified as a(n)

(A) cholinergic blocking drug (e.g., atropine)
(B) α-adrenergic drug
(C) β-adrenergic drug
(D) β-adrenergic blocking drug
(E) α-adrenergic blocking drug

103. The decrease in heart rate that occurred at time 3 was most likely caused by

(A) the direct effect of the drug on the sinoatrial (SA) node
(B) the direct effect of the drug on the ventricular muscle
(C) the occurrence of ventricular extrasystoles
(D) a reflex effect mediated by the chemoreceptors
(E) a reflex effect mediated by the baroreceptors

104. When the arterial O_2 tension drops from 100 mm Hg to 27 mm Hg with an arterial pH of 7.4 and a CO_2 tension of 40 mm Hg, O_2 content in the blood decreases by about

(A) 10%
(B) 25%
(C) 33%
(D) 50%
(E) 75%

105. Given the following data: glomerular capillary hydrostatic pressure = 47 mm Hg, glomerular capillary colloid oncotic pressure = 28 mm Hg, Bowman's capsule hydrostatic pressure = 10 mm Hg, and Bowman's capsule oncotic pressure = 0 mm Hg, what is the effective filtration pressure (EFP)?

(A) 2 mm Hg
(B) 4 mm Hg
(C) 6 mm Hg
(D) 9 mm Hg
(E) 10 mm Hg

106. Epinephrine-forming enzyme activity is increased directly by

(A) hydrocortisone
(B) acetylcholine
(C) norepinephrine
(D) adrenocorticotropic hormone (ACTH)
(E) 11-deoxycortisol

Questions 107–109

A normal 55-year-old man who lives at an altitude of 11,500 feet is seen for an annual physical examination. Findings on examination include a hemoglobin concentration of 18 g/dl, an arterial O_2 tension of 27 mm Hg, and an arterial pH of 7.40.

107. Based on these findings, this man's arterial O_2 content (ml/dl) would be approximately

(A) 8
(B) 12
(C) 15
(D) 18
(E) 24

108. This man's pulmonary artery pressure is likely to be

(A) normal
(B) lower than normal because of the inhibition of chemoreceptors caused by the low CO_2 tension
(C) higher than normal because of the increased cardiac output
(D) higher than normal because of hypoxic pulmonary vasoconstriction (HPV)

109. The hemoglobin saturation in this man's venous blood is likely to be

(A) normal
(B) less than normal because of the decreased arterial hemoglobin saturation
(C) greater than normal because of the increased cardiac output
(D) greater than normal because of the increased hemoglobin concentration

110. Which of the following conditions would simultaneously increase the aortic systolic pressure and decrease the aortic pulse pressure?

(A) Increased heart rate
(B) Increased arterial compliance
(C) Decreased peripheral resistance
(D) Increased stroke volume
(E) Increased elastic constant

DIRECTIONS: Each of the numbered items or incomplete statements in this section is negatively phrased, as indicated by a capitalized word such as NOT, LEAST, or EXCEPT. Select the ONE lettered answer or completion that is BEST in each case.

---

111. All of the following substances can be converted to more biologically active forms after secretion EXCEPT

(A) angiotensinogen
(B) angiotensin I
(C) tetraiodothyronine (T_4)
(D) testosterone

112. β-Adrenergic receptors mediate all of the following responses EXCEPT

(A) ciliary muscle contraction
(B) increased myocardial contractility
(C) vasodilation
(D) insulin secretion
(E) decreased intestinal motility

113. All of the following would increase alveolar O_2 tension EXCEPT

(A) breathing gas containing 24% O_2
(B) breathing gas containing 28% O_2
(C) ascending a mountain
(D) hyperventilating

114. Findings consistent with metabolic acidosis include all of the following EXCEPT

(A) increased excretion of titratable acid
(B) increased excretion of NH_4^+
(C) decreased excretion of $NaHCO_3$
(D) decreased respiratory rate
(E) a negative base excess

115. Cutting the vagus nerve has a major effect on all of the following EXCEPT

(A) the rate of liquid emptying from the stomach
(B) receptive relaxation
(C) migrating motor complex
(D) primary esophageal peristalsis

116. When light strikes the eye, it decreases all of the following EXCEPT the

(A) membrane conductance of rods and cones
(B) amount of synaptic transmitter released by rods and cones
(C) concentration of cyclic guanosine monophosphate (cGMP) in rods and cones
(D) active transport of Na^+ by rods and cones
(E) concentration of all-*trans* retinal in rods and cones

117. A large dose of insulin is administered intravenously to a normal 34-year-old female patient. This is likely to cause an increase in all of the following EXCEPT

(A) plasma epinephrine concentration
(B) plasma K^+ concentration
(C) adrenocorticotropic hormone (ACTH) secretion
(D) growth hormone (GH) secretion
(E) glucagon secretion

118. All of the following are renal responses to severe hemorrhage EXCEPT

(A) increased Na^+ reabsorption by the distal nephron
(B) decreased glomerular filtration rate (GFR)
(C) increased permeability of the collecting duct to water
(D) decreased filtration fraction
(E) increased renin secretion

119. The neural crest gives rise to all of the following tissues and cells EXCEPT

(A) Schwann cells
(B) cartilage and bone of the skull
(C) the neural lobe of the pituitary gland
(D) melanocytes
(E) the adrenal medulla

120. Expansion of the antrum causes an increase in all of the following EXCEPT

(A) secretion of gastrin
(B) secretion of pancreatic enzymes
(C) secretion of gastric acid (HCl)
(D) gastric motility
(E) receptive relaxation

121. Untreated type I diabetes mellitus is associated with all of the following biochemical changes EXCEPT

(A) positive nitrogen balance
(B) ketonemia
(C) ketonuria
(D) low plasma C-peptide concentration
(E) glycosuria

122. Hypophysectomy results in the functional decline of many endocrine organs. All of the following changes are likely to occur after removal of the pituitary gland EXCEPT

(A) atrophy of the thyroid gland
(B) dwarfism (if performed during early adolescence)
(C) deficiency of aldosterone
(D) cessation of menstrual cycles
(E) impaired testosterone secretion

123. The reflex responsible for withdrawing a limb is characterized by all of the following EXCEPT

(A) local sign
(B) irradiation
(C) unmyelinated afferent fibers
(D) monosynaptic reflex
(E) afterdischarge

124. All of the following are substrates for catecholamine-O-methyltransferase (COMT) EXCEPT

(A) epinephrine
(B) norepinephrine
(C) normetanephrine
(D) dihydroxymandelic acid

125. Stimuli for aldosterone secretion include all of the following EXCEPT

(A) angiotensin II
(B) hyperkalemia
(C) hypovolemia
(D) corticotropin
(E) atrial natriuretic peptide (ANP)

126. Growth hormone (GH) secretion is increased by all of the following factors except

(A) insulin administration
(B) arginine administration
(C) somatostatin administration
(D) onset of sleep
(E) exercise

127. In hyperaldosteronemia, all of the following conditions are likely to be observed EXCEPT

(A) decreased hematocrit
(B) fall in plasma oncotic pressure
(C) increased extracellular fluid (ECF) volume
(D) hyperkalemia
(E) metabolic alkalosis

128. Lesions within the basal ganglia produce all of the following signs EXCEPT

(A) hypotonia
(B) hemiballism
(C) tremor
(D) hypokinesia
(E) athetosis

129. All of the following substances are required for thyroxine biosynthesis EXCEPT

(A) active iodide
(B) diiodotyrosine
(C) monoiodotyrosine
(D) thyroglobulin
(E) thyroid peroxidase

130. Rods are able to detect light at much lower intensities than cones for all of the following reasons EXCEPT

(A) the diameter of rods is greater
(B) the rods can more easily detect scattered light within the eye
(C) the number of rods innervating a single ganglion cell is greater
(D) the same light stimulus produces a larger receptor potential in rods
(E) the receptor potential adapts more rapidly in rods

131. Hormones with lipolytic activity include all of the following EXCEPT

(A) glucagon
(B) epinephrine
(C) insulin
(D) cortisol
(E) growth hormone (GH)

DIRECTIONS: Each set of matching questions in this section consists of a list of four to twenty-six lettered options (some of which may be in figures) followed by several numbered items. For each numbered item, select the ONE lettered option that is most closely associated with it. To avoid spending too much time on matching sets with large numbers of options, it is generally advisable to begin each set by reading the list of options. Then, for each item in the set, try to generate the correct answer and locate it in the option list, rather than evaluating each option individually. Each lettered option may be selected once, more than once, or not at all.

Questions 132–137

Match each characteristic with the appropriate substance.

(A) Aldosterone
(B) Antidiuretic hormone (ADH)
(C) Atrial natriuretic peptide (ANP)
(D) Renin
(E) Angiotensin I

132. An octapeptide

133. Regulates plasma $[Na^+]$

134. Controls body Na^+ content

135. Secreted by the zona glomerulosa of the adrenal cortex

136. Substrate for dipeptidyl carboxypeptidase or kininase II

137. Produced by modified smooth muscle cells

Questions 138–142

Match each person described below with the set of blood data that best coincides with that person's condition.

	Arterial P_{O_2} (mm Hg)	Arterial P_{CO_2} (mm Hg)	O_2 Content (ml/dl)	Arterial pH
(A)	98	40	arterial = 10	7.40
(B)	50	65	arterial = 15	7.32
(C)	50	40	arterial = 16	7.38
(D)	105	35	venous = 18	7.45
(E)	50	32	arterial = 10	7.45

138. A healthy 40-year-old man who has been mountain climbing for two days

139. A 30-year-old anemic woman

140. A 73-year-old man who is hypoventilating

141. A 45-year-old woman who is hyperventilating

142. A 56-year-old man with moderately severe obstructive lung disease

Questions 143–146

Match each of the signs of motor disease with the component of the central nervous system (CNS) that is most likely injured.

(A) Posterior cerebellum
(B) Vestibular apparatus
(C) Cerebral cortex
(D) Reticular formation
(E) Basal ganglia

143. Intention tremors

144. Spasticity

145. Rigidity

146. Nystagmus

Questions 147–149

Match each hormone imbalance with its effect on plasma ion concentrations.

(A) Low plasma inorganic $[HPO_4{}^{2-}]$
(B) Low plasma $[Ca^{2+}]$
(C) Both
(D) Neither

147. Excessive parathyroid hormone (PTH) secretion

148. Deficient PTH secretion

149. Vitamin D intoxication

Questions 150–153

Match each site of secretory activity described below with the appropriate lettered region of the nephron.

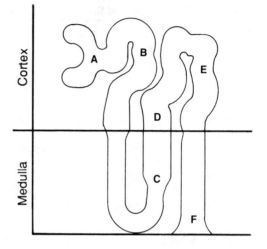

150. Primary site of H^+ secretion

151. Site of K^+ secretion

152. Primary site of $NH_4{}^+$ secretion

153. Site of para-aminohippuric acid (PAH) secretion

Questions 154–156

For each of the following nutrients, choose the substance that is most important for its absorption from the intestine.

(A) Intrinsic factor
(B) Ferritin
(C) Bile salts
(D) Vitamin D
(E) Trypsin

154. Iron

155. Ca^{2+}

156. Cholesterol

Questions 157–160

For each patient described below, select the set of arterial blood values that coincides with that patient's acid-base disorder.

Patient	P_{CO_2} (mm Hg)	$[HCO_3^-]$ (mmol/L)	pH	$[H^+]$ (nmol/L)
(A)	25	15.0	7.40	40.0
(B)	30	16.0	7.31	49.0
(C)	40	25.0	7.41	38.9
(D)	45	35.0	7.51	30.9
(E)	55	27.5	7.37	42.7

157. Patient with metabolic alkalosis and partial respiratory compensation

158. Patient with metabolic acidosis and partial respiratory compensation

159. Patient with highest CO_2 content

160. Patient with highest alveolar ventilation

Questions 161–164

Each of the mechanisms described below is used to alter the force of contraction in certain types of muscle. For each mechanism, select the muscle type or types that alter their contractile force in the manner described.

(A) Cardiac muscle
(B) Skeletal muscle
(C) Smooth and cardiac muscle
(D) Smooth and skeletal muscle
(E) Smooth, cardiac, and skeletal muscle

161. Alteration in the amount of Ca^{2+} released from the sarcoplasmic reticulum (SR)

162. Recruitment of additional muscle fibers

163. Summation of contractions

164. Alteration in preload

Questions 165–169

The following information was obtained from a healthy 24-year-old man who was studied in a renal laboratory:

Inulin Concentration (mg/ml)	Glucose Concentration	Urine Flow Rate	Hematocrit Ratio
Urine = 150	Urine = 1 mg/ml	1.2 ml/min	0.40
Renal arterial plasma = 1.50	Plasma = 90 mg/dl		
Renal venous plasma = 1.20			

Using this data, match each of the following measurements of renal function with the appropriate lettered value.

(A) 0.20
(B) 108 mg/min
(C) 120 ml/min
(D) 600 ml/min
(E) 1000 ml/min

165. Glomerular filtration rate (GFR)

166. Renal blood flow

167. Filtration fraction

168. Renal plasma flow

169. Filtered load of glucose

Questions 170–174

Match each disorder with its associated acid-base state.

(A) Metabolic acidosis
(B) Respiratory acidosis
(C) Both
(D) Neither

170. Hypoventilation

171. Increased [total CO_2]

172. Decreased [total CO_2]

173. Decreased P_{CO_2}

174. Increased $[HCO_3^-]/S \cdot P_{CO_2}$ ratio

Questions 175–178

Match each of the clinical conditions listed below with the digestive system malfunction most likely to cause the condition.

(A) Blockage of the bile duct
(B) Malabsorption of vitamin B_{12}
(C) Malabsorption of water
(D) Malabsorption of carbohydrates

175. Circulatory collapse

176. Malnutrition and steatorrhea

177. Jaundice

178. Anemia

Questions 179–181

Match each cause of an acid-base disturbance with the characteristic set of body fluid changes.

	Plasma pH	Plasma [HCO_3^-] (mEq/L)	Urine pH
(A)	7.27	37	acid
(B)	7.31	16	acid
(C)	7.40	15	alkaline
(D)	7.40	24	acid
(E)	7.55	22	alkaline

179. Hyperventilation

180. Chronic respiratory tract obstruction

181. Diabetic ketoacidosis

ANSWERS AND EXPLANATIONS

1–3. The answers are: 1-A [Chapter 10 II B 2 a; III A], **2-D** [Chapter 10 II B 1 e (2)], **3-D** [Chapter 10 II B 2 a–b]. Cell A is most likely found in the sinoatrial (SA) node. SA cells are characterized by relatively low membrane potentials, spontaneous depolarization (which indicates pacemaker activity), and a relatively fast rate of depolarization.

Cells A and B could exhibit pacemaker activity, because they are capable of diastolic (spontaneous) depolarization (i.e., a decrease in the diastolic membrane potential). The decrease in membrane potential is caused by the decline in K^+ conductance (G_K). Less K^+ leaves the cell during diastole, while Na^+ continues to leak into the cell. The membrane potential declines until threshold is reached, at which time an all-or-none action potential is generated.

Pacemaker cells, such as A and B, are slow fibers. Slow fibers lack Na^+ channels which, in fast fibers, open once the threshold potential is reached, allowing the rapid influx of Na^+. Slow fibers have "slow channels," which apparently limit the rate of ion entry and prolong phase 0 (i.e., depolarization).

4–5. The answers are: 4-C [Chapter 16 IV A 1], **5-B** [Chapter 18 IV B 2 a (1) (b); Figure 18-9]. Lung compliance is calculated as the change in volume per unit change in distending pressure. Since alveolar pressure is zero at the beginning and end of inspiration, the transmural (distending) pressure for the lung is zero minus the interpleural pressure. Only the change in interpleural pressure between the beginning and end of inspiration is given. This difference divided into the tidal volume gives the lung compliance during dynamic conditions, or 150 and 60 ml/cm H_2O. Note that compliance is expressed as volume/pressure.

The change in this patient's lung compliance when she alters her respiratory rate is termed frequency-dependent compliance. Frequency-dependent compliance occurs in the presence of high airway resistance, which causes some acini not to fill completely at rapid rates of respiration, as the result of long time constants. Thus, frequency-dependent compliance indicates the presence of high airway resistance, which is synonymous with obstructive lung disease.

6. The answer is C [Chapter 24 II A; Chapter 26 II B; Figures 24-1, 26-1C]. The Na^+-dependent reabsorption of glucose is blocked by phlorizin, a phenolic glycoside. Thus, this competitive inhibitor virtually blocks the secondary active transport of glucose by the proximal tubule, which leads to glycosuria. Therefore, following phlorizin administration, the clearance of glucose becomes equal to the clearance of inulin. Curve C depicts the clearance curve for a filtered substance that is neither reabsorbed nor secreted, such as inulin.

7. The answer is C [Chapter 41 II B]. Receptive relaxation occurs when food enters the stomach. Although a small amount of receptive relaxation occurs as part of the esophageal reflex, the relaxation of the orad stomach necessary to accommodate the food entering it during a meal is brought about by a vagovagal reflex initiated by the presence of food in the stomach. Chewing is entirely dependent on a nonvagal reflex involving stretch receptors and motor efferents in the jaw muscles. Swallowing is coordinated by a swallowing center in the brain stem and does not involve vagal reflexes. Although gastric emptying is modified to some extent by vagovagal reflexes responding to gastric distention and the presence of chyme in the intestine, local reflexes and hormones are primarily responsible for regulating the rate of gastric emptying. Intestinal contractions are controlled almost entirely by local reflexes and the presence of circulating hormones.

8. The answer is C [Chapter 17 II D 3]. O_2 delivery to the tissues is the product of arterial O_2 content and cardiac output, which normally equals 1 L/min. Of the factors listed, only a 50% reduction in normal hemoglobin concentration would reduce O_2 delivery by half. This would produce a proportional change in O_2 content and an equivalent change in the O_2 delivery. Because of the nonlinear relationship between O_2 tension and hemoglobin saturation, changing the arterial O_2 tension, ventilation rate, or inspired O_2 tension would have a relatively small effect on O_2 delivery.

9. The answer is A [Chapter 6 II C 2]. The auditory ossicles amplify the pressure of the sound stimulus so that sound can pass from air to the fluid environment of the inner ear. Without amplification of the stimulus, the sound reaching the inner ear would be too weak for detection because 99.9% of the sound is normally reflected at the air–fluid interface.

10. The answer is D [Chapter 10 IV B 3 b]. The hexaxial reference system consists of reference lines generated from the three standard (bipolar) limb leads and the three augmented limb leads. The three bipolar limb leads are separated by 60°; thus they lie at 0°, 60°, 120°, 180°, and 240°. The augmented limb leads bisect the angles provided by the bipolar leads; therefore, the hexaxial reference system provides reference lines at every 30°. The hexaxial reference system is in the frontal plane, whereas the precordial (V) leads lie in the horizontal plane.

11. The answer is B [Chapter 25 II B 1 b; Figure 25-3; Table 25-1]. The filtration fraction is that fraction of the plasma flowing through the kidneys that is filtered into Bowman's capsule. Normally about one-fifth of the plasma entering the 2 million glomerular capillaries is filtered. The filtration fraction is the ratio of the glomerular filtration rate (GFR) [125 ml/min] to the renal plasma flow (625 ml/min). Both the GFR and renal plasma flow are regulated in parallel at the afferent arteriole (e.g., constriction decreases both) and inversely at the efferent arteriole (e.g., constriction augments GFR and reduces renal plasma flow). As a result, only changes in efferent arteriolar resistance, not those in afferent arteriolar resistance, affect the ratio of the GFR to the renal plasma flow. Since fluid movement across the glomerulus is governed by Starling's forces, it is proportional to the permeability and surface area of the filtering membrane and to the balance between the hydrostatic and oncotic forces. Ureteral obstruction results in an increase in the hydrostatic pressure in Bowman's capsule, reducing the hydrostatic pressure gradient and, therefore, the GFR and the filtration fraction. An increase in the plasma oncotic pressure contributes to a decrease in GFR as do decreases in the glomerular capillary hydrostatic pressure and the glomerular filtration area.

12. The answer is B [Chapter 49 I D 3 b (2)]. Norepinephrine can be converted to epinephrine by methylation with the epinephrine-forming enzyme, phenylethanolamine-N-methyltransferase (PNMT), which is highly localized in the cytosol of the adrenomedul-lary chromaffin cells. The methyl group donor in this reaction is S-adenosylmethionine. Very high local concentrations of cortisol from the adrenal cortex reach the adrenomedullary chromaffin cells via the adrenal portal system. PNMT is inducible by glucocorticoids. The secretion of adrenomedullary catecholamines is stimulated by acetylcholine (ACh) from the preganglionic nerve endings, which innervate the chromaffin cells. Thus, the adrenal medulla is a functional extension of the nervous system. The synthesis of epinephrine in the adrenal medulla depends on PNMT, cortisol, corticotropin (adrenocorticotropic hormone, ACTH), and the corticotropin-releasing hormone.

13. The answer is B [Chapter 38 IV C; V A; Figures 37-6, 38-1]. In respiratory acidosis, there is a primary increase in plasma CO_2 tension (P_{CO_2}). The renal compensation for respiratory acidosis is increased HCO_3^- reabsorption, which increases the plasma $[HCO_3^-]$. With partial compensation, the pH would not return to normal. Only one set of data (*B*) indicates hypercapnia with a decrease in pH. Note that the pH and the $[HCO_3^-]/S \cdot P_{CO_2}$ ratio are below normal in this patient, which is consistent with acidosis.

14–16. The answers are: 14-D [Chapter 19 I D 2], **15-E** [Chapter 20 IV A 2 c (2)], **16-A** [Chapter 13 II A; Chapter 17 II D 2 b]. The major drive for respiration comes from the medullary chemoreceptors, which respond to local H^+ concentration, which, in turn, depends on the CO_2 tension in the surrounding tissues. During the period of increased ventilation, the CO_2 tension in the body is reduced, thus eliminating the respiratory drive until CO_2 again reaches a threshold value at the end of apnea. The medullary chemoreceptors do not respond to changes in O_2 tension and, thus, are not the cause of the apnea.

During positive-pressure ventilation, the intrathoracic pressure increases, which reduces the pressure gradient between the peripheral tissues and the right side of the heart. Thus, venous return and cardiac output decline during positive-pressure ventilation, and the tissues remove more O_2 from each unit of blood as the blood flows through systemic capillaries. Consequently, during positive-pressure ventilation, venous blood contains less O_2 than normal and the mixed venous O_2 tension decreases. The mixed venous O_2 tension would not be less than 40 mm Hg because of the hypoxia that would occur from breathing this gas mixture, which contains only 0.9% less O_2 than air; the hyperventilation would more than make up for the slight reduction in O_2 tension. Hyperventilation causes a decrease in CO_2 tension and an increased pH, both of which would reduce the P_{50} of hemoglobin, not increase it.

The arteriovenous O_2 content difference is calculated by dividing the O_2 consumption by the cardiac output: 240 ml/min ÷ 6 L/min = 40 ml O_2/L of blood. Because the measurements in the question are given in ml/dl, the correct answer is 4 ml/dl.

17. The answer is A [Chapter 22 III B 1, 3 a; Chapter 38 IV A 1 b]. Severe diarrhea causes volume depletion and electrolyte loss, which defines a state of dehydration. It is important to remember that volume depletion refers to effective circulating blood volume. Because intestinal secretions are rich in K^+ and HCO_3^-, diarrhea causes K^+ depletion and HCO_3^- loss, which lead to hypokalemia and metabolic acidosis, respectively. Vomiting, excessive sweating, and water deprivation can lead to volume depletion but not acidosis. In fact, vomiting causes loss of high concentrations of gastric H^+ and Cl^-, which leads to metabolic alkalosis and hypochloremia. Daily sweat production can exceed 10 L in subjects exercising in a hot climate. Severe sweating is associated with a significant loss of K^+, which may contribute to heat stroke.

18. The answer is A [Chapter 53 VIII C 2]. Calcitriol promotes renal tubular reabsorption of both Ca^{2+} and HPO_4^{2-}, leading to hypercalcemia and hyperphosphatemia. Calcitonin has the opposite effects, promoting the excretion of both electrolytes and, thus, leading to hypocalcemia and hypophosphatemia. Aldosterone favors increased reabsorption of Na^+ and increased excretion of K^+, leading to hypokalemia and alkalemia; the $[Na^+]$ usually remains within normal limits because of a commensurate increase in water reabsorption. Progesterone has an anti-aldosterone–like effect, in that it promotes Na^+ excretion.

19. The answer is C [Chapter 1 III B 3 a (2)]. The initial shrinkage of the red blood cell results from a higher osmotic pressure in the solution than within the cell. However, since the red blood cell eventually returns to its initial volume, the solution must be isotonic. The particles producing the higher osmotic pressure in the solution have reflection coefficients of less than 1 and, thus, are able to diffuse across the membrane. Although these particles initially cause water to flow out of the cell, they eventually reach diffusional equilibrium across the cell membrane and, thus, are unable to keep the water from returning to the cell.

20. The answer is E [Chapter 10 IV B 2]. The terms chest leads, V leads, and precordial leads are synonymous. To determine the axis of any lead, a line is drawn from the electrode site to the electrical zero reference point for the system which, theoretically, lies within the heart. Thus, the axis for each V lead runs from the surface of the chest wall to the heart, which is essentially a horizontal direction.

21. The answer is C [Chapter 29 II A–B]. Spironolactone is a competitive aldosterone antagonist that interferes with the aldosterone-stimulated Na^+ reabsorption in the distal tubular cell and in the collecting duct. Inhibition of Na^+ reabsorption is associated with a marked decrease in urinary excretion of K^+ and H^+. Spironolactone is effective as a diuretic in normal subjects or in patients on a low-Na^+ diet, but not in adrenalectomized patients.

22. The answer is A [Chapter 7 II A 2 b]. Fatigue-resistant muscle fibers can contract for long periods of time without fatiguing because their rich capillary supply and a high concentration of mitochondria enable them to generate the amount of adenosine triphosphate (ATP) required for their activity. However, these muscles cannot contract as rapidly [and thus have lower myosin–adenosine triphosphatase (ATPase) activity] or produce as much force as the non-fatigue resistant muscles. Fast-twitch fatigable muscle fibers utilize anaerobic metabolic pathways to provide ATP and therefore have much higher concentrations of glycolytic enzymes than fatigue-resistant fibers.

23. The answer is B [Chapter 16 VIII C]. The alveolar ventilation in patient B is greater than in patient A. Alveolar ventilation is the minute ventilation minus the dead space ventilation, or the respiratory rate times the difference between the tidal volume and the dead space. Thus, patient B has an alveolar ventilation of 10 • (400 − 150) = 2500 ml/min, whereas patient A has an alveolar ventilation of

$20 \cdot (200 - 150) = 1000$ ml/min. Dead space ventilation in patient A is 3000 ml/min, whereas in patient B it is 1500 ml/min.

24. The answer is A [Chapter 38 IV D; V B; Figure 38-1]. The variable that shows the greatest degree of change in this patient is arterial CO_2 tension (P_{CO_2}), which is decreased by 75% compared to a 46% decrease in $[HCO_3^-]$. These findings are consistent with respiratory alkalosis, which, in this case, is due to hyperventilation brought on by early salicylate toxicity. The resultant hypocapnia reduces tubular H^+ secretion, so that HCO_3^- reabsorption is attenuated, causing a compensatory loss of urinary HCO_3^-. The arterial pH of this patient can be calculated using the Henderson-Hasselbalch equation as

$$pH = pK + \log \frac{[HCO_3^-]}{S \cdot P_{CO_2}}$$
$$= 6.1 + \log \frac{13 \text{ mmol/L}}{0.3 \text{ mmol/L}}$$
$$= 6.1 + \log 43.3$$
$$= 6.1 + 1.6$$
$$= 7.7$$

Note that the increased ratio of $[HCO_3^-]/S \cdot P_{CO_2}$ is consistent with alkalotic states. This condition must be differentiated from metabolic alkalosis, which is associated with increases in both arterial $[HCO_3^-]$ and CO_2 tension. Metabolic acidosis is associated with decreases in $[HCO_3^-]$, CO_2 tension, and pH.

25. The answer is D [Chapter 41 II F]. Gastric motility is controlled primarily by enterogastric reflexes that are elicited when chyme enters the small intestine. The reflexes are both neural and hormonal and act to inhibit gastric contractions. They prevent food from entering the intestine too rapidly. Although distention of the antrum will elicit neuronal and hormonal excitatory reflexes, these reflexes are overcome by the inhibitory enterogastric reflexes.

26. The answer is D [Chapter 50 II F 1 c (8)]. During the postovulatory phase of the menstrual cycle, the thickened, secretory endometrium depends on the continued presence of estradiol and progesterone. If fertilization does not occur, the corpus luteum remains functional for 13–14 days and then undergoes regression (luteolysis). After the regressing corpus luteum loses its ability to produce adequate amounts of estradiol and progesterone, the innermost layer (adluminal, or stratum functionale, layer) of the endometrium becomes ischemic, degenerates, becomes necrotic, and is sloughed into the uterine cavity. The loss of proliferated endometrium (stratum functionale) is accompanied by bleeding known as menstruation (menses). The discontinuation of oral contraceptives after 3 weeks also permits menstruation.

27. The answer is B [Chapter 10 IV C 2 c]. A positive QRS complex recorded from a unipolar electrode indicates that the electrical vector is directed toward the electrode. With positive QRS complexes in leads aV_R and aV_F, the vector must lie in the right axis deviation (RAD) quadrant. A mean electrical axis between 0° and 90° would indicate that the vector lies in the normal quadrant, whereas left ventricular hypertrophy would place the vector in the left axis deviation (LAD) quadrant.

28. The answer is E [Chapter 22 III C 1; Figure 22-1; Table 22-2]. As a result of the hemorrhage, this patient has undergone the loss of water and electrolytes in isotonic concentration leading to dehydration [reduced extracellular fluid (ECF) volume]. In this patient, there will be no net water movement across the cell membranes because both NaCl and water are administered as 1 L of isotonic (isosmotic) NaCl. Because the administered NaCl will remain initially in the extracellular space, the only effects will be a 1 L increase in the ECF volume and a dilution of the plasma protein concentration, the latter effect accounting for the decrease in the plasma colloid oncotic pressure. A 0.9% NaCl solution is isotonic because it maintains the normal red blood cell volume. Furthermore, this 0.9% NaCl solution contains dissociated Na^+ and Cl^-, each in a concentration of 150 mEq/L. Because NaCl is an electrolyte, this same solution is equivalent to a concentration of 300 mOsm/L, assuming 100% dissociation of NaCl. Thus, a 150 mmol/L NaCl solution is isotonic and isosmotic to human plasma. Edema is ruled out in this patient because it represents the retention of both water and NaCl as an isotonic solution, resulting in isotonic overhydration. The infusion of a large volume of isotonic saline at a rate that exceeds urinary excretion results in edema.

29. The answer is A [Chapter 3 III B 2 b; Table 3-1]. The pacemaker cells of the heart contain muscarinic receptors, which are activated by acetylcholine (ACh). When the ACh receptor on these cells is activated, K^+ conductance is increased, which causes the membrane to hyperpolarize and slows pacemaker activity.

30. The answer is C [Chapter 49 I H 1 c (2)]. Epinephrine is a potent hyperglycemic agent for several reasons. It stimulates α-adrenergic receptors on the pancreatic beta cell, inhibiting insulin secretion and, therefore, subsequent facilitated transport of glucose by muscle and adipose tissue. Epinephrine also promotes hepatic and muscle glycogenolysis by activating cyclic adenosine 3',5'-monophosphate (cAMP)-dependent phosphorylase; glycogenolysis in muscle leads to an increase in the plasma level of lactate, which provides the liver with an important glyconeogenic substrate. The lipolytic effect of epinephrine mobilizes free fatty acids, which enhances gluconeogenesis. Additionally, catecholamines directly inhibit peripheral glucose uptake, partly due to the suppression of glucose transporters.

31. The answer is D [Chapter 16 III B 1 a]. The interpleural pressure during the expiratory maneuver would be 15 cm H_2O. Before the Valsalva maneuver, the transmural pressure across the lungs (i.e., the transpulmonary pressure) is equal to the alveolar pressure minus the interpleural pressure, or $0 - (-5) = 5$ cm H_2O. Because lung volume remained constant, the transmural pressure must also remain constant during the Valsalva maneuver, so that when the alveolar pressure increases to 20 cm H_2O, the interpleural pressure must equal 15 cm H_2O in order to maintain the difference of 5 cm H_2O between the inside and outside of the lungs.

32–34. The answers are: 32-B [Chapter 11 II A 3], **33-E** [Chapter 11 II A 4 a (2)], **34-D** [Chapter 11 II A 4 a (3) (b)]. The ventricular volume curve is distinctive because of the decline during systole and the rise during diastole, which can be timed from the relationship with the electrocardiogram (EKG) tracing.

Rapid filling begins with the opening of the atrioventricular (AV) valves early in diastole. Diastole normally begins shortly after the end of the T wave and is indicated by the occurrence of the second heart sound (S_2, not shown).

The duration of diastole (interval C) is largely determined by the heart rate; obviously, when the heart rate increases, the duration of the cardiac cycle shortens. The period of systole also shortens with an increase in heart rate, but is not affected as much as diastole. At very high heart rates, the stroke volume is reduced because the abbreviated diastole renders the ventricular filling time inadequate. Thus, it is important to terminate ventricular tachycardia because of the reduced cardiac output that results from the rapid rate.

35. The answer is E [Chapter 7 VI A 2 a, C 1]. The flocculonodular lobe is also called the vestibulocerebellum to indicate its association with the vestibular system. Lesions within the flocculonodular lobe, like those within the vestibular system, result in a loss of balance (i.e., ataxia). The inability to make rapid, alternating movements (dysdiadochokinesia) or to smoothly reach toward a target is a sign of posterior cerebellar disease. Resting tremors or spontaneous movements are signs of basal ganglia disease. A loss of muscle strength is related to lower motor neuron or muscle disease.

36. The answer is D [Chapter 22 I A–C; Table 22-1; Figure 22-1]. Assuming complete dissociation, this 0.15 mol/L solution of sodium chloride (NaCl) is equivalent to a solution of 300 mOsm/L (0.15 mol/L of Na^+ + 0.15 mol/L of Cl^-). Because this NaCl solution is osmotically balanced with body fluids, there is no shift of water between the major fluid compartments. Thus, the extracellular fluid (ECF) volume increases with no change in the osmolar concentration of the ECF or intracellular fluid (ICF) compartment.

37. The answer is B [Chapter 38 IV B; Figure 37-6; Table 37-1]. The severe loss of gastric fluid in this patient would produce metabolic alkalosis due to the loss of HCl (a noncarbonic acid). She also would exhibit hypovolemia due to the fluid loss. The development of alkalosis is sensed by the chemoreceptors controlling ventilation, resulting in hypoventilation, and an increase in the arterial CO_2 tension (hypercapnia), which reduces the pH toward normal. The kidneys would be expected to excrete the excess bicarbonate, raising the urinary pH.

38. The answer is A [Chapter 41 III B 2]. The formation of HCl by the parietal cells is a two-step process. First, Cl^- is transported into the parietal cell canaliculi. The negative potential developed by the flow of Cl^- allows K^+ to flow into the canaliculi. The K^+ is then actively transported out of the canaliculi in exchange for H^+. Since both K^+ and H^+ are transported against their concentration gradients, an active transport system (H^+-K^+-ATPase) is required.

39. The answer is A [Chapter 49 II C 3 e (1)–(3); Figure 49-2]. A deficiency of 17α-hydroxylase leads to reduction in the 17α-hydroxylation of pregnenolone and progesterone, resulting in hypogonadism and elevated blood gonadotropin levels. This enzyme deficiency also leads to increased production of 11-deoxycorticosterone (11-DOC). This mineralocorticoid causes Na^+ retention, extracellular volume expansion, and hypertension—effects that suppress renin and aldosterone secretion. The 17α-hydroxylase defect also affects the gonads, preventing testicular and adrenal androgen synthesis in males and ovarian estrogen synthesis in females and, thus, resulting in a female phenotype regardless of genotypic sex. These patients require not only cortisol to suppress adrenocorticotropic hormone (ACTH) secretion but also sex steroid treatment consistent with the genotypic sex.

40. The answer is C [Chapter 41 III D; V; Chapter 42 VII B 2]. A significant amount of protein digestion occurs in the stomach because of the action of hydrochloric acid (HCl), which begins to break proteins apart, and pepsin, a protease enzyme secreted by gastric chief cells. These protein digestion products then act as secretagogues that stimulate the secretion of pancreatic proteases.

41. The answer is C [Chapter 33 IV C 2 a; V B 2]. From the data given, this patient's arterial CO_2 tension (P_{CO_2}) is determined to be 50 mm Hg (normal = 40 mm Hg). The Henderson equation is used to estimate P_{CO_2}, but $[H^+]$ must be determined first using the Henderson-Hasselbalch equation. Given a pH of 7.5, $[H^+]$ is calculated as:

$$[H^+] = \text{antilog } (9 - pH)$$
$$= \text{antilog } (9 - 7.5)$$
$$= \text{antilog } 1.5 = 31.6 \text{ nmol/L}$$

Substituting to solve for P_{CO_2},

$$P_{CO_2} = \frac{[H^+]\,[HCO_3^-]}{24}$$
$$= \frac{(31.6 \text{ nmol/L})\,(38 \text{ mmol/L})}{24}$$
$$= 50 \text{ mm Hg}$$

This patient's increased bicarbonate concentration ($[HCO_3^-]$) indicates a metabolic alkalosis. The elevated bicarbonate levels suppress respiratory drive, leading to compensatory elevation of CO_2 tension. Note that unit analysis cannot be used in this equation.

42. The answer is D [Chapter 10 IV A 3, B 2, 3 b]. The largest deflection in the electrocardiogram (EKG) occurs when the electrical vector is parallel to the lead axis. Lead aV_L lies at $-30°$, and therefore should exhibit the largest deflection of any lead.

43–46. The answers are: 43-D [Chapter 22 I B 1 a; II B 1 a (1); Table 22-1], **44-B** [Chapter 22 II A 1, B 2], **45-D** [Chapter 22 II A 1, B], **46-B** [Chapter 22 I A 2]. The man's plasma volume is equal to the Evans blue space, which is 3 L. Because his plasma volume represents 60% of his blood volume, the total blood volume is calculated by dividing the Evans blue space by 0.6, or:

$$\text{Blood volume (L)} = \frac{\text{plasma volume (L)}}{(1 - \text{hematocrit})}$$
$$\text{Blood volume} = \frac{3\text{ L}}{0.6} = 5 \text{ L}$$

The interstitial fluid volume cannot be calculated directly by the dilution principle because there is no substance that is confined to the interstitial fluid space. Because the interstitial fluid space constitutes part of the extracellular fluid (ECF) volume, it can be determined by subtracting the plasma volume from the ECF. The ECF volume of this man is equal to the inulin space. Therefore, the interstitial fluid volume equals the inulin space minus the Evans blue space (8 L − 3 L), or 5 L.

The ECF volume is determined directly by the dilution of such substances as inulin, mannitol, sucrose, thiosulfate, radiosodium, and radiochloride. Thus, the volume of the ECF is equal to the inulin space, which is given as 8 L.

The mathematical relationship between lean body mass (LBM) and total body water (TBW) is:

$$\text{TBW (L)} = 0.7 \text{ LBM (kg)}$$

Therefore,

$$\text{LBM (kg)} = \frac{\text{TBW (L)}}{0.7}$$

Because the total body water is given as the tritiated water space, the calculation is:

$$\text{LBM (kg)} = \frac{35 \text{ L}}{0.7} = 50 \text{ L}$$

Because 50 L of water weighs 50 kg, the 50 kg LBM must be subtracted from the 60 kg body weight, leaving 10 kg (i.e., the weight of the body fat).

47. The answer is A [Chapter 16 VIII C 1–2]. This patient has an alveolar-to-arterial O_2 tension difference of 35 mm Hg. To calculate this, the alveolar O_2 tension (P_{AO_2}) must be determined using the alveolar gas equation. Given an arterial CO_2 tension of 30 mm Hg, the alveolar O_2 tension is determined as:

$$
\begin{aligned}
P_{AO_2} &= (760 - 47) \cdot 0.21 - 30 \\
&= 150 - 30 = 120 \text{ mm Hg}
\end{aligned}
$$

Thus, the alveolar-to-arterial O_2 tension difference is $120 - 85$, or 35 mm Hg. This patient's low arterial CO_2 tension and high pH indicate that he has respiratory alkalosis; thus, he has been hyperventilating and does not have metabolic alkalosis. Respiratory alkalosis causes a decrease in plasma bicarbonate levels due to the slope of the blood buffer line. Lastly, chronic obstructive pulmonary disease (COPD) leads to hypoxia and CO_2 retention, not hyperventilation.

48. The answer is C [Chapter 6 I B 1 c (2)]. The axial length of the eye in an individual with normal eyesight (i.e., an emmetrope) is equal to the eye's focal length, which is equal to the reciprocal of its power. Because the power of the eye is 60 diopters (D), the axial length is 0.0167 meters (16.7 mm). The power of a lens required to focus an image at a given distance from the lens can be determined using the lens formula:

$$P = \frac{1}{o} + \frac{1}{i} = \frac{1}{f} \text{ , where}$$

o, i, and f are the object, image, and focal distances respectively (in meters) and P is the

power of the lens (in diopters). Because the axial length is 0.0167 meters and the object is placed 0.25 meters from the lens, the total power of the lens must be 64 diopters in order to focus the object. Thus, the converging power of the lens must increase by 4 diopters.

49. The answer is A [Chapter 2 III E 1–2]. During attacks of hyperkalemic periodic paralysis, large amounts of K^+ are released from skeletal muscle fibers, causing a rise in extracellular K^+ concentration. When the extracellular K^+ concentration increases, the membrane potential depolarizes, causing inactivation of the Na^+ channels.

50–55. The answers are: 50-D [Chapter 11 IV A; Figure 11-7B], **51-B** [Chapter 11 IV D; Figure 11-7C], **52-A** [Chapter 11 II A 3 b (1)], **53-D** [Figure 11-7], **54-C** [Figure 11-7], **55-C** [Figure 11-7A]. Loop ACDI shows the effect of an increased afterload. The increased ventricular pressure reflects an increase in arterial resistance, which increases the afterload. The maximal pressure–volume point for loop ACDI lies on the same line as that for the normal pressure–volume loop, which indicates that there is no change in contractility. The stroke volume represents the difference in volume between the vertical lines for any loop. In this case, the increased afterload decreases the stroke volume. The ventricular end-diastolic volume (VEDV) is indicated by point A; all three loops start at the same VEDV. Preload is a synonym for VEDV.

An increase in contractility causes the maximal pressure–volume relationship to move to the left and assume a steeper slope. Stroke volume increases and there is a smaller volume of blood that remains in the ventricles at the end of systole.

Mitral valve closure occurs at the onset of systole (point A) and represents the beginning of the isovolumic contraction period. During this interval, as the name implies, the ventricular volume remains constant and ventricular pressure rises until it exceeds the pressure in the aorta.

Ventricular filling occurs during diastole (i.e., the interval between points A and H) and is the increase in volume that occurs in preparation for the next contraction. During steady-state conditions, the ventricular filling equals the stroke volume that is ejected during the next systole. Changes in venous return, the duration of diastole, the afterload, and ventricular contractility can all influence ventricular filling and the stroke volume in a complex fashion.

The slope of the line from point G to point A has the units of mm Hg/ml, which are compatible with elastance. Compliance is the reciprocal of elastance and has units of ml/mm Hg. The line from point G to point A occurs during diastole; therefore, the slope is a measure of the diastolic elastance of the ventricles.

Points D, E, and F represent the ventricular end-systolic volume (VESV). Stroke volume is represented as the width of each loop (i.e., A–I, A–H, or A–G). The VEDV occurs at point A for all three loops. The diastolic reserve volume is the volume of blood that the ventricle can hold from the end-diastolic point to the maximal volume—this volume is not shown on the graph. The residual volume represents a minimal volume of blood that lies between the trabeculae carneae and the papillary muscles. This volume is a part of the VESV and is never ejected from the ventricles.

56. The answer is D [Chapter 26 III D 2 a, 4 b; Figure 36-1]. The primary effect of carbonic anhydrase inhibitors such as acetazolamide is to inhibit both H^+ secretion and $NaHCO_3$ reabsorption, making the urine alkaline. NH_4^+ excretion is reduced as a result of the diminished H^+ secretion. Carbonic anhydrase inhibitors restrict H^+ secretion by inhibiting the intracellular hydration of CO_2, a primary source of intracellular H^+. The decline in H^+ secretion inhibits the Na^+-H^+ exchange at the luminal membrane of the proximal tubule, which is the primary site of $NaHCO_3$ reabsorption. HCO_3^- reabsorption is also inhibited because only a limited amount of HCO_3^- can be reabsorbed by the distal segment. The elevated intraluminal HCO_3^- augments Na^+ and K^+ excretion and results in $NaHCO_3$ diuresis. Carbonic anhydrase inhibitors also block the dehydration of H_2CO_3 formed in the tubular lumen. Chronic doses of such drugs can lead to hyperchloremic acidosis (metabolic acidosis).

57. The answer is A [Chapter 7 III B 1 c (2)]. Reciprocal innervation is a neuronal innervation pattern in which, when one motoneuron is excited, its antagonistic motoneuron is inhibited. Inhibition of flexors during an extension is an example of reciprocal innervation. This pattern of innervation allows movement to occur unimpeded by the activity of antagonistic muscles.

58. The answer is D [Chapter 46 I A]. In humans, there is a diurnal variation in the secretory patterns of adrenocorticotropic hormone (ACTH) and cortisol (hydrocortisone) and in the excretion of 17-hydroxycorticoids. In individuals who sleep regularly from about 11:00 P.M. to 7:30 A.M., the peak of this circadian rhythm occurs between 6 A.M. and 8 A.M., while the nadir is observed between midnight and 2 A.M. Thus, the peak plasma level of glucocorticoid is entrained to the activity cycle with maximal ACTH-cortisol secretion appearing about 1 hour after awakening. Changes in sleep periods or in longitude cause phase shifts in the ACTH-cortisol secretory pattern; however, the rhythm itself persists. There is an abrogation of the pituitary-adrenocortical rhythm in patients with hypercortisolism.

59. The answer is C [Chapter 17 II C, D 2 b]. This patient's O_2 consumption is 520 ml/min/m^2. O_2 consumption can be determined using the Fick principle, an important concept that has wide applicability in physiology and medicine. To solve for O_2 consumption ($\dot{V}O_2$), the Fick principle is expressed as:

$$\dot{V}O_2 = \dot{Q} \bullet (CaO_2 - C\bar{v}O_2), \text{ where}$$

$\dot{Q}$ = blood flow and $CaO_2 - C\bar{v}O_2$ = the arteriovenous O_2 difference.

In this problem, the cardiac index equals the blood flow. Because arterial hemoglobin saturation is assumed to be 100%, the arterial O_2 content (CaO_2) equals the O_2 capacity, which equals hemoglobin • 1.34, or 16 ml/dl (160 ml/L). Using these values in the above formula gives:

$$\begin{aligned} \dot{V}O_2 &= 6.5 \bullet (160 - 80) \\ &= 520 \text{ ml/min} \end{aligned}$$

It is important to use similar units for blood flow and O_2 content. In addition, converting O_2 content to ml/L before calculating is recommended.

60. The answer is B [Chapter 50 II E 2 a–c, F 1; Figure 50-2]. The menstrual cycle consists of four phases. These phases (with synonymous terms) are: menses (4–5 days), the preovulatory phase (also called follicular, estrogenic, or proliferative phase; 10–12 days), the ovulatory phase, and the postovulatory phase (also called luteal, progestational, or secretory phase; 14 days).

61. The answer is B [Chapter 33 V B 2]. From the data given, this patient's arterial bicarbonate concentration ($[HCO_3^-]$) is determined to be 14.7 mmol/L (mEq/L). Arterial $[HCO_3^-]$ is easily estimated using the Henderson equation, which is stated as:

$$[HCO_3^-] = 24 \frac{P_{CO_2}}{[H^+]}$$

where $[HCO_3^-]$ is expressed in mmol/L, $[H^+]$ in nmol/L, and P_{CO_2} in mm Hg. Substituting,

$$[HCO_3^-] = 24 \frac{30}{49}$$
$$= 14.7 \text{ mmol/L}$$

The acid-base disturbance in this case is an almost completely compensated metabolic acidosis (pH 7.32).

62. The answer is E [Chapter 43 III A, C]. Although acidosis can result from loss of fluid from both the intestine and the colon, only the colon secretes K^+. Thus, excessive fluid loss from the colon will result in both acidosis and hypokalemia.

63–64. The answers are: 63-B [Figure 11-2A], **64-A** [Chapter 11 II C 3 a (1)]. The systolic gradient that occurs between the ventricular and aortic pressures is diagnostic of aortic valve stenosis. Normally, the aortic valve provides a negligible resistance and the aortic pressure is nearly identical to the ventricular pressure during the phase of rapid ventricular ejection. Pulmonary valve stenosis would produce similar tracings upon measurement of the right ventricular and pulmonary pressures; however, the pressures are proportionately reduced for the right-sided events because of the low resistance of the pulmonary circulation.

Semilunar (e.g., aortic or pulmonary) valve stenosis represents an impediment to the ejection of blood from the ventricle and results in an systolic ejection murmur. An ejection murmur is diamond-shaped (i.e., it is a crescendo–decrescendo sound that has maximal intensity in midsystole when the pressure gradient is largest).

65–67. The answers are: 65-B [Chapter 20 II C 1 d; Table 20-1], **66-E** [Chapter 18 V A 2 b (3) (b)], **67-A** [Chapter 18 V A 2 b; Chapter 20 II C 1 d]. A right-to-left anatomic shunt causes mixed venous blood to enter the systemic circulation without being oxygenated, leading to hypoxia and an increased alveolar–arterial O_2

tension difference. Ventilation–perfusion imbalance also leads to hypoxia but is not the cause in this patient. With ventilation–perfusion imbalance, arterial O_2 tension should exceed 500 mm Hg when the patient breathes 100% O_2 at sea level, and the alveolar O_2 tension should be 673 mm Hg (which can be calculated using the alveolar gas equation). A left-to-right anatomic shunt does not cause hypoxia, because oxygenated blood (from the left side of the circulation) enters the right ventricle or the pulmonary artery. Administration of 100% O_2 will completely correct hypoxia resulting from diffusion abnormalities or hypoventilation.

The fraction of the cardiac output that represents shunted blood can be calculated using the shunt equation:

$$\frac{\dot{Q}s}{\dot{Q}_T} = \frac{(C_{iO_2} - C_{aO_2})}{(C_{iO_2} - C_{\bar{v}O_2})}, \text{ where}$$

$$
\begin{aligned}
\dot{Q}s &= \text{the shunted flow} \\
\dot{Q}_T &= \text{cardiac output} \\
C_{iO_2} &= \text{pulmonary capillary } O_2 \\
&\quad \text{content} \\
C_{aO_2} \text{ and } C_{\bar{v}O_2} &= \text{arterial and venous } O_2 \\
&\quad \text{content, respectively}
\end{aligned}
$$

The calculation usually is sufficiently accurate only if hemoglobin O_2 is used and the dissolved O_2 is disregarded. In the pulmonary capillaries, hemoglobin should be 100% saturated while the patient is breathing 100% O_2 so that:

$$
\begin{aligned}
C_{iO_2} &= O_2 \text{ capacity} \\
&= 18 \text{ g hemoglobin/dl} \cdot 1.34 \text{ ml } O_2/\text{g} \\
&= 24.1 \text{ ml/dl}
\end{aligned}
$$

Arterial O_2 content can be determined as:

$$
\begin{aligned}
C_{aO_2} &= C_{iO_2} \cdot \text{hemoglobin saturation} \\
&= 24.1 \text{ ml/dl} \cdot 0.85 \\
&= 20.5 \text{ ml/dl}
\end{aligned}
$$

Venous O_2 content can be determined if it is remembered that the normal venous O_2 tension also is 40 mm Hg, which is equivalent to 75% hemoglobin saturation, giving:

$$
\begin{aligned}
C_{\bar{v}O_2} &= 24.1 \cdot 0.75 \\
&= 18.1 \text{ ml/dl}
\end{aligned}
$$

Thus,

$$
\begin{aligned}
\frac{\dot{Q}s}{\dot{Q}_T} &= \frac{24.1 - 20.5}{24.1 - 18.1} \\
&= \frac{3.6}{6.0} = 0.6
\end{aligned}
$$

To determine the site of a right-to-left shunt, the O_2 tension must be measured in the left ventricle and atrium rather than the right (i.e., the unsaturated venous blood is used to locate the point where the O_2 tension drops). Thus, if there were an interatrial septal defect with right-to-left flow, the O_2 tension in the left atrium would be lower than the O_2 tension in the pulmonary veins. A ventricular septal defect would show a drop in O_2 tension in the left ventricle compared with that in the left atrium. If the O_2 tension were low and equal in the pulmonary veins, left atrium, and left ventricle, then the shunt would necessarily be within the lungs, which could indicate the presence of an arteriovenous anastomosis.

68. The answer is A [Chapter 7 II A 2 b, C]. Small motoneurons usually fire before large motoneurons during the performance of a movement. The small motoneurons innervate small, fatigue-resistant muscle fibers that generate long-lasting muscle twitches. Because the duration of the twitch is longer than in other muscle fibers, the frequency of firing required for tetanus is lower. Slow-twitch fibers cannot produce a large amount of force. Because they have a rich capillary supply, slow-twitch fibers do not need to rely on glycolysis as a source of adenosine triphosphate (ATP) and, therefore, are more resistant to fatigue.

69. The answer is D [Chapter 16 VIII C 1]. The alveolar CO_2 tension is directly proportional to the rate at which CO_2 is produced by metabolism and inversely proportional to alveolar ventilation. The increased minute ventilation would not alter the metabolic rate significantly, so any changes in CO_2 tension must be related to alveolar ventilation. Alveolar ventilation is the difference between minute ventilation and dead space ventilation. During control conditions, alveolar ventilation is $(0.5 \text{ L} \cdot 15) - (0.2 \text{ L} \cdot 15) = 4.5$ L/min, which is unchanged by the alteration in respiratory pattern $[30 \cdot (0.35 \text{ L} - 0.2 \text{ L})]$. An increase in alveolar ventilation would cause a decrease in CO_2 tension, which would lead to respiratory alkalosis (i.e., increased arterial pH).

70. The answer is C [Chapter 28 II A 1]. A decrease in plasma osmolality (osmolarity) leads to a decrease in antidiuretic hormone (ADH) secretion. This results in an increase in free-water clearance and diuresis. These osmoreceptors are found in the vicinity of the supraoptic nucleus of the hypothalamus. ADH augments the water permeability of the cortical collecting duct and the water and urea

permeabilities of the medullary collecting duct. It increases renal water reabsorption, resulting in the excretion of a small volume of hypertonic urine. The major stimuli for ADH are an increase in the plasma osmolality and a decrease in the effective circulating blood volume. Aldosterone promotes Na^+ reabsorption, which leads to water retention.

71. The answer is B [Chapter 37 II B 3 b; Chapter 38 IV A; V C; Figure 38-1]. This diabetic patient has metabolic acidosis resulting from ketoacidosis. The ketoacidosis is attributable mainly to the formation of β-hydroxybutyric acid from the partial oxidation of fatty acids. The high concentration of ketoacids in the form of anions is responsible for the increased anion gap; however, it is H^+ retention, not anion accumulation, that is responsible for the acidosis. Osmotic diuresis accounts for this patient's dehydration, which is not only water loss but also increased renal excretion of Na^+, K^+ Cl^-, and glucose. Na^+ and K^+ also are lost when they are excreted in association with the excess quantities of organic anions. In this case, the decline in plasma bicarbonate ($[HCO_3^-]$) is the primary change, and the decreased CO_2 tension (i.e., hyperventilation) is the compensatory response. Complete compensation is evidenced by the normal pH of 7.39 and the normal ratio of $[HCO_3^-]/S \cdot P_{CO_2}$. It is possible to have an acidosis with a normal blood pH, or $[H^+]$, because secondary changes diminish the extent of the acid-base imbalance.

72–74. The answers are: 72-C, 73-A, 74-E [Chapter 16 VII C 4 a]. This patient's functional residual capacity (FRC) is 2.7 L. The dilution test measures the FRC, the lung volume at the beginning of the test. Because the volume of gas in the spirometer is kept constant, the degree of dilution produced by the lungs after equilibration must be determined by calculating the ratio of the initial fraction of helium to the fraction of helium following equilibration (F_1/F_2), which equals $0.05 \div 0.03$, or 1.67. Thus, the volume of the spirometer plus the lung volume is 1.67 times the volume of the spirometer, or $1.67 \cdot 4$ (6.7 L). Subtracting the volume of the spirometer leaves the lung volume at the start of the test (i.e., FRC), or $6.7 - 4$ (2.7 L).

This patient's residual volume (RV) is 0.7 L. Because FRC is the sum of RV and expiratory reserve volume (ERV), RV is determined as: $RV = FRC - ERV$, or $2.7 - 2.0 = 0.7$ L.

This patient's total lung capacity (TLC) is 5.7 L. TLC is the sum of vital capacity (VC) and RV, or 5 + 0.7 = 5.7 L.

75. The answer is B [Chapter 49 I H 1 c (1), I 2]. Pheochromocytoma is an adrenomedullary tumor characterized by the hypersecretion of catecholamines, usually norepinephrine. Catecholamines block insulin release (hypoinsulinemia) and increase both gluconeogenesis and fatty acid mobilization, which account for glucose intolerance, fasting hyperglycemia, and glycosuria. The breakdown of muscle glycogen leads to elevated plasma pyruvate or lactate levels. The metabolic features of pheochromocytoma resemble hyperthyroidism and include tremor, weight loss, heat intolerance, and increased basal metabolic rate. The cardinal sign is hypertension, which may be persistent or paroxysmal. Patients usually complain of attacks that may be precipitated by emotion or physical exercise, which consist of pounding headaches, sweating, pallor, pain or "tightness" in the chest, apprehension, paresthesia, nausea, and vomiting. In this patient, the attacks were provoked by mechanical pressure that was exerted on the tumor by changes in body position.

76. The answer is B [Chapter 7 III B 1 d (2); Figure 7-4]. Unloading (i.e., the reduction in Ia afferent activity that accompanies muscle shortening) can be prevented if the intrafusal and extrafusal muscle fibers are coactivated so that tension on the intrafusal muscle fibers is maintained during shortening. Therefore, gamma motoneurons, which innervate intrafusal muscle fibers, can prevent unloading. Alpha motoneurons innervate extrafusal muscle fibers; Ib afferent fibers innervate Golgi tendon organs; and the free nerve endings of C fibers contain pain, temperature, and mechanical receptors.

77. The answer is D [Chapter 16 II B 1 b (1) (b), 2 b (1)]. The man can lie under 68 cm of water. Calculation of this depth requires converting the pressure given in mm Hg to a pressure expressed in cm H_2O: 50 mm Hg • 1.36 cm H_2O (mm Hg) = 68 cm H_2O. This man's alveolar pressure is equal to the atmospheric pressure when he is under water, whereas the pressure outside his chest wall is 68 cm H_2O higher. To expand his lungs, this man must generate a pressure slightly more negative than −68 cm H_2O. Under these conditions, the man is undergoing negative-pressure

breathing, which, if prolonged, can lead to pulmonary edema.

78. The answer is E [Chapter 38 IV D; V B; Figures 37-6, 38-1]. The most likely diagnosis is respiratory alkalosis resulting from anxiety-induced hyperventilation. The key determinants of the cause and compensation of the acid-base disorder are the arterial CO_2 tension and $[HCO_3^-]$. The greater reduction in CO_2 tension than in $[HCO_3^-]$ indicates that the primary disturbance is respiratory alkalosis. That both the arterial $[HCO_3^-]$ and $[H^+]$ change in the same direction further supports the diagnosis of a respiratory acid-base imbalance. Both obstructive and restrictive lung diseases are common causes of respiratory acidosis.

79. The answer is C [Chapter 2 III F; Chapter 3 II C 1]. The flow of Ca^{2+} into the presynaptic nerve terminal stimulates the release of neurotransmitter from synaptic vesicles by exocytosis. Ca^{2+} enters the nerve terminal down its electrochemical gradient through channels opened by membrane depolarization. Propagation of an action potential is accomplished by generating new action potentials along the nerve fiber. The generation of action potentials is dependent on electrically excitable Na^+ and K^+ channels.

80. The answer is C [Chapter 42 V C]. Vitamin B_{12} is absorbed in the terminal ileum in association with intrinsic factor. Bile salts are also absorbed in the terminal ileum. If bile salts cannot be absorbed and recirculated, lipid absorption will be compromised because the liver is unable to synthesize enough bile to absorb all the fats ingested during a meal.

81. The answer is D [Chapter 25 II A 2]. The simplest flow (Q) equation that summarizes the relationship between the hydrostatic pressure gradient (ΔP) and resistance (R) is:

$$Q = \frac{\Delta P}{R}$$

The pressure gradient is the hydrostatic pressure difference between the renal artery and renal vein. The resistance to blood flow can be determined by dividing this hydrostatic pressure gradient by the renal blood flow:

$$R = \frac{\Delta P}{Q}$$

82. The answer is C [Chapter 6 II C 2 a (2); Figure 6-11]. The middle ear amplifies the pressure of the sound stimulus so that an adequate signal can reach the inner ear, even though 99.9% of the sound is reflected at the air–fluid interface. Because the middle ear cannot amplify all sounds equally well, some sound frequencies are detected at lower strengths than others. Fortunately, the sounds that are heard at the lowest intensities (e.g., between 500 and 5000 Hz) are the sounds that are used for speech.

83–84. The answers are: 83-E [Chapter 11 II A 4; Figure 11-7], **84-B** [Chapter 11 II A 3 b (2)]. Opening of the mitral valve signals the onset of ventricular filling, which occurs when ventricular pressure and volume are at their lowest point (i.e., point E on the pressure–volume loop).

By definition, isovolumic ("equal volume") contraction must be represented by a vertical line on the pressure–volume loop. Isovolumic contraction is the first period of ventricular systole after the mitral valve closes; at the onset of isovolumic contraction, the ventricular pressure rises to reach the aortic diastolic pressure. Isovolumic contraction ends when the semilunar valve opens and ventricular ejection begins. The onset of ejection is signaled by the point where ventricular volume begins to decrease.

85. The answer is C [Chapter 26 III D 4 a–c]. The kidney excretes H^+ as NH_4^+ and titratable acid. Titratable acid exists mainly in the form of $H_2PO_4^-$. The titratable acid is excreted mainly as NaH_2PO_4, which is also called acid phosphate, monosodium phosphate, or monobasic phosphate. The normal kidney excretes almost twice as much acid combined with NH_3 than it excretes titratable acid. The rate of NH_4^+ excretion increases during metabolic acidosis.

86. The answer is D [Chapter 48 IV A 3 d (2)]. This woman's inability to lactate and to menstruate following parturition stems from pituitary failure related to the massive bleeding she experienced during delivery. Most likely, this woman has postpartum pituitary necrosis (Sheehan's syndrome) resulting from ischemia of the hypophysial portal system, which supplies 90% of the blood to the anterior pituitary gland. The resultant pituitary failure may be complete (panhypopituitarism) or partial. In this case, the acidophils responsible for production of prolactin, follicle-stimulating hormone (FSH), and luteinizing hormone (LH) were affected, resulting in this patient's inability to lactate and to menstruate. This patient also would exhibit hypoglycemia because of low plasma growth hormone (GH) levels and low plasma adrenocorticotropic hormone (ACTH) levels, which would lead to decreased secretion of cortisol (a potent hyperglycemic hormone) and decreased synthesis of epinephrine (a hyperglycemic hormone that depends on cortisol). The decreased levels of these hyperglycemic hormones would cause insulin sensitivity.

Because aldosterone secretion does not depend mainly on ACTH secretion, this patient should be able to regulate Na^+ balance. She also should have normal antidiuretic hormone (ADH) activity, because the supraoptic nuclei and pars nervosa are not perfused by the hypophysial portal system. Thus, this patient should demonstrate normal water balance.

87. The answer is B [Chapter 37 II B 3; Chapter 38 IV A; V C; Figures 37-6; 38-1]. The decreases in arterial pH, $[HCO_3^-]$, and CO_2 tension are consistent with metabolic acidosis. There also is a decreased $[HCO_3^-]/S \cdot P_{CO_2}$ ratio (i.e., 12.2) and a widened anion gap (i.e., 47 mEq/L). The primary disturbance is the marked reduction in $[HCO_3^-]$, and the compensatory response in hyperventilation (as indicated by the hypocapnia). That the $[HCO_3^-]$ decreases and the $[H^+]$ increases is evidence of metabolic acid-base imbalance. The hyperglycemia and dehydration (as evidenced by dry skin) support a diagnosis of diabetes mellitus.

88. The answer is C [Chapter 18 IV A 1 b]. The lung is a passive structure whose transmural pressure varies as a function of volume and phase of respiration. At residual volume (RV), the transpulmonary pressure may be negative at the base of the lung, because interpleural pressure could be positive owing to the hydrostatic effects of lung weight and the decreased retractile force at low lung volume. The transpulmonary pressure always is greater at the apex than at the base when standing because of the hydrostatic effects of gravity. Thus, the airways and alveoli at the apex of the lung always are more distended than those at the base of the lung of a person who is standing.

89–92. The answers are: 89-C, 90-B, 91-D, 92-A [Chapter 16 IV A 1–2; VIII C; Chapter 18 V A 2; Chapter 38 IV A]. This patient's blood data are almost consistent with a ventilation–perfusion abnormality. Although the CO_2 tension is normal, the O_2 tension is decreased, as are pH and $[HCO_3^-]$. These latter two findings indicate a metabolic acidosis that is most likely due to accumulation of lactic acid caused by the decreased O_2 tension and cardiac output. Ventilation–perfusion abnormalities are the most common cause of clinical hypoxia. A variety of factors affecting the lungs and cardiovascular system may cause these abnormalities.

Given a predicted dead space of 150 ml, this patient's calculated alveolar ventilation is 7.2 L/min. Alveolar ventilation is the volume/ min that is effective in gas exchange, which is calculated as follows: (tidal volume − dead space) • respiratory rate. Thus, in this patient, the alveolar ventilation equals $(0.6 - 0.15) • 16 = 7.2$ L/min.

This patient's O_2 tension value while breathing 40% O_2 (120 mm Hg) indicates an increase in the alveolar-to-arterial O_2 tension difference. Applying the alveolar gas equation, the alveolar O_2 tension is calculated as

$$\text{alveolar } P_{O_2} = (760 - 47) • 0.40 - \frac{43}{0.8}$$
$$= 230 \text{ mm Hg}$$

Thus, there is a difference of 110 mm Hg in alveolar-to-arterial O_2 tension in this patient $(230 - 120 = 110)$. Normally, the alveolar-to-arterial O_2 tension is 5–10 mm Hg. The large difference in this patient results from a physiologic shunt secondary to the ventilation–perfusion abnormality. Changes in the hemoglobin concentration do not affect the arterial O_2 tension, which depends solely on the amount of O_2 that is in physical solution in the blood.

This patient's respiratory compliance is 0.03 L/mm Hg. The compliance of the respiratory system is calculated as the change in volume (i.e., tidal volume) divided by the change in distending pressure. Because the respirator is delivering intermittent positive pressure (i.e., 20 mm Hg in order to distend the lungs to accept the delivered volume), the pressure in the lungs at the end of expiration is zero (i.e., atmospheric). Thus, compliance of the respiratory system is 0.6 L/20 mm Hg = 0.03 L/mm Hg. Normal respiratory compliance is about 0.1 L/mm Hg. The decreased compliance ("stiff lungs") in this patient could be a result of the pulmonary edema.

93. The answer is B [Chapter 42 VII C 2 d (3)]. Although fats can be absorbed all along the intestine, from the point at which the bile duct enters the duodenum until the bile salts are absorbed from the terminal ileum, almost all of the digested lipids are absorbed by the time the chyme reaches the midjejunum. Little, if any, lipid absorption occurs in the ileum.

94. The answer is D [Chapter 9 II C; Chapter 13 II A]. The data can be used to calculate the cardiac output using the Fick principle:

$$Q = \frac{\dot{V}_{O_2}}{Ca_{O_2} - C\bar{v}_{O_2}}, \text{where}$$

$$\begin{aligned}
Q &= \text{cardiac output} \\
\dot{V}_{O_2} &= O_2 \text{ consumption (ml/min)} \\
Ca_{O_2} &= \text{arterial } O_2 \text{ content} \\
C\bar{v}_{O_2} &= \text{mixed venous } O_2 \text{ content}
\end{aligned}$$

Mixed venous blood from the right ventricle or pulmonary artery must be used to calculate the cardiac output. For this calculation, as in all others, the units must be consistent so every volume can be expressed in ml:

Cardiac output (Q) = 210 (ml/min)/(0.18 ml O_2/ml blood) − (0.11 ml O_2/ml blood) = 210/0.07 = 3000 ml/min.

The normal cardiac output is approximately 5 L/min (5000 ml/min). Therefore, choices B ("cardiac output is approximately 1470 ml/ min") and E ("cardiac output is extremely high") are incorrect. The amount of O_2 transferred to the tissues is given by the difference between the arterial and venous O_2 content, not the sum. The arterial O_2 content is equal to or only slightly less than the pulmonary venous oxygen content. (Thebesian venous flow into the left ventricle may account for the slightly lower arterial O_2 content.) Stroke volume is equal to the cardiac output divided by the heart rate (3000/75 = 40 ml). In this patient, the stroke volume of 40 ml is less than the normal stroke volume of 70 ml, leading to a lower than normal cardiac output.

95. The answer is D [Chapter 42 VII E 3]. Although intrinsic factor is secreted by parietal cells in the stomach, it is not able to bind vitamin B_{12} in the stomach because the vitamin is bound to another protein, called protein R. When the vitamin reaches the intestine, the R protein is removed, and intrinsic factor is able to bind to the vitamin B_{12}. The B_{12}–intrinsic factor complex is absorbed in the terminal ileum.

96. The answer is C [Chapter 50 III B 2 b (2)].
The corpus luteum is the initial source of
plasma progesterone and 17α-hydroxypro-
gesterone, which peak at 3–4 weeks postconception. At 6–8 weeks postconception, the
progesterone reaches its lowest point, while
the 17α-hydroxyprogesterone continues to
decline. Since the placenta cannot synthesize
17α-hydroxyprogesterone, the secondary rise
in progesterone reflects placental (trophoblast)
function, and the 17α-hydroxyprogesterone
curve is a correlate of corpus luteal function
of pregnancy. The theca interna and adrenal
glands do not secrete progesterone. The decidua is not the source of any hormones but is
a specialized region of the endometrium that
develops into the maternal component of the
placenta.

97–99. The answers are: 97-B, 98-C, 99-C
[Chapter 16 VI D; VIII B 2 a (2) (b) (ii), C 1;
Chapter 37 IV A 1; Chapter 38 IV C, V A; Figure 38-1]. The increased CO_2 tension (P_{CO_2})
in this patient indicates that alveolar ventilation is inadequate. Minute ventilation may be
normal or increased while alveolar ventilation
is reduced, because the total dead space increases secondary to ventilation–perfusion
abnormalities caused by the pulmonary disease. (It is important to remember that total
dead space equals alveolar dead space plus
anatomic dead space.) The normal arterial pH
in this patient indicates that renal compensation has occurred. Renal compensation may
require 1–2 weeks, meaning that the abnormality is chronic, ruling out acute respiratory
failure. Anemia and carbon monoxide poisoning are not viable alternatives, since arterial
O_2 tension is normal in these two conditions.
An anatomic shunt causes hypoxia and typically results in a lowered CO_2 tension as a
result of the increased ventilation that results
from this type of hypoxia.

Smoking causes inflammation and edema in
the airways (bronchitis), which can lead to
infection of the airways and consequent mucus production. The resultant airway narrowing increases airway resistance and the work
of breathing. If respiratory work increases sufficiently, there will be respiratory muscle fatigue and a decrease in alveolar ventilation,
which results in hypercapnia and hypoxia.
These latter events cause pulmonary artery
vasoconstriction and pulmonary hypertension.

This patient has respiratory acidosis, by
definition, because of the hypercapnia. Respiratory acidosis is compensated by renal reten-

tion of HCO_3^- (normal = 24 mEq/L). Metabolic or lactic acidosis is not present because
of the increased HCO_3^- levels. Respiratory
alkalosis is produced by a decrease in CO_2
tension.

100–103. The answers are: 100-D [Chapter
14 III C 2], **101-B** [Chapter 14 III C 1], **102-B**
[Chapter 14 III C 1 b (1)], **103-E** [Chapter 14
III C 2]. Stimulation of the right vagus nerve
can cause marked cardiac slowing. From the
record, it is obvious that the major effect of
the intervention at time 1 was a slowing of the
heart rate. Although vagal stimulation can
cause marked cardiac slowing, the ventricular
end-diastolic volume (VEDV) increases so that
stroke volume increases sufficiently to compensate for the reduced heart rate. Cardiac
output must have remained relatively constant
because there is little change in arterial pressure and total peripheral resistance. (Recall
that cardiac output equals arterial pressure
divided by total peripheral resistance.) Electrical stimulation of the superior cervical ganglion or administration of a cholinergic blocking drug would increase the heart rate.
Electrical stimulation of the lumbar sympathetic nerve roots would cause a significant
rise in resistance in the lower portion of the
body. The administration of a β-adrenergic
blocking drug would reduce the heart rate, but
it would also reduce cardiac contractility and
therefore, cardiac output.

There is obviously a rise in heart rate, arterial pressure, and total resistance at time 2.
The only factor that is listed that could cause
this sequence of changes is stimulation of the
portion of the sympathetic system that controls
the heart (i.e., the superior cervical ganglion).

An α-adrenergic drug was most likely administered at time 3. Cholinergic blocking
drugs and β-adrenergic agonists are incorrect
choices because both drugs would increase,
rather than decrease, the heart rate. A
β-adrenergic blocking drug would not cause
such a rise in arterial pressure, although it
would produce a drop in heart rate. An
α-adrenergic blocking drug would cause a
decrease, rather than an increase, in peripheral resistance.

All adrenergic drugs would increase the
heart rate as a direct effect on the sinoatrial
(SA) node, isoproterenol having the greatest
effect and norepinephrine the least. A drug
such as epinephrine, which has both α- and
β-adrenergic properties, can increase contractility and peripheral resistance so that a large

increase in arterial pressure occurs. The increased arterial pressure then affects the baroreceptors and results in a reflex slowing of cardiac rate. When activated, chemoreceptors lead to a rise in heart rate as well as an increase in contractility. Ventricular extrasystoles have no effect on peripheral resistance. The effect of the drug on ventricular muscle would not alter heart rate.

104. The answer is D [Chapter 17 II B 3 a]. A drop in arterial O_2 tension from 100 mm Hg to 27 mm Hg would decrease the O_2 content of the blood by about 50%. The key to answering this question is to recognize that 27 mm Hg is the normal value for the P_{50}, which is the O_2 tension at a hemoglobin saturation of 50%. Since hemoglobin is nearly 100% saturated at an O_2 tension of 100 mm Hg, then O_2 content must decrease by 50%.

105. The answer is D [Chapter 25 I A 1 a; Figure 25-1]. The oncotic and hydrostatic pressures within the capillary and within the interstitium contribute to the regulation of fluid exchange between the plasma and the interstitial fluid (or Bowman's capsule). To determine whether there is a net reabsorption or filtration, it is necessary to compare the magnitude and direction of these two different pressures as

Outward Forces (mm Hg)

Capillary hydrostatic pressure	=	47
Bowman's capsule oncotic pressure	=	0
		47

Inward Forces (mm Hg)

Bowman's capsule hydrostatic pressure	=	10
Capillary oncotic pressure	=	28
		38

Because the outwardly directed forces exceed the inwardly directed forces, there will be an effective filtration pressure (EFP) of 9 mm Hg. Note that the oncotic pressure in Bowman's capsule is assigned a value of zero because the fluid in this region is an ultrafiltrate of plasma.

106. The answer is A [Chapter 49 I D 3 b (2)]. Hydrocortisone (cortisol) directly increases epinephrine-forming enzyme activity. Epinephrine is synthesized in the adrenal medulla and certain brain neurons from norepinephrine by the action of the enzyme phenylethanolamine-N-methyltransferase

(PNMT). The adrenal cortex and adrenal medulla are related both anatomically and functionally. Venous blood from the sinusoids of the cortex enters the adrenal portal system and perfuses the adrenal medulla before entering the systemic circulation. Therefore, the chromaffin cells are exposed to a high concentration of cortisol. Adrenocorticotropic hormone (ACTH) indirectly activates the epinephrine-forming enzyme, because it stimulates the secretion of cortisol.

107–109. The answers are: 107-B [Chapter 17 II B 3, C–D], **108-D** [Chapter 18 III C 1], **109-B** [Chapter 20 II D]. This man's arterial O_2 content is 12 ml/dl. To calculate this, it is important to recognize that the normal P_{50} is 27 mm Hg, meaning that this man's arterial hemoglobin saturation must be 50%. One reason that this man's hemoglobin saturation is low is his exposure to a low barometric pressure at an altitude of 11,500 feet. The barometric pressure at this altitude is about 490 mm Hg, providing an alveolar O_2 tension of about 50 mm Hg. The O_2 capacity is calculated from the hemoglobin concentration as follows: 18 g/dl • 1.34 ml/g = 24.12 ml/dl. Disregarding the small amount of O_2 that is dissolved in plasma, the arterial O_2 content would equal the O_2 capacity • saturation, or 24.12 • 0.5 (12 ml/dl).

This man's pulmonary artery pressure is likely to be higher than normal because of hypoxic pulmonary vasoconstriction (HPV). At high altitudes there would be an increased resistance to flow through the pulmonary vasculature. This increased resistance is the result of smooth muscle contraction in the walls of the pulmonary blood vessels in response to an unknown mediator brought on by hypoxia. The cardiac output in people who live for long periods at high altitudes is in the normal range, although it increases upon initial exposure to such altitudes. Chemoreceptors have no significant effect on the pulmonary circulation.

Venous hemoglobin saturation is decreased at high altitudes, because the hemoglobin starts in the lungs somewhat unsaturated because of the low O_2 tension. Residents at high altitudes have an increased hemoglobin concentration as a compensation for the low O_2 tension in the tissues. If hemoglobin concentration rises sufficiently, the O_2 content may be in the normal range, but the hemoglobin saturation would be reduced secondary to the low O_2 tension. Although cardiac output is normal in people acclimated to high altitude, the venous hemoglobin saturation *is* above normal in people with increased cardiac output. The fact that less O_2 is removed from each unit of blood when O_2 delivery is increased by the higher cardiac output accounts for this phenomenon.

110. The answer is A [Chapter 9 IV B 1, C]. An increased heart rate would simultaneously increase the aortic systolic pressure and decrease the aortic pulse pressure, according to the elastic modulus:

$$E = V \cdot dP/dV, \text{ where}$$
$$E = \text{elastic modulus}$$
$$V = \text{arterial volume}$$
$$dP = \text{pulse pressure}$$
$$dV = \text{arterial uptake during diastole}$$

An increase in heart rate increases cardiac output and raises the arterial volume, which is proportional to the mean aortic pressure. If stroke volume remains constant, then the arterial uptake during diastole will also remain essentially unchanged so that the pulse pressure must decrease to maintain the equation valid. An increased arterial compliance decreases the value of the elastic constant (compliance is the reciprocal of elastance), which lowers systolic pressure. Decreased peripheral resistance decreases the systolic pressure and increases the pulse pressure. An increased stroke volume or an increased elastic constant cause an increase in systolic pressure as well as pulse pressure.

111. The answer is A [Chapter 29 III B 5; Chapter 50 I C 3 a; Chapter 52 I B 1; V B]. Angiotensin I, tetraiodothyronine (T_4), and testosterone all can be considered prohormones. Angiotensinogen is a liver-derived globulin that has no effect on aldosterone secretion. Angiotensinogen is converted by renin into angiotensin I, an inactive decapeptide that is rapidly converted into the active octapeptide, angiotensin II, by angiotensin converting enzyme (ACE), found mainly in pulmonary endothelial cells. Most of the circulating triiodothyronine (T_3) is formed extrathyroidally from T_4 by 5'-deiodinase, an enzyme found mainly in the liver and kidney. On a molar basis, T_3 has 3–5 times the bioactivity of T_4. Testosterone can be reduced to the more potent androgen dihydrotestosterone in an irreversible reaction catalyzed by a 5α-reductase, an enzyme found in the prostate, seminal vesi-

cles, epididymis, and skin. Dihydrotestosterone has twice the bioactivity of testosterone.

112. The answer is A [Chapter 49 I G; Table 49-1]. The ciliary muscle is innervated by both postganglionic sympathetic and postganglionic parasympathetic fibers. Only β-adrenergic, not α-adrenergic, receptors are found in this smooth muscle. β-Adrenergic stimulation causes relaxation of the ciliary muscle, which, in turn, increases the tension on the lens, causing the lens to become thinner and adapted for far vision. The parasympathetic innervation of the ciliary muscle causes it to contract, which, in turn, decreases the tension on the lens, causing the lens to become thick and adapted for near vision (accommodation). The heart contains only β-adrenergic receptors, whereas vascular smooth muscle, pancreatic beta cells, and intestinal smooth muscle contain both α- and β-adrenergic receptors. Stimulation of either type of receptor in the intestine causes inhibition of peristalsis.

113. The answer is C [Chapter 16 VIII C 2 a–b]. Ascending a mountain decreases the O_2 tension, because of the reduction in the barometric pressure. Air contains about 21% O_2; therefore, administering supplemental O_2 would increase the O_2 tension. Hyperventilation would reduce the CO_2 tension, resulting in a higher O_2 tension, according to the alveolar gas equation.

114. The answer is D [Chapter 38 IV A 2]. Metabolic acidosis is associated with a low arterial pH, a reduced plasma [HCO_3^-], and a compensatory increase in alveolar ventilation and renal excretion of H^+. Renal excretion of titratable acid is limited by the amount of filtered buffers (e.g., phosphate buffer) and by the ability of the kidney to lower urinary pH. Thus, the net acid excretion by the kidney is largely controlled by factors that regulate the urinary secretion of NH_3 and the urinary excretion of NH_4^+. There is a concomitant, but limited, increase in titratable acid. Base excess is negative in metabolic acidosis.

115. The answer is C [Chapter 40 IV A 3 (c) (1); Chapter 41 II B, F 1 a; Chapter 42 III B 3]. The migrating motor complex (MMC) is initiated by motilin (a hormone released from the endocrine cells in the small intestine) and is not affected by vagotomy. Cutting the vagus nerve prevents receptive relaxation from occurring and thus increases the rate of liquid emptying from the stomach. Primary esophageal peristalsis is coordinated by the swallowing center and thus is prevented by vagotomy.

116. The answer is E [Chapter 6 I C 2; Figure 6-4]. When light strikes the eye, it causes the photoisomerization of 11-*cis* retinal to all-*trans* retinal. The conformational change leads to the activation of rhodopsin, which activates a G protein called transducin. Transducin activates a cyclic guanosine monophosphate (cGMP) phosphodiesterase, which decreases the concentration of cGMP. The decrease in cGMP concentration causes Na^+ channels to close, which causes the membrane to hyperpolarize and synaptic transmitter release to decrease. Closing the Na^+ channels causes less Na^+ to enter the cell; therefore, Na^+-K^+-adenosine triphosphatase (ATPase) pumps less Na^+ out of the cell.

117. The answer is B [Chapter 51 I D 4 a, 6 a, b]. Insulin is stimulatory to K^+ uptake by cells, and high concentrations of exogenous insulin cause extracellular hypokalemia. This hypokalemic action of insulin is due to the increased K^+ uptake by muscle and liver. Although the primary stimulus to the alpha cell is hypoglycemia, hypoglycemia-induced catecholamine release undoubtedly plays a role. Hypoglycemia also is a potent stimulus for the release of growth hormone (GH, somatotropin) and adrenocorticotropic hormone (ACTH, corticotropin). Among the stimuli for ACTH release are pain, anxiety, pyrogens, and hypoglycemia.

118. The answer is D [Chapter 25 II B 1 b; Chapter 29 III B 1 a, 3 a–b]. Because hemorrhage decreases both glomerular filtration rate (GFR) and the renal plasma flow commensurately, there is no change in the filtration fraction, which is the ratio of GFR to renal plasma flow. Severe hemorrhage results in a state of hypovolemia, which, in turn, is a major stimulus for the secretions of both aldosterone and antidiuretic hormone (ADH). Bodily responses to hypovolemia take precedence over the regulation of solute concentration. The decrease in blood volume initiates a neurocirculatory reflex arc composed of afferent and efferent limbs. The afferent limb is activated by a reduction in plasma volume. The reduced plasma volume leads to a decrease in central blood volume, decreased atrial and pulmo-

nary venous pressures, and a resultant reduction in the activity of low-pressure baroreceptors in these regions. There also is a concomitant reduction in the firing rate of the high-pressure baroreceptors in the carotid sinus and aortic arch. The decreased firing rates of these receptors cause an increase in ADH secretion. Another afferent input originates in the low-pressure baroreceptors in the kidney. Stimulation of the juxtaglomerular cells by hypovolemia evokes the secretion of renin, which forms angiotensin I, a decapeptide converted to angiotensin II by angiotensin converting enzyme (ACE) in the pulmonary circulation. Angiotensin II is also a stimulator of aldosterone secretion. Thus, aldosterone enhances Na^+ reabsorption, and ADH enhances water reabsorption by increasing the permeability of the collecting duct to water.

119. The answer is C [Chapter 48 I B]. The neural lobe of the pituitary gland, or neurohypophysis, is derived from the neural tube (neural ectoderm). The neural crest gives rise to a wide variety of cells, including neurons with perikarya outside the central nervous system (CNS), parafollicular cells of the thyroid gland, fibroblasts, dentin-producing cells, vascular smooth muscle cells, melanocytes, and Schwann cells. Also derived from the neural crest are the cartilage and bone of the skull and the adrenal medulla.

120. The answer is E [Chapter 41 II B 2, F 1 a, 2 a; III A 2 b (2), B 3 a; Chapter 42 III C 2]. Receptive relaxation occurs when the proximal stomach (fundus and corpus) is stretched by the presence of food. Distention of the antrum causes gastrin to be released from G cells. Gastrin enhances gastric acid (HCl) and pancreatic enzyme secretion as well as gastric motility. In addition, expansion of the antrum initiates a vagovagal reflex that enhances antral contractions.

121. The answer is A [Chapter 51 I D 3]. Untreated diabetes mellitus is associated with negative nitrogen balance. The disease reflects a state of severe insulin deficiency combined with a decreased concentration of plasma C-peptide. Since insulin and C-peptide are secreted in equimolar amounts, C-peptide directly reflects pancreatic beta cell secretory activity and, therefore, endogenous insulin secretion. The most characteristic feature of untreated diabetes mellitus is fasting hyperglycemia due, in part, to the lack of insulin and, in larger part, to the unopposed action of the counter-regulatory (insulin-antagonizing) hormones [glucagon, cortisol, epinephrine, growth hormone (GH)]. These four hormones cause the liver to change from a glucose-utilizing to a glucose-producing organ, leading to hyperglycemia and glycosuria. The hormones also increase the rate of lipolysis, which leads to a fatty acid-induced decrease in glucose uptake. Cortisol and glucagon promote muscle proteolysis, leading to hyperaminoacidemia, hyperaminoaciduria, elevated blood urea nitrogen (BUN), and negative nitrogen balance. This hormonal imbalance switches liver fatty acid metabolism from oxidation, reesterification, or both to ketogenesis. The sum of these effects is an increase in blood fatty acids, amino acids, and ketones due to the gradual loss of muscle and adipose tissue. The ketosis leads to ketonemia and, eventually, ketonuria.

122. The answer is C [Chapter 29 III A 1 b (1); Chapter 48 IV A 4 b]. Aldosterone secretion is not significantly depressed following hypophysectomy and patients can regulate Na^+ and K^+ balance. However, hypophysectomy without hormone replacement therapy is incompatible with human life. The pituitary gland is essential for cellular differentiation, somatic growth, adaptation to stress, and reproduction. If the pituitary gland is removed from young, rapidly growing animals, dwarfism occurs as a result of the absence of growth hormone (GH). Atrophy of the gonads, thyroid gland, and adrenal cortex also occurs in the absence of pituitary tropic hormones.

123. The answer is D [Chapter 7 III A 1, 3]. The withdrawal reflex (i.e., the withdrawal of a limb from a painful or irritating stimulus) is produced by a polysynaptic, not a monosynaptic, reflex. The only spinal cord monosynaptic reflex is the stretch reflex. The afferent fibers of the withdrawal reflex include both small myelinated (Aδ) and unmyelinated (C) fibers. In contrast, the afferent fibers used in the stretch reflex are large myelinated fibers. The withdrawal reflex can initiate withdrawal of only the injured area of the body skin (local sign), or it can produce withdrawal of other, non-injured areas of the body (irradiation). Moreover, the reflex contraction of the muscles producing the withdrawal can outlast the stimulus (afterdischarge). The monosynaptic stretch reflex normally involves only the muscle fibers that are stretched. Spread of activity

to other muscles or continuation of muscular activity after the stimulus is withdrawn is a pathological sign.

124. The answer is C [Chapter 49 I F 2 b (2)]. Catecholamine-O-methyltransferase (COMT) catalyzes the metabolic inactivation of dopamine, norepinephrine, and epinephrine. Normetanephrine is deaminated catecholamine produced by the action of COMT on norepinephrine. Dihydroxymandelic acid also is a substrate for COMT. COMT is found in the soluble fraction of cells, especially the liver and kidney, and it is more important in the metabolic degradation of circulating catecholamines.

125. The answer is E [Chapter 29 III A 1 a, 2, 3; Chapter 30 IV A]. Atrial natriuretic peptide (ANP) is secreted by the atrial monocytes in response to increased blood volume. ANP prevents angiotensin formation by inhibiting renin release. The cardiac hormones inhibit angiotensin II and adrenocorticotropic hormone corticotropin (ACTH)—two stimuli for aldosterone secretion. Angiotensin II is the primary stimulus for aldosterone secretion. Angiotensin II synthesis is regulated by renin release from the juxtaglomerular cells. The principal stimuli for renin secretion are decreased perfusion pressure in the afferent arterioles, decreased $[Na^+]$ in the macula densa area of the nephron, and norepinephrine secretion from the sympathetic neurons innervating the juxtaglomerular cells. All actions of renin are mediated through the generation of angiotensin II. Other stimuli for aldosterone include ACTH, a high plasma $[K^+]$, and a low plasma $[Na^+]$. The stimulatory effect of ACTH on aldosterone secretion is powerful but short-lived and is not a major factor in the control of aldosterone production.

126. The answer is C [Chapter 48 IV A 2 b (1)]. Somatostatin is the growth hormone (GH)-inhibiting hormone produced in the arcuate nucleus of the tuberoinfundibular neural tract. Somatostatin is released from the median eminence and enters the hypothalamic-hypophysial portal system that perfuses the adenohypophysis. This neuropeptide also is produced in duodenum and the pancreatic delta cells. The plasma GH level is elevated in response to any form of stress as well as to exercise and to deep sleep (stages III and IV). Both insulin- and arginine-induced hypoglycemia are potent stimuli for GH secretion and can be used as provocative tests of pituitary GH reserve.

127. The answer is D [Chapter 29 II A–D]. Aldosterone acts on the connecting tubule and collecting tubules to increase the reabsorption of Na^+ and the secretion of K^+ and H^+. Thus, excess amounts of this hormone lead to hypokalemia (kaliuresis), alkalemia (alkalosis), and hypertension. By promoting Na^+ retention, excess aldosterone secretion leads to an increase in blood volume via an increase in plasma volume. This effect, in turn, brings about a decline in hematocrit and in plasma oncotic pressure.

128. The answer is A [Chapter 7 V C]. Hypotonia is a sign of cerebellar disease. Lesions within the basal ganglia produce spontaneous, wild, flinging movements (hemiballismus); tremor; slow, twisting movements (athetosis); or lack of spontaneous movements or difficulty initiating movement (hypokinesia).

129. The answer is C [Chapter 52 IV A 1–4]. Thyroid hormone biosynthesis occurs in four steps: (1) active uptake of inorganic iodide (I^-), (2) oxidation of inorganic iodide to active iodide (I^+, known as iodinium), (3) formation of iodotyrosines by the iodination of tyrosine residues within the matrix of thyroglobulin, and (4) coupling (condensation) of iodotyrosines to form iodothyronines. All four steps require thyroid-stimulating hormone (TSH); steps 2, 3, and 4 are catalyzed by the membrane-bound enzyme thyroid peroxidase. Diiodotyrosine and monoiodotyrosine have no biologic activity; however, the iodothyronines—T_3 and T_4 (thyroxine)—are biologically active. Therefore, thyroxine synthesis requires TSH, active iodide, thyroglobulin, thyroid peroxidase, and the coupling of two molecules of diodotyrosine.

130. The answer is E [Chapter 6 I C 3]. Rods adapt to a light stimulus more slowly than cones and, therefore, are capable of summating successive light stimuli that, individually, would be too weak to be detected. The ability of rods to detect low levels of light is enhanced because rods can detect scattered light within the eye, they have larger receptor fields than cones (allowing light from a wider area of the eye to stimulate the ganglion cells), and they produce a larger receptor potential than cones do in response to a light stimulus.

131. The answer is C [Chapter 51 I D 2]. Insulin is a lipogenic as well as antilipolytic hormone. It decreases the activity of the intracellular lipase, triglyceride lipase ("hormone-sensitive lipase"), by inhibiting the formation of cyclic adenosine 3′,5′-monophosphate (cAMP). Insulin activates the extracellular lipase, lipoprotein lipase, which is responsible for the hydrolysis of plasma lipoproteins. The hydrolysis of triglycerides in the adipocyte delivers long-chain fatty acids and glycerol into the circulation. Lipolysis in adipocytes is mediated by triglyceride lipase. Catecholamines play a primary role in lipolysis, whereas glucagon, growth hormone (GH), cortisol, and adrenocorticotropic hormone (ACTH), have lesser lipolytic effects. Cortisol and T_3 modulate the sensitivity of adipocytes to the lipolytic effects of catecholamines.

132–137. The answers are: 132-B [Chapter 28 I A–B], **133-B** [Chapter 28 III A], **134-A** [Chapter 29 II A 4], **135-A** [Chapter 29 I A 1], **136-E** [Chapter 29 III B 5 c (1)], **137-D** [Chapter 29 III B 1 a]. Antidiuretic hormone (ADH), also known as arginine vasopressin, is an octapeptide synthesized mainly in the supraoptic nucleus of the ventral diencephalon. ADH also can be classified as a nonapeptide, if the single cystine moiety is counted as two cysteine residues. ADH is stored in the pars nervosa, from which it is secreted.

ADH regulates plasma osmolality, which normally is about 300 mOsm/kg. It promotes water reabsorption mainly from the tubular fluid in the renal collecting ducts by increasing the water permeability of these cells. ADH allows humans to elaborate a small volume, hypertonic urine in order to conserve water. Because ADH promotes free-water reabsorption, it determines the plasma $[Na^+]$.

Aldosterone, the most potent endogenous mineralocorticoid, acts primarily on the renal collecting ducts to promote Na^+ reabsorption and K^+ and H^+ excretion. It has similar effects on sweat, salivary, and intestinal glands. Thus, aldosterone controls the Na^+ content of the body. In turn, the Na^+ content determines the volume of the various fluid compartments. Aldosterone also increases Na^+ reabsorption by the connecting segment of the nephron. Aldosterone is synthesized in and secreted from the outermost layer of the adrenal cortex, called the zona glomerulosa.

Dipeptidyl carboxypeptidase, also known as angiotensin converting enzyme (ACE) or kininase II, is located mainly on the endothelial surface of pulmonary capillaries. ACE cata-lyzes the conversion of angiotensin I (an inac-tive decapeptide) to angiotensin II (an active octapeptide). In addition, ACE simultaneously inactivates the nonapeptide, bradykinin. Thus, ACE leads to the increased formation of an-giotensin II (a vasoconstrictor) and the de-creased formation of bradykinin (a vasodila-tor). Angiotensin II functions as a vasoconstrictor and aldosterone-inhibiting hormone.

Renin is a proteolytic enzyme produced by the juxtaglomerular cells of the afferent arteri-ole. These cells are modified smooth muscle cells that have acquired secretory function and, thus, are described as myoepithelial cells. Juxtaglomerular cells function as low-pressure baroreceptors that are stimulated by a decrease in renal perfusion pressure caused by a decrease in systemic blood volume or pres-sure.

138–142. The answers are: 138-E [Chapter 20 II D 1 a], **139-A** [Chapter 20 II C 3], **140-B** [Chapter 16 VIII C 1 b (2)], **141-D** [Chapter 16 VIII C 1 b (1)], **142-C** [Chapter 20 II C 1 c]. While mountain climbing, the arterial O_2 ten-sion would be reduced, with arterial CO_2 ten-sion reduced by reflex hyperventilation. Be-cause this is acute exposure to high altitude, arterial pH would be increased; not enough time has passed for the kidneys to correct the blood pH. Only one set of blood data (E) coin-cides with these conditions.

Although the arterial O_2 tension is normal in anemia, the arterial O_2 content is reduced in proportion to the decrease in hemoglobin concentration. Because the chemoreceptors are not stimulated, ventilation does not change and, thus, arterial CO_2 tension and pH are normal. Only one set of blood data (A) coincides with these conditions.

Hypoventilation is synonymous with hyper-capnia and, thus, with an increase in arterial CO_2 tension. Only one set of blood data (B) reflects this change.

During hyperventilation, arterial CO_2 ten-sion decreases and arterial O_2 tension in-creases. Only one set of blood data (D) coin-cides with these conditions.

Patients with chronic obstructive lung dis-ease typically have a reduced arterial O_2 ten-sion owing to the ventilation–perfusion abnor-mality that is present. In early stages of the disease, these patients have a normal-to-low CO_2 tension and an arterial pH in the normal range, because the kidneys can maintain the

correct HCO_3^-/H_2CO_3 ratio. Only one set of blood data (C) coincides with these condi-tions.

143–146. The answers are: 143-A [Chapter 7 VI C 3 c], **144-C** [Chapter 7 IV A 2 b], **145-E** [Chapter 7 V C 4], **146-B** [Chapter 7 IV B 3 a]. The posterior cerebellum (cerebrocerebellum) coordinates the motor programs required for the smooth movement of a limb toward its target. This coordination is lacking in individ-uals with lesions of the cerebrocerebellum, causing the limb to oscillate as it approaches a target. No tremor is present when the limb is resting.

Elimination of inhibitory control over the neurons within the reticular formation that are responsible for activating the spinal motoneu-rons during a movement produces spasticity [i.e., continuous contraction of antigravity (or physiologic extensor) muscles]. Loss of inhibi-tion can occur when lesions damage the in-hibitory upper motor neurons in the cerebral cortex or their axons within the internal cap-sule.

Rigidity is a major sign of Parkinson's dis-ease and results from lesions to the dopamin-ergic neurons within the substantia nigra nu-cleus of the basal ganglia. Rigidity differs from spasticity in that both the extensor and flexor muscles at a joint are activated.

Nystagmus is the spontaneous release of the vestibulo-ocular reflex, in which the eye moves slowly toward the edge of the socket and then moves quickly back to the center. Damage to the vestibular apparatus, vestibular nerve, or vestibular nucleus can cause nystag-mus. Under normal physiologic conditions, the vestibulo-ocular reflex enables the eye to maintain visual fixation while the head is moving.

147–149. The answers are: 147-A, 148-B, 149-D [Chapter 53 VI C, VIII]. Parathyroid hormone (PTH) is the hypercalcemic hormone of the body. Excessive amounts of PTH would produce hypercalcemia and hypophos-phatemia together with hypercalciuria and hyperphosphaturia. The major regulator of PTH synthesis and secretion is serum ionized calcium concentration ($[Ca^{2+}]$). PTH main-tains the normal plasma total calcium concen-tration at about 5 mEq/L by interacting with the kidney, bone, and intestine. PTH stimu-lates bone resorption, and it decreases the maximal tubular transport capacity (Tm) for phosphate by decreasing proximal tubular reabsorption of phosphate, resulting in phosphate diuresis. When PTH secretion is high, the fraction of filtered phosphate that is reabsorbed may fall from the normal 80%–95% to 5%–20%. PTH increases the maximal tubular transport capacity for Ca^{2+} by increasing distal tubular reabsorption of Ca^{2+}. Although PTH stimulates renal Ca^{2+} reabsorption, the urinary Ca^{2+} is greater than normal in states of excess PTH because of the increased filtered load of Ca^{2+}.

There is an inverse relationship between PTH secretion and plasma $[Ca^{2+}]$: when plasma $[Ca^{2+}]$ falls PTH increases, and when plasma $[Ca^{2+}]$ rises PTH secretion is suppressed. Thus, a decrease in PTH secretion is caused by an increase in plasma $[Ca^{2+}]$. This increase in $[Ca^{2+}]$ is associated with a decrease in $[HPO_4^{2-}]$. Inorganic phosphate has no direct influence on PTH secretion; rather, it is the phosphate-induced decrease in plasma $[Ca^{2+}]$ that stimulates PTH secretion. Conversely, a decrease in plasma inorganic $[HPO_4^{2-}]$ increases plasma $[Ca^{2+}]$ and indirectly inhibits PTH secretion.

Overall, the physiologic action of active vitamin D_3 (calcitriol) is to increase extracellular $[Ca^{2+}]$ and $[HPO_4^{2-}]$. The effects of calcitriol are exerted primarily on the intestine and bone and, to a lesser extent, the kidney. Calcitriol increases renal tubular reabsorption of Ca^{2+} and phosphate. It also is the principal mediator of PTH-induced intestinal Ca^{2+} and phosphate absorption. Thus, vitamin D_3 toxicity leads to hypercalcemia directly by bone resorption and renal Ca^{2+} reabsorption and indirectly by mediating PTH-induced intestinal Ca^{2+} absorption. This hypercalcemia suppresses PTH secretion, leading to increased plasma $[HPO_4^{2-}]$ and increased renal phosphate reabsorption. This is in contradistinction to hyperparathyroidism, which causes an increase in $[Ca^{2+}]$, a decrease in $[HPO_4^{2-}]$, and an increase in plasma PTH level. Thus, hyperparathyroidism must be distinguished from other causes of hypercalcemia.

150–153. The answers are: 150-B [Chapter 26 III D 1 b (1)], **151-E** [Chapter 26 III C 1 a], **152-B** [Chapter 26 III D 2 b (2) (a)], **153-B** [Chapter 26 II C 1]. Most H^+ secretion by the nephron occurs in the proximal tubule by Na^+-H^+ exchange. This active H^+ efflux is linked through a countertransport mechanism to Na^+ influx across the luminal membrane. Most of this secreted H^+ is not excreted but is reabsorbed in the form of H_2O. Most impor-

tant, any secreted H^+ that combines with HCO_3^- in the lumen forms H_2CO_3, which is dehydrated into CO_2 and H_2O, both of which are reabsorbed. Thus, the secreted H^+ that combines with HCO_3^- does not contribute to the urinary excretion of acid.

K^+ secretion occurs mainly in the cortical collecting duct. The excreted K^+ is derived mainly from K^+ secretion in this region of the nephron. Secretion of K^+ involves active pumping across the peritubular membrane followed by passive diffusion across the luminal membrane into the tubular lumen.

The cells of the proximal tubule are the major site of ammonia production. Addition of NH_4^+ to the tubular fluid occurs primarily in the proximal tubule. Thus, the accumulation of ammonia in the lumen involves both nonionic diffusion of NH_3 and transport of NH_4^+.

The proximal tubule is the major site for the active secretion of para-aminohippuric acid (PAH) in its anionic form.

154–156. The answers are: 154-B, 155-D, 156-C [Chapter 42 VII C 2, E 4–5]. Iron is absorbed from the duodenum and proximal jejunum by a membrane-bound carrier protein on the luminal surface of the enterocyte. Once inside the cell, iron combines with an iron-binding protein to form a complex called ferritin. Before being extruded from the serosal surface of the cell, the iron dissociates from ferritin.

Ca^{2+} is absorbed from the duodenum by a membrane-bound carrier on the luminal surface of the enterocyte that is formed in response to the presence of vitamin D. Once inside the enterocyte, the Ca^{2+} is extruded from the serosal surface of the cell by an active transport system.

Cholesterol must be dissolved in micelles before it can be absorbed. Micelles are small spherical globules formed from bile salts. The polar, water-soluble end of the bile salt faces outward, and the lipid-soluble tail portion of the bile salt faces inward. Cholesterol dissolves in the interior of the micelle, and, when the micelle makes contact with the intestinal membrane, the cholesterol diffuses from the micelle into the enterocyte.

157–160. The answers are: 157-D, 158-B, 159-D, 160-A [Chapter 37 I C, E 3; IV A 2, B; Chapter 38 IV A, B, D; V B; VI]. The blood data for patient D include high pH ($\downarrow$ $[H^+]$), high $[HCO_3^-]$, and a compensatory increase in CO_2 tension—findings that coincide with partially compensated metabolic alkalosis. That $[H^+]$ and $[HCO_3^-]$ change in opposite directions indicates a primary metabolic disturbance, and the pH of 7.5 indicates alkalosis. The increment in CO_2 tension shows that partial respiratory compensation has occurred.

The blood data for patient B include low pH ($\uparrow [H^+]$), low $[HCO_3^-]$, low CO_2 content, and low CO_2 tension—findings that coincide with a compensated metabolic acidosis. When metabolic acidosis is compensated by a respiratory alkalosis, the pH tends to return to normal, the CO_2 content drops further, and the CO_2 tension decreases.

Patient D has the highest CO_2 content (i.e., $[HCO_3^-] + S \cdot P_{CO_2}$). This patient has partially compensated metabolic alkalosis indicated by a 42% increase in $[HCO_3^-]$ with only a 13% compensatory increase in CO_2 tension. The CO_2 content is equal to 36.35 mmol/L (35 mmol/L + 1.35 mol/L).

The blood data for patient A include a low pH ($\uparrow [H^+]$), CO_2 tension, and a compensatory decrease in $[HCO_3^-]$—findings that coincide with respiratory alkalosis, a condition caused by high alveolar ventilation. Note that the change in the alternate variable ($[HCO_3^-]$) is in the same direction as the change in the primary variable (CO_2 tension). In this patient, the pH of 7.4 together with hypocapnia and decreased $[HCO_3^-]$ indicates full compensation for a respiratory alkalosis (i.e., a $[HCO_3^-]/S \cdot P_{CO_2}$ ratio of 20:1) and thus the highest alveolar ventilation.

161–164. The answers are: 161-C [Chapter 4 III C 3; IV B 2], **162-D** [Chapter 4 II C 2 b; Table 4-1], **163-D** [Chapter 4 III C 1], **164-E** [Chapter 4 II C 2 c (2); III C 2; Table 4-1]. The amount of Ca^{2+} released from the sarcoplasmic reticulum (SR) of the smooth and cardiac muscle cells is normally varied to control contractile force. In skeletal muscle, maximal amounts of Ca^{2+} enter the cell with each contraction.

Skeletal and smooth muscles are able to recruit additional fibers when more force is required. The heart must contract in a coordinated fashion so that all of its muscle fibers are recruited at the same time.

The heart must relax between contractions; thus, it cannot use repetitive stimulation as a mechanism for increased force production.

Smooth, cardiac, and skeletal muscles all are able to influence the force of contraction by varying the initial length (preload) of their sarcomeres.

165–169. The answers are: 165-C [Chapter 24 I A 1; II A 1 a], **166-E** [Chapter 26 II C 3], **167-A** [Chapter 25 II B 1 a], **168-D** [Chapter 26 C 3], **169-B** [Chapter 26 I B 1 a]. The use of inulin clearance (C_{in}) to measure glomerular filtration rate (GFR) is valid because all of the filtered inulin is excreted in the urine without being reabsorbed or secreted by the renal tubules. Thus, GFR is equal to C_{in}:

$$GFR = C_{in} = \frac{U_{in} \cdot \dot{V}}{P_{in}}, \text{ where}$$

U_{in} and P_{in} = urinary and plasma inulin concentrations, respectively, and $\dot{V}$ = the urinary volume/minute. Substituting,

$$GFR = C_{in} = \frac{150 \text{ mg/ml} \cdot 1.2 \text{ ml/min}}{1.5 \text{ mg/ml}}$$

$$= \frac{180 \text{ mg/min}}{1.5 \text{ mg/ml}}$$

$$= 120 \text{ ml/min}$$

Note that with a unit analysis, both mg/ml terms cancel out.

Renal plasma flow (RPF) is calculated from the clearance of para-aminohippuric acid (PAH); however, there are no PAH data provided for this man. Thus, it is necessary to determine RPF from the filtration fraction (FF), defined as the ratio of GFR to renal plasma flow:

$$FF = \frac{GFR}{RPF}$$

The filtration fraction can also be calculated from the arterial and venous inulin concentrations as:

$$\text{fraction of inulin filtered} = \frac{A_{in} - V_{in}}{A_{in}}, \text{ where}$$

A_{in} and V_{in} are the arterial and venous inulin concentrations, respectively. Substituting,

$$FF = \frac{1.50 \text{ mg/ml} - 1.20 \text{ mg/ml}}{1.50 \text{ mg/ml}}$$

$$= \frac{0.3}{1.50} = 0.20$$

With the filtration fraction and GFR determined, the renal plasma flow is easily determined as:

$$RPF = \frac{GFR}{FF} = \frac{120}{0.2} \ ml/min = 600 \ ml/min$$

Because the filtration fraction equals the ratio of the GFR to renal plasma flow, the filtration fraction also can be calculated using the extraction of a substance such as inulin to determine the renal plasma flow. If the urinary concentration of inulin is 150 mg/ml and the urine flow rate is 1.2 ml/min, the urinary excretion rate of inulin is 180 mg/min. If, when the excretion rate was measured, the inulin concentration was 1.50 mg/ml in renal arterial plasma and 1.20 mg/ml in renal venous plasma, each milliliter of plasma traversing the kidneys must have contributed 0.30 mg to the 180 mg that was excreted. Hence, the renal plasma flow must have been 600 ml/min (180 mg/min divided by 0.3 mg/ml). Thus, with the renal plasma flow determined, the filtration fraction can be calculated as:

$$\frac{GFR}{RPF} = \frac{120 \ ml/min}{600 \ ml/min} = 0.2$$

Once renal plasma flow is determined, renal blood flow (RBF) can also be discerned. Renal blood flow is calculated by dividing the renal plasma flow by the term (1 − hematocrit), or

$$RBF = \frac{RPF}{1 - hematocrit} = \frac{600 \ ml/min}{1 - 0.40} =$$

$$\frac{600 \ ml/min}{0.60} = 1000 \ ml/min$$

The quantity (or amount) of the substance filtered per unit time is termed the filtered load, or amount filtered. It is equal to the product of the GFR (or C_{in}) and the plasma concentration of that substance:

$$\begin{aligned} Filtered \ load &= GFR \cdot P_G \\ &= 120 \ ml/min \cdot 0.9 \ mg/ml \\ &= 108 \ mg/min \end{aligned}$$

Note that the ml terms cancel out with a unit analysis and that it is necessary to express the plasma glucose concentration (P_G) in mg/ml and not in the unit of mg/dl given in the original data.

170–174. The answers are: 170-B, 171-B, 172-A, 173-A, 174-D [Chapter 37 I A–E; IV; Chapter 38 IV A, C]. Hypoventilation, or decreased alveolar ventilation, is the cause of respiratory acidosis. In metabolic acidosis, the increased [H^+] in the arterial blood is a stimu-

lus for hyperventilation and a decline in CO_2 tension.

The main components of total CO_2 content ([total CO_2]) are [HCO_3^-] and dissolved CO_2; an increase in either factor will increase the CO_2 content. In respiratory acidosis, [total CO_2] increases via an increase in CO_2 tension, and in metabolic alkalosis, [total CO_2] increases via an increase in [HCO_3^-]. The compensatory responses to respiratory acidosis and metabolic alkalosis increase the CO_2 content further by elevating the [HCO_3^-] and S · P_{CO_2}, respectively.

[Total CO_2] is decreased when either [HCO_3^-] or dissolved CO_2 is decreased. Thus, in respiratory alkalosis, [total CO_2] decreases due to a decline in arterial CO_2 tension, and in metabolic alkalosis, [total CO_2] decreases due to a decline in [HCO_3^-]. The renal compensation to respiratory alkalosis is increased [HCO_3^-] excretion, which lowers [total CO_2] further. Similarly, the respiratory compensation to metabolic acidosis is hyperventilation, which also reduces the CO_2 content further.

The reduced [HCO_3^-] and increased [H^+] of metabolic acidosis lead to a compensatory hyperventilation, which results in decreased arterial CO_2 tension (hypocapnia). Respiratory acidosis is due to hypoventilation, which results in increased arterial CO_2 tension (hypercapnia).

The normal [HCO_3^-]/S · P_{CO_2} ratio is 20:1. This ratio is decreased in acidotic states, by either the increased CO_2 tension of respiratory acidosis or the decreased [HCO_3^-] of metabolic acidosis. This ratio is increased in alkalotic conditions, by either the decreased CO_2 tension of respiratory alkalosis or the elevated [HCO_3^-] of metabolic alkalosis.

175–178. The answers are: 175-C, 176-A, 177-A, 178-B [Chapter 42 V C 3–4, F 1 d 2; VII A 2 e, D 1 a, E 3 d (2)]. If water is not properly absorbed in the small intestine, rapid dehydration and consequent circulatory collapse will result.

Any blockage of the bile duct prevents bile from being secreted and thus prevents the proper absorption of fat. Malabsorption of fat results in malnutrition and steatorrhea. In addition, when bile is not secreted, bilirubin cannot be secreted, and the resulting accumulation of bilirubin in the soft tissues leads to jaundice.

Vitamin B_{12} is required for the normal production of red blood cells (erythropoiesis). Thus, failure to absorb vitamin B_{12}, caused either by lack of intrinsic factor or by intestinal disease, will lead to pernicious anemia.

Carbohydrate malabsorption usually results in watery diarrhea and intestinal gas.

179–181. The answers are: 179-C, 180-A, 181-B [Chapter 37 I A, B; IV; Chapter 38 IV A, C, D; Figure 38-1]. Hyperventilation is defined as alveolar ventilation in excess of the body's need for CO_2 elimination, which results in decreased arterial CO_2 tension—the underlying factor in respiratory alkalosis. Although CO_2 tension is not given, it can be calculated using the mathematical relationship: $[HCO_3^-]/S \cdot P_{CO_2} = 20/1$. Applying this equation, it is clear that the values in set C indicate a decrease in CO_2 tension:

$$\frac{15}{S \cdot P_{CO_2}} = \frac{20}{1}, \text{ or } S \cdot P_{CO_2} = 0.75 \text{ mmol/L}$$

Because

$$S \cdot P_{CO_2} = 0.75 \text{ mmol/L}$$

and

$$S = 0.03 \text{ mmol/L/mm Hg},$$

then

$$P_{CO_2} = \frac{0.75 \text{ mmol/L}}{0.03 \text{ mmol/L/mm Hg}} = 25 \text{ mm Hg}$$

A CO_2 tension of 25 mm Hg is significantly lower than normal (40 mm Hg). In acute respiratory alkalosis, there is a decrease in urinary H^+ excretion and an increase in urinary

HCO_3^- excretion. In this case, there has been complete renal compensation, as evidenced by the return of the $[HCO_3^-]/S \cdot P_{CO_2}$ ratio to normal (20:1).

In chronic respiratory tract obstruction (e.g., due to tracheal stenosis, foreign body, or tumor), alveolar ventilation is insufficient to excrete CO_2 at a rate required by the body, which leads to increased arterial CO_2 tension—the underlying factor in respiratory acidosis. The kidney increases H^+ secretion, resulting in the addition of HCO_3^- to the extracellular fluid (ECF). The values in set A coincide with these changes. In this case, there is partial compensation of the respiratory acidosis as evidenced by the increase in $[HCO_3^-]$. In chronic respiratory acidosis, the respiratory centers become less sensitive to hypercapnia and acidosis and rely on the associated hypoxemia as the primary drive to ventilation. Correction of the low O_2 tension by the administration of O_2 will diminish respiratory drive, resulting in hypoventilation, a further increase in CO_2 tension, and possibly CO_2 narcosis. For this reason, O_2 must be given with extreme caution to patients with chronic hypercapnia.

Metabolic acidosis exhibits the characteristics of increased $[H^+]$ (decreased pH), a reduced $[HCO_3^-]$, and a compensatory hyperventilation resulting in hypocapnia. The kidney responds to the increased H^+ load by increasing the secretion and excretion of NH_4^+ and the excretion of titratable acid ($H_2PO_4^-$). The values in set B coincide with these changes.

Index